W9-CRX-885

PLEASE STAMP DATE DUE, BOTH BELOW AND ON CARD

TE DUE	DATE DUE	DATE

S tring Theory in a Nutshell

String Theory in a Nutshell

Elias Kiritsis

PRINCETON UNIVERSITY PRESS · PRINCETON AND OXFORD

Published by Princeton University Press,
41 William Street, Princeton, New Jersey 08540

In the United Kingdom: Princeton University Press,
3 Market Place, Woodstock, Oxfordshire OX20 1SY

British Library Cataloging-in-Publication Data is available

Library of Congress Cataloging-in-Publication Data
Kiritsis, Elias.
   String theory in a nutshell / Elias Kiritsis.
     p.   cm.
   Includes bibliographical references and index.
   ISBN-13: 978-0-691-12230-4 (acid-free paper)
   ISBN-10: 0-691-12230-X (acid-free paper)
1. String models.   I. Title.
   QC794.6.S85 K565 2007
   530.14—dc22   2006050586

This book has been composed in Scala and Scala Sans

Printed on acid-free paper. ∞

press.princeton.edu

Printed in the United States of America

10 9 8 7 6 5 4 3 2 1

To my parents who gave me everything they had
and
to Takoui who made my life brighter

Contents

Preface

In the fall of 2003, it was suggested to me that a new textbook on the discipline that goes under the name of string theory was needed in order to accommodate several advances that have happened since Polchinski's books appeared.

Although I wrote a short textbook in 1997, I have not yet learned my lesson. Therefore, I embarked on a direct confrontation with deadlines that, although not as spectacular as the ones in Polchinski's book, managed to wreak havoc on my academic and personal schedule. I have not even put sufficient credence in a colleague's statement "it is never too late to stop writing a book." I can only hope now that the result was worth the effort.

This is a textbook on the collection of ideas categorized under the name of "string theory." This is a large domain that has as its central goal the unification of all interactions including gravity. There have been several surges of progress in different directions in the past twenty years, and this book tries to give the student of the field some of the salient ideas.

It is *not* the purpose of this book to provide deep insights into the theory. I can only leave this to more competent colleagues. Its purpose is to provide the fastest possible introduction to the basic formalism and structure of string theory and its main properties and ramifications that are on a reasonably solid footing today.

This effort is unlike writing a textbook about the Standard Model of the fundamental interactions. It is less clear here what will turn out to be just mathematics, what will transform into real physics, and what will be neither of the above. However, the scope and deep interest of the endeavor, namely, to understand the basic mysteries of the universe at the most fundamental level, has driven more than a generation of bright physicists and has provided breakthroughs in our theoretical understanding of both gravitational and gauge theories. It is this interest that drives researchers today, together with the hope that the theory will eventually be seriously confronted with experimental data.

There are several current and past areas of research in string theory that have not been treated in this book. The reasons were varied. They include the subjects of the

Green-Schwarz quantization of superstrings, the Berkovits quantization, string field theory, the whole issue of cosmological backgrounds in string theory, strings and branes at finite temperature and the issue of the Hagedorn transition, as well as the more recent studies of tachyon condensation and the development of the formalism to investigate compactifications with fluxes and moduli stabilization. For all these subjects a brief guide into the bibliography is given.

At the end of each chapter a section of bibliography is provided. It purpose is to guide the student to the starting points of the subject's literature. In particular (s)he is guided to specialized reviews and original papers when they are deemed to have a pedagogical value. Obviously, the choice of references reflects the knowledge and taste of the author, and I apologize in advance for possible omissions or errors.

A serious effort was made to provide many exercises after each chapter. There are 460 exercises in this book and they are of several types. The simplest guide the student to complete calculations that are sketched but not done in detail in the text. Such exercises are typically easy although sometimes they can be labor intensive. Exercises are given where the student is invited to work on other issues that are not directly dealt within the text, but which are nevertheless useful in understanding the issues involved. Finally, there are exercises that initiate an exploration of areas not treated in this book. Some exercises are hard and they have been the subject of full-blown research articles in the past. The intervention of the instructor in the choice of exercises is therefore important.

A website dedicated to this book will be in service: <http://hep.physics.uoc.gr/~kiritsis/string-book/>. It will collect information that might be useful to readers, including the omnipresent corrections of errors and misprints. The author welcomes suggestions for corrections at the address kiritsis@physics.uoc.gr

I am indebted to many people who have indirectly contributed to the present effort. They include my teachers who gave me the first tools to do science, colleagues who shared their knowledge with me, collaborators who shared knowledge and patience, and students who pushed me several times to clarify my understanding of subjects. They are too many to mention by name. I would like to thank, however, P. Anastasopoulos, M. Bianchi, R. Casero, U. Gursoy, F. Nitti, A. Paredes, S. Wadia, and especially L. Alvarez-Gaume and B. Pioline for reading the manuscript at various stages and providing corrections and constructive criticism. Extra thanks go to P. Anastasopoulos for his help with the figures. I would also like to thank the Marie Curie program of the European Commission for funding my research in the past and during the period this book was written. Finally and most importantly, I would like to thank my life partner Takoui for understanding and support during this difficult endeavor.

Elias Kiritsis
January 2006

Abbreviations

ADM	Arnowitt-Deser-Misner
AdS	Anti–de Sitter
ALE	Asymptotically locally Euclidean
BCFT	Boundary conformal field theory
BF	Breitenlohner-Freedman (bound)
BFSS	Banks-Fischler-Shenker-Susskind
BH	Bekenstein-Hawking
BPS	Bogomolnyi-Prasad-Sommerfield
BRST	Becchi-Rouet-Stora-Tyutin
BTZ	Banados-Teitelboim-Zanelli
CFT	Conformal field theory
CP	Chan-Paton
CP	Charge conjugation $\times$ parity
CP^n	Complex projective space
CY	Calabi-Yau (manifold)
DBI	Dirac-Born-Infeld
DD	Dirichlet-Dirichlet (boundary conditions)
DDF	Di Vecchia–Del Giudice–Fubini
DLCQ	Discrete light-cone quantization
DN	Dirichlet-Neumann (boundary conditions)
DNT	Dirac-Nepomechie-Teitelboim
EFT	Effective field theory
FI	Fayet-Iliopoulos
GH	Gibbons-Hawking
GS	Green-Schwarz
GSO	Gliozzi-Scherk-Olive

$\mathcal{H}_2$	Upper half plane
HMS	Hypermultiplet moduli space
HRG	Holographic renormalization group
HW	Highest weight
KK	Kaluza-Klein
KN	Kerr-Newman
MW	Majorana-Weyl
MQM	Matrix quantum mechanics
NN	Neumann-Neumann (boundary conditions)
NS	Neveu-Schwarz
NSR	Neveu-Schwarz-Ramond
OPE	Operator product expansion
PB	Poisson bracket
QFT	Quantum field theory
R	Ramond
RG	Renormalization group
RN	Reissner-Nordström
RP_2	Real projective plane
RS	Randall-Sundrum
SCFT	SuperConformal field theory
SM	Standard Model
ST	String theory
SUGRA	Supergravity
SUSY	Supersymmetry
SYM	Super Yang-Mills
vev(s)	vacuum expectation value(s)
VMS	Vector moduli space
WZ	Wess-Zumino
WZW	Wess-Zumino-Witten

S tring Theory in a Nutshell

1 | Introduction

1.1 Prehistory

The quest in physics has been historically dominated by unraveling the simplicity of the physical laws, moving more and more toward the elementary. Although this is not guaranteed to succeed indefinitely, it has been vindicated so far. The other organizing tendency of the human mind is toward "unification": finding a unique framework for describing seemingly disparate phenomena.

The physics of the late nineteenth and twentieth centuries is a series of discoveries and unifications. Maxwell unified electricity and magnetism. Einstein developed the general theory of relativity that unified the principle of relativity and gravity. In the late 1940s, there was a culmination of two decades' efforts in the unification of electromagnetism and quantum mechanics. In the 1960s and 1970s, the theory of weak and electromagnetic interactions was also unified. Moreover, around the same period there was also a wider conceptual unification. Three of the four fundamental forces known were described by gauge theories. The fourth, gravity, is also based on a local invariance, albeit of a different type, and so far stands apart.[1] The combined theory, containing the quantum field theories of the electroweak and strong interactions together with the classical theory of gravity, formed the Standard Model of fundamental interactions. It is based on the gauge group $SU(3) \times SU(2) \times U(1)$. Its spin-1 gauge bosons mediate the strong and electroweak interactions. The matter particles are quarks and leptons of spin $\frac{1}{2}$ in three copies (known as generations and differing widely in mass), and a spin-0 particle, the Higgs boson, still experimentally elusive, that is responsible for the spontaneous breaking of the electroweak gauge symmetry.

---

[1] Today, we have some intriguing evidence that even gravity may be a strong-coupling facet of an extra underlying four-dimensional gauge theory.

The Standard Model has been experimentally tested and has survived thirty years of accelerator experiments.[2] This highly successful theory, however, is not satisfactory:

• A classical theory, namely, gravity, described by general relativity, must be added to the Standard Model in order to agree with experimental data. This theory is not renormalizable at the quantum level. In other words, new input is needed in order to understand its high-energy behavior. This has been a challenge to the physics community since the 1930s and (apart from string theory) very little has been learned on this subject since then.

• The three SM interactions are not completely unified. The gauge group is semisimple. Gravity seems even further from unification with the gauge theories. A related problem is that the Standard Model contains many parameters that look *a priori* arbitrary.

• The model is unstable as we increase the energy (hierarchy problem of mass scales) and the theory loses predictivity as one starts moving far from current accelerator energies and closer to the Planck scale. Gauge bosons are protected from destabilizing corrections because of gauge invariance. The fermions are equally protected due to chiral symmetries. The real culprit is the Higgs boson.

Several attempts have been made to improve on the problems above.

The first attempts focused on improving on unification. They gave rise to the grand unified theories (GUTs). All interactions were collected in a simple group SU(5) in the beginning, but also SO(10), E_6, and others. The fermions of a given generation were organized in the (larger) representations of the GUT group. There were successes in this endeavor, including the prediction of $\sin^2 \theta_W$ and the prediction of light right-handed neutrinos in some GUTs. However, there was a need for Higgs bosons to break the GUT symmetry to the SM group and the hierarchy problem took its toll by making it technically impossible to engineer a light electroweak Higgs.

The physics community realized that the focus must be on bypassing the hierarchy problem. A first idea attacked the problem at its root: it attempted to banish the Higgs boson as an elementary state and to replace it with extra fermionic degrees of freedom. It introduced a new gauge interaction (termed technicolor) which bounds these fermions strongly; one of the techni-hadrons should have the right properties to replace the elementary Higgs boson as responsible for the electroweak symmetry breaking. The negative side of this line of thought is that it relied on the nonperturbative physics of the technicolor interaction. Realistic model building turned out to be difficult and eventually this line of thought was mostly abandoned.

A competing idea relied on a new type of symmetry, supersymmetry, that connects bosons to fermions. This property turned out to be essential since it could force the bad-mannered spin-0 bosons to behave as well as their spin-$\frac{1}{2}$ partners. This works well, but supersymmetry stipulated that each SM fermion must have a spin-0 superpartner with equal mass. This being obviously false, supersymmetry must be spontaneously broken at

---

[2] With the exception of the neutrino sector that was suspected to be incomplete and is currently the source of interesting discoveries.

an energy scale not far away from today's accelerator energies. Further analysis indicated that the breaking of global supersymmetry produced superpartners whose masses were correlated with those of the already known particles, in conflict with experimental data.

To avoid such constraints global supersymmetry needed to be promoted to a local symmetry. As a supersymmetry transformation is in a sense the square root of a translation, this entailed that a theory of local supersymmetry must also incorporate gravity. This theory was first constructed in the late 1970s, and was further generalized to make model building possible. The flip side of this was that the inclusion of gravity opened the Pandora's box of nonrenormalizability of the theory. Hopes that (extended) supergravity might be renormalizable soon vanished.

In parallel with the developments above, a part of the community resurrected the old idea of Kaluza and Klein of unifying gravity with the other gauge interactions. If one starts from a higher-dimensional theory of gravity and compactifies the theory to four dimensions, one ends with four-dimensional gravity plus extra gauge interactions. Although gravity in higher dimensions is more singular in the UV, physicists hoped that at least at the classical level one would get a theory that is very close to the Standard Model. Although progress was made, a stumbling block turned out to be obtaining a four-dimensional chiral spectrum of fermions as in the SM.

Although none of the directions above provided a final and successful theory, the ingredients were very interesting ideas that many felt would form a part of the ultimate theory.

1.2 The Case for String Theory

String theory has been the leading candidate over the past two decades for a theory that consistently unifies all fundamental forces of nature, including gravity. It gained popularity because it provides a theory that is UV finite.[3]

The basic characteristic of the theory is that its elementary constituents are extended strings rather than pointlike particles as in quantum field theory. This makes the theory much more complicated than QFT, but at the same time it imparts some unique properties.

One of the key ingredients of string theory is that it provides a finite theory of quantum gravity, at least in perturbation theory. To appreciate the difficulties with the quantization of Einstein gravity, we look at a single-graviton exchange between two particles (Fig. 1.1a). Then, the amplitude is proportional to E^2/M_P^2, where E is the energy of the process and M_P is the Planck mass, $M_P \sim 10^{19}$ GeV. It is related to the Newton constant as

$$M_P^2 = \frac{1}{16\pi G_N}. \tag{1.2.1}$$

---

[3] Although there is no rigorous proof to all orders that the theory is UV finite, there are several all-orders arguments as well as rigorous results at low-loop order. In closed string theory, amplitudes must be carefully defined via analytic continuation, standard in S-matrix theory. When open strings are present, there are divergences. However, they are interpreted as IR divergences (due to the exchange of massless states) in the dual closed string channel. They are subtracted in the "Wilsonian" S-matrix elements.

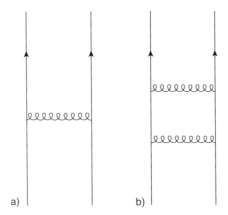

Figure 1.1 Gravitational interaction between two particles via graviton exchange.

Therefore, the gravitational interaction is irrelevant in the IR ($E \ll M_P$) but strongly relevant in the UV. In particular, this implies that the two-graviton exchange diagram (Fig. 1.1b)) is proportional to the dimensionless ratio

$$\frac{E^2}{M_P^4} \int_0^\Lambda d\tilde{E} \; \tilde{E} \sim \frac{\Lambda^2 E^2}{M_P^4}, \tag{1.2.2}$$

where E is the outgoing particle energy and $\tilde{E}$ is the internal particle energy. This is strongly UV divergent. It is known that Einstein gravity coupled to matter is nonrenormalizable in perturbation theory. Supersymmetry makes the UV divergence softer but the nonrenormalizability persists.

There are two ways out of this:

- There is a nontrivial UV fixed point that governs the high-energy behavior of quantum gravity. To date, no credible example of this possibility has been offered.

- There is new physics at $E \sim M_P$ (or even lower) and Einstein gravity is the IR limit of a more general theory, valid at and beyond the Planck scale. You could consider the analogous situation with the Fermi theory of weak interactions. There, a nonrenormalizable current-current interaction with similar problems occurred, but today we know that this is the IR limit of the standard weak interaction mediated by the $W^\pm$ and Z^0 gauge bosons. So far, there is no consistent field theory that can make sense at energies beyond M_P and contains gravity. Good reviews of the ultraviolet problems of Einstein gravity can be found in [1,2].

Strings provide a theory that induces new physics at the string scale M_s which in perturbation theory (the string coupling g_s being weak) is much lower than the Planck scale M_P. It is still true, however, that the string perturbation theory becomes uncontrollable when the energies approach the Planck scale.

There are two important reasons why closed string theory does not have UV divergences. One is the fact that string dynamics and interactions are inextricably linked to the geometry of two-dimensional surfaces. For example, for closed strings, by decomposing the string

Figure 1.2 Open and closed Riemann surfaces.

Feynman diagrams, it is obvious that there is essentially a universal three-point interaction in the theory. This interaction is dictated by the two-dimensional geometry of closed Riemann surfaces as is obvious from figure 1.2. The other related reason is the presence of an infinite tower of excitations with masses in multiples of the string scale. Their interactions are carefully tuned to become soft at distances larger than the string length ℓ_s but still longer than the Planck length ℓ_P.

For open strings the situation is subtler. There, UV divergences are present, but are interpreted as IR closed-string divergences in the dual closed-string channel. This UV-IR open-closed string duality is at the heart of many of the recent developments in the field.

Another key ingredient of string theory is that it unifies gravity with gauge interactions. It does this in several different ways. The simplest is via the traditional KK approach. Super-string theory typically is defined in ten dimensions. Standard four-dimensional vacua can be obtained via compactification on a six-dimensional compact manifold. However, gauge symmetry can also arise from D-branes that sometimes are part of the vacuum (as in orientifolds). There is even gauge symmetry coming from a nongeometrical part of the theory as happens in the heterotic string. The unified origin of gravity and gauge symmetry extends even further to other interactions. For example, the Yukawa interactions, crucial for giving mass to the SM particles, are also intimately related to the gauge interactions.

Unlike earlier Kaluza-Klein approaches to unification, string theory is capable of providing, upon appropriate compactifications, chiral matter in four dimensions. This happens via a subtle interplay between anomaly-related interactions and the process of compactification.

Another characteristic ingredient of string theory is that the presence of space-time fermions in the theory implies the appearance of space-time supersymmetry at least at high energies. Supersymmetry, consequently, is an important ingredient of the theory.

There are good reasons to believe that the theory is unique, although there are many possible vacua that could be stable. Recent understanding of nonperturbative dualities has strengthened the belief in the uniqueness of this string theory structure. It has also pointed out new corners in the overall theory, which many call M-theory, that are still uncharted.

Despite all this, string theory, after thirty-five years of research, has many important questions still unanswered. Physicists feel that the fundamental definition of the theory is not known. How it fits the real world in detail is also not known presently.

It may be that string theory will turn into an "intellectual classical black hole." It may also be that it is the correct description of physics at short distances. Time and experiment will show.

1.3 A Stringy Historical Perspective

In the 1960s, physicists tried to make sense of a large amount of experimental data relevant to the strong interaction. There were lots of particles (or "resonances") and the situation could best be described as chaotic. Some regularities were observed:

Almost linear Regge behavior. It was noticed that a relatively large number of resonances could be nicely put on (almost) straight lines by plotting their mass versus their spin

$$m^2 = \frac{J}{\alpha'} + \alpha_0, \tag{1.3.1}$$

with $\alpha' \sim 1 \text{ GeV}^{-2}$; this relation was checked up to $J = 11/2$.

s-t duality. If we consider a scattering amplitude of two hadrons $\to$ two hadrons $(1, 2 \to 3, 4)$, then it can be described by the Mandelstam invariants

$$s = -(p_1 + p_2)^2, \quad t = -(p_2 + p_3)^2, \quad u = -(p_1 + p_3)^2, \tag{1.3.2}$$

with $s + t + u = \sum_i m_i^2$. We are using a metric with signature $(-+++)$. Such an amplitude depends on the flavor quantum numbers of hadrons (for example SU(3)). Consider the flavor part, which is cyclically symmetric in flavor space. For the full amplitude to be symmetric, it must also be cyclically symmetric in the momenta p_i. This symmetry amounts to the interchange $t \leftrightarrow s$. Thus, the amplitude should satisfy $A(s, t) = A(t, s)$. Consider a t-channel contribution due to the exchange of a spin-J particle of mass M. Then, at high energy

$$A_J(s, t) \sim \frac{(-s)^J}{t - M^2}. \tag{1.3.3}$$

Thus, this partial amplitude increases with s and its behavior becomes worse for large values of J. If one sews amplitudes of this form together to make a loop amplitude, then there are uncontrollable UV divergences for $J > 1$. Any finite sum of amplitudes of the form (1.3.3) has this bad UV behavior. Moreover, such a finite sum has no s-channel poles. However, if one allows an infinite number of terms then it is conceivable that the UV behavior might be different.

A proposal for such a dual amplitude was made by Veneziano [3],

$$A(s, t) = \frac{\Gamma(-\alpha(s))\Gamma(-\alpha(t))}{\Gamma(-\alpha(s) - \alpha(t))}, \tag{1.3.4}$$

where Γ is the standard Euler Γ-function and

$$\alpha(s) = \alpha(0) + \alpha' s. \tag{1.3.5}$$

By using the standard properties of the Γ-function, it can be checked that the amplitude (1.3.4) has an infinite number of s, t-channel poles:

$$A(s, t) = -\sum_{n=0}^{\infty} \frac{(\alpha(s) + 1) \cdots (\alpha(s) + n)}{n!} \frac{1}{\alpha(t) - n}. \tag{1.3.6}$$

In this expansion, the $s \leftrightarrow t$ interchange symmetry of (1.3.4) is not manifest. The poles in (1.3.6) correspond to the exchange of an infinite number of particles of mass $M^2 = \frac{(n - \alpha(0))}{\alpha'}$ and high spins. It can also be checked that the high-energy behavior of the Veneziano amplitude is softer than any local quantum field theory amplitude, and the infinite number of poles is crucial for this.

It was subsequently realized by Nambu, Goto, Nielsen, and Susskind that such amplitudes came out of theories of relativistic strings. However, string theories had several shortcomings in explaining the dynamics of strong interactions.

- All of them seemed to predict a particle with negative mass squared, the tachyon.

- Several of them seemed to contain a massless spin-2 particle that was impossible to get rid of.

- All of them seemed to require a space-time dimension of 26 in order not to break Lorentz invariance at the quantum level.

- They contained only bosons.

At the same time, experimental data from SLAC showed that at even higher energies hadrons have a pointlike structure; this opened the way for quantum chromodynamics as the correct theory that describes strong interactions.

However, some work continued in the context of "dual models" and in the mid-1970s several interesting breakthroughs were made.

- It was understood by Neveu, Schwarz, and Ramond how to include space-time fermions in string theory.

- Gliozzi, Scherk, and Olive also understood how to get rid of the omnipresent tachyon. In the process, the constructed theory gained space-time supersymmetry.

- Scherk and Schwarz, and independently Yoneya, proposed that closed string theory, always having a massless spin-2 particle, naturally describes gravity and that the scale α' should be related to the Planck scale. Moreover, the theory can be defined in four dimensions using the Kaluza-Klein idea, namely, considering the extra dimensions to be compact and small.

However, the new big impetus for string theory came in 1984. After a general analysis of gauge and gravitational anomalies [4], it was realized that anomaly-free theories in higher dimensions are very restricted. Green and Schwarz showed in [5] that open superstrings in ten dimensions are anomaly-free if the gauge group is O(32). $E_8 \times E_8$ was also anomaly-free but could not appear in open string theory. In [6] it was shown that another supersymmetric string exists in ten dimensions, a hybrid of the superstring and the bosonic string, which can realize the $E_8 \times E_8$ or O(32) gauge symmetries.

Since the early 1980s, the field of string theory has been continuously developing. There was much heterotic model building in the late 1980s; the matrix model approach to two-dimensional string theory was developed in the early 1990s, followed by the study of stringy black holes. In the mid-1990s, nonperturbative dualities between different supersymmetric string theories were uncovered. This development gave rise to the hope that the theory is unique, and led to the name M-theory. D-branes were discovered and studied. They turned out to be crucial for the construction of controllable models for the identification of black-hole microstates and the microscopic explanation of the black-hole entropy. This, moreover, led to the formulation of AdS/CFT correspondence and its generalizations.

String theory is a continuously evolving subject and this book gives only a brief introduction to some of the best understood topics.

1.4 Conventions

Unless otherwise stated we use natural units in which $\hbar = c = 1$. The string length ℓ_s is kept explicitly throughout the book. It is related to the Regge slope α' and the string (mass) scale M_s by

$$\ell_s = \sqrt{\alpha'} = \frac{1}{M_s}.$$

In the literature, most of the time $\alpha' = 2$ in closed-string theory, $\alpha' = 1/2$ in open-string theory, and sometimes $\alpha' = 1$ in CFT.

The fundamental string tension is $T = \frac{1}{2\pi\ell_s^2}$. We denote by T_p the D_p-brane tension, by T_{M_2, M_5} the respective M$_2$- and M$_5$-brane tensions, and by $\tilde{T}_5$ the NS$_5$-brane tension.

We use X^μ for the space-time coordinates of the string and x^μ for their zero modes. By convention, the left-moving part of the string is the holomorphic part, with conformal dimensions $(\Delta, 0)$. The right-moving part is the antiholomorphic part with dimensions $(0, \bar{\Delta})$. The right-moving part is taken as the nonsupersymmetric side of the heterotic string.

$F_{L,R}$ is the world-sheet (left-moving or right-moving) fermion number. The operator that we use is $(-1)^{F_{L,R}}$. In the NS sector, it counts the number of fermion oscillators modulo 2. Its action is explained in sections 4.12 on page 71 and 7.7.1 on page 174. It should be distinguished from the "space-time fermion number" operators $\mathbf{F}_{L,R}$. $\mathbf{F}_L = 0$ in the left-moving NS sector and 1 in the left-moving Ramond sector. A similar definition holds for $\mathbf{F}_R$. Note that, in the heterotic string, $\mathbf{F}_L$ is indeed the space-time fermion number. In a type-II string the space-time fermion number is $\mathbf{F}_L + \mathbf{F}_R$ modulo 2.

We are using the "mostly plus" convention for the signature of the space-time metric. Our curvature conventions are such that the n-sphere S^n has positive scalar curvature. When there is no risk of confusion, we use for the volume element of the metric $\sqrt{-\det g} \leftrightarrow \sqrt{\det g} \leftrightarrow \sqrt{g}$.

Our conventions on the two-dimensional geometry are spelled out in appendix A, on page 503. Those on differential forms, the ϵ-density, the related E-tensor, and the Hodge dual are found in appendix B on page 505.

We use a unified notation for extended supersymmetry in diverse dimensions. Generically, n supersymmetries in d dimensions are denoted by $\mathcal{N} = n_d$. In two, six, and ten dimensions, because of the existence of Majorana-Weyl spinors, we may define extended supersymmetries with p left-handed and q right-handed MW supercharges. In this case, we use the notation $\mathcal{N} = (p, q)_d$. However, sometimes, even in such dimensions, if the chirality of the supersymmetry is not important for our purposes, we might still use the $\mathcal{N} = n_d$ notation.

For symplectic groups we use the notation Sp(2N) with $\mathrm{Sp}(2) \sim \mathrm{SU}(2)$. For the groups SO(2n), the subscript $\pm$ on the spinor indicates its eigenvalue ± 1 under the appropriate generalization of γ^5.

We also call the heterotic string based on the $\mathrm{Spin}(32)/\mathbb{Z}_2$ lattice "the O(32) heterotic string," for simplicity.

Bibliography

The guide to the associated literature presented in this book has been compiled with pedagogy as its main motivation. This book is intended as a textbook, and an appropriately chosen bibliography is a crucial complement. Review articles have been favored here but also original papers when they are deemed to have pedagogical value.

Most of the papers and reviews after 1991 have appeared in the electronic physics archives, are referred to as "arXiv:hep-th/*yymmnnn*," and are available from the central archive site http://arXiv.org/ and its mirrors worldwide.

There are several books and lecture notes on string theory. The first benchmark is that by Green, Schwarz, and Witten (GSW) [7]. It is a reference two-volume set. It summarizes the older literature on string theory, and describes in detail string compactifications up to the mid-1980s. The best and most detailed exposition of the Green-Schwarz approach to the superstring is presented here in detail. There is a balance between covariant and light-cone methods used in the quantization.

The second benchmark is Polchinski's two-volume set [8]. It focuses on the modern approach to string theory via covariant quantization and the use of CFT methods. It also contains a description of D-branes and their roles in nonperturbative string dualities.

There are several other books with different characteristics. Johnson's recent book [9] provides an in-depth look at D-branes and their effects in string theory, while at the same time providing an introduction to the basics. Ortin's book [10] provides a coherent and in-depth exposition of geometric aspects of the theory and its many interesting classical solutions. Szabo's book [11] is a short introduction to string theory (120 pages in all) which covers the very basics. Last but not least is the recent book of Zwiebach [12], which is written at a more introductory level and is addressed to advanced undergraduates. It discusses several interesting subjects in string theory at an accessible level.

There have been several other books and good reviews over the years which are limited by their scope or date of appearance. There is, however, some merit in consulting them since they may have other advantages, like the in-depth description of special subjects. These are [14–21]. We should also note the relatively short but up-to-date introduction to string theory by Danielsson in [21].

Marolf's resource letter [22] is an excellent source of various articles and reviews on string theory. It includes a wide spectrum of sources, from popular science to specialized reviews. The review article of Seiberg and Schwarz [23] provides a useful overview of the field, its achievements, and its goals.

2 | Classical String Theory

As in field theory, there are two approaches to discussing classical and quantum string theory. One is the first-quantized approach, which treats the dynamics of a single string. The dynamical variables are the space-time coordinates of the string. This is an on-shell approach. The other is the second-quantized or field theory approach. Here the dynamical variables are functionals of the string coordinates, or string fields. They allow an off-shell formulation. Unfortunately, although there is an elegant formulation of open string field theory, the closed string field theory approaches are complicated and difficult to use. In this book we will follow the first-quantized approach.

2.1 The Point Particle

Before discussing strings, it is useful to first consider the relativistic point particle. This will introduce the (known) basic concepts and will serve as a benchmark in our discussion of strings. We will use the first-quantized path integral language. Point particles, when traveling from one point in space-time to another, classically follow an extremal path. The natural action is proportional to the length of the world-line between some initial and final points:

$$S = \int \mathcal{L}\, d\tau = -m \int_{S_i}^{S_f} ds = -m \int_{\tau_0}^{\tau_1} d\tau \sqrt{-\eta_{\mu\nu}\dot{x}^\mu \dot{x}^\nu}, \tag{2.1.1}$$

where we consider a flat space-time, $\eta_{\mu\nu} = \mathrm{diag}(-1,+1,\ldots,+1)$. The coordinates x^μ parametrize the space in which the point particle is moving. It is called the *target space*. On the other hand, the coordinate τ parametrizes the path of the particle. It is known as a *world-line* coordinate.

From now on we will use the notation $\eta_{\mu\nu}\dot{x}^\mu \dot{x}^\nu \equiv \dot{x} \cdot \dot{x} \equiv \dot{x}^2$. The momentum conjugate to $x^\mu(\tau)$ is

$$p_\mu = \frac{\delta \mathcal{L}}{\delta \dot{x}^\mu} = \frac{m\dot{x}_\mu}{\sqrt{-\dot{x}^2}}, \tag{2.1.2}$$

and the Lagrange equations coming from varying the action (2.1.1) with respect to $x^\mu(\tau)$ read

$$\partial_\tau \left(\frac{m\dot{x}_\mu}{\sqrt{-\dot{x}^2}} \right) = 0. \tag{2.1.3}$$

Equation (2.1.2) gives the following mass-shell constraint:

$$p_\mu p^\mu + m^2 = 0. \tag{2.1.4}$$

The canonical Hamiltonian is given by

$$H_{\text{canonical}} = \frac{\partial \mathcal{L}}{\partial \dot{x}^\mu} \dot{x}^\mu - \mathcal{L} = 0. \tag{2.1.5}$$

Inserting (2.1.2) into (2.1.5), we see that $H_{\text{canonical}}$ vanishes identically. Thus, the constraint (2.1.4) completely governs the dynamics of the system. This is always true in theories with a (local) time reparametrization invariance.

We can add the constraint (2.1.4) to the Hamiltonian using a Lagrange multiplier N. The system will then be described by

$$H = \frac{N}{2m}(p^2 + m^2), \tag{2.1.6}$$

from which the equations of motion follow:

$$\dot{x}^\mu = \{x^\mu, H\} = \frac{N}{m} p^\mu = \frac{N\dot{x}^\mu}{\sqrt{-\dot{x}^2}}, \quad \dot{x}^2 = -N^2, \tag{2.1.7}$$

where $\{\ \}$ is the Poisson bracket. We are describing timelike trajectories for N real. The choice $N = 1$ corresponds to a choice of scale for the parameter τ, the proper time.

The square root in (2.1.1) is an undesirable feature. For the free particle it is not a problem, but, as we will see later, it will be a problem for the string. Moreover, the action we used above is ill defined for massless particles. Classically, there exists an alternative action which does not contain the square root and in addition allows the generalization to the massless case. Consider the following action:

$$S = \tfrac{1}{2} \int d\tau \, e(\tau) \left[e^{-2}(\tau) \dot{x} \cdot \dot{x} - m^2 \right]. \tag{2.1.8}$$

The auxiliary variable $e(\tau)$ can be viewed as an einbein on the world-line. The associated metric would be $g_{\tau\tau} = e^2$, and (2.1.8) could be rewritten as

$$S = \tfrac{1}{2} \int d\tau \sqrt{\det g_{\tau\tau}} \, (g^{\tau\tau} \partial_\tau x \cdot \partial_\tau x - m^2). \tag{2.1.9}$$

The action is invariant under reparametrizations of the world-line. An infinitesimal reparametrization is given by

$$\delta x^\mu(\tau) = x^\mu(\tau + \xi(\tau)) - x^\mu(\tau) = \xi(\tau)\dot{x}^\mu + \mathcal{O}(\xi^2), \quad \delta e = \xi \dot{e} + \dot{\xi} e + \mathcal{O}(\xi^2). \tag{2.1.10}$$

Varying e in (2.1.8) leads to

$$\delta S = -\tfrac{1}{2} \int d\tau \left(\frac{\dot{x}^2}{e^2(\tau)} + m^2 \right) \delta e(\tau). \tag{2.1.11}$$

Setting $\delta S = 0$ gives us the equation of motion for e:

$$e^{-2}\dot{x}^2 + m^2 = 0 \rightarrow e = \frac{1}{m}\sqrt{-\dot{x}^2}. \qquad (2.1.12)$$

Varying x gives

$$\delta S = \frac{1}{2}\int d\tau\, e(\tau)\left(e^{-2}(\tau)\, 2\dot{x}^\mu\right)\partial_\tau \delta x^\mu. \qquad (2.1.13)$$

After partial integration, we find the equation of motion

$$\partial_\tau(e^{-1}\dot{x}^\mu) = 0. \qquad (2.1.14)$$

Substituting (2.1.12) into (2.1.14), we find the same equations as before (cf. eq. (2.1.3)). If we substitute (2.1.12) directly into the action (2.1.8), we find (2.1.1), which establishes the classical equivalence of both actions.

We will now derive the propagator for the point particle. By definition, it is given by the functional integral

$$\langle x'|x\rangle = \hat{N}\int_{x(0)=x}^{x(1)=x'}\mathcal{D}e\mathcal{D}x^\mu\,\exp\left[-\frac{1}{2}i\int_0^1\left(\frac{1}{e}(\dot{x})^2 - e\,m^2\right)d\tau\right], \qquad (2.1.15)$$

where we have put $\tau_0 = 0$, $\tau_1 = 1$, and $\hat{N}$ is a normalization constant to be specified.

Under reparametrizations of the world-line, the einbein transforms as a covariant vector. To first order, this means

$$\delta e = \partial_\tau(\xi e). \qquad (2.1.16)$$

This is the local reparametrization invariance of the path. Since we are integrating over e, the presence of the invariance implies that (2.1.15) will give an infinite result. Therefore, we need to gauge-fix the reparametrization invariance (2.1.16). We can gauge-fix e to be constant. However, (2.1.16) now indicates that we cannot fix more. To see what this constant may be, notice that the length of the path of the particle is

$$L = \int_0^1 d\tau\sqrt{\det g_{\tau\tau}} = \int_0^1 d\tau\, e, \qquad (2.1.17)$$

so the best we can do is to fix $e = L$.

This is the simplest example of leftover (Teichmüller) parameters in the metric after gauge fixing. The e integration contains an integral over the constant mode (reparametrization invariant) as well as the rest of the modes (reparametrization parameters). The rest of the modes therefore makes up the "gauge volume." We must remove it via gauge fixing. To make the path integral converge, we also rotate to Euclidean time, $\tau \rightarrow -i\tau$. Thus, we arrive at the gauge-fixed path integral

$$\langle x|x'\rangle = \hat{N}\int_0^\infty dL\int_{x(0)=x}^{x(1)=x'}\mathcal{D}x^\mu\exp\left[-\frac{1}{2}\int_0^1\left(\frac{1}{L}\dot{x}^2 + Lm^2\right)d\tau\right]. \qquad (2.1.18)$$

We now expand around a classical path

$$x^\mu(\tau) = x^\mu + (x'^\mu - x^\mu)\tau + \delta x^\mu(\tau). \qquad (2.1.19)$$

The boundary conditions become $\delta x^\mu(0) = \delta x^\mu(1) = 0$. The measure for the fluctuations δx^μ is

$$\| \delta x \|^2 = \int_0^1 d\tau\, e(\delta x^\mu)^2 = L \int_0^1 d\tau\, (\delta x^\mu)^2, \tag{2.1.20}$$

so that

$$\mathcal{D}x^\mu \sim \prod_\tau \sqrt{L}\, d\delta x^\mu(\tau), \tag{2.1.21}$$

where the product runs over a discretization of the interval $[0,1]$. Then

$$\langle x|x'\rangle = \hat{N} \int_0^\infty dL \int \prod_\tau \sqrt{L}\, d\delta x^\mu(\tau)\, e^{-(x'-x)^2/2L - m^2 L/2} e^{-(1/2L)\int_0^1 d\tau\, (\delta \dot{x}^\mu)^2}. \tag{2.1.22}$$

The Gaussian integral involving $\delta \dot{x}^\mu$ can be evaluated directly:

$$\int \prod_\tau \sqrt{L}\, d\delta x^\mu(\tau) e^{-(1/2L)\int_0^1 (\delta \dot{x}^\mu)^2} \sim \left[\det\left(-\frac{1}{L^2}\partial_\tau^2\right) \right]^{-D/2}. \tag{2.1.23}$$

We must compute the determinant of the operator $-\partial_\tau^2/L^2$. The determinant is given as the product of all its eigenvalues. The associated eigenvalue problem is

$$-\frac{1}{L^2}\partial_\tau^2 \psi(\tau) = \lambda \psi(\tau) \tag{2.1.24}$$

with the boundary conditions $\psi(0) = \psi(1) = 0$. Note that there is no zero-mode problem here because of the boundary conditions. The solution is

$$\psi_n(\tau) = C_n \sin(n\pi\tau), \quad \lambda_n = \frac{n^2}{L^2}, \quad n = 1, 2, \ldots, \tag{2.1.25}$$

and thus

$$\det\left(-\frac{1}{L^2}\partial_\tau^2\right) = \prod_{n=1}^\infty \frac{n^2}{L^2}. \tag{2.1.26}$$

Obviously the determinant is infinite and we have to regularize it. We will use ζ-function regularization in which[1]

$$\prod_{n=1}^\infty L^{-2} = L^{-2\zeta(0)} = L, \quad \prod_{n=1}^\infty n^a = e^{-a\zeta'(0)} = (2\pi)^{a/2}. \tag{2.1.27}$$

Adjusting the normalization factor we finally obtain

$$\langle x|x'\rangle = \frac{1}{2(2\pi)^{D/2}} \int_0^\infty dL\, L^{-D/2} e^{-(x'-x)^2/2L - m^2 L/2}$$

$$= \frac{1}{(2\pi)^{D/2}} \left(\frac{|x-x'|}{m}\right)^{(2-D)/2} K_{(D-2)/2}(m|x-x'|). \tag{2.1.28}$$

---

[1] The Riemann ζ-function is defined as $\zeta(s) = \sum_{n=1}^\infty n^{-s}$.

This is the free propagator of a scalar particle in D dimensions, with K_ν the modified Bessel function. To obtain the more familiar expression, we have to pass to momentum space

$$|p\rangle = \int d^D x \, e^{ip \cdot x} |x\rangle, \tag{2.1.29}$$

$$
\begin{aligned}
\langle p|p'\rangle &= \int d^D x \, e^{-ip \cdot x} \int d^D x' \, e^{ip' \cdot x'} \langle x|x'\rangle \\
&= \frac{1}{2} \int d^D x' e^{i(p'-p) \cdot x'} \int_0^\infty dL \, e^{-(L/2)(p^2 + m^2)} = (2\pi)^D \delta(p - p') \frac{1}{p^2 + m^2},
\end{aligned} \tag{2.1.30}
$$

just as expected.

Here we should make one more comment. The momentum space amplitude $\langle p|p'\rangle$ can also be computed directly if we insert in the path integral $e^{ip \cdot x(0)}$ for the initial state and $e^{-ip' \cdot x(1)}$ for the final state. Thus, amplitudes are given by path-integral averages of the quantum-mechanical wave functions of free particles.

In exercise 2.6 on page 26 you are invited to generalize the present discussion to a charged particle moving in an electromagnetic field.

2.2 Relativistic Strings

We now use the ideas of the previous section to construct actions for strings. In the case of point particles, the action was proportional to the length of the world-line between some initial point and final point. For strings, it will be related to the surface area of the "world-sheet" swept by the string as it propagates through space-time. The Nambu-Goto action is defined as

$$S_{NG} = -T \int dA. \tag{2.2.1}$$

The constant factor T is the string tension and makes the action dimensionless; its dimensions must be [length]$^{-2}$ or [mass]^2. It is related to the conventional α' parameter (Regge slope) as

$$T = \frac{1}{2\pi\alpha'}. \tag{2.2.2}$$

We will also introduce for future convenience the string length ℓ_s and string scale M_s as

$$\ell_s = \sqrt{\alpha'} = \frac{1}{M_s}, \quad T = \frac{1}{2\pi\ell_s^2} = \frac{M_s^2}{2\pi}. \tag{2.2.3}$$

Suppose $\xi^\alpha, \alpha = 0, 1 \sim (\tau, \sigma)$, are coordinates on the world-sheet and $G_{\mu\nu}$ is the metric of the space-time in which the string propagates (target space). $X^\mu(\xi^\alpha)$ are the coordinates of the string describing its embedding in space-time. Then, $G_{\mu\nu}$ induces a metric on the world-sheet:

$$ds^2 = G_{\mu\nu}(X)dX^\mu dX^\nu = G_{\mu\nu} \frac{\partial X^\mu}{\partial \xi^\alpha} \frac{\partial X^\nu}{\partial \xi^\beta} d\xi^\alpha d\xi^\beta = \hat{G}_{\alpha\beta} d\xi^\alpha d\xi^\beta, \tag{2.2.4}$$

where the induced metric is

$$\hat{G}_{\alpha\beta} = G_{\mu\nu} \partial_\alpha X^\mu \partial_\beta X^\nu. \tag{2.2.5}$$

This metric can be used to calculate the surface area. If the space-time is flat Minkowski space, then $G_{\mu\nu} = \eta_{\mu\nu}$ and the Nambu-Goto action becomes

$$S_{NG} = -T \int \sqrt{-\det \hat{G}_{\alpha\beta}} \, d^2\xi = -T \int \sqrt{(\dot{X} \cdot X')^2 - (\dot{X}^2)(X'^2)} \, d^2\xi, \tag{2.2.6}$$

where $\dot{X}^\mu = \frac{\partial X^\mu}{\partial \tau}$ and $X'^\mu = \frac{\partial X^\mu}{\partial \sigma}$. The equations of motion are

$$\partial_\tau \left(\frac{\delta \mathcal{L}}{\delta \dot{X}^\mu} \right) + \partial_\sigma \left(\frac{\delta \mathcal{L}}{\delta X'^\mu} \right) = 0. \tag{2.2.7}$$

The equations of motion must be supplemented by boundary conditions. They depend on the type of string we study. In the case of closed strings, the world-sheet is a tube. If we let σ run from 0 to $\bar{\sigma} = 2\pi$, we must impose the periodicity condition

$$X^\mu(\sigma + \bar{\sigma}) = X^\mu(\sigma). \tag{2.2.8}$$

For open strings, the world-sheet is a strip. In this case we put by convention $\bar{\sigma} = \pi$. Two kinds of boundary conditions are frequently used[2]:

Neumann: $\left. \dfrac{\delta \mathcal{L}}{\delta X'^\mu} \right|_{\sigma = 0 \text{ or } \bar{\sigma}} = 0;$ (2.2.9)

Dirichlet: $\left. \dfrac{\delta \mathcal{L}}{\delta \dot{X}^\mu} \right|_{\sigma = 0 \text{ or } \bar{\sigma}} = 0.$ (2.2.10)

As we shall see at the end of this section, Neumann conditions imply that no momentum flows off the ends of the string. It is appropriate for free string end points. The Dirichlet condition implies that the end points of the string are fixed in space-time.

The momentum conjugate to X^μ is

$$\Pi^\mu = \frac{\delta \mathcal{L}}{\delta \dot{X}^\mu} = -T \frac{(\dot{X} \cdot X')X'^\mu - (X')^2 \dot{X}^\mu}{[(X' \cdot \dot{X})^2 - (\dot{X})^2(X')^2]^{1/2}}. \tag{2.2.11}$$

The matrix $\frac{\delta^2 \mathcal{L}}{\delta \dot{X}^\mu \delta \dot{X}^\nu}$ has two zero eigenvalues, with eigenvectors $\dot{X}^\mu$ and X'^μ. This signals the occurrence of two constraints that follow directly from the definition of the conjugate momenta. They are

$$\Pi \cdot X' = 0, \quad \Pi^2 + T^2 X'^2 = 0. \tag{2.2.12}$$

The canonical Hamiltonian

$$H = \int_0^{\bar{\sigma}} d\sigma \, (\dot{X} \cdot \Pi - \mathcal{L}) \tag{2.2.13}$$

vanishes identically, just as in the case of the point particle. Again, the dynamics is governed solely by the constraints.

The square root in the Nambu-Goto action makes the treatment of the quantum theory quite complicated. We can simplify the action by introducing an intrinsic fluctuating

---

[2] One could also impose an arbitrary linear combination of the two boundary conditions. This is the subject of exercise 2.14 on page 27.

metric on the world-sheet. In this way we obtain the Polyakov[3] action for strings moving in flat space-time

$$S_P = -\frac{T}{2} \int d^2\xi \sqrt{-\det g} \; g^{\alpha\beta} \partial_\alpha X^\mu \partial_\beta X^\nu \; \eta_{\mu\nu}. \tag{2.2.14}$$

As is well known from field theory, varying the action with respect to the metric yields the stress-tensor[4] on the world-sheet:

$$T_{\alpha\beta} \equiv -\frac{4\pi}{\sqrt{-\det g}} \frac{\delta S_P}{\delta g^{\alpha\beta}} = -\frac{1}{\ell_s^2} \left[\partial_\alpha X \cdot \partial_\beta X - \tfrac{1}{2} g_{\alpha\beta} g^{\gamma\delta} \partial_\gamma X \cdot \partial_\delta X \right]. \tag{2.2.15}$$

Setting this variation to zero and solving for $g_{\alpha\beta}$, we obtain

$$g_{\alpha\beta} = \lambda \partial_\alpha X \cdot \partial_\beta X. \tag{2.2.16}$$

where λ is an arbitrary function. This ambiguity reflects the invariance of (2.2.14) under Weyl rescalings.

The world-sheet metric $g_{\alpha\beta}$ is classically proportional to the induced metric. If we substitute this back into the action, we find the Nambu-Goto action. The two actions are therefore equivalent, at least classically. Whether this is also true quantum-mechanically is not clear in general. However, they can be shown to be equivalent in the critical dimension.

From now on, we will follow the Polyakov approach to the quantization of string theory. By varying (2.2.14) with respect to X^μ, we obtain the equations of motion:

$$\frac{1}{\sqrt{-\det g}} \partial_\alpha (\sqrt{-\det g} \, g^{\alpha\beta} \partial_\beta X^\mu) = 0. \tag{2.2.17}$$

The world-sheet action in the Polyakov approach consists of D two-dimensional scalar fields X^μ coupled to the dynamical two-dimensional metric. We are therefore considering a theory of two-dimensional (quantum) gravity coupled to matter. One could ask whether there are other terms that can be added to (2.2.14). It turns out that there are only two terms with at most two derivatives: the cosmological term

$$\lambda_1 \int \sqrt{-\det g} \tag{2.2.18}$$

and the Gauss-Bonnet term

$$\lambda_2 \int \sqrt{-\det g} \, R^{(2)}, \tag{2.2.19}$$

where $R^{(2)}$ is the two-dimensional scalar curvature associated with $g_{\alpha\beta}$. This gives the Euler number of the world-sheet, which is a topological invariant. Therefore this term cannot influence the local dynamics of the string, but it will give factors that weight various topologies differently.

It is not difficult to prove that (2.2.18) has to be zero classically. In fact the classical equations of motion for $\lambda_1 \neq 0$ imply that $g_{\alpha\beta} = 0$, which gives trivial dynamics. We will not consider it further. For the open string, there are other possible terms, which are defined on the boundary of the world-sheet.

---

[3] This action, in terms of a dynamical world-sheet metric, was first written down by Brink, di Vecchia, and Howe [24] and Deser and Zumino [25], for both the bosonic string and the superstring. It is colloquially known as the Polyakov action, after Polyakov's study of quantum effects, including the conformal anomaly [34].

[4] Note that our definition in two dimensions is different by a factor of 4π from the standard definition.

We now discuss the symmetries of the Polyakov action in a Minkowski target-space background:

- Poincaré invariance:

$$\delta X^\mu = \omega^\mu{}_\nu X^\nu + \alpha^\mu, \quad \delta g_{\alpha\beta} = 0, \tag{2.2.20}$$

where $\omega_{\mu\nu} = -\omega_{\nu\mu}$;

- local two-dimensional reparametrization invariance:

$$\delta g_{\alpha\beta} = \xi^\gamma \partial_\gamma g_{\alpha\beta} + \partial_\alpha \xi^\gamma g_{\beta\gamma} + \partial_\beta \xi^\gamma g_{\alpha\gamma} = \nabla_\alpha \xi_\beta + \nabla_\beta \xi_\alpha,$$
$$\delta X^\mu = \xi^\alpha \partial_\alpha X^\mu,$$
$$\delta(\sqrt{-\det g}) = \partial_\alpha(\xi^\alpha \sqrt{-\det g}); \tag{2.2.21}$$

- Weyl invariance:

$$\delta X^\mu = 0, \quad \delta g_{\alpha\beta} = 2\Lambda g_{\alpha\beta}. \tag{2.2.22}$$

Weyl invariance implies that the stress-tensor is traceless. This is in fact true in general. Consider an action $S(g_{\alpha\beta}, \phi^i)$ in arbitrary space-time dimensions. We assume that it is invariant under the scale transformations

$$\delta g_{\alpha\beta} = 2\Lambda(x)g_{\alpha\beta}, \quad \delta\phi^i = d_i \Lambda(x)\phi^i, \tag{2.2.23}$$

where d_i is the conformal weight of the field ϕ_i.

The variation of the action under infinitesimal conformal transformations is

$$0 = \delta S = \int d^2\xi \left(2\frac{\delta S}{\delta g^{\alpha\beta}} g^{\alpha\beta} + \sum_i d_i \frac{\delta S}{\delta \phi_i} \phi_i \right) \Lambda. \tag{2.2.24}$$

Using the equations of motion for the fields ϕ_i, i.e., $\frac{\delta S}{\delta \phi_i} = 0$, we find

$$T^\alpha_\alpha \sim \frac{\delta S}{\delta g^{\alpha\beta}} g^{\alpha\beta} = 0, \tag{2.2.25}$$

which follows without the use of the equations of motion, if and only if $d_i = 0$. This is the case for the bosonic string, described by the Polyakov action, but not for fermionic extensions.

Just as we could gauge-fix $e(\tau)$ for the point particle using reparametrization invariance, we can reduce $g_{\alpha\beta}$ to $\eta_{\alpha\beta} = \text{diag}(-1, +1)$. This is called the conformal gauge. First, we choose a parametrization that makes the metric conformally flat, i.e.,

$$g_{\alpha\beta} = e^{2\Lambda(\xi)} \eta_{\alpha\beta}. \tag{2.2.26}$$

It can be proven that in two dimensions, this is always possible for world-sheets with trivial topology. We will discuss the subtle issues that appear for nontrivial topologies later on.

Using the Weyl symmetry, we can further reduce the metric to $\eta_{\alpha\beta}$. We also work with "light-cone coordinates"

$$\xi^+ = \tau + \sigma, \quad \xi^- = \tau - \sigma. \tag{2.2.27}$$

The metric becomes

$$ds^2 = -d\xi^+ d\xi^-. \tag{2.2.28}$$

The components of the metric are

$$g_{++} = g_{--} = 0, \quad g_{+-} = g_{-+} = -\frac{1}{2} \tag{2.2.29}$$

and

$$\partial_\pm = \frac{1}{2}(\partial_\tau \pm \partial_\sigma). \tag{2.2.30}$$

The Polyakov action in conformal gauge is

$$S_P = 2T \int d^2\xi \, \partial_+ X^\mu \partial_- X^\nu \eta_{\mu\nu}. \tag{2.2.31}$$

By going to conformal gauge, we have not completely fixed all reparametrizations. In particular, the reparametrizations

$$\xi^+ \longrightarrow f(\xi^+), \quad \xi^- \longrightarrow g(\xi^-) \tag{2.2.32}$$

rescale the metric (2.2.28) by a factor $\partial_+ f \partial_- g$, so they can be compensated by the transformation of $d^2\xi$.

Notice that in two dimensions we have exactly enough symmetry to completely fix the metric. A metric on a d-dimensional world-sheet has $d(d+1)/2$ independent components. Using reparametrizations, d of them can be fixed. Weyl invariance removes one more component. The number of remaining components is

$$\frac{d(d+1)}{2} - d - 1. \tag{2.2.33}$$

This is zero in the case $d = 2$ (strings). It is equivalent to the fact that two-dimensional gravity has no propagating degrees of freedom. An analogous treatment of higher-dimensional extended objects is much more difficult.[5]

We will derive the equations of motion from the Polyakov action in conformal gauge (eq. (2.2.31)). By varying X^μ, we obtain (after partial integration)

$$\delta S = T \int d^2\xi (\delta X^\mu \partial_+ \partial_- X_\mu) - T \int_{\tau_0}^{\tau_1} d\tau \, X'_\mu \delta X^\mu. \tag{2.2.34}$$

Using periodic boundary conditions for the closed string and Neumann boundary conditions

$$X'^\mu \big|_{\sigma=0,\bar\sigma} = 0 \tag{2.2.35}$$

for the open string, we find the equations of motion

$$\partial_+ \partial_- X^\mu = 0. \tag{2.2.36}$$

Even after gauge fixing, the equations of motion for the metric have to be imposed. From (2.2.15) they are

$$T_{\alpha\beta} = 0, \tag{2.2.37}$$

or

$$T_{10} = T_{01} = \tfrac{1}{2}\dot{X} \cdot X' = 0, \quad T_{00} = T_{11} = \frac{1}{4}(\dot{X}^2 + X'^2) = 0, \tag{2.2.38}$$

---

[5] Three-dimensional gravity has also no propagating degrees of freedom. However, the membrane theory is an interacting theory even in the light-cone gauge, as described in section 14.1.1 on page 471.

which can also be written as

$$(\dot{X} \pm X')^2 = 0. \tag{2.2.39}$$

These are known as the Virasoro constraints. They are the analog of the Gauss law in the string case. They are also the generalizations of the Hamiltonian constraint (2.1.5) of the point-particle case.

In light-cone coordinates, the components of the stress-tensor are

$$T_{++} = \tfrac{1}{2}\partial_+ X \cdot \partial_+ X, \quad T_{--} = \tfrac{1}{2}\partial_- X \cdot \partial_- X, \quad T_{+-} = T_{-+} = 0. \tag{2.2.40}$$

This last expression is equivalent to $T^\alpha_\alpha = 0$ and it is trivially satisfied. Energy-momentum conservation $\nabla^\alpha T_{\alpha\beta} = 0$ becomes

$$\partial_- T_{++} + \partial_+ T_{-+} = \partial_+ T_{--} + \partial_- T_{+-} = 0. \tag{2.2.41}$$

Using (2.2.40), this states that

$$\partial_- T_{++} = \partial_+ T_{--} = 0, \tag{2.2.42}$$

which leads for closed strings to an infinite number of conserved charges

$$Q_f = \int_0^{\bar{\sigma}} f(\xi^+) T_{++}(\xi^+)\, d\sigma, \tag{2.2.43}$$

corresponding to the symmetries (2.2.32). Similar remarks apply to T_{--}. To convince ourselves that Q_f is indeed conserved, we need to calculate

$$0 = \int d\sigma\, \partial_-(f(\xi^+)T_{++}) = \partial_\tau Q_f + f(\xi^+)T_{++}\big|_0^{\bar{\sigma}}. \tag{2.2.44}$$

For closed strings, the boundary term is not there since the closed string world-sheet has no boundaries. In the case of open strings, a linear combination of T_{++} and T_{--} is conserved.

Of course, there are other conserved charges in the theory, namely, those associated with Poincaré invariance:

$$P^\alpha_\mu = -T\sqrt{-\det g}\, g^{\alpha\beta} \partial_\beta X_\mu, \tag{2.2.45}$$

$$J^\alpha_{\mu\nu} = -T\sqrt{-\det g}\, g^{\alpha\beta}(X_\mu \partial_\beta X_\nu - X_\nu \partial_\beta X_\mu). \tag{2.2.46}$$

We have $\partial_\alpha P^\alpha_\mu = 0 = \partial_\alpha J^\alpha_{\mu\nu}$ because of the equation of motion for X^μ. The associated charges are

$$P_\mu = \int_0^{\bar{\sigma}} d\sigma\, P^\tau_\mu, \quad J_{\mu\nu} = \int_0^{\bar{\sigma}} d\sigma\, J^\tau_{\mu\nu}. \tag{2.2.47}$$

These are conserved on shell, e.g.,

$$\frac{dP_\mu}{d\tau} = T\int_0^{\bar{\sigma}} d\sigma\, \partial_\tau^2 X_\mu = T\int_0^{\bar{\sigma}} d\sigma\, \partial_\sigma^2 X_\mu = T(\partial_\sigma X_\mu(\sigma = \bar{\sigma}) - \partial_\sigma X_\mu(\sigma = 0)). \tag{2.2.48}$$

In the second equality we used the equation of motion for X, (2.2.36). This expression automatically vanishes for the closed string. For open strings, it vanishes only for Neumann boundary conditions. We have then shown that these boundary conditions imply that there is no momentum flow off the ends of the string. The nonconservation for Dirichlet boundary conditions will be interpreted in chapter 8. Similar remarks apply to angular momentum.

2.3 Oscillator Expansions

We will now solve the equations of motion for the bosonic string,

$$\partial_+\partial_- X^\mu = 0, \tag{2.3.1}$$

and then impose the boundary conditions.

The most general solution to equation (2.3.1) can be written as

$$X^\mu(\tau,\sigma) = X_L^\mu(\tau+\sigma) + X_R^\mu(\tau-\sigma), \tag{2.3.2}$$

with left-moving and right-moving components

$$X_L^\mu(\tau+\sigma) = \frac{x^\mu}{2} + \frac{\ell_s^2\,p^\mu}{2}(\tau+\sigma) + \frac{i\ell_s}{\sqrt{2}}\sum_{k\neq 0}\frac{\alpha_k^\mu}{k}e^{-ik(\tau+\sigma)},$$

$$X_R^\mu(\tau-\sigma) = \frac{x^\mu}{2} + \frac{\ell_s^2\,\bar{p}^\mu}{2}(\tau-\sigma) + \frac{i\ell_s}{\sqrt{2}}\sum_{k\neq 0}\frac{\bar{\alpha}_k^\mu}{k}e^{-ik(\tau-\sigma)}. \tag{2.3.3}$$

The α_k^μ and $\bar{\alpha}_k^\mu$ are arbitrary Fourier modes. The function $X^\mu(\tau,\sigma)$ must be real, so we know that x^μ, p^μ, and $\bar{p}^\mu$ must also be real. We may also derive the following reality condition for the α's:

$$(\alpha_k^\mu)^* = \alpha_{-k}^\mu \qquad \text{and} \qquad (\bar{\alpha}_k^\mu)^* = \bar{\alpha}_{-k}^\mu. \tag{2.3.4}$$

We may now impose boundary conditions. Since these differ in the closed and open string cases we will discuss them below separately. We will first consider the case of the closed string.

2.3.1 Closed strings

The propagation of a closed string sweeps out a cylinder in space-time (figure 2.1). The closed string noncompact coordinates satisfy the periodicity condition

$$X^\mu(\tau,\sigma+2\pi) = X^\mu(\tau,\sigma). \tag{2.3.5}$$

If we impose it on the general solution (2.3.3), we obtain that the wave number k must be an integer.

Thus, for closed strings

$$X_L^\mu(\tau+\sigma) = \frac{x^\mu}{2} + \frac{\ell_s^2\,p^\mu}{2}(\tau+\sigma) + \frac{i\ell_s}{\sqrt{2}}\sum_{n\in\mathbb{Z}-\{0\}}\frac{\alpha_n^\mu}{n}e^{-in(\tau+\sigma)},$$

$$X_R^\mu(\tau-\sigma) = \frac{x^\mu}{2} + \frac{\ell_s^2\,\bar{p}^\mu}{2}(\tau-\sigma) + \frac{i\ell_s}{\sqrt{2}}\sum_{n\in\mathbb{Z}-\{0\}}\frac{\bar{\alpha}_n^\mu}{n}e^{-in(\tau-\sigma)}. \tag{2.3.6}$$

If we define

$$\alpha_0^\mu = \frac{\ell_s}{\sqrt{2}}\,p^\mu, \quad \bar{\alpha}_0^\mu = \frac{\ell_s}{\sqrt{2}}\,\bar{p}^\mu, \tag{2.3.7}$$

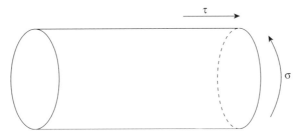

Figure 2.1 The parametrization of the cylinder.

we can write

$$\partial_+ X_L^\mu = \frac{\ell_s}{\sqrt{2}} \sum_{n \in \mathbb{Z}} \alpha_n^\mu e^{-in(\tau+\sigma)}, \tag{2.3.8}$$

$$\partial_- X_R^\mu = \frac{\ell_s}{\sqrt{2}} \sum_{n \in \mathbb{Z}} \bar{\alpha}_n^\mu e^{-in(\tau-\sigma)}. \tag{2.3.9}$$

The periodicity condition also imposes $p^\mu = \bar{p}^\mu$. We may now calculate the center-of-mass position of the string:

$$X_{CM}^\mu \equiv \frac{1}{2\pi} \int_0^{2\pi} d\sigma \, X^\mu(\tau, \sigma) = x^\mu + \ell_s^2 \, p^\mu \tau. \tag{2.3.10}$$

Thus, x^μ is the center-of-mass position at $\tau = 0$ and is moving as a free particle. In the same way we can calculate the center-of-mass momentum, or just the momentum of the string. From (2.2.47) and (2.2.3) we obtain

$$p_{CM}^\mu = T \int_0^{2\pi} d\sigma \dot{X}^\mu = \frac{1}{\sqrt{2}(2\pi \ell_s)} \int_0^{2\pi} d\sigma \sum_{n \in \mathbb{Z}} (\alpha_n^\mu e^{-in(\tau+\sigma)} + \bar{\alpha}_n^\mu e^{-in(\tau-\sigma)})$$

$$= \frac{1}{\sqrt{2}\ell_s} (\alpha_0^\mu + \bar{\alpha}_0^\mu) = p^\mu. \tag{2.3.11}$$

We observe that the variables that describe the classical motion of the string are the center-of-mass position x^μ and momentum p^μ plus an infinite collection of "oscillator" variables α_n^μ and $\bar{\alpha}_n^\mu$. This reflects the fact that the string can move as a whole, but it can also vibrate in various modes. The oscillator variables represent the vibrational degrees of freedom.

A similar calculation can be done for the angular momentum of the string:

$$J^{\mu\nu} = T \int_0^{2\pi} d\sigma \, (X^\mu \dot{X}^\nu - X^\nu \dot{X}^\mu) = l^{\mu\nu} + E^{\mu\nu} + \bar{E}^{\mu\nu}, \tag{2.3.12}$$

where we have separated the contributions to the angular momentum coming from the center-of-mass motion and the vibrations,

$$l^{\mu\nu} = x^\mu p^\nu - x^\nu p^\mu, \tag{2.3.13}$$

$$E^{\mu\nu} = -i \sum_{n=1}^\infty \frac{1}{n} (\alpha_{-n}^\mu \alpha_n^\nu - \alpha_{-n}^\nu \alpha_n^\mu), \tag{2.3.14}$$

$$\bar{E}^{\mu\nu} = -i \sum_{n=1}^\infty \frac{1}{n} (\bar{\alpha}_{-n}^\mu \bar{\alpha}_n^\nu - \bar{\alpha}_{-n}^\nu \bar{\alpha}_n^\mu). \tag{2.3.15}$$

If a target-space coordinate is compact, then there is more structure. We will assume that X parametrizes a circle of radius R. Then, the periodicity condition (2.3.5) is replaced by

$$X(\tau, \sigma + 2\pi) = X(\tau, \sigma) + 2\pi n R, \tag{2.3.16}$$

since a translation by an integer multiple of the length of the circle in target space is immaterial. The term $(p - \bar{p})\sigma$ in $X(\sigma, \tau)$ is known as the winding contribution. The reason is that as we go once around a closed string, $X(\sigma + 2\pi) = X(\sigma) + \pi \ell_s^2 (p - \bar{p})$, the string coordinate winds (extends) by $2\pi (p - \bar{p})$. This has a literal meaning when the coordinate is compact. In the case of open strings we will see the interpretation further on. Going through the same procedure as above we find that

$$p - \bar{p} = \frac{2}{\ell_s^2} n R, \quad n \in \mathbb{Z}. \tag{2.3.17}$$

On the other hand, the total momentum evaluated from (2.3.11), is $(p + \bar{p})/2$ and must be quantized as usual,

$$\frac{p + \bar{p}}{2} = \frac{m}{R}, \quad m \in \mathbb{Z}. \tag{2.3.18}$$

Thus, from (2.3.7)

$$\alpha_0^\mu = \frac{1}{\sqrt{2}} \left(m \frac{\ell_s}{R} + n \frac{R}{\ell_s} \right), \quad \bar{\alpha}_0^\mu = \frac{1}{\sqrt{2}} \left(m \frac{\ell_s}{R} - n \frac{R}{\ell_s} \right). \tag{2.3.19}$$

2.3.2 Open strings

We will now derive the oscillator expansion (2.3.3) in the case of the open string. As mentioned earlier, there are two basic boundary conditions we could impose on an end point: free end points correspond to the Neumann boundary conditions (2.2.9) while fixed end points correspond to Dirichlet boundary conditions (2.2.10).

The first simple case corresponds to open strings with free end points. In this case all coordinates have Neumann-Neumann (NN) boundary conditions:

$$X'^\mu(\tau, \sigma)|_{\sigma = 0, \pi} = 0.$$

If we substitute the solutions of the wave equation (2.2.36) we obtain the following condition:

$$X'^\mu|_{\sigma = 0} = \frac{\ell_s^2}{2} (p^\mu - \bar{p}^\mu) + \frac{\ell_s}{\sqrt{2}} \sum_{k \neq 0} e^{-ik\tau} (\alpha_k^\mu - \bar{\alpha}_k^\mu) = 0, \tag{2.3.20}$$

from which we can draw the following conclusion:

$$p^\mu = \bar{p}^\mu \quad \text{and} \quad \alpha_k^\mu = \bar{\alpha}_k^\mu. \tag{2.3.21}$$

We observe that the left-movers and right-movers get mixed by the boundary condition. This is known from the dynamics of everyday strings, since a left-moving wave, reflects back into a right-moving wave upon hitting the free end point. The boundary condition at the other end, $\sigma = \pi$, implies that k is an integer. *We redefine also p^μ by multiplying by* *a factor of 2* in (2.3.6) so that, as shown below, it represents the total momentum of the center of mass of the string. Therefore, the solution becomes:

$$\text{NN: } X^\mu(\tau,\sigma) = x^\mu + 2\ell_s^2\, p^\mu\tau + i\sqrt{2}\ell_s \sum_{n\in\mathbb{Z}-\{0\}} \frac{\alpha_n^\mu}{n}e^{-in\tau}\cos(n\sigma). \tag{2.3.22}$$

If we define $\alpha_0^\mu = \sqrt{2}\ell_s p^\mu$, we may write

$$\partial_\pm X^\mu = \sqrt{2}\ell_s \sum_{n\in\mathbb{Z}} \alpha_n^\mu e^{-in(\tau\pm\sigma)}. \tag{2.3.23}$$

We should also mention here that the condition $p^\mu = \bar{p}^\mu$ implies that no winding is allowed. This is in accordance with the common intuition that an open string with free ends cannot wind a circle in a stable fashion.

We may now calculate the center-of-mass position of the string:

$$X_{CM}^\mu \equiv \frac{1}{\pi}\int_0^\pi d\sigma\, X^\mu(\tau,\sigma) = x^\mu + 2\ell_s^2\, p^\mu\tau, \tag{2.3.24}$$

x^μ is again the center-of-mass position at $\tau = 0$ and is moving as a free particle. The center-of-mass momentum, or just the momentum is given in (2.2.47). We obtain

$$p_{CM}^\mu = T\int_0^\pi d\sigma\, \dot{X}^\mu = \frac{1}{\sqrt{2\pi}\ell_s}\int_0^\pi d\sigma \sum_{n\in\mathbb{Z}} \alpha_n^\mu e^{-in\tau}\cos n\sigma$$

$$= \frac{1}{\sqrt{2}\ell_s}\alpha_0^\mu = p^\mu, \tag{2.3.25}$$

which justifies our earlier rescaling of p^μ in the open case.

In some cases, namely, when we later describe D_p-branes with $p < 9$, some coordinates may have Dirichlet conditions on both end points (DD).

Imposing the Dirichlet boundary condition

$$\dot{X}^I(\tau,\sigma)|_{\sigma=0} = 0$$

at one end point we obtain

$$\dot{X}^I|_{\sigma=0} = \frac{\ell_s^2}{2}(p^I + \bar{p}^I) + \frac{\ell_s}{\sqrt{2}}\sum_{k\neq 0} e^{-ik\tau}(\bar{\alpha}_k^I + \alpha_k^I) = 0, \tag{2.3.26}$$

and therefore

$$p^I = -\bar{p}^I \quad \text{and} \quad \alpha_k^I = -\bar{\alpha}_k^I. \tag{2.3.27}$$

Only winding is allowed, but no momentum. Consequently, in the noncompact directions where $p = \bar{p}$, no momentum is allowed at all for DD boundary conditions.

Imposing the Dirichlet condition at the other end $\sigma = \pi$ implies that k is an integer. Consequently we may write

$$\text{DD: } X^I(\tau,\sigma) = x^I + w^I\sigma - \sqrt{2}\ell_s \sum_{n\in\mathbb{Z}} \frac{\alpha_n^\mu}{n}e^{-in\tau}\sin(n\sigma). \tag{2.3.28}$$

We will investigate the interpretation of the winding term w^I here. Note that from the solution (2.3.28)

$$X^I(\tau,0) = x^I, \quad X^I(\tau,\pi) \equiv y^I = x^I + \pi\, w^I. \tag{2.3.29}$$

Thus, the two end points are fixed at x^I, y^I with $y^I - x^I = \pi w^I$. Consequently πw^I is the distance between the two fixed end points.

An open string in a given direction may have an end point fixed (say at $\sigma = 0$) and the other free (at $\sigma = \pi$). This is the DN case.

The Dirichlet boundary condition may be imposed as above with the result that

$$p^I = -\bar{p}^I \qquad \text{and} \qquad \alpha_k^I = -\bar{\alpha}_k^I.$$

From

$$X'^I|_{\sigma=\pi} = \frac{\ell_s^2}{2}(p^I - \bar{p}^I) + \frac{\ell_s}{\sqrt{2}} \sum_{k \neq 0} e^{-ik(\tau - \pi)}((-1)^{2k}\alpha_k^I - \bar{\alpha}_k^I) = 0, \tag{2.3.30}$$

the Neumann condition at $\sigma = \pi$ now implies that $p^I = 0$ and that k must be half integer. Therefore,

$$\text{DN:} \quad X^I(\tau, \sigma) = x^I - \sqrt{2}\ell_s \sum_{k \in \mathbb{Z}+1/2} \frac{\alpha_k^I}{k} e^{-ik\tau} \sin(k\sigma). \tag{2.3.31}$$

2.3.3 The Virasoro constraints

In the Hamiltonian picture, we have equal-τ Poisson brackets (PB) for the dynamical variables, the X^μ fields, and their conjugate momenta:

$$\{X^\mu(\sigma, \tau), \dot{X}^\nu(\sigma', \tau)\}_{PB} = \frac{1}{T}\delta(\sigma - \sigma')\eta^{\mu\nu}. \tag{2.3.32}$$

The other brackets $\{X, X\}$ and $\{\dot{X}, \dot{X}\}$ vanish. We can derive from (2.3.32) the PB for the oscillators and center-of-mass position and momentum:

$$\{\alpha_m^\mu, \alpha_n^\nu\} = \{\bar{\alpha}_m^\mu, \bar{\alpha}_n^\nu\} = -im\delta_{m+n,0}\,\eta^{\mu\nu},$$
$$\{\bar{\alpha}_m^\mu, \alpha_n^\nu\} = 0, \quad \{x^\mu, p^\nu\} = \eta^{\mu\nu}. \tag{2.3.33}$$

For the open-string case, the $\bar{\alpha}$'s are absent.

The Hamiltonian

$$H = \int d\sigma(\dot{X}\Pi - \mathcal{L}) = \frac{T}{2}\int d\sigma(\dot{X}^2 + X'^2) \tag{2.3.34}$$

can also be expressed in terms of oscillators. In the case of closed strings it is given by

$$H_{\text{closed}} = \tfrac{1}{2}\sum_{n \in \mathbb{Z}}(\alpha_{-n} \cdot \alpha_n + \bar{\alpha}_{-n} \cdot \bar{\alpha}_n), \tag{2.3.35}$$

while for open strings with NN or DD boundary conditions it is

$$H_{\text{NN}} = \ell_s^2\, p^2 + \sum_{n=1}^{\infty} \alpha_{-n} \cdot \alpha_n, \tag{2.3.36}$$

$$H_{\text{DD}} = \frac{(x - y)^2}{(2\pi\ell_s)^2} + \sum_{n=1}^{\infty} \alpha_{-n} \cdot \alpha_n. \tag{2.3.37}$$

Finally, for DN boundary conditions the Hamiltonian is

$$H_{\text{DN}} = \tfrac{1}{2}\sum_{n \in \mathbb{Z}+1/2} \alpha_{-n} \cdot \alpha_n. \tag{2.3.38}$$

In the previous section we saw that the Virasoro constraints in the conformal gauge were $T_{--} = \frac{1}{2}(\partial_- X)^2 = 0$ and $T_{++} = \frac{1}{2}(\partial_+ X)^2 = 0$. We define the Virasoro operators as the Fourier modes of the stress-tensor. For the closed string they become

$$L_m = 2T \int_0^{2\pi} d\sigma \; T_{--} e^{im(\tau-\sigma)}, \quad \bar{L}_m = 2T \int_0^{2\pi} d\sigma \; T_{++} e^{im(\tau+\sigma)}, \tag{2.3.39}$$

or, expressed in oscillators,

$$L_m = \frac{1}{2} \sum_{n \in \mathbb{Z}} \alpha_{m-n} \cdot \alpha_n, \quad \bar{L}_m = \frac{1}{2} \sum_{n \in \mathbb{Z}} \bar{\alpha}_{m-n} \cdot \bar{\alpha}_n. \tag{2.3.40}$$

They satisfy the reality conditions

$$L_m^* = L_{-m} \quad \text{and} \quad \bar{L}_m^* = \bar{L}_{-m}. \tag{2.3.41}$$

If we compare these expressions with (2.3.35), we observe that the Hamiltonian can be written in terms of Virasoro modes as

$$H = L_0 + \bar{L}_0. \tag{2.3.42}$$

The Hamiltonian generates time translations. $H = 0$ is one of the classical constraints. Another operator $\bar{L}_0 - L_0$, is the generator of translations in σ, as can be shown with the help of the basic Poisson brackets (2.3.32). The constraint $\bar{L}_0 - L_0 = 0$ implies that there is no preferred point on the string.

For a noncompact coordinate (2.3.35) becomes

$$L_0 = \frac{\ell_s^2}{4} p^2 + \sum_{n=1}^{\infty} \alpha_{-n} \alpha_n, \quad \bar{L}_0 = \frac{\ell_s^2}{4} p^2 + \sum_{n=1}^{\infty} \bar{\alpha}_{-n} \bar{\alpha}_n, \tag{2.3.43}$$

while for a compact one

$$L_0 = \frac{1}{4}\left(m\frac{\ell_s}{R} + n\frac{R}{\ell_s}\right)^2 + \sum_{n=1}^{\infty} \alpha_{-n} \alpha_n, \quad \bar{L}_0 = \frac{1}{4}\left(m\frac{\ell_s}{R} - n\frac{R}{\ell_s}\right)^2 + \sum_{n=1}^{\infty} \bar{\alpha}_{-n} \bar{\alpha}_n, \tag{2.3.44}$$

where we have used (2.2.3),(2.3.7), and (2.3.19).

In the case of open strings, the α's and $\bar{\alpha}$'s are identified. The Virasoro modes are defined as

$$L_m = 2T \int_0^{\pi} d\sigma \, \{ T_{--} e^{im(\tau-\sigma)} + T_{++} e^{im(\tau+\sigma)} \}. \tag{2.3.45}$$

They can be expressed in oscillators as

$$L_m = \frac{1}{2} \sum_{n \in \mathbb{Z}} \alpha_{m-n} \cdot \alpha_n. \tag{2.3.46}$$

The Hamiltonian is

$$H = L_0.$$

With the help of the Poisson brackets for the oscillators, we can derive the brackets for the Virasoro constraints. They form an algebra known as the classical Virasoro algebra:

$$\{L_m, L_n\}_{PB} = -i(m-n)L_{m+n},$$
$$\{\bar{L}_m, \bar{L}_n\}_{PB} = -i(m-n)\bar{L}_{m+n},$$
$$\{L_m, \bar{L}_n\}_{PB} = 0. \tag{2.3.47}$$

In the open-string case, $L_m = \bar{L}_m$ so the $\bar{L}$'s are effectively absent.

Bibliography

Further details on the subject of this section can be found in the GSW [7] and Polchinski [8] books. The book by Zwiebach [12] describes this section in detail and moreover discusses many classical string solutions. Details on the calculation of determinants for the point particles and strings can be also found in the review [26].

Exercises

2.1. For an arbitrary einbein on the world-line of a point particle, find an explicit diffeomorphism of the form (2.1.16) on page 12 that maps it to a constant. Verify that you cannot change the physical length variable, as expected.

2.2. Use ζ-function regularization of the determinant (2.1.26) on page 13 to produce (2.1.27).

2.3. Consider the world-line action of a point particle in an arbitrary space-time metric $G_{\mu\nu}$. Derive the equations of motion for the path and show that these are equivalent to the geodesic equations.

2.4. Consider the world-line action of a point particle in an arbitrary space-time metric $G_{\mu\nu}$. Fix the world-line diffeomorphisms by choosing the static gauge $X^0 = \tau$. Derive the equations of motion.

2.5. Derive the nonrelativistic limit of the relativistic particle action (2.1.1) in an arbitrary space-time metric $G_{\mu\nu}$. Show that the action can be written as a sum of a kinetic and a potential contribution.

2.6. Consider a point particle of charge e moving in a nontrivial metric and an electromagnetic potential A_μ. Show that the electromagnetic coupling is described by the addition of

$$\Delta S = e \int d\tau\, A_\mu \dot{x}^\mu \tag{2.1E}$$

to the action. Derive the equations of motion. If one coordinate is cyclic, derive the associated conserved momentum.

2.7. By considering all possible terms that can appear in the world-sheet action for a string propagating in flat space-time, show that the area term (2.2.1) on page 14 is the most relevant one in the IR.

2.8. Consider, in analogy with the point-particle, the nonrelativistic limit of the Nambu-Goto action. Show that the action is again a sum of kinetic and a potential piece. Show that the kinetic energy comes only from the transverse motion of the string.

2.9. Show that according to the classical string equations of motion, the endpoints of the relativistic open string move with the speed of light.

2.10. Show that the addition of a two-dimensional cosmological term (2.2.18) on page 16 to the Polyakov action leads to $g_{\alpha\beta} = 0$.

2.11. Derive the Poincaré currents (2.2.45) and (2.2.46) on page 19 and show that they are conserved.

2.12. Use the oscillator expansions (2.3.6) and the Poisson brackets (2.3.32) on page 24 to derive (2.3.33).

2.13. Generalize the discussion (2.3.16)–(2.3.19) on page 22 to a string moving on an n-torus with constant metric G_{IJ}.

2.14. Consider an open string with boundary conditions

$$(X'^{\mu} + \lambda \dot{X}^{\mu})\big|_{\sigma=0,\pi} = 0 \tag{2.2E}$$

that are linear combinations of Dirichlet and Neumann boundary conditions. Find the classical solution that satisfies these boundary conditions. What is its interpretation? Study the limits $\lambda \to 0$ and $\lambda \to \infty$.

2.15. Calculate the flow of momentum at a Dirichlet end point.

2.16. Derive the Hamiltonians (2.3.35), (2.3.36)–(2.3.38).

2.17. Use the Poisson brackets (2.3.33) to derive the classical Virasoro algebra (2.3.47) on page 26.

3 Quantization of Bosonic Strings

There are several ways to quantize relativistic strings. Some rely on operator methods and others on path integrals.

Covariant canonical quantization. The classical unconstrained variables of the string motion become operators. We then impose the constraints in the quantum theory as conditions on states in the Hilbert space. This procedure preserves manifest Lorentz invariance and is known as the old covariant approach. It leads to a Hilbert space with indefinite metric, as in the Gupta-Bleuler approach to QED.

Light-cone quantization. In this case the constraints are solved at the level of the classical theory. This is easily done in the light-cone gauge. We then quantize by replacing the left-over variables by operators. Manifest Lorentz invariance is however lost. Its presence has to be checked *a posteriori*.

Path integral quantization. This can be combined with techniques of the Becchi-Rouet-Stora-Tyutin (BRST) symmetry. It has manifest Lorentz invariance, but it works in an extended Hilbert space that also contains ghost fields. It is the analog of the Faddeev-Popov method for gauge theories.[1]

All three methods of quantization agree whenever they can be compared. Each one has some advantages, depending on the nature of the questions we ask in the quantum theory. All three will be presented in some detail.

3.1 Covariant Canonical Quantization

The usual way to perform the canonical quantization is to replace all fields by operators and replace the Poisson brackets by commutators

$$\{\,,\,\}_{\mathrm{PB}} \longrightarrow -i[\,,\,].$$

---

[1] The BRST quantization can also be implemented in the operator approach.

The Virasoro constraints are then operator constraints whose positive frequency components must annihilate physical states.

Using the canonical prescription, the commutators for the oscillators and center-of-mass
position and momentum become

$$[x^\mu, p^\nu] = i\eta^{\mu\nu}, \tag{3.1.1}$$

$$[\alpha_m^\mu, \alpha_n^\nu] = m\delta_{m+n,0}\,\eta^{\mu\nu}. \tag{3.1.2}$$

There is a similar expression for the $\bar{\alpha}$'s in the case of closed strings, while α_n^μ and $\bar{\alpha}_n^\mu$
commute. The reality condition (2.3.4) now becomes a hermiticity condition on the oscil-
lators. If we absorb the factor m in (3.1.2) in the oscillators, $\alpha_m^\mu \to \sqrt{m}\,\alpha_m^\mu$, we can write
the commutation relation as

$$[\alpha_m^\mu, \alpha_n^{\nu\dagger}] = \delta_{m,n}\eta^{\mu\nu}, \quad m, n > 0, \tag{3.1.3}$$

which is just the harmonic oscillator commutation relation for an infinite set of oscillators.

We must define a Hilbert space on which the operators act. Our system is an infinite
collection of harmonic oscillators so the answer is the usual Fock space. The negative
frequency modes α_m, $m < 0$, are raising operators and the positive frequency modes are
the lowering operators. We define the ground state of our Hilbert space as the state that
is annihilated by all lowering operators. This does not yet define the state completely:
we also have to consider the center-of-mass operators x^μ and p^μ. Again this is known
from elementary quantum mechanics: if we diagonalize p^μ, then the states will be also
characterized by the momentum. We denote the ground state by $|p^\mu\rangle$ and we have

$$\alpha_m^\nu |p^\mu\rangle = 0 \quad \forall \ m > 0. \tag{3.1.4}$$

We can build other states by acting on this ground state with the negative frequency modes[2]

$$|p^\mu\rangle, \quad \alpha_{-1}^\nu|p^\mu\rangle, \quad \alpha_{-1}^\sigma\alpha_{-1}^\nu\alpha_{-2}^\rho|p^\mu\rangle, \quad \text{etc.} \tag{3.1.5}$$

There seems to be a problem, however: because of the Minkowski metric in the commutator for the oscillators, we obtain

$$|\ \alpha_{-1}^0|p^\mu\rangle\ |^2 = \langle p^\mu|\alpha_1^0\alpha_{-1}^0|p^\mu\rangle = -1, \tag{3.1.6}$$

which means that there are negative norm states. But we still have to impose the classical
constraints $L_m = 0$. Imposing these constraints will help us avoid the states with negative
norm.

Before we go further, however, we have to face a typical ambiguity when quantizing
a classical system. The classical variables are functions of coordinates and momenta. In
the quantum theory, coordinates and momenta are noncommuting operators. A specific
ordering prescription has to be made in order to define them as operators in the quantum
theory. In particular, we would like their eigenvalues on physical states to be finite; we will
therefore have to pick a normal-ordering prescription as in quantum field theory. Normal
ordering puts all positive frequency modes to the right of the negative frequency modes.

---

[2] We consider here for simplicity the case of the open string.

The Virasoro operators in the quantum theory are now defined by their normal-ordered expressions

$$L_m = \tfrac{1}{2} \sum_{n \in \mathbb{Z}} : \alpha_{m-n} \cdot \alpha_n : . \tag{3.1.7}$$

Because of the commutation relations (3.1.2), only L_0 is sensitive to normal ordering,

$$L_0 = \tfrac{1}{2}\alpha_0^2 + \sum_{n=1}^{\infty} \alpha_{-n} \cdot \alpha_n. \tag{3.1.8}$$

The commutator of two oscillators is a constant. Therefore the general form of the quantum version of L_0 will differ from the normal-ordered one by a constant. We therefore include a normal-ordering constant a in all expressions containing L_0. Otherwise stated, we replace L_0 by $L_0 - a$.

We may now calculate the algebra of the L_m's. Because of the normal ordering this has to be done with great care. The Virasoro algebra becomes

$$[L_m, L_n] = (m - n)L_{m+n} + \frac{c}{12}m(m^2 - 1)\delta_{m+n,0}, \tag{3.1.9}$$

where c is known as the central charge and in this case $c = d$, the dimension of the target space or the number of free scalar fields on the world-sheet.

We can now see that we cannot impose the classical constraints $L_m = 0$ as operator constraints $L_m|\phi\rangle = 0$ because

$$0 = \langle\phi|[L_m, L_{-m}]|\phi\rangle = 2m \, \langle\phi|L_0|\phi\rangle + \frac{d}{12}m(m^2 - 1)\langle\phi|\phi\rangle \neq 0.$$

This is analogous to a similar phenomenon that occurs in gauge theory. There, one follows the Gupta-Bleuler approach, which makes sure that the constraints vanish "weakly" (their expectation value on physical states vanishes). Here, the maximal set of constraints we can impose on physical states is

$$L_{m>0}|\text{phys}\rangle = 0, \quad (L_0 - a)|\text{phys}\rangle = 0, \tag{3.1.10}$$

and, in the case of closed strings, equivalent expressions for the $\bar{L}$'s. This is consistent with the weak vanishing of the classical constraints because $\langle\text{phys}'|L_n - a\delta_{n,0}|\text{phys}\rangle = 0$ for all $n \in \mathbb{Z}$ using the hermiticity conditions $L_n^\dagger = L_{-n}$.

Thus, the physical states in the theory are the states in the Hilbert space that also satisfy (3.1.10). It turns out that apart from physical states, there are so-called spurious states in the Hilbert space. They satisfy $|\text{spurious}\rangle = L_{-n}|\text{any}\rangle$. They are orthogonal to all physical states. There are even states in the Hilbert space which are both physical and spurious. Such states must decouple from the physical Hilbert space since they can be shown to correspond to null states.

There is a detailed analysis of the physical spectrum of string theory, which culminates with the famous "no-ghost" theorem: The theorem states that if the space-time dimension is $d = 26$, the physical spectrum defined by (3.1.10) and $a = 1$ contains only positive norm states.

We will now analyze the L_0 condition. If we substitute the expression for L_0 in (3.1.10) with $p^2 = -m^2$ and $a = 1$ we obtain the mass-shell condition

$$\ell_s^2 m^2 = 4(N - 1), \tag{3.1.11}$$

where N is the level-number operator:

$$N = \sum_{m=1}^{\infty} \alpha_{-m} \cdot \alpha_m. \tag{3.1.12}$$

For closed strings we can deduce a similar expression for $(\bar{L}_0 - 1)$. From $L_0 = \bar{L}_0$ it follows that $\bar{N} = N$. We observe that the state with the minimal mass, $N = 0$, is a tachyon.

3.2 Light-cone Quantization

In this approach we must first solve the classical constraints. This will leave us with a smaller number of classical variables. Then we will quantize them.

There is a gauge in which the solution of the Virasoro constraints is simple. This is the light-cone gauge. The light-cone coordinates are defined as

$$X^{\pm} = X^0 \pm X^1.$$

Remember that we still have some residual invariance after choosing the conformal gauge:

$$\xi'^+ = f(\xi^+), \qquad \xi'^- = g(\xi^-).$$

This invariance can be used to set

$$X^+ = x^+ + \ell_s^2 p^+ \tau. \tag{3.2.1}$$

This gauge can indeed be reached because, according to the gauge transformations, the transformed coordinates σ' and τ' have to satisfy the wave equation in terms of the old coordinates and X^+ clearly does so.

Imposing now the classical Virasoro constraints (2.2.39) on page 19, we can solve for X^- in terms of the transverse coordinates X^i. This means that we can eliminate both X^+ and X^- using the constraints. We will then have to work only with the transverse directions. Therefore, after solving the constraints, we are left with all positions and momenta of the string, but only the transverse oscillators.

Solving the constraints (2.2.39), the light-cone oscillators can be expressed as

$$\alpha_n^+ = \bar{\alpha}_n^+ = \frac{\ell_s}{\sqrt{2}} p^+ \delta_{n,0},$$

$$\alpha_n^- = \frac{\sqrt{2}}{\ell_s p^+} \left\{ \sum_{m \in \mathbb{Z}} : \alpha_{n-m}^i \alpha_m^i : -2a\delta_{n,0} \right\}, \tag{3.2.2}$$

and a similar expression for $\bar{\alpha}^-$.

We can now quantize. We replace x^μ, p^μ, α_n^i, and $\bar{\alpha}_n^i$ with operators. The index i takes values in the $d - 2$ transverse directions. However, we have given up the manifest Lorentz covariance of the theory. Since this theory in the light-cone gauge originated from a manifest Lorentz-invariant theory in d dimensions, we would expect that after fixing the gauge this invariance is still present. However, it turns out that in the quantum theory this is true only if $d = 26$ and $a = 1$. To put it differently, the Poincaré algebra closes if $d = 26$.

3.3 Spectrum of the Bosonic String

We now analyze the spectrum both of closed and of open strings. In the light-cone gauge we have solved almost all of the Virasoro constraints. However, we still have to impose $(L_0 - a)|\text{phys}\rangle = 0$ and a similar one $(\bar{L}_0 - \bar{a})|\text{phys}\rangle$ for the closed string. In exercise 3.5 on page 47 you are asked to show that only $a = \bar{a}$ gives a nontrivial closed-string spectrum consistent with Lorentz invariance. In particular this implies that $L_0 = \bar{L}_0$ on physical states.

The states are constructed in a fashion similar to that of section 3.1. One starts from the state $|p^\mu\rangle$, which is the vacuum for the transverse oscillators, and then creates more states by acting with the negative frequency modes of the transverse oscillators.

For the closed string, the ground state is $|p^\mu\rangle$, for which we have the mass-shell condition $\ell_s^2 m^2 = -4a$.

The first excited level will be (imposing $L_0 = \bar{L}_0$)

$$\alpha^i_{-1}\bar{\alpha}^j_{-1}|p^\mu\rangle. \tag{3.3.1}$$

We can decompose this into irreducible representations of the transverse rotation group $SO(d-2)$ in the following manner:

$$\alpha^i_{-1}\bar{\alpha}^j_{-1}|p^\mu\rangle = \alpha^{[i}_{-1}\bar{\alpha}^{j]}_{-1}|p^\mu\rangle + \left[\alpha^{\{i}_{-1}\bar{\alpha}^{j\}}_{-1} - \frac{1}{d-2}\delta^{ij}\alpha^k_{-1}\bar{\alpha}^k_{-1}\right]|p^\mu\rangle$$

$$+ \frac{1}{d-2}\delta^{ij}\alpha^k_{-1}\bar{\alpha}^k_{-1}|p^\mu\rangle. \tag{3.3.2}$$

These states can be interpreted as a spin-2 particle G_{ij} (graviton), an antisymmetric tensor B_{ij}, and a scalar Φ.

Lorentz invariance requires physical states to be representations of the little group of the Lorentz group $SO(d-1, 1)$, which is $SO(d-1)$ for massive states and $SO(d-2)$ for massless states. Thus, we conclude that states at this first excited level must be massless, since the representation content is such that they cannot be assembled into $SO(d-1)$ representations. Their mass-shell condition is

$$\ell_s^2 m^2 = 4(1-a),$$

from which we can derive the value of the normal-ordering constant, $a = 1$. This constant can also be expressed in terms of the target-space dimension d via ζ-function regularization. In exercise 3.1 on page 46 you are asked to show that $a = \frac{d-2}{24}$. We conclude that Lorentz invariance requires that $a = 1$ and $d = 26$. These are the values that we will assume henceforth.

What about the next level? It turns out that higher excitations, which are naturally tensors of SO(24), can be uniquely combined in representations of SO(25). This is consistent with Lorentz invariance for massive states and can be shown to hold for all higher-mass excitations.

Now consider the open string. Again the ground state is tachyonic with mass

$$\ell_s^2 m^2 = -1. \tag{3.3.3}$$

The first excited level is

$$\alpha_{-1}^i |p^\mu\rangle,$$

which is again massless and is the vector representation of SO(24), as it should be for a massless vector in 26 dimensions. The second-level excitations are given by

$$\alpha_{-2}^i |p^\mu\rangle, \quad \alpha_{-1}^i \alpha_{-1}^j |p^\mu\rangle,$$

which are tensors of SO(24). These two parts uniquely combine into a symmetric traceless SO(25) massive tensor.

In the case of the open string we see that at level n with mass-shell condition

$$\ell_s^2 m^2 = (n - 1), \tag{3.3.4}$$

we always have a state described by a symmetric tensor of rank n and we can conclude that the maximal spin at level n can be expressed in terms of the mass as

$$j^{\text{max}} = \ell_s^2 m^2 + 1.$$

This is the so-called leading Regge trajectory. Its presence gave an extra impetus to "old" string theory as a candidate theory of the strong interactions.

We have seen, by studying the spectrum, that a consistent quantization of the bosonic string giving a Lorentz-invariant theory is possible only in 26 space-time dimensions. This dimension is called the critical dimension. String theories can also be defined in less than 26 dimensions and are called noncritical. They are not Lorentz invariant. Their consistent backgrounds turn out to be different than Minkowski space. We will discuss examples in chapter 6 on page 144.

3.4 Unoriented Strings

The bosonic closed string we have discussed has a well-defined notion of world-sheet orientation, since the left- and right-movers are distinct. It is therefore an *oriented* string theory. Moreover, the theory is invariant under the world-sheet parity symmetry Ω that interchanges right- and left-movers:

$$\Omega: \sigma \to 2\pi - \sigma, \quad \tau \to \tau. \tag{3.4.1}$$

Ω is an involution, $\Omega^2 = 1$. Therefore, the Ω eigenvalues are ± 1. Applying this transformation to the mode expansion of the closed string (2.3.6) on page 20 we obtain by definition

$$\Omega \, \alpha_k^\mu \, \Omega^{-1} = \bar{\alpha}_k^\mu, \quad \Omega \, \bar{\alpha}_k^\mu \, \Omega^{-1} = \alpha_k^\mu, \tag{3.4.2}$$

while the momentum remains invariant. It remains to specify its action on the ground state. The argument involves tadpole cancellation, a constraint that will be developed in section 5.3 on page 133. Consistency of interactions requires that the closed string tachyon is invariant,

$$\Omega |p^\mu\rangle = |p^\mu\rangle, \tag{3.4.3}$$

as you will be required to show in exercise 6.14 on page 154 in a later chapter.[3]

We may obtain a truncation of the bosonic string theory by throwing away all states that are not invariant under Ω. The fact that Ω is a symmetry guarantees that this is consistent with the interactions. The string theory thus obtained is an unoriented closed string theory.

In particular, the closed string tachyon is kept. From the massless states $\alpha^i_{-1} \bar\alpha^j_{-1} |p^\mu\rangle$, the graviton and dilaton have $\Omega = 1$ and are kept, while the two-index antisymmetric tensor has $\Omega = -1$ and is projected out.

In the case of the open string, the world-sheet parity transformations is

$$\Omega : \ \sigma \to \pi - \sigma, \quad \tau \to \tau, \tag{3.4.4}$$

and acts on the oscillators as

$$\text{(NN):} \quad \Omega \, \alpha^\mu_k \, \Omega^{-1} = (-1)^k \, \alpha^\mu_k, \quad \text{(DD):} \quad \Omega \, \alpha^\mu_k \, \Omega^{-1} = (-1)^{k+1} \, \alpha^\mu_k, \tag{3.4.5}$$

while the vacuum is invariant. It also transforms ND boundary conditions to DN ones.

Before discussing the open string spectrum, it is appropriate to introduce, in the next section, a new degree of freedom for the open strings.

3.4.1 Open strings and Chan-Paton factors

Open strings are allowed to carry charges at the end points. These are known as Chan-Paton factors and give rise to non-Abelian gauge groups of the type Sp(2N) or O(N) in the unoriented case and U(N) in the oriented case.

To see how this comes about, we will attach charges labeled by an index $i = 1, 2, \ldots, N$ at the two end points of the open string as in figure 3.1. This is a new degree of freedom that does not transform under space-time transformations and has trivial world-sheet dynamics. It is just an index.

Now the ground state is labeled, apart from the momentum, by the end-point charges $|p, i, j\rangle$, where i labels one end of the string and j the other. Therefore, in this case we have N^2 tachyons. Moreover, the whole spectrum comes in N^2 copies. In particular, the massless states are $\alpha^\mu_{-1}|p, i, j\rangle$ and they give a collection of N^2 vectors. We can use the N^2 Hermitian matrices λ^a_{ij}, generators of the U(N) Lie algebra, and normalized to

$$\text{Tr}[\,\lambda^a \, \lambda^b\,] = \delta^{ab} \tag{3.4.6}$$

to define an alternative basis for the states

$$|p; a\rangle = \sum_{ij} \lambda^a_{ij} \, |p, i, j\rangle. \tag{3.4.7}$$

---

[3] For another view see also exercise 5.15 on page 143.

Figure 3.1 Chan-Paton labels at the end points of the string.

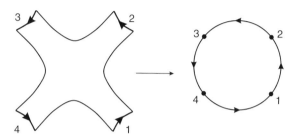

Figure 3.2 a) The world-sheet of the four-point tree-level interaction of four open strings. b) Its conformal transform to a disk with four insertion points on the boundary.

When such strings interact and join two end points, the natural prescription is that this is possible only if the respective CP indices are equal. This guarantees that all string amplitudes will be proportional to traces of products of λ-matrices associated with the external states. For example, for the four-point amplitude shown in figure 3.2 the trace factor will be $\mathrm{Tr}[\lambda_1^a \lambda_2^b \lambda_3^c \lambda_4^d]$. Consequently, the amplitudes will be invariant under global U(N) transformations.

The presence of N^2 massless vectors indicates that the theory has a local U(N) gauge invariance. This can be verified by explicit analysis of the scattering amplitudes. There is a quick argument to that effect: The U(N) symmetry is by construction a world-sheet symmetry, under which the string coordinates are neutral. As is standard in string theory, global world-sheet symmetries transform into local space-time symmetries. Therefore, we expect the U(N) to become a gauge symmetry in space-time. However, the orthodox way to show this is to do exercise 5.11 on page 143 when you reach chapter 5.

Other groups can be generated in unoriented open strings. Here we will also have to project by the transformation Ω that interchanges the two string end points. Ω also reverses the orientation of the string itself. The most general (unitary) action of Ω will interchange the CP labels and will permute them by a unitary transformation γ_Ω. We may therefore write

$$\Omega: \ |p\,;ij\rangle \to \epsilon\,(\gamma_\Omega)_{ii'} \, |p\,;,j'i'\rangle \, (\gamma_\Omega^{-1})_{j'j}, \tag{3.4.8}$$

where the $N \times N$ unitary matrix γ_Ω describes the action of Ω on one end point of the string and ϵ is a phase that we will determine from consistency conditions in the following. Ω^2 is given by

$$\Omega^2: \ |p\,;ij\rangle \to \epsilon^2\,(\gamma_\Omega(\gamma_\Omega^T)^{-1})_{ii'} \, |p\,;i'j'\rangle \, (\gamma_\Omega^T \gamma_\Omega^{-1})_{j'j}, \tag{3.4.9}$$

and $\Omega^2 = 1$ therefore implies that

$$\gamma_\Omega = \zeta \gamma_\Omega^T, \quad \epsilon^2 \zeta^2 = 1. \tag{3.4.10}$$

Unless ζ is a phase, the only solution to (3.4.10) is $\gamma_\Omega = 0$. Taking the determinant of (3.4.10) we obtain

$$\zeta^N = 1. \tag{3.4.11}$$

Therefore, at best, the phase ζ is an N-th root of unity.

We will now derive further constraints on ϵ and ζ, and eventually classify the solutions of (3.4.10). The strongest constraints can be obtained by considering the massless vector states

$$A^\mu = \alpha^\mu_{-1} \sum_{ij} |p; ij\rangle \lambda_{ij}. \tag{3.4.12}$$

The vectors that will survive the Ω projection have eigenvalue 1. Their CP wave-functions λ must satisfy

$$\lambda = -\epsilon \, \gamma_\Omega \, \lambda^T \, \gamma_\Omega^{-1}, \tag{3.4.13}$$

where we used (3.4.5). The gauge group is the space of solutions of (3.4.13) where λ is Hermitian.

We may also take the transpose of (3.4.13), use (3.4.10), and match back to (3.4.13) to obtain

$$\epsilon^2 = 1, \quad \zeta^2 = 1, \tag{3.4.14}$$

where the second relation follows from the first and (3.4.10). Therefore, both ϵ and ζ are signs. So far, such constraints on the phases are true for any state of the open string. However, for the vectors there is a further constraint: that their CP matrices form a Lie algebra. This is necessary for the consistency of their interactions. Using, (3.4.13) we obtain for the commutator of two matrices

$$[\lambda_1, \lambda_2] = \epsilon^2 \gamma_\Omega \, [\lambda_1^T, \lambda_2^T] \, \gamma_\Omega^{-1} = -\epsilon^2 \gamma_\Omega \, [\lambda_1, \lambda_2]^T \, \gamma_\Omega^{-1}. \tag{3.4.15}$$

For the commutator to also satisfy (3.4.13) we must have

$$\epsilon^2 = \epsilon \quad \Longrightarrow \quad \epsilon = 1. \tag{3.4.16}$$

We may now proceed to solve (3.4.13) for the two possible cases in (3.4.10), namely, the cases of symmetric and antisymmetric unitary matrices γ_Ω.

(i) In the symmetric case, $\zeta = 1$, we may diagonalize γ_Ω. Since γ_Ω is unitary, the N eigenvalues are phases, $e^{i\theta_i}$. Solving directly (3.4.13) with $\epsilon = 1$ from (3.4.16) we obtain up to a sign

$$\lambda_{ii} = 0, \quad \lambda_{i<j} = iR_{ij}e^{(i/2)(\theta_i - \theta_j)}, \quad R_{ij} \in \mathbb{R}^+. \tag{3.4.17}$$

In exercise 3.9 on page 47 you are required to show that this is a subgroup of U(N), isomorphic to SO(N) for any of the values of the θ_i. The θ_i provide different embeddings of the SO(N) inside the space of Hermitian matrices. Therefore, without loss of generality, when $\zeta = 1$ we will, from now on, take $\gamma_\Omega = 1$.

(ii) Consider now the antisymmetric case $\zeta = -1$. Then, from (3.4.11) N must be even. By a choice of basis, the matrix γ_Ω can be written in the 2×2 block basis as

$$
\gamma_\Omega = \begin{pmatrix} 0 & e^{i\theta_1} & 0 & 0 & \cdots \\ -e^{i\theta_1} & 0 & 0 & 0 & \cdots \\ 0 & 0 & 0 & e^{i\theta_2} & \cdots \\ 0 & 0 & -e^{i\theta_2} & 0 & \cdots \\ \vdots & \vdots & \vdots & \vdots & \ddots \end{pmatrix}, \quad \theta_i \in [0, 2\pi].
\tag{3.4.18}
$$

We parametrize now λ in (3.4.13) in terms of $(N/2)^2$ 2×2 matrices λ_{ab}, with

$$
(\lambda_{aa})^\dagger = \lambda_{aa}, \quad (\lambda_{ab})^\dagger = \lambda_{ba}.
\tag{3.4.19}
$$

The general solution of (3.4.13) is then

$$
\lambda_{aa} = \begin{pmatrix} B_a & E_a \\ E_a^* & -B_a \end{pmatrix}, \quad B_a \in \mathbb{R}, \quad \lambda_{ab} = \begin{pmatrix} C_{ab} & D_{ab}^* \, e^{i(\theta_a - \theta_b)} \\ D_{ab} & -C_{ab}^* \, e^{i(\theta_a - \theta_b)} \end{pmatrix}.
\tag{3.4.20}
$$

These matrices generate the algebra of Sp(2N) for arbitrary values of the phases θ_i. The θ-independent generators of the algebra are

$$
J_{aa}^i = \sigma^i, \quad J_{ab}^\mu = \sigma^\mu, \quad \sigma^0 = i\mathbf{1}_2.
\tag{3.4.21}
$$

We may therefore choose a canonical form for γ_Ω by taking $\theta_i = \frac{\pi}{2}$ and rearranging the basis so that

$$
\gamma_\Omega = i \begin{pmatrix} \mathbf{0}_{N/2} & \mathbf{1}_{N/2} \\ -\mathbf{1}_{N/2} & \mathbf{0}_{N/2} \end{pmatrix}, \quad \gamma_\Omega^T = -\gamma_\Omega,
\tag{3.4.22}
$$

where the subscripts indicate the dimension of the matrix.

Once we have fixed the signs ϵ and ζ, we may discuss the other states in the open string spectrum. The tachyon matrices in particular must solve

$$
\lambda = \gamma_\Omega \, \lambda^T \, \gamma_\Omega^{-1}.
\tag{3.4.23}
$$

This has an extra minus sign compared to the vectors.

In the symmetric case, $\zeta = 1$ this gives a real symmetric matrix, which transforms as the two-index symmetric traceless tensor of SO(N) plus a singlet. In the antisymmetric case, the corresponding matrix transforms into the two-index skew-traceless antisymmetric representation of Sp(2N) plus a singlet.

3.5 Path Integral Quantization

In this section we will use the path integral approach to quantize the string, starting from the Polyakov action. Consider the bosonic string partition function[4]

$$
Z = \int \frac{\mathcal{D}g \, \mathcal{D}X^\mu}{V_{\text{gauge}}} e^{-S_p(g, X^\mu)}.
\tag{3.5.1}
$$

---

[4] As usual, to make the path integral well defined we must Wick-rotate to Euclidean space.

The measures are defined from the norms:

$$\|\delta g\| = \int d^2\xi \sqrt{g} g^{\alpha\beta} g^{\delta\gamma} \delta g_{\alpha\gamma} \delta g_{\beta\delta},$$
$$\|\delta X^\mu\| = \int d^2\xi \sqrt{g} \delta X^\mu \delta X^\nu \eta_{\mu\nu}.$$

The action is Weyl invariant, but the measures are not. This implies that generically in the quantum theory the Weyl factor will couple to the rest of the fields. We can use conformal reparametrizations to rescale our metric,

$$g_{\alpha\beta} = e^{2\phi} h_{\alpha\beta}.$$

The variation of the metric under reparametrizations and Weyl rescalings can be decomposed into a traceless and a pure trace part,

$$\delta g_{\alpha\beta} = \nabla_\alpha \xi_\beta + \nabla_\beta \xi_\alpha + 2\Lambda g_{\alpha\beta} = (\hat{P}\xi)_{\alpha\beta} + 2\tilde{\Lambda} g_{\alpha\beta}, \tag{3.5.2}$$

where $(\hat{P}\xi)_{\alpha\beta} = \nabla_\alpha \xi_\beta + \nabla_\beta \xi_\alpha - (\nabla_\gamma \xi^\gamma) g_{\alpha\beta}$. We also introduce $\tilde{\Lambda} = \Lambda + \frac{1}{2}\nabla_\gamma \xi^\gamma$. The integration measure can be written as

$$\mathcal{D}g = \mathcal{D}(\hat{P}\xi)\mathcal{D}(\tilde{\Lambda}) = \mathcal{D}\xi \mathcal{D}\Lambda \left|\frac{\partial(\hat{P}\xi, \tilde{\Lambda})}{\partial(\xi, \Lambda)}\right|, \tag{3.5.3}$$

where the Jacobian (Faddeev-Popov determinant) is

$$\left|\frac{\partial(\hat{P}\xi, \tilde{\Lambda})}{\partial(\xi, \Lambda)}\right| = \left|\det\begin{pmatrix} \hat{P} & * \\ 0 & 1 \end{pmatrix}\right| = \left|\det\hat{P}\right| = \sqrt{\det\hat{P}\hat{P}^\dagger}. \tag{3.5.4}$$

The $*$ here stands for an operator that does not affect the value of the determinant.

There are two sources of Weyl noninvariance in the path integral: the Faddeev-Popov determinant and the X^μ measure. The Weyl factor of the metric decouples in the quantum theory only if $d = 26$. This is the way that the critical dimension is singled out in the path integral approach. If $d \neq 26$, the Weyl factor has to be kept. This is described in more detail in chapter 6 on page 144.

Here, we will assume that we are in the critical dimension. We can factor out the integration over the reparametrizations and the Weyl group, in which case the partition function becomes

$$Z = \int \mathcal{D}X^\mu \sqrt{\det\hat{P}\hat{P}^\dagger} \, e^{iS_p(\hat{h}_{\alpha\beta}, X^\mu)}, \tag{3.5.5}$$

where $\hat{h}_{\alpha\beta}$ is some fixed reference metric that can be chosen at will. We can now use the so-called Faddeev-Popov trick: we can exponentiate the determinant using anti-commuting ghost variables c^α and $b_{\alpha\beta}$, where $b_{\alpha\beta}$ is a symmetric and traceless tensor:

$$\sqrt{\det\hat{P}\hat{P}^\dagger} = \int \mathcal{D}c\mathcal{D}b \, e^{i\int d^2\sigma \sqrt{g} g^{\alpha\beta} b_{\alpha\gamma} \nabla_\beta c^\gamma}. \tag{3.5.6}$$

If we now choose $h_{\alpha\beta} = \eta_{\alpha\beta}$ the partition function becomes:

$$Z = \int \mathcal{D}X\mathcal{D}c\mathcal{D}b \, e^{i(S_p[X] + S_{gh}[c,b])}, \tag{3.5.7}$$

where

$$S_p[X] = 2T \int d^2\xi \, \partial_+ X^\mu \partial_- X_\mu, \tag{3.5.8}$$

$$S_{gh}[b, c] = \frac{1}{\pi} \int (b_{++}\partial_- c^+ + b_{--}\partial_+ c^-). \tag{3.5.9}$$

Apart from the ghosts, we have arrived at the action in the conformal gauge, (2.2.31). This was the starting point for the old covariant quantization. The presence of the ghosts signals that there is a better (and more general) way to quantize. We will pursue this further in the next few sections.

3.6 Topologically Nontrivial World-sheets

We have seen in the previous section that gauge-fixing the diffeomorphisms and Weyl rescalings gives rise to a Faddeev-Popov determinant. Subtleties arise when this determinant is zero, and we will discuss the appropriate treatment here.

As already mentioned, under the combined effect of reparametrizations and Weyl rescalings, the metric transforms as in (3.5.2). The operator $\hat{P}$ maps vectors to traceless symmetric tensors. Those reparametrizations satisfying

$$(\hat{P}\xi^*)_{\alpha\beta} = 0 \tag{3.6.1}$$

do not affect the metric. Equation (3.6.1) is called the conformal Killing equation, and its solutions are the conformal Killing vectors. These are the zero modes of $\hat{P}$. When a surface admits conformal Killing vectors then there are reparametrizations that cannot be fixed by fixing the conformal class of the metric. They have to be fixed separately.

Now define the natural inner product for vectors and tensors:

$$(V_\alpha, W_\alpha) = \int d^2\xi \sqrt{\det g} g^{\alpha\beta} V_\alpha W_\beta \tag{3.6.2}$$

and

$$(T_{\alpha\beta}, S_{\alpha\beta}) = \int d^2\xi \sqrt{\det g} g^{\alpha\gamma} g^{\beta\delta} T_{\alpha\beta} S_{\gamma\delta}. \tag{3.6.3}$$

The decomposition (3.5.2) separating the traceless part from the trace is orthogonal. The Hermitian conjugate with respect to this product maps traceless symmetric tensors $T_{\alpha\beta}$ to vectors:

$$(\hat{P}^\dagger T)_\alpha = -2\nabla^\beta T_{\alpha\beta}. \tag{3.6.4}$$

The zero modes of $\hat{P}^\dagger$ are the solutions of

$$\hat{P}^\dagger T^* = 0 \tag{3.6.5}$$

and correspond to symmetric traceless tensors, which cannot be written as $(\hat{P}\xi)_{\alpha\beta}$ for any vector field ξ. Indeed, if (3.6.5) is satisfied, then for all ξ^α, $0 = (\xi, \hat{P}^\dagger t^*) = (\hat{P}\xi, t^*)$. Thus, zero modes of $\hat{P}^\dagger$ correspond to deformations of the metric that cannot be compensated for by reparametrizations and Weyl rescalings. Such deformations do not correspond to gauge transformations. They are called Teichmüller deformations. We have already seen an example of this in the point-particle case. The length of the path in (2.1.17) on page 12, was a Teichmüller parameter, since it could not be changed by diffeomorphisms.

The following table gives the number of conformal Killing vectors and zero modes of $\hat{P}^{\dagger}$, depending on the topology of the closed string world-sheet. The genus g is equal to the number of handles of a closed oriented surface.

Genus	No. of zeros of $\hat{P}$	No. of zeros of $\hat{P}^{\dagger}$
0	3	0
1	1	1
≥ 2	0	$3g - 3$

The results described above are important for the calculation of loop corrections to scattering amplitudes.

The perturbative expansion of a given amplitude takes the form

$$S(g_s) = \sum_\chi g_s^\chi \, S_\chi = \sum_{g=0}^\infty g_s^{2-2g} \, S_g,$$ (3.6.6)

where g_s is the string coupling constant, χ is the Euler number of contributing surfaces, and in the second equation we assumed that only closed Riemann surfaces contribute.

3.7 BRST Primer

The general way to covariantly quantize theories with local symmetries is the BRST formalism. We will take here a brief look at this formalism in general. We will subsequently apply it to the quantization of the string.

Consider a theory with fields ϕ_i, which has a certain gauge symmetry. The gauge transformations will satisfy an associative algebra[5]

$$[\delta_\alpha, \delta_\beta] = f_{\alpha\beta}{}^\gamma \delta_\gamma,$$ (3.7.1)

with the $f_{\alpha\beta}{}^\gamma$ being field independent.

We can now fix the gauge by imposing some appropriate gauge conditions

$$F^A(\phi_i) = 0.$$ (3.7.2)

Using again the Faddeev-Popov trick, we can write the path integral as

$$\int \frac{\mathcal{D}\phi}{V_{\text{gauge}}} \, e^{-S_0} \sim \int \mathcal{D}\phi \, \delta(F^A(\phi)) \, \mathcal{D}b_A \, \mathcal{D}c^\alpha \, e^{-S_0 - \int b_A (\delta_\alpha F^A) c^\alpha}$$

$$\sim \int \mathcal{D}\phi \, \mathcal{D}B_A \, \mathcal{D}b_A \, \mathcal{D}c^\alpha \, e^{-S_0 - i \int B_A F^A(\phi) - \int b_A (\delta_\alpha F^A) c^\alpha}$$

$$= \int \mathcal{D}\phi \, \mathcal{D}B_A \, \mathcal{D}b_A \, \mathcal{D}c^\alpha \, e^{-S},$$ (3.7.3)

---

[5] This is not the most general algebra possible, but it is sufficient for our purposes.

where

$$S = S_0 + S_1 + S_2, \quad S_1 = i \int B_A F^A(\phi), \quad S_2 = \int b_A (\delta_\alpha F^A) c^\alpha. \qquad (3.7.4)$$

The antighost B_A is a Lagrange multiplier. Note that the index α associated with the ghost c_α is in one-to-one correspondence with the parameters of the gauge transformations in (3.7.1). The index A associated with the ghost b_A and the antighost B_A are in one-to-one correspondence with the gauge-fixing conditions.

The full gauge-fixed action S is invariant under the *BRST transformation*,

$$\begin{aligned}
\delta \, \phi_i &= -i\epsilon c^\alpha \delta_\alpha \phi_i, \\
\delta \, b_A &= -\epsilon B_A, \\
\delta \, c^\alpha &= -\tfrac{1}{2}\epsilon c^\beta c^\gamma f_{\beta\gamma}{}^\alpha, \\
\delta \, B_A &= 0.
\end{aligned} \qquad (3.7.5)$$

In these transformations, ϵ must be a Grassmann variable. The first transformation is just the original gauge transformation on ϕ_i, but with the gauge parameter replaced by the ghost c_α.

The extra terms in the action due to the ghosts and gauge fixing in (3.7.3) can be written in terms of a BRST transformation:

$$\delta(b_A F^A) = \epsilon[B_A F^A(\phi) + b_A c^\alpha \delta_\alpha F^A(\phi)]. \qquad (3.7.6)$$

The concept of the BRST symmetry is important for the following reason. When we gauge-fix and introduce the ghosts, the theory is no longer invariant under the original symmetry. The BRST symmetry is a global remnant of the original symmetry, which remains intact.

Consider now a small change in the gauge-fixing condition δF, and look at the change induced in a physical amplitude

$$\epsilon \delta_F \langle \psi | \psi' \rangle = -i \langle \psi | \delta(b_A \delta F^A) | \psi' \rangle = \langle \psi | \{Q_B, b_A \delta F^A\} | \psi' \rangle, \qquad (3.7.7)$$

where Q_B is the conserved charge generating the BRST variation. The amplitude should not change under variation of the gauge condition and we conclude that $(Q_B^\dagger = Q_B)$

$$Q_B |\text{phys}\rangle = 0. \qquad (3.7.8)$$

Therefore, *all physical states must be BRST invariant.*

Next, we have to check whether this BRST charge stays conserved, or, equivalently, whether it commutes with the change in the Hamiltonian under a variation of the gauge condition. The conservation of the BRST charge is equivalent to the statement that our original gauge symmetry is intact. Because of this, we do not want to compromise its conservation in the quantum theory just by changing our gauge-fixing condition:

$$\begin{aligned}
0 = [Q_B, \delta H] &= [Q_B, \delta_B(b_A \delta F^A)] \\
&= [Q_B, \{Q_B, b_A \delta F^A\}] = [Q_B^2, b_A \delta F^A].
\end{aligned} \qquad (3.7.9)$$

This should be true for an arbitrary change in the gauge condition and we conclude

$$Q_B^2 = 0, \qquad (3.7.10)$$

that is, the BRST charge has to be nilpotent for our description of the quantum theory to be consistent. If, for example, there is an anomaly in the gauge symmetry at the quantum level this will show up as a failure of the nilpotency of the BRST charge in the quantum theory. This would imply that the quantum theory as it stands is inconsistent: we have fixed a classical symmetry that is not a symmetry at the quantum level.

The nilpotency of the BRST charge has strong consequences. Consider the state $Q_B|\chi\rangle$. This state will be annihilated by Q_B whatever $|\chi\rangle$ is, so it is physical. However, this state is orthogonal to all physical states including itself and therefore it is a *null state*. Thus, it should be ignored when we discuss quantum dynamics. Two states related by

$$|\psi'\rangle = |\psi\rangle + Q_B|\chi\rangle$$

have the same inner products and are indistinguishable. This is the remnant, in the gauge-fixed version, of the original gauge symmetry. The Hilbert space of physical states is then the cohomology of Q_B, i.e., physical states are the BRST closed states modulo the BRST exact states:

$$Q_B|\text{phys}\rangle = 0 \qquad \text{and} \qquad |\text{phys}\rangle \neq Q_B|\text{something}\rangle. \tag{3.7.11}$$

3.8 BRST in String Theory and the Physical Spectrum

We are now ready to apply this formalism to the bosonic string. We will get rid of the antighost B by explicitly solving the gauge-fixing condition as we did before, by setting the two-dimensional metric to be equal to some fixed reference metric. Expressed in the world-sheet light-cone coordinates, we obtain the following BRST transformations:

$$
\begin{aligned}
\delta_B X^\mu &= i\epsilon(c^+\partial_+ + c^-\partial_-)X^\mu, \\
\delta_B c^\pm &= \pm i\epsilon(c^+\partial_+ + c^-\partial_-)c^\pm, \\
\delta_B b_\pm &= \pm i\epsilon(T_\pm^X + T_\pm^{gh}).
\end{aligned}
\tag{3.8.1}
$$

We used the shorthand notation $T_\pm^X = T_{\pm\pm}(X)$, and so on. The action containing the ghost terms is

$$S_{gh} = \frac{1}{\pi}\int d^2\xi \, (b_{++}\partial_- c^+ + b_{--}\partial_+ c^-). \tag{3.8.2}$$

The stress-tensor for the ghosts has the nonvanishing terms

$$
\begin{aligned}
T_{++}^{gh} &= i(2b_{++}\partial_+ c^+ + \partial_+ b_{++} c^+), \\
T_{--}^{gh} &= i(2b_{--}\partial_- c^- + \partial_- b_{--} c^-),
\end{aligned}
\tag{3.8.3}
$$

and its conservation becomes

$$\partial_- T_{++}^{gh} = \partial_+ T_{--}^{gh} = 0. \tag{3.8.4}$$

The equations of motion for the ghosts are

$$\partial_- b_{++} = \partial_+ b_{--} = \partial_- c^+ = \partial_+ c^- = 0. \tag{3.8.5}$$

We have to impose again the appropriate periodicity (closed strings) or boundary (open strings) conditions on the ghosts, and then we can expand the fields in Fourier modes again:

$$c^+ = \sum \bar{c}_n e^{-in(\tau+\sigma)}, \quad c^- = \sum c_n e^{-in(\tau-\sigma)},$$
$$b_{++} = \sum \bar{b}_n e^{-in(\tau+\sigma)}, \quad b_{--} = \sum c_n e^{-in(\tau-\sigma)}.$$

The Fourier modes satisfy the following anticommutation relations:

$$\{b_m, c_n\} = \delta_{m+n,0}, \quad \{b_m, b_n\} = \{c_m, c_n\} = 0. \tag{3.8.6}$$

We can define the Virasoro operators for the ghost system as the expansion modes of the stress-tensor. We find

$$L_m^{gh} = \sum_n (m-n) : b_{m+n} c_{-n} : , \quad \bar{L}_m^{gh} = \sum_n (m-n) : \bar{b}_{m+n} \bar{c}_{-n} : . \tag{3.8.7}$$

From this we can compute the algebra of Virasoro operators:

$$[L_m^{gh}, L_n^{gh}] = (m-n)L_{m+n}^{gh} + \tfrac{1}{6}(m - 13m^3)\delta_{m+n,0}. \tag{3.8.8}$$

The total Virasoro operators for the combined system of X^μ fields and ghost then become

$$L_m = L_m^X + L_m^{gh} - a\delta_{m,0}, \tag{3.8.9}$$

where the constant term is due to normal ordering of L_0. The algebra of the combined system can then be written as

$$[L_m, L_n] = (m-n)L_{m+n} + A(m)\delta_{m+n}, \tag{3.8.10}$$

with

$$A(m) = \frac{d}{12}m(m^2 - 1) + \frac{1}{6}(m - 13m^3) + 2am. \tag{3.8.11}$$

This term vanishes, if and only if $d = 26$ and $a = 1$. This is exactly the same result we obtained from requiring Lorentz invariance after quantization in the light-cone gauge or the absence of negative norm states in the old covariant quantization.

The need for $d = 26$ can also be shown using the BRST formalism. Invariance under BRST transformation induces, via Noether's theorem, a BRST current

$$j_B = cT^X + \tfrac{1}{2} : cT^{gh} := cT^X + : bc\partial c : , \tag{3.8.12}$$

and the BRST charge becomes

$$Q_B = \int d\sigma j_B.$$

The anomaly now shows up in Q_B^2: the BRST charge is nilpotent if and only if $d = 26$.

We can express the BRST charge in terms of the X^μ Virasoro operators and the ghost oscillators as

$$Q_B = \sum_n c_n L_{-n}^X + \sum_{m,n} \frac{m-n}{2} : c_m c_n b_{-m-n} : -c_0, \tag{3.8.13}$$

where the c_0 term comes from the normal ordering of L_0^X. In the case of closed strings there is of course a similar expression for $\bar{Q}_B$, and the BRST charge is $Q_B + \bar{Q}_B$.

We will now find the physical spectrum in the BRST framework. According to our previous discussion, the physical states must be annihilated by the BRST charge, and not be of the form $Q_B|\text{any}\rangle$.

We must describe our extended Hilbert space that includes the ghosts. As far as the X^μ oscillators are concerned the situation is the same as in the previous sections. We therefore concentrate on the ghost Hilbert space. The full Hilbert space will be a tensor product of the two.

First, we must describe the ghost vacuum state. This should be annihilated by the positive ghost oscillator modes

$$b_{n>0}|\text{ghost vacuum}\rangle = c_{n>0}|\text{ghost vacuum}\rangle = 0. \tag{3.8.14}$$

However, there is a subtlety because of the presence of the zero modes b_0 and c_0 which, according to (3.8.6), satisfy

$$b_0^2 = c_0^2 = 0, \quad \{b_0, c_0\} = 1. \tag{3.8.15}$$

These anticommutation relations are the same as those of the Clifford algebra in two space-time dimensions in light-cone coordinates. The simplest representation of this algebra is therefore two dimensional and is realized by $b_0 = (\sigma^1 + i\sigma^2)/\sqrt{2}$ and $c_0 = (\sigma^1 - i\sigma^2)/\sqrt{2}$ where σ^i are the Pauli matrices. Thus, in this representation, there should be two states: a "spin-up" and a "spin-down" state, satisfying[6]

$$b_0|\downarrow\rangle = 0, \quad b_0|\uparrow\rangle = |\downarrow\rangle, \quad c_0|\uparrow\rangle = 0, \quad c_0|\downarrow\rangle = |\uparrow\rangle, \quad ((\uparrow|)^\dagger = |\downarrow\rangle. \tag{3.8.16}$$

Because of this, we have to impose one more condition on the physical states. This condition asks that the zero mode b_0 annihilates the physical states:

$$b_0|\text{phys}\rangle = 0. \tag{3.8.17}$$

This is known as the "Siegel gauge" and although its imposition seems mysterious at this level, it is needed.

To see why, consider the simpler case of the massive point particle. This corresponds to keeping the zero modes only in our expansions, and using the identification $L_0 = H = p^2 + m^2$. We tensor the X^μ Hilbert space (which in the particle case is given by the states $|p^\mu\rangle$) with that of the ghosts. We obtain two types of states $|\vec{p}; \downarrow\rangle$ and $|\vec{p}; \uparrow\rangle$. The BRST charge is here $Q_B = c_0 H = c_0 L_0$.

Acting with it on the two possible types of states we obtain

$$Q_B|\vec{p}; \downarrow\rangle = (p^2 + m^2)|\vec{p}; \uparrow\rangle, \quad Q_B|\vec{p}; \uparrow\rangle = 0, \tag{3.8.18}$$

[6] There is an SL$(2, \mathbb{C})$-invariant ghost vacuum $|0\rangle$ satisfying the standard requirements
$$b_{n \geq -1}|0\rangle = 0, \quad c_{n \geq 2}|0\rangle = 0, \quad L_{n \geq -1}^{gh}|0\rangle = 0.$$
The states $|\downarrow\rangle$ and $|\uparrow\rangle$ are then given as
$$|\downarrow\rangle = c_1|0\rangle, \quad |\uparrow\rangle = c_0 c_1|0\rangle$$
and therefore satisfy (3.8.16). The nontrivial inner product in this language is $\langle 0|c_{-1}c_0c_1|0\rangle \neq 0$, reflecting the presence of the three c zero modes on the sphere.

which indicates that $|\vec{p}; \downarrow\rangle$ is BRST closed on shell while $|\vec{p}; \uparrow\rangle$ is closed for any p. We now have to factor exact states, that is, states of the form given in (3.8.18).

It is obvious that the states $|\vec{p}; \uparrow\rangle$ for $p^2 + m^2 \neq 0$ are BRST exact. Therefore the physical states (closed/exact) are $|\vec{p}; \downarrow\rangle$ and $|\vec{p}; \uparrow\rangle$ both with $p^2 + m^2 = 0$. We have obtained two copies of the expected spectrum (a single state satisfying the mass-shell condition).

Note that, since the states $|\vec{p}; \uparrow\rangle$ are exact for all off-shell values of $\vec{p}$, and therefore null, their amplitudes are zero except on mass shell. Thus, their amplitudes are proportional to $\delta(p^2 + m^2)$ which is an unacceptable behavior for physical amplitudes (which must have poles and branch cuts at best in the p^2-plane). We conclude that this copy must be projected out, and this is the purpose of the Siegel gauge condition $b_0|\text{phys}\rangle = 0$.

We therefore impose also (3.8.17) which implies that the correct ghost vacuum is $| \downarrow\rangle$. We can now create states from this vacuum by acting with the negative modes of the ghosts b_m, c_n. We cannot act with c_0 since the new state does not satisfy the Siegel condition (3.8.17).

We are now ready to describe the physical states in the open string. Note that since Q_B in (3.8.13) has "level" zero,[7] we can impose BRST invariance on physical states level by level. We will decompose the BRST charge in levels as

$$Q_B = Q_0 + Q_1 + \cdots, \quad Q_0 = c_0(L_0 - 1), \quad Q_1 = c_1 L_{-1} + c_{-1} L_1 + c_0(c_{-1}b_1 + b_{-1}c_1). \qquad (3.8.19)$$

At level zero, there is only one state, the total vacuum $| \downarrow, p^\mu\rangle$. The relevant part of Q_B is Q_0 and

$$0 = Q_B| \downarrow, p\rangle = (L_0^X - 1)c_0| \downarrow, p\rangle. \qquad (3.8.20)$$

BRST invariance gives the same mass-shell condition, namely, $L_0^X - 1 = 0$, that we obtained in previous quantization schemes. This state cannot be a BRST exact state; it is therefore physical: it is the tachyon.

At the first level, the possible operators are α_{-1}^μ, b_{-1}, and c_{-1}. The most general state of this form is then

$$|\psi\rangle = (\zeta \cdot \alpha_{-1} + \xi_1 c_{-1} + \xi_2 b_{-1})| \downarrow, p\rangle, \qquad (3.8.21)$$

which has 28 parameters: a 26-vector ζ_μ and two more constants ξ_1, ξ_2. The BRST condition demands

$$0 = Q_B|\psi\rangle = [c_0 \ell_s^2 p^2 (\zeta \cdot \alpha_{-1} + \xi_1 c_{-1} + \xi_2 b_{-1}) + \sqrt{2}\ell_s(p \cdot \zeta c_{-1} + \xi_2 p \cdot \alpha_{-1})]| \downarrow, p\rangle. \qquad (3.8.22)$$

This holds only if $p^2 = 0$ (the state is massless), $p \cdot \zeta = 0$, and $\xi_2 = 0$. So there are only 26 parameters left. Next we have to make sure that this state is not Q_B exact: a general Q_B-exact state is $Q_B|\chi\rangle$ with $|\chi\rangle$ of the same form as (3.8.21), but with parameters $\zeta'^\mu, \xi'_{1,2}$.

---

[7] By level here we mean total mode number. Thus, L_0 and $L_{-n}L_n$ both have level zero.

The state χ must also satisfy $p^2 = 0$ so that $Q_B|\chi\rangle$ satisfies the Siegel condition. Therefore, the most general Q_B-exact state at this level with $p^2 = 0$ will be

$$Q_B|\chi\rangle = \sqrt{2}\ell_s(p \cdot \zeta' c_{-1} + \xi'_2 p \cdot \alpha_{-1})|\downarrow, p\rangle.$$

This means that the c_{-1} part in (3.8.21) is BRST exact and that the polarization has the equivalence relation $\zeta_\mu \sim \zeta_\mu + \xi'_2 p_\mu$. This leaves us with the 24 physical degrees of freedom we expect for a massless vector particle in 26 dimensions.

We have seen explicitly in our analysis that the exact states $Q_B|\text{phys}\rangle$ correspond to gauge transformations of the physical states.

It is clear that the same procedure can be applied to the higher levels. In the case of the closed string we have to include the barred operators, and of course we have to use $Q_B + \bar{Q}_B$.

Bibliography

Side reading can be found in [7] where the different quantization procedures are described in detail. The connection of the light-cone quantization to the old covariant quantization is discussed and is accompanied by the construction of the physical DDF operators. The book [12] treats the light-cone quantization in considerable detail. The no-ghost theorem was proved in [27,28]. It is also presented in [7] and [8].

Constrained quantization is described in [29]. The BRST quantization in general is described in two good review papers [30,31]. In the context of gauge theory and string theory it is discussed in [32,33].

The Polyakov approach is described in [34]. The lectures in [35] are also very useful. A detailed, rigorous, and well-explained account of the Polyakov approach on arbitrary open and closed Riemann surfaces can be found in [36]. A review on various aspects of Riemann surfaces entering string perturbation theory can be found in [37].

The various gauge groups that can appear in open strings due to CP factors as well as the relevant tree-level constraints are analyzed in [38].

For a field theory of strings, the reader may consult [39] for a description of open string field theory. Reviews on open string field theory can be found in [40,41]. A review of the current status of closed string field theory is [42].

Exercises

3.1. Using the commutation relations of the oscillators, calculate carefully the commutator of the Virasoro operators in (3.1.9) on page 30. Show that the normal ordering constant a is formally equal to

$$a = \frac{D-2}{2} \sum_{n=1}^{\infty} n. \tag{3.1E}$$

Regularize this divergent sum using the zeta function regularization

$$\sum_{n=1}^{\infty} n = \zeta(-1) = -\frac{1}{12}. \qquad (3.2E)$$

3.2. Redo the previous calculation using appropriate expectation values of the commutator, in order to unambiguously compute the central term.

3.3. Show that the Virasoro algebra (3.1.9) on page 30 satisfies the Jacobi identity.

3.4. Solve the light-cone constraints to find (3.2.2) on page 31.

3.5. Show that only when the left and right intercepts are the same in the closed string, $a = \bar{a}$, do we obtain a nontrivial spectrum consistent with Lorentz invariance.

3.6. Consider the massive states of the bosonic open string in $d = 26$ at the second level. Show that they form representations of massive little group O(25).

3.7. Consider the generators of the Lorentz symmetry introduced at the classical level in (2.3.12) on page 21. Normal-order them and then compute their commutators with the Virasoro generators. Use this to evaluate their commutation relations. Show that they represent faithfully the Poincaré algebra when $d = 26$.

3.8. Derive (3.4.5) on page 34.

3.9. Consider the Hermitian matrices in (3.4.17). Show that they form an SO(N) algebra, for any of the the values of the phases θ_i. Similarly show that the matrices (3.4.20) in the antisymmetric case generate an Sp(2N) group.

3.10. Consider matrices λ_V satisfying the condition (3.4.13) on page 36 for the vectors and matrices λ_R satisfying the opposite sign conditions (3.4.23) on page 37. Show that λ_R form a representation of the algebra of the λ_V. Identify these representations for γ_Ω both symmetric and antisymmetric.

3.11. Consider an n-index traceless symmetric tensor in two dimensions. Show that it has two independent components.

3.12. Consider the two-dimensional sphere with the round metric. Find explicitly the six conformal Killing vectors. Show that there are no Teichmüller parameters here.

3.13. Consider the two-dimensional torus. Find explicitly the two conformal Killing vectors. Find also the two (real) Teichmüller parameters.

3.14. Calculate the Euler characteristic of the disk, by picking first a flat metric and then a round metric where the disk is a hemisphere.

3.15. Show that the Euler characteristic of a torus with n handles is $\chi = 2 - 2n$.

3.16. Derive the gauge-fixed form of the point-particle path integral including the b and c ghosts in analogy with the string case. Use this to set up the BRST quantization of the free particle.

3.17. Derive the BRST variations (3.8.1) on page 42. Use them to show that classically $Q_B^2 = 0$.

3.18. Derive (3.8.6) and (3.8.7) on page 43 and calculate the central charge of the ghost stress-tensor, verifying (3.8.8).

3.19. Use the BRST transformations (3.8.1) on page 42 to derive the BRST current in (3.8.12).

3.20. Show by direct calculation that for the bosonic string $Q_B^2 = 0$ when $d = 26$.

3.21. Consider the closed bosonic string. Redo the BRST analysis and find the physical states at the zeroth and first (massless) levels.

3.22. Consider the ghost zero-mode algebra (3.8.15) on page 44. We have used the representation of this algebra that had the minimal dimension, namely, two. Pick now a higher-dimensional representation of this Clifford algebra. Rediscuss the space of physical states and the necessary gauge conditions. Is this a better choice?

4 Conformal Field Theory

We have seen so far that the world-sheet quantum theory that describes the bosonic string is a conformally invariant quantum field theory in two dimensions. In this chapter we will give a basic introduction to such conformal field theories and their application in string theory. We will assume Euclidean signature in two dimensions. Whenever world-sheet scalars appear, we kept the string length ℓ_s explicit in the formulas. Comparisons can be made with a major part of the literature by setting $\ell_s = \sqrt{2}$, corresponding to $\alpha' = 2$.

4.1 Conformal Transformations

Under general coordinate transformations $x \to x'$, the metric transforms as

$$g_{\mu\nu} \to g'_{\mu\nu}(x') = \frac{\partial x^\alpha}{\partial x'^\mu} \frac{\partial x^\beta}{\partial x'^\nu} g_{\alpha\beta}(x). \tag{4.1.1}$$

The group of conformal transformations, in any dimension, is then defined as the sub-group of these coordinate transformations that leave the metric invariant up to a rescaling[1]

$$g_{\mu\nu}(x) \to g'_{\mu\nu}(x') = \Omega(x)g_{\mu\nu}(x). \tag{4.1.2}$$

These are precisely the coordinate transformations that preserve the angle between two vectors, hence the name conformal transformations. For flat space note that the Poincaré group is a subgroup of the conformal group (with $\Omega(x) = 1$).

We will examine the generators of these transformations. Under infinitesimal coordinate transformations $x^\mu \to x'^\mu = x^\mu + \epsilon^\mu$, we obtain

$$\delta g_{\mu\nu} = -(\partial^\lambda \epsilon_\mu g_{\lambda\nu} + \partial^\lambda \epsilon_\nu g_{\lambda\mu}) - \epsilon^\lambda \partial_\lambda g_{\mu\nu}. \tag{4.1.3}$$

For it to be a conformal transformation around the flat metric we must have

$$\partial_\mu \epsilon_\nu + \partial_\nu \epsilon_\mu = \frac{2}{d}(\partial \cdot \epsilon)\delta_{\mu\nu}. \tag{4.1.4}$$

---

[1] Not to be confused with Weyl rescalings of the metric that do not act on the coordinates.

The proportionality factor can be found by contracting both sides with $\delta^{\mu\nu}$. If we act on both sides of this equation with ∂^μ we obtain

$$\Box \epsilon_\nu + \left(1 - \frac{2}{d}\right)\partial_\nu(\partial \cdot \epsilon) = 0. \tag{4.1.5}$$

If we now act on both sides of (4.1.4) with $\Box = \partial_\mu \partial^\mu$ we obtain

$$\partial_\mu \Box \epsilon_\nu + \partial_\nu \Box \epsilon_\mu = \frac{2}{d}\delta_{\mu\nu}\Box(\partial \cdot \epsilon). \tag{4.1.6}$$

With these two equations, we can write the constraints on the parameter as follows:

$$\left[\delta_{\mu\nu}\Box + (d-2)\partial_\mu\partial_\nu\right]\partial \cdot \epsilon = 0. \tag{4.1.7}$$

We can already see in (4.1.7) that $d = 2$ will be a special case. Indeed for $d > 2$, (4.1.7) implies that the parameter ϵ can be at most quadratic in x. This is because it is cubic in derivatives and nondegenerate. We can identify the following possibilities for ϵ:

$$
\begin{aligned}
\epsilon^\mu &= a^\mu, & &\text{translations,} \\
\epsilon^\mu &= \omega^\mu{}_\nu x^\nu, & &\text{rotations } (\omega_{\mu\nu} = -\omega_{\nu\mu}), \\
\epsilon^\mu &= \lambda x^\mu, & &\text{scale transformations,} \\
\epsilon^\mu &= b^\mu x^2 - 2x^\mu(b \cdot x), & &\text{special conformal transformations.}
\end{aligned}
\tag{4.1.8}
$$

The finite scale and special conformal transformations are obtained by exponentiating the infinitesimal ones:

$$D: \quad x^\mu \to \lambda x^\mu, \quad K_\mu: \quad x^\mu \to \frac{x^\mu + x^2\, a^\mu}{1 + 2x \cdot a + x^2\, a^2}. \tag{4.1.9}$$

Together with the generators of the Poincaré group, namely, translations P_μ and rotations $J_{\mu\nu}$, they form the conformal group. The generators are

$$P_\mu = -i\partial_\mu, \quad J_{\mu\nu} = i(x_\mu \partial_\nu - x_\nu \partial_\mu), \quad K_\mu = -i\left[x^2\partial_\mu - 2x_\mu(x \cdot \partial)\right], \quad D = -ix \cdot \partial, \tag{4.1.10}$$

with commutation relations

$$
\begin{aligned}
&[J_{\mu\nu}, P_\rho] = -i(\eta_{\mu\rho}P_\nu - \eta_{\nu\rho}P_\mu), \\
&[P_\mu, K_\nu] = 2iJ_{\mu\nu} - 2i\eta_{\mu\nu} D, \\
&[J_{\mu\nu}, J_{\rho\sigma}] = -i\left[\eta_{\mu\rho}J_{\nu\sigma} - \eta_{\mu\sigma}J_{\nu\rho} - \eta_{\nu\rho}J_{\mu\sigma} + \eta_{\nu\sigma}J_{\mu\rho}\right], \\
&[J_{\mu\nu}, K_\rho] = -i(\eta_{\mu\rho}K_\nu - \eta_{\nu\rho}K_\mu), \\
&[D, K_\mu] = iK_\mu, \quad [D, P_\mu] = -iP_\mu, \quad [J_{\mu\nu}, D] = 0.
\end{aligned}
\tag{4.1.11}
$$

You are asked to verify this in exercise 4.1 on page 118.

We have a total of

$$d + \tfrac{1}{2}d(d-1) + 1 + d = \tfrac{1}{2}(d+2)(d+1)$$

parameters. In a space of signature (p, q) with $d = p + q$, the Lorentz group is $O(p, q)$. The conformal group is then $O(p+1, q+1)$. The generators can be relabeled as

$$M_{\mu\nu} = J_{\mu\nu}, \quad M_{\mu,d} = \frac{1}{2}(K_\mu - P_\mu), \quad M_{\mu,d+1} = \frac{1}{2}(K_\mu + P_\mu), \quad M_{d,d+1} = D. \tag{4.1.12}$$

You are invited to verify this in exercise 4.2 on page 118.

4.1.1 *The case of two dimensions*

We will now investigate the special case $d = 2$. The restriction that ϵ can be at most of second order no longer applies, and (4.1.4) in Euclidean space ($g_{\mu\nu} = \delta_{\mu\nu}$) reduces to

$$\partial_1\epsilon_1 = \partial_2\epsilon_2, \qquad \partial_1\epsilon_2 = -\partial_2\epsilon_1. \tag{4.1.13}$$

This can be further simplified by going to complex coordinates $z, \bar{z} = x^1 \pm ix^2$. If we define the complex parameters $\epsilon, \bar{\epsilon} = \epsilon_1 \pm i\epsilon_2$, the equations for the parameters become

$$\partial\bar{\epsilon} = 0, \qquad \bar{\partial}\epsilon = 0, \tag{4.1.14}$$

where we used the shorthand notation $\bar{\partial} = \partial_{\bar{z}}$. This means that ϵ can be an arbitrary function of z, but it is independent of $\bar{z}$, and vice versa for $\bar{\epsilon}$. Globally, this means that conformal transformations in two dimensions consist of the analytic coordinate transformations[2]

$$z \to f(z) \quad \text{and} \quad \bar{z} \to \bar{f}(\bar{z}). \tag{4.1.15}$$

We can expand the infinitesimal transformation parameter as

$$\epsilon(z) = -\sum a_n z^{n+1}.$$

The generators corresponding to these transformations are then

$$\ell_n = -z^{n+1}\partial_z, \tag{4.1.16}$$

i.e., ℓ_n generates the transformation with $\epsilon = -z^{n+1}$. The generators satisfy the following algebra:

$$[\ell_m, \ell_n] = (m - n)\ell_{m+n}, \qquad [\bar{\ell}_m, \bar{\ell}_n] = (m - n)\bar{\ell}_{m+n}, \tag{4.1.17}$$

and $[\bar{\ell}_m, \ell_n] = 0$. Therefore, the conformal group in two dimensions is infinite dimensional.

An interesting subalgebra of this algebra is spanned by the generators $\ell_{0,\pm 1}$ and $\bar{\ell}_{0,\pm 1}$. These are the only generators that are globally well defined on the Riemann sphere $S^2 = \mathbb{C} \cup \infty$. They form the algebra $O(3,1) \sim SL(2,\mathbb{C})$. They generate the following transformations:

Generator	Infinitesimal Transformation	Finite Transformation	
ℓ_{-1}	$z \to z - \epsilon$	$z \to z + \alpha$	Translations
ℓ_0	$z \to z - \epsilon z$	$z \to \lambda z$	Scaling
ℓ_1	$z \to z - \epsilon z^2$	$z \to \frac{z}{1-\beta z}$	Special conformal

with equivalent expressions for the barred generators. From this, it is immediately clear that the generator $i(\ell_0 - \bar{\ell}_0)$ generates a rescaling of the phase or, in other words, it generates rotations in the z-plane. Dilatations are generated by $\ell_0 + \bar{\ell}_0$. The transformations generated by $\ell_{0,\pm 1}$ can be summarized by the expression

$$z \to \frac{az + b}{cz + d}, \tag{4.1.18}$$

---

[2] You may find some useful two-dimensional complex geometry facts and conventions in appendix A on page 503.

where $a, b, c, d \in \mathbb{C}$ and $ad - bc = 1$. This is the group $\mathrm{PSL}(2, \mathbb{C}) = \mathrm{SL}(2,\mathbb{C})/\mathbb{Z}_2$, where the $\mathbb{Z}_2$ fixes the freedom to replace all parameters a, b, c, d by minus themselves, leaving the transformation (4.1.18) unchanged. We will call this finite-dimensional subgroup of the conformal group the *restricted conformal group*.

4.2 Conformally Invariant Field Theory

A two-dimensional theory will be called conformally invariant if the trace of its stress-tensor vanishes in the quantum theory in flat space. Such a theory has the following properties:

(i) It contains an (infinite) set of fields $\{A_i\}$. In particular, this set will contain all the derivatives of local fields.

(ii) There exists a subset $\{\phi_j\} \subset \{A_i\}$, called quasi-primary fields, that transforms under *restricted* conformal transformations

$$z \to f(z) = \frac{az + b}{cz + d}, \quad \bar{z} \to \bar{f}(\bar{z}) = \frac{\bar{a}\bar{z} + \bar{b}}{\bar{c}\bar{z} + \bar{d}}, \tag{4.2.1}$$

as tensors of weight $(\Delta, \bar{\Delta})$, namely,

$$\Phi(z, \bar{z}) = \left(\frac{\partial f}{\partial z}\right)^\Delta \left(\frac{\partial \bar{f}}{\partial \bar{z}}\right)^{\bar{\Delta}} \Phi\left(f(z), \bar{f}(\bar{z})\right). \tag{4.2.2}$$

Such tensors are described in appendix A on page 503. All fields that are not derivatives of other fields are quasi-primary.

(iii) Finally there are the so-called primary fields, which transform as in (4.2.2) for all conformal transformations; $\Delta, \bar{\Delta}$ are real valued ($\bar{\Delta}$ is not the complex conjugate of Δ).

The expression

$$\Phi(z, \bar{z}) \, dz^\Delta \, d\bar{z}^{\bar{\Delta}}$$

is invariant under conformal transformations; $(\Delta, \bar{\Delta})$ are the conformal weights of the primary field.

(iv) The theory is covariant under conformal transformations. Consequently, the correlation functions satisfy

$$\left\langle \prod_{i=1}^N \Phi_i(z_i, \bar{z}_i) \right\rangle = \prod_{i=1}^N \left(\frac{\partial f}{\partial z}\right)_{z \to z_i}^{\Delta_i} \left(\frac{\partial \bar{f}}{\partial \bar{z}}\right)_{\bar{z} \to \bar{z}_i}^{\bar{\Delta}_i} \left\langle \prod_{j=1}^N \Phi_j\left(f(z_j), \bar{f}(\bar{z}_j)\right) \right\rangle. \tag{4.2.3}$$

As we shall see later on, the conformal anomaly spontaneously breaks the invariance of the full conformal group. On the sphere, the unbroken subgroup is the restricted conformal group and (4.2.3) is thus valid only for $\mathrm{SL}(2, \mathbb{C})$. However, there will be Ward identities that will encode the full conformal covariance of the theory.

Infinitesimally, under $z \to z + \epsilon(z)$ and $\bar{z} \to \bar{z} + \bar{\epsilon}(\bar{z})$, from (4.2.2), a primary field transforms as

$$\delta_{\epsilon, \bar{\epsilon}} \Phi(z, \bar{z}) \equiv \Phi(z, \bar{z}) - \Phi'(z + \epsilon(z), \bar{z} + \bar{\epsilon}(\bar{z})) = \left[(\Delta \partial \epsilon + \epsilon \partial) + (\bar{\Delta} \bar{\partial} \bar{\epsilon} + \bar{\epsilon} \bar{\partial})\right] \Phi(z, \bar{z}). \tag{4.2.4}$$

Covariance of the two-point function $G^{(2)}(z_i, \bar{z}_i) = \langle \Phi(z_1, \bar{z}_1)\Phi(z_2, \bar{z}_2) \rangle$ under all conformal transformations would imply

$$\delta_{\epsilon,\bar{\epsilon}} G^{(2)}(z_i, \bar{z}_i) = \langle \delta_{\epsilon,\bar{\epsilon}} \Phi_1, \Phi_2 \rangle + \langle \Phi_1, \delta_{\epsilon,\bar{\epsilon}} \Phi_2 \rangle = 0. \tag{4.2.5}$$

The combination of (4.2.4) and (4.2.5) yields the following differential equation for the two-point function:

$$\left[\left(\epsilon(z_1)\partial_{z_1} + \Delta_1 \partial \epsilon(z_1) + \epsilon(z_2)\partial_{z_2} + \Delta_2 \partial \epsilon(z_2) \right) + (\text{barred terms}) \right] G^{(2)}(z_i, \bar{z}_i) = 0. \tag{4.2.6}$$

As we will show in section 4.6 on page 59 this equation is valid only for the conformal transformations left unbroken by the vacuum. For the sphere, these are the restricted conformal transformations forming the SL(2,$\mathbb{C}$) group.

We can now use the series expansion of $\epsilon(z)$ to analyze this equation. If we first take $\epsilon(z) = 1$ and $\bar{\epsilon}(\bar{z}) = 1$ (translations), then (4.2.6) tells us that $G^{(2)}(z_i, \bar{z}_i)$ depends only on $z_{12} = z_1 - z_2, \bar{z}_{12} = \bar{z}_1 - \bar{z}_2$. This is not very surprising because in a translationally invariant theory we would expect the correlation functions to depend only on the relative distance. If we next use $\epsilon(z) = z, \bar{\epsilon}(\bar{z}) = \bar{z}$ (rotational invariance), we find $G^{(2)} \sim 1/(z_{12}^{\Delta_1+\Delta_2} \bar{z}_{12}^{\bar{\Delta}_1+\bar{\Delta}_2})$ and if we finally use $\epsilon(z) = z^2$ (special conformal transformation) we find the restriction $\Delta_1 = \Delta_2 = \Delta$ and $\bar{\Delta}_1 = \bar{\Delta}_2 = \bar{\Delta}$.

The conclusion is that the two-point function is completely fixed up to a constant:

$$G^{(2)}(z_i, \bar{z}_i) = \frac{C_{12}}{z_{12}^{2\Delta} \bar{z}_{12}^{2\bar{\Delta}}}. \tag{4.2.7}$$

This constant can be set to 1, by normalizing the operators.

A similar analysis can be done for the three-point function and it turns out to be also completely determined up to a constant. You are invited to explore this in exercise 4.5 on page 119.

The next correlation function, however, the four-point function, is not fully determined. Conformal invariance restricts it, using the procedure outlined above, to have the following form:

$$G^{(4)}(z_i, \bar{z}_i) = f(x, \bar{x}) \prod_{i<j} z_{ij}^{-(\Delta_i+\Delta_j)+\Delta/3} \prod_{i<j} \bar{z}_{ij}^{-(\bar{\Delta}_i+\bar{\Delta}_j)+\bar{\Delta}/3}, \tag{4.2.8}$$

where $\Delta = \sum_{i=1}^4 \Delta_i$, $\bar{\Delta} = \sum_{i=1}^4 \bar{\Delta}_i$. The function f is arbitrary, but depends only on the cross ratio $x = z_{12}z_{23}/z_{13}z_{24}$ and $\bar{x}$.

The general N-point function of quasi-primary fields on the sphere

$$G^N(z_1, \bar{z}_1, \ldots z_N, \bar{z}_N) = \left\langle \prod_{i=1}^N \Phi_i(z_i, \bar{z}_i) \right\rangle \tag{4.2.9}$$

satisfies the following constraints coming from SL(2,$\mathbb{C}$) covariance:

$$\sum_{i=1}^N \partial_i \, G^N = 0, \tag{4.2.10}$$

$$\sum_{i=1}^N (z_i \partial_i + \Delta_i) \, G^N = 0, \tag{4.2.11}$$

$$\sum_{i=1}^N (z_i^2 \partial_i + 2z_i \Delta_i) \, G^N = 0, \tag{4.2.12}$$

and similar ones with $z_i \rightarrow \bar{z}_i$, $\Delta_i \rightarrow \bar{\Delta}_i$. These are the Ward identities reflecting SL(2,$\mathbb{C}$) invariance of the correlation functions on the sphere.

4.3 Radial Quantization

We will now study the Hilbert space of a conformally invariant theory. We start from the two-dimensional Euclidean plane with coordinates $\sigma^0 = \tau$ and $\sigma^1 = \sigma$. (Note that we will eventually go from a two-dimensional Euclidean space to Minkowski space by means of a Wick rotation, $\tau \rightarrow i\tau$.) To avoid IR problems, we will compactify the space direction, $\sigma = \sigma + 2\pi$, and the two-dimensional space becomes a cylinder. Next, we make the conformal transformation

$$z = e^{\tau+i\sigma}, \qquad \bar{z} = e^{\tau-i\sigma}, \tag{4.3.1}$$

which maps the cylinder onto the complex plane (topologically a sphere) as shown in figure 4.1.

Surfaces of equal time on the cylinder will become circles of equal radius on the complex plane. This means that the infinite past ($\tau = -\infty$) gets mapped onto the origin of the plane ($z = 0$) and the infinite future becomes $z = \infty$. Time reversal becomes $z \rightarrow 1/z^*$ on the complex plane, and parity $z \rightarrow z^*$.

We already saw that ℓ_0 was the generator of dilatations on the cylinder, $z \rightarrow \lambda z$, so $\ell_0 + \bar{\ell}_0$ will move us in the radial direction on the plane, which corresponds to the time direction on the cylinder. This means that the dilatation operator is the Hamiltonian of our system,[3]

$$H = \ell_0 + \bar{\ell}_0. \tag{4.3.2}$$

An integral over the space direction σ will become a contour integral around the origin of the complex plane. This enables us to use all the powerful techniques developed in complex analysis.

Infinitesimal coordinate transformations are generated by the stress-tensor, which is traceless in the case of a conformal field theory (CFT),

$$T_\mu{}^\mu = 0. \tag{4.3.3}$$

This is true only in flat space. We will discuss the general case when the world-sheet is curved in section 4.9 on page 64.

Condition (4.3.3) in complex coordinates means that the stress-tensor has nonvanishing components T_{zz} and $T_{\bar{z}\bar{z}}$, while $T_{z\bar{z}} = 0$ since $T_{z\bar{z}}$ is the trace of the stress-tensor. This can be shown by expressing them back in Euclidean coordinates, $z = x^0 + ix^1$,

$$T_{z\bar{z}} = T_{\bar{z}z} = \tfrac{1}{4}(T_{00} + T_{11}) = \tfrac{1}{4}T_\mu{}^\mu, \tag{4.3.4}$$

$$T_{zz} = \tfrac{1}{4}(T_{00} - T_{11} + 2iT_{01}), \quad T_{\bar{z}\bar{z}} = \tfrac{1}{4}(T_{00} - T_{11} - 2iT_{01}). \tag{4.3.5}$$

---

[3] Since we are in Euclidean space, the term Hamiltonian may appear bizarre. The proper name should be transfer operator, which upon Wick rotation becomes the Hamiltonian. Similarly, the exponential of the transfer operator gives the transfer matrix, which would become the time evolution operator upon Wick rotation.

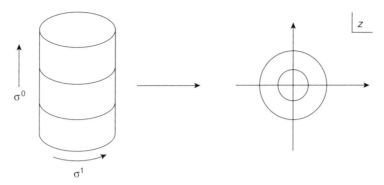

Figure 4.1 The map from the cylinder to the compactified complex plane.

The conservation law $\partial^\mu T_{\mu\nu} = 0$ gives us, together with the traceless condition,

$$\partial_z T_{\bar z\bar z} = 0 \quad \text{and} \quad \partial_{\bar z} T_{zz} = 0, \tag{4.3.6}$$

which implies that the two nonvanishing components of the stress-tensor are holomorphic and antiholomorphic, respectively:

$$T(z) \equiv T_{zz} \quad \text{and} \quad \bar T(\bar z) \equiv T_{\bar z\bar z}. \tag{4.3.7}$$

Thus, we can construct an infinite number of conserved currents, because if $T(z)$ is conserved, then $\epsilon(z)T(z)$ is also conserved, for every holomorphic function $\epsilon(z)$.

These currents produce the following conserved charges:

$$Q_\epsilon = \frac{1}{2\pi i} \oint dz\, \epsilon(z) T(z), \quad Q_{\bar\epsilon} = \frac{1}{2\pi i} \oint d\bar z\, \bar\epsilon(\bar z) \bar T(\bar z). \tag{4.3.8}$$

Such charges are the generators of the infinitesimal conformal transformations

$$z \to z + \epsilon(z), \qquad \bar z \to \bar z + \bar\epsilon(\bar z).$$

The variation of fields under these transformations is given, as usual, by the commutator of the fields with the generators

$$\delta_{\epsilon,\bar\epsilon}\Phi(z,\bar z) = [Q_\epsilon + Q_{\bar\epsilon}, \Phi(z,\bar z)]. \tag{4.3.9}$$

We know that expectation values of products of operators are well defined in a quantum theory only if the operators are time ordered. The analog of this in radial quantization on the complex plane is radial ordering. The radial-ordering operator R is defined as

$$R(A(z)B(w)) = \begin{cases} A(z)B(w), & |z| > |w|, \\ (-1)^F B(w)A(z), & |z| < |w|. \end{cases} \tag{4.3.10}$$

For fermionic operators, $F = 1$ above.

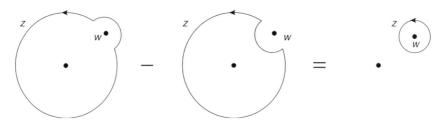

Figure 4.2 Rearrangement of contours relevant for commutators.

With the help of this ordering, we can write an equal-time commutator of an operator with a spatial integral over another operator as a contour integral over the radially ordered product of the two operators:

$$\left[\int d\sigma\, B, A \right] = \oint dz\, R(B(z)A(w)) \tag{4.3.11}$$

as shown in figure 4.2.

In general, the radially ordered product of two operators can be expanded in terms of a complete set of orthonormal local operators

$$\Phi_i(z, \bar{z})\Phi_j(w, \bar{w}) = \sum_k C_{ijk}(z - w)^{\Delta_k - \Delta_i - \Delta_j}(\bar{z} - \bar{w})^{\bar{\Delta}_k - \bar{\Delta}_i - \bar{\Delta}_j} \Phi_k(w, \bar{w}), \tag{4.3.12}$$

where the numerical constants C_{ijk} can be shown to coincide with the constants in the three-point function $\langle \Phi_i \Phi_j \Phi_k \rangle$. This is true in any quantum field theory. Here however, because of conformal invariance, there is no mass scale that appears in the OPE. This type of expansion can be thought of as a way to encode the correlation functions, since knowledge of (4.3.12) determines them completely in a positive (unitary) theory, and vice versa.

We may now use (4.3.11) to rewrite (4.3.9) as

$$\delta_{\epsilon, \bar{\epsilon}} \Phi(z, \bar{z}) = \frac{1}{2\pi i} \oint \left(dz \epsilon(z) R(T(z)\Phi(w, \bar{w})) + d\bar{z} \bar{\epsilon}(\bar{z}) R(\bar{T}(\bar{z})\Phi(w, \bar{w})) \right)$$

$$= \left[(\Delta \partial \epsilon(w) + \epsilon(w)\partial) + (\bar{\Delta}\bar{\partial}\bar{\epsilon}(\bar{w}) + \bar{\epsilon}(\bar{w})\bar{\partial}) \right] \Phi(w, \bar{w}), \tag{4.3.13}$$

where the last line is the desired result copied from (4.2.4). This equality will hold only if T and $\bar{T}$ have the following short-distance singularities with Φ:

$$R(T(z)\Phi(w, \bar{w})) = \frac{\Delta}{(z - w)^2} \Phi(w, \bar{w}) + \frac{1}{z - w} \partial_w \Phi(w, \bar{w}) + \cdots, \tag{4.3.14}$$

$$R(\bar{T}(\bar{z})\Phi(w, \bar{w})) = \frac{\bar{\Delta}}{(\bar{z} - \bar{w})^2} \Phi(w, \bar{w}) + \frac{1}{\bar{z} - \bar{w}} \partial_{\bar{w}} \Phi(w, \bar{w}) + \cdots, \tag{4.3.15}$$

where the ellipses denote regular terms. From now on we shall drop the R symbol and assume that the operator product expansion is always radially ordered. The OPE with the stress-tensor can be used as a definition of a quasiprimary field of weight $(\Delta, \bar{\Delta})$ instead of (4.2.2).

We will describe here the general Ward identities for insertions of the stress-tensor. Consider the correlation function

$$F^N(z, z_i, \bar{z}_i) = \left\langle T(z) \prod_{i=1}^{N} \Phi_i(z_i, \bar{z}_i) \right\rangle, \tag{4.3.16}$$

where Φ_i are primary fields. Viewed as a function of z, F^N is meromorphic with poles when $z \to z_i$. The residues of these poles can be calculated with the help of (4.3.15). On the sphere, a meromorphic function is uniquely specified by its poles and residues. Therefore, we obtain

$$F^N(z, z_i, \bar{z}_i) = \sum_{i=1}^{N} \left(\frac{\Delta_i}{(z - z_i)^2} + \frac{\partial_{z_i}}{z - z_i} \right) \left\langle \prod_{i=1}^{N} \Phi_i(z_i, \bar{z}_i) \right\rangle. \tag{4.3.17}$$

This Ward identity expresses correlation functions of primary fields with an insertion of the stress-tensor in terms of the correlator of the primary fields themselves. Multiple insertions can also be handled using in addition (4.5.1).

4.4 Mode Expansions

In this section we would like to write the analog of the Fourier expansions on the cylinder that are convenient in many cases. These are known as mode expansions of the conformal operators.

We will first write the mode expansion for the stress-tensor as

$$T(z) = \sum_{n \in \mathbb{Z}} z^{-n-2} L_n, \quad \bar{T}(\bar{z}) = \sum_{n \in \mathbb{Z}} \bar{z}^{-n-2} \bar{L}_n. \tag{4.4.1}$$

The exponent $-n-2$ is chosen such that for the scale change $z \to \frac{z}{\lambda}$, under which $T(z) \to \lambda^2 T\left(\frac{z}{\lambda}\right)$, we have $L_{-n} \to \lambda^n L_{-n}$. L_{-n} and $\bar{L}_{-n}$ then have scaling dimension n. If we consider a theory on a closed string world-sheet, the transformation from the Euclidean space cylinder to the complex plane is given by

$$w = \tau + i\sigma \to z = e^w. \tag{4.4.2}$$

For a holomorphic field Φ with conformal weight Δ, we would write

$$\Phi_{\text{cyl}}(w) = \sum_{n \in \mathbb{Z}} \phi_n e^{-nw} = \sum_{n \in \mathbb{Z}} \phi_n e^{in(i\tau - \sigma)} = \sum_{n \in \mathbb{Z}} \phi_n z^{-n}. \tag{4.4.3}$$

When going to the plane and using (4.2.2) on page 52 this becomes, for primary fields,

$$\Phi(z) = \sum_{n \in \mathbb{Z}} \phi_n z^{-n-\Delta}. \tag{4.4.4}$$

Nonprimary fields also have an inhomogeneous piece in (4.2.2). This is relevant for the stress-tensor and we will discuss it in the next section.

The mode expansion for the stress-tensor can be inverted by

$$L_n = \oint \frac{dz}{2\pi i} z^{n+1} T(z), \quad \bar{L}_n = \oint \frac{d\bar{z}}{2\pi i} \bar{z}^{n+1} \bar{T}(\bar{z}). \tag{4.4.5}$$

4.5 The Virasoro Algebra and the Central Charge

The stress-tensor $T_{\alpha\beta}$ is conserved. In exercise 4.12 on page 119 you are invited to show that in such a case the stress-tensor has a scaling dimension that is exactly 2. In particular, $T(z)$ has conformal weight (2,0) and $\bar{T}(\bar{z})$ (0,2). They are obviously quasi-primary fields since they cannot be derivatives of other fields.[4] From these properties we can write the most general OPE between two stress-tensors compatible with conservation (holomorphicity) and conformal invariance:

$$T(z)T(w) = \frac{c/2}{(z-w)^4} + 2\frac{T(w)}{(z-w)^2} + \frac{\partial T(w)}{z-w} + \cdots . \tag{4.5.1}$$

The fourth-order pole can only be proportional to a constant from dimension counting. This constant has to be positive in a unitary theory since $\langle T(z)T(w)\rangle = c/2(z-w)^4$. There can be no third-order pole since the OPE has to be symmetric under $z \leftrightarrow w$. Finally the rest of the singular terms are fixed by the fact that T has conformal weight (2,0). We have a similar OPE for $\bar{T}$ with $z \to \bar{z}$ and $c \to \bar{c}$ and

$$T(z)\bar{T}(\bar{w}) = \text{regular}. \tag{4.5.2}$$

Comparing (4.5.1) with (4.3.15) on page 56 we can conclude that $T(z)$ itself is not a primary field due to the presence of the most singular term. The constant c is called the (left) central charge and $\bar{c}$ the right central charge. Invariance under two-dimensional world-sheet parity requires $c = \bar{c}$.

The operator product expansions of $T(z)T(w)$ and $\bar{T}(\bar{z})\bar{T}(\bar{w})$ can now be written in terms of the modes. We have

$$[L_n, L_m] = \left(\oint \frac{dz}{2\pi i} \oint \frac{dw}{2\pi i} - \oint \frac{dw}{2\pi i} \oint \frac{dz}{2\pi i} \right) z^{n+1} T(z) w^{m+1} T(w)$$

$$= \oint \frac{dw}{2\pi i} \oint_{C_w} \frac{dz}{2\pi i} z^{n+1} w^{m+1} \left(\frac{c/2}{(z-w)^4} + \frac{2T(w)}{(z-w)^2} + \frac{\partial T(w)}{z-w} + \cdots \right)$$

$$= \oint \frac{dw}{2\pi i} \left(\frac{c}{12}(n+1)n(n-1)w^{n-2}w^{m+1} \right.$$

$$\left. + 2(n+1)w^n w^{m+1} T(w) + w^{n+1} w^{m+1} \partial T(w) \right). \tag{4.5.3}$$

The residue of the first term comes from $\frac{1}{3!}\partial_z^3 z^{n+1}\big|_{z=w} = \frac{1}{6}(n+1)n(n-1)w^{n-2}$. We integrate the last term by parts and combine it with the second term. This gives $(n-m)w^{n+m+1}T(w)$. Performing the w integration leads to the Virasoro algebra

$$[L_n, L_m] = (n-m)L_{n+m} + \frac{c}{12}(n^3 - n)\delta_{n+m,0}. \tag{4.5.4}$$

---

[4] The case where they are derivatives of currents is a degenerate case that will not interest us further. Its consequences are explored in exercise 4.17 on page 120.

The analogous calculation for $\bar{T}(\bar{z})$ yields

$$[\bar{L}_n, \bar{L}_m] = (n - m)\bar{L}_{n+m} + \frac{\bar{c}}{12}(n^3 - n)\delta_{n+m,0}. \tag{4.5.5}$$

Since $T\bar{T}$ has no singularities in its OPE,

$$[L_n, \bar{L}_m] = 0. \tag{4.5.6}$$

Equations (4.5.4)–(4.5.6) are generalizations of the centerless algebras in (4.1.17).

Therefore, every conformally invariant theory realizes the conformal algebra, and its spectrum decomposes into its representations. For $c = \bar{c} = 0$, it reduces to the classical algebra. In a diffeomorphism invariant theory $c = \bar{c}$.

4.6 The Hilbert Space

To describe the Hilbert space of a CFT, we will use the standard formalism of in- and out-states of quantum field theory adapted to our coordinate system. For quasi-primary fields $A(z, \bar{z})$, the in-states are defined as

$$|A_{\mathrm{in}}\rangle = \lim_{\tau \to -\infty} A(\tau, \sigma)|0\rangle = \lim_{z \to 0} A(z, \bar{z})|0\rangle. \tag{4.6.1}$$

For the out-states, we need a description in the neighborhood of $z \to \infty$. If we define $z = \frac{1}{w}$, then $z = \infty$ corresponds to the point $w = 0$. The map $f : w \to z = \frac{1}{w}$ is a conformal transformation, under which $A(z, \bar{z})$ transforms as

$$\tilde{A}(w, \bar{w}) = A(f(w), \bar{f}(\bar{w}))(\partial f(w))^{\Delta}(\bar{\partial}\bar{f}(\bar{w}))^{\bar{\Delta}}. \tag{4.6.2}$$

Substituting $f(w) = \frac{1}{w}$, we find

$$\tilde{A}(w, \bar{w}) = A\left(\frac{1}{w}, \frac{1}{\bar{w}}\right)(-w^{-2})^{\Delta}(-\bar{w}^{-2})^{\bar{\Delta}}. \tag{4.6.3}$$

It is natural to define

$$\langle A_{\mathrm{out}}| = \lim_{w, \bar{w} \to 0} \langle 0|\tilde{A}(w, \bar{w}). \tag{4.6.4}$$

We would like $\langle A_{\mathrm{out}}|$ to be the Hermitian conjugate of $|A_{\mathrm{in}}\rangle$. Hermitian conjugation of operators of weight $(\Delta, \bar{\Delta})$ is defined by

$$[A(z, \bar{z})]^{\dagger} = A\left(\frac{1}{\bar{z}}, \frac{1}{z}\right)\bar{z}^{-2\Delta}z^{-2\bar{\Delta}}. \tag{4.6.5}$$

This definition finds its justification in the continuation from Euclidean space back to Minkowski space. The missing factor of i in Euclidean time evolution $A(\sigma, \tau) = e^{\tau H} A(\sigma, 0) e^{-\tau H}$ must be compensated for in the definition of the adjoint by a Euclidean time reversal, which is implemented on the plane by $z \to 1/\bar{z}$. With the definition (4.6.5), we find

$$
\langle A_{\text{out}} | = \lim_{w \to 0} \langle 0 | \tilde{A}(w, \bar{w}) = \lim_{z \to 0} \langle 0 | A \left(\frac{1}{z}, \frac{1}{\bar{z}} \right) \bar{z}^{-2\Delta} z^{-2\bar{\Delta}}
$$

$$
= \lim_{z \to 0} \langle 0 | \, [A(z, \bar{z})]^{\dagger} = | A_{\text{in}} \rangle^{\dagger}. \tag{4.6.6}
$$

The fact that the stress-tensor is a Hermitian operator can be expressed using (4.6.5) in the following way:

$$
T^{\dagger}(z) = \sum_{m \in \mathbb{Z}} \frac{L_m^{\dagger}}{\bar{z}^{m+2}} \equiv \sum_{m \in \mathbb{Z}} \frac{L_m}{\bar{z}^{-m-2}} \frac{1}{\bar{z}^4}, \tag{4.6.7}
$$

where we have expanded in a Fourier series. In terms of the oscillator modes,

$$
L_m^{\dagger} = L_{-m}, \tag{4.6.8}
$$

and analogously $\bar{L}_m^{\dagger} = \bar{L}_{-m}$.

These conditions can also be derived from the Hermiticity of T in Minkowski space. The conditions on the vacuum follow from the regularity of

$$
T(z)|0\rangle = \sum_{m \in \mathbb{Z}} L_m z^{-m-2} |0\rangle \tag{4.6.9}
$$

at $z = 0$. Only positive powers of z are allowed, so we must demand

$$
L_m |0\rangle = 0, \quad m \geq -1. \tag{4.6.10}
$$

The same condition for $\lim_{w \to 0} \langle 0 | \tilde{T}(w)$ gives

$$
\langle 0 | L_m = 0, \quad m \leq 1. \tag{4.6.11}
$$

Equation (4.6.10) states that the in-vacuum is SL(2,$\mathbb{C}$) invariant, along with extra conditions for $m > 1$. The rest of the Virasoro operators create nontrivial states out of the vacuum. The only operators that annihilate both $\langle 0 |$ and $|0\rangle$ are generated by $L_{\pm 1,0}$ and $\bar{L}_{\pm 1,0}$ and constitute the SL(2,$\mathbb{C}$) subgroup of the conformal group.

If we consider holomorphic fields with mode expansion as in (4.4.4), conformal invariance and the SL(2,$\mathbb{C}$) invariance of the vacuum imply

$$
\Phi_{n > -\Delta} |0\rangle = 0. \tag{4.6.12}
$$

4.7 The Free Boson

The action for a noncompact free boson in two dimensions as we encountered it in string theory is

$$
S = \frac{1}{4\pi \ell_s^2} \int d^2\xi \, \partial_\alpha X \partial^\alpha X = \frac{1}{2\pi \ell_s^2} \int d^2z \, \partial X \bar{\partial} X. \tag{4.7.1}
$$

We use the conventions of appendix A on page 503.

The field $X(z, \bar{z})$ has the propagator

$$\langle X(z, \bar{z})X(w, \bar{w})\rangle = -(2\pi \ell_s^2)\,\Box^{-1} = -\frac{\ell_s^2}{2}\log\left(|z-w|^2\mu^2\right). \tag{4.7.2}$$

This is obtained by taking the massless limit of the massive scalar propagator in two dimensions; μ is an IR cutoff. Equation (4.7.2) can be obtained by starting with the massive propagator, with mass μ, and taking the limit $\mu \to 0$, keeping terms that do not vanish in the limit. You are invited to do this calculation explicitly in exercise 4.9 on page 119.

For the theory to be well defined, the dependence on μ should disappear from correlation functions. Note that X itself is not a conformal field since its correlation functions are IR divergent. Its derivative $\partial_z X$, however, is well behaved. The OPE of the derivative with itself is

$$\partial_z X(z)\partial_w X(w) = \partial_z \partial_w \langle XX\rangle + :\partial_z X \partial_w X:$$
$$= -\frac{\ell_s^2}{2(z-w)^2} + :\partial_z X \partial_w X:, \tag{4.7.3}$$

and $\partial_z X$ is a conformal field of weight $(1, 0)$. Note that μ has disappeared. Note also that the first line of (4.7.3) serves as the definition of normal ordering.

We will now calculate the OPE of $\partial_z X$ with the stress-tensor. According to the action (4.7.1) the stress-tensor for the free boson is given by

$$T(z) = -\frac{1}{\ell_s^2}:\partial X \partial X: = -\frac{1}{\ell_s^2}\lim_{z\to w}\left[\partial_z X \partial_w X + \frac{\ell_s^2}{2(z-w)^2}\right], \tag{4.7.4}$$

$$\bar{T}(\bar{z}) = -\frac{1}{\ell_s^2}:\bar{\partial} X \bar{\partial} X: = -\frac{1}{\ell_s^2}\lim_{\bar{z}\to \bar{w}}\left[\partial_{\bar{z}} X \partial_{\bar{w}} X + \frac{\ell_s^2}{2(\bar{z}-\bar{w})^2}\right]. \tag{4.7.5}$$

Using Wick's theorem, we can calculate

$$T(z)\,\partial X(w) = -\frac{1}{\ell_s^2}:\partial X(z)\partial X(z):\partial X(w)$$

$$= -\frac{2}{\ell_s^2}\partial X(z)\langle \partial X(z)\partial X(w)\rangle + \cdots$$

$$= \frac{\partial X(z)}{(z-w)^2} + \cdots$$

$$= \frac{\partial X(w)}{(z-w)^2} + \frac{1}{z-w}\partial^2 X(w) + \cdots, \tag{4.7.6}$$

where the ellipses indicate terms that are not singular as $z \to w$. Similarly we find $\bar{T}\partial X = $ regular. Therefore, ∂X is a $(1, 0)$ primary field. In the same way we find that $\bar{\partial} X$ is a $(0, 1)$ primary field.

There are many other (quasi-)primary fields constructed out of products of derivatives of X.

We will consider, however, another interesting class of operators, the "vertex" operators $V_p(z) =: e^{ipX(z, \bar{z})}:$. The OPE with the stress-tensor is

$$T(z)V_p(w, \bar{w}) = -\frac{1}{\ell_s^2}:\partial X(z)\partial X(z):\sum_{n=0}^{\infty}\frac{i^n p^n}{n!}:X^n(w, \bar{w}):. \tag{4.7.7}$$

For all terms in the expansion there can be either one or two contractions. We obtain

$$
\begin{aligned}
T(z) V_p(w, \bar{w}) &= -\frac{1}{\ell_s^2} \left[ip\partial\langle XX\rangle \right]^2 e^{ipX(w,\bar{w})} - \frac{1}{\ell_s^2} (2ip) : \partial X(z)\partial\langle XX\rangle e^{ipX(w,\bar{w})} : + \cdots \\
&= \frac{\ell_s^2\, p^2}{4(z-w)^2} e^{ipX(w,\bar{w})} + \frac{ip\partial X(z)}{z-w} e^{ipX(w,\bar{w})} + \cdots \\
&= \frac{\ell_s^2\, p^2}{4(z-w)^2} V_p(w, \bar{w}) + \frac{1}{z-w}\partial V_p(w, \bar{w}) + \cdots.
\end{aligned}
\tag{4.7.8}
$$

Therefore, the vertex operator V_p is a conformal field of weight $\left(\frac{\ell_s^2\, p^2}{4}, \frac{\ell_s^2\, p^2}{4} \right)$.

Consider now a correlation function of vertex operators

$$
G^N = \left\langle \prod_{i=1}^N V_{p_i}(z_i, \bar{z}_i) \right\rangle = \exp\left[-\frac{1}{2} \sum_{i,j=1}^N p_i p_j \langle X(z_i, \bar{z}_i) X(z_j, \bar{z}_j)\rangle \right],
\tag{4.7.9}
$$

where the second step in the above formula is due to the fact that we have a free (Gaussian) field theory.

Using the propagator (4.7.2) we can see that the IR divergences cancel only if

$$
\sum_i p_i = 0.
\tag{4.7.10}
$$

This is a charge- (momentum-)conservation condition.

For the two-point function we obtain

$$
\begin{aligned}
\langle V_p(z, \bar{z}) V_{-p}(w, \bar{w}) \rangle &= \langle : e^{ipX(z,\bar{z})} :: e^{-ipX(w,\bar{w})} : \rangle \\
&= e^{-(p^2 \ell_s^2/2) \log |z-w|^2} = \frac{1}{|z-w|^{\ell_s^2 p^2}},
\end{aligned}
\tag{4.7.11}
$$

which confirms that $\ell_s^2\, p^2 = 4\Delta = 4\bar{\Delta}$.

In this theory, the operator $i\partial X$ is a U(1) current, which is chirally conserved. It is associated to the symmetry of the action under $X \to X + \epsilon$. There is also the antichiral current $i\bar{\partial}X$ associated with the same translation symmetry because X is free. The zero modes of the currents are the left- and right-moving charge operators. We will normalize the currents as

$$
J = i\frac{\sqrt{2}}{\ell_s}\partial X, \quad \bar{J} = i\frac{\sqrt{2}}{\ell_s}\bar{\partial}X.
\tag{4.7.12}
$$

From

$$
J(z) V_p(w, \bar{w}) = \frac{p\ell_s}{\sqrt{2}} \frac{V_p(w, \bar{w})}{(z-w)} + \text{finite},
\tag{4.7.13}
$$

we can tell that the operator V_p carries charge $p\ell_s/\sqrt{2}$. The charge-conservation condition (4.7.10) is precisely due to the U(1) invariance of the theory. In the case of string theory, this type of U(1) invariance is essentially momentum conservation.

We will calculate the value of $c, \bar{c}$ for the free-boson theory. With the stress-tensor $T(z) = -\frac{1}{\ell_s^2} : \partial X \partial X :$ we can calculate the OPE

$$
\begin{aligned}
T(z)T(w) &= \frac{1}{\ell_s^4} \left\{ 2 \left(\partial\partial\langle XX\rangle \right)^2 + 4 : \partial X(z)\partial X(w) : \partial\partial\langle XX\rangle + \cdots \right\} \\
&= \frac{1/2}{(z-w)^4} + \frac{2}{(z-w)^2} T(w) + \frac{1}{z-w}\partial T(w) + \cdots.
\end{aligned}
\tag{4.7.14}
$$

We conclude that a single free boson has central charge $c = \bar{c} = 1$. In the bosonic string theory we have d free bosons; consequently the central charge is $c = \bar{c} = d$.

4.8 The Free Fermion

We will now analyze the conformal field theory that describes a free massless fermion. In two dimensions, with Minkowski signature, it is possible to have spinors that are both Majorana and Weyl, and these will have only one component. The γ-matrices can be represented by the Pauli matrices, i.e., $\gamma^1 = \sigma^1$, $\gamma^2 = \sigma^2$, so that the chirality projectors are $\frac{1}{2}(1 \pm \sigma^3)$. The Dirac operator becomes

$$\slashed{\partial} = \sigma^1 \partial_1 + \sigma^2 \partial_2 = \begin{pmatrix} 0 & \partial_1 - i\partial_2 \\ \partial_1 + i\partial_2 & 0 \end{pmatrix} = 2 \begin{pmatrix} 0 & \partial \\ \bar{\partial} & 0 \end{pmatrix}. \tag{4.8.1}$$

The action for a Majorana spinor $\chi = \begin{pmatrix} \psi \\ \bar{\psi} \end{pmatrix}$ and its conjugate $\chi^\dagger = (\psi, \bar{\psi})$ is

$$S = -\frac{1}{2\pi} \int d^2x \, \chi^\dagger \gamma^1 \slashed{\partial} \chi = -\frac{1}{2\pi} \int d^2z (\psi \bar{\partial} \psi + \bar{\psi} \partial \bar{\psi}). \tag{4.8.2}$$

The equations of motion are

$$\bar{\partial}\psi = \partial\bar{\psi} = 0, \tag{4.8.3}$$

which means that the left and right chiralities are represented by a holomorphic and an antiholomorphic spinor, respectively.

The operator product expansions of ψ and $\bar{\psi}$ with themselves can be found either by transforming the action into momentum space or by explicitly writing down the most general power expression with the correct conformal dimension. They are given by

$$\psi(z)\psi(w) = \frac{1}{z-w} + \cdots, \quad \bar{\psi}(\bar{z})\bar{\psi}(\bar{w}) = \frac{1}{\bar{z}-\bar{w}} + \cdots. \tag{4.8.4}$$

Up to a constant factor, the only expressions with conformal dimension $(2, 0)$ and $(0, 2)$, respectively, are

$$T(z) = -\frac{1}{2} : \psi(z)\partial\psi(z) :, \quad \bar{T}(\bar{z}) = -\frac{1}{2} : \bar{\psi}(\bar{z})\bar{\partial}\bar{\psi}(\bar{z}) :. \tag{4.8.5}$$

This stress-tensor has the correct operator product expansion

$$T(z)T(w) = \frac{1/4}{(z-w)^4} + \frac{2}{(z-w)^2} T(w) + \frac{1}{z-w} \partial T(w), \tag{4.8.6}$$

and a similar expression for $\bar{T}(\bar{z})$, so that the central charge of the free-fermion theory is $c = \bar{c} = \frac{1}{2}$.

As we will see later in section 4.12 on page 71, there are several important ingredients in this theory that are crucial for string theory.

4.9 The Conformal Anomaly

For the purposes of this section, we will assume that we are on an arbitrary Riemann surface and we will restore the tensor indices of $T \to T_{zz}$ for convenience.

Using (4.3.9) on page 55 we may evaluate the variation of the stress-tensor under the conformal transformation $z \to z + \epsilon(z)$,

$$\delta_\epsilon T_{zz}(z) = \frac{1}{2\pi i} \oint_{C_z} dw \, \epsilon(w) \, T_{ww}(w) T_{zz}(z) = \frac{c}{12} \partial_z^3 \epsilon(z) + 2\partial_z \epsilon(z) T_{zz}(z) + \epsilon(z)\partial_z T_{zz}(z). \tag{4.9.1}$$

This comes from the short-distance properties of the OPE, and is therefore insensitive to the global geometry of the surface.

The two last terms are the standard terms implying that T_{zz} is a tensor of weight $(2,0)$. However the first piece, proportional to the central charge, spoils this property. We will now show that a nonzero c signals the presence of a conformal anomaly, breaking the conformal invariance on nonflat world-sheets.

The infinitesimal transformation (4.9.1) can be integrated to a finite transformation as

$$T_{zz}(z) = \left(\frac{\partial z'}{\partial z}\right)^2 T_{z'z'}(z') + \frac{c}{12}\{z', z\}, \quad \{f(z), z\} \equiv \frac{f'''}{f'} - \frac{3}{2}\left(\frac{f''}{f'}\right)^2. \tag{4.9.2}$$

The term multiplying $T(z')$ is straightforward, due to the fact that T is a tensor of rank 2. The inhomogeneous piece can be determined from its infinitesimal limit in (4.9.1) and the cocycle property for finite conformal transformations. You are invited to do this in exercise 4.19 on page 120. We may now use the metric to modify T_{zz} so that it is a tensor. It is not difficult to show that

$$\hat{T}_{zz} = T_{zz} - \frac{c}{24}\left[2\partial_z(g^{z\bar{z}}\partial_z g_{z\bar{z}}) - (g^{z\bar{z}}\partial_z g_{z\bar{z}})^2\right] \tag{4.9.3}$$

is a true tensor transforming as

$$\hat{T}_{zz} = (f')^2 \hat{T}_{z'z'}(f(z)). \tag{4.9.4}$$

We will now show the existence of a nonzero trace (conformal anomaly) on curved world-sheets. On general grounds, since the β-functions are zero in a CFT, the trace of the stress tensor can be proportional only to a curvature invariant of dimension 2. There is only one such scalar, the scalar curvature, so we must have[5]

$$T_\alpha^\alpha = A \, R^{(2)}, \tag{4.9.5}$$

with A a constant and $R^{(2)}$ the Ricci scalar of the world-sheet. We can translate this equation into complex coordinates as

$$T_{z\bar{z}} = \frac{A}{2}g_{z\bar{z}}R^{(2)}. \tag{4.9.6}$$

---

[5] There could also be a constant with dimensions of [mass]^2 on the right-hand side of equation (4.9.5). Such a constant can always be canceled by adding an appropriate cosmological constant on the world-sheet.

Taking the covariant derivative of (4.9.6), and using the stress-tensor conservation $\nabla^{\bar{z}} T_{\bar{z}z} + \nabla^{z}\hat{T}_{zz} = 0$ and (A.10), we obtain

$$g^{z\bar{z}}\partial_{\bar{z}}\hat{T}_{zz} \equiv \nabla^{z}\hat{T}_{zz} = -\nabla^{\bar{z}}T_{\bar{z}z} = -\frac{A}{2}\nabla^{\bar{z}}(g_{z\bar{z}}R^{(2)}) = -\frac{A}{2}\partial_{z}R^{(2)}. \tag{4.9.7}$$

Since T_{zz} is holomorphic, the only contribution to the left-hand side of (4.9.7) comes from the nonholomorphic correction in (4.9.3). Substituting in (4.9.7) we find equality if and only if $A = c/12$. We have therefore proven the conformal anomaly formula: the stress tensor of a two-dimensional CFT with central charge c, on a curved surface satisfies[6]

$$T^{\alpha}{}_{\alpha} = -\frac{c}{12}R^{(2)}. \tag{4.9.8}$$

In a generic nonconformally invariant theory, the trace can be a more general function of its various fields with the β-functions as proportionality coefficients. In a CFT, it is proportional only to the scalar curvature. This implies that in a CFT with $c \neq 0$, the theory depends on the conformal factor of the metric, but in the very specific form implied by (4.9.8).

Equation (4.9.8) implies that the variation of the effective action $Z \sim \int \mathcal{D}X e^{-S(X)}$ under an infinitesimal Weyl transformation $\delta g^{\alpha\beta} = -g^{\alpha\beta}\delta\phi$ is

$$\delta \log Z = -\delta S = \frac{1}{4\pi}\int d^2\xi\,\sqrt{g}\,T_{\alpha\beta}\delta g^{\alpha\beta} = \frac{c}{48\pi}\int d^2\xi\,\sqrt{g}\,R^{(2)}\,\delta\phi. \tag{4.9.9}$$

We can integrate (4.9.9) to obtain the dependence of the quantum theory on the conformal factor. Let $\hat{g}_{\alpha\beta} = e^{\phi}g_{\alpha\beta}$. Then

$$\int [DX]_{\hat{g}}e^{-S[\hat{g}_{\alpha\beta},X]} = e^{cS_L[g_{\alpha\beta},\phi]}\int [DX]_{g}e^{-S[g_{\alpha\beta},X]}, \tag{4.9.10}$$

where X is a generic set of fields and

$$S_L[g_{\alpha\beta},\phi] = \frac{1}{96\pi}\int d^2\xi\,\sqrt{g}\,g^{\alpha\beta}\partial_{\alpha}\phi\partial_{\beta}\phi + \frac{1}{48\pi}\int d^2\xi\,\sqrt{g}\,R^{(2)}\,\phi. \tag{4.9.11}$$

This is the Liouville action. In critical string theory, the ghost system cancels the central charge of the string coordinates and the full theory is independent of the scale factor.

When the world-sheet has boundaries, then there are extra invariants with the correct dimension to potentially contribute to the conformal anomaly. The most general expression for the variation of the effective action is

$$\delta \log Z = \frac{c}{48\pi}\int_{\Sigma} d^2\xi\,\sqrt{g}\,R^{(2)}\,\delta\phi + \int_{\partial\Sigma} d\sigma\,(B_1 + B_2\,K + B_3 n^a\nabla_a)\delta\phi, \tag{4.9.12}$$

where Σ denotes the two-dimensional world-sheet, $\partial\Sigma$ is its boundary, n^{α} and t^{α} are the normal and tangent unit vectors to the boundary, respectively, and $K = -t^{\alpha}n_{\beta}\nabla_{\alpha}t^{\beta}$ is the geodesic curvature of the boundary.

---

[6] To compare with other books, one should remember that we are using the slightly modified definition for the stress tensor $T_{\alpha\beta} = -(4\pi/\sqrt{g})(\delta S/\delta g^{\alpha\beta})$ as well as specific conventions for the curvature that can affect the sign.

The contribution proportional to B_1 can be canceled by a boundary cosmological constant while the B_3-term can be canceled by a term proportional to the geodesic curvature. Finally, the WZ consistency condition on the Weyl anomaly,

$$[\delta_1, \delta_2] \log Z = 0, \tag{4.9.13}$$

implies that $B_2 = \frac{c}{24\pi}$.

4.10 Representations of the Conformal Algebra

In CFT, the spectrum decomposes into representations of the generic symmetry algebra, namely, two copies of the Virasoro algebra.[7] We will describe here only the left algebra with operators L_m to avoid repetition.

The Cartan subalgebra of the Virasoro algebra is generated by L_0. The positive modes are raising operators and the negative ones are lowering operators. Highest-weight (HW) representations are constructed by starting from a state that is annihilated by all raising operators. The representation is then generated by acting on the HW state by the lowering operators.

Suppose Φ is a primary field (operator) of left-moving dimension Δ. From the operator product expansion with the stress-tensor (4.3.15) on page 56, we find

$$[L_n, \Phi(w)] = \oint \frac{dz}{2\pi i} z^{n+1} T(z)\Phi(w) = \Delta(n+1)w^n \Phi(w) + w^{n+1}\partial\Phi(w). \tag{4.10.1}$$

In most CFTs, there is a one-to-one correspondence between states and local operators.[8] The state associated with this operator is

$$|\Delta\rangle \equiv \Phi(0)|0\rangle. \tag{4.10.2}$$

Since $[L_n, \Phi(0)] = 0$, $n > 0$, it follows that

$$L_{m>0}|\Delta\rangle = L_{m>0}\Phi(0)|0\rangle = [L_m, \Phi(0)]|0\rangle + \Phi(0)L_{m>0}|0\rangle = 0. \tag{4.10.3}$$

Primary fields are therefore in one-to-one correspondence with HW states. Each primary field generates a representation of the Virasoro algebra. Also, $L_0|\Delta\rangle = \Delta|\Delta\rangle$. More generally, in-states $|\Delta, \bar{\Delta}\rangle$, defined by (4.10.2) with Φ of conformal dimension $(\Delta, \bar{\Delta})$ also satisfy

$$L_0|\Delta, \bar{\Delta}\rangle = \Delta|\Delta, \bar{\Delta}\rangle, \quad \bar{L}_0|\Delta, \bar{\Delta}\rangle = \bar{\Delta}|\Delta, \bar{\Delta}\rangle, \quad L_{n>0}, \bar{L}_{n>0}|\Delta, \bar{\Delta}\rangle = 0. \tag{4.10.4}$$

The rest of the states in the representation generated by $|\Delta\rangle$ are of the form

$$|\chi\rangle = L_{-n_1} L_{-n_2} \cdots L_{-n_k}|\Delta\rangle, \tag{4.10.5}$$

where all $n_i > 0$, and are called descendants. They are L_0 eigenstates with eigenvalues $\Delta + \sum_k n_k$. This type of representation is called a Verma module.

---

[7] There are exceptional cases where the CFT spectrum forms reducible but indecomposable representations. Such theories are known as logarithmic CFTs. A review can be found in [43,44].

[8] This correspondence may fail in CFTs with a continuous spectrum of dimensions. A prime example in this class is Liouville theory [45].

We have seen that we can have a one-to-one correspondence with HW states $|\Delta\rangle$ and primary fields $\Phi_\Delta(z)$ given by (4.10.2). A similar statement can be made for descendants. Consider the state $L_{-1}|\Delta\rangle$. It is not difficult to show that the operator that creates this state out of the vacuum is

$$(L_{-1}\Phi)(z) \equiv \oint_{C_z} \frac{dw}{2\pi i}\, T(w)\Phi_\Delta(z), \tag{4.10.6}$$

using (4.4.5). For the general state (4.10.5) we have to use nested contours

$$\Phi_\chi(z) = \prod_{i=1}^{k} \oint \frac{dw_i}{2\pi i}\, (w_i - z)^{-n_i+1} T(w_i)\Phi_\Delta(z). \tag{4.10.7}$$

Therefore, a general correlation function of descendant operators can be written in terms of multiple contour integrals of a correlation function of the associated primary fields and several insertions of the stress-tensor. Conformal Ward identities express such a correlation function in terms of the one with primary fields only. Therefore, knowledge of the correlators of primary fields determines all correlators of the CFT.

We will also discuss the quasi-primary fields. We have seen that on the sphere L_{-1} is the translation operator

$$[L_{-1}, O(z, \bar{z})] = \partial_z O(z, \bar{z}). \tag{4.10.8}$$

The quasiprimary states are the HW states of the global conformal group. Consider the part generated by $L_{\pm1}, L_0$. The raising operator is L_1, while L_{-1} is the lowering operator. The HW states are annihilated by L_1. The rest of the representation is generated by acting several times with L_{-1}. Therefore, the descendant (non-quasi-primary) states are derivatives of quasi-primary ones.

An interesting function of a conformal representation generated by a primary operator of dimension Δ, is the character

$$\chi_\Delta(q) \equiv \text{Tr}[q^{L_0-c/24}], \tag{4.10.9}$$

where the trace is taken over the whole representation.

There is an extra shift of L_0 in (4.10.9) proportional to the central charge. The reason is as follows: Characters appear when discussing the partition function on the torus. The torus can be thought of as the cylinder with the two end points identified (with a twist). Going from the sphere to the cylinder, and using the anomalous transformation law for the stress tensor (4.9.2) there is a shift of $L_0 \to L_0 - \frac{c}{24}$.

In exercise 4.25 on page 120 you are invited to show that a representation without null vectors has

$$\chi_\Delta(q) = \frac{q^{\Delta-c/24}}{\prod_{n=1}^{\infty}(1-q^n)} \tag{4.10.10}$$

as a character.

There is a special representation, which is called the vacuum representation. If one starts with the unit operator, the state associated with it via (4.10.2) is the vacuum state. The rest of the representation is generated by the negative Virasoro modes. Note, however, that from (4.10.1) L_{-1} acts as a z derivative. However the z derivative of the unit operator is zero. This is equivalent to the statement that L_{-1} annihilates the vacuum state. For $c \geq 1$ the vacuum character is given by

$$\chi_0(q) = \frac{q^{-c/24}}{\prod_{n=2}^{\infty} (1 - q^n)}. \tag{4.10.11}$$

The term with $n = 1$ is missing here since L_{-1} does not generate any states out of the vacuum.

In a positive (unitary) theory the norms of states have to be positive. The norm of the state $L_{-n}|0\rangle$, $n > 0$, is

$$\| L_{-n}|0\rangle \|^2 = \langle 0|L_{-n}^{\dagger}L_{-n}|0\rangle = \langle 0| \left[\frac{c}{12}(n^3 - n) + 2nL_0 \right] |0\rangle$$
$$= \frac{c}{12}(n^3 - n), \tag{4.10.12}$$

where we have used the commutation relations of the Virasoro algebra and the SL(2,$\mathbb{C}$) invariance of the vacuum. Unitarity demands this to be positive. For large enough n, this means $c \geq 0$ (if $c = 0$, the Hilbert space is one dimensional and spanned by $|0\rangle$). A more detailed investigation shows that for $c \geq 1$ we cannot obtain other direct constraints from unitarity.

When $0 < c < 1$, unitarity implies that c must be of the form

$$c = 1 - \frac{6}{m(m+1)}, \quad m = 2, 3, \ldots. \tag{4.10.13}$$

An example of a CFT with $m = 3$ is the Ising model, with $m = 4$, the tricritical Ising model; and with $m = 5$ the three-state Potts model; $m = 2$ is the trivial theory with $c = 0$.

Generically, the Verma modules described above correspond to irreducible representations of the Virasoro algebra. However, in special cases, it may happen that the Verma module contains "null" states (states of zero norm that are orthogonal to any other state). Then, the irreducible representation is obtained by factoring out the null states. Such representations are called degenerate.

We give here an example of a null state. Consider the Ising model, $m = 3$ above, with $c = 1/2$. This is essentially the conformal field theory of a Majorana fermion that we discussed earlier. Consider the primary state with $\Delta = 1/2$, $|1/2\rangle$, corresponding to the fermion and the following descendant state:

$$|\chi\rangle = \left(L_{-2} - \frac{3}{4}L_{-1}^2 \right) |1/2\rangle. \tag{4.10.14}$$

You are asked in exercise 4.27 on page 121 to show that $|\chi\rangle$ is a null state. This amounts to showing that its norm is zero and that it is orthogonal to every other state in the representation. This is equivalent to showing that $L_n|\chi\rangle = 0$ for every positive n. Therefore $|\chi\rangle = 0$. This implies that not all states in the Verma module are independent. The irreducible representation is obtained by appropriately subtracting such equivalences.

4.11 Affine Current Algebras

So far we have seen that in any CFT there is a holomorphic stress-tensor T of weight $(2,0)$. There may also be conserved currents J_μ which are also chirally conserved:

$$\partial^\mu J_\mu = 0, \quad \epsilon^{\mu\nu}\partial_\mu J_\nu = 0. \tag{4.11.1}$$

In complex coordinates, J_z generates a chiral symmetry and has weight $(1,0)$. Similarly $J_{\bar{z}}$ generates an (anti)chiral symmetry and has weight $(0,1)$. The conservation equations (4.11.1) translate into holomorphicity conditions

$$\partial_{\bar{z}} J_z = 0, \quad \partial_z J_{\bar{z}} = 0. \tag{4.11.2}$$

Consider the whole set of such holomorphic currents $J^a(z)$ present in a theory. We can write the most general OPE of these currents compatible with their scaling dimension, chiral conservation, and conformal invariance as

$$J^a(z)J^b(w) = \frac{G^{ab}}{(z-w)^2} + \frac{if^{ab}{}_c J^c(w)}{z-w} + \text{finite}, \tag{4.11.3}$$

where $f^{ab}{}_c$ is antisymmetric in the upper indices and G^{ab} is symmetric. Using associativity of the operator products, it can be shown that the f^{ab}_c also satisfy a Jacobi identity and $f^{abc} = f^{ab}{}_d G^{dc}$ is totally antisymmetric. Therefore they must be the structure constants of a Lie algebra g with invariant Killing metric G^{ab}.

Expanding $J^a(z) = \sum_n J^a_n z^{-n-1}$, we can translate (4.11.3) into commutation relations for the modes of the currents

$$[J^a_m, J^b_n] = m\, G^{ab}\, \delta_{m+n,0} + if^{ab}{}_c J^c_{m+n}. \tag{4.11.4}$$

This algebra $\hat{g}$ is an infinite-dimensional generalization of the Lie algebra g and is known as an affine or current algebra. Clearly, the subalgebra of the zero modes J^a_0 constitutes a Lie algebra (g) with structure constants $f^{ab}{}_c$.

A conformal field of weight $(1,0)$ is necessarily Virasoro primary in a positive theory. Thus, the OPE with the stress-tensor should be

$$T(z)J^a(w) = \frac{J^a(w)}{(z-w)^2} + \frac{\partial J^a(w)}{z-w}, \tag{4.11.5}$$

and $\bar{T}(\bar{z})J^a(w) = \text{regular}$.

This type of algebra is realized, as we shall see, in many CFTs. The prototype is the chiral model with a Wess-Zumino term. This is a theory in two dimensions, where the basic field $g(x)$ is in a matrix representation of a group G. The action is

$$S = \frac{1}{4\lambda^2} \int_{M_2} d^2\xi \, \text{Tr}(\partial_\mu g \partial^\mu g^{-1}) + \frac{ik}{8\pi} \int_{B;\, \partial B = M_2} d^3\xi \, \text{Tr}(\epsilon_{\alpha\beta\gamma} U^\alpha U^\beta U^\gamma), \tag{4.11.6}$$

where $U_\mu = g^{-1}\partial_\mu g$. The second term in the action is integrated over a three-dimensional manifold B whose boundary is the two-dimensional space M_2 on which the CFT is defined. This is the WZ term. It has the special property that its variation is a total derivative.

It therefore gives a two-dimensional instead of a three-dimensional contribution to the equations of motion.

There is a consistency condition that has to be imposed, however. Consider another three-manifold with the same boundary. We would like the theory to be the same. For this, the action difference between the two prescriptions must be an integer multiple of 2π, so the two path integrals are the same. This provides a quantization condition[9] on the coupling k.

The theory above has two different couplings, λ and k. It can be shown that when $\lambda^2 = 4\pi/k$ then the theory is conformally invariant (this is called the WZW model). In this case, it can be directly verified (see exercise 4.31 on page 121) that the classical equations imply that the matrix currents $J = g^{-1}\partial g$ and $\bar{J} = \bar{\partial}gg^{-1}$ are chirally conserved,

$$\bar{\partial}J = \partial\bar{J} = 0.$$

This is a reflection of the symmetry of the action (4.11.6) under $g \to h_1 g h_2$, where $h_{1,2}$ are arbitrary G elements. Therefore, the currents J generate a G_L affine algebra while the currents $\bar{J}$ generate a G_R affine algebra.

An interesting feature of this theory (which turns out to be generic) is that the stress-tensor can be written as a bilinear in terms of the currents. This is known as the affine-Sugawara form.

Consider the group G to be simple. Then by a change of basis in (4.11.3) we can set $G^{ab} = k\delta^{ab}$. Normalize the long roots to have squares equal to 2. Then the (2,0) operator

$$T_G(z) = \frac{1}{2(k + \tilde{h})} : J^a(z)J^a(z) : \tag{4.11.7}$$

satisfies the Virasoro algebra with central charge

$$c_G = \frac{kD_G}{k + \tilde{h}}. \tag{4.11.8}$$

$\tilde{h}$ is the dual Coxeter number of the group G while D_G is the dimension of the group G. In the case of SU(N), we have $\tilde{h} =$ N; for SO(N), $\tilde{h} =$ N-2 etc. With this normalization, k should be a positive integer in order to have a positive theory. It is called the level of the affine algebra, $\hat{g}_k$.

In this type of theory, the affine symmetry is "larger" than the Virasoro symmetry since we can construct the Virasoro operators out of the current operators. In particular, the spectrum will form representations of the affine algebra. To describe such representations we will use a procedure similar to the case of a Virasoro algebra. The representation is generated by a set of states $|R_i\rangle$ that transform in the representation R of the zero-mode subalgebra J_0^a and are annihilated by the positive modes of the currents

$$J_{m>0}^a|R_i\rangle = 0, \quad J_0^a|R_i\rangle = (T_R^a)_{ij}|R_j\rangle. \tag{4.11.9}$$

[9] There is another way to argue for the quantization of k. If we demand positivity (unitarity) of the quantum theory then we obtain the same quantization condition.

The rest of the affine representation is generated from the states $|R_i\rangle$ by the action of the negative modes of the currents. The states $|R_i\rangle$ are generated as usual, out of the vacuum, by local operators $R_i(z, \bar{z})$. Conditions (4.11.9) translate into the following OPE:

$$J^a(z) R_i(w, \bar{w}) = \frac{(T_R^a)_{ij}}{(z - w)} R_j(w, \bar{w}) + \cdots . \tag{4.11.10}$$

This is the definition of *affine primary* fields which play the same role as the primary fields in the case of the conformal algebra.

The conformal weight of affine primaries can be calculated from the affine-Sugawara form of the stress-tensor and is given by

$$\Delta_R = \frac{C_R}{k + \tilde{h}}, \tag{4.11.11}$$

where C_R is the quadratic Casimir of the representation R. For example, the spin-j representation of SU(2) has $\Delta_j = j(j + 1)/(k + 2)$.

We have seen so far that the irreducible representations of the affine algebra $\hat{g}_k$ are in one-to-one correspondence with those of the finite Lie algebra g. This is not the end of the story, however. It turns out that not all representations of the finite algebra can appear, but only the so-called integrable ones. In the case of SU(2) this implies $j \leq k/2$. For SU(N)$_k$, the integrable representations are those with at most k columns in their Young tableau.

Nonintegrable representations are not unitary, and they can be shown to decouple from the correlation functions.

4.12 Free Fermions and O(N) Affine Symmetry

Free fermions and bosons can be used to realize particular representations of current algebras. It will be useful for our later purposes to consider the CFT of N free Majorana-Weyl fermions ψ^i:

$$S = -\frac{1}{2\pi} \int d^2z \, \psi^i \bar{\partial} \psi^i. \tag{4.12.1}$$

Clearly, this model exhibits a global O(N) symmetry, $\psi^i \to \Omega_{ij}\psi_j$, $\Omega^T\Omega = 1$, which leads to the chirally conserved Hermitian ($J_m^{ij\dagger} = J_{-m}^{ij}$) currents

$$J^{ij}(z) = i : \psi^i(z)\psi^j(z) :, \quad i < j. \tag{4.12.2}$$

Using the OPE

$$\psi^i(z)\psi^j(w) = \frac{\delta^{ij}}{z - w} \tag{4.12.3}$$

and Wick's theorem, we can calculate

$$J^{ij}(z)J^{kl}(w) = \frac{G^{ij,kl}}{(z - w)^2} + i f^{ij,kl}{}_{mn} \frac{J^{mn}(w)}{(z - w)} + \cdots , \tag{4.12.4}$$

where $G^{ij,kl} = (\delta^{ik}\delta^{jl} - \delta^{il}\delta^{jk})$ is the Killing O(N) metric and

$$2f^{ij,kl}{}_{mn} = (\delta^{ik}\delta^{ln} - \delta^{il}\delta^{kn})\delta^{jm} + (\delta^{jl}\delta^{kn} - \delta^{jk}\delta^{ln})\delta^{im} - (m \leftrightarrow n) \tag{4.12.5}$$

are the structure constants of O(N) in a basis where the long roots have length squares equal to 2. Thus, N free fermions realize the O(N) current algebra at level $k = 1$.

We can construct the affine-Sugawara stress-tensor

$$T(z) = \frac{1}{2(N-1)} \sum_{i<j}^{N} : J^{ij}(z)J^{ij}(z) : . \tag{4.12.6}$$

As discussed previously, $T(z)$ will satisfy an operator product expansion

$$T(z)T(w) = \frac{c_G/2}{(z-w)^4} + \frac{2T(w)}{(z-w)^2} + \frac{\partial T(w)}{z-w}, \tag{4.12.7}$$

where

$$c_G = \frac{kD}{k+\tilde{h}}. \tag{4.12.8}$$

For SO(N), one has $\tilde{h} = N - 2$ and $D = \frac{1}{2}N(N-1)$. With $k = 1$, this gives

$$c_G = \frac{N(N-1)/2}{1+N-2} = \frac{N}{2}, \tag{4.12.9}$$

i.e., each fermion contributes $\frac{1}{2}$ to the central charge. This is expected since the central charge of the tensor product of two noninteracting theories is the sum of the two central charges. Moreover, if we use the explicit form of the currents in terms of the fermions we can directly evaluate the normal-ordered product in (4.12.6) with the result

$$T(z) = -\frac{1}{2} \sum_{i=1}^{N} : \psi^i \partial \psi^i : , \tag{4.12.10}$$

which is also the stress-tensor that follows from the free-fermion action (4.12.1).

Since N free fermions realize the O(N)$_1$ affine symmetry, we should be able to classify the spectrum into irreducible representations of the O(N)$_1$ current algebra. From the representation theory of current algebra we learn that at level 1 there exist the following integrable (unitary) representations: the unit (vacuum) representation constructed by acting on the vacuum with the negative current modes, the vector V representation, and the spinor representation(s).

If N is odd there is a single spinor representation of dimension $2^{(N-1)/2}$. When N is even, there are two inequivalent spinor representations of dimension $2^{N/2-1}$: the spinor S and the conjugate spinor C. From now on we will assume N to be even because this is the case of most interest in what follows. Applying (4.11.11) to our case we find that the conformal weight of the vector is

$$\Delta_V = \frac{(N-1)/2}{1+N-2} = \frac{1}{2}. \tag{4.12.11}$$

The candidate affine primary fields for the vector are the fermions themselves. Their conformal weight is 1/2 and they transform as a vector under the global O(N) symmetry. This can be verified by computing the OPE

$$J^{ij}(z)\psi^k(w) = i\frac{T^{ij}_{kl}}{z-w}\psi^l(w) + \cdots , \tag{4.12.12}$$

where $T^{ij}_{kl} = (\delta^{il}\delta^{jk} - \delta^{ik}\delta^{jl})$ are the representation matrices of the vector representation.

Comparing (4.12.12) with (4.11.10) we verify that ψ^i are the affine primaries of the vector representation.

The conformal weights of the spinor and conjugate spinor are equal and, from (4.11.11), we obtain $\Delta_S = \Delta_C = N/16$. Operators with such a conformal weight do not seem to exist in the free-fermion theory in the way it has been described so far.

Notice, however, that the action (4.12.1) has a $\mathbb{Z}_2$ symmetry

$$\psi^i \to -\psi^i. \tag{4.12.13}$$

This symmetry commutes with the O(N) symmetry.

On the cylinder, because the S^1 is noncontractible, we have the option of putting two distinct boundary conditions on the fermions. Both preserve the O(N) invariance:

Neveu-Schwarz: $\psi^i(\sigma + 2\pi) = -\psi^i(\sigma)$;

Ramond: $\psi^i(\sigma + 2\pi) = \psi^i(\sigma)$.

The mode expansion of a periodic holomorphic field on the cylinder is

$$\psi(\tau + i\sigma) = \sum_n \psi_n e^{-n(\tau + i\sigma)}, \tag{4.12.14}$$

where n is an integer. Thus, in the Ramond (R) sector ψ is integer moded. In the Neveu-Schwarz (NS) sector, ψ is antiperiodic so its Fourier expansion is like (4.12.14) but now n is half integer. When we go from the cylinder to the sphere, $z = e^{\tau + i\sigma}$, (4.2.2) on page 52 implies that the mode expansion becomes

$$\psi^i(z) = \sum_n \psi_n^i z^{-n-\Delta} = \sum_n \psi_n^i z^{-n-1/2}. \tag{4.12.15}$$

We observe that in the NS sector (half-integer n) the field $\psi^i(z)$ is single valued (invariant under $z \to e^{2\pi i} z$). In the R sector, it has a $\mathbb{Z}_2$ branch cut. To summarize

- $n \in \mathbb{Z}$ (Ramond),

- $n \in \mathbb{Z} + \frac{1}{2}$ (Neveu-Schwarz).

The OPE (4.12.3) implies the following anticommutation relations for the fermionic modes:

$$\{\psi_m^i, \psi_n^j\} = \delta^{ij} \delta_{m+n,0}, \tag{4.12.16}$$

in both the NS and the R sectors.

We now focus on the NS sector. Here the fermionic oscillators are half-integrally moded and (4.12.16) shows that $\psi^i_{-n-1/2}$, $n \le 0$, are creation operators, while $\psi^i_{n+1/2}$ are annihilation operators. Consequently, the vacuum satisfies

$$\psi_{n>0}^i |0\rangle = 0 \tag{4.12.17}$$

and the full spectrum is generated by acting on the vacuum with the negative frequency oscillators.

We expect to obtain here the vacuum and the vector representation. In particular, the primary states of the vector can be identified with

$$|i\rangle = \psi_{-1/2}^i |0\rangle \tag{4.12.18}$$

and the rest of the representation is constructed from the above states by acting with the negative current modes.

At this point it is useful to introduce the fermion number operator F and the fermion parity $(-1)^F$. It essentially counts the number of fermionic modes modulo 2. It satisfies

$$\{(-1)^F, \psi_n^i\} = 0 \tag{4.12.19}$$

and the NS vacuum has eigenvalue 1: $(-1)^F|0\rangle = |0\rangle$. Using (4.12.19) we can calculate that the vector primary states (4.12.18) have $(-1)^F = -1$. Since the currents contain an even number of fermion modes we can state the following:

- All states of the vacuum (unit) representation have $(-1)^F = 1$. The first nontrivial states correspond to the currents themselves:

$$J_{-1}^{ij}|0\rangle = i\psi_{-1/2}^i \psi_{-1/2}^j |0\rangle. \tag{4.12.20}$$

- All states of the vector representation have $(-1)^F = -1$. The first nontrivial states below the primaries are

$$J_{-1}^{jk}|i\rangle = i\left[\delta^{ik}\psi_{-3/2}^j - \delta^{ij}\psi_{-3/2}^k + \psi_{-1/2}^i \psi_{-1/2}^j \psi_{-1/2}^k\right]|0\rangle. \tag{4.12.21}$$

We will now calculate the characters (multiplicities) in the NS sector. We will first calculate the trace of $q^{L_0-c/24}$ in the full NS sector. This is not difficult to do since every negative moded fermionic oscillator $\psi_{-n-1/2}^i$ contributes $1 + q^{n+1/2}$. The first term corresponds to the oscillator being absent, while the second corresponds to it being excited. Since the oscillators are fermionic, their square is zero and therefore no more terms can appear. Putting everything together, we obtain

$$\mathrm{Tr}_{NS}[q^{L_0-c/24}] = q^{-N/48}\prod_{n=1}^{\infty}(1 + q^{n-1/2})^N. \tag{4.12.22}$$

Using identities (C.8) and (C.10) from Appendix C on page 507, we can write this in terms of ϑ-functions as

$$\mathrm{Tr}_{NS}[q^{L_0-c/24}] = \left[\frac{\vartheta_3}{\eta}\right]^{N/2}, \tag{4.12.23}$$

where $\vartheta_i = \vartheta_i(0|\tau)$. In order to separate the contributions of the unit and vector representations, we also need to calculate the same trace but with $(-1)^F$ inserted. Then $\psi_{-n-1/2}^i$ contributes $1 - q^{n+1/2}$ and

$$\mathrm{Tr}_{NS}[(-1)^F q^{L_0-c/24}] = q^{-N/48}\prod_{n=1}^{\infty}(1 - q^{n-1/2})^N = \left[\frac{\vartheta_4}{\eta}\right]^{N/2}. \tag{4.12.24}$$

Now we can project onto the vector or the unit representation:

$$\chi_O = \mathrm{Tr}_{NS}\left[\frac{(1 + (-1)^F)}{2} q^{L_0-c/24}\right] = \frac{1}{2}\left(\left[\frac{\vartheta_3}{\eta}\right]^{N/2} + \left[\frac{\vartheta_4}{\eta}\right]^{N/2}\right), \tag{4.12.25}$$

$$\chi_V = \mathrm{Tr}_{NS}\left[\frac{(1 - (-1)^F)}{2} q^{L_0-c/24}\right] = \frac{1}{2}\left(\left[\frac{\vartheta_3}{\eta}\right]^{N/2} - \left[\frac{\vartheta_4}{\eta}\right]^{N/2}\right). \tag{4.12.26}$$

It turns out that sometimes inequivalent current algebra representations have the same conformal weight and multiplicities, and therefore the same characters. This will happen for the spinors. To distinguish them we will define a refined character (the *affine* character), where we insert an arbitrary affine group element in the trace. By an adjoint action (that leaves the trace invariant) we can bring this element into the Cartan torus. In this case the group element can be written as an exponential of the Cartan generators $g = e^{2\pi i \sum_i v_i J_0^i}$. We will consider

$$\chi_R(v_i) = \text{Tr}_R\left[q^{L_0 - c/24} e^{2\pi i \sum_i v_i J_0^i}\right], \tag{4.12.27}$$

where i runs over the Cartan subalgebra and J_0^i are the zero modes of the Cartan currents. The Cartan subalgebra of O(N) for N even is generated by $J_0^{12}, J_0^{34}, \dots J_0^{N/2-1,N/2}$ and has dimension N/2. We will calculate the affine characters of the unit and vector representations. Consider the contribution of the fermions ψ^1 and ψ^2. By going to the basis $\psi^\pm = \psi^1 \pm i\psi^2$ it is easy to see that the J_0^{12} eigenvalues of $\psi_n^\pm$ are ± 1. Putting everything together and using the ϑ-function product formulas from appendix C on page 507 we obtain

$$\chi_O(v_i) = \frac{1}{2}\left[\prod_{i=1}^{N/2} \frac{\vartheta_3(v_i)}{\eta} + \prod_{i=1}^{N/2} \frac{\vartheta_4(v_i)}{\eta}\right], \tag{4.12.28}$$

$$\chi_V(v_i) = \frac{1}{2}\left[\prod_{i=1}^{N/2} \frac{\vartheta_3(v_i)}{\eta} - \prod_{i=1}^{N/2} \frac{\vartheta_4(v_i)}{\eta}\right]. \tag{4.12.29}$$

We will now turn to the Ramond sector and construct the associated Hilbert space. Here the fermions are integrally moded. For ψ_n^i with $n \neq 0$ the same discussion as before applies. We separate creation and annihilation operators, and the vacuum should be annihilated by the annihilation operators. However, an important difference here is the presence of anticommuting zero modes

$$\{\psi_0^i, \psi_0^j\} = \delta^{ij}. \tag{4.12.30}$$

A similar situation occurred when discussing the ghost system in (3.8.15) on page 44. Equation (4.12.30) is the O(N) Clifford algebra. It is realized by the Hermitian O(N) γ-matrices. Consequently, the "vacuum" must be a (Dirac) spinor $\hat{S}$ of O(N) with $2^{N/2}$ components. We label the R vacuum by $|\hat{S}_\alpha\rangle$ and we have

$$\psi_{m>0}^i|\hat{S}_\alpha\rangle = 0, \quad \psi_0^i|\hat{S}_\alpha\rangle = \frac{1}{\sqrt{2}}\gamma_{\alpha\beta}^i|\hat{S}_\beta\rangle. \tag{4.12.31}$$

The origin of the $\sqrt{2}$ above is due to the unconventional normalization of the Clifford algebra in (4.12.30).

Consider the chirality operator

$$\gamma^{N+1} = i^{N(N-1)/2}\prod_{i=1}^N (\sqrt{2}\,\psi_0^i), \quad \{\gamma^{N+1}, \psi_0^i\} = 0, \quad [\gamma^{N+1}]^2 = 1. \tag{4.12.32}$$

This matrix plays a similar role to that of γ^5 for four-dimensional spinors. It can be used to define Weyl spinors. Thus, we obtain the spinor $S = (1 + \gamma^{N+1})/2\,\hat{S}$ and the conjugate spinor $C = (1 - \gamma^{N+1})/2\,\hat{S}$. In the Ramond sector

$$(-1)^F = \gamma^{N+1} (-1)^{\sum_{n=1}^\infty \psi_{-n}^i \psi_n^i} \tag{4.12.33}$$

and with this definition

$$(-1)^F |S\rangle = |S\rangle, \quad (-1)^F |C\rangle = -|C\rangle. \tag{4.12.34}$$

Acting with the negative moded fermionic oscillators we construct the full spectrum of the Ramond sector. The R vacuum has the correct conformal weight, namely, N/16. This is the subject of exercise 4.37 on page 122.

We will now compute the multiplicities in the Ramond sector. Every fermionic oscillator ψ^i_{-n} with $n > 0$ will give a contribution $1 + q^n$. There will also be the multiplicity $2^{N/2}$ from the S and C ground states. Thus,

$$\mathrm{Tr}_R[q^{L_0 - c/24}] = 2^{N/2}\, q^{N/16 - N/48} \prod_{n=1}^{\infty} (1 + q^n)^N = \left[\frac{\vartheta_2}{\eta}\right]^{N/2}. \tag{4.12.35}$$

If we consider the trace with $(-1)^F$ inserted, we will obtain 0 since, for any state, there is another one of opposite $(-1)^F$ eigenvalue related by the zero modes. The fact that $\mathrm{Tr}[(-1)^F] = 0$ translates into the statement that the full R spectrum is nonchiral. Indeed, both C and S appear. Therefore,

$$\chi_S = \chi_C = \frac{1}{2}\left[\frac{\vartheta_2}{\eta}\right]^{N/2}. \tag{4.12.36}$$

The affine character does distinguish between the C and S representations:

$$\chi_S(\nu_i) = \frac{1}{2}\left[\prod_{i=1}^{N/2} \frac{\vartheta_2(\nu_i)}{\eta} + e^{-i\pi N/4} \prod_{i=1}^{N/2} \frac{\vartheta_1(\nu_i)}{\eta}\right], \tag{4.12.37}$$

$$\chi_C(\nu_i) = \frac{1}{2}\left[\prod_{i=1}^{N/2} \frac{\vartheta_2(\nu_i)}{\eta} - e^{-i\pi N/4} \prod_{i=1}^{N/2} \frac{\vartheta_1(\nu_i)}{\eta}\right]. \tag{4.12.38}$$

For $\nu_i = 0$ they reduce to (4.12.36).

Finally, we expect the R vacua corresponding to the C and S representations to be created out of the NS vacuum $|0\rangle$ by affine primary fields $\hat{S}_\alpha(z)$:

$$|\hat{S}_\alpha\rangle = \lim_{z \to 0} \hat{S}_\alpha(z)|0\rangle. \tag{4.12.39}$$

These are known as spin fields. We have the OPEs

$$\psi^i(z)\hat{S}_\alpha(w) = \frac{\gamma^i_{\alpha\beta}}{\sqrt{2}} \frac{\hat{S}_\beta(w)}{\sqrt{z - w}} + \cdots, \tag{4.12.40}$$

$$J^{ij}(z)\hat{S}_\alpha(w) = \frac{i}{4}[\gamma^i, \gamma^j]_{\alpha\beta} \frac{\hat{S}_\beta(w)}{(z - w)} + \cdots, \tag{4.12.41}$$

$$\hat{S}_\alpha(z)\hat{S}_\beta(w) = \frac{C_{\alpha\beta}}{(z - w)^{N/8}} + \frac{\gamma^i_{\alpha\beta}}{\sqrt{2}} \frac{\psi^i(w)}{(z - w)^{N/8 - 1/2}} + \frac{i}{4}[\gamma^i, \gamma^j]_{\alpha\beta} \frac{J^{ij}(w)}{(z - w)^{N/8 - 1}} + \cdots, \tag{4.12.42}$$

where $C_{\alpha\beta}$ is the charge conjugation matrix. (4.12.40)–(4.12.42) follow from group theory and the basic properties of the spin fields.

The theory has a (chiral) symmetry that we will call with hindsight space-time fermion parity,[10] F_L. It is a $\mathbb{Z}_2$ symmetry under which all NS states are even and all R states are

---

[10] Sometimes it is also referred as space-time fermion number modulo 2.

odd. It is a generalization of the symmetry in O(N) representation theory, under which spinors are odd, while tensors are even. It implies in particular that correlators with an odd number of Ramond state insertions are zero.

4.13 Superconformal Symmetry

We have seen that the conformal symmetry of a CFT is encoded in the OPE of the stress-tensor T which is a chiral (2,0) operator. We have also encountered other chiral operators. They include chiral fermions (1/2,0) and currents (1,0). They generate extra symmetries.

Here we will study symmetries whose conserved chiral currents have spin 3/2. They are associated with fermionic symmetries known as supersymmetries.

World-sheet supersymmetry plays the role of the gauge symmetry in the superstrings. It is therefore of prime importance. In the next subsections we will analyze the structure of the symmetries obtained by marrying conformal invariance and (extended) supersymmetry.

4.13.1 $\mathcal{N} = (1,0)_2$ superconformal symmetry

Consider the theory of a free scalar and a Majorana fermion[11] with action

$$S = \frac{1}{2\pi\ell_s^2} \int d^2z \, \partial X \bar\partial X - \frac{1}{2\pi\ell_s^2} \int d^2z (\psi\bar\partial\psi + \bar\psi\partial\bar\psi) \tag{4.13.1}$$

with two-point functions

$$\langle X(z,\bar z) X(0,0)\rangle = -\frac{\ell_s^2}{2}\log|z|^2, \quad \langle \psi(z)\psi(0)\rangle = \frac{\ell_s^2}{z}, \quad \langle \bar\psi(\bar z)\bar\psi(0)\rangle = \frac{\ell_s^2}{\bar z}. \tag{4.13.2}$$

The action is invariant under a left-moving supersymmetry transformation

$$\delta X = \epsilon(z)\psi, \quad \delta\psi = \epsilon(z)\partial X, \quad \delta\bar\psi = 0, \tag{4.13.3}$$

and a right-moving one

$$\delta X = \bar\epsilon(\bar z)\bar\psi, \quad \delta\bar\psi = \bar\epsilon(\bar z)\bar\partial X, \quad \delta\psi = 0. \tag{4.13.4}$$

ϵ and $\bar\epsilon$ are anticommuting. In our conventions, this is $\mathcal{N} = (1,1)_2$ supersymmetry.

The associated conservation laws can be written as $\partial\bar G = \bar\partial G = 0$ and the conserved chiral currents are

$$G(z) = i\frac{\sqrt 2}{\ell_s^2}\psi\partial X, \quad \bar G(\bar z) = i\frac{\sqrt 2}{\ell_s^2}\bar\psi\bar\partial X. \tag{4.13.5}$$

In the following we will focus on the left-moving part of the theory. We can obtain the OPE

$$G(z)G(w) = \frac{1}{(z-w)^3} + 2\frac{T(w)}{z-w} + \cdots,$$
$$T(z)G(w) = \frac{3}{2}\frac{G(w)}{(z-w)^2} + \frac{\partial G(w)}{z-w} + \cdots, \tag{4.13.6}$$

---

[11] In anticipation of the superstring case, we take the fermions to have dimensions of space-time length. In this way they will be worthy supersymmetric partners of the string coordinates. Compared to section 4.8 on page 63 this simply amounts to $\psi \to \psi/\ell_s$.

where $T(z)$ is the total stress-tensor of the theory satisfying (4.5.1) on page 58 with $c = 3/2$,

$$T(z) = -\frac{1}{\ell_s^2} : \partial X \partial X : -\frac{1}{2\ell_s^2} : \psi \partial \psi : . \tag{4.13.7}$$

(4.13.6) implies that $G(z)$ is a primary field of dimension 3/2. The algebra generated by T and G is the $\mathcal{N} = (1, 0)_2$ superconformal algebra. It encodes the presence of conformal invariance and one supersymmetry. The most general such algebra can be written down using conformal invariance and associativity.

To do this, it is customary to define $\hat{c} = 2c/3$. Then, the superconformal algebra, apart from (4.5.1), contains the following OPEs:

$$G(z)G(w) = \frac{\hat{c}}{(z-w)^3} + 2\frac{T(w)}{z-w} + \cdots ,$$
$$T(z)G(w) = \frac{3}{2}\frac{G(w)}{(z-w)^2} + \frac{\partial G(w)}{z-w} + \cdots . \tag{4.13.8}$$

Introducing the modes of the supercurrent $G(z) = \sum_r G_r/z^{r+3/2}$ we obtain the following (anti)commutation relations:

$$\{G_r, G_s\} = \frac{\hat{c}}{2}\left(r^2 - \frac{1}{4}\right)\delta_{r+s,0} + 2L_{r+s},$$
$$[L_m, G_r] = \left(\frac{m}{2} - r\right)G_{m+r}, \tag{4.13.9}$$

together with those of the Virasoro algebra.

This algebra has the symmetry (external automorphism) $G \to -G$ and $T \to T$. Consequently, NS or R boundary conditions are possible for the supercurrent. In the explicit realization (4.13.5), they correspond to the respective boundary conditions for the fermion.

In the NS sector, the supercurrent modes are half integral and $G_r|0\rangle = 0$ for $r > 0$. Primary states are annihilated by the positive modes of G and T and the superconformal representation is generated by the action of the negative modes of G and T. The generic character is

$$\chi_{\mathcal{N}=1}^{\text{NS}} = \text{Tr}[q^{L_0 - c/24}] = q^{\Delta - c/24}\prod_{n=1}^{\infty}\frac{1 + q^{n-1/2}}{1 - q^n}. \tag{4.13.10}$$

In the Ramond sector, G is integrally moded and has in particular a zero mode G_0, which according to (4.13.9) satisfies

$$\{G_0, G_0\} = 2G_0^2 = 2L_0 - \frac{\hat{c}}{8}. \tag{4.13.11}$$

Primary states are again annihilated by the positive modes. In a unitary theory, (4.13.11) indicates that for any state $\Delta \geq \hat{c}/16$. When the right-hand side of (4.13.11) is nonzero, the state is doubly degenerate and G_0 maps one degenerate state to the other. There is no degeneracy when $\Delta = \hat{c}/16$ since from (4.13.11) $G_0^2 = 0$, which implies that $G_0 = 0$ on such a state. As in the case of free fermions, we can introduce the operator $(-1)^F$, which anticommutes with G and counts fermion number modulo 2.

The pairing of degenerate states in the R sector due to G_0 can be stated as follows: the trace of $(-1)^F$ in the R sector has nonzero contributions only from the ground states with $\Delta = \hat{c}/16$. This trace is called the Witten index of a supersymmetric theory. It is also known as the *elliptic genus* of the $\mathcal{N} = (1, 0)_2$ superconformal field theory. It is the CFT generalization of the Dirac index.

The generic character in the Ramond sector is given by

$$\chi^R_{\mathcal{N}=1} = \text{Tr}[q^{L_0 - c/24}] = q^{\Delta - c/24} \prod_{n=1}^{\infty} \frac{1 + q^n}{1 - q^n}. \tag{4.13.12}$$

The $\mathcal{N} = (1, 0)_2$ and $\mathcal{N} = (1, 1)_2$ superconformal theories have an elegant formulation in superspace. We will describe the $\mathcal{N} = (1, 1)_2$ case here. Along with the coordinates $z, \bar{z}$, we introduce two anticommuting variables $\theta, \bar{\theta}$ and the covariant derivatives

$$D_\theta = \frac{\partial}{\partial \theta} + \theta \partial_z, \quad \bar{D}_{\bar{\theta}} = \frac{\partial}{\partial \bar{\theta}} + \bar{\theta} \partial_{\bar{z}}. \tag{4.13.13}$$

The fields X and $\psi, \bar{\psi}$ can now be described by a function in superspace, the scalar superfield

$$\hat{X}(z, \bar{z}, \theta, \bar{\theta}) = X + \theta \psi + \bar{\theta} \bar{\psi} + \theta \bar{\theta} F, \tag{4.13.14}$$

where F is an auxiliary field with no dynamics. The action (4.13.1) becomes

$$S = \frac{1}{2\pi \ell_s^2} \int d^2 z \int d\theta d\bar{\theta} \, D_\theta \hat{X} \, \bar{D}_{\bar{\theta}} \hat{X}. \tag{4.13.15}$$

The equations of motion set $F = 0$.

Relevant supersymmetric vertex operators are constructed out of $e^{ip\hat{X}}$ as well as the covariant derivatives $D_\theta \hat{X}$, $\bar{D}_{\bar{\theta}} \hat{X}$.

4.13.2 $\mathcal{N} = (2, 0)_2$ superconformal symmetry

There are further generalizations of superconformal symmetry. The next simplest case is the $\mathcal{N} = (2, 0)_2$ superconformal algebra which contains, apart from the stress-tensor, two supercurrents $G^\pm$ and a U(1) current J. Its OPEs, apart from the Virasoro one, are

$$G^+(z)G^-(w) = \frac{2c}{3} \frac{1}{(z-w)^3} + \left(\frac{2J(w)}{(z-w)^2} + \frac{\partial J(w)}{z-w} \right) + \frac{2}{z-w} T(w) + \cdots, \tag{4.13.16}$$

$$G^+(z)G^+(w) = \text{regular}, \quad G^-(z)G^-(w) = \text{regular}, \tag{4.13.17}$$

$$T(z)G^\pm(w) = \frac{3}{2} \frac{G^\pm(w)}{(z-w)^2} + \frac{\partial G^\pm(w)}{z-w} + \cdots, \tag{4.13.18}$$

$$J(z)G^\pm(w) = \pm \frac{G^\pm(w)}{z-w} + \cdots, \tag{4.13.19}$$

$$T(z)J(w) = \frac{J(w)}{(z-w)^2} + \frac{\partial J(w)}{z-w} + \cdots, \tag{4.13.20}$$

$$J(z)J(w) = \frac{c/3}{(z-w)^2} + \cdots. \tag{4.13.21}$$

The global symmetry of the $\mathcal{N} = (2, 0)_2$ superconformal algebra is a continuous O(2) rotation of the two supercharges (in a real basis $G^1 = G^+ + G^-$, $G^2 = i(G^+ - G^-)$). The SO(2) part is an internal automorphism. The extra $\mathbb{Z}_2$ transformation $G^1 \to G^1$,

$G^2 \to -G^2$ is an external automorphism. We can use the symmetry to impose various boundary conditions. Using the external automorphism, one of the supercurrents has periodic boundary conditions and the other antiperiodic ones. This provides an inequivalent algebra, the twisted $\mathcal{N} = (2,0)_2$ algebra.

More interesting for our purposes is to use the SO(2)$\sim$U(1) symmetry in order to impose boundary conditions

$$G^{\pm}(e^{2\pi i}z) = e^{\mp 2\pi i\alpha} G^{\pm}(z), \tag{4.13.22}$$

while T, J are single valued. For $\alpha = 0$ we have the NS sector, where both supercharges are half-integrally moded. For $\alpha = \pm 1/2$ we obtain the Ramond sector, where both supercharges are integrally moded. Since the U(1) symmetry we used is an internal automorphism, the algebras obtained for the various boundary conditions, labeled by α, are isomorphic. We can write this isomorphism (known as "spectral flow") explicitly:

$$J_n^{\alpha} = J_n - \alpha \frac{c}{3}\delta_{n,0}, \quad L_n^{\alpha} = L_n - \alpha J_n + \alpha^2 \frac{c}{6}\delta_{n,0}, \tag{4.13.23}$$

$$G_{r+\alpha}^{\alpha,+} = G_r^+, \quad G_{r-\alpha}^{\alpha,-} = G_r^-, \tag{4.13.24}$$

where $n \in \mathbb{Z}$ and $r \in \mathbb{Z} + \frac{1}{2}$. The spectral flow provides a continuous interpolation between the NS and R sectors.

In the NS sector, HW irreducible representations are generated by a HW state $|\Delta, q\rangle$, annihilated by the positive modes of $T, J, G^{\pm}$ and characterized by the eigenvalues Δ of L_0 and q of J_0. The SL(2,$\mathbb{C}$)-invariant vacuum has $\Delta = q = 0$. The rest of the states of the representations are generated by the action of the negative modes of the superconformal generators.

In the R sector, HW states are again annihilated by the positive modes. Here, however, we also have the zero modes of the supercurrents $G_0^{\pm}$ satisfying

$$(G_0^{\pm})^2 = 0, \quad \{G_0^+, G_0^-\} = 2\left(L_0 - \frac{c}{24}\right). \tag{4.13.25}$$

Unitarity implies that $\Delta \geq c/24$ in the R sector. When $\Delta > c/24$ both $G_0^{\pm}$ act nontrivially and the lowest level consists of four states. When $\Delta = c/24$, then $G_0^{\pm}$ are null and the HW vector is a singlet.

We introduce the $(-1)^F$ operator in a way analogous to the $\mathcal{N} = (1,0)_2$ case. The trace of $(-1)^F$ in the R sector (elliptic genus) obtains contributions only from states with $\Delta = c/24$.

Using (4.13.24) we deduce that

$$J_0^{R\pm} = J_0^{NS} \mp \frac{c}{6}, \quad L_0^{R\pm} - \frac{c}{24} = L_0^{NS} \mp \frac{1}{2}J_0^{NS}. \tag{4.13.26}$$

Thus, the positivity condition $L_0^R - c/24 \geq 0$ translates in the NS sector to $2\Delta - |q| \geq 0$. The Ramond ground states correspond to NS states with $2\Delta = |q|$, known as chiral states. They are generated from the vacuum by the chiral field operators. Because of charge conservation, their OPE at short distances is regular. It can be written as a ring, the *chiral ring*,

$$O_{q_1}(z)O_{q_2}(z) = O_{q_1+q_2}(z). \tag{4.13.27}$$

The chiral ring captures some nontrivial information about the $\mathcal{N} = (2, 0)_2$ superconformal theory.

From (4.13.26) we can deduce that the unit operator ($\Delta = q = 0$) in the NS sector is mapped, under spectral flow, to an operator with ($\Delta = c/24, q = \pm c/6$) in the R sector. This is the maximal-charge ground state in the R sector, and applying the spectral flow once more we learn that there must be a chiral operator with ($\Delta = c/6, q = \pm c/3$) in the NS sector. As we will see later on, this operator plays an important role for space-time supersymmetry in string theory.

Similarly, the relation

$$\{G^-_{3/2}, G^+_{-3/2}\} = 2L_0 - 3J_0 + \frac{2c}{3} \tag{4.13.28}$$

when acted on a chiral primary gives $|q| \leq \frac{c}{3}$. Thus, this chiral operator has the maximal U(1) charge of the chiral ring.

$\mathcal{N} = (2, 2)_2$ superconformal theories can be realized as nonlinear σ-models on manifolds with SU(N) holonomy. The six-dimensional case corresponds to Calabi-Yau (CY) manifolds, which are Ricci flat. The central charge of the $\mathcal{N} = (2, 2)_2$ algebra in the CY case is $c = 9$. The ($\Delta = 3/2, q = \pm 3$) state, mentioned above, can be interpreted as the unique (3,0) form of the CY manifold.

As we will see in section 9.2, this symmetry will be relevant for heterotic superstring vacua with $\mathcal{N} = 1_4$ space-time supersymmetry or type-II vacua with $\mathcal{N} = 2_4$ space-time supersymmetry.

4.13.3 $\mathcal{N} = (4, 0)_2$ superconformal symmetry

Another extended superconformal algebra that is useful in string theory is the "short" $\mathcal{N} = (4, 0)_2$ superconformal algebra. It contains, apart from the stress-tensor, four supercurrents and three currents that form the current algebra of SU(2)$_k$. The four supercurrents transform as two conjugate spinors under SU(2)$_k$. The Virasoro central charge c is related to the level k of the SU(2) current algebra as $c = 6k$.

The algebra is defined in terms of the usual Virasoro OPE, the statement that J^a, G^α, $\bar{G}^\alpha$ are primary with the appropriate conformal weight, and the following OPEs:

$$J^a(z)J^b(w) = \frac{k}{2}\frac{\delta^{ab}}{(z-w)^2} + i\epsilon^{abc}\frac{J^c(w)}{(z-w)} + \cdots, \tag{4.13.29}$$

$$J^a(z)G^\alpha(w) = \frac{1}{2}\sigma^a_{\beta\alpha}\frac{G^\beta(w)}{(z-w)} + \cdots, \quad J^a(z)\bar{G}^\alpha(w) = -\frac{1}{2}\sigma^a_{\alpha\beta}\frac{\bar{G}^\beta(w)}{(z-w)} + \cdots, \tag{4.13.30}$$

$$G^\alpha(z)\bar{G}^\beta(w) = \frac{4k\delta^{\alpha\beta}}{(z-w)^3} + 2\sigma^a_{\beta\alpha}\left[\frac{2J^a(w)}{(z-w)^2} + \frac{\delta J^a(w)}{(z-w)}\right] + 2\delta^{\alpha\beta}\frac{T(w)}{(z-w)} + \cdots, \tag{4.13.31}$$

$$G^\alpha(z)G^\beta(w) = \text{regular}, \quad \bar{G}^\alpha(z)\bar{G}^\beta(w) = \text{regular}. \tag{4.13.32}$$

As in the $\mathcal{N} = (2, 0)_2$ case, there are various boundary conditions we can impose. We will be interested in NS and R boundary conditions here. There is a spectral flow, similar to the $\mathcal{N} = (2, 0)_2$ one, that interpolates between NS and R boundary conditions.

In the NS sector, primary states are annihilated by the positive modes and are characterized by their conformal weight Δ and $SU(2)_k$ spin j. As usual, for unitarity we have $j \leq k/2$.

The representations saturating the above bounds are called "massless," since they would correspond to massless states in the appropriate string context. In the particular case of $k = 1$, relevant for string compactification, the $\mathcal{N} = (4,0)_2$ superconformal algebra can be realized in terms of a σ-model on a four-dimensional Ricci-flat, Kähler manifold with $SU(2)$ holonomy. In the compact case this is the K3 class of manifolds. In the NS sector, the two massless representations have $(\Delta, j) = (0,0)$ and $(1/2, 1/2)$, while in the R sector, $(\Delta, j) = (1/4, 0)$ and $(1/4, 1/2)$.

The trace in the R sector of $(-1)^F$ obtains contributions from ground states only and provides the elliptic genus of the $\mathcal{N} = (4,0)_2$ superconformal theory.

4.14 Scalars with Background Charge

There is a variant of the free massless scalar field with a nontrivial coupling to the two-dimensional scalar curvature. Consider the action

$$S = \frac{1}{4\pi \ell_s^2} \int d^2\xi \sqrt{g} g^{\alpha\beta} \partial_\alpha X \partial_\beta X + \frac{Q}{\sqrt{2}(4\pi \ell_s)} \int d^2\xi \sqrt{g} \, R^{(2)} X. \tag{4.14.1}$$

The curvature coupling has the correct scaling dimension to preserve conformal invariance, at least classically.

The stress tensor can be directly calculated to be

$$T_{\alpha\beta} \equiv -\frac{4\pi}{\sqrt{g}} \frac{\delta S}{\delta g^{\alpha\beta}} = -\frac{1}{\ell_s^2} \left[\partial_\alpha X \partial_\beta X - \frac{1}{2} g_{\alpha\beta} (\partial X)^2 \right] + \frac{Q}{\sqrt{2}\ell_s} \left[\nabla_\alpha \nabla_\beta - g_{\alpha\beta} \Box \right] X, \tag{4.14.2}$$

where we have used

$$\delta R = \left[R_{\mu\nu} + g_{\mu\nu} \Box - \nabla_\mu \nabla_\nu \right] \delta g^{\mu\nu}, \tag{4.14.3}$$

valid in any dimension. The equation of motion for X is

$$\Box X = \frac{\ell_s Q}{2\sqrt{2}} R^{(2)}. \tag{4.14.4}$$

We may now calculate the classical piece of the energy momentum trace from (4.14.2) using (4.14.4):

$$T^\alpha{}_\alpha \big|_{\text{classical}} = -\frac{Q}{\ell_s \sqrt{2}} \Box X = -\frac{Q^2}{4} R^{(2)}, \tag{4.14.5}$$

where in the second equality we have used the equation of motion.

Combining this with the conformal anomaly of a massless scalar

$$T^\alpha{}_\alpha \big|_{\text{quantum}} = -\frac{1}{12} R^{(2)} \tag{4.14.6}$$

from (4.7.14) on page 62 and (4.9.8) on page 65, we find that the theory is conformally invariant with central charge

$$c = 1 + 3Q^2. \tag{4.14.7}$$

The holomorphic component of the stress tensor is

$$T = -\frac{1}{\ell_s^2} : \partial X \partial X : + \frac{Q}{\ell_s \sqrt{2}} \partial^2 X. \tag{4.14.8}$$

The propagator on the sphere in flat coordinates is unaffected by the background charge

$$\langle X(z, \bar{z}) X(w, \bar{w}) \rangle = -\frac{\ell_s^2}{2} \log |z - w|^2. \tag{4.14.9}$$

The OPE of the vertex operator $V_p =: e^{ipX} :$ with the stress tensor is now

$$T(z) V_p = \Delta_p \frac{V_p(w)}{(z - w)^2} + \frac{\partial_w V_p(w)}{z - w} + \cdots, \quad \Delta_p = \frac{(\ell_s p)^2}{4} + i\frac{(\ell_s p) Q}{2\sqrt{2}}. \tag{4.14.10}$$

In order to obtain real conformal weights $p \to p - i\frac{Q}{\sqrt{2}\ell_s}$. Then $U_p =: e^{ipX} e^{(Q/\sqrt{2})X/\ell_s} :$ has conformal dimension $\Delta = \frac{(\ell_s p)^2}{4} - \frac{Q^2}{8}$. Its conjugate operator is U_{-p}. For the vertex operators V_{p_i} the correlator $\langle \prod_{i=1}^n V_{p_i} \rangle$ is nonzero if

$$i\sqrt{2}\ell_s \sum_{i=1}^n p_i = Q\chi, \tag{4.14.11}$$

where χ is the Euler number of the surface, defined in (5.0.1) on page 128 or alternatively in (6.1.2) on page 145. You are invited to derive this in exercise 4.42 on page 123 by considering the integration in the path integral over the zero mode of the scalar X.

This condition can be formulated in terms of the charges of the "anomalous" U(1) symmetry generated by the standard U(1) chiral current in (4.7.12) on page 62, normalized as $J(z)J(w) = (z - w)^{-2} + \cdots$. It can be verified by direct computation of the OPE that the U(1) current satisfies the "anomalous" OPE

$$T(z)J(w) = \frac{A}{(z - w)^3} + \frac{J(w)}{(z - w)^2} + \frac{\partial_w J(w)}{(z - w)} + \cdots, \quad A = iQ. \tag{4.14.12}$$

The coefficient A is the anomaly in the (chiral) conservation of the U(1) current on a curved surface. Indeed, we can use (4.14.4) to derive

$$\partial_{\bar{z}} J = \frac{A}{8} R^{(2)}. \tag{4.14.13}$$

If we define the charge q of a primary operator V_q as

$$J(z) V_q(w) = q \frac{V_q(w)}{z - w} + \cdots, \tag{4.14.14}$$

the charge neutrality condition (4.14.11) reads

$$\sum_{i=1}^n q_i = -\frac{A}{2} \chi. \tag{4.14.15}$$

On the sphere we may introduce the standard Fourier modes and write (4.14.12) and (4.14.14) as

$$[L_m, J_n] = -n J_{m+n} + \frac{A}{2} m(m + 1)\delta_{m+n,0}, \quad [J_m, V_q(w)] = q \, w^m V_q(w). \tag{4.14.16}$$

The stress-tensor and the current were defined so that their modes are Hermitian. However, the presence of the anomaly term in (4.14.16) necessarily modifies the Hermiticity conditions of the U(1) current zero mode:

$$L_m^\dagger = L_{-m}, \quad J_m^\dagger = J_{-m} - A\delta_{m,0}. \tag{4.14.17}$$

You are invited in exercise 4.43 on page 123 to show that this is the only set of Hermiticity conditions for J compatible with the commutation relations (4.14.16). If we therefore define the in-vacuum as chargeless, $J_0|0\rangle = 0$, the out-vacuum carries a nonzero charge

$$\langle 0|J_0 = A\langle 0|. \tag{4.14.18}$$

Evaluating in two ways the correlator $\langle 0|J_0 \prod_{i=1}^n V_{q_i}|0\rangle$ we obtain the neutrality condition (4.14.15) on the sphere ($\chi = 2$).

Because of the modified neutrality condition, this system is known as the Coulomb gas with a charge at infinity.[12]

4.15 The CFT of Ghosts

We have seen that in the covariant quantization of the string we had to introduce an anticommuting ghost system containing the b ghost with conformal weight 2 and the c ghost[13] with conformal weight -1. Here, anticipating further applications, we will describe in general the CFT of such ghost systems. We will treat the holomorphic part only to avoid repetition.

The field b will have conformal weight $\Delta = \lambda$ while $c, \Delta = 1 - \lambda$. We will also allow them to be anticommuting ($\epsilon = 1$) or commuting ($\epsilon = -1$). They are governed by the free action

$$S_\lambda = \frac{1}{\pi} \int d^2z\, b\bar\partial c, \tag{4.15.1}$$

from which we obtain the OPEs

$$c(z)b(w) = \frac{1}{z-w}, \quad b(z)c(w) = \frac{\epsilon}{z-w}. \tag{4.15.2}$$

The equations of motion $\bar\partial b = \bar\partial c = 0$ imply that the fields are holomorphic. Their conformal weights determine their mode expansions on the sphere and their Hermiticity properties

$$c(z) = \sum_{n\in\mathbb{Z}} z^{-n-(1-\lambda)}\, c_n, \quad c_n^\dagger = c_{-n}, \tag{4.15.3}$$

$$b(z) = \sum_{n\in\mathbb{Z}} z^{-n-\lambda}\, b_n, \quad b_n^\dagger = \epsilon b_{-n}. \tag{4.15.4}$$

Their (anti)commutation relations are

$$c_m b_n + \epsilon b_n c_m = \delta_{m+n,0}, \quad c_m c_n + \epsilon c_n c_m = b_m b_n + \epsilon b_n b_m = 0. \tag{4.15.5}$$

---

[12] To compare with most of the literature you will have to put $\ell_s = \sqrt{2}$ in this section.

[13] c here is a two-dimensional field, not to be confused with the central charge.

Due to the $\mathbb{Z}_2$ symmetry $b \to -b, c \to -c$, we can introduce the analog of NS and R sectors (corresponding to antiperiodic and periodic boundary conditions on the cylinder):

$$\text{NS: } b_n, \quad n \in \mathbb{Z} - \lambda, \quad c_n, \quad n \in \mathbb{Z} + \lambda, \tag{4.15.6}$$

$$\text{R: } b_n, \quad n \in \frac{1}{2} + \mathbb{Z} - \lambda, \quad c_n, \quad n \in \frac{1}{2} + \mathbb{Z} + \lambda. \tag{4.15.7}$$

The stress-tensor is fixed by the conformal properties of the bc system to be

$$T = -\lambda b \partial c + (1 - \lambda)(\partial b)c. \tag{4.15.8}$$

Under this stress-tensor, b and c transform as primary fields with conformal weights $(\lambda,0)$ and $(0,1 - \lambda)$. T satisfies the Virasoro algebra with central charge

$$c = -2\epsilon(6\lambda^2 - 6\lambda + 1) = \epsilon(1 - 3Q^2), \quad Q = \epsilon(1 - 2\lambda). \tag{4.15.9}$$

We have already encountered two special cases of this system. The first is $\lambda = 2$ and $\epsilon = 1$, which corresponds to the reparametrization ghosts with $c = -26$. The second is $\lambda = 1/2$ and $\epsilon = 1$, which corresponds to a complex (Dirac) fermion or equivalently to two Majorana fermions with $c = 1$.

There is a classical U(1) symmetry in (4.15.1): $b \to e^{i\theta}b, c \to e^{-i\theta}c$. The associated U(1) current is

$$J(z) = - : b(z)c(z) := \sum_{n \in \mathbb{Z}} z^{-n-1}J_n, \tag{4.15.10}$$

where the normal ordering is chosen with respect to the standard SL(2,$\mathbb{C}$)-invariant vacuum $|0\rangle$, in which $\langle c(z)b(w)\rangle = 1/(z - w)$. It generates a U(1) current algebra

$$J(z)J(w) = \frac{\epsilon}{(z - w)^2} + \cdots . \tag{4.15.11}$$

The ghosts b, c are affine primary

$$J(z)b(w) = -\frac{b(w)}{z - w} + \cdots , \quad J(z)c(w) = \frac{c(w)}{z - w} \cdots . \tag{4.15.12}$$

A direct computation of the OPE gives

$$T(z)J(w) = \frac{Q}{(z - w)^3} + \frac{J(w)}{(z - w)^2} + \frac{\partial_w J(w)}{z - w} + \cdots . \tag{4.15.13}$$

Note the appearance of the central term in (4.15.13), which makes it different from the standard TJ OPE in (4.11.5) on page 69 and similar to (4.14.12). Translating into commutation relations, we obtain

$$[L_m, J_n] = -nJ_{m+n} + \frac{Q}{2}m(m + 1)\delta_{m+n,0}. \tag{4.15.14}$$

As described in detail in section 4.14, the central term implies an anomaly in the current algebra. The U(1) charge conservation condition is modified to $\sum_i q_i = -Q$ (Q is a background charge for the system). The anomaly is related to an index theorem involving the zero-mode structure of the bc system. It translates into

$$\text{no. of zero modes of } c - \text{no. of zero modes of } b = -\frac{\epsilon}{2}Q\chi, \tag{4.15.15}$$

where $\chi = 2(1 - g)$ is the Euler number of a genus g surface.

According to (4.6.12) on page 60 we obtain (NS sector)

$$b_{n>-\lambda}|0\rangle = c_{n>\lambda-1}|0\rangle = 0. \tag{4.15.16}$$

Consequently, for the standard reparametrization ghosts ($\lambda = 2$) the state with lowest L_0 eigenvalue is not the vacuum but $c_1|0\rangle$ with L_0 eigenvalue equal to -1.

4.16 CFT on the Disk

So far, our discussion of CFT on surfaces without boundary (closed Riemann surfaces) is suitable for the description of closed string propagation. In order to describe the propagation of open strings, we must introduce boundaries on the two-dimensional surfaces.

The simplest example is given by the tree-level propagation of an open string. It gives rise to an infinitely long strip, with coordinates $-\infty < \tau < \infty$ and $0 \leq \sigma \leq \pi$. As is customary, we have rescaled σ to the range $[0, \pi]$ by a scale transformation.

By a conformal transformation $z = e^{\tau+i\sigma}$, the strip can be mapped to the upper half plane $\mathcal{H}_2$ as shown in figure 4.3. The $\sigma = 0$ boundary of the strip is mapped to the positive real axis, while the $\sigma = \pi$ boundary of the strip is mapped to the negative real axis. Upon a further conformal mapping $w = \frac{z+i}{z-i}$, the upper half plane is mapped to the unit disk (the interior of the unit circle) $|w| \leq 1$. We introduce polar coordinates as $w = re^{i\theta}$. In these coordinates the boundary is at $r = 1$. The original $\sigma = 0$ boundary of the strip is now at $0 \leq \theta \leq \pi$ while the $\sigma = \pi$ boundary of the strip is now at $-\pi \leq \theta \leq 0$.

A useful way to describe such surfaces is via involutions (orbifolds) of closed Riemann surfaces, in this case the sphere. Let the sphere coordinate be z. We identify points on the sphere as

$$z' = \frac{1}{\bar{z}}. \tag{4.16.1}$$

We obtain the interior of the unit disk D_2 as the fundamental region (See figure 4.4b). The boundary is at the fixed points of the involution, ie. at $|z| = 1$. The upper half plane can be obtained from the sphere by means of the involution

$$z' = \bar{z}. \tag{4.16.2}$$

The boundary (fixed manifold) is now on the real axis. In this parametrization, the conformal Killing group is PSL(2,$\mathbb{R}$), which consists of the PSL(2,$\mathbb{C}$) transformations that preserve the boundary. The Euler number of the disk is 1.

There are advantages in constructing the disk (and other) surfaces as involutions of the sphere. Many data on the disk, like propagators, can be obtained from those on the sphere by the method of images as we will discuss below.

4.16.1 Free massless bosons on the disk

Having a surface with a boundary, we must specify the boundary conditions. We will indicate this in the simplest (and useful case) of a free massless world-sheet scalar X. As

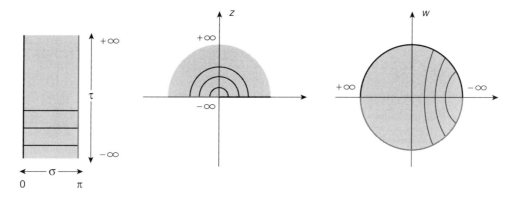

Figure 4.3 From left to right: the infinite strip conformally mapped to the upper half plane and further to the unit disk.

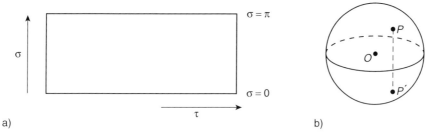

a)   b)

Figure 4.4 a) Parametrization of the strip. b) The disk involution.

discussed in section 2.3.2 on page 22, a Neumann boundary condition on a given boundary amounts to

$$J - \bar{J}|_{\text{boundary}} = i \frac{\sqrt{2}}{\ell_s} (\partial X - \bar{\partial} X)\Big|_{\text{boundary}} = 0, \tag{4.16.3}$$

while a Dirichlet boundary condition becomes

$$J + \bar{J}|_{\text{boundary}} = i \frac{\sqrt{2}}{\ell_s} (\partial X + \bar{\partial} X)\Big|_{\text{boundary}} = 0. \tag{4.16.4}$$

Thus, on $\mathcal{H}_2$, a NN string world-sheet involves a scalar satisfying $J = \bar{J}$ on the whole real axis. This condition implies that there is no charge (momentum) flowing out of the boundary. Momentum therefore is conserved. A DD string, on the other hand, is described by a scalar satisfying $J = -\bar{J}$ on the boundary, and charge (momentum) conservation is broken.

Finally, a DN string involves a $J = \bar{J}$ boundary condition on the positive real axis and a $J = -\bar{J}$ boundary condition on the negative real axis. In this case, the boundary condition jumps discontinuously at $z = 0$ and this can interpreted as due to the presence of a defect (insertion of a boundary field) at $z = 0$. This field turns out to be the boundary analog of a $\mathbb{Z}_2$ orbifold twist field, (described in section 4.21 on page 107).

Since $T = \frac{1}{2} : J^2 :$ and $\bar{T} = \frac{1}{2} : \bar{J}^2 :$, it is obvious that in all the cases above, the stress tensor satisfies a Neumann boundary condition at the boundary:

$$T = \bar{T}\Big|_{\text{boundary}}. \tag{4.16.5}$$

This condition is the minimal necessary condition for a boundary to preserve conformal invariance present in the bulk CFT.

We may now describe the calculation of some simple correlation functions on the disk topology. We will use for concreteness the $\mathcal{H}_2$ representation.

Let us fist consider the massless scalar propagator with NN boundary conditions. By splitting X_{NN} into left- and right-moving pieces $X_{NN}(z, \bar{z}) = X_{NN}(z) + \bar{X}_{NN}(\bar{z})$ we may evaluate

$$\langle X_{NN}(z)X_{NN}(w)\rangle = -\frac{\ell_s^2}{2}\log(z-w), \quad \langle X_{NN}(z)\bar{X}_{NN}(\bar{w})\rangle = -\frac{\ell_s^2}{2}\log(z-\bar{w}), \tag{4.16.6}$$

which follow from (2.3.21) and (2.3.22) on page 23 and the commutation relations of the oscillators.

The full two-point function is the sum of the images of the sphere propagator under the involution (4.16.2)

$$\langle X_{NN}(z, \bar{z})X_{NN}(w, \bar{w})\rangle = -\frac{\ell_s^2}{2}\left(\log|z-w|^2 + \log|z-\bar{w}|^2\right). \tag{4.16.7}$$

(4.16.7) also follows from (4.16.6).

For DD boundary conditions we must antisymmetrize under the involution. Using (2.3.27) and (2.3.28) we obtain

$$\langle X_{DD}(z)X_{DD}(w)\rangle = -\frac{\ell_s^2}{2}\log(z-w), \quad \langle X_{DD}(z)\bar{X}_{DD}(\bar{w})\rangle = +\frac{\ell_s^2}{2}\log(z-\bar{w}), \tag{4.16.8}$$

from which

$$\langle X_{DD}(z, \bar{z})X_{DD}(w, \bar{w})\rangle = -\frac{\ell_s^2}{2}\left(\log|z-w|^2 - \log|z-\bar{w}|^2\right) \tag{4.16.9}$$

follows.

4.16.2 Free massless fermions on the disk

We must first impose the various boundary conditions on the fermion. We introduce also a right-moving fermion that we will label $\bar{\psi}$.

We will start with the NN boundary conditions. A guideline for possible boundary conditions is the fact that for the string, the $\mathcal{N} = (1, 1)_2$ superconformal symmetry must remain intact in the presence of boundaries. We work here with cylinder coordinates. The supercurrents $G = \psi J$ and $\bar{G} = \bar{\psi}\bar{J}$ in the R sector where the fermions are periodic must satisfy

$$G = \bar{G}, \quad \text{R sector}, \tag{4.16.10}$$

at both boundaries. This guarantees that the supercurrent has integer modes.

In the NS sector one of the boundaries flips sign, in order to obtain half-integer modes,

$$G + \bar{G}\big|_{\sigma=0}, \quad G - \bar{G}\big|_{\sigma=\pi}, \quad \text{NS sector.} \tag{4.16.11}$$

This implies the appropriate boundary conditions on the fermions

NN R sector : $\quad \psi - \bar{\psi}\big|_{\sigma=0} = \psi - \bar{\psi}\big|_{\sigma=\pi} = 0,$ $\tag{4.16.12}$

NN NS sector : $\quad \psi + \bar{\psi}\big|_{\sigma=0} = \psi - \bar{\psi}\big|_{\sigma=\pi} = 0.$ $\tag{4.16.13}$

As for the bosons, the DD boundary condition is essentially the same with $\bar{\psi} \to -\bar{\psi}$:

DD NS sector : $\quad \psi - \bar{\psi}\big|_{\sigma=0} = \psi + \bar{\psi}\big|_{\sigma=\pi} = 0,$ $\tag{4.16.14}$

DD R sector : $\quad \psi + \bar{\psi}\big|_{\sigma=0} = \psi + \bar{\psi}\big|_{\sigma=\pi} = 0.$ $\tag{4.16.15}$

We will introduce the mode expansions on the cylinder.

NS: $\quad \psi(\sigma,\tau) = \sum_{n\in\mathbb{Z}} b_{n+1/2} e^{(n+1/2)(\tau+i\sigma)}, \quad \bar{\psi}(\sigma,\tau) = \sum_{n\in\mathbb{Z}} \bar{b}_{n+1/2} e^{(n+1/2)(\tau-i\sigma)},$ $\tag{4.16.16}$

R: $\quad \psi(\sigma,\tau) = \sum_{n\in\mathbb{Z}} b_n e^{n(\tau+i\sigma)}, \quad \bar{\psi}(\sigma,\tau) = \sum_{n\in\mathbb{Z}} \bar{b}_n e^{n(\tau-i\sigma)}.$ $\tag{4.16.17}$

Then the open string boundary conditions (4.16.12)–(4.16.15) give the following identifications:

NN: $\quad \bar{b}_{n+1/2} = -b_{n+1/2}, \quad \bar{b}_n = b_n,$ $\tag{4.16.18}$

DD: $\quad \bar{b}_{n+1/2} = b_{n+1/2}, \quad \bar{b}_n = -b_n.$ $\tag{4.16.19}$

There is also the possibility of DN boundary conditions:

DN NS sector: $\quad \psi + \bar{\psi}\big|_{\sigma=0} = \psi + \bar{\psi}\big|_{\sigma=\pi} = 0,$ $\tag{4.16.20}$

DN R sector: $\quad \psi - \bar{\psi}\big|_{\sigma=0} = \psi + \bar{\psi}\big|_{\sigma=\pi} = 0.$ $\tag{4.16.21}$

The oscillator expansions here are inverted between the NS and R sectors. In particular, the NS sector has (4.16.17) as an oscillator expansion while the R sector has (4.16.16). Imposing the boundary conditions (4.16.20),(4.16.21), we obtain (4.16.19) with the difference that integral oscillators refer to the NS sector and half integral to the R sector.

When we have N fermions with DN boundary conditions, the NS ground state transforms as an O(N) spinor, with conformal weight N/16, while the R vacuum is the SL(2,$\mathbb{R}$)-invariant vacuum.

We have now all the ingredients to discuss the two-point functions on the disk. In the NS sector we obtain,

$$\langle \psi_{NN}(z)\psi_{NN}(w)\rangle = \frac{1}{z-w}, \quad \langle \psi_{NN}(z)\bar{\psi}_{NN}(\bar{w})\rangle = -\frac{1}{z-\bar{w}}, \tag{4.16.22}$$

$$\langle \psi_{DD}(z)\psi_{DD}(w)\rangle = \frac{1}{z-w}, \quad \langle \psi_{DD}(z)\bar{\psi}_{DD}(\bar{w})\rangle = \frac{1}{z-\bar{w}}, \tag{4.16.23}$$

$$\langle \psi_{DN}(z)\psi_{DN}(w)\rangle = \frac{z+w}{2\sqrt{zw}(z-w)}, \quad \langle \psi_{DN}(z)\bar{\psi}_{DN}(\bar{w})\rangle = -\frac{z+\bar{w}}{2\sqrt{z\bar{w}}(z-\bar{w})}, \tag{4.16.24}$$

while the R sector propagators are the subject of exercise 4.53 on page 124.

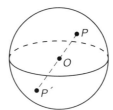

Figure 4.5 Real projective plane RP$_2$.

4.16.3 The projective plane

The projective plane RP$_2$ is another simple two-dimensional surface which is nonorientable and therefore relevant for unoriented strings.

It can be obtained from the disk by pairwise identifying opposite points on its boundary. Its Euler number is the same as that of the disk, namely, 1. It can also be obtained from the sphere by a $\mathbb{Z}_2$ involution that changes the orientation,

$$z' = -\frac{1}{\bar{z}}. \tag{4.16.25}$$

This identifies diametrically opposite points on the sphere (see figure 4.5). There are no fixed points here, hence no boundary. The real projective plane RP$_2$ can be thought of as a disk with its boundary replaced by a cross cap, namely, an S^1 with antidiametric points identified.

Using again the method of images we obtain for the scalar propagator

$$\langle X(z, \bar{z})X(w, \bar{w})\rangle_{\mathrm{RP}_2} = -\frac{\ell_s^2}{2}\left(\log|z-w|^2 + \log|1+z\bar{w}|^2\right). \tag{4.16.26}$$

4.17 CFT on the Torus

Consider the next simplest closed Riemann surface after the sphere, the torus. It has genus $g = 1$, Euler number $\chi = 0$, and is therefore flat. Using conformal symmetry, we can pick a constant metric so that the volume is normalized to 1. Pick coordinates $\sigma_1, \sigma_2 \in [0, 1]$. Then the volume is 1 if the determinant of the metric is 1. We can parametrize the metric, which is also a symmetric and positive-definite matrix, by a single complex number $\tau = \tau_1 + i\tau_2$, with positive imaginary part $\tau_2 \geq 0$, as follows:

$$g_{ij} = \frac{1}{\tau_2}\begin{pmatrix} 1 & \tau_1 \\ \tau_1 & |\tau|^2 \end{pmatrix}. \tag{4.17.1}$$

The line element is

$$ds^2 = g_{ij}d\sigma_i d\sigma_j = \frac{1}{\tau_2}|d\sigma_1 + \tau\, d\sigma_2|^2 = \frac{dw\, d\bar{w}}{\tau_2}, \tag{4.17.2}$$

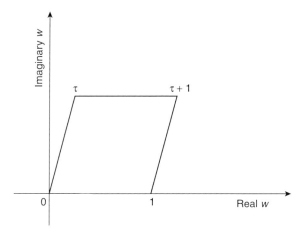

Figure 4.6 The torus as a quotient of the complex plane.

where

$$w = \sigma_1 + \tau\sigma_2, \quad \bar{w} = \sigma_1 + \bar{\tau}\sigma_2 \tag{4.17.3}$$

are the complex coordinates of the torus. This is the reason the parameter τ is known as the complex structure (or modulus) of the torus. It cannot be changed by infinitesimal diffeomorphisms or Weyl rescalings and is thus the complex Teichmüller parameter of the torus. The periodicity properties of σ_1, σ_2 translate to

$$w \rightarrow w + 1, \quad w \rightarrow w + \tau. \tag{4.17.4}$$

Therefore we may think of the torus as the points of the complex plane w identified under two translation vectors corresponding to the complex numbers 1 and τ, as suggested in Fig. 4.6.

The volume form is

$$d\sigma_1 \wedge d\sigma_2 = \frac{i}{2}\frac{dw \wedge d\bar{w}}{\tau_2} \equiv \frac{d^2w}{\tau_2}, \quad \int \frac{d^2w}{\tau_2} = 1. \tag{4.17.5}$$

Although τ is invariant under infinitesimal diffeomorphisms, it does transform under some topologically nontrivial transformations. Consider, instead of the parallelogram in figure 4.6 defining the torus, the one in figure 4.7a. Obviously, they are equivalent, due to the periodicity conditions (4.17.4). However, the second one corresponds to a modulus $\tau + 1$. We conclude that the transformation

$$T : \tau \rightarrow \tau + 1 \tag{4.17.6}$$

leaves the torus invariant. Consider another equivalent choice of a parallelogram, which is depicted in figure 4.7b. It is characterized by complex numbers τ and $\tau + 1$. To bring it to the original form (one side on the real axis) and preserve its orientation, we have to scale both sides down by a factor of $\tau + 1$. It will then correspond to an equivalent torus with modulus $\tau/(\tau + 1)$. We have obtained a second modular transformation

$$X : \tau \rightarrow \frac{\tau}{\tau + 1}. \tag{4.17.7}$$

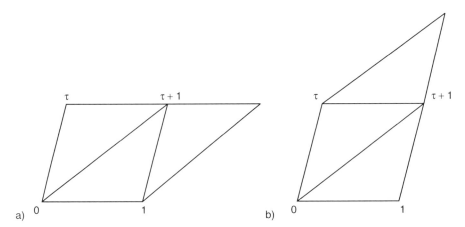

Figure 4.7 a) The modular transformation $\tau \to \tau + 1$. b) The modular transformation $\tau \to \tau/(\tau + 1)$.

It can be shown that taking products of these transformations generates the full modular group of the torus. A convenient set of generators is given by T in (4.17.6) and

$$T^{-1}XT^{-1} \equiv S : \tau \to -\frac{1}{\tau}, \quad S^2 = 1, \quad (ST)^3 = 1. \tag{4.17.8}$$

The most general transformation is of the form

$$\tau' = \frac{a\tau + b}{c\tau + d} \leftrightarrow A = \begin{pmatrix} a & b \\ c & d \end{pmatrix}, \tag{4.17.9}$$

where the matrix A has integer entries and determinant 1. Such matrices form the group $SL(2,\mathbb{Z})$. Since changing the sign of the matrix does not affect the modular transformation in (4.17.9) the modular group is $PSL(2,\mathbb{Z}) = SL(2,\mathbb{Z})/\mathbb{Z}_2$.

As mentioned above, the modulus takes values in the upper half plane $\mathcal{H}_2$ ($\tau_2 \geq 0$), which is the Teichmüller space of the torus. However, to find the moduli space of truly inequivalent tori we have to quotient this by the modular group. It can be shown that the fundamental domain $\mathcal{F} = \mathcal{H}_2/PSL(2,\mathbb{Z})$ of the modular group is the area contained in between the lines $\tau_1 = \pm 1/2$ and above the unit circle with center at the origin. It is shown in figure 4.8.

There is an interesting construction of the torus starting from the cylinder. Consider a cylinder of length $2\pi\tau_2$ and circumference 1. Take one end, rotate it by an angle $2\pi\tau_1$, and glue it to the other end. This produces a torus with modulus $\tau = \tau_1 + i\tau_2$. This construction gives a very useful relation between the path integral of a CFT on the torus and a trace over the Hilbert space. First, the propagation along the cylinder is governed by the Hamiltonian (transfer matrix) $H = L_0^{\text{cyl}} + \bar{L}_0^{\text{cyl}}$. The rotation around the cylinder is implemented by the "momentum" operator $P = L_0^{\text{cyl}} - \bar{L}_0^{\text{cyl}}$. Gluing together the two ends gives a trace in the Hilbert space. From (4.9.2) on page 64

$$L_0^{\text{cyl}} = L_0 - \frac{c}{24}, \quad \bar{L}_0^{\text{cyl}} = \bar{L}_0 - \frac{\bar{c}}{24}, \tag{4.17.10}$$

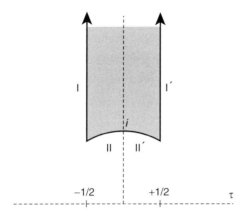

Figure 4.8 The moduli space of the torus.

where $L_0, \bar{L}_0$ are the operators on the sphere. Putting everything together, we obtain

$$Z = \int e^{-S} = \mathrm{Tr}\left[e^{-2\pi \tau_2\, H}\, e^{2\pi i \tau_1\, P}\right] = \mathrm{Tr}\left[e^{2\pi i \tau L_0^{\text{cyl}}}\, e^{-2\pi i \bar{\tau} \bar{L}_0^{\text{cyl}}}\right]$$

$$= \mathrm{Tr}\left[q^{L_0 - c/24}\, \bar{q}^{\bar{L}_0 - \bar{c}/24}\right], \tag{4.17.11}$$

where $q = \exp[2\pi i \tau]$. The trace includes also possible continuous parts of the spectrum. This is a very useful relation. It provides the correct normalization of the path integral.

4.18 Compact Scalars

In section 4.7 we described the CFT of a noncompact real scalar field on the torus. Here we will consider a compact scalar field X taking values on a circle of radius R. Consequently, the values X and $X + 2\pi m R$, $m \in \mathbb{Z}$, will be considered equivalent.

We will evaluate the path integral of the theory on the torus. The action is

$$S = \frac{1}{4\pi \ell_s^2} \int d^2\sigma \sqrt{g}\, g^{ij} \partial_i X \partial_j X$$

$$= \frac{1}{4\pi \ell_s^2} \int_0^1 d\sigma_1 \int_0^1 d\sigma_2\, \frac{1}{\tau_2} |\tau \partial_1 X - \partial_2 X|^2$$

$$= -\frac{1}{4\pi \ell_s^2} \int d^2\sigma\, X \square\, X, \tag{4.18.1}$$

where the Laplacian is given by

$$\square = \frac{1}{\tau_2} |\tau \partial_1 - \partial_2|^2. \tag{4.18.2}$$

We wish to evaluate the path integral

$$Z(R) = \int \mathcal{D}X\, e^{-S} \tag{4.18.3}$$

on the torus.

As usual, we will have to find the classical solutions of finite action (instantons) and calculate the fluctuations around them. The field X should be periodic on the torus and is a map from the torus (topologically $S^1 \times S^1$) to the circle S^1. Such maps are classified by two integers that specify how many times X winds around each of the two cycles of the torus. The equation of motion $\Box X = 0$ has the following instanton solutions:

$$X_{\text{class}} = 2\pi R(n\sigma_1 + m\sigma_2), \quad m, n \in \mathbb{Z}. \tag{4.18.4}$$

They have the correct periodicity properties

$$X_{\text{class}}(\sigma_1 + 1, \sigma_2) = X(\sigma_1, \sigma_2) + 2\pi nR, \quad X_{\text{class}}(\sigma_1, \sigma_2 + 1) = X(\sigma_1, \sigma_2) + 2\pi mR, \tag{4.18.5}$$

and the following classical action:

$$S_{m,n} = \frac{\pi R^2}{\tau_2 \ell_s^2} |m - n\tau|^2. \tag{4.18.6}$$

Thus, we can separate $X = X_{\text{class}} + \chi$, and the path integral can be written as

$$Z(R) = \sum_{m,n\in\mathbb{Z}} \int \mathcal{D}\chi \, e^{-S_{m,n} - S(\chi)} = \sum_{m,n\in\mathbb{Z}} e^{-S_{m,n}} \int \mathcal{D}\chi \, e^{-S(\chi)}. \tag{4.18.7}$$

What remains to be done is the path integral over χ. There is always the constant zero mode that we can separate, $\chi(\sigma_1, \sigma_2) = \chi_0 + \delta\chi(\sigma_1, \sigma_2)$, with $0 \le \chi_0 \le 2\pi R$. The field $\delta\chi$ can be expanded in the eigenfunctions of the Laplacian

$$\Box\psi_i = -\lambda_i \, \psi_i. \tag{4.18.8}$$

It is not difficult to see that these eigenfunctions and the associated eigenvalues are

$$\psi_{m_1,m_2} = e^{2\pi i(m_1\sigma_1 + m_2\sigma_2)}, \quad \lambda_{m_1,m_2} = \frac{4\pi^2}{\tau_2}|m_1\tau - m_2|^2. \tag{4.18.9}$$

The eigenfunctions satisfy

$$\int d^2\sigma \, \psi_{m_1,m_2}\psi_{n_1,n_2} = \delta_{m_1+n_1,0}\delta_{m_2+n_2,0}, \tag{4.18.10}$$

so we expand

$$\delta\chi = \sum_{m_1,m_2\in\mathbb{Z}}' A_{m_1,m_2}\psi_{m_1,m_2}, \tag{4.18.11}$$

where the prime implies omission of the constant mode $(m_1, m_2) = (0, 0)$. Reality implies $A^*_{m_1,m_2} = A_{-m_1,-m_2}$. The action becomes

$$S(\chi) = \frac{1}{4\pi \ell_s^2} \sum_{m_1,m_2\in\mathbb{Z}}' \lambda_{m_1,m_2} |A_{m_1,m_2}|^2. \tag{4.18.12}$$

We can specify the measure from the norm

$$\|\delta X\|^2 = \frac{1}{\ell_s^2} \int d^2\sigma \sqrt{\det G} \, (d\chi)^2 = \sum_{m_1,m_2}' \frac{|dA_{m_1,m_2}|^2}{\ell_s^2} \tag{4.18.13}$$

to be

$$\int \mathcal{D}\chi = \int_0^{2\pi R} \frac{d\chi_0}{\ell_s} \int \prod_{m_1,m_2}' \frac{dA_{m_1,m_2}}{2\pi \ell_s}. \tag{4.18.14}$$

Putting everything together we obtain

$$\int \mathcal{D}\chi\, e^{-S(\chi)} = \frac{2\pi R}{\ell_s \prod'_{m_1,m_2} \lambda^{1/2}_{m_1,m_2}} = \frac{2\pi R}{\ell_s \sqrt{\det' \Box}}. \tag{4.18.15}$$

Using the explicit form of the eigenvalues, the determinant of the Laplacian can be calculated using ζ-function regularization,

$$\det' \Box = 4\pi^2 \tau_2 \eta^2(\tau) \bar{\eta}^2(\bar{\tau}), \tag{4.18.16}$$

where η is the Dedekind function defined in (C.10). Collecting all terms in (4.18.7) we obtain

$$Z(R) = \frac{R}{\ell_s \sqrt{\tau_2} |\eta|^2} \sum_{m,n \in \mathbb{Z}} e^{-(\pi R^2/\tau_2 \ell_s^2)|m - n\tau|^2}. \tag{4.18.17}$$

This is the Lagrangian form of the partition function. We mentioned in the previous chapter that the partition function on the torus can also be written in Hamiltonian form as in (4.17.11). To do this we have to perform a Poisson resummation (see appendix A) on the integer m. We obtain

$$Z(R) = \sum_{\tilde{m},n \in \mathbb{Z}} \frac{q^{P_L^2/2} \bar{q}^{P_R^2/2}}{\eta \bar{\eta}} \tag{4.18.18}$$

with

$$P_L = \frac{1}{\sqrt{2}} \left(\tilde{m} \frac{\ell_s}{R} + n \frac{R}{\ell_s} \right), \quad P_R = \frac{1}{\sqrt{2}} \left(\tilde{m} \frac{\ell_s}{R} - n \frac{R}{\ell_s} \right), \tag{4.18.19}$$

so that

$$\Delta = \frac{1}{4} \left(\tilde{m} \frac{\ell_s}{R} + n \frac{R}{\ell_s} \right)^2, \quad \bar{\Delta} = \frac{1}{4} \left(\tilde{m} \frac{\ell_s}{R} - n \frac{R}{\ell_s} \right)^2. \tag{4.18.20}$$

This is in agreement with (2.3.44) on page 25. From now on we drop the tilde from $\tilde{m}$.

The vacuum amplitude we just calculated is in the form (4.17.11) and from it we can read off the spectrum of conformal weights and multiplicities of the theory. Before we do this, however, let us return to the sphere and discuss the current-algebra structure of the theory. As in section 4.7 there is a holomorphic and an antiholomorphic U(1) current:

$$J(z) = i \frac{\sqrt{2}}{\ell_s} \partial X, \quad \bar{J}(\bar{z}) = i \frac{\sqrt{2}}{\ell_s} \bar{\partial} X, \tag{4.18.21}$$

satisfying the U(1) current algebra

$$J(z)J(w) = \frac{1}{(z-w)^2} + \text{finite}, \quad \bar{J}(\bar{z})\bar{J}(\bar{w}) = \frac{1}{(\bar{z}-\bar{w})^2} + \text{finite}. \tag{4.18.22}$$

We can write the stress-tensor in the affine-Sugawara form

$$T(z) = -\frac{1}{\ell_s^2} (\partial X)^2 = \frac{1}{2} : J^2 :, \quad \bar{T}(\bar{z}) = -\frac{1}{\ell_s^2} (\bar{\partial} X)^2 = \frac{1}{2} : \bar{J}^2 :. \tag{4.18.23}$$

The spectrum can be decomposed into affine HW representations as discussed in section 4.11 on page 69. An affine primary field is specified by its charges Q_L and Q_R under the left- and right-moving current algebras. From (4.18.23) we obtain that its conformal weights are given by

$$\Delta = \frac{1}{2}Q_L^2, \quad \bar{\Delta} = \frac{1}{2}Q_R^2. \tag{4.18.24}$$

The rest of the representation is constructed by acting on the affine primary state with the negative current modes J_{-n} and $\bar{J}_{-n}$. We can compute the character of such a representation ($c = \bar{c} = 1$):

$$\chi_{Q_L,Q_R}(q,\bar{q}) = \text{Tr}[q^{L_0 - 1/24}\bar{q}^{\bar{L}_0 - 1/24}] = \frac{q^{Q_L^2/2}\,\bar{q}^{Q_R^2/2}}{\eta\bar{\eta}}. \tag{4.18.25}$$

A comparison with (4.18.18) shows that the spectrum contains an infinite number of affine U(1) representations labeled by m, n with $Q_L = P_L$ and $Q_R = P_R$. For $m = n = 0$ we have the vacuum representation whose HW state is the standard vacuum. The other HW states, labeled by m, n satisfy

$$J_0|m,n\rangle = P_L\,|m,n\rangle, \quad \bar{J}_0|m,n\rangle = P_R\,|m,n\rangle. \tag{4.18.26}$$

In the operator picture, they are created out of the vacuum by the vertex operators (we split $X(z,\bar{z}) \sim X_L(z) + X_R(\bar{z})$ as usual)

$$V_{m,n} =: \exp[i\,p_L\,X_L + i\,p_R\,X_R]:, \tag{4.18.27}$$

$$J(z)V_{m,n}(w,\bar{w}) = \frac{\ell_s p_L}{\sqrt{2}}\frac{V_{m,n}(w,\bar{w})}{z-w} + \cdots, \quad \bar{J}(\bar{z})V_{m,n}(w,\bar{w}) = \frac{\ell_s p_R}{\sqrt{2}}\frac{V_{m,n}(w,\bar{w})}{\bar{z}-\bar{w}} + \cdots. \tag{4.18.28}$$

Their correlators are given by the Gaussian formula

$$\left\langle \prod_{i=1}^N V_{m_i,n_i}(z_i,\bar{z}_i) \right\rangle = \prod_{i<j}^N z_{ij}^{(\ell_s^2/2)p_L^i p_L^j}\,\bar{z}_{ij}^{(\ell_s^2/2)p_R^i p_R^j}, \tag{4.18.29}$$

where $z_{ij} = z_i - z_j$. Using the Gaussian formula

$$: e^{iaX(z)} :: e^{ibX(w)} := (z-w)^{(\ell_s^2/2)ab} : e^{iaX(z)+ibX(w)} :$$

$$= (z-w)^{(\ell_s^2/2)ab}\left[: e^{i(a+b)X(w)} : +\mathcal{O}(z-w)\right], \tag{4.18.30}$$

we obtain the following OPE rule for U(1) representations:

$$[V_{m_1,n_1}] \cdot [V_{m_2,n_2}] \sim [V_{m_1+m_2,n_1+n_2}], \tag{4.18.31}$$

compatible with U(1) charge conservation. Under the $U(1)_L \times U(1)_R$ transformation associated with the group element $e^{i\theta_L + i\theta_R}$ the oscillators are invariant but the states $|m,n\rangle$ pick up a phase $e^{i(m+n)\theta_L + i(m-n)\theta_R}$.

In the canonical representation, the momentum operator is taking values m/R as required by the usual (point-particle) quantum mechanical quantization condition on a circle of radius R. The existence of the extra spatial dimension of the string allows for the possibility of X winding around the circle n times. This is precisely the interpretation of the integer n in (4.18.19). It has no point-particle (one-dimensional) analog.

4.18.1 Modular invariance

We return to the torus to take another look at the partition function. For string theory purposes we would like it to be invariant under the full diffeomorphism group. In particular it should be invariant under the large transformations, namely, the modular transformations. This is important in string theory since modular invariance is at the very heart of finiteness of closed string theory. Moreover it is essential for the cancellation of space-time anomalies.

It suffices to prove invariance under the two transformations T and S, since these generate the modular group. We will use the Lagrangian representation of the partition function (4.18.17). It is not difficult to verify, using the formulas of appendix C, that $\sqrt{\tau_2}\eta\bar{\eta}$ is by itself modular invariant. Thus, we need to consider only the instanton sum. Under $\tau \to \tau + 1$ we can change the summation $(m, n) \to (m + n, n)$, and the full sum is invariant. Under $\tau \to -1/\tau$ we can change the summation $(m, n) \to (-n, m)$ and again the sum is invariant. We conclude that the torus partition function of the compact boson is modular invariant. It is interesting to note that invariance under the T transformation in the Hamiltonian representation (4.17.11) in general implies

$$\Delta - \bar{\Delta} - \frac{c - \bar{c}}{24} = \text{integer} \tag{4.18.32}$$

for the whole spectrum. In particular, for the vacuum state $\Delta = \bar{\Delta} = 0$, it implies that $c - \bar{c} = 0 \bmod (24)$. In our case $c = \bar{c} = 1$ and from (4.18.19) and (4.18.24) $P_L^2/2 - P_R^2/2 = mn \in \mathbb{Z}$.

4.18.2 Decompactification

Another comment concerns the torus partition function of a noncompact boson. This can be obtained by taking the limit $R \to \infty$. We expect that the partition function in this limit will diverge. The divergence is expected to be linear in the volume. Therefore, the free energy per unit volume is finite. From (4.18.17) we note that as R gets large, the only term that is not exponentially suppressed is the one with $m = n = 0$, so

$$Z_{\text{noncompact}} = \lim_{R \to \infty} Z(R) = \frac{R}{\ell_s \sqrt{\tau_2}\, \eta\, \bar{\eta}} = \frac{V}{(2\pi\ell_s)\sqrt{\tau_2}\, \eta\, \bar{\eta}}, \tag{4.18.33}$$

where $V = 2\pi R$ is the volume of the (very large) circle. This is the properly normalized noncompact partition function evaluated also directly as

$$Z_{\text{noncompact}} = V \int_{-\infty}^{\infty} \frac{dp}{2\pi} \frac{q^{\ell_s^2 \frac{p^2}{4}} \bar{q}^{\ell_s^2 \frac{p^2}{4}}}{\eta\, \bar{\eta}} = \frac{V}{(2\pi\ell_s)\sqrt{\tau_2}\, \eta\, \bar{\eta}}. \tag{4.18.34}$$

4.18.3 The torus propagator

We will derive here the torus propagator for the boson. This will be useful later on for one-loop calculations in string theory. The propagator can be directly written in terms of the nonzero eigenvalues and the associated eigenfunctions (4.18.9) of the Laplacian (4.18.6):

$$\Delta(\sigma_1, \sigma_2) \equiv \langle \delta\chi(\sigma_1, \sigma_2)\delta\chi(0,0)\rangle = -\frac{\ell_s^2}{2}\sum_{m,n}{}' \frac{1}{|m\tau - n|^2}e^{2\pi i(m\sigma_1 + n\sigma_2)}. \tag{4.18.35}$$

The sum is conditionally convergent and has to be regularized using ζ-function regularization. We obtain

$$\square\,\Delta(\sigma_1, \sigma_2) = \frac{2\pi^2\ell_s^2}{\tau_2}\left[\delta(\sigma_1)\delta(\sigma_2) - 1\right], \tag{4.18.36}$$

so that the integral over the torus gives zero, in accordance with the fact that we have omitted the zero mode. It can also be expressed in complex coordinates in terms of ϑ-functions as

$$\Delta(\sigma_1, \sigma_2) = -\frac{\ell_s^2}{2}\log G(z, \bar{z}), \quad G = e^{-2\pi(\mathrm{Im}z^2)/\tau_2}\left|\frac{\vartheta_1(z)}{\vartheta_1'(0)}\right|^2. \tag{4.18.37}$$

4.18.4 Marginal deformations

In a CFT, there is a special class of operators, known as marginal operators, with $(\Delta, \bar{\Delta}) = (1, 1)$. For such an operator $\phi_{1,1}$, the density $\phi_{1,1}dzd\bar{z}$ is conformally invariant. If we perturb our action by $g\int \phi_{1,1}$, we can expect that the theory remains conformally invariant, at least classically.

In the quantum theory however, short-distance singularities typically spoil conformal invariance. When conformal invariance persists, $\phi_{1,1}$ is called an exactly marginal operator. In this way, perturbing by $\phi_{1,1}$ we obtain a continuous family of CFTs parameterized by the coupling g. The central charge cannot change during a marginal perturbation.

Our present example exhibits an occurrence of this phenomenon. There is a $(1,1)$ operator, namely, $\phi = \partial X \bar{\partial} X \sim J\bar{J}$. By adding this to the action (4.18.1), it is easy to see that the effect of the perturbation is to change the effective radius R. The theory, being again a free field theory, remains conformally invariant. In this case the operation seems trivial, however, marginal operators exist in more complicated CFTs. They are responsible for generating a continuous manifold of CFTs, parameterized by the associated couplings, known as the moduli of the CFT.

The exactly marginal operators $\phi_{(1,1)}$ are vertex operators for string theory massless states ar zero momentum. The associated moduli are interpreted as the vevs of such scalar fields. The fact that they can be continuously deformed, means that such scalars (also called moduli scalars) have a flat potential. In particular they are massless.

The occurrence of such massless scalars in string theory poses phenomenological problems as we will see later.

4.18.5 Multiple compact scalars

The above discussion can easily be generalized to the case of N free compact scalar fields X^i, $i = 1, 2, \ldots, N$. We will scale them to become dimensionless angles with values in

$[0, 2\pi]$. They parametrize an N-dimensional torus. The most general quadratic action is

$$S = \frac{1}{4\pi\ell_s^2} \int d^2\sigma \sqrt{g} g^{\alpha\beta} G_{ij}\partial_\alpha X^i \partial_\beta X^j + \frac{1}{4\pi\ell_s^2} \int d^2\sigma \epsilon^{\alpha\beta} B_{ij}\partial_\alpha X^i \partial_\beta X^j, \tag{4.18.38}$$

where $g_{\alpha\beta}$ is the torus metric (4.17.1), $\epsilon^{\alpha\beta}$ is the usual ϵ-symbol, $\epsilon^{12} = 1$; G_{ij} is a constant symmetric positive-definite matrix that plays the role of metric in the space of the X^i (target-space N-torus). The constant matrix B_{ij} is antisymmetric. It is the constant background value of the two-index antisymmetric tensor over the N-torus. It is the analog of the θ-term in four-dimensional gauge theories. Now G, B have space-time dimension [length]^2.

An analogous calculation of the path integral produces

$$Z_{N,N}(G, B) = \frac{\sqrt{\det G}}{\ell_s^N (\sqrt{\tau_2}\eta\bar\eta)^N} \sum_{\vec{m},\vec{n}} e^{[\pi(G_{ij}+B_{ij})]/\tau_2\ell_s^2(m_i+n_i\tau)(m_j+n_j\bar\tau)}. \tag{4.18.39}$$

This partition function reduces for N=1, $G = R^2$, $B = 0$ to (4.17.11). Using a multiple Poisson resummation on the m_i, it can be transformed into the Hamiltonian representation

$$Z_{N,N}(G, B) = \frac{\Gamma_{d,d}(G, B)}{\eta^N \bar\eta^N} = \sum_{\vec{m},\vec{n}\in\mathbb{Z}^N} \frac{q^{P_L^2/2} \bar{q}^{P_R^2/2}}{\eta^N \bar\eta^N}, \tag{4.18.40}$$

where

$$P_{L,R}^2 \equiv P_{L,R}^i G_{ij} P_{L,R}^j, \tag{4.18.41}$$

$$P_L^i = \frac{G^{ij}}{\sqrt{2}} \left(\ell_s m_j + \frac{(B_{jk} + G_{jk})}{\ell_s} n_k \right), \quad P_R^i = \frac{G^{ij}}{\sqrt{2}} \left(\ell_s m_j + \frac{(B_{jk} - G_{jk})}{\ell_s} n_k \right). \tag{4.18.42}$$

The theory has a left-moving and a right-moving U(1)^N current algebra generated by the currents

$$J^i(z) = i\sqrt{2}\,\partial X^i, \quad \bar{J}^i = i\sqrt{2}\,\bar\partial X^i, \tag{4.18.43}$$

$$\langle X^i(z)X^j(w)\rangle = -\frac{\ell_s^2}{2} G^{ij} \log(z-w) \Rightarrow J^i(z)J^j(w) = \frac{\ell_s^2 G^{ij}}{(z-w)^2} + \cdots, \tag{4.18.44}$$

and similarly for $\bar{J}^i$. The stress-tensor is again of the affine-Sugawara form

$$T(z) = -\frac{G_{ij}}{\ell_s^2}\,\partial X^i \partial X^j = \frac{G_{ij}}{2\ell_s^2}\, : J^i J^j : . \tag{4.18.45}$$

Affine primaries are characterized by charges $Q_{L,R}^i$ and

$$\Delta = \frac{G_{ij}}{2\ell_s^2} Q_L^i Q_L^j, \quad \bar\Delta = \frac{G_{ij}}{2\ell_s^2} Q_R^i Q_R^j. \tag{4.18.46}$$

Comparing this with (4.18.40) we obtain $Q_{L,R}^i = \ell_s P_{L,R}^i$.

In exercise 4.60 on page 124 you are invited to show that (4.18.40) is modular invariant.

4.18.6 Enhanced symmetry and the string Brout-Englert-Higgs effect

Something special happens to the CFT of the single compact boson when the radius is $R = \ell_s$. The conformal weights of the primaries are now given by

$$\Delta = \frac{1}{4}(m+n)^2, \quad \bar{\Delta} = \frac{1}{4}(m-n)^2. \tag{4.18.47}$$

Notice that the two states with $m = n = \pm 1$ are $(1,0)$ operators. For generic R the only $(1,0)$ (chiral) operator is the U(1) current $J(z)$. Now we have two additional ones. The current algebra becomes larger if we also include these operators. Similarly, the states with $m = -n = \pm 1$ are $(0,1)$ operators and the right-moving current algebra is also enhanced. We will discuss only the left-moving part, since the right-moving part behaves in a similar way. The two operators that become $(1,0)$ can be written as vertex operators (4.18.27)

$$J^{\pm}(z) = \frac{1}{\sqrt{2}} : e^{\pm 2iX(z)/\ell_s} : . \tag{4.18.48}$$

Define also

$$J^3(z) = \frac{1}{\sqrt{2}}J(z) = \frac{i}{\ell_s}\partial X(z). \tag{4.18.49}$$

They satisfy the following OPEs, which can be computed directly using $\langle X(z)X(0)\rangle = -\frac{\ell_s^2}{2}\log z$:

$$J^3(z)J^{\pm}(w) = \pm\frac{J^{\pm}(w)}{z-w} + \cdots, \quad J^+(z)J^+(w) = \cdots, \quad J^-(z)J^-(w) = \cdots, \tag{4.18.50}$$

$$J^+(z)J^-(w) = \frac{1/2}{(z-w)^2} + \frac{J^3(w)}{z-w} + \cdots, \quad J^3(z)J^3(w) = \frac{1/2}{(z-w)^2} + \cdots. \tag{4.18.51}$$

This is the SU(2) current algebra at level $k = 1$. This is not too surprising, since the central charge of SU(2)$_k$ is given by (4.11.8) to be $c = 3k/(k+2)$. It indeed becomes $c = 1$ when $k = 1$. This realization of current algebra at level 1 in terms of free bosons is known as the Halpern-Frenkel-Kač-Segal construction.

We have seen before that SU(2)$_1$ has two integrable affine representations, the vacuum representation with $j = 0$ and the $j = 1/2$ representation with conformal weight $\Delta = 1/4$ (from (4.11.11) on page 71). The primary state of the $j = 0$ representation is the vacuum. The primary operators of the $j = 1/2$ representation transform as a two-component spinor of SU(2)$_L$ and a two-component spinor of SU(2)$_R$ with conformal weights $(1/4,1/4)$. They are represented by the four vertex operators $V_{m,n}$ with $(m,n) = (0,\pm 1)$ and $(\pm 1,0)$. They have the correct conformal weight and OPEs with the currents (4.18.48).

This phenomenon generalizes to the N-dimensional toroidal models. The U(1) charges $p^i_{L,R}$ take values on a 2N-dimensional lattice of signature (N,N) that depends on G_{ij}, B_{ij}. For special values of G,B this lattice coincides with the root lattice of a Lie group G with rank N. Then, some vertex operators become extra chiral currents and along with the N abelian currents J^i form an affine algebra G at level $k = 1$.

When the toroidal CFT acquires enhanced current algebra symmetry then the associated bosonic string theory acquires enhanced gauge symmetry. Consider the bosonic string with one of the 26 dimensions (say X^{25}) compactified on a circle of radius R.

Then the massless states are again similar, but with a slightly different interpretation. There are now 25 noncompact dimensions, so we have 25-dimensional Lorentz invariance.

The analogues of the tachyon state are now $|p^\mu; m, n\rangle$ where m, n specify the momentum and winding along the circle.

The massless states are

$$\alpha^\mu_{-1}\bar{\alpha}^\nu_{-1}|p^\mu; 0, 0\rangle, \quad \alpha^\mu_{-1}\bar{\alpha}^{25}_{-1}|p^\mu; 0, 0\rangle, \quad \alpha^{25}_{-1}\bar{\alpha}^\mu_{-1}|p^\mu; 0, 0\rangle, \quad \alpha^{25}_{-1}\bar{\alpha}^{25}_{-1}|p^\mu; 0, 0\rangle, \tag{4.18.52}$$

which from the point of view of the 25 noncompact dimensions are the graviton, antisymmetric tensor, dilaton, two U(1) gauge fields, and a scalar. Note that the scalar state is generated by $\partial X^{25}\bar{\partial} X^{25}$, which is the perturbation that changes the radius. We may say that the expectation value of this scalar is the radius R. There are other massive states. Among them we have

$$|A^\pm_\mu\rangle = \bar{\alpha}^\mu_{-1}|p^\mu; \pm 1, \pm 1\rangle, \tag{4.18.53}$$

which are massive vector bosons with mass $m^2 = \left(\frac{1}{R} - \frac{R}{\ell_s^2}\right)^2$ and

$$|\bar{A}^\pm_\mu\rangle = \alpha^\mu_{-1}|\pm 1, \mp 1\rangle \tag{4.18.54}$$

with the same mass as above. As we vary R, the mass changes and at $R = \ell_s$ the vectors become massless. At that point, the string theory acquires an $SU(2)_L \times SU(2)_R$ gauge symmetry. Moving away from $R = \ell_s$, $SU(2)_L \times SU(2)_R$ gauge symmetry is spontaneously broken to $U(1)_L \times U(1)_R$. This is the usual Brout-Englert-Higgs effect. The scalar whose expectation value is the radius plays the role of the Higgs scalar (although there is no potential for R here).

4.18.7 T-duality

The theory of a single scalar, compactified on a circle of radius R, has another remarkable property.

As we have seen, the primaries have

$$H = L_0 + \bar{L}_0 = \frac{1}{2}\left(m^2\frac{\ell_s^2}{R^2} + n^2\frac{R^2}{\ell_s^2}\right), \quad P = L_0 - \bar{L}_0 = mn. \tag{4.18.55}$$

It is obvious that the this spectrum is invariant under

$$R \to \frac{\ell_s^2}{R}, \quad m \leftrightarrow n. \tag{4.18.56}$$

This corresponds to the following transformation of the U(1) charges:

$$P_L \to P_L, \quad P_R \to -P_R. \tag{4.18.57}$$

Only the right charge changes sign. The action on the respective currents is analogous,

$$J(z) \to J(z), \quad \bar{J}(\bar{z}) \to -\bar{J}(\bar{z}). \tag{4.18.58}$$

In terms of the boson, the previous equation reads

$$\partial_\sigma X \leftrightarrow \partial_\tau X, \tag{4.18.59}$$

and from (2.3.2)

$$X^\mu(\tau, \sigma) = X_L^\mu(\tau + \sigma) + X_R^\mu(\tau - \sigma) \rightarrow X_L^\mu(\tau + \sigma) - X_R^\mu(\tau - \sigma). \tag{4.18.60}$$

It can be verified that not only the spectrum but also the interactions respect this property. This is a peculiar property since it implies that a CFT cannot distinguish a circle of radius R from another of radius ℓ_s^2/R. This is, strictly speaking, not a symmetry of the two-dimensional theory. It states that two *a priori* different theories are in fact equivalent. However, in the context of string theory it will become a true symmetry and is known under the name T-duality. Notice that the presence of winding modes is essential for the presence of T-duality. Therefore it can appear in string theory but not in point-particle field theory.

There is an interesting interpretation of T-duality in string theory. Start from the CFT with $R = \ell_s$. We have seen that at this point there is an enhanced symmetry $SU(2)_L \times SU(2)_R$. Then, at this point, the duality transformation (4.18.58) is an $SU(2)_R$ Weyl reflection, which is an obvious symmetry of the CFT. This explains the self-duality at $R = \ell_s$.

We now move infinitesimally away from $R = \ell_s$ by perturbing the CFT with the marginal operator $\epsilon \int J^3 \bar{J}^3 \sim \epsilon \int \partial X \bar{\partial} X$. Because of the self-duality of the unperturbed theory, the ϵ perturbation and $-\epsilon$ perturbation give identical theories. This is the infinitesimal version of $R \rightarrow \ell_s^2/R$ duality around $R = \ell_s$. We can further extend this duality on the whole line. In this sense, the duality is a consequence of the $SU(2)$ symmetry at $R = \ell_s$. At this point the duality transformation is an $SU(2)_R$ transformation.

Consider again the bosonic string with one dimension compactified. At $R = \ell_s$ the $SU(2)_L$ transformation is a gauge transformation. Away from $R = \ell_s$ the gauge symmetry is broken and the duality symmetry is a discrete remnant of the original gauge symmetry.

We can generalize the T-duality symmetry to the N-dimensional toroidal models; here the duality transformations form an infinite discrete group, unlike the one-dimensional case where the group is $\mathbb{Z}_2$.

First observe that the partition function (4.18.39) is invariant under shifts of B_{ij}/ℓ_s^2 by any antisymmetric matrix with integer entries. Also, by construction, the theory is invariant under GL(N) rotations of the scalars G_{ij} and B_{ij}. However, since the rotations also act on m_i, n_i they must rotate them again to integer values. Therefore, the GL(N) matrix must have integer entries and such matrices form the discrete group GL(N,$\mathbb{Z}$). Finally, there are transformations such as the radius inversion, which leave the spectrum invariant. Together, all of these transformations combine into an infinite discrete group O(N,N,$\mathbb{Z}$). It is described by 2N×2N integer-valued matrices of the form

$$\Omega = \begin{pmatrix} A & B \\ C & D \end{pmatrix}, \tag{4.18.61}$$

where A, B, C, D are $N \times N$ matrices. Define also the O(N,N)-invariant metric

$$
L = \begin{pmatrix} 0 & 1_N \\ 1_N & 0 \end{pmatrix},
\tag{4.18.62}
$$

where 1_N is the N-dimensional unit matrix. Ω belongs to O(N,N,$\mathbb{Z}$) if it has integer entries and satisfies

$$
\Omega^T L \Omega = L.
\tag{4.18.63}
$$

Define the dimensionless matrix $E_{ij} = \frac{G_{ij} + B_{ij}}{\ell_s^2}$. Then the duality transformations are

$$
E \rightarrow (AE + B)(CE + D)^{-1}, \qquad \begin{pmatrix} \vec{m} \\ \vec{n} \end{pmatrix} \rightarrow \Omega \begin{pmatrix} \vec{m} \\ \vec{n} \end{pmatrix}.
\tag{4.18.64}
$$

In the special (but useful) case N=2 we can parametrize

$$
G_{ij} = \ell_s^2 \frac{T_2}{U_2} \begin{pmatrix} 1 & U_1 \\ U_1 & U_1^2 + U_2^2 \end{pmatrix}, \qquad B_{ij} = \ell_s^2 \begin{pmatrix} 0 & T_1 \\ -T_1 & 0 \end{pmatrix},
\tag{4.18.65}
$$

with $T_2, U_2 \geq 0$, and T, U dimensionless. Defining the complex parameters $T = T_1 + iT_2$, $U = U_1 + iU_2$, the lattice sum (4.18.40) becomes

$$
\Gamma_{2,2}(T, U) = \sum_{\vec{m}, \vec{n}} \exp\left[-\frac{\pi \tau_2}{T_2 U_2} | - m_1 U + m_2 + T(n_1 + Un_2)|^2 + 2\pi i\tau(m_1 n_1 + m_2 n_2) \right].
$$
$$
\tag{4.18.66}
$$

The duality group O(2,2,$\mathbb{Z}$) acts on T and U with independent PSL(2,$\mathbb{Z}$) transformations (4.17.9) as well as with the exchange $T \leftrightarrow U$.

4.19 Free Fermions on the Torus

In section 4.12 we analyzed the CFT of N free Majorana-Weyl fermions. We will now consider the partition function of this theory on the torus. The action was given in (4.12.1 on page 71). To do the path integral, we have to choose boundary conditions for the fermions around the two-cycles of the torus. For each cycle we have the choice between periodic and antiperiodic boundary conditions. In total we have four possible sectors. The fermionic path integral will give a power of the fermionic determinant defined with the appropriate boundary conditions (also known as spin structures):

$$
\int \mathcal{D}\psi \, e^{-S} = (\det \partial)^{N/2}.
\tag{4.19.1}
$$

This can be computed by finding the appropriate eigenvalues and taking the ζ-regularized product. We will first consider antiperiodic boundary conditions on both cycles (A,A).

Then the eigenvalues are

$$\lambda_{AA} \sim \left(\left(m_1 + \frac{1}{2} \right) \tau + \left(m_2 + \frac{1}{2} \right) \right), \quad m_{1,2} \in \mathbb{Z}. \tag{4.19.2}$$

A calculation of the regularized product gives

$$(\det \partial)_{AA} = \frac{\vartheta_3(\tau)}{\eta(\tau)}. \tag{4.19.3}$$

For (A, P) boundary conditions we obtain

$$\lambda_{AP} \sim \left(\left(m_1 + \frac{1}{2} \right) \tau + m_2 \right), \quad m_{1,2} \in \mathbb{Z}, \tag{4.19.4}$$

$$(\det \partial)_{AP} = \frac{\vartheta_4(\tau)}{\eta(\tau)}. \tag{4.19.5}$$

For (P, A) boundary conditions we have

$$\lambda_{PA} \sim \left(m_1 \tau + \left(m_2 + \frac{1}{2} \right) \right), \quad m_{1,2} \in \mathbb{Z}, \tag{4.19.6}$$

$$(\det \partial)_{PA} = \frac{\vartheta_2(\tau)}{\eta(\tau)}. \tag{4.19.7}$$

Finally for (P, P) boundary conditions the determinant vanishes, since these boundary conditions now allow zero modes. By coupling to constant gauge fields (which act as sources for the zero modes) it can be seen that the determinant here is proportional to $\vartheta_1(\tau)$, which indeed is identically zero.

We can summarize the above results as follows. Let $a = 0, 1$ indicate A, P boundary conditions, respectively, around the first cycle and $b = 0, 1$ indicate A, P around the second. Then

$$(\det \partial)[^a_b] = \frac{\vartheta[^a_b](\tau)}{\eta(\tau)}. \tag{4.19.8}$$

The (P, P) spin structure is known as the odd spin structure, the rest as even spin structures. From appendix C we can see that modular transformations permute the various boundary conditions since they permute the various cycles. To construct something that is modular invariant, we will have to sum over all boundary conditions.[14] Including also the right-moving fermions we can write the full partition function as

$$Z_N^{\text{fermionic}} = \frac{1}{2} \sum_{a,b=0}^{1} \left| \frac{\vartheta[^a_b]}{\eta} \right|^N. \tag{4.19.9}$$

It can be checked directly that it is modular invariant. To expose the spectrum, we can express the partition function in terms of the characters (4.12.28), (4.12.29), (4.12.37), and (4.12.38) on page 76 as

$$Z_N^{\text{fermionic}} = |\chi_O|^2 + |\chi_V|^2 + |\chi_S|^2 + |\chi_C|^2, \tag{4.19.10}$$

from which we see that all $O(N)_1$ integrable representations participate.

---

[14] Since ϑ_1 is modular covariant by itself, it would seem that we have the option to exclude it. It turn out that consistency with higher-genus amplitudes requires its inclusion in the torus partition function.

The two-point functions of the fermions in the even spin structures can be fixed in terms of their pole structure and periodicity properties. They are given by the Szegö kernel

$$\langle \psi^i(z)\psi^j(0)\rangle = \delta^{ij} \, S[^a_b](z), \quad S[^a_b](z) = \frac{\vartheta[^a_b](z)\vartheta'_1(0)}{\vartheta_1(z)\vartheta[^a_b](0)}. \tag{4.19.11}$$

We will discuss the zero modes in the odd spin structure further. Each real fermion has a zero mode, and the path integral vanishes. The first nonzero correlation function must contain N fermions so that they soak up all the zero modes. The integral over the zero modes gives a completely antisymmetric tensor, which we normalize to the invariant ϵ-tensor. The rest of the contribution is given by the partition function in the absence of zero modes. Since the oscillators are integrally moded and since there is a $(-1)^F$ insertion, the nonzero-mode contribution is

$$q^{-N/24} \prod_{n=1}^{\infty} (1 - q^n)^N = \eta^N = \left[\frac{1}{2\pi} \frac{\partial_\nu \vartheta_1(\nu)|_{\nu=0}}{\eta} \right]^{N/2}. \tag{4.19.12}$$

Thus,

$$\left\langle \prod_{k=1}^{N} \psi^{i_k}(z_k) \right\rangle_{\text{odd}} = \epsilon^{i_1,\dots,i_N} \eta^N. \tag{4.19.13}$$

4.20 Bosonization

In this section we will indicate an equivalence between boson and fermion theories in two dimensions.

Consider two Majorana-Weyl fermions $\psi^i(z)$ with

$$\psi^i(z)\psi^j(w) = \frac{\delta^{ij}}{z - w} + \cdots . \tag{4.20.1}$$

We can change basis to[15]

$$\psi = \frac{1}{\sqrt{2}}(\psi^1 + i\psi^2), \quad \bar{\psi} = \frac{1}{\sqrt{2}}(\psi^1 - i\psi^2). \tag{4.20.2}$$

The theory contains a U(1) current algebra generated by the (1,0) current:

$$J(z) =: \psi\bar{\psi} :, \quad J(z)J(w) = \frac{1}{(z-w)^2} + \cdots , \tag{4.20.3}$$

$$J(z)\psi(w) = \frac{\psi(w)}{z - w} + \cdots , \quad J(z)\bar{\psi}(w) = -\frac{\bar{\psi}(w)}{z - w} + \cdots . \tag{4.20.4}$$

Equation (4.20.4) states that $\psi, \bar{\psi}$ are affine primaries with charges 1 and -1. The stress-tensor is

$$T(z) = -\frac{1}{2} : \psi^i \partial \psi^i := \frac{1}{2} : J^2 : . \tag{4.20.5}$$

It has central charge $c = 1$.

---

[15] In this section $\bar{\psi}$ does not denote a right-moving fermion. Both ψ and $\bar{\psi}$ are left-movers.

We can represent the same operator algebra using a single chiral boson $X(z)$; namely,

$$J(z) = i\frac{\sqrt{2}}{\ell_s}\partial X, \quad \psi =: e^{i\sqrt{2}X/\ell_s} :, \quad \bar{\psi} =: e^{-i\sqrt{2}X/\ell_s} :. \tag{4.20.6}$$

Applying these definitions to (4.20.5) they produce the correct stress-tensor of the scalar, namely, $T = -\frac{1}{\ell_s^2} : \partial X^2 :$. This chiral operator construction suggests that two Majorana-Weyl fermions and a chiral boson might give equivalent theories. However, the full theories contain also right-moving parts. When these are included, we are considering on the one hand a Dirac fermion and on the other a scalar. For the scalar theory, however, we have to specify the radius R. To do this we start from the partition function of the torus for a Dirac fermion (4.19.9) for N=2.

Applying a Poisson resummation to the ϑ-functions we obtain

$$|\vartheta[_b^a]|^2 = \frac{1}{\sqrt{2\tau_2}} \sum_{m,n\in\mathbb{Z}} \exp\left[-\frac{\pi}{2\tau_2}|n - b + \tau(m - a)|^2 + i\pi mn\right]$$

$$= \frac{1}{\sqrt{2\tau_2}} \sum_{m,n\in\mathbb{Z}} \exp\left[-\frac{\pi}{2\tau_2}|n + \tau m|^2 + i\pi(m + a)(n + b)\right]. \tag{4.20.7}$$

The second equation is valid when $a, b \in \mathbb{Z}$. Then,

$$Z = \frac{1}{2}\sum_{a,b=0}^{1}\left|\frac{\vartheta[_b^a]}{\eta}\right|^2$$

$$= \frac{1}{2\sqrt{2\tau_2}} \sum_{a,b=0}^{1}\sum_{m,n\in\mathbb{Z}} \exp\left[-\frac{\pi}{2\tau_2}|n + \tau m|^2 + i\pi(m + a)(n + b)\right]. \tag{4.20.8}$$

Summation over b gives a factor of 2 and sets $m + a$ to be even. Thus, $m = 2\tilde{m} + a$. Summing over a resets m to be an arbitrary integer. Thus,

$$Z_{\text{Dirac}} = \frac{1}{\sqrt{2\tau_2}} \sum_{m,n\in\mathbb{Z}} \exp\left[-\frac{\pi}{2\tau_2}|n + \tau m|^2\right] \tag{4.20.9}$$

and, comparing with (4.18.17), we see that it is the same as that of a boson with radius $R = \ell_s/\sqrt{2}$.

To summarize, a Dirac fermion is equivalent to a compact boson with radius $R = \ell_s/\sqrt{2}$.

4.20.1 " Bosonization" of bosonic ghost system

We will describe here the bosonization of the bosonic ghost systems since it is very convenient in the superstring case. We consider the ghost system of section 4.15 with $\epsilon = -1$.

We first bosonize the U(1) current by introducing a new boson ϕ:

$$J(z) = -\partial\phi, \quad \langle\phi(z)\phi(w)\rangle = -\log(z - w). \tag{4.20.10}$$

The stress-tensor that gives the OPE (4.15.13) on page 85 is

$$\hat{T} = \frac{1}{2} : J^2 : +\frac{1}{2}Q\partial J = \frac{1}{2}(\partial\phi)^2 - \frac{Q}{2}\partial^2\phi, \quad Q = 2\lambda - 1. \tag{4.20.11}$$

The boson ϕ has "background charge" because of the derivative term in its stress-tensor. It is described by the following action:

$$S_Q = \frac{1}{2\pi} \int d^2z \left[\partial\phi\bar\partial\phi - \frac{Q}{2}\sqrt{g}R^{(2)}\phi \right], \tag{4.20.12}$$

where $R^{(2)}$ is the two-dimensional scalar curvature. Using (6.1.2) on page 145 we see that there is a background charge of $-Q\chi/2$, where $\chi = 2(1 - g)$ is the Euler number of the surface. This implies in practice that a correlator is nonzero if the ϕ charges add up to $-Q\chi/2$.

A direct computation shows that $\hat{T}$ has central charge $\hat{c} = 1 + 3Q^2$. The original central charge of the theory was $c = \hat{c} - 2$, as can be seen from (4.15.9) on page 85. Thus, we must also add an auxiliary Fermi system with $\lambda = 1$, composed of a dimension-one field $\eta(z)$ and a dimension-zero field $\xi(z)$. This system has central charge -2. The stress-tensor of the original system can be written as

$$T = \hat{T} + T_{\eta\xi}. \tag{4.20.13}$$

Exponentials of the scalar ϕ have the following OPEs with the stress-tensor and the U(1) current:

$$T(z) : e^{q\phi(w)} := \left[-\frac{q(q + Q)}{2(z - w)^2} + \frac{1}{z - w}\partial_w \right] : e^{q\phi(w)} : + \cdots , \tag{4.20.14}$$

$$J(z) : e^{q\phi(w)} := \frac{q}{z - w} : e^{q\phi(w)} : \cdots \rightarrow [J_0, : e^{q\phi(w)} :] = q : e^{q\phi(w)} :. \tag{4.20.15}$$

In terms of the new variables we can express the original b, c ghosts as

$$c(z) = e^{\phi(z)}\eta(z), \quad b(z) = e^{-\phi(z)}\partial\xi(z). \tag{4.20.16}$$

Finally, the spin fields of b, c that interpolate between NS and R sectors are given by $e^{\pm\phi/2}$ with conformal weight $-(1 \pm 2Q)/8$. Note that the zero mode of the field ξ does not enter the definition of b, c. Thus, the bosonized Hilbert space provides two copies of the original Hilbert space since any state $|\rho\rangle$ has a degenerate partner $\xi_0|\rho\rangle$.

4.21 Orbifolds

The notion of the orbifold arises when we consider a manifold M that has a discrete symmetry group G. We may consider a new manifold $\tilde{M} \equiv M/G$, which is obtained from the old one by modding out by the symmetry group G. If G is freely acting (M has no fixed points under the G action) then M/G is a smooth manifold. On the other hand, if G has fixed points, then M/G is no longer a smooth manifold but has conical singularities at the fixed points. These are known as orbifold singularities. We will now provide simple examples of the above.

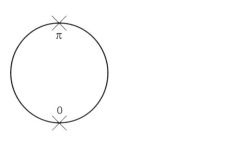

Figure 4.9 The orbifold $S^1/\mathbb{Z}_2$.

Consider the real line $\mathbb{R}$. It has a $\mathbb{Z}_2$ symmetry $x \to -x$. This symmetry has one fixed point, namely, $x = 0$. The orbifold $\mathbb{R}/\mathbb{Z}_2$ is the half-line with an orbifold point (singularity) at the boundary $x = 0$. On the other hand the real line $\mathbb{R}$ has another discrete infinite symmetry group, namely, translations $x \to x + 2\pi\lambda$. This symmetry is freely acting, and the resulting orbifold is a smooth manifold, namely a circle of radius λ.

Orbifolds are interesting in the context of CFT and string theory, since they provide spaces for string compactification that are richer than tori, but admit an exact CFT description. Moreover, although their classical geometry can be singular, strings propagate smoothly on them. In other words, the correlation functions of the associated CFT are finite.

We will describe here some simple examples of orbifolds in order to indicate the important issues. They will be useful later on, in order to break supersymmetry in string theory.

We first consider an example of a non-freely-acting orbifold. Start from a circle of radius R, parametrized by $x \in [0, 2\pi]$, and mod out the symmetry $x \to -x$. There are two fixed points under the symmetry action, $x = 0$ and $x = \pi$. The resulting orbifold is a line segment with the fixed points at the boundaries (figure 4.9).

It is not difficult to construct the CFT of the orbifold. Every operator in the original Hilbert space has a well-defined behavior under the $\mathbb{Z}_2$ orbifold transformation, $X \to -X$, and for the vertex operators, $V_{m,n} \to V_{-m,-n}$.

The orbifold construction indicates that we should keep only the operators invariant under the orbifold transformation. Thus, the orbifold theory contains the $\mathbb{Z}_2$ invariant operators. Their correlators are the same as in the original theory. In particular, the invariant vertex operators are $V_{m,n}^{+} = \frac{1}{\sqrt{2}}(V_{m,n} + V_{-m,-n})$.

However, this is not the end of the story. What we have constructed so far is the "untwisted sector." An indication that we must have more can be seen from the torus partition function. We will start from (4.18.18) in the Hamiltonian representation. In order to keep only the invariant states we will have to insert a projector in the trace. This projector is $(1 + g)/2$, where g is the nontrivial orbifold group element acting on states as

$$g\left[\prod_{i=1}^{N}\alpha_{-n_i}\prod_{j=1}^{\bar{N}}\bar{\alpha}_{-\bar{n}_j}|m,n\rangle\right] = (-1)^{N+\bar{N}}\prod_{i=1}^{N}\alpha_{-n_i}\prod_{j=1}^{\bar{N}}\bar{\alpha}_{-\bar{n}_j}|-m,-n\rangle. \tag{4.21.1}$$

Thus,

$$Z(R)^{\text{invariant}} = \frac{1}{2}Z(R) + \frac{1}{2}\text{Tr}[g\, q^{L_0 - 1/24}\bar{q}^{\bar{L}_0 - 1/24}]. \tag{4.21.2}$$

To evaluate the second trace we note that $\langle m_1, n_1 | m_2, n_2 \rangle \sim \delta_{m_1+m_2}\delta_{n_1+n_2}$, which implies that only states with $m = n = 0$ contribute to the g trace. These are pure oscillator states (the vacuum module) and every oscillator is weighted with a factor of -1 due to the action of g. We obtain

$$\frac{1}{2}\text{Tr}[g\, q^{L_0 - 1/24}\,\bar{q}^{\bar{L}_0 - 1/24}] = \frac{1}{2}(q\bar{q})^{-1/24}\prod_{n=1}^{\infty}\frac{1}{(1+q^n)(1+\bar{q}^n)} = \left|\frac{\eta}{\vartheta_2}\right| \tag{4.21.3}$$

and

$$Z(R)^{\text{invariant}} = \frac{1}{2}Z(R) + \left|\frac{\eta}{\vartheta_2}\right|. \tag{4.21.4}$$

A simple look at the modular properties of the ϑ-functions indicates that this partition function is not modular invariant. Something is missing.

The missing states are the *twisted* states. Because of the $\mathbb{Z}_2$ symmetry, there is another boundary condition possible for the field X, namely, $X(\sigma + 2\pi) = -X(\sigma)$. This is again a periodicity condition, since now X and $-X$ are identified. In this sector (which is similar to the Ramond sector for the fermions) the momentum and winding are forced to be zero by the boundary condition and the oscillators are half-integrally moded. Imposing the boundary condition above on the solution of the Laplace equation, we obtain the following mode expansion in the twisted sector:

$$X(\sigma, \tau) = x_0 + \frac{i\ell_s}{\sqrt{2}}\sum_{n\in\mathbb{Z}}\left(\frac{\alpha_{n+1/2}}{n+1/2}e^{i(n+1/2)(\sigma+\tau)} + \frac{\bar{\alpha}_{n+1/2}}{n+1/2}e^{-i(n+1/2)(\sigma-\tau)}\right). \tag{4.21.5}$$

The zero mode x_0 is forced to lie at the two fixed points: $x_0 = 0, \pi$. This indicates the presence of two ground states $|H^{0,\pi}\rangle$ in this sector. They are primaries under the Virasoro algebra and invariant under the orbifold transformation. They satisfy

$$\alpha_{n+1/2}|H^{0,\pi}\rangle = \bar{\alpha}_{n+1/2}|H^{0,\pi}\rangle = 0, \quad n \geq 0. \tag{4.21.6}$$

Their conformal weight can be computed in the same way as we did for the spin fields in the fermionic case. It is $\Delta = \bar{\Delta} = 1/16$. The rest of the states are generated by the action of the negative moded oscillators on the ground states. However, not all of the states are invariant. To pick the invariant states we will have to do a trace with our projector in the twisted sector:

$$Z^{\text{twisted}} = \frac{1}{2}\text{Tr}[(1+g)q^{L_0-1/24}\,\bar{q}^{\bar{L}_0-1/24}]$$

$$= \frac{1}{2}\frac{1}{(q\bar{q})^{1/48}}\left[\prod_{n=1}^{\infty}\frac{1}{(1-q^{n-1/2})(1-\bar{q}^{n-1/2})} + \prod_{n=1}^{\infty}\frac{1}{(1+q^{n-1/2})(1+\bar{q}^{n-1/2})}\right]$$

$$= \left|\frac{\eta}{\vartheta_4}\right| + \left|\frac{\eta}{\vartheta_3}\right|. \tag{4.21.7}$$

The full partition function

$$Z^{\text{orb}}(R) = Z^{\text{untwisted}} + Z^{\text{twisted}} = \frac{1}{2}Z(R) + \left|\frac{\eta}{\vartheta_2}\right| + \left|\frac{\eta}{\vartheta_4}\right| + \left|\frac{\eta}{\vartheta_3}\right| \tag{4.21.8}$$

is now modular invariant. In fact, the four different parts in (4.21.8) can be interpreted as the result of performing the path integral on the torus, with the four different boundary conditions around the two cycles, as in the case of the fermions. We will introduce the notation $Z[^h_g]$ where h, g take values $0, 1$; $h = 0$ labels the untwisted sector, $h = 1$ the twisted sector; $g = 0$ implies no projection, while $g = 1$ implies a projection. In this notation the orbifold partition function can be written as

$$Z^{\text{orb}} = \frac{1}{2} \sum_{h,g=0}^{1} Z[^h_g], \tag{4.21.9}$$

with $Z[^0_0] = Z(R)$ and

$$Z[^h_g] = 2 \left| \frac{\eta}{\vartheta[^{1-h}_{1-g}]} \right|, \quad (h, g) \neq (0, 0). \tag{4.21.10}$$

They transform as follows under modular transformations:

$$\tau \to \tau + 1 : Z[^h_g] \to Z[^h_{h+g}], \tag{4.21.11}$$

$$\tau \to -\frac{1}{\tau} : Z[^h_g] \to Z[^g_h], \tag{4.21.12}$$

and we conclude that (4.21.9) is modular invariant.

Notice also that the whole twisted sector does not depend on the radius. This is a general characteristic of non-freely-acting orbifolds. As we will see later, the situation is different for freely acting orbifolds.

The twisted ground states are generated from the SL(2,$\mathbb{C}$)-invariant vacuum by the twist operators $H^{0,\pi}(z, \bar{z})$. Correlation functions of twist operators are more difficult to compute, but this calculation can be done. The following schematic OPEs can be established:

$$[H^0] \cdot [H^0] \sim \sum_{n,m} C^{2m,2n}[V^+_{2m,2n}] + C^{2m,2n+1}[V^+_{2m,2n+1}], \tag{4.21.13}$$

$$[H^\pi] \cdot [H^\pi] \sim \sum_{n,m} C^{2m,2n}[V^+_{2m,2n}] - C^{2m,2n+1}[V^+_{2m,2n+1}], \tag{4.21.14}$$

$$[H^0] \cdot [H^\pi] \sim \sum_{n,m} C^{2m+1,2n}[V^+_{2m+1,2n}]. \tag{4.21.15}$$

Here $[V^+_{m,n}]$ stands for the whole U(1) representation generated from the primary vertex operator $V^+_{m,n} = (V_{m,n} + V_{-m,-n})/\sqrt{2}$ by the action of the U(1) current modes. The OPE coefficients are given by

$$C_{m,n} = 2^{-2(\Delta_{m,n}+\bar{\Delta}_{m,n})+1/2}, \quad C_{0,0} = 1, \tag{4.21.16}$$

and

$$\Delta_{m,n} = \frac{1}{4}\left(m\frac{\ell_s}{R} + n\frac{R}{\ell_s}\right)^2, \quad \bar{\Delta}_{m,n} = \frac{1}{4}\left(m\frac{\ell_s}{R} - n\frac{R}{\ell_s}\right)^2. \tag{4.21.17}$$

Notice that the two U(1) currents ∂X and $\bar{\partial}X$ of the original theory have been projected out. Consequently, in the orbifold theory we do not expect to have the continuous U(1)$_\text{L}$ × U(1)$_\text{R}$ invariance any longer. This is already obvious in the twisted OPEs, which show that the charges m, n are no longer conserved. There remains, however, a residual $\mathbb{Z}_2 \times \mathbb{Z}_2$ symmetry,

$$(H^0, H^\pi, V^+_{m,n}) \to (-H^0, H^\pi, (-1)^m V^+_{m,n}), \tag{4.21.18}$$

$$(H^0, H^\pi, V^+_{m,n}) \to (H^\pi, H^0, (-1)^n V^+_{m,n}). \tag{4.21.19}$$

When these transformations are combined with the extra symmetry that changes the sign of the twist fields[16]

$$(H^0, H^\pi, V^+_{m,n}) \to (-H^0, -H^\pi, V^+_{m,n}), \tag{4.21.20}$$

they generate the dihedral group D_4, which is the invariance group of the orbifold.

The orbifold theory depends also on a continuous parameter, the radius R. Moreover, we also have here the duality symmetry $R \to \ell_s^2/R$, since from (4.21.8)

$$Z^{\text{orb}}(R) = Z^{\text{orb}}(1/R). \tag{4.21.21}$$

The orbifold above can be easily generalized in various directions. First we can consider other starting CFTs, such as higher-dimensional tori or interacting CFTs. Moreover, the symmetry we use can be a bigger abelian or nonabelian discrete group. We will not delve further in this direction for the moment.

We will now discuss a simple example of a freely acting orbifold group. We start again from the theory of a scalar on a circle of radius R. However, here we will use a $\mathbb{Z}_2$ subgroup of the U(1) symmetry that acts as $|m, n\rangle \to (-1)^m |m, n\rangle$ and leaves the oscillators invariant. The geometrical action is a half-lattice shift: $X \to X + \pi R$. Thus, this orbifold is a translation orbifold. Although the result of such an orbifolding is trivial (it halves the radius of the circle), the orbifolding procedure is pedagogically illuminating. Moreover, the conformal blocks of this trivial translation orbifold can be ingredients in nontrivial freely acting orbifolds.

We will calculate the partition function using the same method as above. It will be written again in the form (4.21.9) with $Z[^0_0] = Z(R)$. $Z[^0_1]$ must include the group element in the trace:

$$Z[^0_1] = \sum_{m,n \in \mathbb{Z}} (-1)^m \frac{\exp\left[\frac{i\pi\tau}{2}\left(\frac{m}{R} + nR\right)^2 - \frac{i\pi\bar{\tau}}{2}\left(\frac{m}{R} - nR\right)^2\right]}{\eta\bar{\eta}}. \tag{4.21.22}$$

The computation of $Z[^1_0]$ can be made by noting that the twisted boundary condition is identical to the periodicity condition for a circle of half the radius, so that $n \to n + 1/2$. The same result is obtained by performing a $\tau \to -1/\tau$ transformation on $Z[^0_1]$. Both methods give

$$Z[^1_0] = \sum_{m,n \in \mathbb{Z}} \frac{\exp\left[\frac{i\pi\tau}{2}\left(\frac{m}{R} + \left(n + \frac{1}{2}\right)R\right)^2 - \frac{i\pi\bar{\tau}}{2}\left(\frac{m}{R} - \left(n + \frac{1}{2}\right)R\right)^2\right]}{\eta\bar{\eta}}. \tag{4.21.23}$$

---

[16] This is similar to the space-time fermion number symmetry for the free fermions.

Finally $Z[^1_1]$ can be obtained from $Z[^1_0]$ by a $\tau \to \tau + 1$ transformation or by inserting the group element in the trace:

$$Z[^1_1] = \sum_{m,n \in \mathbb{Z}} (-1)^m \frac{\exp\left[\frac{i\pi\tau}{2}\left(\frac{m}{R} + \left(n + \frac{1}{2}\right)R\right)^2 - \frac{i\pi\bar\tau}{2}\left(\frac{m}{R} - \left(n + \frac{1}{2}\right)R\right)^2\right]}{\eta\bar\eta}. \tag{4.21.24}$$

We can summarize the above by

$$Z[^h_g] = \sum_{m,n \in \mathbb{Z}} (-1)^{gm} \frac{\exp\left[\frac{i\pi\tau}{2}\left(\frac{m}{R} + \left(n + \frac{h}{2}\right)R\right)^2 - \frac{i\pi\bar\tau}{2}\left(\frac{m}{R} - \left(n + \frac{h}{2}\right)R\right)^2\right]}{\eta\bar\eta} \tag{4.21.25}$$

or, in the Lagrangian representation, by

$$Z[^h_g] = \frac{R}{\sqrt{\tau_2}\eta\bar\eta} \sum_{m,n, \in \mathbb{Z}} \exp\left[-\frac{\pi R^2}{\tau_2}\left|m + \frac{g}{2} + \left(n + \frac{h}{2}\right)\tau\right|^2\right]. \tag{4.21.26}$$

Summing up the contributions as in (4.21.9) we obtain, not to our surprise, the partition function for a boson compactified on a circle of radius $R/2$. This is what we would have expected from the geometrical action of the orbifold element. Note also that here the twisted sectors have a nontrivial dependence on the radius. This is a generic feature of freely acting orbifolds.

Although this orbifold example looks trivial, it can be combined with other projections to make nontrivial orbifold CFTs. In the exercises, you are invited to work out such examples.

We will comment here on the most general orbifold group of a toroidal model. The generic symmetry of a d-dimensional toroidal CFT contains the $U(1)_L^d \times U(1)_R^d$ chiral symmetry. The transformations associated with it are arbitrary lattice translations. They act on a state with momenta m_i and windings n_i as

$$g_{\text{translation}} = \exp\left[2\pi i \sum_{i=1}^{d} (m_i\theta_i + n_i\phi_i)\right], \tag{4.21.27}$$

where θ_i, ϕ_i are rational in order to obtain a discrete group. There are also symmetries that are subgroups of the $O(d,d)$ duality group not broken by the moduli G_{ij} and B_{ij}. These depend on the point of the moduli space. Consequently, the generic element is a combination of a translation and a rotation acting on the left part of the theory and an *a priori* different rotation and translation acting on the right part of the theory.

There are constraints imposed by modular invariance that restrict the choice of orbifold groups. The orbifolding procedure can be viewed as a gauging of a discrete symmetry. It can happen that the discrete symmetry is anomalous. Then, the theory will not be modular invariant. An example is proposed in the exercise 4.67 on page 125.

4.22 CFT on Other Surfaces of Euler Number Zero

We have analyzed so far the torus surface, associated with one-loop diagrams of closed strings. For the open strings, the analog of a one-loop diagram has the topology of a cylinder (or annulus) (figure 4.10). It has no holes but two boundaries. It has Euler number zero, like the torus.

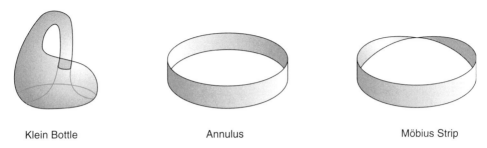

Klein Bottle   Annulus   Möbius Strip

Figure 4.10 The three open and/or unoriented surfaces with Euler number zero.

For unoriented strings there are two more surfaces, the Klein bottle that implements the orientation projection in the closed string sector, and the Möbius strip that does the same in the open sector (figure 4.10). We shall also discuss them in turn.

The cylinder (annulus)
The cylinder C_2 can be described as the region on the plane satisfying

$$0 \leq \mathrm{Re}\, z \leq \pi, \quad z \simeq z + 2\pi it. \tag{4.22.1}$$

It is therefore a strip of width π and length $2\pi t$ with its end points joined. The real number $t \in [0, \infty)$ is a Teichmüler parameter. The cylinder has a single conformal Killing vector associated with the symmetry of rotations in the direction $\mathrm{Im} z$.

The cylinder can be obtained from the torus by a $\mathbb{Z}_2$ involution. We start from a torus with $\tau = it$. We then identify under the involution (see figure 4.11)

$$z' = -\bar{z}. \tag{4.22.2}$$

To match with the previous description, we have to scale-up all lengths by 2π. The fixed lines are then at $\mathrm{Re}\, z = 0, \pi$ and form the two boundaries.

The Klein bottle
The Klein bottle K_2 can be constructed by a freely acting involution on the cylinder,

$$z \simeq z + 2\pi \simeq -\bar{z} + 2\pi it. \tag{4.22.3}$$

It can also be obtained from the plane by modding out by the following action:

$$(\sigma_1, \sigma_2) \simeq (\sigma_1 + 2\pi, \sigma_2) \simeq (-\sigma_1, \sigma_2 + 2\pi t). \tag{4.22.4}$$

It is therefore a cylinder of circumference 2π and length $2\pi t$ with the two ends fitted together with a parity-reversal transformation. Because of this, there are no boundaries. There are, however, two crosscaps. The modulus t is a positive real number. There is a translation symmetry along σ_2.

The Klein bottle can be also obtained by a freely acting $\mathbb{Z}_2$ involution on a torus with $\tau = 2it$,

$$z' = -\bar{z} + 2\pi it. \tag{4.22.5}$$

This involution translates one cycle and inverts the direction of the other one.

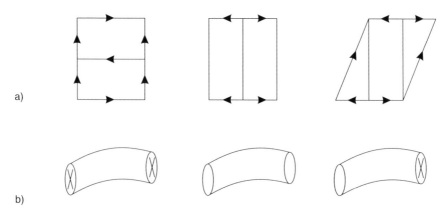

Figure 4.11 a) The representation of the Klein bottle, cylinder, and Möbius strip in terms of parallelograms with identifications. b) A schematic representation of the three surfaces in terms of boundaries and cross caps.

The Möbius Strip

The Möbius strip M_2 can be obtained from a standard strip by pasting it with a parity (Ω) twist:

$$0 \leq z \leq \pi, \quad z \simeq -\bar{z} + \pi + 2\pi it. \tag{4.22.6}$$

The modulus t is a positive real number, and translations along σ^2 are a symmetry.

It can be obtained from a torus with $\tau = 2it$ by modding out by the two involutions

$$z' = -\bar{z}, \quad z' = z + 2\pi \left(it + \frac{1}{2}\right). \tag{4.22.7}$$

The first creates a cylinder while the second creates the crosscap.

As in the torus case, the path integral of a CFT on the cylinder is given by

$$Z_{C_2}^{\mathrm{bc}} = \mathrm{Tr}\left[e^{-2\pi t L_0^{\mathrm{bc}}}\right], \tag{4.22.8}$$

where we must specify boundary conditions (bc) at both boundaries. L_0^{bc} is the open Hamiltonian on the cylinder with the given boundary conditions. The trace is taken in the open string Hilbert space. It is related to the open Hamiltonian in the disk/sphere coordinates (as usual) by $L_0^{\mathrm{cyl}} = L_0^{\mathrm{sphere}} - \frac{c}{24}$.

We will first calculate the cylinder partition function for a noncompact boson with NN boundary conditions. From the form of L_0 in (2.3.36) on page 24 we obtain

$$L_0 = -\frac{1}{24} + \ell_s^2 p^2 + \sum_{n=1}^{\infty} \alpha_{-n}\alpha_n \tag{4.22.9}$$

and

$$Z_{C_2,\mathrm{boson}}^{\mathrm{NN}} = V\,N_1 N_2 \int_{-\infty}^{\infty} \frac{dp}{2\pi} \frac{e^{-2\pi\ell_s^2 p^2 t}}{\eta(it)} = \frac{V\,N_1 N_2}{(2\pi\ell_s)\sqrt{2t}\,\eta(it)}, \tag{4.22.10}$$

where V is the (infinite) volume of the scalar, and N_1, N_2 are the multiplicities of the CP indices at the two boundaries. Thus, $N_1 N_2$ is the overall multiplicity of every state.

For DD boundary conditions, from (2.3.37) on page 24

$$L_0 = -\frac{1}{24} + \left(\frac{\Delta x}{2\pi \ell_s}\right)^2 + \sum_{n=1}^{\infty} \alpha_{-n}\alpha_n, \tag{4.22.11}$$

so that

$$Z_{C_2,\text{boson}}^{\text{DD}} = N_1 N_2 \frac{e^{-[(\Delta x)^2/2\pi \ell_s^2]t}}{\eta(it)}, \tag{4.22.12}$$

where Δx is the distance between the end points of the string in target space. The CP multiplicity is again present. Note, however, that there is no factor of volume as the end points are fixed.

Finally, for DN boundary conditions an appropriate quantum version of (2.3.38) is

$$L_0 = \frac{1}{48} + \sum_{n=0}^{\infty} \alpha_{-n-1/2}\alpha_{n+1/2}. \tag{4.22.13}$$

The intercept is a sum of the standard $-1/24$ as well as the $1/16$, which is the weight of the ground state due to the $\mathbb{Z}_2$ twist. This was shown in the simple $\mathbb{Z}_2$ orbifold in section 4.21. Putting everything together we obtain

$$Z_{C_2,\text{boson}}^{\text{DN}} = N_1 N_2 \frac{e^{-\pi t/24}}{\prod_{n=0}^{\infty}(1 - e^{-\pi t(2n+1)})} = N_1 N_2 \sqrt{\frac{\eta(it)}{\vartheta_3(it)}}. \tag{4.22.14}$$

There is no factor of volume here either.

We proceed to consider the path integral on the Möbius strip. Its Hamiltonian interpretation follows directly from its construction: it is a strip glued up to a parity (Ω) transformation. Therefore

$$Z_{M_2}^{bc} = \frac{1}{2}\text{Tr}\left[\Omega\, e^{-2\pi t L_0^{bc}}\right]. \tag{4.22.15}$$

The factor of $\frac{1}{2}$ is to provide the correct normalization of the projection. The full unoriented open partition function is

$$Z_{\text{unoriented}} = \text{Tr}\left[\frac{(1 + \Omega)}{2}\, e^{-2\pi t L_0^{bc}}\right]. \tag{4.22.16}$$

From the action of Ω on the oscillators (3.4.5) on page 34 we can evaluate the trace in the free-boson NN case as

$$\begin{aligned}
Z_{M_2,\text{boson}}^{\text{NN}} &= V\, CP \int_{-\infty}^{\infty} \frac{dp}{2\pi} \frac{e^{-2\pi \ell_s^2 p^2 t}}{\prod_{n=1}^{\infty}(1 - (-1)^n e^{-2\pi nt})} \\
&= \frac{V\, CP}{(2\pi \ell_s)\sqrt{2t}\prod_{n=1}^{\infty}(1 - e^{-4\pi nt})(1 + e^{-2\pi(2n-1)t})} = \frac{V\, CP}{(2\pi \ell_s)\sqrt{2t}\sqrt{\vartheta_3(2it)\eta(2it)}}.
\end{aligned} \tag{4.22.17}$$

In general, the Ω projection acts on the CP sector. CP stands for the trace of Ω on the CP indices. This is analyzed in detail later, in section 5.3.4 on page 138.

According to (3.4.5) on page 34, for the DD case Ω acts nontrivially on the oscillators, and we obtain

$$Z_{M_2,\text{boson}}^{\text{DD}} = CP\delta(\Delta x)\frac{1}{\prod_{n=1}^{\infty}(1 + (-1)^n e^{-2\pi nt})} = CP\delta(\Delta x)\frac{\sqrt{2}\,\eta(2it)}{\sqrt{\vartheta_4(2it)\vartheta_2(2it)}}, \tag{4.22.18}$$

where CP stands for the appropriate CP trace. Ω interchanges the end points of the string coordinate. We obtain a nonzero contribution in the trace only if they are at the same point. This explains the presence of $\delta(\Delta x)$ in (4.22.18).

Finally, the Ω projection interchanges ND and DN strings, so the Möbius partition function for DN is zero, since no term survives in the trace with an Ω inserted.

For the path integral on the Klein bottle we must trace over the closed string spectrum with an Ω insertion. This is obvious from the definition of the surface. The combination of the torus and the Klein bottle amplitude projects on the $\Omega = 1$ states:

$$Z_{\text{closed}}^{\text{unoriented}} = \text{Tr}\left[\frac{1+\Omega}{2} e^{-2\pi t(L_0 + \bar{L}_0 - c/12)}\right] = \frac{1}{2} Z_{\text{closed}}^{\text{torus}} + Z_{K_2}. \tag{4.22.19}$$

From the action of Ω (3.4.2), we may evaluate the trace for a massless scalar field. Only states that are left-right symmetric survive the trace.

Thus,

$$Z_{K_2} \equiv \frac{1}{2}\left[\Omega\, e^{-2\pi t(L_0 + \bar{L}_0 - c/12)}\right] = \frac{V}{2} \int_{-\infty}^{\infty} \frac{dp}{2\pi} \frac{e^{-\pi \ell_s^2 t p^2}}{\eta(2it)} = \frac{V}{(4\pi \ell_s)\sqrt{t}\, \eta(2it)}. \tag{4.22.20}$$

4.23 CFT on Higher-genus Riemann Surfaces

So far we have analyzed CFT on surfaces of low genus, namely, the Riemann sphere ($g = 0$), the disk, and RP$_2$, as well as surfaces with Euler number zero. Similarly, we can define and analyze CFTs on more complicated surfaces.

A general N-point function on a genus-($g \geq 2$) closed Riemann surface depends on the N (complex) positions of the operators and on $3(g-1)$ complex numbers that are the moduli of the surface. They are the generalizations of the modulus τ of the torus. There is also the notion of a modular group that acts on the moduli. The N-point function must be covariant under the modular group of a genus g surface with N punctures. The partition function, in particular, must be invariant.

There is a set of relations between correlation functions of the same CFT defined on various Riemann surfaces. This is known as factorization. Consider as an example the partition function of a CFT on a genus-2 surface depicted in figure 4.12a. It depends on three complex moduli. In particular, there is a modulus, which we will denote by q, such that as $q \to 0$ the surface develops a long cylinder in between and, at $q = 0$, degenerates into two tori with one puncture each (figure 4.12b). This implies a Hamiltonian degeneration formula for the partition function:

$$\langle 1 \rangle_{g=2} = \sum_i q^{\Delta_i - c/24} \bar{q}^{\bar{\Delta}_i - \bar{c}/24} \langle \phi_i \rangle_{g=1} \langle \phi_i \rangle_{g=1}, \tag{4.23.1}$$

where the sum is over all states of the theory and the one-point functions are evaluated on the once-punctured tori. This happens because as the intermediate cylinder becomes long we can use the cylinder Hamiltonian to describe this part of the theory. Equation (4.23.1) is schematically represented in figure 4.13. This can be generalized to arbitrary correlation functions and arbitrary degenerations.

Factorization is important since it implies perturbative unitarity in the underlying string

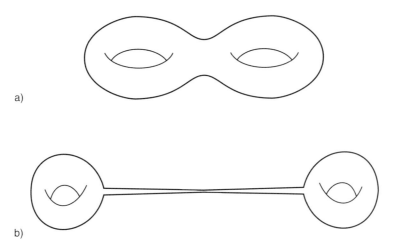

a)

b)

Figure 4.12 a) The double torus. b) The degeneration limit into two tori.

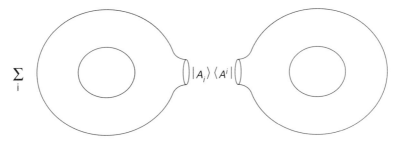

Figure 4.13 The Hamiltonian description of the degeneration of a double torus into a pair of tori.

theory. For example a $g = 2$ amplitude is a two-loop correction to a scattering amplitude and we should be able to construct it, in perturbation theory, by sewing one-loop amplitudes as suggested in figures 4.12 and 4.13.

Bibliography

There are two very good modern books which treat two-dimensional CFT in very great detail, namely, [46,47]. The lecture notes in [26] and [48] are a very good introduction to CFT of closed surfaces. More mathematical aspects of rational CFTs are reviewed in [49,50].

A very good book on classical and affine Lie algebras and their representations is [51]. An excellent review of affine current algebra can be found in [52] . Further issues, like the exact solution of coset theories, as well as the construction of generalized CFTs from affine currents, with generically irrational central charges are reviewed in [53]. The extended chiral symmetries of rational CFTs are reviewed in [54]. The relation between the affine symmetries and current algebra is explained in [55].

The formulation of CFT, presented here, originates in [56]. There, among other things, the technique for solving th minimal models with $c < 1$ is developed. The calculation of the correlators of such theories can be efficiently given in terms of the Coulomb gas [57]. The ADE classification of modular invariants for minimal CFTs can be found in [58].

The application of CFT techniques to the formulation and covariant quantization of string theory as well as the intricacies of the ghost systems can be found in the detailed paper [59].

Coming to more specialized topics, a rather complete description of conformal anomalies in any dimension can be found in the book [60]. A recent classification of superconformal algebras can be found in [61]. A formulation of $\mathcal{N} = (1, 1)_2$ superconformal symmetry in complex superspace can be found in [62]. A similar analysis for $\mathcal{N} = (2, 2)_2$ superconformal theories can be found in [63]. Reviews on $\mathcal{N} = (2, 0)_2$ superconformal symmetry and string compactification can be found in [47, 64,65,66,67,68]. Extended superconformal symmetry including the $\mathcal{N} = (4, 0)_2$ algebra is reviewed in [69]. T-duality for more complicated CFTs is reviewed extensively in [70]. The elliptic genus [71,72] is reviewed in [73]. Finally, progress in noncompact CFTs is reviewed in [74].

Off-critical descriptions are sometimes useful in understanding CFTs and their spectrum. One such tool is associated to the geometry of renormalization group flows in two dimensions and the associated c-theorem [75,76]. Another, is the concept of Landau-Ginsburg models [77], which can be very helpful in the presence of enough supersymmetry [64,65]. The third is the concept of linear σ-models. Again these are two-dimensional gauge theories coupled to matter fields. In the IR, the models flow to nontrivial CFTs. This is also a formalism that can be predictive at least in the presence of enough supersymmetry [78].

Orbifold compactifications were introduced in [79] where a detailed description of the general ideas can be found. The calculation of twist field correlators can be found in [80,81]. A detailed study of orbifolds on various Riemann surfaces can be found in [82,83]. Asymmetric orbifolds are a generalization of the simple orbifolds we described here. There, the action of the orbifold group is not the same for the left- and the right-movers. They are extensively discussed in [84].

Boundary CFT is a subject developed later than CFT on closed Riemann surfaces, because the heterotic (closed) string monopolized the attention of physicists up to 1995. The original papers [85,86] still remain the best introduction to the subject. They are simple, well written and physically motivated. The reviews [87,88,89] and the paper [90] are more formal but introduce with great clarity the main BCFT tools, boundary fields, boundary states, OPE, and fusion algebra. The review [91] is informal but contains a quick introduction to BCFT and its applications to D-branes, as well as remarks on AdS$_3$.

We will not discuss the method of boundary states in this book. This is a concept that can be useful in certain circumstances, in particular if we are interested in the coupling of closed strings to open strings. The paper [92] and the reviews [93,94,95,96] are very good starting points for understanding the boundary state formalism.

The excellent review on open strings [97] is also a good place to find many of the properties of BCFT, with the focus being on open string vacua.

CFTs with indecomposable spectrum are known as logarithmic CFTs. They are reviewed in [43,44], although their relevance for string theory is not yet clear.

Exercises

4.1. Use the infinitesimal conformal transformations (4.1.8) in order to derive the algebra of the conformal group (4.1.11) on page 50.

4.2. Show that the algebra of conformal transformations in a space-time with signature (p,q) is isomorphic to the Lie algebra of $O(p + 1, q + 1)$.

4.3. Derive the infinitesimal transformation of primary fields (4.2.4) on page 52 under conformal transformations.

4.4. Derive the SL(2) Ward identities (4.2.10)–(4.2.12) from (4.2.6) on page 53.

4.5. Solve the Ward identities and show that the most general form allowed for the three-point function is

$$G^{(3)}(z_i, \bar{z}_i) = \frac{C_{123}}{z_{12}^{\Delta_{12}} z_{23}^{\Delta_{23}} z_{31}^{\Delta_{31}} \bar{z}_{12}^{\bar{\Delta}_{12}} \bar{z}_{12}^{\bar{\Delta}_{12}} \bar{z}_{12}^{\bar{\Delta}_{12}}}, \tag{4.1E}$$

where $\Delta_{12} = \Delta_1 + \Delta_2 - \Delta_3$, $\bar{\Delta}_{12} = \bar{\Delta}_1 + \bar{\Delta}_2 - \bar{\Delta}_3$, etc.

4.6. Use the Ward identities to prove 4.2.8 on page 53. What is the generalization to n-point functions?

4.7. We have set the radius of the σ direction in section 4.3 on page 54 to be 2π. Discuss why there is no loss of generality in doing this.

4.8. Show that the constants C_{ijk} that appear in the OPE (4.3.12) on page 56 are the same coefficients appearing in the three-point function $\langle \Phi_i \Phi_j \Phi_k \rangle$.

4.9. Start from the Euclidean massive scalar propagator in two dimensions

$$\Delta_F(x, y) = \frac{\ell_s^2}{2\pi} \int d^2p \, \frac{e^{ip \cdot (x-y)}}{p^2 + \mu^2} \tag{4.2E}$$

and take the $\mu \to 0$ limit to obtain (4.7.2).

4.10. Show that

$$\partial \bar{\partial} \log |z|^2 = \partial \frac{1}{\bar{z}} = \bar{\partial} \frac{1}{z} = 2\pi \delta^{(2)}(z), \tag{4.3E}$$

either by using the Stokes' divergence theorem or by regulating the singularity at $z = 0$ and then removing the regularization.

4.11. Derive the stress tensor of the free massless boson in (4.7.1) and verify (4.7.4), (4.7.5) on page 61.

4.12. Consider the stress-tensor in a two-dimensional quantum field theory. Classically it has dimension two. Show that if it is conserved in the quantum theory, its exact scaling dimension remains two.

4.13. Derive equation (4.7.9) on page 62.

4.14. There is a single primary operator of dimension (4,0) constructed out of products of derivatives of a free massless scalar field. Find it.

4.15. Derive equation (4.7.13) on page 62.

4.16. Use Wick's theorem to show that

$$: e^{iaX(z)} : \ : e^{ibX(w)} : := (z-w)^{(\ell_s^2/2)ab} : e^{iaX(z)+ibX(w)} : . \tag{4.4E}$$

4.17. Consider a stress tensor which is the derivative of a $(1,0)$ current: $T = \partial_z J$. Compute the TT OPE and show that it implies via (4.3.9) on page 55 that T has scaling dimension zero, an obvious inconsistency.

4.18. By calculating the expansions of $T(z)\psi(w)$ and $\bar{T}(\bar{z})\bar{\psi}(\bar{w})$, show that ψ and $\bar{\psi}$ are primary fields of conformal weight $(\frac{1}{2}, 0)$ and $(0, \frac{1}{2})$, respectively.

4.19. Use the infinitesimal variation of the stress tensor (4.9.1) and the cocycle property to prove (4.9.2). To do this, set $T(z) = (f')^2 T(f) + C(f(z))$ and show that for two consecutive holomorphic transformations $z' = f(z)$ and $z = g(w)$, C must satisfy the cocycle condition

$$C(z'(w)) = C(z(w)) + \left(\frac{\partial z}{\partial w}\right)^2 C(z'(z)). \tag{4.5E}$$

(4.9.2) on page 64 implies that C contains at most third derivatives. Show that the only such function satisfying (4.5E) is proportional to the Schwarzian derivative. You can then fix the proportionality factor from (4.9.1).

4.20. Show that $\hat{T}$ defined as in (4.9.3) transforms under conformal transformations as in (4.9.4) on page 64.

4.21. Prove that equation (4.9.7) on page 65 is valid with $A = -c/12$.

4.22. Show that the Liouville action (4.9.11) on page 65 satisfies the cocycle property

$$S_L(g_{\alpha\beta}e^{\chi}, \phi) = S_L(g_{\alpha\beta}, \phi) + \frac{1}{48\pi} \int \sqrt{g}\, g^{\alpha\beta} \partial_\alpha \phi \partial_\beta \chi. \tag{4.6E}$$

This is necessary so that the relation (4.9.10) on page 65 satisfies the abelian group property.

4.23. Find the value of B_2 in (4.9.12) on page 65 by solving the WZ consistency condition (4.9.13).

4.24. Use (4.9.2) on page 64 to show that

$$L_0^{\text{torus}} = L_0^{\text{sphere}} - \frac{c}{24}. \tag{4.7E}$$

4.25. For a generic Virasoro representation without null vectors, calculate the character and show that it is given by

$$\chi_\Delta(q) = \frac{q^{\Delta-c/24}}{\prod_{n=1}^{\infty}(1-q^n)}. \tag{4.8E}$$

From this expression we can read off the multiplicities of states at any given level.

4.26. Show that in a positive (unitary) theory with $c = 0$, the Hilbert space contains only one state: the vacuum $|0\rangle$.

4.27. Show that $|\chi\rangle$ in (4.10.14) on page 68, although being a descendant, is also primary and its norm is zero.

4.28. Let $\psi(z)$ be the conformal field with dimension $1/2$ of the $c = 1/2$ CFT and $\chi(z)$ the null conformal field corresponding to the state in (4.10.14). From

$$\langle \chi(z_1) \prod_{i=2}^{N} \psi(z_i) \rangle = 0, \tag{4.9E}$$

derive a differential equation for correlators of ψ. Use it together with the SL(2) Ward identities to determine the three- and four-point functions.

4.29. Show that a conformal field of weight $(1,0)$ is necessarily primary in a positive theory.

4.30. Derive the current algebra relations (4.11.4) on page 69 from the OPE of the currents (4.11.3).

4.31. Derive the equations of motion from the action (4.11.6) on page 69. Show that they imply $\bar{\partial}J = \partial\bar{J} = 0$ when $\lambda^2 = 4\pi/k$.

4.32. Couple the WZW model to a world-sheet metric and derive the stress tensor. Show that it is traceless and that T is given by a formula similar to (4.11.7) on page 70. Can you justify the difference with (4.11.7)?

4.33. Show that OPE (4.11.10) on page 71 implies

$$[J_m^a, R_i(z, \bar{z})] = z^m (T^a)_{ij} R_j(z, \bar{z}). \tag{4.10E}$$

4.34. Show that the affine-Sugawara form of the stress tensor (4.11.7) on page 70 can be written in mode form as

$$L_n = \frac{1}{2(k + \tilde{h})} \sum_{m \in \mathbb{Z}} : J_{n+m}^a J_{-m}^a : . \tag{4.11E}$$

This implies that for a primary state $|R_i\rangle$, the following state is null:

$$|\chi_i\rangle = \left(L_{-1}\delta_{ij} - \frac{1}{(k + \tilde{h})} T_{ij}^a J_{-1}^a \right) |R_j\rangle. \tag{4.12E}$$

By further considering the correlator $\langle \chi_{i_1}(z_1) \prod_{k=2}^{N} R_{i_k}(z_k) \rangle = 0$ and using the current Ward identities (4.10E) on page 121 derive the Knizhnik-Zamolodchikov equation

$$\left(\partial_{z_i} - \frac{1}{(k + \tilde{h})} \sum_{j \neq i}^{N} \frac{T_i^a \otimes T_j^a}{z_i - z_j} \right) \left\langle \prod_{k=1}^{N} R(z_k) \right\rangle = 0, \tag{4.13E}$$

where T_i^a acts on the primary field $R(z_i)$.

4.35. The coset construction: Consider the affine-Sugawara stress-tensor T_G associated with the group G. Pick a subgroup $H \subset G$ with regular embedding and consider its associated affine-Sugawara stress-tensor T_H constructed out of the H currents, which are a subset of the G currents. Consider also $T_{G/H} = T_G - T_H$. Show that

$$T_{G/H}(z) J^H(w) = \text{regular}, \quad T_{G/H}(z) T_H(w) = \text{regular}. \tag{4.14E}$$

Show also that $T_{G/H}$ satisfies the Virasoro algebra with central charge $c_{G/H} = c_G - c_H$. The interpretation of the above construction is that, roughly speaking, the G-WZW theory can be decomposed into the H-theory and the G/H theory described by the stress-tensor $T_{G/H}$. As an application, show that if you choose $G = SU(2)_m \times SU(2)_1$ and H to be the diagonal subgroup $SU(2)_{m+1}$ then the G/H theory is that of the minimal models with central charge (4.10.13) on page 68. You may want to look at [98]. For a generalization of this construction, see [99,53].

4.36. Using the mode expansion (4.12.15) on page 73, and the commutation relations (4.12.16) and (4.12.31) on page 75 show that the two-point function of fermions in the Ramond vacuum is

$$G_R^{ij}(z,w) = \langle \hat{S} | \psi^i(z) \psi^j(w) | \hat{S} \rangle = \delta^{ij} \frac{z+w}{2\sqrt{zw}} \frac{1}{z-w}. \tag{4.15E}$$

4.37. Show that for any normalized quasiprimary operator ϕ with dimension Δ

$$\langle T(z_1) \phi(z_2) \phi(z_3) \rangle = \frac{\Delta}{z_{12}^2 z_{13}^2 z_{23}^{2\Delta - 2}}. \tag{4.16E}$$

Use it to show that for any normalized quasiprimary state $|X\rangle$ in a CFT generated out of the $SL(2,\mathbb{C})$-invariant vacuum by an operator with conformal weight Δ we have

$$\langle X | T(z) | X \rangle = \frac{\Delta}{z^2}. \tag{4.17E}$$

Now, evaluate $\langle \hat{S} | T(z) | \hat{S} \rangle$ using the propagator of the previous exercise and show that the conformal weight of the Ramond vacuum is N/16.

4.38. Write the action (4.13.15) on page 79 in components by doing the integral over the anticommuting coordinates, and show that it is equivalent to (4.13.1) on page 77.

4.39. Use (4.13.14) on page 79 for several scalar superfields to show that after imposing the equations of motion

$$D_\theta \hat{X}^\mu \, \bar{D}_{\bar\theta} \hat{X}^\nu \, e^{ip \cdot \hat{X}} \Big|_{\theta\bar\theta} = \left(\partial X^\mu + i(p \cdot \psi) \psi^\mu \right) \left(\bar\partial X^\nu + i(p \cdot \bar\psi) \bar\psi^\nu \right) e^{ip \cdot X}. \tag{4.18E}$$

4.40. Use the same procedure as that used in the $\mathcal{N} = (2,0)_2$ superconformal case to show that for unitary $\mathcal{N} = (4,0)_2$ superconformal primaries in the NS sector $\Delta - j \geq 0$, while in the R sector, $\Delta \geq k/4$.

4.41. Compute the OPE of the stress tensor (4.14.8) on page 83 with itself and verify that the central charge is given by (4.14.7). Verify the OPE (4.14.10).

4.42. Derive (4.14.11) on page 83 by considering the integration in the path integral over the zero mode of the scalar X.

4.43. Show that (4.14.17) on page 84 are the only set of Hermiticity conditions for J compatible with the commutation relations (4.14.16).

4.44. Show that the ghost stress tensor satisfies the Virasoro algebra with central charge given in (4.15.9) on page 85.

4.45. Derive the OPE of the BRST current in (3.8.12) on page 43 with itself. Use this result to calculate Q_B^2.

4.46. Use the expressions of (4.20.16) on page 107 to verify by direct computation (4.15.2), (4.15.8), and (4.15.10) on page 85.

4.47. Find the propagator for a massless scalar on $\mathcal{H}_2$ with DN boundary conditions.

4.48. Consider a massless scalar on the disk. Show that the normal ordered operator $: \partial X \bar{\partial} X :$, although finite inside the disk, has a divergence at the boundary. This can be seen by considering a mirror charge. We therefore need to introduce a modified normal ordering that we will label by $\overset{*}{_*}\partial X \bar{\partial} X\overset{*}{_*}$ to remove the divergences on the boundary. Is there any difference between $\overset{*}{_*}\partial X \bar{\partial} X\overset{*}{_*}$ and $\overset{*}{_*}\partial X \partial X\overset{*}{_*}$? Explain.

4.49. Consider an Ising (Majorana) fermion on the disk with boundary condition $\psi = \bar{\psi}$ at the boundary. Calculate the correlator

$$\left\langle \prod_{i=1}^{m} \psi(z_i) \prod_{j=1}^{2n-m} \bar{\psi}(\bar{z}_j) \right\rangle. \tag{4.19E}$$

4.50. Consider a boson on the strip with DN boundary conditions

$$\partial_\tau X^I\big|_{\sigma=0} = 0, \quad \partial_\sigma X^I\big|_{\sigma=\pi} = 0. \tag{4.20E}$$

Derive the mode expansion compatible with (4.20E) and show that there is no winding or momentum allowed and that the oscillators have half-integral frequencies.

4.51. Consider the theory of N real massless scalars X^i on the upper half plane with boundary conditions

$$\partial_z X^i - O^{ij}\partial_{\bar{z}} X^j\big|_{z=\bar{z}} = 0, \tag{4.21E}$$

where O^{ij} is a constant matrix. For which O does the boundary condition preserve conformal invariance? Find the mode expansion of the X^i and calculate their propagator.

4.52. Derive the NS fermion propagators (4.16.22)–(4.16.24) on page 89.

4.53. Derive the fermion propagators in the Ramond sector on the disk.

4.54. Calculate the one-point function of the vertex operator $V_a =: e^{iaX} :$ on RP$_2$.

4.55. Derive the $\mathcal{N} = (1,0)_2$ and $\mathcal{N} = (2,0)_2$ superconformal algebras with background charges.

4.56. Use the same procedure as in (2.1.24)–(2.1.27) on page 13 to calculate the determinant of the scalar Laplacian on the torus and derive (4.18.16) on page 95.

4.57. Consider perturbing a conformal field theory by $g \int \phi_{(d,d)}$. Calculate the β-function of the coupling g and show that

$$\beta(g) = (2 - 2d)g + Cg^2 + \mathcal{O}(g^3),\tag{4.22E}$$

where C is the three-point coupling in $\langle \phi_{(d,d)} \phi_{(d,d)} \phi_{(d,d)} \rangle$.

4.58. Consider a CFT with chiral currents J^a and $\bar{J}^{\bar{a}}$. Consider the class of $(1,1)$ perturbations

$$\delta S = \sum_{a,\bar{a}} g_{a\bar{a}} J^a \bar{J}^{\bar{a}}.\tag{4.23E}$$

What are the conditions for these interactions to be exactly marginal?

4.59. Show that a marginal perturbation cannot change the central charge.

4.60. Show that the multiscalar partition function in (4.18.40) on page 99 is modular invariant.

4.61. Verify that the bosonization relations (4.20.6) on page 106 reproduce the same OPEs as in the fermionic theory.

4.62. Use the OPEs in (4.21.13)–(4.21.15) on page 110 to deduce the following transformation rule for the twist fields under $R \to 1/R$ duality:

$$\begin{pmatrix} H^0 \\ H^\pi \end{pmatrix} \to \frac{1}{\sqrt{2}} \begin{pmatrix} 1 & 1 \\ 1 & -1 \end{pmatrix} \begin{pmatrix} H^0 \\ H^\pi \end{pmatrix}.\tag{4.24E}$$

4.63. Consider further "orbifolding" the $S^1/\mathbb{Z}_2$ orbifold theory by the $\mathbb{Z}_2$ transformation in (4.21.20) on page 111. Show that the resulting theory is the original toroidal theory. In this respect the toroidal theory is no more fundamental than the orbifold one.

4.64. Show that when $R = \ell_s/\sqrt{2}$ the orbifold partition function becomes the square of the Ising model partition function

$$Z^{\text{Ising}} = \frac{1}{2} \left[\left| \frac{\vartheta_2}{\eta} \right| + \left| \frac{\vartheta_3}{\eta} \right| + \left| \frac{\vartheta_4}{\eta} \right| \right],\tag{4.25E}$$

which was computed in (4.19.9) on page 104 with $N = 1$. You will also need (C.14) on page 508.

4.65. Consider the CFT of a two-dimensional torus, which is a direct product of two circles of radii $R_{1,2}$ and coordinates $X_{1,2}$. This theory has, among others, the $\mathbb{Z}_2$ symmetry, which acts simultaneously as $X_1 \to -X_1$ and $X_2 \to X_2 + \pi R_2$. It is a freely acting symmetry. Construct the orbifold partition function. Show that geometrically the orbifold as a manifold is the Klein bottle.

4.66. Consider the CFT of the product of two circles with equal radii. It is invariant under the interchange of the two circles. This transformation forms a $\mathbb{Z}_2$ subgroup of the rotation group O(2). Orbifold by this symmetry and construct the orbifold blocks of the partition function. Is the partition function modular invariant? You will need (C.17) on page 508.

4.67. Redo the freely acting orbifold of a free scalar, but now use the group element $g = (-1)^{m+n}$. It corresponds to a translation of X_L. Show that it is impossible to construct a modular-invariant partition function. Therefore, this is an anomalous symmetry. This is to be expected since it corresponds to a $\mathbb{Z}_2$ subgroup of the chiral $U(1)_L$ symmetry.

4.68. Consider the CFT of a complex massless free boson X. Orbifold by the $\mathbb{Z}_N$ rotation group acting as

$$X \to e^{2\pi i/N} X. \tag{4.26E}$$

Show that the ground states in the kth twisted sector have dimension

$$\Delta_k = \frac{k(N-k)}{2N^2}, \quad k = 1, 2, \ldots, N-1. \tag{4.27E}$$

4.69. Derive the two-point function for a massless scalar on the torus. From this, derive by involution, the scalar propagator, on the Klein bottle, cylinder, and Möbius strip.

4.70. Calculate the Möbius strip traces $\text{Tr}_{NS,R} \left[\Omega (-1)^F q^{L_0} \right]$ and $\text{Tr}_{NS,R} \left[\Omega q^{L_0} \right]$ for a free fermion.

5 | Scattering Amplitudes and Vertex Operators

In the previous chapter we have analyzed in detail CFTs. They will now be used to describe the dynamics of string theory. The question we will address in this chapter is how to compute scattering amplitudes of physical string states.

Consider the tree-level process of two closed strings, which come from far, interact, and eventually move off to infinity (figure 5.1a). By a conformal transformation we can map the string world-sheet to a sphere with four infinitesimal holes (punctures) (figure 5.1b). At each puncture we have to put appropriate boundary conditions that will specify which is the external physical state that participates in the interaction. In the path integral we will have to insert a "vertex operator," namely, the appropriate wave function as we do in the case of the point particle. Then, we will have to take the path-integral average of these vertex operators weighted with the Polyakov action on the sphere. In the operator language, this amplitude, an S-matrix element, will be given by a correlation function of these vertex operators in the two-dimensional world-sheet quantum theory. We will also have to integrate over the positions of these vertex operators. On the sphere there are three conformal Killing vectors, which implies that there are three reparametrizations that have not been fixed. We can fix them by fixing the positions of three vertex operators. The positions of the rest are Teichmüller moduli and should be integrated over.

What is the vertex operator associated with a given physical state? This can be found directly from the two-dimensional world-sheet theory using the state-operator correspondence. The correct vertex operator will produce the appropriate physical state as it comes close to the out vacuum.

Consider further the string diagram in figure 5.2a. This is the string generalization of a one-loop amplitude contribution to the scattering of four particles in figure 5.1a. Again by a conformal transformation it can be deformed into a torus with four punctures (figure 5.2b).

It follows that the closed string zero-, one-, and two-point amplitudes on the sphere are not well defined. This is consistent with the fact that such amplitudes do not exist on shell. The zero-point amplitude at one loop is not well defined either. When we will be talking

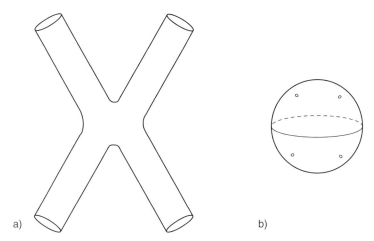

Figure 5.1 a) Tree closed-string diagram describing four-point scattering. b) Its conformal equivalent, the four-punctured sphere.

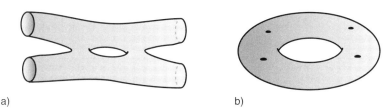

Figure 5.2 a) World-sheet relevant for the one-loop contribution to four-point scattering. b) Its conformal transform where the holes become punctures on a torus.

about the one-loop vacuum amplitude below, we will implicitly consider the one-point dilaton amplitude at zero momentum.

In a similar fashion, closed string g-loop diagrams can be calculated by integrating correlators on the CFT on a genus-g Riemann surface. One final ingredient is the string coupling g_s. A g-loop contribution has to be additionally weighted by a factor $g_s^{-\chi}$, where $\chi = 2(1 - g)$ is the Euler number of the Riemann surface. The perturbative expansion is thus a topological expansion. Notice also that the insertion of a vertex operator creates an infinitesimal hole in the Riemann surface and decreases its Euler number by 1. It is thus accompanied by a factor of $1/g_s$.

We briefly describe here the topological expansion for the case of open strings. A tree-level four-point diagram in this case is shown in figure 5.4 on page 128. By a conformal transformation it can be mapped to a disk with four points marked on the boundary (figure 5.4b). These are the positions of insertion of the appropriate vertex operators. Thus, the open string vertex operators are inserted at the boundary of the surface. Open strings can also emit closed strings. In such amplitudes, closed string emission is represented by the insertion of closed string vertex operators in the interior of the surface.

Here the topological expansion also includes Riemann surfaces with boundary. Moreover, we can consider oriented strings (where the string is oriented) as well as nonoriented

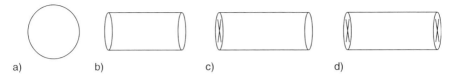

Figure 5.3 a) The disk. b) The cylinder. c) The Möbius strip. d) the Klein bottle.

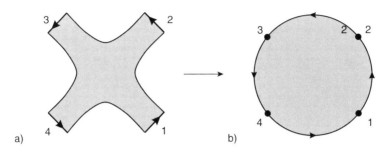

Figure 5.4 a) The four-point open-string tree amplitude. b) Its conformal transform to a disk with four points marked on the boundary.

strings (where the orientation of the string is not preserved by the interactions). In the second case we will have to include nonorientable Riemann surfaces in the topological expansion. Such surfaces are characterized by the number of handles g, the number of boundaries B, and the number of cross caps C that introduce the nonorientability of the surface. A cross cap is $S^1/\mathbb{Z}_2 = \mathrm{RP}^1$. The Euler number is given by

$$\chi = 2(1 - g) - B - C. \tag{5.0.1}$$

In figure 5.3 we show the other four simplest surfaces with boundaries and/or crosscaps: the disk with $(g, B, C) = (0, 1, 0)$, $\chi = 1$, the cylinder with $(g, B, C) = (0, 2, 0)$, $\chi = 0$, as well as two nonorientable surfaces, the Möbius strip with $(g, B, C) = (0, 1, 1)$, $\chi = 0$ and the Klein bottle with $(g, B, C) = (0, 0, 2)$, $\chi = 0$.

As we will see later on, consistent theories of open unoriented strings necessarily include couplings to closed unoriented strings. A quick way to see this is to consider the cylinder diagram (figure 5.3b). If we take time to run upward, then it describes a one-loop diagram of an open string. If, however, we take time to run sideways then it describes the tree-level propagation of a closed string.

5.1 Physical Vertex Operators

We have seen in the previous chapter that to each state in the CFT there corresponds a local operator.[1] This operator creates the corresponding state out of the $\mathrm{SL}(2,\mathbb{C})$-invariant vacuum. Therefore, to all states in string theory (on asymptotically flat space-times),

---

[1] This is true always for unitary and compact CFTs. It is also true for flat noncompact CFTs.

we have corresponding local operators on the world-sheet. However, we need only consider physical states. For this we must translate the physical state conditions on the local operators.

We will work first, for purposes of illustration, in the old covariant approach and consider the case of closed strings. We have seen that physical states had to satisfy $L_0 = \bar{L}_0 = 1$ and they should be annihilated by the positive modes of the Virasoro operators. In CFT language they must be primary fields of conformal weight $(1,1)$. Moreover, from their definition, spurious states correspond to Virasoro descendants.

The Polyakov action in the conformal gauge is

$$S_P = \frac{1}{4\pi\ell_s^2} \int d^2\xi \, \partial X^\mu \bar{\partial} X^\nu \eta_{\mu\nu}. \tag{5.1.2}$$

From it we obtain the two-point function

$$\langle X^\mu(z,\bar{z}) X^\nu(w,\bar{w}) \rangle = -\frac{\ell_s^2}{2} \eta^{\mu\nu} \log |z-w|^2, \tag{5.1.3}$$

the stress-tensor

$$T = -\frac{1}{\ell_s^2} \eta_{\mu\nu} : \partial X^\mu \partial X^\nu :, \tag{5.1.4}$$

and similarly for $\bar{T}$.

The states at level zero $|p\rangle$ correspond to the operators $V_p =: e^{ip^\mu X^\mu}:$. They are primary with conformal weights $\Delta = \bar{\Delta} = \ell_s^2 p^2/4$. In order for them to have dimension $(1,1)$, and be therefore physical, we need $p^2 = -m^2 = 4/\ell_s^2$. This is the tachyon mass-shell condition.

The next set of states $\alpha_{-1}^\mu \bar{\alpha}_{-1}^\nu |p\rangle$ correspond to the operators $: \partial X^\mu \bar{\partial} X^\nu V_p :$. We consider the linear combination $O(\epsilon) = \epsilon_{\mu\nu} : \partial X^\mu \bar{\partial} X^\nu V_p :$ of such operators and compute the OPE with $T, \bar{T}$:

$$T(z)O(w,\bar{w}) = -ip^\mu \epsilon_{\mu\nu} \frac{\ell_s^2}{4} \frac{\bar{\partial} x^\nu V_p}{(z-w)^3} + \left(1 + \frac{\ell_s^2 p^2}{4}\right) \frac{O(w,\bar{w})}{(z-w)^2} + \frac{\partial_w O(w,\bar{w})}{z-w} + \cdots, \tag{5.1.5}$$

and a similar one for $\bar{T}$. In order for O to be a primary $(1,1)$ operator the third-order pole must vanish

$$p^\mu \epsilon_{\mu\nu} = p^\nu \epsilon_{\mu\nu} = 0 \tag{5.1.6}$$

and $p^2 = 0$. These are the mass-shell and transversality conditions for the graviton (ϵ symmetric and traceless), the antisymmetric tensor (ϵ antisymmetric), and the dilaton ($\epsilon_{\mu\nu} \sim \eta_{\mu\nu} - p_\mu \bar{p}_\nu - \bar{p}_\mu p_\nu$ with $\bar{p}^2 = 0$, $p \cdot \bar{p} = 1$). Higher levels work in a similar fashion.

In modern (BRST) covariant quantization the physical state condition translates into $[Q_{\text{BRST}}, V_{\text{phys}}(z,\bar{z})] = 0$, which reduces to the usual conditions on physical states studied earlier. In this context, the physical vertex operators are the ones we found in the old covariant case multiplied by the ghosts $c(z)\bar{c}(\bar{z})$.

N-point scattering amplitudes (S-matrix elements) on the sphere are constructed by calculating the appropriate N-point correlator of the associated vertex operators and integrating it over the positions of the insertions. As we mentioned before, there is a residual

SL(2,$\mathbb{C}$) invariance on the sphere that was not fixed by going to the conformal gauge. This can be used to set the positions of three vertex operators to three fixed points taken conventionally to be $0, 1, \infty$. We also need to insert three c-ghosts at the position of the three vertex operators to soak up the three fermionic zero modes due to the three conformal Killing vectors of the sphere. They play the role of a Faddeev-Popov determinant for the SL(2,$\mathbb{C}$) Killing symmetry. Then, we integrate over the remaining $N - 3$ positions. We will calculate such amplitudes in the next section.

Moving to one-loop diagrams, we have a similar prescription. For an N-point one-loop amplitude we first have to calculate the N-point function of the appropriate vertex operators on the torus. Due to the translational symmetry of the torus ($c, \bar{c}$ zero modes) the correlator depends on $N - 1$ positions as well as on the torus modulus τ. These are Teichmüller moduli, and they should be integrated over. Diffeomorphism invariance implies that the correlator integrated over the $N - 1$ positions should be invariant under modular transformations. Finally we have to integrate over τ in the fundamental domain (Fig. 4.8).

5.2 Calculation of Tree-level Tachyon Amplitudes

5.2.1 The closed string

We will calculate explicitly here the N-point tachyon amplitude in the closed bosonic string. According to our previous discussion, the formula is

$$S_{\text{N,tree}}(p_i) = \frac{8\pi i}{\ell_s^2} g_s^{N-2} \left\langle \prod_{i=1}^{3} : c(z_i) \bar{c}(\bar{z}_i) e^{ip_i \cdot X} : \int \prod_{j=4}^{N} d^2 z_j : e^{ip_j \cdot X} : \right\rangle. \tag{5.2.1}$$

Using the scalar propagator on the sphere (4.7.2) on page 61 and the ghost correlator[2]

$$\langle c(z_1)\bar{c}(\bar{z}_1)c(z_2)\bar{c}(\bar{z}_2)c(z_3)\bar{c}(\bar{z}_3)\rangle = |z_{12}|^2 |z_{13}|^2 |z_{23}|^2, \tag{5.2.2}$$

we obtain

$$S_{\text{N,tree}}(p_i) = \frac{8\pi i}{\ell_s^2} g_s^{N-2} (2\pi)^{26} \delta^{(26)} \left(\sum_i p_i \right) |z_{12}|^2 |z_{13}|^2 |z_{23}|^2 \int \prod_{n=4}^{N} d^2 z_n \prod_{i<j=1}^{N} |z_{ij}|^{\ell_s^2 p_i \cdot p_j}. \tag{5.2.3}$$

The overall normalization used here can be justified by unitarity. Factorizing the four-point amplitude into two three-point amplitudes and an intermediate propagator with canonical residue fixes the normalization.

We should comment on the zero-, one-, and two-point amplitudes on the sphere. They all vanish. The reason is the suppression due to the volume of the sphere symmetry PSL(2,$\mathbb{C}$). In BRST quantization this is because we need at least three c-ghosts to have a nonzero correlator on the sphere. This is obvious from (4.15.15) on page 85.

---

[2] This is obtained from $c(z) = \sum_{n \in \mathbb{Z}} (c_n / z^{n-1})$ and $\langle 0 | c_{-1} c_0 c_1 | 0 \rangle = 1$.

We will evaluate the four-point S-matrix amplitude known also as the Shapiro-Virasoro amplitude. We set $N = 4$ into (5.2.3), use the standard Mandelstam invariants

$$s = -(p_1 + p_2)^2, \quad t = -(p_2 + p_3)^2, \quad u = -(p_1 + p_3)^2, \tag{5.2.4}$$

with $s + t + u = -\frac{16}{\ell_s^2}$, and by an SL(2,$\mathbb{C}$) transformation bring $z_{1,2,3}$ into $0, 1, \infty$ to obtain

$$
\begin{aligned}
S_{4,\text{tree}}(s,t,u) &= \frac{8\pi i}{\ell_s^2} g_s^2 (2\pi)^{26} \delta^{(26)} \left(\sum_i p_i \right) \int d^2 z \, |z|^{\ell_s^2 t/2} \, |1-z|^{\ell_s^2 u/2} \\
&= \frac{8\pi i}{\ell_s^2} g_s^2 (2\pi)^{27} \delta^{(26)} \left(\sum_i p_i \right) \frac{\Gamma\left(-1 - \ell_s^2 \frac{t}{4}\right) \Gamma\left(-1 - \ell_s^2 \frac{u}{4}\right) \Gamma\left(-1 - \ell_s^2 \frac{s}{4}\right)}{\Gamma\left(2 + \ell_s^2 \frac{t}{4}\right) \Gamma\left(2 + \ell_s^2 \frac{u}{4}\right) \Gamma\left(2 + \ell_s^2 \frac{s}{4}\right)}.
\end{aligned}
\tag{5.2.5}
$$

This amplitude has Regge behavior in the large-s, fixed-t limit

$$S_{4,\text{tree}}(s,t,u) \sim s^{2 + \ell_s^2 t/2} \frac{\Gamma\left(-1 - \ell_s^2 \frac{t}{4}\right)}{\Gamma\left(2 + \ell_s^2 \frac{t}{4}\right)}, \tag{5.2.6}$$

while it is exponentially soft for s, t, u all large

$$S_{4,\text{tree}}(s,t,u) \sim e^{-(\ell_s^2/2)(s \log s + t \log t + u \log u)}. \tag{5.2.7}$$

This is a general characteristic of string interactions at high energy.

5.2.2 *The open string*

For the open string, the surface giving rise to tree diagrams is conformally equivalent to the disk D_2 (or its conformal equivalent, the upper half plane $\mathcal{H}_2$). If the strings are unoriented then there is also the projective plane RP_2.

The next important ingredients are the Green's functions on $\mathcal{H}_2$. These were obtained in section 4.16 on page 86. For the NN case it is

$$\langle X^\mu(z,\bar{z}) X^\nu(w,\bar{w}) \rangle_{D_2} = -\frac{\ell_s^2}{2} \eta^{\mu\nu} \left(\log |z - w|^2 + \log |z - \bar{w}|^2 \right). \tag{5.2.8}$$

We can calculate the expectation values of vertex operators, some of them inserted inside the disk (closed string vertex operators) and some on the boundary (open string vertex operators) using the Gaussian formula (4.7.9) on page 62:

$$
\left\langle \prod_{i=1}^m : e^{ip_i \cdot X(z_i,\bar{z}_i)} : \prod_{I=1}^n : e^{iq_I \cdot X(w_I)} : \right\rangle_{D_2} = (2\pi)^{26} \delta^{(26)} \left(\sum p_i + q_I \right)
$$
$$
\times \prod_{i<j}^m |(z_i - z_j)(z_i - \bar{z}_j)|^{\ell_s^2 p_i \cdot p_j} \prod_{I<J}^n |w_I - w_J|^{2\ell_s^2 q_I \cdot q_J} \prod_{I,i} |w_I - z_i|^{\ell_s^2 q_I \cdot p_i} |w_I - \bar{z}_i|^{\ell_s^2 q_I \cdot p_i}, \tag{5.2.9}
$$

where w_I are coordinates on the boundary (the real line).

For the ghosts inserted in the bulk of D_2 we obtain

$$\langle c(z_1)\bar{c}(\bar{z}_1) c(z_2)\bar{c}(\bar{z}_2) c(z_3)\bar{c}(\bar{z}_3) \rangle_{D_2} = |z_{12}(z_1 - \bar{z}_3)(z_2 - \bar{z}_3)|^2. \tag{5.2.10}$$

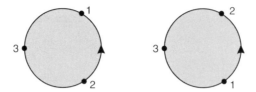

Figure 5.5 The two possible orderings of the three-point open-string tree amplitude.

In the projective plane, the correlators can also be computed using the image method as done in section (4.16.3) on page 90, based on the involution $z' = -\frac{1}{\bar{z}}$. From the scalar propagator,

$$\langle X^\mu(z, \bar{z}) X^\nu(w, \bar{w})\rangle_{\mathrm{RP}_2} = -\frac{\ell_s^2}{2} \eta^{\mu\nu} \left(\log|z - w|^2 + \log|1 + z\bar{w}|^2\right), \tag{5.2.11}$$

we obtain

$$\left\langle \prod_{i=1}^m : e^{ip_i \cdot X(z_i, \bar{z}_i)} : \right\rangle_{\mathrm{RP}_2} = (2\pi)^{26} \delta^{(26)}\left(\sum_i p_i\right) \prod_{i=1}^m |1 + z_i\bar{z}_i|^{\ell_s^2 p_i \cdot p_i/2}$$

$$\times \prod_{\substack{i<j}}^m |(z_i - z_j)(1 + z_i\bar{z}_j)|^{\ell_s^2 p_i \cdot p_j}. \tag{5.2.12}$$

Finally for the ghosts inserted in RP_2 we obtain

$$\langle c(z_1)\bar{c}(\bar{z}_1)c(z_2)\bar{c}(\bar{z}_2)c(z_3)\bar{c}(\bar{z}_3)\rangle_{\mathrm{RP}_2} = |z_{12}(1 + z_1\bar{z}_3)(1 + z_2\bar{z}_3)|^2. \tag{5.2.13}$$

We are now in a position to calculate the scattering of open tachyons and derive the Veneziano amplitude.

$$S_{N,tree}^{D_2} = \frac{2ig_s^{N-2}}{\ell_s^2} \left\langle \prod_{i=1}^3 : c(w_i)e^{ip_i \cdot X} : \prod_{j=4}^N \int dw_j : e^{ip_j \cdot X} : \right\rangle$$

$$= \frac{2ig_s^{N-2}}{\ell_s^2} (2\pi)^{26} \delta^{(26)}\left(\sum_i p_i\right) \int \prod_{n=4}^N dw_n \, |w_{12}w_{13}w_{23}| \prod_{\substack{i<j=1}}^N |w_{ij}|^{2\ell_s^2 p_i \cdot p_j}, \tag{5.2.14}$$

where the integrals are over the real line. Using a PSL(2,$\mathbb{R}$) transformation, we can fix the three points w_1, w_2, w_3 again to $0, 1, \infty$. However here we cannot change the cyclic ordering of the points using PSL(2,$\mathbb{R}$), so we must add the two independent orderings shown in figure 5.5.

For the four-point amplitude we obtain

$$S_{4,tree}^{D_2} = \frac{2ig_s^2}{\ell_s^2}(2\pi)^{26}\delta^{(26)}\left(\sum_i p_i\right) |w_{12}w_{13}w_{23}| \int_{-\infty}^\infty dw_4 \prod_{\substack{i<j=1}}^4 |w_{ij}|^{2\ell_s^2 p_i \cdot p_j} + (p_2 \leftrightarrow p_3)$$

$$= \frac{2ig_s^2}{\ell_s^2}(2\pi)^{26}\delta^{(26)}\left(\sum_i p_i\right)\left[\int_{-\infty}^\infty dw \, |w|^{-\ell_s^2 u - 2}|1 - w|^{-\ell_s^2 t - 2} + (t \leftrightarrow s)\right], \tag{5.2.15}$$

where in the second line we introduced the standard Mandelstam variables, which now satisfy $s + t + u = -\frac{4}{\ell_s^2}$. We may now split the range of integration to the three intervals

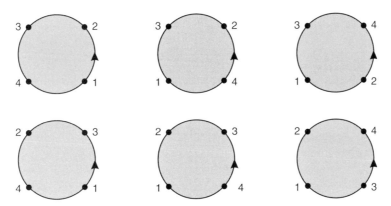

Figure 5.6 The six possible orderings of the four-point open-string tree amplitude.

$(-\infty, 0]$, $[0, 1]$ and $[1, \infty)$. This splitting together with the $(t \leftrightarrow s)$ gives six contributions that correspond to all distinct cyclic orderings of the four vertex operators as shown in figure 5.6.

We obtain

$$S_{4,\text{tree}}^{D_2} = \frac{4 i g_s^2}{\ell_s^2} (2\pi)^{26} \delta^{(26)} \left(\sum_i p_i \right) [A(s,t) + A(t,u) + A(s,u)], \qquad (5.2.16)$$

with

$$A(s,t) = \int_0^1 dw \, w^{-\ell_s^2 s - 2} (1 - w)^{-\ell_s^2 t - 2} = \frac{\Gamma\left(-\ell_s^2 s - 1\right) \Gamma\left(-\ell_s^2 t - 1\right)}{\Gamma\left(\ell_s^2 u + 2\right)}. \qquad (5.2.17)$$

This is the Veneziano amplitude whose behavior we have analyzed in section 1.3 on page 6.

5.3 The One-loop Vacuum Amplitudes

We will study, in some detail, the simplest quantum corrections in string theory, the vacuum one-loop amplitudes from both closed and open strings, both oriented and unoriented. The interpretation of such an amplitude in flat space is that it represents a correction to the vacuum energy. If this is nonzero (and it is in bosonic string theory) then the original flat vacuum is unstable.

We know already that the bosonic string theory vacuum is unstable due to the presence of the tachyon. However, as we will see in later sections, in the superstring the tachyon is absent. There, a nonvanishing one-loop vacuum energy will destabilize the flat vacuum. This is a facet of the cosmological constant problem which is as serious in string theory, as it is in field theory.

The vacuum energy in D dimensions is given as usual by

$$\Lambda = \underbrace{\sum \frac{\hbar \omega_{\vec{p}}}{2}}_{\text{bosons}} - \underbrace{\sum \frac{\hbar \omega_{\vec{p}}}{2}}_{\text{fermions}} = -\frac{i}{2} V_D \int \frac{d^D p}{(2\pi)^D} \sum_i (-1)^{F_i} \log (p^2 + m_i^2), \qquad (5.3.1)$$

where we have kept the $\hbar$ only on the left to indicate the one-loop origin of this vacuum energy. m_i are the space-time masses of fields and F_i stands for the fermion number of the ith particle. We have also added an i since the momentum integral is written over Euclidean space. Upon Wick rotation this extra i makes the vacuum energy real in Minkowski space.

Using the formal identity

$$\log S = -\lim_{\epsilon \to 0} \frac{d}{d\epsilon} S^{-\epsilon} = -\lim_{\epsilon \to 0} \frac{d}{d\epsilon} \left[\epsilon \int \frac{dt}{t^{1-\epsilon}} e^{-2\pi t S} \right] = -\int \frac{dt}{t} e^{-2\pi t S}, \qquad (5.3.2)$$

we may rewrite (5.3.1) as

$$\begin{aligned}
\Lambda &= \frac{i}{2} V_D \sum_i \int \frac{d^D p}{(2\pi)^D} \int_0^\infty \frac{dt}{t} (-1)^{F_i} e^{-2\pi t (p^2 + m_i^2)} \\
&= \frac{i}{2} \frac{V_D}{(4\pi)^D} \int_0^\infty \frac{dt}{t^{1+D/2}} \mathrm{Tr}' \left[(-1)^F e^{-2\pi t \, m^2} \right],
\end{aligned} \qquad (5.3.3)$$

where in the second equality we have performed the momentum integral and Tr' stands for the trace in the nonzero mode sector. We may now proceed in the evaluation of the vacuum energy of the bosonic string theories.

5.3.1 The torus

For closed strings, the one-loop vacuum amplitude comes from the torus. We will calculate it in two distinct ways: by doing the string theory path integral over the torus and by summing up the field theory contributions of all the string modes. The comparison of the two approaches is instructive for what concerns the absence of UV divergences in string theory.

We will first calculate the one-loop vacuum energy of the closed bosonic string by evaluating the one-loop string (torus) diagram. This corresponds to calculating the torus partition function of the underlying CFT and integrating it over the torus moduli space. We will do this computation in the covariant approach. We have seen that the torus partition function for a single noncompact boson is given by $V/(2\pi \ell_s \sqrt{\tau_2} \eta \bar{\eta})$, and we have 26 of these. The b, c ghosts contribute η^2 to the partition function and cancel the contribution of two left-moving oscillators. Similarly the $\bar{b}, \bar{c}$ ghosts contribute $\bar{\eta}^2$. Finally, the integration measure contains an integral over τ_1, which imposes $L_0 = \bar{L}_0$ and integral over the Schwinger parameter τ_2.

There are three extra subtleties. The first is that there is a discrete symmetry, namely, $\sigma_i \to -\sigma_i$ that is not fixed by the standard gauge fixing. To avoid double counting, we must divide by a factor of 2. The second is that on the torus we have two translational Killing symmetries that must be fixed. In the presence of vertex operator insertions this can be done very easily by fixing the position of one of them. In our case, since we have no vertex operators, we can directly divide by the volume of the symmetry $\int d^2 z = \tau_2$. Finally there is an overall factor of i, coming from the Wick rotation to Euclidean space.

Putting everything together, we obtain

$$\Lambda_{\text{torus}} = iV_{26} \int_{\mathcal{F}} \frac{d\tau_1 d\tau_2}{2\tau_2} \frac{\eta^2 \bar\eta^2}{((2\pi \ell_s)\sqrt{\tau_2}\, \eta\, \bar\eta)^{26}} = \frac{iV_{26}}{4(2\pi \ell_s)^{26}} \int_{\mathcal{F}} \frac{d^2\tau}{\tau_2^2} \frac{1}{\tau_2^{12}|\eta|^{48}}, \tag{5.3.4}$$

where $\mathcal{F}$ is the fundamental domain of the moduli space of the torus in figure 4.8 page 93 and according to our conventions $d^2\tau = 2d\tau_1 d\tau_2$.

It is only the transverse oscillations of the string that contribute to the vacuum amplitude.

We will now redo the same calculation by considering the string as a collection of an infinite number of field theory modes and use the one-loop formula (5.3.3). For closed strings (2.3.43) on page 25 implies that for the string states

$$p^2 + m^2 = \frac{2}{\ell_s^2}(L_0 + \bar L_0), \tag{5.3.5}$$

and L_0 is the Hamiltonian on the strip, related to that of the sphere as usual by $L_0^{\text{strip}} = L_0^{\text{sphere}} - \frac{c}{24}$. Moreover, we may impose the constraint $L_0 = \bar L_0$ by doing an unrestricted trace but including

$$\delta(L_0 - \bar L_0) = \int_{-\infty}^{\infty} d\theta\, e^{2\pi i\theta(L_0 - \bar L_0)}. \tag{5.3.6}$$

We therefore obtain

$$\Lambda_B = \frac{iV_{26}}{2} \int \frac{d^{26}p}{(2\pi)^{26}} \int_0^\infty \frac{dt}{t} \int_{-\infty}^\infty d\theta\, \text{Tr}' \left[e^{(-4\pi t/\ell_s^2)(L_0 + \bar L_0)} e^{2\pi i\theta(L_0 - \bar L_0)} \right]. \tag{5.3.7}$$

By redefining $\tau = \tau_1 + i\tau_2$, $2t = \ell_s^2 \tau_2$, $\theta = \tau_1$, we observe that the complex variable τ takes values in the upper half plane $\mathcal{H}_2$. We may therefore rewrite

$$\Lambda_B = \frac{iV_{26}}{4} \int \frac{d^{26}p}{(2\pi)^{26}} \int_{\mathcal{H}_2} \frac{d^2\tau}{\tau_2} \text{Tr}' \left[e^{2\pi i\tau L_0 - 2\pi i\bar\tau \bar L_0} \right], \tag{5.3.8}$$

where the trace is over transverse physical states. Integrating explicitly over the 26-dimensional momentum we obtain

$$\Lambda_B = \frac{iV_{26}}{4(2\pi \ell_s)^{26}} \int_{\mathcal{H}_2} \frac{d^2\tau}{\tau_2^2} \frac{Z_B(\tau, \bar\tau)}{\tau_2^{12}} = \frac{iV_{26}}{4(2\pi \ell_s)^{26}} \int_{\mathcal{H}_2} \frac{d^2\tau}{\tau_2^2} \frac{1}{\tau_2^{12}|\eta(\tau)|^{48}}, \tag{5.3.9}$$

where $Z_B(\tau, \bar\tau) = \eta^{-24}\bar\eta^{-24}$ is the light-cone (mass-generating) partition function of the string.

We have obtained the same result as in (5.3.4) with one important difference. The t-integral now reaches the $t = 0$ boundary, and this is where UV divergences come from. In the first calculation, the symmetries of the theory, namely, the SL(2,$\mathbb{Z}$) symmetry acting on τ, restrict the integration domain to the fundamental domain $\mathcal{F}$. The UV region is therefore absent.

We may therefore view the string result as an interesting (and consistent) way of cutting off the UV divergences of the theory by imposing a cutoff at about the string scale M_s.

Of course, in the case of bosonic strings the vacuum energy is IR divergent due to the presence of the tachyon.

5.3.2 The cylinder

Our next step is to calculate the cylinder contribution to the vacuum amplitude due to open strings.

We will consider for simplicity open strings with NN boundary conditions on all coordinates (therefore, free open strings in 26 dimensions), and N CP factors on their end points.

The amplitude can be written in the Hamiltonian formulation as

$$\Lambda_{C_2} = \int_0^\infty \frac{dt}{2t} \, \mathrm{Tr}_{\mathrm{open}} \left[e^{-2\pi t \, (L_0 - c/24)} \right], \tag{5.3.10}$$

where the trace is in the open string Hilbert space, and L_0 is the Hamiltonian on the sphere, related to that of the strip as usual by $L_0^{\mathrm{strip}} = L_0^{\mathrm{sphere}} - \frac{c}{24}$.

As calculated in (4.22.10) on page 114 each coordinate provides a factor of $V/\sqrt{8\pi^2 \ell_s^2 t} \, \eta(it)$ and the b, c ghosts a factor of $\eta^2(it)$. All three subtleties that we discussed for the torus occur here as well and justify the overall factors in front of (5.3.10). Putting everything together we obtain

$$\Lambda_{C_2} = i N^2 V_{26} \int_0^\infty \frac{dt}{2t} \frac{1}{(8\pi^2 \ell_s^2 t)^{13} \eta(it)^{24}}, \tag{5.3.11}$$

where N^2 is the multiplicity of the CP ground states.

The behavior of the integrand at $t \to \infty$ (the IR region) is dominated by the light open string modes.[3] In the bosonic theory we are considering, the integrand blows up exponentially due to the open string tachyon. However, in the supersymmetric theories without tachyons, there will be no divergence in this region.

The other dangerous region is $t \to 0$, where UV divergences originate. This region was cut off in the case of the closed strings due to modular invariance. This is not the case for open strings. This region will cause divergences. Although from the open string point of view these are UV divergences, from the closed string point of view they are IR divergences, as we now show.

t is the radius of the circle in C_2. Instead of taking this radius to be small, we may do a conformal transformation so that instead we take the length of the cylinder to become large (as $1/t$). This diagram can also be interpreted as a closed string created from the vacuum, propagating for time π/t and then disappearing again.

To see this explicitly, we introduce the dual closed string time variable

$$\ell = \frac{\pi}{t}, \tag{5.3.12}$$

---

[3] This representation of the amplitude in terms of the t variable is known as the open string or "direct channel."

and use the modular transformation of the η-function,

$$\eta(it) = \frac{1}{\sqrt{t}} \, \eta\left(i\frac{\ell}{\pi}\right), \tag{5.3.13}$$

to write[4]

$$\Lambda_{C_2} = \frac{iN^2 V_{26}}{2\pi (8\pi^2 \ell_s^2)^{13}} \int_0^\infty d\ell \, \eta\left(i\frac{\ell}{\pi}\right)^{-24}. \tag{5.3.14}$$

We expand the η-functions for large ℓ,

$$\eta\left(i\frac{\ell}{\pi}\right)^{-24} = e^{2\ell} + 24 + \mathcal{O}\left(e^{-2\ell}\right). \tag{5.3.15}$$

The first term is again due to the closed string tachyon and its divergence is not interesting here (and will be absent in better string theories later). However, the massless contribution also produces a divergence,

$$\mathcal{T}_{C_2} = 24 \frac{iN^2 V_{26}}{2\pi (8\pi^2 \ell_s^2)^{13}} \int_0^\infty d\ell. \tag{5.3.16}$$

This is an IR divergence from the point of view of the closed string. This divergence can be understood as follows: The cylinder amplitude in the $\ell \to \infty$ limit can be factorized into two disk one-point functions and an intermediate closed string propagator as shown in figure 5.7.

As described in exercise 5.8 on page 142, the massless states of the closed string have nonzero one-point functions on the disk at zero momentum. Such nonzero one-point functions are known as tadpoles and exist whenever the vacuum around which we are expanding does not solve the equations of motion.

In our case, there is a source in the vacuum that couples to the massless fields. Such sources will be later related to D-branes and O-planes. We will return to them soon. Here, there is a term in the effective action of the form

$$\delta S \sim \int d^{26}x \sqrt{G} e^{-\phi} \tag{5.3.17}$$

which provides nonzero one-point functions for the metric and the dilaton. Because of this term, flat space is no longer a solution to the equations of motion.

Typically, when such tadpoles are present in a perturbative expansion, perturbation theory is ill defined. The reason is that higher diagrams factorize into nonzero tadpoles connected with on-shell propagators that give infinite contributions.

This is exactly what is happening here. The intermediate on-shell propagator of massless states $1/p^2$ is responsible for the divergence we are finding.

To avoid such divergences we must expand around a vacuum which solves the equations of motion. In the case at hand, with oriented closed plus open bosonic strings, such a vacuum will necessarily be non-Poincaré-invariant, and therefore difficult to deal with.

---

[4] The cylinder amplitude, written in terms of the closed string "time" ℓ, is referred to as the closed-string or "transverse" channel.

Figure 5.7 The UV factorization of the cylinder amplitude into tree-level closed-string vertices and propagators.

5.3.3 The Klein bottle

Let us consider now unoriented closed strings and compute the extra contribution to the vacuum amplitude due to the Klein bottle. This is given as

$$\Lambda_{K_2} = \int_0^\infty \frac{dt}{4t} \operatorname{Tr}\left[\Omega \, e^{-2\pi t(L_0 + \bar{L}_0 - c/12)}\right], \tag{5.3.18}$$

where there is an extra factor of $1/2$ coming from the projection $(1 + \Omega)/2$. Using the scalar contributions computed in (4.22.20) on page 116 we obtain

$$\Lambda_{K_2} = i\frac{V_{26}}{(2\pi \ell_s)^{26}} \int_0^\infty \frac{dt}{4t} \frac{1}{t^{13}\eta(2it)^{24}}. \tag{5.3.19}$$

We can rewrite it in terms of the closed string time variable. For the Klein bottle it is

$$\ell = \frac{\pi}{2t}. \tag{5.3.20}$$

This can be seen starting from the parallelogram representing the Klein bottle described in section 4.22 on page 112, exchange the axes, and rescale to the standard normalization. Using (5.3.13) the result is

$$\Lambda_{K_2} = i\frac{2^{26} V_{26}}{4\pi (8\pi^2 \ell_s^2)^{13}} \int_0^\infty d\ell \, \eta\left(i\frac{\ell}{\pi}\right)^{-24}. \tag{5.3.21}$$

Again we have here a similar divergence due to tree-level massless tadpoles,

$$T_{K_2} = 24i\frac{2^{26} V_{26}}{4\pi (8\pi^2 \ell_s^2)^{13}} \int_0^\infty d\ell. \tag{5.3.22}$$

The tadpole originates in the one-point functions of the massless states on RP_2, as shown in figure 5.8.

5.3.4 The Möbius strip

In the case of unoriented open strings, the one-loop diagram that implements the orientation projection is the Möbius strip.

This contribution can be written in the Hamiltonian language as

$$\Lambda_{M_2} = \int_0^\infty \frac{dt}{4t} \operatorname{Tr}_{\text{open}}\left[\Omega \, e^{-2\pi t \, (L_0 - c/24)}\right], \tag{5.3.23}$$

where the trace is in the open string Hilbert space, and L_0 is the open string Hamiltonian

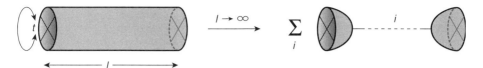

Figure 5.8 The UV factorization of the Klein bottle amplitude into tree-level closed-string vertices and propagators.

on the sphere. As in the Klein bottle case, there is an extra factor of $1/2$ included in (5.3.23) coming from the projection $(1 + \Omega)/2$ in the open sector.

To evaluate (5.3.23) for the bosonic string we must use the result (4.22.17) on page 115 according to which a boson contributes (up to CP factors) $(\sqrt{8\pi^2 \ell_s^2 t}\ \vartheta_3(2it)\eta(2it))^{-1}$, and the b, c ghosts $\vartheta_3(2it)\eta(2it)$.

An extra important ingredient is the action of orientation-reversal Ω on the N CP factors at the end of the string. This was described in (3.4.8). We obtain for the CP trace

$$\sum_{ij}\langle i,j|\Omega|i,j\rangle = \sum_{ij}\langle i,j|j',i'\rangle(\gamma_\Omega)_{ii'}(\gamma_\Omega^{-1})_{j'j} = \mathrm{Tr}[\gamma_\Omega^T \gamma_\Omega^{-1}]. \tag{5.3.24}$$

According to the discussion in section 3.4.1 on page 34, γ_Ω is either symmetric or antisymmetric. We therefore obtain ζN for the trace. Putting everything together we finally obtain

$$\Lambda_{M_2} = i\zeta N V_{26} \int_0^\infty \frac{dt}{4t} \frac{1}{(8\pi^2 \ell_s^2 t)^{13}\vartheta_3(2it)^{12}\eta(2it)^{12}}. \tag{5.3.25}$$

As in the case of the Klein bottle, the Möbius strip can be represented as a cylinder with a boundary on one side and a crosscap on the other side (figure 5.3 on page 128). The length of the cylinder is in this case

$$\ell = \frac{\pi}{4t}. \tag{5.3.26}$$

Performing a modular transformation on the amplitude (5.3.25) we may write it in the transverse channel as

$$\Lambda_{M_2} = 2i\zeta N \frac{2^{13} V_{26}}{4\pi (8\pi^2 \ell_s^2)^{13}} \int_0^\infty d\ell\ \vartheta_3\left(2i\frac{\ell}{\pi}\right)^{-12} \eta\left(2i\frac{\ell}{\pi}\right)^{-12}. \tag{5.3.27}$$

The tadpole here is

$$\mathcal{T}_{M_2} = -48 i\zeta N \frac{2^{13} V_{26}}{4\pi (8\pi^2 \ell_s^2)^{13}} \int_0^\infty d\ell. \tag{5.3.28}$$

It is due to the propagation of closed-string massless states between a disk and an crosscap tadpole as indicated in figure 5.9.

5.3.5 Tadpole cancellation

For the unoriented open and closed bosonic string the total massless tadpole is

$$\mathcal{T} = \mathcal{T}_{K_2} + \mathcal{T}_{C_2} + \mathcal{T}_{M_2} = i\frac{12 V_{26}}{2\pi (8\pi^2 \ell_s^2)^{13}} \left(2^{13} - \zeta N\right)^2 \int_0^\infty d\ell. \tag{5.3.29}$$

Thus, for the choice $N = 2^{13}$ and $\zeta = 1$ (SO(N) gauge group) there is no tadpole and

Figure 5.9 The UV factorization of the Möbius amplitude into tree-level closed-string vertices and propagators.

the vacuum solves the equations of motion. This is the simplest example of a general consistency condition for unoriented open-string vacua that we will meet again later-on.

One remark is in order here. It may seem that the tadpole cancellation condition as described above is ambiguous, since by rescaling closed-string time variables ℓ differently on different surfaces, it can be changed. This is another reflection of the fact that (5.3.29) is an equation for infinite quantities.

Typically in such a case, we must regularize, cancel the tadpole, and then remove the cutoff. It is not known how to do this in unoriented string theory. However, the definition of ℓ on the three surfaces (5.3.12), (5.3.20), and (5.3.26) is specified by the two-dimensional geometry, and this definition guarantees that the geometric structure of string perturbation theory is preserved. According to our discussion in section 4.23 on page 116, this will guarantee the correct factorization of string amplitudes.

5.3.6 UV structure and UV-IR correspondence

We have calculated the various one-loop vacuum amplitudes for open and closed bosonic strings. They are the simplest diagrams where divergences appear because of quantum effects. It is therefore interesting to discuss the mechanism by which UV divergences are absent.

The oriented closed-string amplitude is UV finite because the UV regime $\tau_2 \to 0$ is missing from the integration region. Modular invariance is responsible for this. The fact that this is a geometric symmetry guarantees that similar higher-loop amplitudes have the same property. They should degenerate properly to preserve unitarity. On the other hand, the invariance of the closed-string contributions under the modular transformation $\tau \to -\frac{1}{\tau}$ indicates that the would be UV behavior is similar to the IR one ($\tau_2 \to \infty$).

Consequently, there are two ways of interpreting UV finiteness of the closed string. The first is that of theory with a smart geometric cutoff (imposed effectively by modular invariance) of the order of the string scale,[5] M_s. This is obvious already at one loop.

Another interpretation is that modular invariance turned potential UV divergences into IR divergences. Unlike UV divergences, IR divergences are interpreted in terms of the presence of massless particles and we know how to handle them. Thus, closed strings have (physical) IR divergences only. We should also add that modular invariance is also

---

[5] The string scale cutoff is much smaller than the Planck scale at weak string coupling. This should be related to the fact that string perturbation theory breaks down at energies larger than the string scale and close to the Planck scale.

responsible for the absence of space-time anomalies in closed string theories. This will be discussed in more detail in section 7.9 on page 176. It suffices to say here that anomalies exist only if there are UV divergences in a theory. A finite theory does not have anomalies.

Once we pass to oriented open strings, the one-loop amplitude on the cylinder has potential UV divergences as $t \to 0$, as well as the expected IR divergences as $t \to \infty$. Here, there is no modular invariance to protect the $t \to 0$ region.

However, the analog of the modular transformation, $t \to \frac{1}{t}$ reinterprets the amplitude in terms of a tree-level amplitude of closed strings. Moreover, now the original UV divergences of the open string amplitude are interpreted as IR divergences due to the exchange of massless closed-string modes. Such divergences are, as usual, subtracted from Wilsonian effective action.

Again, UV divergences are mapped to IR divergences, but in a different way from the closed string. Moreover, for this to successfully work at each order of perturbation theory, the whole structure of open Riemann surfaces comes into play. This correspondence of open string UV divergences with closed-string IR divergences is at the heart of gauge theory/gravity correspondence that we will describe in chapter 13 on page 403.

The analog of the constraint of modular invariance in the closed sector is in a sense tadpole cancellation in the open sector. This removes UV divergences that destabilize the perturbation theory. Such divergences, although they have an IR interpretation, cannot be subtracted in a way that respects the perturbative expansion. Some of them, could be subtracted by a thorough reorganization of perturbation theory, by changing the vacuum, provided the new vacuum solves the classical equations of motion.

Bibliography

We have gone through vertex operators and scattering rather fast. We have given only the central ideas and few representative amplitudes. Although such amplitudes capture most of the important characteristics, there are several details we glossed over. The following bibliographic recommendations will guide you further in the literature.

The vertex operators and their derivation are described in detail in chapter 11 of [7]. Their original derivation can be found in [100]. The structure of the amplitudes both on the sphere and on higher-genus surfaces has been discussed in several articles. The tree amplitudes and the associated unitarity issues have been discussed in detail since the seventies. Useful discussions can be found in older reviews like [101,102,103]. The one-loop amplitudes have been constructed first by sewing tree-level amplitudes using unitarity [103]. In [12] there is a clear exposition of string interactions in light-cone gauge and the associated Riemann surfaces.

Relations between open and closed string tree-level amplitudes have been discussed in [104].

The one-loop vacuum energy in field theory has been discussed in [105]. ϑ-functions and related modular geometry are discussed in [106]. For closed string loop amplitudes we mention the review [107] on string perturbation theory as well as [59,108,109]. The sewing construction of higher closed amplitudes and the associated constraints can be found in [110].

The cancellation of open string tadpoles was first discussed in [111–113]. For open string theory, especially at one loop, a good review is [97]. The papers [114–119] are also helpful in studying open string theory on various open surfaces as involutions of closed surfaces. The sewing of open surfaces and the associated constraints have been discussed in [120].

Of particular interest has been high-energy/fixed-angle scattering in string theory since in this regime, the high-energy region is probed. This has been analyzed in [121,122]. It turns out that for such scattering perturbation theory seems unreliable although there are arguments that the eikonal approximation may capture the essential physics . The fast growth of string perturbation theory was pointed out in [123]. Attempts to Borel-resum string perturbation theory were made [124,125].

A complete review of the cosmological constant problem can be found in [126].

Exercises

5.1. Calculate the general holomorphic ghost correlator on the sphere

$$\left\langle \prod_{i=1}^{n+3} c(z_i) \prod_{j=1}^{n} b(w_j) \right\rangle. \tag{5.1E}$$

5.2. Show that the PSL$(2,\mathbb{C})$ transformation

$$z \to \frac{(z - z_1)z_{32}}{(z - z_3)z_{12}} \tag{5.2E}$$

maps (z_1, z_2, z_3) to $(0, 1, \infty)$. Use this change of variables to show that for $N = 4$ (5.2.3) becomes (5.2.5).

5.3. Show that the N-point tachyon amplitude is invariant under SL$(2,\mathbb{C})$ transformations of the z_i coordinates.

5.4. Verify the correctness of the normalization of the closed tachyon amplitudes in (5.2.3) on page 130, using factorization (unitarity).

5.5. Calculate the three-point amplitude between two closed-string tachyons and one closed string massless state.

5.6. Calculate the four-point amplitude of four massless closed-string states. How do you fix the normalization?

5.7. Use the Gaussian formula (4.7.9) on page 62 to prove (5.2.9) and (5.2.12) on page 132.

5.8. Consider the amplitude of a single massless closed string vertex operator on the disk with NN boundary conditions. Momentum conservation implies that we must take it to have zero momentum. Show that fixing properly the PSL$(2,\mathbb{R})$ symmetry does not make this amplitude zero. Calculate this amplitude. This is a nonzero tadpole (one-point function) for the massless closed fields. Its nonzero answer implies that the flat vacuum is unstable.

5.9. Calculate the scattering amplitude of two open and one closed tachyon on the disk.

5.10. Calculate the scattering amplitude of two closed-string tachyons on RP_2.

5.11. Calculate the amplitude of four gauge bosons on the disk. Include the CP factors and pay attention to the ordering of traces. Show that the amplitude factorizes respecting the structure of the nonabelian three-vertex for the $U(N)$ gauge group. This proves that the theory has local $U(N)$ invariance.

5.12. Use (5.3.1) on page 134 to calculate (5.3.11).

5.13. Use the torus propagator in section 4.18.3 on page 97 to calculate the torus amplitude for n tachyons in the closed bosonic string.

5.14. Calculate the scattering amplitude of n open-string tachyons and m closed string tachyons on the cylinder.

5.15. Consider the unoriented bosonic string with the most general action of orientation reversal Ω on the closed-string ground state

$$\Omega|p^\mu\rangle_{\text{closed}} = \epsilon_c\,|p^\mu\rangle_{\text{closed}}. \tag{5.3E}$$

Go through again the tadpole cancellation conditions and show that they can be canceled if and only if $\epsilon_o = 1$, justifying in retrospect equation (3.4.3) on page 34.

6 | Strings in Background Fields

We have described the propagation of strings in flat 26-dimensional Minkowski space. Although this is important, it falls short of our purposes. There are good reasons to believe that we must understand string physics in curved backgrounds. For one reason, we must have some of the dimensions compact. Although there are compact and flat manifolds (tori), they do not seem phenomenologically appealing as we will learn in detail in later chapters. We also know (from cosmological observations) that our space-time is curved at large scales.

It is therefore important to understand string propagation in nontrivial backgrounds. In this chapter we will provide the basic elements of this description.

6.1 The Nonlinear σ-model Approach

The background fields that are typically nontrivial are the massless fields.[1] We will therefore describe string physics when the massless fields $G_{\mu\nu}$, $B_{\mu\nu}$, and Φ have nontrivial VEVs. To achieve this, we would like to build the background from the infinitesimal perturbations corresponding to these fields, namely their vertex operators studied in the previous chapter. A finite background will appear by exponentiating the relevant vertex operators for the massless fields.

Doing this, the Polyakov action becomes[2]

$$S_P = \frac{1}{4\pi \ell_s^2} \int d^2\xi \left[\sqrt{g}g^{\alpha\beta} G_{\mu\nu}(X) + \epsilon^{\alpha\beta} B_{\mu\nu}(X) \right] \partial_\alpha X^\mu \partial_\beta X^\nu$$
$$+ \frac{1}{4\pi} \int d^2\xi \sqrt{g} R^{(2)} \Phi(X), \qquad (6.1.1)$$

---

[1] We are ignoring the tachyon in this discussion. For the issues we discuss here the presence of the tachyon is not important. Eventually we will be dealing with tachyon-free superstring theories and this discussion will apply. Most of the massive excitations are unstable. Moreover, they correspond to higher-derivative terms on the world-sheet. We therefore set all massive backgrounds to zero.

[2] We are discussing here closed oriented strings.

where $R^{(2)}$ is the scalar curvature of the intrinsic word-sheet metric $g_{\alpha\beta}$. The couplings of the metric and the antisymmetric tensor are natural in the action above. The coupling of the dilaton is less obvious.

A related observation is the following: consider the constant part of the dilaton field (VEV) Φ_0. The Euler character of the world-sheet is given by

$$\chi = \frac{1}{4\pi} \int d^2\xi \sqrt{g}\, R^{(2)}. \tag{6.1.2}$$

It is positive ($\chi = 2$) for the sphere, zero for the torus, and negative for the higher-genus Riemann surfaces. We therefore observe that there is a factor $e^{-\chi\Phi_0}$ in front of e^{-S_P}. From this and equation (3.6.6) on page 40 it follows that the string coupling is essentially given by the dilaton VEV

$$g_s = e^{\Phi_0}. \tag{6.1.3}$$

The action (6.1.1) describes what is known as a nonlinear σ-model. It is the generalization of the flat Polyakov action to nontrivial string backgrounds. This is a nonlinear two-dimensional quantum field theory. It can be generically studied in perturbation theory. The true dimensionless coupling constant is the ratio of the length scale of the target space metric $G_{\mu\nu}$ to ℓ_s. It is however customary to talk about perturbation theory for small ℓ_s. This is the pointlike limit, where the size of the string is much smaller than the physical length scales of the metric. This is known as (small) α'-perturbation theory and it is the standard weak-coupling perturbation theory of the σ-model (6.1.1). If the characteristic scales of the metric and other background fields are much larger than ℓ_s, then at the string scale the fields are almost flat. Therefore the perturbation theory is a weak-field perturbation theory.

Note that the G and B couplings of the string σ-model (6.1.1) are classically conformally invariant. This is not the case with the dilaton coupling unless the dilaton is constant. For general backgrounds however, quantum effects break the conformal symmetry of the σ-model. This breaking of the conformal symmetry can be studied generically in α'-perturbation theory.

The breaking of conformal invariance is signaled by a non-vanishing trace of the stress tensor. This was studied, in the special case of a CFT in section 4.9 on page 64. Here, however, the situation is more general.

A direct computation in σ-model perturbation theory gives

$$T_\alpha{}^\alpha = \frac{\beta^\Phi}{12} R^{(2)} + \frac{1}{2\ell_s^2} (\beta^G_{\mu\nu}\, g^{\alpha\beta} + \beta^B_{\mu\nu}\epsilon^{\alpha\beta})\partial_\alpha X^\mu \partial_\beta X_\nu. \tag{6.1.4}$$

The β-functions can be calculated perturbatively in the weak-coupling expansion of the σ-model, $\ell_s \to 0$. To leading nontrivial order,

$$\frac{\beta^G_{\mu\nu}}{\ell_s^2} = R_{\mu\nu} - \frac{1}{4}H_{\mu\rho\sigma}H_\nu{}^{\rho\sigma} + 2\nabla_\mu\nabla_\nu\Phi + \mathcal{O}(\ell_s^2), \tag{6.1.5}$$

$$\frac{\beta^B_{\mu\nu}}{\ell_s^2} = -\frac{1}{2}\nabla^\rho\left[e^{-2\Phi}H_{\mu\nu\rho}\right] + \mathcal{O}(\ell_s^2), \tag{6.1.6}$$

and

$$\beta^\Phi = D - 26 + \frac{3}{2}\ell_s^2\left[4(\nabla\Phi)^2 - 4\square\Phi - R + \frac{1}{12}H^2\right] + \mathcal{O}(\ell_s^4), \tag{6.1.7}$$

where $H_{\mu\nu\rho}$ is the totally antisymmetric field strength of $B_{\mu\nu}$,

$$H_{\mu\nu\rho} = \partial_\mu B_{\nu\rho} + \partial_\nu B_{\rho\mu} + \partial_\rho B_{\mu\nu}, \tag{6.1.8}$$

and D is the space-time dimension. The -26 in the dilaton β-function is the contribution of the reparametrization ghosts.

When G, B, Φ are such that $\beta_{\mu\nu}^G = \beta_{\mu\nu}^B = 0$, then the σ-model describes a CFT with central charge $c = \beta^\Phi$. In exercise 6.3 on page 152 you are asked to show that β^Φ is a constant when the other β-functions vanish.

The conditions for conformal invariance and thus consistent string propagation are given by the equations

$$\beta^\Phi = \beta_{\mu\nu}^G = \beta_{\mu\nu}^B = 0. \tag{6.1.9}$$

These conditions are second-order equations for the background fields and can be obtained by varying an action

$$S_{\text{tree}} = \frac{1}{2\kappa^2}\int d^D x\sqrt{-\det G}e^{-2\Phi}\left[R + 4(\nabla\Phi)^2 - \frac{1}{12}H^2 + 2\frac{26 - D}{3\ell_s^2}\right] + \mathcal{O}(\ell_s^2). \tag{6.1.10}$$

In (6.1.10) the fields that appear are those that couple to the string σ-model. This is known as the "string frame." The characteristic factor of $e^{-2\Phi}$ multiplying the full action is due to the fact that this action comes from the sphere (as suggested by (6.1.2) and (6.1.3)). Corrections to this action from higher loops will be proportional to $e^{\chi\Phi}$.

In the string frame, the kinetic terms of the metric G and the dilaton are not diagonal. They become diagonal in the "Einstein frame," related to the string frame by a conformal rescaling of the metric. We separate the expectation value of the dilaton $\Phi \to \Phi_0 + \Phi$, and define the Einstein metric as

$$G_{\mu\nu}^E = e^{-4\Phi/(D-2)}G_{\mu\nu}. \tag{6.1.11}$$

We also use

$$\hat{R}_{\mu\nu} = R_{\mu\nu} - \frac{1}{2}(\nabla^2\phi)g_{\mu\nu} - \frac{D-2}{2}\left[\nabla_\mu\nabla_\nu\phi + \frac{1}{2}(\nabla\phi)^2 g_{\mu\nu} - \frac{1}{2}\nabla_\mu\phi\nabla_\nu\phi\right] \tag{6.1.12}$$

for $\hat{g}_{\mu\nu} = e^\phi g_{\mu\nu}$ to obtain the action in the Einstein frame:

$$S_E^{\text{tree}} = \frac{1}{2\kappa^2}\int d^D x\sqrt{G^E}\left[R - \frac{4}{D-2}(\nabla\Phi)^2 - \frac{e^{-8\Phi/(D-2)}}{12}H^2 + 2\frac{26-D}{3\ell_s^2}e^{4\Phi/(D-2)}\right]$$
$$+ \mathcal{O}(\ell_s^2). \tag{6.1.13}$$

The gravitational constant is given by

$$\kappa^2 \sim g_s^2\ell_s^{(D-2)}. \tag{6.1.14}$$

We now consider the notion of the (Wilsonian) effective action for the light fields. All other particles have masses of the order of the string scale. We can imagine integrating out the heavy fields. This will induce corrections to the action of the light fields. This is the definition of the low-energy effective action. This effective action contains only the light fields and is valid up to energies of the order of the mass of the heavy fields.[3] At tree level, this procedure can be implemented by considering the full (on-shell) scattering amplitudes of the light fields from string calculations, expanding them in α' and finding the extra interactions induced on the light fields. Since the amplitudes used are on shell, the effective action can be calculated up to terms that vanish by using the equations of motion.

It can be shown that, up to terms that vanish on shell, the σ-model conformal invariance conditions and the string amplitude calculations produce the same low-energy effective action (6.1.10). This is intuitively expected as the α'-perturbation theory is computing the breaking of conformal invariance due to massless fluctuations, an information also contained in the massless scattering amplitudes.

Therefore, the conformal invariance of the σ-model gives the (classical) string equations of motion and by extension, the tree-level S-matrix elements. In the σ-model approach, the off-shell freedom is related to changing renormalization schemes in σ-model perturbation theory.

To summarize, for backgrounds satisfying (6.1.9) to leading order, we have a CFT, which can be used to construct scattering amplitudes, as described in chapter 5. For slowly varying background fields, the higher-order corrections in α' can be neglected, and the scattering amplitudes we construct, describe strings interacting in the background specified by $G_{\mu\nu}$, $B_{\mu\nu}$, Φ in the σ-model (6.1.1). Moreover, an effective action for the massless fields that reproduces their scattering amplitudes is given in (6.1.10).

6.2 The Quest for Conformal Invariance

Why do we need conformal invariance? Going back to the discussion of the classical string in section 2.2 on page 14, we saw that, classically, a string described by the Nambu-Goto action, once transformed to the Polyakov description, gives a classically conformally invariant theory, because of the freedom in (2.2.16) on page 16. In the Polyakov description, the two-dimensional metric is a dynamical field. If the two-dimensional theory that we couple to gravity is conformally invariant with the correct central charge, (that is, all β-functions vanish), then the conformal factor of the two-dimensional metric decouples.

In such a case we must factor out the volume of Weyl transformations from the path integral and we obtain the description of string theory advocated so far. Moreover, the vanishing of the β-functions implies, to leading order in α', second-order equations for the background fields. In this setup, string vacua are in one-to-one correspondence with two-dimensional CFTs with central charge $c = 26$.

However, in a theory of quantum gravity, conformal invariance is imposed dynamically. Integrating over all metrics has the effect of washing out the scale dependence of a cutoff theory. We may thus try to relax a bit our algorithm for constructing string theory vacua.

[3] If all massive fields are unstable, then the S-matrix elements computed from the effective action are valid at higher energies.

The first modest step in this direction is to start from a CFT with $c \neq 26$. In this case only the dilaton β-function is nonzero. Now, the conformal factor of the two-dimensional metric no longer decouples. Since $T_a{}^a \sim \delta_\phi \log Z$, where ϕ is the conformal factor, we can solve this equation to derive the dependence of the effective action on ϕ. This has been done in section 4.9 on page 64 with the (semiclassical) result

$$S_\phi = \frac{26-c}{96\pi} \int d^2\sigma \sqrt{g}\, g^{\alpha\beta} \partial_\alpha \phi \partial_\beta \phi + \frac{26-c}{48\pi} \int d^2\sigma \sqrt{g}\, R^{(2)}\, \phi. \qquad (6.2.1)$$

This action should be added to the CFT action, which, because of conformal invariance, does not obtain any ϕ-dependent corrections. This semiclassical result is almost correct: there is a finite renormalization in the full theory of the overall multiplicative factor $26 - c \to 25 - c$ to account for the extra conformal anomaly of the free ϕ field. Thus, the total action is (after gauge fixing to a fiducial metric $\hat{g}$)

$$S_{\text{string}} = S_{\text{CFT}} + S_L + S_{\text{ghosts}}, \qquad (6.2.2)$$

where

$$S_L = \frac{25-c}{96\pi} \int d^2\sigma \sqrt{\hat{g}}\, \hat{g}^{\alpha\beta} \partial_\alpha \phi \partial_\beta \phi + \frac{25-c}{48\pi} \int d^2\sigma \sqrt{\hat{g}}\, \hat{R}^{(2)}\, \phi. \qquad (6.2.3)$$

We will assume for the moment that $c < 25$. Redefining $X = \ell_s \sqrt{\frac{25-c}{24}}\, \phi$ we can rewrite the action as

$$S_L = \frac{1}{4\pi \ell_s^2} \int d^2\sigma \sqrt{\hat{g}}\, \hat{g}^{\alpha\beta} \partial_\alpha X \partial_\beta X + \frac{Q}{\sqrt{2}(4\pi \ell_s)} \int d^2\sigma \sqrt{\hat{g}}\, \hat{R}^{(2)}\, X \qquad (6.2.4)$$

with $Q = \sqrt{\frac{(25-c)}{3}}$. We observe that the conformal factor behaves as an extra space-time coordinate, with a linear dilaton $\Phi = QX$ in that direction (compare with (6.1.1)). As described in section 4.14 on page 82, this is a CFT with central charge $c_L = 1 + 3Q^2 = 26 - c$. Thus, the total central charge of the original CFT, the Liouville theory, and the ghosts adds up to zero. We are effectively back to the previous case, by appending to the original CFT the Liouville CFT with the appropriate central charge. Moreover, now the string theory is living in a space-time with one more dimension than the original CFT would indicate. This new dimension is spacelike when $c < 25$. It is timelike, on the other hand, when $c > 25$.

The next generalization emerges when two-dimensional gravity is coupled to a nonconformal theory, where $\beta^{\Phi,B,G}$ do not vanish. The conformal factor does not decouple. The general form of the anomaly equation is given in (6.1.4) where we have neglected a non-derivative function of the X's. This can be integrated as in (4.9.9)–(4.9.11) on page 65 to reveal the conformal factor dependence. It will be further renormalized when we promote ϕ to a new quantum field with standard measure. Renaming $\phi = X^D$, the final form of the action will be as in (6.1.1) but in one more dimension. Now, however, the new theory must be conformally invariant. Thus, the new metric, antisymmetric tensor, and dilaton must satisfy (6.1.9) in one more dimension.

6.3 Linear Dilaton and Strings in $D < 26$ Dimensions

In this section we will pursue further string theory in D flat dimensions. As argued in the previous section, the Liouville field does not decouple, but is promoted to an extra space dimension, with a dilaton field that is linear. Therefore, we are describing an effective theory in $D + 1$ dimensions.

The corresponding background is

$$G_{\mu\nu} = \eta_{\mu\nu}, \quad B_{\mu\nu} = 0, \quad \Phi = \frac{Q}{\sqrt{2}} \phi. \tag{6.3.1}$$

$\phi = X^D / \ell_s$ is the Liouville field and $Q = \sqrt{\frac{(25 - D)}{3}}$. It solves all the β-function equations (6.1.9), and provides a CFT with $c = 26$.

However, because of the linear dilaton, the effective string coupling behaves as

$$g_{eff} = g_s \, e^{Q\phi/\sqrt{2}}, \tag{6.3.2}$$

and diverges as $\phi \to \infty$. Perturbation theory is unfortunately unreliable in this background. We may, however, modify this background in order to shield the strong-coupling region.

To do this, we must investigate the low-lying spectrum of the theory. The simplest operators are plane waves $V_{q,\vec{p}} =: e^{i\vec{p}\cdot\vec{x} + iq\phi}$: where $\vec{p}, \vec{x}$ refer to all directions except the Liouville one. We are interested in operators with ϕ dependence only, in order to preserve translation invariance in the other directions. We will therefore set $\vec{p} = 0$.

In $D + 1$ dimensions, the intercept of the L_0 operator was worked out in exercise 3.1 on page 46 to be $(D - 1)/24$. Using also the L_0 eigenvalues in (4.14.10) on page 83, we find that the operators that satisfy the physical state condition $L_0 = \frac{D-1}{24}$ are

$$T_\pm(X) =: \exp\left(\left[\sqrt{\frac{25 - D}{6}} \pm \sqrt{\frac{1 - D}{6}}\right] \phi\right) :. \tag{6.3.3}$$

They are real when $D \le 1$, but complex when $D > 1$. Such a background would have been a tachyon in $D = 25$. However, for $D < 1$ it has a positive mass squared, and it becomes massless in $D = 1$.

We can add $\int T$ to the action in order to shield the strong-coupling region. Since $T_\pm$ satisfies the physical state conditions, it preserves the conformal invariance of the perturbed theory to leading order. Moreover, it turns out that it is T_-, which is good for our purposes.[4] The new action is

$$S_{2d} = \frac{1}{4\pi \ell_s^2} \int d^2\sigma \sqrt{\hat{g}} \, \hat{g}^{\alpha\beta} \partial_\alpha X^\mu \partial_\beta X_\mu + \frac{Q}{4\pi \sqrt{2}} \int d^2\sigma \sqrt{\hat{g}} \, \hat{R}^{(2)} \, \phi$$
$$+ \frac{T_0}{\ell_s^2} \int d^2\sigma \sqrt{\hat{g}} : \exp\left(\left[\sqrt{\frac{25 - D}{6}} - \sqrt{\frac{1 - D}{6}}\right] \phi\right) :. \tag{6.3.4}$$

---

[4] T_+ is not a good operator in the linear dilaton theory. We refer the reader to [45] for a detailed discussion of this issue.

Here $\hat{g}$ is a fiducial background world-sheet metric related to the physical fluctuating metric g as $g = e^\phi \hat{g}$.

Now the region of strong coupling is screened by the potential. Therefore, at $\phi \to -\infty$ we are at very weak coupling, which grows as we move towards larger values of ϕ. Finally, we hit the potential wall which puts an upper bound on the string effective coupling.

The potential term in the action can be interpreted as a renormalized version of the world-sheet cosmological constant term. Indeed, for $g_{\alpha\beta} = e^\phi \hat{g}_{\alpha\beta}$ we have

$$T_0 \int d^2\sigma \sqrt{g} = T_0 \int d^2\sigma \sqrt{\hat{g}}\, e^\phi. \tag{6.3.5}$$

Renormalization effects can be interpreted as changing $e^\phi \to e^{\beta\phi}$ with

$$\beta = \sqrt{\frac{25-D}{6}} - \sqrt{\frac{1-D}{6}}. \tag{6.3.6}$$

The dynamics of this model is not very easy to calculate in the world-sheet description. However, in chapter 14 on page 470 we will see an alternative description of the same theory that can be handled computationally.

We will derive a characteristic datum of the theory, namely the string critical exponent. To define it we consider the fixed (renormalized) area partition function, defined as

$$Z(A) = \int \mathcal{D}\phi \mathcal{D}X\, e^{-S}\, \delta\left(\int d^2\sigma \sqrt{\hat{g}}\, e^{\beta\phi} - A\right) \tag{6.3.7}$$

for fixed-genus Riemann surfaces.

We now shift the field ϕ as $\phi \to \phi + \frac{\epsilon}{\beta}$ where ϵ is a constant. The measure of the path integral is invariant. The action S transforms because of the linear dilaton term,

$$\frac{Q}{4\pi\sqrt{2}} \delta \int d^2\sigma \sqrt{g}\, \hat{R}^{(2)}\, \phi = \frac{Q\epsilon}{4\pi\sqrt{2}\beta} \int d^2\sigma \sqrt{g}\, \hat{R}^{(2)} = \frac{Q\epsilon}{\sqrt{2}\beta}\chi, \tag{6.3.8}$$

where χ is the Euler number. We therefore obtain

$$Z(A) = \int \mathcal{D}\phi \mathcal{D}X\, e^{-S-(Q\epsilon/\sqrt{2}\beta)\chi}\, \delta\left(e^\epsilon \int d^2\sigma \sqrt{\hat{g}}\, e^{\beta\phi} - A\right) = e^{-\epsilon\left((Q/\sqrt{2}\beta)\chi+1\right)} Z(A\, e^{-\epsilon}). \tag{6.3.9}$$

This functional relation can be solved as

$$Z(A) = A^{-(Q/\sqrt{2}\beta)\chi-1} Z(1). \tag{6.3.10}$$

Defining the string exponent γ as

$$Z(A) \sim A^{(\gamma-2)\chi/2-1}, \tag{6.3.11}$$

we finally obtain

$$\gamma = 2 - \frac{\sqrt{2}Q}{\beta} = \frac{1}{12}\left[D - 1 - \sqrt{(25-D)(1-D)}\right]. \tag{6.3.12}$$

We will rederive the string exponent in chapter 14 from a completely different point of view.

6.4 T-duality in Nontrivial Backgrounds

We have seen in section 4.18.7 on page 101, that the CFT of a compact boson is invariant under the inversion of the radius measured in string units. This is a special case of the general T-duality transformation that we will investigate here. Assume that the string background has a nontrivial U(1) isometry. Choosing an adapted compact coordinate θ so that the isometry can be written as $\theta \to \theta + \epsilon$, we may write the metric and antisymmetric tensor as

$$ds^2 = G_{00}\, d\theta^2 + 2G_{0i}\, d\theta\, dx^i + G_{ij}\, dx^i dx^j, \quad B = B_{0i}\, d\theta \wedge dx^i + B_{ij}\, dx^i \wedge dx^j. \tag{6.4.1}$$

All functions including the dilaton Φ depend on the coordinates x^i but not θ. The generalized T-duality transformation generates the new background

$$\tilde{G}_{00} = \frac{1}{G_{00}}, \quad \tilde{G}_{0i} = \frac{B_{0i}}{G_{00}}, \quad \tilde{B}_{0i} = \frac{G_{0i}}{G_{00}}, \quad \tilde{\Phi} = \Phi - \frac{1}{2}\log G_{00}, \tag{6.4.2}$$

$$\tilde{G}_{ij} = G_{ij} - \frac{G_{0i}G_{0j} - B_{0i}B_{0j}}{G_{00}}, \quad \tilde{B}_{ij} = B_{ij} + \frac{G_{0i}B_{0j} - B_{0i}G_{0j}}{G_{00}}. \tag{6.4.3}$$

In exercise 6.10 on page 153 you are instructed to prove that T-duality is a symmetry of bosonic string theory. This proof however, is semiclassical and may fail in curved σ-models. It can be shown that in compact σ-models where the Killing isometry is associated to chiral currents, T-duality is an exact symmetry of the CFT. In WZW models and cosets, its validity relies on nontrivial symmetries of the compact affine current algebras, namely, affine automorphisms. In noncompact σ-models, the situation is more complicated and we will not discuss it further here.

Bibliography

The correct prescription for the coupling of the dilaton to the σ-model is given in [127].

The calculation of β-functions in σ-models is described in detail in [128,129]. Their application to the heterotic string is found in [130]. The σ-model approach is reviewed in [131,132]. Two-loop contributions to the β-functions are discussed in [133] while four-loop contributions in the supersymmetric case are discussed in [134]. The calculation of the effective action from scattering amplitudes is described in detail in [135]. A survey of exact solutions to the string equations in closed string theory can be found in [136].

The linear dilaton theories in $D < 2$ is a large subject and are reviewed extensively in [137,138]. In chapter 14 we will study an alternative description of such theories in terms of the dynamics of matrices.

A pedestrian introduction of T-duality can be found in [12]. It is extensively reviewed in [70,139]. Its action in solvable compact CFTs has been discussed in [140]. It is directly related to the equivalence of axial and vector gauging of chiral U(1) symmetries of CFTs [141,142]. In $\mathcal{N} = (2,2)_2$ supersymmetric σ-models, it corresponds to the duality between chiral and twisted multiplets of the associated supersymmetry, [143].

Exercises

6.1. From (6.1.4) on page 145 find the change of the effective action after a finite conformal transformation.

6.2. By varying the action (6.1.10) on page 146 using

$$\delta R = (R_{\mu\nu} - \nabla_\mu \nabla_\nu)\delta g^{\mu\nu} + g_{\mu\nu}\Box\delta g^{\mu\nu}, \tag{6.1E}$$

derive the equations (6.1.5)–(6.1.7) on page 146.

6.3. Using Bianchi identities show that when $\beta_G = \beta_B = 0$ then β_Φ is constant.

6.4. Consider two-dimensional gravity coupled to a $c = 25$ CFT. Find out what kind of string theory we obtain.

6.5. Consider the string σ-model on a compact manifold. Show that the α'-expansion is a large-volume expansion and the central charge is given by

$$c = D + \mathcal{O}(1/\text{Volume}^{1/D}). \tag{6.2E}$$

6.6. Show that the integral of the three-form $H = dB$ on any compact three-manifold M_3 must satisfy

$$\frac{1}{4\pi^2 \ell_s^2} \int_{M_3} H \in \mathbb{Z}. \tag{6.3E}$$

6.7. Consider the three-dimensional σ-model over the round three-sphere with radius R and antisymmetric tensor flux given by

$$G_{ij} dx^i dx^j = R^2 (d\psi^2 + \sin^2 \psi (d\theta^2 + \sin^2 \theta d\phi^2)), \tag{6.4E}$$

$$B_{ij} dx^i \wedge dx^j = R^2 (\psi - \sin \psi \cos \psi) \sin \theta \, d\theta \wedge d\phi. \tag{6.5E}$$

(a) Show that the quantization condition of the previous exercise implies that

$$\frac{R^2}{\ell_s^2} = k \in \mathbb{Z}. \tag{6.6E}$$

(b) Show that $\beta^B = \beta^G = 0$ and that the central charge is (we ignore ghosts)

$$c = \beta^\Phi = 3 - \frac{6}{k} + \mathcal{O}\left(\frac{1}{k^2}\right). \tag{6.7E}$$

(c) The metric has an SU(2) × SU(2) symmetry that is also respected by the antisymmetric tensor background. Show that this symmetry generates an SU(2)$_L$ × SU(2)$_R$ current algebra. Thus, the model is the SU(2) WZW model described in section 4.11 on page 69. From this we learn the exact central charge

$$c = \frac{3k}{k+2}, \tag{6.8E}$$

which matches with (6.7E). Note that the radius of the S^3 cannot become smaller than the string length.

6.8. Consider strings moving in one space and one time dimension. Consider backgrounds that have a spatial translational isometry. Find all solutions to the conformal invariance conditions.

6.9. Consider the S^3 σ-model without any flux. This is a nonconformal theory. Couple this to two-dimensional gravity. Our arguments in section 6.2 indicate that the gravitationally dressed theory will have a metric of the form

$$ds^2 = F(\phi)d\phi^2 + \phi R^2 d\Omega_3^2, \tag{6.9E}$$

as well as a dilaton $\Phi(\phi)$ where ϕ is the conformal factor. Determine $F(\phi)$ and $\Phi(\phi)$ so that the theory is conformally invariant. What is the new (dressed) geometry in four dimensions?

6.10. One way to prove the T-duality transformation rules (6.4.2),(6.4.3) is the following: gauge the Killing symmetry $\theta \to \theta + \epsilon$ of the σ-model by introducing a two-dimensional gauge field and sending $\partial_\alpha \theta \to \partial_\alpha \theta + A_\alpha$. Add also to the action the term $\delta S = \frac{1}{2\pi} \int \phi \, \epsilon^{\alpha\beta} \partial_\alpha A_\beta$ where ϕ is a new dynamical compact scalar. Integrating out ϕ imposes that A_α is pure gauge and thus can be removed by a gauge transformation. This shows that the new theory is equivalent to the initial one.[5] On the other hand, you may integrate out the gauge fields, and then gauge-fix $\theta \to 0$, to obtain the dual σ-model. The transformation of the dilaton follows from a careful treatment of the (regularized) determinant, coming from integrating out the gauge fields.

6.11. Show that the T-duality transformations (6.4.2),(6.4.3) are symmetries of the leading-order string equations (6.1.5)–(6.1.7), in the following sense: applied to a solution they give back a solution. Hint: Consider the effective action (6.1.10) for backgrounds of the form (6.4.1) that do not depend on θ and show that it is invariant under (6.4.2),(6.4.3). You will need the formulas of appendix E on page 516.

6.12. Consider the KK parametrization of the background in (6.4.1) on page 151:

$$ds^2 = G_{00}(d\theta + A_i dx^i)^2 + g_{ij} dx^i dx^j, \quad B = (d\theta + A_i dx^i) \wedge B_j dx^j + b_{ij} dx^i \wedge dx^j. \tag{6.10E}$$

Show that the T-duality transformations (6.4.2),(6.4.3) simplify to

$$\tilde{G}_{00} = \frac{1}{G_{00}}, \quad \tilde{A}_i = B_i, \quad \tilde{B}_i = A_i, \tag{6.11E}$$

$$\tilde{g}_{ij} = g_{ij}, \quad \tilde{b}_{ij} = b_{ij}, \quad \tilde{\Phi} = \Phi - \frac{1}{2} \log G_{00}. \tag{6.12E}$$

---

[5] You have to be careful with the radius of ϕ for this to be true.

6.13. When a σ-model has N commuting isometries, generalize the transformations (6.4.2),(6.4.3) to an O(N,N;$\mathbb{Z}$) T-duality group.

6.14. Show that if the tachyon state of the bosonic string transforms with a minus sign under the orientation reversal Ω, then the dilaton and the graviton transform with a minus sign while the antisymmetric tensor is invariant. Show that this is incompatible with their effective interactions summarized in (6.1.10).

7 | Superstrings and Supersymmetry

Bosonic strings have two major problems:

- Their spectrum always contains a tachyon. As a consequence, their vacuum is unstable.

- They do not contain space-time fermions.

We have already seen during our study of free-fermion CFTs that they contain states that transform as spinors under the associated orthogonal symmetry. Therefore we should add free fermions on the world-sheet of the string in order to obtain states that transform as spinors in space-time.

These fermions should carry a space-time index, i.e., ψ^μ, $\bar\psi^\mu$, in order for the spinor to be a space-time spinor. However, in such a case, there will be additional negative norm states associated with the modes of ψ^0. In order for these to be removed from the physical spectrum, we need more constraints than the Virasoro constraints alone. The appropriate symmetry algebra, providing all necessary constraints is the $\mathcal{N} = (1,1)_2$ superconformal algebra.

7.1 $\mathcal{N} = (1,1)_2$ World-sheet Superconformal Symmetry

In the bosonic case, we started with two-dimensional gravity coupled to D scalars X^μ on the world-sheet. The theory eventually implied a set of Virasoro constraints on the Hilbert space. Here, we would like to start from the two-dimensional $\mathcal{N} = (1,1)_2$ supergravity coupled to D superfields, each containing a bosonic coordinate X^μ and two fermionic coordinates, one left-moving ψ^μ and one right-moving $\bar\psi^\mu$. The $\mathcal{N} = (1,1)_2$ supergravity multiplet contains the metric and a gravitino χ_a.

The analog of the bosonic Polyakov action is

$$S_P^{II} = \frac{1}{4\pi \ell_s^2} \int d^2\sigma \sqrt{g} \left[g^{ab} \partial_a X^\mu \partial_b X^\mu + \frac{i}{2} \bar\psi^\mu \slashed{\partial} \psi^\mu + \frac{i}{2} (\bar\chi_a \gamma^b \gamma^a \psi^\mu) \left(\partial_b X^\mu - \frac{i}{4} \bar\chi_b \psi^\mu \right) \right]. \quad (7.1.1)$$

It is invariant under a local left-moving supersymmetry

$$\delta g_{ab} = i\bar{\epsilon}(\gamma_a \chi_b + \gamma_b \chi_a), \quad \delta\chi_a = 2\nabla_a \epsilon, \tag{7.1.2}$$

$$\delta X^\mu = i\bar{\epsilon}\psi^\mu, \quad \delta\psi^\mu = \gamma^a\left(\partial_a X^\mu - \frac{i}{2}\chi_a\psi^\mu\right)\epsilon, \quad \delta\bar{\psi}^\mu = 0, \tag{7.1.3}$$

where ϵ is a left-moving Majorana-Weyl spinor. There is a similar right-moving supersymmetry involving a right-moving Majorana-Weyl spinor $\bar{\epsilon}$ and the fermions $\bar{\psi}^\mu$. In our notation we have $\mathcal{N} = (1, 1)_2$ supersymmetry.

The analog of the conformal gauge is the superconformal gauge

$$g_{ab} = e^\phi \delta_{ab}, \quad \chi_a = \gamma_a \zeta, \tag{7.1.4}$$

where ζ is a constant Majorana spinor; ϕ and ζ decouple from the classical action (7.1.1). Apart from the Virasoro operators we also have the supercurrents

$$G_{\text{matter}} = i\frac{\sqrt{2}}{\ell_s^2}\psi^\mu \partial X^\mu, \quad \bar{G}_{\text{matter}} = i\frac{\sqrt{2}}{\ell_s^2}\bar{\psi}^\mu \bar{\partial} X^\mu. \tag{7.1.5}$$

We must also introduce the appropriate ghosts. We still have the usual b, c system with $\lambda = 2$ associated with diffeomorphisms, but now we also need a commuting set of ghosts β, γ with $\epsilon = -1$, $\lambda = \frac{3}{2}$ associated with the local supersymmetry. Superconformal invariance will be present at the quantum level, provided the ghost central charge cancels the matter central charge. Each bosonic and fermionic coordinate contributes $3/2$ to the central charge. Since we have D of them, the matter central charge is $c_{\text{matter}} = \frac{3}{2}D$. The b, c system contributes -26 to the central charge while the β, γ system contributes $+11$. The total central charge vanishes, provided that $D = 10$. This is the critical dimension for the superstring.

The classical constraints imply the vanishing of $T, G, \bar{T}, \bar{G}$. Consequently, we have enough constraints to remove the negative norm states.

The BRST current is

$$j_{\text{BRST}} = \gamma G_{\text{matter}} + cT_{\text{matter}} + \frac{1}{2}\left(cT_{\text{ghost}} + \gamma G_{\text{ghost}}\right), \tag{7.1.6}$$

where

$$G_{\text{matter}} = i\frac{\sqrt{2}}{\ell_s^2}\psi^\mu \partial X^\mu, \quad T_{\text{matter}} = -\frac{1}{\ell_s^2}\partial X^\mu \partial X^\mu - \frac{1}{2\ell_s^2}\psi^\mu \partial\psi^\mu, \tag{7.1.7}$$

$$G_{\text{ghost}} = -i\left(c\,\partial\beta - \frac{1}{2}\gamma\,b + \frac{3}{2}\partial c\,\beta\right), \quad T_{\text{ghost}} = T_{bc} - \frac{1}{2}\gamma\,\partial\beta - \frac{3}{2}\partial\gamma\,\beta. \tag{7.1.8}$$

G_{ghost} and T_{ghost} satisfy the OPEs of the $\mathcal{N} = (1, 1)_2$ superconformal algebra (4.5.1) on page 58, and (4.13.8) on page 78 with the correct central charge.

The BRST charge is

$$Q = \frac{1}{2\pi i}\left[\oint dz\,j_{\text{BRST}} + \oint d\bar{z}\,\bar{j}_{\text{BRST}}\right]. \tag{7.1.9}$$

It is nilpotent for $D = 10$ and can be used in the standard way to define physical states.

7.2 Closed (Type-II) Superstrings

We will first consider the closed (type-II) superstring case. We will work in a physical gauge and derive the spectrum. The analog of the light-cone gauge in the supersymmetric case is[1]

$$X^+ = x^+ + p^+\tau, \quad \psi^+ = \bar{\psi}^+ = 0. \tag{7.2.1}$$

As in the bosonic case, we can explicitly solve the constraints by expressing $X^-, \psi^-, \bar{\psi}^-$ in terms of the transverse modes. Then, the physical states can be constructed out of the transverse bosonic and fermionic oscillators. However, all zero modes are present.

As we mentioned before, we have two left-moving sectors corresponding to NS and R boundary conditions for ψ^μ and G and another two sectors corresponding to $\overline{\text{NS}}$ and $\overline{\text{R}}$ boundary conditions for $\bar{\psi}^\mu$ and $\bar{G}$.

For the moment we will discuss only the left sector to avoid repetition. We will introduce as usual the modes L_n and G_r of the superconformal generators.

In the light-cone gauge we have solved the constraints, apart from those associated with the zero modes. In the NS sector, G is half-integrally moded and the only zero mode is L_0. There is also a normal-ordering constant, which can be calculated either by demanding Lorentz invariance of the physical spectrum, as we have done for the bosonic string, or by realizing that in the covariant formulation the lowest "energy" state is not the usual vacuum $|0\rangle$ but $c_1\gamma_{-1/2}|0\rangle$. Both approaches result in a normal-ordering constant equal to $a = \frac{1}{2}$, and therefore $L_0 - \frac{1}{2}$ should be zero on physical states. The state $|p\rangle$ is a physical state with $p^2 = -m^2 = 2/\ell_s^2$. It is again a tachyon. The next states are of the form $\psi^i_{-1/2}|p\rangle$ and satisfy $L_0 = \frac{1}{2}$ if $p^2 = 0$. These states are massless. However, we would prefer not to have a tachyonic state. Since the tachyon has $(-1)^{F_L} = 1$ we would like to impose the extra constraint (GSO projection): physical states in the NS sector should have odd fermion number.

In the R sector we have two zero modes: L_0 and G_0. The L_0 constraint is the same as in the NS sector. A quick look at the expression for G_0 in (7.1.5) is enough to convince us that there can be no normal-ordering constant, and that G_0 should be zero on physical states. On the other hand we know from the superconformal algebra

$$0 = \{G_0, G_0\} = 2\left(L_0 - \frac{D-2}{16}\right). \tag{7.2.2}$$

Compatibility with the L_0 constraint implies that $D = 10$. The R ground states are spinors of O(9,1). Consequently, these states satisfy the L_0 constraint. Also remember that $G_0 = \psi_0^\mu \alpha_0^\mu + 2\sum_{n\neq 0}^\infty \psi_n^i \alpha_{-n}^i$. As shown in section 4.12 on page 71, the operator ψ_0^μ is represented by the ten-dimensional Γ-matrices Γ^μ and α_0^μ by p_μ. The other terms in G_0 do not contribute to the ground states. $G_0 = 0$ implies the Dirac equation $\not{p} \equiv \Gamma^\mu p_\mu = 0$.

---

[1] There is a subtlety here concerning the super-light-cone gauge. If ψ^+, for example, has NS boundary conditions, then it can be set to zero. If it has R boundary conditions, then it can be set to zero except for its zero mode. A similar remark applies to $\bar{\psi}^+$.

Therefore, the potentially massless states in the R sector are a spinor S and a conjugate spinor C of O(9,1) satisfying the massless Dirac equation. Under $(-1)^{F_L}$, S has eigenvalue 1 and C has -1. All other states are built on these ground-states and are massive. So far, there is no *a priori* reason to impose also a GSO projection in the R sector. As we will see later on, one-loop modular invariance will force us to do so. Anticipating this fact, we will also fix the fermion parity in the R sector. Since $(-1)^F = +$ or $-$ is a matter of convention in the R sector, we will allow both possibilities. We will only keep the S or C spinor ground states, but not both.

A similar discussion applies to the right-moving sector. Combining the two we have overall four sectors:

• (NS-NS): These are bosons since they transform in tensor representations of the rotation group. The projection here is $(-1)^{F_L} = (-1)^{F_R} = -1$. The lowest states allowed by the physical constraints and the GSO projection are of the form $\psi^i_{-1/2}\bar{\psi}^j_{-1/2}|p\rangle$, they are massless and correspond to a symmetric traceless tensor (the graviton), an antisymmetric tensor and a scalar. The tachyon is removed!

• (NS-R): These are fermions. The GSO projection on the NS side is $(-1)^{F_L} = -1$ and by convention we keep the S representation in the R sector: $(-1)^{F_R} = 1$. The lowest-lying states, $\psi^i_{-1/2}|p, \bar{S}\rangle$ are massless space-time fermions and contain a C Majorana-Weyl gravitino and an S fermion.

• (R-NS): Here the GSO projection in NS is $(-1)^{F_R} = -1$, but in the R sector we have two physically distinct options: keep the S spinor (type IIB) or the C spinor (type IIA). Again the lowest-lying states $\bar{\psi}^i_{-1/2}|p, S$ or $C\rangle$ are massless space-time fermions.

• (R-R): In the IIA case the massless states are $|S, \bar{C}\rangle$, which decomposes into a vector and a three-index antisymmetric tensor, as will be shown in section 7.2.1. In type IIB they are $|S, \bar{S}\rangle$, which decomposes into a scalar, a two-index antisymmetric tensor, and a self-dual four-index antisymmetric tensor.

There are also the bosonic oscillators for us to use but, since the intercept is $-1/2$, they are not involved in the massless states. They do, however, contribute to the massive spectrum.

Both type-IIA and -IIB theories have two gravitini and are thus expected to have $\mathcal{N} = 2_{10}$ local supersymmetry. In type IIB the gravitini have the same space-time chirality, while the two spin-$\frac{1}{2}$ fermions have opposite chirality. Thus, the theory is chiral. The type-IIA theory is nonchiral since the two gravitini and the spin-$\frac{1}{2}$ fermions come in both chiralities.

In the light-cone gauge, the leftover constraints are essentially the linearized equations of motion. In the NS-NS sector the constraints are

$$L_0 = \bar{L}_0, \quad L_0 - \frac{1}{2} = 0; \qquad (7.2.3)$$

and, as we have seen already in the bosonic case, it gives the mass-shell condition. This corresponds to the free Klein-Gordon equation. The R-NS sector contains space-time fermions and the constraints are as in (7.2.3), plus the $G_0 = 0$ constraint, which provides as we showed above, the Dirac equation both for massless and massive states. Its square, from

(7.2.2), gives the Klein-Gordon equation; the independent equations are thus $G_0 = 0$ and $L_0 = \bar{L}_0$. Similar remarks apply for the NS-R sector. Finally in the R-$\bar{\text{R}}$ sector the states are bi-spinors and they satisfy two Dirac equations: $G_0 = \bar{G}_0 = 0$.

R-R massless states are special for two reasons. First, they can always be written in terms of higher-form gauge potentials. They are also always coupled to other perturbative states via derivatives. Therefore, no perturbative states are charged under them. As we will see later on, they are at the heart of nonperturbative duality conjectures. A more detailed discussion of their properties can be found in the next section.

We will examine more closely the space-time meaning of the operators $(-1)^{F_{L,R}}$. In the left-moving NS sector

$$(-1)^{F_L} = \exp\left[i\pi \sum_{r \in \mathbb{Z}+1/2} \psi_r^i \psi_{-r}^i \right]. \tag{7.2.4}$$

In the Ramond sector

$$(-1)^{F_L} = \prod_{\mu=0}^{9} \psi_0^\mu \exp\left[i\pi \sum_{n=1}^\infty \psi_n^i \psi_{-n}^i \right] = \Gamma^{11} \exp\left[i\pi \sum_{n=1}^\infty \psi_n^i \psi_{-n}^i \right], \tag{7.2.5}$$

where Γ^{11} is the analog of γ^5 in ten dimensions. Similar relations hold for the right-moving sector. We can deduce from the form of the supercurrents (7.1.5) that the zero modes satisfy

$$\{(-1)^{F_L}, G_0\} = 0, \quad \{(-1)^{F_R}, \bar{G}_0\} = 0. \tag{7.2.6}$$

These generalize the field theoretic relation

$$\{\Gamma^{11}, \partial\!\!\!/\} = 0. \tag{7.2.7}$$

Note that in string theory this equation holds also for the massive Dirac operator G_0.

At the massless level, and all massive levels as well, there is an equal number of on-shell fermionic and bosonic degrees of freedom. This is necessary if the theory has (at least one) space-time supersymmetry.

7.2.1 Massless R-R states

We will now consider in more detail the massless R-R states of type-IIA,B string theory. They have unusual properties and play a central role in nonperturbative duality symmetries.

We start by describing in detail the Γ-matrix conventions in flat ten-dimensional Minkowski space.

The (32×32)-dimensional Γ-matrices satisfy

$$\{\Gamma^\mu, \Gamma^\nu\} = 2\eta^{\mu\nu}, \quad \eta^{\mu\nu} = (-++\cdots+). \tag{7.2.8}$$

The Γ-matrix indices are raised and lowered with the flat Minkowski metric $\eta^{\mu\nu}$:

$$\Gamma_\mu = \eta_{\mu\nu} \Gamma^\nu, \quad \Gamma^\mu = \eta^{\mu\nu} \Gamma_\nu. \tag{7.2.9}$$

We shall use the Majorana representation where the Γ-matrices are real, Γ^0 is antisymmetric, and the rest symmetric. Also

$$\Gamma^0 \Gamma_\mu^\dagger \Gamma^0 = \Gamma_\mu, \quad \Gamma^0 \Gamma_\mu \Gamma^0 = \Gamma_\mu^T. \tag{7.2.10}$$

The Majorana spinors S_α are real: $S_\alpha^* = S_\alpha$;

$$\Gamma_{11} = \Gamma_0 \cdots \Gamma_9, \quad (\Gamma_{11})^2 = 1, \quad \{\Gamma_{11}, \Gamma^\mu\} = 0. \tag{7.2.11}$$

Γ_{11} is symmetric and real. This is the reason why, in ten dimensions, the Weyl condition $\Gamma_{11} S = \pm S$ is compatible with the Majorana condition.[2] We use the convention that for the Levi-Cività tensor, $\epsilon^{01\cdots9} = 1$. We will define the antisymmetrized products of Γ-matrices

$$\Gamma^{\mu_1 \cdots \mu_k} = \Gamma^{[\mu_1} \cdots \Gamma^{\mu_k]} = \frac{1}{k!} \left(\Gamma^{\mu_1} \cdots \Gamma^{\mu_k} \pm \text{permutations} \right). \tag{7.2.12}$$

We can derive by straightforward computation the following identities among Γ-matrices:

$$\Gamma_{11} \Gamma^{\mu_1 \cdots \mu_k} = \frac{(-1)^{[(k+1)/2]}}{(10-k)!} \epsilon^{\mu_1 \cdots \mu_{10}} \Gamma_{\mu_{k+1} \cdots \mu_{10}}, \tag{7.2.13}$$

$$\Gamma^{\mu_1 \cdots \mu_k} \Gamma_{11} = \frac{(-1)^{[k/2]}}{(10-k)!} \epsilon^{\mu_1 \cdots \mu_{10}} \Gamma_{\mu_{k+1} \cdots \mu_{10}}, \tag{7.2.14}$$

with $[x]$ denoting the integer part of x. Then

$$\Gamma^\mu \Gamma^{\nu_1 \cdots \nu_k} = \Gamma^{\mu \nu_1 \cdots \nu_k} + k \, \eta^{\mu[\nu_1} \Gamma^{\nu_2 \cdots \nu_k]}, \tag{7.2.15}$$

$$\Gamma^{\nu_1 \cdots \nu_k} \Gamma^\mu = \Gamma^{\nu_1 \cdots \nu_k \mu} + k \, \eta^{\mu[\nu_k} \Gamma^{\nu_1 \cdots \nu_{k-1}]}, \tag{7.2.16}$$

with square brackets defined in appendix B on page 505. The invariant Lorentz scalar product of two spinors χ, ϕ is $\chi_\alpha^* (\Gamma^0)_{\alpha\beta} \phi_\beta$.

Now consider the ground states of the R-R sector. On the left, we have a Majorana spinor S_α satisfying $\Gamma_{11} S = S$ by convention. On the right, we have another Majorana spinor $\tilde{S}_\alpha$ satisfying $\Gamma_{11} \tilde{S} = \xi \tilde{S}$, where $\xi = 1$ for the type-IIB string and $\xi = -1$ for the type-IIA string. The total ground state is the product of the two. To represent it, it is convenient to define the following bispinor field:

$$F_{\alpha\beta} = S_\alpha (\Gamma^0)_{\beta\gamma} \tilde{S}_\gamma, \quad S_\alpha \tilde{S}_\beta = F_{\alpha\gamma} \Gamma^0_{\gamma\beta}. \tag{7.2.17}$$

With this definition, $F_{\alpha\beta}$ is real and the trace $F_{\alpha\beta} \delta^{\alpha\beta}$ is Lorentz invariant. The chirality conditions on the spinor translate into

$$\Gamma_{11} F = F, \quad F \Gamma_{11} = -\xi F, \tag{7.2.18}$$

where we have used the fact that Γ_{11} is symmetric and anticommutes with Γ^0.

We can now expand the bispinor F into the complete set of antisymmetrized Γ's:

$$F_{\alpha\beta} = \sum_{k=0}^{10} \frac{(-1)^k}{k!} F_{\mu_1 \cdots \mu_k} (\Gamma^{\mu_1 \cdots \mu_k})_{\alpha\beta}, \tag{7.2.19}$$

where the $k = 0$ term is proportional to the unit matrix and the tensors $F_{\mu_1 \cdots \mu_k}$ are real.

---

[2] In a space with signature (p, q) the Majorana and Weyl conditions are compatible, provided $|p - q|$ is a multiple of 8.

We can now translate the first of the chirality conditions in (7.2.18) using (7.2.14) to obtain the following equation:

$$F^{\mu_1 \cdots \mu_k} = \frac{(-1)^{[(k+1)/2]}}{(10-k)!} \epsilon^{\mu_1 \cdots \mu_{10}} F_{\mu_{k+1} \cdots \mu_{10}}. \tag{7.2.20}$$

The second chirality condition implies

$$F^{\mu_1 \cdots \mu_k} = \xi \frac{(-1)^{[k/2]+1}}{(10-k)!} \epsilon^{\mu_1 \cdots \mu_{10}} F_{\mu_{k+1} \cdots \mu_{10}}. \tag{7.2.21}$$

Compatibility between (7.2.20) and (7.2.21) implies that type-IIB theory ($\xi = 1$) contains tensors of odd rank (the independent ones being $k = 1, 3$ and $k = 5$ satisfying a self-duality condition) and type-IIA theory ($\xi = -1$) contains tensors of even rank (the independent ones having $k = 0, 2, 4$). The number of independent tensor components adds up in either case to $16 \times 16 = 256$.

As mentioned in section 7.2, the mass-shell conditions imply that the bispinor field (7.2.8) obeys two massless Dirac equations coming from G_0 and $\bar{G}_0$:

$$(p_\mu \Gamma^\mu) F = F(p_\mu \Gamma^\mu) = 0. \tag{7.2.22}$$

To convert these to equations for the tensors, we use the gamma identities (7.2.15) and (7.2.16). After some straightforward algebra one finds

$$p^{[\mu} F^{\nu_1 \cdots \nu_k]} = p_\mu F^{\mu \nu_2 \cdots \nu_k} = 0, \tag{7.2.23}$$

which are the Bianchi identity and the free massless equation for an antisymmetric tensor field strength. Using the language of forms, which is reviewed in appendix B on page 505, we may write

$$dF = d\,{}^\star F = 0. \tag{7.2.24}$$

Solving the Bianchi identity locally allows us to express the k-index field strength as the exterior derivative of a $(k-1)$-form potential

$$F_{\mu_1 \cdots \mu_k} = k\, \partial_{[\mu_1} C_{\mu_2 \cdots \mu_k]}, \tag{7.2.25}$$

or in form notation

$$F_{(k)} = dC_{(k-1)}. \tag{7.2.26}$$

Consequently, the type-IIA theory has a vector (C_μ) and a three-index tensor potential ($C_{\mu\nu\rho}$), in addition to a constant nonpropagating zero-form field strength (F), while the type-IIB theory has a zero-form (C), a two-form ($C_{\mu\nu}$), and a four-form potential ($C_{\mu\nu\rho\sigma}$), the latter with self-dual field strength. The number of physical transverse degrees of freedom adds up in both cases to $64 = 8 \times 8$.

It is not difficult to see that in the perturbative string spectrum there are no states charged under the R-R forms. First, couplings of the form $\langle s|R\bar{R}|s \rangle$ are not allowed by the separately conserved left and right (space-time) fermion number symmetries $\mathbf{F}_{R,L}$. Second, the R-R vertex operators contain the field strengths rather than the potentials and equations of motion and Bianchi identities enter on an equal footing. If there were electric states in perturbation theory we would also have magnetic states.

R-R forms have another peculiarity. There are various ways to deduce that their couplings to the dilaton are exotic. The dilaton dependence of an F^{2m} term at the kth order of perturbation theory is $e^{2(k-1)\Phi}e^{2m\Phi}$ instead of the usual $e^{2(k-1)\Phi}$ term for NS-NS fields. For example, at tree level, the quadratic terms are dilaton independent in the string frame. In this frame, the gauge transformation properties of R-R fields as well as their Bianchi identities are standard. In exercise 7.34 on page 186, you are invited to derive this property.

7.3 Type-I Superstrings

We now consider open superstrings. There are two possibilities: oriented and unoriented open superstrings. Unoriented open strings are obtained by identifying open strings by the operation that exchanges the two end points. The bosonic case was discussed earlier in section 2.3.2 on page 22. We may add CP factors at the end points. In the oriented case we have obtained a U(N) gauge group, while in the unoriented case we obtained O(N) or Sp(2N) gauge groups. Moreover, we had shown that only the O(2^{13}) open and closed bosonic unoriented theory was free of tadpoles (but not of tachyons).

As it might be expected, in the superstring case, the GSO projection will remove the tachyonic ground state and the lowest bosonic states will be a collection of massless vectors. Here also, as in the bosonic case, tadpoles will be present, due to the massless states. Their cancellation will force us to consider the supersymmetric closed plus open unoriented theory.

We start by studying the spectrum in the closed sector. To obtain the unoriented spectrum, we must project out using the orientation reversal Ω transformation. This is a symmetry only in the IIB string.

To study the spectrum, we must find out how Ω acts on the various closed string oscillators as well its action on the ground states. For the oscillators we have the standard action

$$\Omega\,\alpha_k^\mu\,\Omega^{-1} = \bar\alpha_k^\mu, \quad \Omega\,\bar\alpha_k^\mu\,\Omega^{-1} = \alpha_k^\mu, \quad \Omega\,\psi_r^\mu\,\Omega^{-1} = \bar\psi_r^\mu, \quad \Omega\,\bar\psi_r^\mu\,\Omega^{-1} = -\psi_r^\mu. \tag{7.3.1}$$

The extra minus sign in the fermion transformation is inserted in order for the product $\psi_r\bar\psi_r$ to be orientation invariant. This is mostly convenience, since it can be reabsorbed into the transformation of the vacuum.

A crucial ingredient is the transformation of the vacua, under Ω. The NS-R and R-NS sectors transform into each other. Therefore, in the unoriented theory, one copy remains. At the massless level we obtain one Majorana-Weyl gravitino and a Majorana-Weyl fermion of the opposite chirality. In the NS-NS and R-R vacua we have instead

$$\Omega|p^\mu\rangle_{\text{NS-NS}} = \epsilon_{\text{NS}}|p^\mu\rangle_{\text{NS-NS}}, \quad \Omega|p^\mu; S_\alpha; \tilde{S}_\beta\rangle_{\text{R-}\bar{\text{R}}} = \epsilon_{\text{R}}|p^\mu; S_\beta; \tilde{S}_\alpha\rangle_{\text{R-}\bar{\text{R}}}, \tag{7.3.2}$$

where $\epsilon_{\text{NS}}^2 = \epsilon_{\text{R}}^2 = 1$ from $\Omega^2 = 1$.

Consistency of the interactions in the NS sector imposes $\epsilon_{\text{NS}} = 1$. To show this, assume the opposite, $\epsilon_{\text{NS}} = -1$. In that case the graviton and dilaton are noninvariant and are projected out. The NS-NS antisymmetric tensor is kept. Since there is a three-point vertex in the IIB theory where two antisymmetric tensors couple to a graviton, this implies that in the four-point amplitude involving four antisymmetric tensors there will be graviton intermediate states. This is inconsistent with the projection. Therefore it is necessary that $\epsilon_{\text{NS}} = 1$. This choice is also compatible with supersymmetry. There is no ten-dimensional supermultiplet that contains a two-index antisymmetric tensor but no graviton.

Similarly, in the R$\bar{\text{R}}$ sector, supersymmetry will imply that $\epsilon_{\text{R}} = -1$. You are asked in exercise 7.8 on page 184 to show, using (7.3.2), that the IIB bispinor (7.2.17) transforms under Ω as[3]

$$\Omega \, F \, \Omega^{-1} = -\epsilon_{\text{R}} \, \Gamma^0 \, F^T \, \Gamma^0. \tag{7.3.3}$$

Using the Γ-matrix properties, we find that the IIB forms transform as

$$F_{\mu_1\mu_2\cdots\mu_k} \to \epsilon_{\text{R}} \, (-1)^{k+1} F_{\mu_k\mu_{k-1}\cdots\mu_1} = \epsilon_{\text{R}} \, (-1)^{[(k+1)/2]+1} F_{\mu_1\mu_2\cdots\mu_k}, \quad k \text{ odd.} \tag{7.3.4}$$

Therefore, for $\epsilon_{\text{R}} = -1$, the two-index antisymmetric tensor survives, but the scalar and the four-index self-dual antisymmetric tensor are projected out. Therefore, in total, we have the graviton, a scalar, an antisymmetric tensor, as well as a Majorana-Weyl gravitino and a Majorana-Weyl fermion. This is the content of the (chiral) $\mathcal{N} = 1_{10}$ supergravity multiplet.

As we will see in section 7.9, pure $\mathcal{N} = 1_{10}$, supergravity in ten dimensions is anomalous. There must be an extra sector that cancels the anomalies. As in the bosonic case, this extra sector will be the unoriented open string.

To describe its spectrum we must investigate how Ω acts on the open string oscillators as well as the NS and R vacua. For the bosonic oscillators, this was derived in (3.4.5) on page 34. We reproduce them here:

$$\text{NN}: \quad \Omega \, \alpha_k^\mu \, \Omega^{-1} = (-1)^k \, \alpha_k^\mu, \quad \text{DD}: \quad \Omega \, \alpha_k^\mu \, \Omega^{-1} = (-1)^{k+1} \, \alpha_k^\mu. \tag{7.3.5}$$

A similar argument can be applied for the mode expansion of the fermions to obtain

$$\text{NN}: \quad \Omega \psi_r \Omega^{-1} = (-1)^r \psi_r, \quad \text{DD}: \quad \Omega \psi_r \Omega^{-1} = (-1)^{r+1} \psi_r. \tag{7.3.6}$$

You are asked to derive this in exercise 7.12 on page 184. In particular, in the NS sector of the open string with NN boundary conditions

$$\Omega \psi_{n+1/2} \Omega^{-1} = i^{2n+1} \psi_{n+1/2}. \tag{7.3.7}$$

We first consider the NS sector. We parametrize the action on the CP factors as usual

$$\Omega: \; |p\,; ij\rangle_{\text{NS}} \to \epsilon_{\text{NS}} \, (\gamma_\Omega)_{ii'} \, |p\,; j'i'\rangle \, (\gamma_\Omega)_{j'j}^{-1}, \tag{7.3.8}$$

following the discussion of section 3.4.1 on page 34.

It is convenient to first consider the massless vector states,

$$A^\mu = \psi_{-1/2}^\mu \sum_{ij} |p\,; ij\rangle_{\text{NS}} \, \lambda_{ij}. \tag{7.3.9}$$

---

[3] We have added the extra sign ϵ_{R} that originates in the transformation of the ground state. We also treat the left and right spinors S_α and $\tilde{S}_\alpha$ as communting with each other.

The action of Ω^2 on the massless states is

$$\Omega^2: \ \psi^\mu_{-1/2}|p\,;ij\rangle_{\rm NS} \to -\epsilon^2_{\rm NS} \, (\gamma_\Omega[(\gamma_\Omega)^T]^{-1})_{ii'} \, \psi^\mu_{-1/2}|p\,;i'j'\rangle_{\rm NS} \, ((\gamma_\Omega)^T(\gamma_\Omega)^{-1})_{j'j}. \tag{7.3.10}$$

Imposing that $\Omega^2 = 1$ therefore implies that

$$\gamma_\Omega = \zeta \, (\gamma_\Omega)^T, \quad \epsilon^2_{\rm NS}\zeta^2 = -1. \tag{7.3.11}$$

To have nontrivial solutions of the first equation in (7.3.11), ζ must be a phase.

Demanding that Ω gives an eigenvalue one when acting on (7.3.9) and using (7.3.7) we obtain

$$\lambda = -i\epsilon_{\rm NS} \, \gamma_\Omega \, \lambda^T \, (\gamma_\Omega)^{-1}. \tag{7.3.12}$$

This condition selects the vectors that are invariant under the Ω projection.

The determinant of (7.3.11) gives $\zeta^N = 1$. Taking the transpose of (7.3.12), using (7.3.11) and matching back to (7.3.12), we obtain

$$\epsilon^2_{\rm NS} = -1, \quad \zeta^2 = 1, \tag{7.3.13}$$

where the second relation follows from the first and (7.3.11).

We now ask that the invariant vectors form a Lie algebra. Using (7.3.12) we obtain for the commutator of two matrices

$$[\lambda_1, \lambda_2] = -\epsilon^2_{\rm NS} \, \gamma_\Omega \, [\lambda^T_1, \lambda^T_2] \, (\gamma_\Omega)^{-1} = - \, \gamma_\Omega \, [\lambda_1, \lambda_2]^T \, \gamma_\Omega. \tag{7.3.14}$$

For the commutator to also satisfy (7.3.12) we must have[4]

$$\epsilon_{\rm NS} = -i. \tag{7.3.15}$$

(7.3.13) allows two values for ζ. As shown in section 3.4.1 on page 34, $\zeta = 1$ would give an SO(N) gauge group, but $\zeta = -1$ a symplectic one.

For the higher levels in the NS sector, the Ω projection is now completely fixed. You are invited in exercise 7.10 on page 184 to find the physical spectrum in the first massive sector of the unoriented open superstring.

We now move to the R sector and write the action of Ω on the ground states as

$$\Omega|p; i, j; S_\alpha\rangle_{\rm R} = \epsilon_{\rm R}(\gamma_\Omega)_{ii'}|p; j', i'; S_\alpha\rangle_{\rm R}(\gamma_\Omega)^{-1}_{j'j}. \tag{7.3.16}$$

Note that we have taken the same γ_Ω matrix as in the NS sector. This is required for consistency of the interactions of the fermions with those of the NS sector, notably the vectors. It is only in this case, that the fermions will transform into representations of the gauge group. Another way to see this is to note that the R ground states are given by the fermionic spinors (transforming with a phase) acting on the NS vacuum.

Imposing $\Omega^2 = 1$ on the physical states we obtain (using $\zeta^2 = 1$)

$$\epsilon^2_{\rm R} = 1. \tag{7.3.17}$$

The Ω-invariant massless fermions $\lambda_{ij}|p; i, j; S_\alpha\rangle_{\rm R}$ must satisfy

$$\lambda = \epsilon_{\rm R} \, \gamma_\Omega \, \lambda^T \, (\gamma_\Omega)^{-1}. \tag{7.3.18}$$

---

[4] In the covariant formulation this i emerges directly. It comes from the transformation of the ghost oscillators acting on the disk vacuum.

If we want to impose supersymmetry, then we must choose $\epsilon_R = -1$. Comparison with (7.3.12) shows that in this case we obtain the same gauge group representations for bosons and fermions. We end up, at the massless level, with the ten-dimensional SYM multiplet with gauge group SO(N) ($\zeta = 1$) or Sp(2N) ($\zeta = -1$).

As we will see in section 7.6.3, in order for the R tadpoles to cancel we must choose $\epsilon_R \zeta = -1$ and N = 32.

To obtain the supersymmetric type-I theory, we must choose also $\epsilon_R = -1$. To cancel the tadpoles, $\zeta = 1$ and we obtain the vector supermultiplet of SO(32). This is the anomaly-free supersymmetric type-I superstring theory.

7.4 Heterotic Superstrings

We have used either the Virasoro algebra (bosonic strings) or the $\mathcal{N} = (1, 1)_2$ superconformal algebra (superstrings) to remove negative-norm states from string theories. Also, the closed theories were left-right symmetric, in the sense that a similar algebra is acting on both the left and the right. We might however envisage the possibility of using a Virasoro algebra on the right and the superconformal algebra on the left.

Consider a string theory where we have on the left (holomorphic) side a number of bosonic coordinates and an equal number of left-moving world-sheet fermions. The left constraint algebra will be that of the superstring and, the absence of Weyl anomaly will imply that the number of left-moving coordinates must be 10. In the right-moving sector, we will include just a number of bosonic coordinates. The constraint algebra will be the Virasoro algebra and the Weyl anomaly cancellation implies that the number of right-moving coordinates is 26. Together, we have ten left+right bosonic coordinates $X^\mu(z, \bar{z})$, ten left-moving fermions $\psi^\mu(z)$ and an extra 16 right-moving coordinates $\phi^I(\bar{z})$, $I = 1, 2, \ldots, 16$. The X^μ are noncompact, but the ϕ^I are necessarily compact (for reasons of modular invariance) and must take values in some 16-dimensional lattice L_{16}.

To remove the tachyon, we will also impose the usual GSO projection on the left, namely, $(-1)^{F_L} = -1$. Here, we will have two sectors, generated by the left-moving fermions, the NS sector (space-time bosons) and the R sector (space-time bosons). Also the non-compact space-time dimension is ten, the ϕ^I being compact ("internal") coordinates.

As we describe below, the new nontrivial element of the heterotic string is the presence of the 16-dimensional internal lattice. The contribution of the left-moving compact bosons ϕ^I can be obtained by extracting the right-moving part of the toroidal CFT (4.18.40) on page 99:

$$Z_{\text{compact}}(\bar{q}) = \sum_{L_{16}} \frac{\bar{q}^{\vec{p}_R^2/2}}{\bar{\eta}^{16}} = \frac{\bar{\Gamma}_{16}(\bar{q})}{\bar{\eta}^{16}}, \tag{7.4.1}$$

where $\vec{p}_R$ is a lattice vector.

For the heterotic partition function to be modular invariant, the lattice sum $\bar{\Gamma}_{16}$ must be invariant under $\tau \to \tau + 1$. This implies that the lattice must be even ($\vec{p}_R^2 = $ even integer for all lattice vectors). It must also transform as

$$\tau \to -\frac{1}{\tau} : \bar{\Gamma}_{16} \to \bar{\tau}^8 \, \bar{\Gamma}_{16}. \tag{7.4.2}$$

This implies that the lattice is self-dual (the dual of the lattice coincides with the lattice itself). There are two Euclidean 16-dimensional lattices that satisfy the above requirements:

$E_8 \times E_8$ lattice

This is the root lattice of the group $E_8 \times E_8$. The roots of E_8 are composed of the roots of O(16), $\vec{\epsilon}_{ij}$, as well as the spinor weights of O(16), $\vec{\epsilon}_\alpha^s$. The O(16) roots $\vec{\epsilon}_{ij}$ are eight-dimensional vectors with a ± 1 in position i, a ± 1 in position j and zero elsewhere. For the spinor weights, $\vec{\epsilon}_\alpha^s = (\zeta_1, \zeta_2, \ldots, \zeta_8)/2$, $\alpha = 1, 2, \ldots, 128$ with $\zeta_i = \pm 1$, and $\sum_i \zeta_i = 0 \bmod (4)$.

The roots have squared length equal to 2. A general lattice vector can be written as $\sum_{i<j} n_{ij} \vec{\epsilon}_{ij} + \sum_\alpha m_\alpha \vec{\epsilon}_\alpha^s$, with $n_{ij}, m_\alpha \in \mathbb{Z}$. The lattice sum can be written in terms of ϑ-functions as

$$\bar{\Gamma}_{E_8 \times E_8} = (\bar{\Gamma}_8)^2 = \left[\frac{1}{2} \sum_{a,b=0,1} \bar{\vartheta} \begin{bmatrix} a \\ b \end{bmatrix}^8 \right]^2 = 1 + 2 \cdot 240\, \bar{q} + \mathcal{O}(\bar{q}^2). \tag{7.4.3}$$

Combining it with the oscillators, coming from the $\bar{\eta}$-functions, we observe that there are $2 \cdot 240 + 16 = 2 \cdot 248$ states with $\bar{L}_0 = 1$, which make the adjoint representation of $E_8 \times E_8$. This right-moving theory realizes the current algebra of $E_8 \times E_8$, both of them at level 1. The only integrable representation is the vacuum representation, and the first non-trivial states above the vacuum are generated by the current modes $\bar{J}_{-1}^a$.

Spin(32)/$\mathbb{Z}_2$ lattice

This is the root lattice of O(32) augmented by one of the two spinor weights. The roots of O(32) are $\vec{\epsilon}_{ij}$, which are 16-dimensional vectors with a ± 1 in position i, a ± 1 in position j and zero elsewhere. The spinor weights are $\vec{\epsilon}_\alpha^s = (\zeta_1, \zeta_2, \ldots, \zeta_{16})/2$, $\alpha = 1, 2, 3, \ldots, 2^{16}$, with $\zeta_i = \pm 1$, and $\sum_i \zeta_i = 0 \bmod (4)$. The roots have squared length equal to 2. The generic lattice vector is $\sum_{i<j} n_{ij} \vec{\epsilon}_{ij} + \sum_\alpha m_\alpha \vec{\epsilon}_\alpha^s$, with $n_{ij}, m_\alpha \in \mathbb{Z}$. The lattice sum can be written as

$$\bar{\Gamma}_{O(32)/\mathbb{Z}_2} = \frac{1}{2} \sum_{a,b=0,1} \bar{\vartheta} \begin{bmatrix} a \\ b \end{bmatrix}^{16} = 1 + 480\, \bar{q} + \mathcal{O}(\bar{q}^2). \tag{7.4.4}$$

This theory has a O(32) right-moving current algebra at level 1. The integrable representations present are the vacuum and the spinor. The states at $\bar{L}_0 = 1$ come from the current modes $\bar{J}_{-1}^a$. The spinor ground-states have $\bar{L}_0 = 2$.

Both right-moving current algebra theories can also be constructed from 32 free right-moving fermions $\bar{\psi}^i$, $i = 1, 2, \ldots, 32$. We start from the O(32) theory. The currents

$$\bar{J}^{ij} = i \bar{\psi}^i \bar{\psi}^j \tag{7.4.5}$$

form the level-one O(32) current algebra. In the R sector, all fermions are periodic, in which case O(32) invariance is not broken and we obtain the two spinors S, C of O(32). We impose a GSO-like projection $(-1)^F = 1$ where $(-1)^F$ is the fermion number of the 32 fermions. This keeps the vacuum representation in the NS sector and one of the spinors in the R sector.

To obtain the $E_8 \times E_8$ theory we split the fermions in two groups of 16. We then consider separate periodic or antiperiodic conditions for the two groups. In this case the $O(32)$ invariance is broken to $O(16) \times O(16)$. In the R sector, however, we obtain one of the spinors of $O(16)$ with $\bar{L}_0 = 1$. This spinor combines with the adjoint of $O(16)$ to make the adjoint of E_8. Note that this projection is opposite from the GSO projection on the left-moving fermions ψ^μ.

We can now describe the massless spectrum of the heterotic string theory (light-cone gauge). In the NS sector the constraints are

$$L_0 = \frac{1}{2}, \quad \bar{L}_0 = 1. \tag{7.4.6}$$

Taking also into account the GSO projection, we find that there is no tachyon and the massless states are $\psi^i_{-1/2}\bar{\alpha}^j_{-1}|p\rangle$, which gives the graviton, antisymmetric tensor and dilaton, and $\psi^i_{-1/2}\bar{J}^a_{-1}|p\rangle$, which gives vectors in the adjoint of $G = E_8 \times E_8$ or $O(32)$.

In the R sector the independent constraints are

$$G_0 = 0, \quad \bar{L}_0 = 1, \tag{7.4.7}$$

which, together with the GSO condition, give a Majorana-Weyl gravitino, a Majorana-Weyl fermion, and a set of Majorana-Weyl fermions in the adjoint of the gauge group G. The theory has $\mathcal{N} = 1_{10}$ supersymmetry in ten dimensions and contains at the massless level the supergravity multiplet, and a vector supermultiplet in the adjoint of G. Moreover, the theory is chiral.

There is another interesting heterotic theory we can construct in ten dimensions. This can be obtained as a $\mathbb{Z}_2$ orbifold of the $E_8 \times E_8$ theory. The first symmetry we will use is $(-1)^{F_R}$ that transforms space-time fermions with a minus sign. In each of the two E_8's there is also a symmetry that leaves the vector of the $O(16)$ subgroup invariant and changes the sign of the $O(16)$ spinor. We will call this symmetry generator $\mathcal{S}_i$, $i = 1, 2$ acting on the first, respectively second E_8's. The $\mathbb{Z}_2$ element by which we will orbifold is $(-1)^{F_R} \mathcal{S}_1 \mathcal{S}_2$.

In the sector of the left-moving world-sheet fermions only $(-1)^{F_R}$ acts nontrivially. The twisted blocks are

$$Z_{\text{fermions}}[^h_g] = \frac{1}{2} \sum_{a,b=0}^{1} (-1)^{a+b+ab+ag+bh+gh} \frac{\vartheta^4[^a_b]}{\eta^4}. \tag{7.4.8}$$

On each of the E_8's the nontrivial projection is $\mathcal{S}_i$, which gives the following orbifold blocks

$$\bar{Z}_{E_8}[^h_g] = \frac{1}{2} \sum_{\gamma,\delta=0}^{1} (-1)^{\gamma g+\delta h} \frac{\bar{\vartheta}^8[^\gamma_\delta]}{\bar{\eta}^8}. \tag{7.4.9}$$

The total partition function is

$$Z_{O(16)\times O(16)}^{\text{heterotic}} = \frac{1}{2} \sum_{h,g=0}^{1} \frac{\bar{Z}_{E_8}[^h_g]^2}{(\sqrt{\tau_2}\eta\bar{\eta})^8} \frac{1}{2} \sum_{a,b=0}^{1} (-1)^{a+b+ab+ag+bh+gh} \frac{\vartheta^4[^a_b]}{\eta^4}. \tag{7.4.10}$$

This is a heterotic ten-dimensional vacuum with gauge group $O(16)\times O(16)$ and no space-time supersymmetry. Exercises 7.6 and 7.31 on page 186 explore this theory further.

7.5 Superstring Vertex Operators

In analogy with the bosonic string, the vertex operators must be primary states of the superconformal algebra. We first describe the left-moving part of the superstring and therefore use chiral superfield language (see (4.13.13),(4.13.14) on page 79) where

$$\hat{X}^{\mu}(z,\theta) = X^{\mu}(z) + \theta \psi^{\mu}(z). \tag{7.5.1}$$

The left-moving vertex operators can be written in the form

$$\int dz \int d\theta \ V(z,\theta) = \int dz \int d\theta \ (V_0(z) + \theta V_{-1}(z)) = \int dz \ V_{-1}. \tag{7.5.2}$$

The conformal weight of V_0 is $\frac{1}{2}$ while that of V_{-1} is 1. The integral of V_{-1} has conformal weight zero. For the massless space-time bosons the vertex operator is[5]

$$V^{\text{boson}}(\epsilon, p, z, \theta) = \epsilon_{\mu} : D\hat{X}^{\mu} \ e^{ip\cdot\hat{X}} :, \tag{7.5.3}$$

$$V_0^{\text{boson}}(\epsilon, p, z) = \epsilon_{\mu} \psi^{\mu} e^{ip\cdot X}, \quad V_{-1}^{\text{boson}}(\epsilon, p, z) = \epsilon_{\mu} : (\partial X^{\mu} + ip \cdot \psi \ \psi^{\mu}) e^{ip\cdot X} :, \tag{7.5.4}$$

where $\epsilon \cdot p = 0$.

We would like to present the vertex operators in the covariant quantization. The reason is that the fermion vertex has a simple form only in the modern covariant quantization. It is useful to use the bosonization of the β-γ system described in section 4.20.1 on page 106 in terms of a scalar field ϕ with background charge $Q = 2$, and the η-ξ system as

$$\gamma(z) = e^{\phi(z)}\eta(z), \quad \beta(z) = e^{-\phi(z)}\partial\xi(z). \tag{7.5.5}$$

η has dimension one while ξ has dimension zero. In general, the vertex operator $: e^{q\phi} :$ has conformal weight $-q - \frac{q^2}{2}$. The spin fields of β, γ that interpolate between NS and R sectors are given by $e^{\pm\phi/2}$ with conformal weight $-5/8$ and $3/8$. It should be noted that the zero mode of the ξ field does not appear in the bosonization of the β-γ system. It introduces an extra redundancy in the new Hilbert space.

There is a subtlety in the case of fermionic strings having to do with the β-γ system. As we have seen, in the bosonized form, the presence of the background charge alters the charge neutrality condition.[6] This is related to the existence of supermoduli and super-Killing spinors. Therefore, depending on the correlation function and surface we must have different representatives for the vertex operators of a given physical state with different ϕ-charges. The precise condition, derived in section 4.20.1 is that the sum of ϕ charges must be equal to the Euler number of the surface, χ.

Different representatives are constructed as follows. Consider a physical vertex operator with ϕ charge q, V_q. It is BRST invariant, $[Q_{\text{BRST}}, V_q] = 0$. We can construct another physical vertex operator representing the same physical state but with charge $q + 1$. It is $V_{q+1} = [Q_{\text{BRST}}, \xi V_q]$ and has charge $q + 1$ since Q_{BRST} carries charge 1. Since it is a BRST

---

[5] See also exercise 4.39 on page 122.

[6] The notion of the charge neutrality condition was described in section 4.14 on page 82. Its interpretation in terms of zero modes and the index can be found in section 4.15 on page 84.

commutator, V_{q+1} is also BRST invariant. However, we have seen that states that are BRST commutators of physical states are spurious. This is not the case here since the ξ field appears in the commutator and its zero mode lies outside the ghost Hilbert space. The different ϕ charges are usually called "pictures" in the literature.

In the covariant setup, the massless NS vertex operator becomes[7]

$$V_{-1}^{\text{boson}}(\epsilon, p, z) = e^{-\phi(z)} \epsilon \cdot \psi \, e^{ip \cdot X}. \tag{7.5.6}$$

We can construct another equivalent vertex operator in the zero picture as

$$V_0^{\text{boson}}(\epsilon, p, z) = [Q_{\text{BRST}}, \xi(z) e^{-\phi(z)} \epsilon \cdot \psi \, e^{ip \cdot X}] = \epsilon_\mu (\partial X^\mu + ip \cdot \psi \psi^\mu) \, e^{ip \cdot X}. \tag{7.5.7}$$

The space-time fermion vertex operators can only be constructed in the covariant formalism. For the massless states ($p^2 = 0$), in the canonical $-\frac{1}{2}$ picture they are of the form

$$V_{-1/2}^{\text{fermion}}(u, p, z) = u^\alpha(p) : e^{-\phi(z)/2} \, S_\alpha(z) \, e^{ip \cdot X} :. \tag{7.5.8}$$

S_α is the spin field of the fermions ψ^μ forming an $O(10)_1$ current algebra, with weight $5/8$ (see section 4.12 on page 71). The total conformal weight of $V_{-1/2}$ is 1. Finally, u^α is a spinor wave function, satisfying the massless Dirac equation $\slashed{p} u = 0$.

The $\frac{1}{2}$ picture for the fermion vertex can be computed to be

$$V_{1/2}^{\text{fermion}}(u, p) = [Q_{\text{BRST}}, \xi(z) V_{-1/2}^{\text{fermion}}(u, p, z)]$$

$$= u^\alpha(p) \, e^{\phi/2} \, (\Gamma^\mu)_{\alpha\beta} \, S^\beta \, \partial X^\mu \, e^{ip \cdot X} + \cdots, \tag{7.5.9}$$

where the ellipsis involves terms that do not contribute to four-point amplitudes. The ten-dimensional space-time supersymmetry charges can be constructed from the fermion vertex at zero momentum,

$$Q_\alpha = \frac{1}{2\pi i} \oint dz \; : e^{-\phi(z)/2} \, S_\alpha(z). \tag{7.5.10}$$

It transforms fermions into bosons and vice versa,

$$[Q_\alpha, V_{-1/2}^{\text{fermion}}(u, p, z)] = V_{-1}^{\text{boson}}(\epsilon^\mu = u^\beta \gamma_{\beta\alpha}^\mu, p, z), \tag{7.5.11}$$

$$[Q_\alpha, V_0^{\text{boson}}(\epsilon, p, z)] = V_{-1/2}^{\text{fermion}}(u^\beta = ip^\mu \epsilon^\nu (\gamma_{\mu\nu})_\alpha^\beta, p, z). \tag{7.5.12}$$

There are various pictures for the supersymmetry charges also.

In the open string, the vertex operators described above are the whole story.

In the type-II closed strings the vertex operators are products of two copies of the supersymmetric chiral vertex operators.

For the heterotic string theory, the left-moving part is supersymmetric and the vertex operators are as above. The right-moving part is nonsupersymmetric and the vertex oper-

---

[7] We are using the same symbol for the vertex operator in both the old covariant and the BRST formulation.

ators are those discussed in chapter 5 on page 126. In particular, for the massless states they are currents multiplied by exponentials. We have only left-moving pictures here.

The massless vertex operators are as follows:

- Graviton/antisymmetric tensor/dilaton,

$$V_0 = \epsilon_{\mu\nu}(\partial X^\mu + ip \cdot \psi \psi^\mu)\bar{\partial}X^\nu e^{ip \cdot X}.$$ (7.5.13)

- Gravitino/fermion

$$V_{-1/2} = u_\nu^\alpha(p) : e^{-\phi(z)/2} \, S_\alpha(z)\bar{\partial}X^\nu \, e^{ip \cdot X}.$$ (7.5.14)

- Gauge boson

$$V_0^a = \epsilon_\mu(\partial X^\mu + ip \cdot \psi \psi^\mu)\bar{J}^a e^{ip \cdot X}.$$ (7.5.15)

where $\bar{J}^a$ is an antiholomorphic $O(32)$ or $E_8 \times E_8$ current.

- Gaugino

$$V_{-1/2}^a = u^\alpha(p) : e^{-\phi(z)/2} \, S_\alpha(z)\bar{J}^a \, e^{ip \cdot X} : .$$ (7.5.16)

7.6 One-loop Superstring Vacuum Amplitudes

We will now study the one-loop vacuum amplitude (or partition function) of the various supersymmetric superstring theories namely, type IIA/B, type I, and heterotic. The important difference from the similar calculation of the bosonic string is that we must include the contributions of the world-sheet fermions and impose the proper GSO projections. In the presence of unbroken supersymmetry in flat space, the one-loop vacuum amplitudes will vanish. This will no longer be the case, however, in vacua where supersymmetry is completely broken. Flat space is destabilized in such vacua. This is a form of the cosmological constant problem.

7.6.1 The type-IIA/B superstring

For the bosonic part of the action we have eight transverse oscillators and, in analogy with the case of the bosonic string, we will get a contribution of $((2\pi \ell_s)\sqrt{\tau_2}\eta\bar{\eta})^{-8}$. Here, however, we also have the contribution of the world-sheet fermions. We will consider first the IIB case. In the NS-NS sector the two GSO projections imply that we have the vector on both sides. So the contribution to the partition function is $\chi_V \bar{\chi}_V$ where χ_V was defined in (4.12.26) on page 74. From the R-$\bar{\text{R}}$ sector we have projected out the S representation so the contribution is $\chi_C \bar{\chi}_C$. In the R-NS and NS-R sectors we obtain $-\chi_C \bar{\chi}_V$ and $-\chi_V \bar{\chi}_C$, respectively. The minus sign is there since space-time fermions contribute to the one-loop amplitude with a minus sign relative to space-time bosons. The total is

$$\Lambda_{\text{IIB}} = \frac{iV_{10}}{4(2\pi \ell_s)^{10}} \int_{\mathcal{F}} \frac{d^2\tau}{\tau_2^2} \frac{(\chi_V - \chi_C)(\bar{\chi}_V - \bar{\chi}_C)}{(\sqrt{\tau_2}\eta \, \bar{\eta})^8}.$$ (7.6.1)

Using the formulas (4.12.26) and (4.12.37) on page 76 for the SO(8) characters, we can write the partition function as

$$\Lambda_{\text{IIB}} = \frac{iV_{10}}{4(2\pi\ell_s)^{10}} \int_{\mathcal{F}} \frac{d^2\tau}{\tau_2^2} \frac{1}{(\sqrt{\tau_2}\eta\bar{\eta})^8} \frac{1}{2} \sum_{a,b=0}^{1} (-1)^{a+b} \frac{1}{2} \sum_{\bar{a},\bar{b}=0}^{1} (-1)^{\bar{a}+\bar{b}} \frac{\vartheta^4[^a_b] \bar{\vartheta}^4[^{\bar{a}}_{\bar{b}}]}{\eta^4 \bar{\eta}^4}. \tag{7.6.2}$$

$a = 0$ labels the NS sector, $a = 1$ the R sector, and similarly for the right-movers.

In the type-IIA case, the only difference is that $\bar{\chi}_C$ should be substituted by $\bar{\chi}_S$. The vacuum amplitude becomes

$$\Lambda_{\text{IIA}} = \frac{iV_{10}}{4(2\pi\ell_s)^{10}} \int_{\mathcal{F}} \frac{d^2\tau}{\tau_2^2} \frac{1}{(\sqrt{\tau_2}\eta\bar{\eta})^8} \frac{1}{2} \sum_{a,b=0}^{1} (-1)^{a+b} \frac{1}{2} \sum_{\bar{a},\bar{b}=0}^{1} (-1)^{\bar{a}+\bar{b}+\bar{a}\bar{b}} \frac{\vartheta^4[^a_b] \bar{\vartheta}^4[^{\bar{a}}_{\bar{b}}]}{\eta^4 \bar{\eta}^4}. \tag{7.6.3}$$

The integrants in (7.6.2),(7.6.3) are modular invariant. They are also identically zero. This implies that, at each mass level, there is an equal number of bosonic and fermionic degrees of freedom. This is consistent with the presence of space-time supersymmetry.

7.6.2 The heterotic superstring

For the heterotic superstring, the ten space-time coordinates as well as the right-moving fermions are the same as in the type-II strings. The only difference is that the right-moving fermions of the type-II string here are replaced by the $E_8 \times E_8$ or $\text{Spin}(32)/\mathbb{Z}_2$ lattice. Therefore, the vacuum amplitude is

$$\Lambda_{\text{het}} = \frac{iV_{10}}{4(2\pi\ell_s)^{10}} \int_{\mathcal{F}} \frac{d^2\tau}{\tau_2^2} \frac{1}{2} \frac{\left(\sum_{a,b=0}^{1} (-1)^{a+b+ab} \vartheta^4[^a_b]\right)}{(\sqrt{\tau_2}\eta\bar{\eta})^8 \eta^4} \frac{\bar{\Gamma}_{16}}{\bar{\eta}^{16}}, \tag{7.6.4}$$

where the lattice was defined in (7.4.1) on page 165.

7.6.3 The type-I superstring

Here, we have four Riemann surfaces contributing to the vacuum amplitude. The torus contribution is that of the IIB string, calculated in the previous section. It now has an extra factor of $1/2$ coming from the projector $(1 + \Omega)/2$ that will project on the unoriented states.

The other closed surface is the Klein bottle. This is the trace of Ω in the IIB string. By including the $1/2$ of the GSO projection and the $1/2$ of the Ω projection we obtain

$$\Lambda_{K_2} = i\frac{V_{10}}{(2\pi\ell_s)^{10}} \int_0^\infty \frac{dt}{8t} \frac{\sum_{a,b=0}^{1} (-1)^{a+b} \vartheta^4[^a_b](2it)}{t^5 \, \eta(2it)^{12}}$$

$$= i\frac{2^2 V_{10}}{\pi(2\pi\ell_s)^{10}} \int_0^\infty d\ell \frac{\sum_{a,b=0}^{1} (-1)^{a+b} \vartheta^4[^a_b]\left(2i\frac{\ell}{\pi}\right)}{\eta\left(2i\frac{\ell}{\pi}\right)^{12}}. \tag{7.6.5}$$

where in the second line we have translated the amplitude to the transverse channel. The only sectors that contribute are the NS-NS and the R-R sectors. The total closed string amplitude is $\frac{1}{2}\Lambda_{\text{IIB}} + \Lambda_{K_2}$.

The Klein bottle amplitude has divergences, due to massless tadpoles, as in the bosonic case discussed in section 5.3. They are obtained by keeping the long distance contributions in the second line of (7.6.5),

$$\mathcal{T}_K^{NS} = -\mathcal{T}_K^R = i\frac{V_{10}}{2^4\pi^{11}\ell_s^{10}} \int_0^\infty d\ell. \tag{7.6.6}$$

We will now evaluate the cylinder amplitude. The bosonic pieces are the same as in the bosonic string, the fermionic piece is the standard one, and we obtain (remembering an extra $1/2$ factor from the Ω projection)

$$\Lambda_{C_2} = iV_{10}N^2 \int_0^\infty \frac{dt}{8t} \frac{\sum_{a,b=0}^1 (-1)^{a+b} \vartheta^4[{}^a_b](it)}{(8\pi^2\ell_s^2 t)^5 \eta^{12}(it)}$$

$$= i\frac{V_{10}N^2}{8\pi(8\pi^2\ell_s^2)^5} \int_0^\infty d\ell \frac{\sum_{a,b=0}^1 (-1)^{a+b} \vartheta^4[{}^a_b]\left(i\frac{\ell}{\pi}\right)}{\eta^{12}\left(i\frac{\ell}{\pi}\right)}. \tag{7.6.7}$$

The associated tadpoles are

$$\mathcal{T}_C^{NS} = -\mathcal{T}_C^R = i\frac{N^2 V_{10}}{2^{14}\pi^{11}\ell_s^{10}} \int_0^\infty d\ell. \tag{7.6.8}$$

Finally to calculate the Möbius strip we need (7.3.5)–(7.3.17). This action of Ω may be defined for a general character

$$\chi_\Delta(q) = \mathrm{Tr}[q^{L_0-c/24}] = q^{\Delta-c/24} \sum_{n=0}^\infty a_n q^n \tag{7.6.9}$$

as passage to the careted character

$$\hat{\chi}_\Delta(q) = \mathrm{Tr}[\Omega\, q^{L_0-c/24}] = q^{\Delta-c/24} \sum_{n=0}^\infty a_n (-1)^n q^n = e^{-i\pi(\Delta-c/24)}\chi_\Delta(-q). \tag{7.6.10}$$

As we have seen, there is a sign ambiguity in the definition of the careted character, corresponding to the transformation properties of the vacuum.

The amplitude can then be calculated along the lines of section 4.22 on page 112 and using the results of section 7.3 and exercise 4.70 on page 125. In the NS sector we obtain[8]

$$\Lambda_{M_2}^{NS} = -i\zeta NV_{10} \int_0^\infty \frac{dt}{8t} \frac{\sum_{b=0}^1 (-1)^b \hat{\vartheta}^4[{}^0_b](it)}{(8\pi^2\ell_s^2 t)^5 \hat{\eta}^{12}(it)}$$

$$= -i\zeta NV_{10} \int_0^\infty \frac{dt}{8t} \frac{\vartheta_2(2it)^4 \vartheta_4(2it)^4}{(8\pi^2\ell_s^2 t)^5 \eta(2it)^8 \vartheta_3(2it)^4}$$

$$= -2i\zeta N\frac{2^5 V_{10}}{8\pi(8\pi^2\ell_s^2)^5} \int_0^\infty d\ell \frac{\vartheta_2\left(2i\frac{\ell}{\pi}\right)^4 \vartheta_4\left(2i\frac{\ell}{\pi}\right)^4}{\eta\left(2i\frac{\ell}{\pi}\right)^8 \vartheta_4\left(2i\frac{\ell}{\pi}\right)^4}, \tag{7.6.11}$$

---

[8] When going to the transverse channel via a modular inversion care is necessary, since the transformation mixes contributions from the NS and R sectors. In particular, $\mathrm{Tr}_{NS}[(-1)^F] \leftrightarrow \mathrm{Tr}_R[1]$.

while in the R sector

$$
\Lambda_{M_2}^R = -i\epsilon_R \zeta N V_{10} \int_0^\infty \frac{dt}{8t} \frac{\sum_{b=0}^1 (-1)^b \, \hat{\vartheta}^4[^1_b](it)}{(8\pi^2 \ell_s^2 t)^5 \, \hat{\eta}^{12}(it)}
$$

$$
= -i\epsilon_R \zeta N V_{10} \int_0^\infty \frac{dt}{8t} \frac{\vartheta_2(2it)^4 \vartheta_4(2it)^4}{(8\pi^2 \ell_s^2 t)^5 \, \eta(2it)^8 \vartheta_3(2it)^4}
$$

$$
= -2i\epsilon_R \zeta N \frac{2^5 V_{10}}{8\pi (8\pi^2 \ell_s^2)^5} \int_0^\infty d\ell \, \frac{\vartheta_2\left(2i\frac{\ell}{\pi}\right)^4 \vartheta_4\left(2i\frac{\ell}{\pi}\right)^4}{\eta\left(2i\frac{\ell}{\pi}\right)^8 \vartheta_4\left(2i\frac{\ell}{\pi}\right)^4}. \tag{7.6.12}
$$

We have also used the duplication formulas for the ϑ-functions in appendix C on page 508.
The associated tadpoles are

$$
\mathcal{T}_M^{NS} = -i\zeta \frac{N V_{10}}{2^8 \pi^{11} \ell_s^{10}} \int_0^\infty d\ell, \quad \mathcal{T}_M^R = -i\epsilon_R \zeta \frac{N V_{10}}{2^8 \pi^{11} \ell_s^{10}} \int_0^\infty d\ell. \tag{7.6.13}
$$

We may now collect the contributions of all surfaces to the tadpoles,

$$
\mathcal{T}^{NS} = \left[N^2 - 2^6 \zeta N + (2^5)^2\right] \frac{i V_{10}}{16\pi (2\pi \ell_s)^{10}} \int_0^\infty d\ell \tag{7.6.14}
$$

and

$$
\mathcal{T}^R = -\left[N^2 + 2^6 \epsilon_R \zeta N + (2^5)^2\right] \frac{i V_{10}}{16\pi (2\pi \ell_s)^{10}} \int_0^\infty d\ell. \tag{7.6.15}
$$

The NS tadpole divergence is related to a nonzero graviton and dilaton one-point function on the disk as in the bosonic case. There is also a similar divergence coming from the R sector that in the transverse channel should be interpreted as a nonzero one-point function of a R-R massless field on the disk.

We would like to know what kind of R-R form gives such a tadpole. As we mentioned already, the type-IIB string theory has p-form gauge potentials of even degree. The only one that could give such a constant tadpole would be a ten-form. A ten-form in ten dimensions has a vanishing field strength and trivial dynamics. As we will verify later, it has minimal couplings, like $\mu \int C_{10}$ that would generate the requisite tadpoles. In fact since its field strength vanishes identically, there is no kinetic term for this form. Its equation of motion states that $\mu = 0$. μ can be thought of as the charge of the vacuum under the ten-form, in analogy with the usual electromagnetic minimal coupling.

A non zero charge μ would imply a tree-level closed string contribution of the form $\mu^2/0$ due to its zero propagator. This is the divergence we are finding in the R sector. It is proportional to the (charge)^2 of the ten-form. Consistency and finiteness of the theory imposes that we take this charge to be zero. This implies in particular that

$$
N = 32, \quad \epsilon_R \zeta = -1. \tag{7.6.16}
$$

In the NS sector, the tadpoles are due to the graviton and the dilaton. This is due to an nontrivial energy and dilaton charge of the vacuum. We must therefore either cancel tadpoles separately in the NS and R sectors or find another vacuum without tadpoles. Canceling the NS tadpoles we obtain

$$
N = 32, \quad \zeta = 1. \tag{7.6.17}
$$

There are two possibilities here.

(a) If we preserve supersymmetry, $\zeta = 1 = -\epsilon_R$ and we obtain the type-I theory. In this case the open sector provides an SO(32) SYM multiplet at the massless level.

(b) We may break supersymmetry in the open sector by choosing $\zeta = 1 = \epsilon_R$. The bosons form the adjoint of Sp(32), while the fermions form the antisymmetric representation. This is known as the Sugimoto "vacuum." Although the R tadpoles are cancelled, there still remain NS tadpoles (graviton and dilaton sources).

As we will see in the later section 8.7 on page 201, there is an interesting and useful interpretation of the tadpole cancellation condition. The Klein bottle tadpoles are due to a nondynamical plane (nine-plane here) with charge 32. It is known as an orientifold plane, or O_9 plane. It has negative tension and negative brane charge. Its energy is cancelled by another object, a D-brane. We will learn more on this in the next chapter.

7.7 Closed Superstrings and T-duality

In section 4.18.7 on page 101 we saw that T-duality was a symmetry in the bosonic string. We will investigate here the action of T-duality in the closed superstring theories since the presence of world-sheet fermions is responsible for some important subtleties.

7.7.1 The type-II string theories

Consider the compactification of the x^9 coordinate on a circle. We have seen in section 4.18.7 that T-duality flips the sign of the anti-chiral U(1) current $\partial x^9 \to \partial x^9$, $\bar{\partial} x^9 \to -\bar{\partial} x^9$ and is a symmetry of the bosonic part of the theory. The right-moving world-sheet supercharge contains the combination $\bar{\psi}^9 \bar{\partial} x^9$. In order to preserve the $\mathcal{N} = (1, 1)_2$ super-conformal invariance, we must also flip at the same time the sign of $\bar{\psi}^9$ while keeping ψ^9 invariant. This is implemented by a right-moving fermion parity transformation, $(-1)^{F_R}$.

In the right-moving R sector, the parity transformation of the fermions changes the chirality of the R spinor.

To study the effect in detail, we must focus on the action of fermion parity on the spin fields. The parity action must anticommute with Γ^9 and commute with the other Γ-matrices. Then, up to a phase,[9] it acts as

$$\tilde{S}_\alpha \to \Gamma^9 \Gamma^{11} \tilde{S}_\alpha. \tag{7.7.1}$$

This has the effect of flipping the GSO projection on the right-hand side. Remember that Γ^{11}, because of the existing GSO projection gives a $\pm$ sign in (7.7.1), but Γ^9 changes the chirality. Thus, the T-dual of the IIA string of radius R is the IIB string at radius $1/R$, and vice versa.

To see the action on the R-R forms, we must investigate the action of the parity transformation 7.7.1 on the decomposition (7.2.19) on page 160 that defines the R-R forms.

---

[9] This phase has been fixed using the reality properties of the spinors.

As argued above, the Γ^{11} action is trivial giving only a sign. However, from (7.2.15) on page 160 we may observe that the action of Γ^9 gives an extra 9 index to the R-R forms if there isn't one, and removes one if there is. You are asked to work this out in detail in exercise 7.7 on page 184.

It is obvious that T-dualities in different directions anticommute instead of commuting in the R-R sector, since

$$\Gamma^\mu \Gamma^{11} \Gamma^\nu \Gamma^{11} + \Gamma^\nu \Gamma^{11} \Gamma^\mu \Gamma^{11} = 0. \tag{7.7.2}$$

In particular, doing the T-duality transformation twice, we obtain unity in the NS sector but -1 in the R sector. This is summarized in the statement that the square of T-duality is $(-1)^{\mathbf{F}_R}$, where $\mathbf{F}_R$ is the right-moving space-time fermion number[10] whose value is 0 in the right-moving NS sector and 1 in the associated R sector.

7.7.2 The heterotic string

T-duality in the heterotic string is simpler. The reason is that the theory is supersymmetric on one side only. The change of parity in the R sector consequently gives an isomorphic theory. Thus, T-duality is a symmetry in the heterotic theory. Here the subtlety lies in the way T-duality acts on the Wilson lines, and a glimpse of this will be seen in section 11.1 on page 322.

7.8 Supersymmetric Effective Actions

For the supersymmetric string theories we would like to find the effective field theories that describe the dynamics of the massless fields as we have done in chapter 6 for the bosonic string. A straightforward approach would be the one we used in the bosonic case, namely, either extracting them from scattering amplitudes or requiring Weyl invariance of the associated σ-model in general background fields. In the presence of supersymmetry, however, and at the two-derivative level, these effective actions are uniquely fixed. They were constructed during the late 1970s/early 1980s, as supergravity theories.

First, we would like to obtain the low-energy effective action at the leading order in the α' expansion. When only bosonic fields are present, we just have to keep terms of up to two derivatives. In the presence of fermions, however, we would like to give equal weight to the kinetic terms $\phi \Box \phi$ for bosons and $\bar{\psi} \partial\!\!\!/ \psi$ for fermions. We will give weight 0 to bosons, weight $\frac{1}{2}$ to fermions and 1 to a derivative. Then, both kinetic terms have the same weight, namely 2. These weights are also respected by supersymmetry (SUSY) as can be verified directly from the generic SUSY transformations

$$\delta_\epsilon \phi \sim \phi^m \, \psi \epsilon, \quad \delta_\epsilon \psi \sim \partial \phi^m \, \epsilon + \phi^m \, \psi^2 \epsilon. \tag{7.8.1}$$

The effective actions in the leading order must have weight 2, and this is true for all supergravity actions.

---

[10] Not to be confused with the world-sheet fermion number $F_{L,R}$.

In appendix H on page 525 we present the bosonic parts of the relevant $\mathcal{N} = 1_{10}$ and $\mathcal{N} = 2_{10}$ (IIA,B) supergravity actions in ten dimensions, as well as the supersymmetry transformations of fermions. These two pieces of data are the most useful practically. We also present the eleven-dimensional supergravity as it is closely related to the IIA theory.

7.9 Anomalies

An anomaly is the breakdown of a classical symmetry in the quantum theory. Two types of symmetries can have anomalies, global or local (gauge) symmetries. In the following we will be interested in anomalies of local symmetries. If a local symmetry has an anomaly, this implies that longitudinal degrees of freedom no longer decouple. This signals a breakdown of unitarity. In two dimensions, anomalies are not fatal. The example of the chiral Schwinger model, a U(1) gauge theory coupled to a massless fermion, indicates that one can include the extra degrees of freedom and obtain a consistent theory. However, we do not yet know how to implement this procedure in more than two dimensions. Therefore, we will require the absence of anomalies.

Consider the physical effective action of a theory containing gauge fields as well as a metric $\Gamma^{\text{eff}}[A_\mu, g_{\mu\nu}, \dots]$. The gauge current and the energy-momentum tensor are

$$J^\mu = \frac{\delta \Gamma^{\text{eff}}}{\delta A_\mu}, \quad T^{\mu\nu} = \frac{1}{\sqrt{-g}} \frac{\delta \Gamma^{\text{eff}}}{\delta g_{\mu\nu}}. \tag{7.9.1}$$

The variation of the effective action under a gauge transformation $\delta_\Lambda A = D\Lambda + [a, \Lambda]$ is

$$\delta_\Lambda \Gamma^{\text{eff}} = \text{Tr} \int D_\mu \Lambda \frac{\delta \Gamma^{\text{eff}}}{\delta A_\mu} = \text{Tr} \int \Lambda D_\mu \frac{\delta \Gamma^{\text{eff}}}{\delta A_\mu} = \int \text{Tr} \, [\Lambda \, D_\mu J^\mu], \tag{7.9.2}$$

where we have used integration by parts. Consequently, iff $D_\mu J^\mu \neq 0$ there is an anomaly in the gauge symmetry. Similar remarks apply to the invariance under diffeomorphisms:

$$\delta_{\text{diff}} \Gamma^{\text{eff}} = \int (\nabla^\mu \epsilon^\nu + \nabla^\nu \epsilon^\mu) \frac{\delta \Gamma^{\text{eff}}}{\delta g_{\mu\nu}} = -\int \epsilon^\mu \nabla_\nu T^{\mu\nu}. \tag{7.9.3}$$

Thus, a gravitational anomaly implies the nonconservation of the stress-tensor in the quantum theory.

Anomalies in field theory appear due to UV problems. Consider the triangle graph in four dimensions (figure 7.1a). It is superficially linearly divergent, and gauge invariance reduces this to a logarithmic divergence. If the fermions going around the loop are nonchiral, we can regularize the diagram using Pauli-Villars regularization and we can easily show that the graph vanishes when one of the gauge field polarizations is longitudinal. There is no anomaly in this case. However, if the fermions are chiral, Pauli-Villars (or any other regularization) will break gauge invariance, which will not be recovered when the regulator mass is taken to infinity.

In ten dimensions, the leading graph that can give a contribution to anomalies is the hexagon diagram depicted in figure 7.1b. The external lines can be either gauge bosons

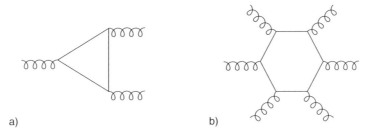

Figure 7.1 a) The anomalous triangle diagram in four dimensions. b) The anomalous hexagon diagram in ten dimensions.

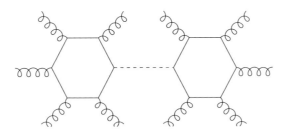

Figure 7.2 Two-loop diagram with physical external legs in which longitudinal modes propagate.

or gravitons. It can be shown that only the completely symmetric part of the graph gives a nontrivial contribution to the anomaly. Nonsymmetric contributions can be canceled by local counterterms. If the diagram were nonzero when one of the external lines is longitudinal, then this will imply that the unphysical polarizations will propagate in the two-loop diagram in figure 7.2.

We will consider the linearized approximation, which is relevant for the leading hexagon diagram: $F = F_0 + A^2$, $F_0 = dA$ and $A \rightarrow A + d\Lambda$. Here, Λ is the gauge parameter matrix (zero-form). The anomaly due to the hexagon diagram with gauge fields in the external lines can have the following general form

$$\delta\Gamma|_{\text{gauge}} \sim \int d^{10}x \; \left[c_1 \text{Tr}[\Lambda F_0^5] + c_2 \text{Tr}[\Lambda F_0]\text{Tr}[F_0^4] + c_3 \text{Tr}[\Lambda F_0](\text{Tr}[F_0^2])^2 \right], \tag{7.9.4}$$

where powers of forms are understood as wedge products. For comparison, the similar expression in four dimensions is proportional to $\text{Tr}[\Lambda F_0^2]$. The three different coefficients c_i correspond to the three group invariants $\text{Tr}[T^6]$, $\text{Tr}[T^4]\text{Tr}[T^2]$, and $(\text{Tr}[T^2])^3$ of a given group generator T in a symmetric group trace. There is a similar result for the gravitational anomaly, where the role of F is played by the $O(D)$ two-form $R_{\mu\nu}^{ab} = e_\rho^a e_\sigma^b R^{\rho\sigma}{}_{\mu\nu}$. The matrix valued two-form R is obtained by multiplying R^{ab} with the $O(D)$ adjoint matrices $T^{ab} \sim [\Gamma^a, \Gamma^b]$. It can be written in terms of the spin connection one-form ω as $R = d\omega + \omega^2$.

Considering the anomaly diagram with graviton external lines we obtain

$$\delta\Gamma|_{\text{grav}} \sim \int d^{10}x \; \left[d_1 \text{Tr}[\Theta R_0^5] + d_2 \text{Tr}[\Theta R_0]\text{Tr}[R_0^4] + d_3 \text{Tr}[\Theta R_0](\text{Tr}[R_0^2])^2 \right]. \tag{7.9.5}$$

Finally, by considering some of the external lines to be gauge bosons and some to be gravitons we obtain the mixed anomaly

$$\delta\Gamma|_{\text{mixed}} \sim \int d^{10}x \left[e_1 \text{Tr}[\Lambda F_0]\text{Tr}[R_0^4] + e_2 \text{Tr}[\Theta R_0]\text{Tr}[F_0^4] \right.$$
$$\left. + e_3 \text{Tr}[\Theta R_0](\text{Tr}[F_0^2])^2 + e_4 \text{Tr}[\Lambda F_0](\text{Tr}[R_0])^2 \right]. \tag{7.9.6}$$

There is also another potential term $\text{Tr}[\Lambda F_0]\text{Tr}[F_0^2]\text{Tr}[R_0^2]$, but it can be removed by a local counterterm.

There is a geometric construction that provides the full anomaly from the leading linearized piece. First, the anomaly satisfies the so-called Wess-Zumino consistency condition, which reflects the group structure of gauge transformations. Let $G(\Lambda) = \delta\Gamma/\delta\Lambda$. Then

$$\delta_{\Lambda_1} G(\Lambda_2) - \delta_{\Lambda_2} G(\Lambda_1) = G([\Lambda_1, \Lambda_2]). \tag{7.9.7}$$

The field strengths transform as follows under gauge transformations and diffeomorphisms:

$$\delta F = [F, \Lambda], \quad \delta R = [R, \Theta]. \tag{7.9.8}$$

It is straightforward to show that the traces $\text{Tr}[R^m]$ and $\text{Tr}[F^m]$ are gauge invariant and closed:

$$d\,\text{Tr}[R^m] = d\,\text{Tr}[F^m] = 0. \tag{7.9.9}$$

In order to construct the anomaly D-form in D dimensions we start with the most general gauge-invariant and closed $(D+2)$-form $I^{D+2}(R, F)$, which can be written as a linear combination of products of even traces of F, R. Since I^{D+2} is closed, it can be written (locally) as

$$I^{D+2}(R, F) = d\Omega^{D+1}(\omega, A), \tag{7.9.10}$$

where the $(D+1)$-form Ω^{D+1} is no longer gauge invariant, but changes under gauge transformations as

$$\delta_\Lambda \Omega^{D+1}(\omega, A) = d\Omega^D(\omega, A, \Lambda). \tag{7.9.11}$$

This is required by the fact that I^{D+2} is gauge invariant. Finally, the D-dimensional anomaly is the piece of Ω^D linear in Λ.

Except for the irreducible part of the gauge anomaly proportional to $\text{Tr}\,\Lambda F_0^5$ and $\text{Tr}[\Theta R_0^5]$, the rest can be canceled, if it appears in a suitable linear combination. This is known as the Green-Schwarz mechanism of anomaly cancellation.

Assume that the reducible part of the anomaly factorizes as follows:

$$\delta\Gamma|_{\text{reduc}} \sim \int d^{10}x\, (\text{Tr}[\Lambda F_0] - \text{Tr}[\Theta R_0]) \left(a_1 \text{Tr}[F_0^4] + a_2 \text{Tr}[R_0^4] \right.$$
$$\left. + a_3 (\text{Tr}[F_0^2])^2 + a_4 (\text{Tr}[R_0^2])^2 + a_5 \text{Tr}[F_0^2]\text{Tr}[R_0^2] \right). \tag{7.9.12}$$

In $\mathcal{N} = 1_{10}$ supergravity, the field strength of the antisymmetric tensor is shifted by the gauge Chern-Simons form, as described in appendix H.5 on page 530. We must also add the gravitational Chern-Simons form in order to cancel gravitational anomalies,

$$\hat{H} = dB - \frac{\kappa^2}{g^2}\Omega_3^{YM}(A) + \frac{\kappa^2}{g^2}\Omega_3^{GR}(\omega). \tag{7.9.13}$$

This addition makes the B-form transform under gauge transformations and diffeomorphisms to keep $\hat{H}$ invariant. Since

$$\delta_\Lambda \Omega_3^{YM}(A) = d\,\mathrm{Tr}[\Lambda\,dA], \quad \delta_\Theta \Omega_3^{GR}(\omega) = d\mathrm{Tr}[\Theta\,d\omega], \tag{7.9.14}$$

the antisymmetric tensor must transform as

$$\delta B = \frac{\kappa^2}{g^2}\mathrm{Tr}[\Lambda\,F_0 - \Theta\,R_0]. \tag{7.9.15}$$

Therefore, the presence of the term

$$\Gamma_{GS} \sim \int d^{10}x\; B\; \left(a_1 \mathrm{Tr}[F_0^4] + a_2 \mathrm{Tr}[R_0^4] + a_3 (\mathrm{Tr}[F_0^2])^2 \right.$$
$$\left. + a_4 (\mathrm{Tr}[R_0^2])^2 + a_5 \mathrm{Tr}[F_0^2]\mathrm{Tr}[R_0^2]\right) \tag{7.9.16}$$

in the effective action is instrumental for the cancellation of the reducible anomaly.

This mechanism can also work in other dimensions, provided there exist appropriate antisymmetric tensors in the theory. There are also generalizations of this mechanism in theories with more than one antisymmetric tensor. Such theories can be obtained by compactifying superstring theories down to six dimensions.

What kind of fields can contribute to the anomalies? First of all they have to be massless and second they must be chiral. Chirality exists in even dimensions, and fields that can be chiral are (spin-1/2) fermions, (spin-3/2) gravitini, and (anti-)self-dual antisymmetric tensors $B_{\mu_1\cdots\mu_{D/2-1}}$. Their field strength $F = dB$ is a $D/2$-form that is (anti-)self-dual

$$F_{\mu_1\cdots\mu_{D/2}} = \pm\frac{i}{(D/2)!}E_{\mu_1\cdots\mu_D}F^{\mu_{D/2+1}\cdots\mu_D}. \tag{7.9.17}$$

For gravitational anomalies to appear, we must also have chiral representations of the Lorentz group $O(1,D-1)$. They exist in $D = 4k + 2$ dimensions. For gauge anomalies, we must have chiral representations of the gauge group G. This can happen in even dimensions and when the gauge group admits complex representations.

We will now give the contributions to the anomalies coming from the various chiral fields. As we argued before, the anomaly is completely characterized by a closed, gauge-invariant $(D+2)$-form. By an orthogonal transformation we can bring the $D \times D$ antisymmetric matrix R_0 to the following block-diagonal form:

$$R_0 = \begin{pmatrix} 0 & x_1 & 0 & 0 & & \cdots & \\ -x_1 & 0 & 0 & 0 & & \cdots & \\ 0 & 0 & 0 & x_2 & & \cdots & \\ 0 & 0 & -x_2 & 0 & & \cdots & \\ & \cdots & \cdots & \cdots & \cdots & \cdots & \\ & & \cdots & & & 0 & x_{D/2} \\ & & \cdots & & & -x_{D/2} & 0 \end{pmatrix}. \tag{7.9.18}$$

Then $\text{Tr}[R_0^{2m}] = 2(-1)^m \sum_i x_i^{2m}$. The contribution to the gravitational anomaly of a spin-$\frac{1}{2}$ fermion is given by

$$\hat{I}_{1/2}(R) = \prod_{i=1}^{D/2} \left(\frac{x_i/2}{\sinh (x_i/2)} \right). \tag{7.9.19}$$

In the previous formula, we have to expand it in a series that contains forms of various orders and pick the $(D+2)$-form piece. Similarly, we have the following contributions for the chiral gravitini:

$$I_{3/2}(R) = \hat{I}_{1/2}(R) \left(-1 + 2 \sum_{i=1}^{D/2} \cosh (x_i) \right) \tag{7.9.20}$$

and self-dual tensors

$$I_A(R) = -\frac{1}{8} \prod_{i=1}^{D/2} \left(\frac{x_i}{\tanh (x_i)} \right). \tag{7.9.21}$$

The gravitini and self-dual tensors do not contribute to gauge or mixed anomalies, since they cannot be charged under the gauge group. However the spin 1/2 fermions can transform nontrivially and their total contribution to anomalies is given by

$$I_{1/2}(R, F) = \text{Tr}[e^{iF}]\hat{I}_{1/2}(R). \tag{7.9.22}$$

Assuming $D = 10$ and expanding the formulas above we obtain

$$\begin{aligned}
I_{1/2}(R, F)\big|_{12\text{-form}} = &-\frac{\text{Tr}[F^6]}{720} + \frac{\text{Tr}[F^4]\text{Tr}[R^2]}{24 \cdot 48} \\
&-\frac{\text{Tr}[F^2]}{256} \left(\frac{\text{Tr}[R^4]}{45} + \frac{(\text{Tr}[R^2])^2}{36} \right) \\
&+ \frac{n}{64} \left(\frac{\text{Tr}[R^6]}{5670} + \frac{\text{Tr}[R^2]\text{Tr}[R^4]}{4320} + \frac{(\text{Tr}[R^2])^3}{10368} \right),
\end{aligned}$$

where n is the total number of spin-1/2 fermions,

$$\begin{aligned}
I_{3/2}(R)\big|_{12\text{-form}} = &-\frac{495}{64} \left(\frac{\text{Tr}[R^6]}{5670} + \frac{\text{Tr}[R^2]\text{Tr}[R^4]}{4320} + \frac{(\text{Tr}[R^2])^3}{10368} \right) \\
&+ \frac{\text{Tr}[R^2]}{384} \left(\text{Tr}[R^4] + \frac{(\text{Tr}[R^2])^2}{4} \right),
\end{aligned}$$

$$I_A(R)\big|_{12\text{-form}} = \hat{I}_{1/2}(R)\big|_{12\text{-form}} - I_{3/2}(R)\big|_{12\text{-form}}. \tag{7.9.23}$$

The anomaly contributions $I_{1/2}$ and $I_{3/2}$ given above correspond to Weyl fermions. Since in ten dimensions we also have Majorana-Weyl fermions, their contribution to anomalies is half of that given above.

We are now in a position to examine which ten-dimensional theories are free of anomalies.

The theory of $\mathcal{N} = 1_{10}$ supergravity without matter contains a Majorana-Weyl gravitino and a Majorana-Weyl spin $\frac{1}{2}$ fermion of opposite chirality. It can easily be checked from the formulas above that this is anomalous.

Type-IIA supergravity is nonchiral and thus trivially anomaly-free. Type IIB, however, is chiral and contains two Majorana-Weyl gravitini contributing $I_{3/2}$ to the anomaly, two

Majorana-Weyl fermions of the opposite chirality contributing $-I_{1/2}$, and a self-dual tensor contributing I_A. The anomaly can be seen to vanish from (7.9.23).

We will now consider $\mathcal{N} = 1_{10}$ supergravity coupled to vector multiplets. The gaugini have the same chirality as the gravitino. The total anomaly is

$$2I^{N=1} = I_{3/2}(R) - I_{1/2}(R) + I_{1/2}(R, F). \tag{7.9.24}$$

We should first require that the irreducible anomaly corresponding to the traces of R^6 and F^6 cancels. The $\text{Tr}[R^6]$ cannot be written as a product of lower traces, since the group $O(9,1)$ has an independent Casimir of order 6. Thus, the coefficient of the $\text{Tr}[R^6]$ term in (7.9.24) must vanish. This implies that $n = 496$. Since the gaugini are in the adjoint representation of the gauge group, their number n is the dimension of the gauge group. We obtain that a necessary (but not sufficient) condition for anomaly cancellation is $\dim G = 496$. Inserting $n = 496$ in (7.9.24) we obtain

$$96 I^{\text{total}} = -\frac{\text{Tr}[F^6]}{15} + \frac{\text{Tr}[R^2]\text{Tr}[F^4]}{24} + \frac{\text{Tr}[R^2]\text{Tr}[R^4]}{8} + \frac{(\text{Tr}[R^2])^3}{32}$$
$$- \frac{\text{Tr}[F^2]}{960} \left(4\text{Tr}[R^4] + 5(\text{Tr}[R^2])^2\right). \tag{7.9.25}$$

The only hope for canceling the leftover anomaly is to be able to use the Green-Schwarz mechanism. It requires that we should be able to factorize I^{total}. This will happen iff

$$\text{Tr}[F^6] = \frac{1}{48}\text{Tr}[F^2]\text{Tr}[F^4] - \frac{1}{14400}(\text{Tr}[F^2])^3. \tag{7.9.26}$$

Then

$$96 I^{\text{total}} = \left(\text{Tr}[R^2] - \frac{1}{30}\text{Tr}[F^2]\right) X_8, \tag{7.9.27}$$

with

$$X_8 = \frac{\text{Tr}[F^4]}{24} - \frac{(\text{Tr}[F^2])^2}{720} - \frac{\text{Tr}[F^2]\text{Tr}[R^2]}{240} + \frac{\text{Tr}[R^4]}{8} + \frac{(\text{Tr}[R^2])^2}{32}, \tag{7.9.28}$$

and the rest of the anomaly can be canceled via the Green-Schwarz mechanism. The only nontrivial condition that remains is (7.9.26).

Consider first the gauge group to be $O(N)$. Then the following formulas apply:

$$\text{Tr}[F^6] = (N - 32)\text{tr}[F^6] + 15\text{tr}[F^2]\text{tr}[F^4], \tag{7.9.29}$$

$$\text{Tr}[F^4] = (N - 8)\text{tr}[F^4] + 3(\text{tr}[F^2])^2, \quad \text{Tr}[F^2] = (N - 2)\text{tr}[F^2], \tag{7.9.30}$$

where Tr stands for the trace in the adjoint and tr for the trace in the fundamental of $O(N)$.

In exercise 7.29 on page 186 you are required to show that the factorization condition (7.9.26) and $\dim G = 496$ are satisfied by G=O(32). Thus, the type-I and heterotic string theories with G=O(32) are anomaly-free.

Consider now $G = E_8 \times E_8$, which also has dimension 496. E_8 has no independent Casimirs of order 4 and 6,

$$\text{Tr}[F^6] = \frac{1}{7200}(\text{Tr}[F^2])^3, \quad \text{Tr}[F^4] = \frac{1}{100}(\text{Tr}[F^2])^2. \tag{7.9.31}$$

Hence (7.9.26) is satisfied. Consequently, the $E_8 \times E_8$ heterotic string is also anomaly-free.

The presence of the Green-Schwarz term (7.9.16) necessary for the cancellation of the reducible anomaly, can be verified by a one-loop computation in the heterotic string.

From (7.9.13) we obtain that ($G = O(32)$)

$$\frac{g^2}{\kappa^2}\, d\hat{H} = -\text{tr}[R^2] + \frac{1}{30}\text{Tr}[F^2] = -\text{tr}[R^2] + \text{tr}[F^2]. \tag{7.9.32}$$

Integrating (7.9.32) over any closed four-dimensional submanifold, we obtain the important constraint to be satisfied by the background fields:

$$\int \text{tr}[R^2] = \frac{1}{30}\int \text{Tr}[F^2]. \tag{7.9.33}$$

Bibliography

The $\mathcal{N} = (1,0)_2$ superconformal invariance, the superghost system, and the covariant quantization of superstring theory are described in detail and with considerable clarity in [59]. The superstring vertex operators, pictures and picture changing, as well a discussion of space-time supersymmetry can also be found in this reference. GSW [7] and Polchinski [8] offer extra insights.

A clear discussion of the R-R spectrum in type-II strings, can be found in [144]. The Γ-matrix conventions used in [144] are different from those used here. The properties of spinors in various dimensions as well as basic supersymmetry properties are explained in detail in [145].

A rather detailed and easy to follow description of the heterotic string theory can be found in the original papers [6]. There are several other nonsupersymmetric superstring theories that are discussed in [146]. The most well-known example is the $O(16) \times O(16)$ theory [147].

Opens superstrings and their one-loop partition functions are described in the review [97]. The tadpole cancellation conditions are interpreted in [148] for ten dimensions and in [117] for less. The Sugimoto solution is in [149].

The effective actions of the ten-dimensional superstring theories are the three ten-dimensional supergravities: types IIA,B and $\mathcal{N} = 1_{10}$. An extended discussion appears in [7]. The most useful ingredients, namely, the bosonic action and the supersymmetric variations of the fermions are given in appendix H. There, the interested reader will find also suggestions for further reading in that direction.

Higher derivative corrections to the tree-level effective action are generated when we integrate out the massive modes of the string. Such terms are suppressed by powers of the string length ℓ_s. In the σ-model approach they come from higher σ-model loop corrections. Such corrections for the supersymmetric strings are discussed in [134]. They may also be extracted from tree-level string amplitude calculations by matching them to the effective field theory. A extensive discussion can be found in [135].

We have only discussed here the NSR quantization of the superstring. Despite its many advantages, this formulation of superstring theory has also drawbacks. The major drawback is that it is not known how to practically handle backgrounds with nontrivial R-R fields. The reason is that such fields involve finite perturbations by the spin fields of the world-sheet fermions of the form $F_{\alpha\beta} S^\alpha \tilde{S}^\beta$. This is a nonlocal interaction with respect to the basic fermionic fields. Since there are physically very interesting backgrounds involving R-R fields, it is of importance to seek alternatives.

There are two other world-sheet formulations of superstrings. The first is known as the Green-Schwarz (GS) formulation. It trades the world-sheet fermions (which transform as space-time vectors) with world-sheet anticommuting scalars that transform as space-time spinors. This is a formulation that is manifestly space-time supersymmetric. The analogue of the world-sheet $\mathcal{N} = (1,1)_2$ superconformal invariance, used to decouple the ghosts in the NSR formalism is played here by another fermionic symmetry known as κ-symmetry. Having world-sheet fields with spinor indices, now the R-R fields couple locally to the world-sheet fields. There is a price to pay: it is not yet known how to covariantly quantize the theory. Until now its efficiency has been put to use in the light-cone gauge. This approach is described in very much detail in GSW [7].

There is a related approach, under the name of Berkovits formulation which seeks to use spinor variables as in the GS formulation but impose a nonlinear constraint on them à la BRST. This is a formulation that is still under active exploration. The original idea can be found in [150]. It evolved out of the hybrid formalism [151] that treats some of the space-time dimensions à la NSR and some à la Berkovits. Beyond tree level, the formulation can be found in [152].

The original reference on gravitational anomalies [4] remains a very pedagogical paper. An extensive review on anomalies in general can be found in [153]. Anomalies are also extensively discussed in chapters 10 and 13 of [7]. The original reference, which prompted the second lifespan of superstring theory, can be found in [5] and is definitely worth reading. The Green-Schwarz counterterms have been calculated by a direct one-loop calculation in [154]. In the same reference a direct link between modular invariance and anomaly cancellation was found.

Exercises

7.1. Verify that G_{ghost} and T_{ghost} satisfy the OPEs of the $\mathcal{N} = (1,0)_2$ superconformal algebra (4.5.1) on page 58, (4.13.8) on page 78 with the correct central charge.

7.2. Show that the BRST charge (7.1.9) on page 156 satisfies $Q^2 = 0$ when $c = 15$.

7.3. Show that at the massless level of the type-IIA,B theories there is an equal number of on-shell fermionic and bosonic degrees of freedom. Find also the physical states at the next (massive) level in both type-IIA and type-IIB theory. Show that they combine into $SO(9)$ representations, as they should, and that there is again an equal number of fermionic and bosonic degrees of freedom.

7.4. Show that Z^{IIB}, Z^{IIA} are modular invariant. Using (C.18) on page 508, show that they also are identically zero.

7.5. Consider the 0A and 0B closed string theories described by the partition functions

$$Z_{0B} = \frac{|\chi_O|^2 + |\chi_V|^2 + |\chi_C|^2 + |\chi_S|^2}{(\sqrt{\tau_2}\eta\,\bar{\eta})^8}, \quad Z_{0A} = \frac{|\chi_O|^2 + |\chi_V|^2 + \chi_C\bar{\chi}_S + \chi_S\bar{\chi}_C}{(\sqrt{\tau_2}\eta\,\bar{\eta})^8}. \tag{7.1E}$$

Show that they are modular invariant. Derive the massless spectrum, and show that there is a tachyon and no space-time supersymmetry.

7.6. Show that (7.4.10) on page 167 is modular invariant. Show also that it describes a ten-dimensional theory with gauge group $O(16) \times O(16)$ and find the massless spectrum.

7.7. Consider type-IIA/B theory with x^9 compactified on a circle. Use the T-duality action (7.7.1) on page 174 and the discussion in section 7.2.1 on page 159 to derive the following T-duality transformations of R-R fields

$$\tilde{C}^{(p)}_{\mu_1\cdots\mu_{p-1}9} = C^{(p-1)}_{\mu_1\cdots\mu_{p-1}}, \quad \tilde{C}^{(p)}_{\mu_1\cdots\mu_p} = C^{(p+1)}_{\mu_1\cdots\mu_p9}. \tag{7.2E}$$

When G_{i9} and B_{i9} are nontrivial the T-duality relations can be found in [155].

7.8. Use the properties of Γ-matrices in section 7.2.1 on page 159 to prove (7.3.3) and (7.3.4).

7.9. Consider the phase ϵ_R in (7.3.2) on page 162 to be $\epsilon_R = 1$ In this case the zero and self-dual four-forms survive in the R-R sector. Is this choice consistent with interactions?

7.10. Find the representations appearing in the first massive level of the unoriented open superstring both in the NS and the R sector, after imposing the tadpole cancellation conditions $\zeta = 1$, $\epsilon_R\zeta_R = 1$.

7.11. In this exercise you asked to investigate the definition of open strings as twisted sectors of the unoriented closed string. A direct transcription of the twisted boundary condition gives

$$X(\sigma + 2\pi) = X(2\pi - \sigma). \tag{7.3E}$$

Using the fact that the world-volume fields satisfy the two-dimensional Laplace equation $(\partial_\tau^2 - \partial_\sigma^2)X = 0$, show that the solutions with such a boundary condition can be written as a sum of open string coordinates satisfying NN boundary conditions and integral modding as well as open string coordinates satisfying NN boundary conditions and half-integral modding.

7.12. Use the fermionic mode expansions (4.16.16),(4.16.17) on page 89 to derive (up to an overall phase) (7.3.6) on page 163.

7.13. To construct the partition functions of ten-dimensional heterotic theories we need in general the characters of O(8) for the left-moving fermions and the characters of a rank 16, level one current algebra for the internal right-moving part (the bosonic contribution is always the same). Consider first the case G=O(32). Write the most general partition function as linear combinations of the characters, and then impose the following constraints:

- Normalization of the vacuum contribution to 1.

- Modular invariance.

- Correct spin-statistics relation.

- Absence of tachyons.

How many theories do you find? How many are supersymmetric?
Repeat the procedure above for G = $E_8 \times$ O(16) and O(16) $\times$ O(16).

7.14. Consider the heterotic string with the 32 internal fermions describing the 16-dimensional lattice. Putting the same boundary conditions to all 32 of them show that their partition function reproduces (7.4.1),(7.4.3) on page 166.

7.15. Show that the lattice generated by the spectrum of free massless bosons describing T^n (described in section 8.7.10) is even and self-dual. What is its signature?

7.16. Rewrite the BRST current in terms of the boson ϕ and the η-ξ system.

7.17. Calculate the rest of the terms in the 1/2-picture fermion vertex $V_{1/2}^{\text{fermion}}$ in (7.5.9) on page 169.

7.18. By manipulating commutators of the BRST charge show that it does not matter what representatives (pictures) we pick for the vertex operators, provided they satisfy the charge neutrality condition.

7.19. Derive the supersymmetric variation equations for the massless superstring vertex operators (7.5.11) and (7.5.12) on page 169.

7.20. Calculate the disk amplitude between three massless gauge boson states. You may choose two vertex operators in the -1 picture and one in the 0 picture.

7.21. Calculate the disk amplitude between a massless gauge boson and two massless gaugini.

7.22. Calculate the sphere amplitude between three gravitons in the type-II string. Calculate the same amplitude in the heterotic string.

7.23. Calculate the disk amplitude for four gaugini.

7.24. Use the dimensional reduction results of appendix E on page 516 to obtain (H.14) from (H.9) on page 526.

7.25. It is obvious that the IIA effective action is invariant under gauge transformations of the two-form $B_2 \rightarrow B_2 + d\Lambda_1$ and three-form $C_3 \rightarrow C_3 + d\Lambda_2$. Show that it is also invariant under A gauge transformations $A \rightarrow A + d\epsilon$ if the three-form transforms as $C_3 \rightarrow C_3 + \epsilon H_3$.

7.26. By redefining the three-form as $C_3' = C_3 + A \wedge B_2$, the new field strength becomes $\hat{G}_4 = dC_3' - da \wedge B_2$. What are the new gauge transformations?

7.27. Show that the IIB action (H.22) on page 529 is SL(2,$\mathbb{R}$) invariant.

7.28. Show that the IIB action is invariant under the a, B_2, C_2, and C_4 gauge transformations. Due to the Chern-Simons terms in F_5 some of them must be modified.

7.29. Show that the factorization condition (7.9.26) on page 181 and $\dim G = 496$ are satisfied by $G=O(32)$. This is at the heart of the anomaly freedom of the type-I and heterotic string theories with gauge group $G=O(32)$.

7.30. Verify that $E_8 \times E_8$ satisfies (7.9.26) on page 181. Thus, the $E_8 \times E_8$ heterotic string is also anomaly-free. Check also that the groups $E_8 \times U(1)^{248}$ and $U(1)^{496}$ are anomaly-free. No known ten-dimensional string theory corresponds to these groups.

7.31. Consider the nonsupersymmetric $O(16) \times O(16)$ heterotic string in ten dimensions. It is a chiral theory with fermionic content transforming as (V, V), $(\bar{S}, 1)$, and $(1, \bar{S})$ under the gauge group. S stands for the 128-dimensional spinor representation of $O(16)$. Use

$$\mathrm{tr}_S[F^6] = 16\mathrm{tr}[F^6] - 15\mathrm{tr}[F^2]\mathrm{tr}[F^4] + \frac{15}{4}(\mathrm{tr}[F^2])^3, \tag{7.4E}$$

$$\mathrm{tr}_S[F^4] = -8\mathrm{tr}[F^4] + 6(\mathrm{tr}[F^2])^2, \quad \mathrm{tr}_S[F^2] = 16\mathrm{tr}[F^2], \tag{7.5E}$$

with tr_S the trace in the spinor representation space and tr the trace in the fundamental representation space to show that the theory is anomaly-free. What is the Green-Schwarz counterterm? Are there any other chiral, nonsupersymmetric, anomaly-free theories in ten dimensions?

7.32. Consider the $E_8 \times E_8$ heterotic string in ten dimensions. This theory has a symmetry $\mathcal{I}$ that interchanges the two E_8 factors. Consider the $\mathbb{Z}_2$ orbifold of this theory with respect to the symmetry transformation $g = (-1)^F \cdot \mathcal{I}$. Construct the modular-invariant partition function (you will need the duplication formulae for the ϑ-functions that you can find in appendix C on page 507). What is the gauge group and the massless spectrum? Is this theory supersymmetric? Chiral? Anomaly-free?

7.33. Use the results obtained in the last section to derive the anomaly formulas relevant for six-dimensional theories coupled to gravity. These will be applicable to any six-dimensional theory that does not contain self-dual two-form gauge fields.

7.34. Consider the type-II theory in a linear dilaton background. Derive the equations of motion of the R-R forms. Generalize the result, to find the dilaton couplings of the R-R forms.

7.35. Consider an $\mathcal{N} = 1_6$ supersymmetric theory in six dimensions with N_H hypermultiplets, N_V vector multiplets and N_T tensor multiplets. Show that the cancellation of the gravitational anomalies implies

$$N_H - N_V - 29N_T = 273. \tag{7.6E}$$

7.36. Show that the open nonsupersymmetric Sugimoto theory is anomaly-free.

8 | D-branes

We have already seen that the spectrum of the type-II string contains several massless forms coming from the R-R sector. Their standard $\left(-\frac{1}{2}, -\frac{1}{2}\right)$ vertex operators however, represent field strengths instead of gauge potentials. This implies that, in string perturbation theory, the closed-string states are not minimally coupled to them. It turns out, however, that objects that are minimally charged under the R-R forms exist and the purpose of this section is to uncover them.

8.1 Antisymmetric Tensors and *p*-branes

We will use the language of forms and we will represent a rank-p antisymmetric tensor $A_{\mu_1\mu_2\cdots\mu_p}$ by the associated p-form[1]

$$A_p \equiv \frac{1}{p!} A_{\mu_1\mu_2\cdots\mu_p} dx^{\mu_1} \wedge \cdots \wedge dx^{\mu_p}. \tag{8.1.1}$$

Such p-forms transform under generalized gauge transformations:

$$A_p \to A_p + d\, \Lambda_{p-1}, \tag{8.1.2}$$

where d is the exterior derivative and Λ_{p-1} is a $(p-1)$-form that serves as the parameter of gauge transformations. The familiar case of (abelian) gauge fields corresponds to $p = 1$. The gauge-invariant field strength is

$$F_{p+1} = dA_p, \tag{8.1.3}$$

satisfying the free Maxwell equations

$$d\,{}^\star F_{p+1} = 0. \tag{8.1.4}$$

---

[1] Differential forms are introduced in appendix B.

The natural objects, charged under a $(p + 1)$-form A_{p+1}, are p-branes. A p-brane is an extended object with p spatial dimensions. For example, point particles correspond to $p = 0$ and strings to $p = 1$. The natural coupling of A_{p+1} to a p-brane is given by

$$iQ_p \int_{\text{world-volume}} A_{p+1} = iQ_p \int A_{\mu_0 \ldots \mu_p} dx^{\mu_0} \wedge \cdots \wedge dx^{\mu_p}, \tag{8.1.5}$$

which generalizes the $A_\mu \dot{x}^\mu$ coupling in the case of electromagnetism. Note also that for $p = 1$, this is precisely the σ-model coupling of the fundamental string to the NS antisymmetric tensor in (6.1.1) on page 144.

The world-volume of a p-brane is $(p + 1)$ dimensional. The charge density Q_p is the usual electric charge for $p = 0$ and the string tension for $p = 1$. For the p-branes we will be considering, the (electric) charges will be related to their tensions (mass per unit volume).

In analogy with electromagnetism, we can also introduce magnetic charges. First, we must define the analog of the magnetic field: the magnetic (dual) form. This is done by first dualizing the field strength and then rewriting it as the exterior derivative of another form[2]:

$$d\tilde{A}_{D-p-3} = \tilde{F}_{D-p-2} = {}^\star F_{p+2} = {}^\star dA_{p+1}, \tag{8.1.6}$$

where D is the the dimension of space-time. Thus, the dual (magnetic) form couples to $(D - p - 4)$-branes that play the role of magnetic monopoles with "magnetic charges" $\tilde{Q}_{D-p-4}$.

There is a generalization of the Dirac quantization condition for general p-form charges. Consider an electric p-brane with charge Q_p and a magnetic $(D - p - 4)$-brane with charge $\tilde{Q}_{D-p-4}$. Normalize the forms so that the kinetic term is $\frac{1}{2} \int F_{p+2} {}^\star F_{p+2}$. Integrating the field strength ${}^\star F_{p+2}$ on a $(D - p - 2)$-sphere surrounding the p-brane we obtain the total flux $\Phi = Q_p$. We can also write

$$\Phi = \int_{S^{D-p-2}} {}^\star F_{p+2} = \int_{S^{D-p-3}} \tilde{A}_{D-p-3}, \tag{8.1.7}$$

where we have used (8.1.6) and we have integrated around the "Dirac string." When the magnetic brane circles the Dirac string it picks up a phase $e^{i\Phi \tilde{Q}_{D-p-4}}$, as can be seen from (8.1.5). Unobservability of the Dirac string implies the Dirac-Nepomechie-Teitelboim (DNT) quantization condition

$$\Phi \tilde{Q}_{D-p-4} = Q_p \tilde{Q}_{D-p-4} = 2\pi k, \quad k \in \mathbb{Z}. \tag{8.1.8}$$

8.2 Open Strings and T-duality

We will consider open strings with X^9 compactified on a circle of radius R. One of the possibilities here is to turn on a constant gauge field (also known as a Wilson line) in the direction of the circle

$$A_9 = \frac{\chi}{2\pi R}, \tag{8.2.1}$$

---

[2] This is guaranteed by (8.1.4).

with χ constant. This could be removed by a gauge transformation $A_9 \to A_9 + \partial_{x^9}\epsilon$ with $\epsilon = \frac{\chi}{2\pi R}x^9$ but this is not an acceptable gauge transformation on the circle. Thus, the gauge field (8.2.1) is nontrivial. In particular, it is expected to affect the spectrum.

To illustrate the effect, consider first for simplicity a charged point particle with action

$$S = \int d\tau \left(\frac{1}{2}\dot{x}^M \dot{x}_M - \frac{m^2}{2} + qA_M\dot{x}^M \right). \tag{8.2.2}$$

The canonical momentum has an extra contribution as usual

$$p_9 = \dot{x}^9 + \frac{q\chi}{2\pi R}. \tag{8.2.3}$$

Since we are on a circle, p_9 is quantized: $p_9 = n/R$. The Hamiltonian is

$$H = \frac{1}{2}\left[p^\mu p_\mu + \left(\frac{2\pi n - q\chi}{2\pi R} \right)^2 + m^2 \right]. \tag{8.2.4}$$

We observe that the Kaluza-Klein masses are shifted from those of the standard case to

$$m_n^2 = m^2 + \left(\frac{n}{R} - \frac{q\chi}{2\pi R} \right)^2. \tag{8.2.5}$$

We will now return to the string and include also the U(N) CP factors. We can diagonalize the constant background gauge field to

$$A_9 = \frac{1}{2\pi R}\text{diag}(\chi_1, \chi_2, \ldots, \chi_N), \tag{8.2.6}$$

under the U(1)^N subgroup of the U(N).

For a string in the state $|ij\rangle$, one end has charge $+1$ under U(1)$_i$ and the other -1 under U(1)$_j$. Thus, the mass formula (3.3.4) on page 33 for the open superstring is modified to

$$m^2 = \frac{\tilde{N} - \frac{1}{2}}{\ell_s^2} + \left(\frac{n}{R} - \frac{\chi_i - \chi_j}{2\pi R} \right)^2, \tag{8.2.7}$$

where $\tilde{N}$ is the oscillator number. In particular for the zero-mode vectors ($\tilde{N} = \frac{1}{2}$, $n = 0$) we obtain a mass

$$m_{ij}^2 = \left(\frac{\chi_i - \chi_j}{2\pi R} \right)^2. \tag{8.2.8}$$

Generically, we have N massless vectors (those for $i = j$). If various groups of χs are equal, we obtain massless vectors for $\prod_i U(N_i)$ with $\sum_i N_i = N$. This spectrum can also be obtained from the low-energy effective action.

If we take $R \to \infty$, the circle decompactifies, the KK states become a continuum, and we obtain the ten-dimensional open string. Consider, however, the $R \to 0$ limit. Open strings with free end points do not have winding modes, (see sections 2.3.2 and 2.3.3 on page 24). Therefore, it seems that all the masses of the KK modes move off to infinity and they decouple. As far as the center-of-mass motion of the string is concerned, we end up with a string in nine dimensions.

Several arguments indicate that this cannot be the whole story. Open strings are always coupled to closed strings, and closed strings when $R \to 0$ become again ten dimensional in the T-dual coordinates. Since the major difference of an open string from its closed counterpart is the presence of end points, we should expect that the bulk of the open strings move again in ten dimensions.[3] The correct setup turns out to be given by defining the $R \to 0$ limit in the open string using T-duality. That is, we should first T-dualize and then take $R \to \infty$ instead.

We found in section 4.18.7 on page 101 that T-duality acts on the currents as

$$\partial_\sigma X \leftrightarrow \partial_\tau X. \tag{8.2.9}$$

Thus, the Neumann boundary condition on the X^9 coordinate becomes a Dirichlet condition. This will fix the end points but not the interior of the open strings, as we argued earlier. To see where this position is fixed, we consider the T-dual of the oscillator expansion (2.3.22), which was given in (2.3.28) on page 23,

$$\tilde{X}^9(\tau, \sigma) = \frac{p^9 \sigma}{\pi T} - \frac{1}{\sqrt{\pi T}} \sum_{k \in \mathbb{Z}} \frac{\alpha_k^\mu}{k} e^{-ik\tau} \sin(k\sigma). \tag{8.2.10}$$

Assume first that there are no Wilson lines. Then, one end point of the string is at $\tilde{x}^9(\sigma = 0) = 0$ while the other is at

$$\tilde{X}^9(\sigma = \pi) = \frac{p^9}{T} = 2\pi \frac{\ell_s^2}{R} n = 2\pi \tilde{R} n, \tag{8.2.11}$$

where $\tilde{R}$ is the dual radius. Since $\tilde{X}^9(\sigma = 0) - \tilde{X}^9(\sigma = \pi)$ is a multiple of the radius of the dual circle, both end points are constrained to be at the origin.

Adding now the Wilson lines we obtain

$$X^9(\sigma = \pi) - X^9(\sigma = 0) = \tilde{R} \left(2\pi n + \chi_j - \chi_i\right) \sim \tilde{R} \left(-\chi_i + \chi_j\right). \tag{8.2.12}$$

Thus, the ith end point of the dual open string in the ninth direction is fixed at points related to the Wilson lines as

$$\tilde{X}_i^9 = -\chi_i \tilde{R} = -2\pi \ell_s^2 A_{9,ii}, \tag{8.2.13}$$

and the mass of the stretched string is given by (see (2.3.37) on page 24)

$$m^2 = \left(\frac{\delta x}{2\pi \ell_s^2}\right)^2 + \frac{\tilde{N} - \frac{1}{2}}{\ell_s^2}. \tag{8.2.14}$$

The extra contribution is the classical energy due to the stretching of the string.

This generalizes to several compact dimensions. When p dimensions become smaller than the string length, T-duality implies that the weakly-coupled description involves the string with Dirichlet boundary conditions on p coordinates. This breaks translation invariance.

---

[3] For gauge theory strings, this is the case. Both closed and open strings are flux tubes, but the open strings end on quarks.

8.3 D-branes

In the previous section we have argued that open strings with Dirichlet boundary conditions on some of the space-time coordinates are required in order to interpret physics in a way compatible with T-duality. This gives rise to the concept of D-branes. In this section, we will give further motivation for their nature and role in string theory.

We will consider again the case of generic Wilson lines along x^9 as in the previous section. The massless states are

$$A_{\mu,ii} \sim \psi^{\mu}_{-1/2}|p; ii\rangle, \quad A_{9,ii} \sim \psi^{9}_{-1/2}|p; ii\rangle. \tag{8.3.1}$$

The analogous $i \neq j$ states are massive with masses given in (8.2.8) as represented in figure 8.1. Note that the momentum p in (8.3.1) is nine dimensional in the T-dual picture. From the original point of view, the U(N) gauge symmetry is broken to U(1)^N by the Wilson lines. From the T-dual point of view, the strings end on N different hyperplanes, at the positions given in (8.2.13). There is a nine-dimensional massless vector A^{μ}_i living on each hyperplane. There is also an extra massless scalar A^9_i.

In the T-dual description, $A_{9,ii}$ represents the collective coordinate of the ith hyperplane as suggested (when A^9_i is constant) by (8.2.13). In fact, we expect, in the more general case of x dependence, $A_{9,ii}$ to have the same interpretation, namely, the position of a (fluctuating) nine-plane inside ten dimensions. The hyperplanes can curve and fluctuate and $A_{9,ii}$ describes these fluctuations.

Therefore we conclude that the hyperplanes are dynamical objects. Their fluctuations are described by the open strings that are constrained to end on them by the Dirichlet boundary conditions (see figure 8.2). Their massless spectrum exhausts the collective excitations of the brane.

Such hyperplanes are known as Dirichlet branes, or D-branes for short. The particular object we have here is a D$_8$-brane. This is because it is a hyperplane transverse to the ninth direction and has eight spatial dimensions. More generally, by T-dualizing p coordinates we will obtain a D$_{9-p}$-brane, having Dirichlet boundary conditions on p coordinates. Extending the concept, we may say that the original open strings are moving on a D$_9$-brane that fills all ten-dimensional space.

As we mentioned earlier, T-duality interchanges Neumann and Dirichlet boundary conditions. Thus, if applied to a direction transverse to a D$_p$ brane it transforms it into a D$_{p+1}$-brane. On the other hand applied, to a longitudinal direction produces a D$_{p-1}$-brane.

We will now derive the massless fluctuations of a single D$_p$-brane in the closed superstring theory.[4] We have NN boundary conditions on the $p+1$ coordinates x^{μ} and their fermionic partners and DD boundary conditions on the $9-p$ coordinates x^I and their fermionic partners. In the NS sector, the massless bosonic states are a $(p+1)$-vector A_{μ} corresponding to the state $b^{\mu}_{-1/2}|p\rangle$ and $9-p$ scalars X^I corresponding to the states $b^I_{-1/2}|p\rangle$. The X^I represent the position coordinates of the D$_p$-brane in transverse space.

---

[4] Only type-IIA, type-IIB, and type-I theories have D-branes. The heterotic theory has none.

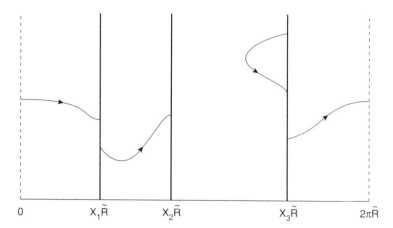

Figure 8.1 Hyperplanes and the associated open strings with Dirichlet boundary conditions.

Figure 8.2 An open string with its end points stuck on a D-brane.

These are the states we would obtain by reducing a ten-dimensional vector to $p + 1$ dimensions. Similarly, from the R sector we obtain world-volume fermions that are the reduction of a ten-dimensional gaugino to $p + 1$ dimensions. In total, we obtain the reduction of a ten-dimensional U(1) vector supermultiplet to $p + 1$ dimensions.

A further comment concerns supersymmetry. We consider the case of the type-II string with $\mathcal{N} = 2_{10}$ supersymmetry. The D-branes constructed in that theory will have only $\mathcal{N} = 1_{10}$ supersymmetry. The reason is that, as we have seen, their fluctuations are described by open superstrings that have half of the supersymmetry of closed strings. Viewed as states of the type-II theory, D-branes preserve half of the space-time supersymmetry and in this sense are (extended) BPS states as will become obvious in chapter 11. In agreement with our discussion in section 8.1, the IIB theory contains stable D$_p$-branes with p odd, and the IIA theory with p even.

We may analyze in more detail the supersymmetry preserved by D-branes. It is simpler to start from the D$_9$-branes that correspond to the standard open strings in ten dimensions.

Let Q_α and $\tilde{Q}_\alpha$ be the space-time supercharges of the type-IIB string theory coming from the left- and right-moving sectors. They are Majorana-Weyl spinors, with 16 real components each. The open string boundary condition preserves only the sum of the supercharges $Q_\alpha + \tilde{Q}_\alpha$ leaving therefore half of the space-time supersymmetry unbroken. By T-dualizing, we may obtain the supercharges preserved by other D-brane solutions. You are asked in exercise 8.10 on page 214 to show that a D_p-brane preserves $Q + \delta^\perp \tilde{Q}$, where

$$\delta^\perp \equiv \prod_{m \notin D} \delta^m, \quad \delta^\mu = \Gamma^\mu \Gamma^{11}. \tag{8.3.2}$$

The product is over all directions transverse to the D-brane.

A special mention is reserved for a D-brane with all boundary conditions being Dirichlet. A string starting and ending on this brane will have DD boundary conditions on all coordinates including time. Since this an object localized in time, its natural interpretation is as an instanton in Euclidean time. It is known as the D_{-1}-instanton. Its world-volume is pointlike. It couples electrically to a zero-form and magnetically to an eight-form. We will have more things to say about D-instantons in section 8.4.1.

To conclude this section, we recapitulate. A D-brane is a (lower-dimensional) defect in ten-dimensional space-time. Its intrinsic dynamics is described by open strings with end points glued on the brane. The brane also interacts with the closed strings that propagate over the whole ten-dimensional space (sometimes referred to as the "bulk"). As it propagates, a closed string may hit the brane, in which case it opens up and its end points stick to the brane. The opposite process is also possible: a pair of open strings leaves the brane, becoming a closed string, and propagating through the bulk. The interplay of open and closed string dynamics is at the center of many fascinating new effects that we will slowly uncover in the rest of this book.

8.4 D-branes and R-R Charges

According to our discussion in section 8.1, the minimally charged objects with respect to $(p + 1)$-form gauge fields are p-branes. As we will show in this section, the D-branes we have described in the previous section, are in fact minimally charged under the type-II R-R forms.

To investigate this issue, we may calculate the one-point function of the appropriate R-R form on the disk, in the presence of the D_p-brane. This calculation is relatively straightforward, but it poses a normalization problem: we do not know *a priori* the relation between the closed and open string coupling. Equivalently, we do not know the relative normalization of open and closed string amplitudes.

A calculation that does not have this ambiguity is the tree-level interaction of two parallel D_p-branes via the exchange of a closed string depicted schematically in figure 8.3. For this interpretation, time runs horizontally. However, if we take time to run vertically, then, the same diagram can be interpreted as a (one-loop) vacuum fluctuation of open strings

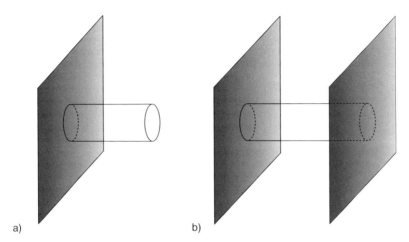

Figure 8.3 a) Emission of a closed string from a D-brane. b) Two D-branes interacting via the tree-level exchange of a closed string.

with their end points attached to the D-branes. In this second picture we can calculate this diagram to be

$$\mathcal{A} = 2iV_{p+1} \int \frac{d^{p+1}k}{(2\pi)^{p+1}} \int_0^\infty \frac{dt}{2t} e^{-2\pi t \ell_s^2 k^2 - t|Y|^2/2\pi \ell_s^2} \frac{1}{\eta^8(it)} \frac{1}{2} \sum_{a,b} (-1)^{a+b+ab} \frac{\vartheta^4[{}^a_b](it)}{\eta^4(it)}$$

$$= 2iV_{p+1} \int_0^\infty \frac{dt}{2t} (8\pi^2 \ell_s^2 t)^{-(p+1)/2} e^{-t|Y|^2/2\pi \ell_s^2} \frac{1}{\eta^{12}(it)} \frac{1}{2} \sum_{a,b=0}^1 (-1)^{a+b+ab} \vartheta^4[{}^a_b](it), \qquad (8.4.1)$$

where $|Y|^2$ is the distance between the D-branes in the transverse space. The only difference from the calculation of a similar cylinder amplitude in section 7.6 on page 170 is that now there is a volume factor as well as a momentum integral in $p + 1$ directions, an extra factor of 2 due to the two possible orientations of the string, and there is the Dirichlet-related shift in the L_0 proportional to the distance squared (see (2.3.37) on page 24).

Of course the total result is zero, because of the ϑ-identity (C.15) on page 508. This reflects the fact that flat D-branes are BPS states and exert no static force on each other. Put differently, the force due to massless graviton and dilaton exchange (from the NS-NS sector) is canceled by the force coming from the massless $(p + 1)$-form coming from the R-R sector. This cancellation is due to the remaining $\mathcal{N} = 1_{10}$ supersymmetry. It persists at higher oscillator levels.

Our purpose is to disentangle the contributions of the various intermediate massless states in the closed string channel. This can be obtained by taking the leading $t \to 0$ behavior of the integrand. In order to do this, we have to perform a modular transformation $t \to 1/t$ in the ϑ- and η-functions. We obtain

$$\mathcal{A}|_{\text{massless}}^{\text{closed string}} = 8i(1 - 1)V_{p+1} \int_0^\infty \frac{dt}{t} (8\pi^2 \ell_s^2 t)^{-(p+1)/2} t^4 e^{-t|Y|^2/2\pi \ell_s^2}$$

$$= 2i\pi (1 - 1)V_{p+1}(4\pi^2 \ell_s^2)^{3-p} G_{9-p}(|Y|), \qquad (8.4.2)$$

where

$$G_d(|Y|) = \frac{1}{4\pi^{d/2}} \int_0^\infty \frac{dt}{t^{(4-d)/2}} e^{-t|Y|^2} \tag{8.4.3}$$

is the massless scalar propagator in d dimensions. The (1) (-1) comes from the NS-NS (R-R) sector, respectively.

Now consider the effective action of the R-R forms C, coupled to p-branes,

$$S = \frac{\alpha_p}{2} \int F_{p+2} \wedge {}^\star F_{p+2} + iT_p \int_{\text{branes}} C_{p+1}, \tag{8.4.4}$$

with $F_{p+2} = dC_{p+1}$. Using this action, the same amplitude for an exchange of C_{p+1} between two D-branes at distance $|Y|$ in the transverse space of dimension $10-(p+1) = 9-p$ is given by

$$\mathcal{A}|_{\text{field theory}} = -i\frac{(T_p)^2}{\alpha_p} V_{p+1} G_{9-p}(|Y|), \tag{8.4.5}$$

where the factor of volume is present since the R-R field can be absorbed or emitted at any point on the world-volume of the D-brane. Matching with the string calculation we obtain

$$\frac{T_p^2}{\alpha_p} = 2\pi (2\pi \ell_s)^{2(3-p)}. \tag{8.4.6}$$

We now return to the DNT quantization condition (8.1.8). With our normalization of the R-R forms and $D = 10$ it becomes

$$\frac{T_p T_{6-p}}{\alpha_p} = 2\pi n. \tag{8.4.7}$$

From (8.4.6) we can verify directly that D-branes satisfy this quantization condition for the minimum quantum $n = 1$.

Therefore, D-branes, describe nonperturbative extended BPS states of the type-II string carrying the minimum R-R charge.

We now proceed to describe a uniform normalization of the D-brane tensions. Our starting point is the type-IIA ten-dimensional effective action (11.4.1 on page 329). The gravitational coupling κ_{10} is given in terms of the string length as

$$2\kappa_{10}^2 = (2\pi)^7 \ell_s^8 g_s^2. \tag{8.4.8}$$

We will also normalize all forms so that their kinetic terms are

$$\frac{1}{4\kappa_{10}^2} \int F_{p+2} \wedge {}^\star F_{p+2} = \frac{1}{(2\kappa_{10}^2)2(p+2)!} \int d^{10}x \sqrt{-g} \, F_{\mu_1 \cdots \mu_{p+2}} F^{\mu_1 \cdots \mu_{p+2}},$$

where $F_{p+2} = dC_{p+1}$. This corresponds to $\alpha_p = 1/(2\kappa_{10}^2)$. We will also define the tensions of various p-branes via their (Euclidean) world-volume action of the form

$$S_p = -T_p \int d^{p+1}\xi \, e^{-\Phi} \sqrt{\det \hat{G}} - iT_p \int C_{p+1}, \tag{8.4.9}$$

where $\hat{G}$ is the metric induced on the world-volume,

$$\hat{G}_{\alpha\beta} = G_{\mu\nu} \frac{\partial X^\mu}{\partial \xi^\alpha} \frac{\partial X^\nu}{\partial \xi^\beta}, \tag{8.4.10}$$

and

$$\int C_{p+1} = \frac{1}{(p+1)!} \int d^{p+1}\xi \, C_{\mu_1 \cdots \mu_{p+1}} \frac{\partial X^{\mu_1}}{\partial \xi^{\alpha_1}} \cdots \frac{\partial X^{\mu_{p+1}}}{\partial \xi^{\alpha_{p+1}}} \epsilon^{\alpha_1 \cdots \alpha_{p+1}}. \tag{8.4.11}$$

The dilaton dependence will be explained in the next section. The DNT quantization condition in (8.4.7) becomes

$$2\kappa_{10}^2 T_p T_{6-p} = 2\pi n, \tag{8.4.12}$$

while (8.4.6) and (8.4.8) finally give

$$T_p = \frac{1}{(2\pi)^p \, \ell_s^{p+1} \, g_s}. \tag{8.4.13}$$

There is a consistency condition on the D-brane tension that comes from T-duality. Consider a D_p-brane with its world-volume wrapped around a p-dimensional orthogonal torus with radii R_i. Its mass is the tension times the volume of the torus

$$M = T_p \prod_{i=1}^{p} (2\pi R_i). \tag{8.4.14}$$

T-dualize one of the compact directions. In section 6.4 we found that the dilaton trans forms as

$$R'_p = \frac{\ell_s^2}{R_p}, \quad e^{-\Phi'} R'_p = e^{-\Phi} R_p, \quad g'_s = g_s \frac{\ell_s}{R_p}, \quad T'_p = T_p \frac{R_p}{\ell_s}, \tag{8.4.15}$$

so that

$$M = M' = 2\pi \frac{\ell_s^2}{R_p} T_p \frac{R_p}{\ell_s} \prod_{i=1}^{p-1} (2\pi R_i) = T_{p-1} \prod_{i=1}^{p-1} (2\pi R_i), \tag{8.4.16}$$

from which we have following recursion relation:

$$T_{p-1} = (2\pi \ell_s) \, T_p. \tag{8.4.17}$$

This is indeed satisfied by (8.4.13).

8.4.1 D-instantons

We note that the previous discussion also applies to $p = -1$. Its natural interpretation works with a Euclidean signature.[5] This corresponds to a 0-dimensional object (localized

---

[5] We may also consider it in Minkowski space. It is then a simple example of an S-brane, namely, a defect localized in real time. However, in such a case it is the source of imaginary long-range R-R fields.

in time) which is charged under the IIB axionic scalar. Thus, D_{-1} has the properties of a space-time instanton, and is known as the D-instanton. It gives rise to nonperturbative instantonic effects in string theory. By T-dualizing along transverse dimensions we may generate higher-dimensional Euclidean D_p-branes. Such branes may wrap a $(p + 1)$-dimensional compact manifold to produce a (pointlike) instanton configuration.

Observe that the D-brane tension, unlike that of the fundamental string, depends on the string coupling constant. In particular, in string perturbation theory where $g_s \ll 1$, the D-brane energy density is very large. Consequently, D-branes can be treated as heavy, semiclassical objects just as we do for solitons.

Soliton masses in field theory, scale as $1/g^2$ where g is a relevant field theory coupling. The power of the coupling for D-branes is different, and this reflects their stringy nature.

We may consider Euclidean D-branes as instanton configurations in string theory. Unlike the standard field-theory instanton action that scales as $1/g^2$, the one due to instantonic D-branes scales as $1/g$. In field theory, the e^{-1/g^2} nature of instanton effects indicates that the nth order of perturbation theory diverges as $n!$. In string theory, $e^{-1/g}$ effects indicate that the nth order of perturbation theory diverges as $(2n)!$. This divergence of closed string perturbation theory was known before the advent of the D-brane concept. It has been used to argue about the existence of $e^{-1/g}$ effects in string theory.

8.5 D-brane Effective Actions

The low-energy dynamics of D-branes can be encoded in the effective action. It contains the interactions of the brane fluctuations, which determine the internal dynamics of the brane as well as the interaction of the brane with the massless closed string sector.

The direct way to calculate the world-volume effective action is to calculate the scattering amplitudes of the massless states of the world-volume theory and their interaction with the closed string massless fields. The leading contribution comes from the disk diagram (tree level) and is thus weighted with a factor of $e^{-\Phi}$. The calculation is similar to the calculation of the effective action in the ten-dimensional open oriented string theory. There are however, shortcuts using general principles of string theory that will lead us to the correct form of the effective action with less calculation. Needless to say, the result has been tested against amplitude calculations. In the previous section, we have calculated the minimal coupling of the appropriate R-R form to the D-brane. Here we will provide the rest of the couplings.

8.5.1 The Dirac-Born-Infeld action

There are two possible starting points. The first is the fact that a D_9 brane is essentially standard open+closed string theory in ten-dimensions. This is a theory that has been analyzed, and the nonlinear effective action for the massless open string gauge-boson has been calculated (remember that there are no transverse scalars for a D_9 brane). The result is the Born-Infeld action for the gauge field[6]

$$S_{BI} = -T_9 \int d^{10}x \, e^{-\Phi} \sqrt{-\det\left(\eta_{\mu\nu} + 2\pi \ell_s^2 F_{\mu\nu}\right)}. \tag{8.5.1}$$

where derivatives of $F_{\mu\nu}$ have been neglected in (8.5.1). This action is reliable for slowly varying fields, $\ell_s^3 \partial_\mu F_{\nu\rho} << 1$.

Another starting point is the D_0-brane of the type-IIA theory. This is a particle. Apart from the minimal coupling to the R-R one-form, its low-energy effective action should be that of massive relativistic particle, namely,

$$S_{D_0} = -T_0 \int d\tau \, e^{-\Phi} \sqrt{-G_{\mu\nu}\dot{x}^\mu \dot{x}^\nu}. \tag{8.5.2}$$

The world-volume gauge boson has no dynamics in one dimension and has been suppressed. Note that $ds^2 = G_{\mu\nu}\dot{x}^\mu \dot{x}^\nu d\tau^2$ is the induced metric on the world-line. In flat space, and picking a static gauge $\tau = x^0$ the action (8.5.2) becomes

$$S_{D_0} = -T_0 \int d\tau \, e^{-\Phi} \sqrt{1 - \dot{x}^I \dot{x}^I}. \tag{8.5.3}$$

x^I are the spatial dimensions. They are at the same time the nine massless transverse scalars on the world-volume. T-duality can be used to bridge the two actions (8.5.1) and (8.5.3). The idea is that turning on a world-volume magnetic field on a two-plane on D_p, and T-dualizing along one of the two directions, using (8.2.13) we obtain a D_{p-1}-brane without magnetic field but tilted on the two-plane. Comparing the two actions we can relate (8.5.1) and (8.5.3) and this is the subject of exercise 8.13. The T-duality between magnetized and intersecting branes is described in detail in section 9.16.

The arguments above allow us to write the action that has the correct couplings to the metric and dilaton as well as the R-R form as

$$S_p = -T_p \int d^{p+1}\xi \, e^{-\Phi} \sqrt{-\det(\hat{G}_{\alpha\beta} + 2\pi\alpha' F_{\alpha\beta})} + iT_p \int C_{p+1}, \tag{8.5.4}$$

where α, β are world-volume indices, the induced metric $\hat{G}_{\alpha\beta}$ and R-R $(p+1)$-form are defined in (8.4.10) and (8.4.11).

The next question concerns the coupling of the NS-NS antisymmetric tensor. This can be found as follows. The world-sheet coupling of the string to $B_{\mu\nu}$ and the open string vector A_μ can be summarized in

$$S_B = \frac{i}{4\pi\ell_s^2} \int_{M_2} d^2\xi \, \epsilon^{\alpha\beta} B_{\mu\nu} \partial_\alpha x^\mu \partial_\beta x^\nu + i \int_{B_1} ds \, A_\mu \partial_s x^\mu, \tag{8.5.5}$$

where M_2 is the two-dimensional world-sheet with one-dimensional boundary B_1. Under a gauge transformation $\delta B_{\mu\nu} = \partial_\mu \Lambda_\nu - \partial_\nu \Lambda_\mu$, the above action changes by a boundary term,

$$\delta S_B = -\frac{i}{2\pi\ell_s^2} \int_{B_1} ds \, \Lambda_\mu \partial_s x^\mu. \tag{8.5.6}$$

---

[6] We are in Minkowski space.

To reinstate gauge invariance, the vector A_μ has to transform as $\delta A_\mu = -\frac{1}{2\pi \ell_s^2} \Lambda_\mu$. Thus, the gauge-invariant combination is

$$\mathcal{F}_{\mu\nu} = (2\pi \ell_s^2) F_{\mu\nu} + B_{\mu\nu} = 2\pi \ell_s^2 (\partial_\mu A_\nu - \partial_\nu A_\mu) + B_{\mu\nu} \,. \tag{8.5.7}$$

We can pull it back on the brane-world volume as

$$\mathcal{F}_{\alpha\beta} = (2\pi \ell_s^2) F_{\alpha\beta} + \hat{B}_{\alpha\beta} \,, \tag{8.5.8}$$

where $\hat{B}_{\alpha\beta} = B_{\mu\nu} \frac{\partial X^\mu}{\partial \xi^\alpha} \frac{\partial X^\nu}{\partial \xi^\beta}$ is the induced antisymmetric tensor on the world-volume.

We may now summarize the leading-order D_p-brane action as

$$S_p = -T_p \int d^{p+1}\xi \, e^{-\Phi} \sqrt{-\det(\hat{G}_{\alpha\beta} + \mathcal{F}_{\alpha\beta})} + iT_p \int C_{p+1} \,. \tag{8.5.9}$$

8.5.2 Anomaly-related terms

As we saw in section 8.4, the minimal coupling to the R-R form comes from the disk diagram (although we computed it by factorizing a cylinder amplitude). Since it involves the coupling to a R-R form, according to the discussion in section 7.2.1 on page 159, it does not carry any dilaton dependence. This is a $\mathcal{CP}$-odd term, since it involves the ϵ-tensor on the world-volume (see equation (8.4.11)).

There are however additional $\mathcal{CP}$-odd couplings that involve q-forms with $q < p$. They are higher order in derivatives. Their appearance is due to a combination of T-duality covariance[7] and cancellation of anomalies.[8] Equivalently, they guarantee the gauge invariances of the bulk R-R forms that are nonstandard due to the supergravity Chern-Simons terms (see for example (H.15) on page 527).

We will present the result here. It involves the roof genus $\hat{I}_{1/2}(R)$ in (7.9.19) on page 180 and the Chern character. Equation (8.5.9) is extended to[9]

$$S_p = -T_p \int d^{p+1}\xi \, e^{-\Phi} \sqrt{-\det(\hat{G} + \mathcal{F})} + iT_p \int C \wedge \mathrm{Tr}[e^{\mathcal{F}}] \wedge \mathcal{G}, \tag{8.5.10}$$

where C stands for a formal sum of all R-R forms, and the integration in the second term picks up the $(p+1)$-form in the sum. The gravitational form $\mathcal{G}$ is

$$\mathcal{G} = \sqrt{\hat{I}_{1/2}(\mathcal{T})/\hat{I}_{1/2}(\mathcal{N})} = 1 - \frac{(2\pi \ell_s)^4}{48} [p_1(\mathcal{T}) - p_1(\mathcal{N})] + \cdots \tag{8.5.11}$$

where $\mathcal{N}$ and $\mathcal{T}$ are the normal and tangent bundles of the brane, respectively, and p_1 is the first Pontryagin class. You are asked to derive or check some parts of this result in exercises 8.23–8.25 on page 216.

---

[7] See exercise 8.23 on page 216.

[8] The anomaly cancellation arguments here are related to anomaly inflow. More information on this can be found in the references at the end of this chapter.

[9] A word is in order here concerning the reality properties of $\mathcal{CP}$-odd terms in the action, proportional to the ϵ-density. In Minkowski space the $\mathcal{CP}$-odd action is real, and therefore e^{iS} a phase. In Euclidean space, S is purely imaginary, and e^{-S} also a phase.

As an explicit example, we will consider the action of the D1-string of type-IIB theory. The relevant forms that couple here is the R-R two-form C_2 as well as the R-R scalar (zero-form) C_0. The action is

$$S_1 = -\frac{1}{2\pi \ell_s^2} \left[\int d^2\xi \, e^{-\Phi} \sqrt{-\det(\hat{G} + \mathcal{F})} - i \int (C_2 + C_0 \mathcal{F}) \right]. \tag{8.5.12}$$

8.6 Multiple Branes and Nonabelian Symmetry

So far, we have discussed a single D_p-brane, interacting with the background type-II fields. We have seen already that the dynamics of multiple D-branes is richer. In section 8.3 we noted that if N parallel branes are at the same point in transverse space, their gauge symmetry is enhanced to U(N). The different branes are labeled by N distinct CP factors.

Moving them apart is equivalent to turning on constant Wilson lines in the T-dual theory. This generically breaks the gauge group $U(N) \rightarrow U(1)^N$ by the standard Brout-Englert-Higgs effect. Thus, the action of displacing the branes is equivalent (after T-duality) to the Higgs effect via (8.2.13).

For N coincident supersymmetric D_p-branes of type-II theory, the world-volume theory is the U(N) super Yang-Mills (SYM) theory with 16 supercharges. It is the dimensional reduction of the $\mathcal{N} = 1_{10}$, U(N) SYM theory. The $(p+1)$-dimensional SYM theory, apart from the vector boson in the adjoint of U(N) also contains $9-p$ scalars X^I also in the adjoint. In the single-brane case, the scalars X^I had the interpretation of collective transverse coordinates of the brane. We are interested in the geometrical interpretation of the scalars in the nonabelian case.

The action to lowest order is[10] (see exercise 8.21 on page 215)

$$S_{N,p} = \frac{T_p(2\pi \ell_s^2)^2}{4} \int d^{p+1}\xi \, e^{-\Phi} \, \text{Tr} \left[F_{\mu\nu}^2 + 2(D_\mu X^I)^2 + \sum_{I,J} [X^I, X^J]^2 \right], \tag{8.6.1}$$

$$F_{\mu\nu} = \partial_\mu A_\nu - \partial_\nu A_\mu + [A_\mu, A_\nu], \quad D_\mu X^I = \partial_\mu X^I + [A_\mu, X^I]. \tag{8.6.2}$$

The Yang-Mills coupling constant is

$$g_{YM,p}^2 = \frac{1}{(2\pi \ell_s^2)^2 \, T_p} = (2\pi \ell_s)^{p-4} \ell_s g_s. \tag{8.6.3}$$

Both A_μ and X^I are U(N) matrices. The potential in (8.6.1) is positive definite. Thus, configurations with $[X^I, X^J] = 0$ are minima. In section 9.1 we will investigate this effect further. The condition $[X^I, X^J] = 0$ implies that X^I belong to the Cartan subalgebra of U(N). Thus, at the minimum of the potential, the commuting matrices X^I can be simultaneously diagonalized. Their eigenvalues x_i^I, $i = 1, 2, \ldots, N$, can be naturally interpreted as the positions of the N D_p-branes. If all of them are equal, the U(N) symmetry is unbroken (see figure 8.4). In exercise 8.22 you are invited to calculate the masses of the non-Cartan

---

[10] We ignore here the fermions and the leading term coming from the tension.

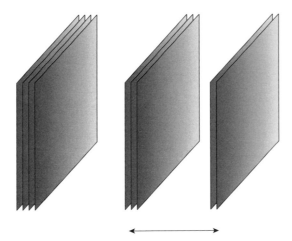

Figure 8.4 A stack of coincident D-branes realizing unbroken nonabelian gauge symmetry, and the Brout-Englert-Higgs effect due to brane separation in transverse space.

gauge bosons A_μ^{ij} with $i \neq j$ and show that they are proportional to the distance in transverse space between the ith and jth branes.

The analogue of the DBI action (8.5.10) in the nonabelian case is much harder to determine. One reason is that in the the nonabelian case, derivatives of field strengths can be traded with commutators using the equations of motion. Substituting in (8.5.10) X's and F's that are $N \times N$ matrices, replacing X-derivatives with gauge-covariant derivatives, and taking an overall symmetric trace gives the correct answer to order $\mathcal{O}(F^4)$. The all-order effective action is not known.

8.7 T-duality and Orientifolds

In this section we will describe the effect of T-duality on unoriented strings. This will lead to the concept of orientifolds. They are a direct generalization of the orientation reversal projection used to define the unoriented strings.

We will start from the closed sector. The unoriented closed strings are defined as an equivalence class, where a string and its reversed cousin are considered to be equivalent. This is identical to the statement that the orientation reversal Ω is gauged. It is also equivalent to the statement that we "orbifold" by the action of Ω, which translates into keeping states in the closed Hilbert space that have $\Omega = +1$. The Ω action on the string coordinates can be described as follows: If we split the coordinates into right-moving and left-moving pieces then the orientation reversal interchanges the two.

We will label now the string coordinates on which T-duality will act with letters from the beginning of the Latin alphabet, keeping Greek indices for the directions unaffected by T-duality. As discussed earlier, T-duality acts as

$$\text{T}: X_L^\mu(z) + X_R^\mu(\bar{z}) \rightarrow X_L^\mu(z) + X_R^\mu(\bar{z}), \quad X_L^a(z) + X_R^a(\bar{z}) \rightarrow X_L^a(z) - X_R^a(\bar{z}). \tag{8.7.1}$$

Combined with the orientifold projection

$$\Omega: X_L^i(z) \to X_L^i(\bar{z}), \quad X_R^i(\bar{z}) \to X_R^i(z), \quad i = \mu, a, \tag{8.7.2}$$

we obtain that the dual coordinates $X'^a(z, \bar{z}) = X_L^a(z) - X_R^a(\bar{z})$ transform as

$$\Omega: X'^a(z, \bar{z}) \to -X'^a(\bar{z}, z). \tag{8.7.3}$$

We may rewrite this action in operator form

$$T_i \, \Omega \, T_i^{-1} = \mathcal{I}_i \, \Omega, \tag{8.7.4}$$

where T_i is T-duality along the ith direction and $\mathcal{I}_i$ is the reflection $x^i \to -x^i$. Therefore, on the dual coordinates, the orientation reversal acts as the space reflection $\mathcal{I}$ and as worldsheet parity interchanging z and $\bar{z}$. The gauging of $\mathcal{I}$ gives the standard orbifold. In the dual theory, this is accompanied at the same time by an orientation reversal. This is an *orientifold*.

The generic states of the orientifold are projections of the original states. To see this effect in detail, we separate the wave function of a state into the zero-mode part depending on (x^μ, x^a) and an oscillator part. We can take this second part to have a well-defined eigenvalue under Ω, $+1$ or -1. Then, the wave function of the zero-mode part is symmetric in the first case under $(x^\mu, x^a) \to (x^\mu, -x^a)$, and antisymmetric in the second, so that the total eigenfunction is invariant. In particular for the NS-NS massless states we obtain

$$g_{\mu\nu}(x^\mu, -x^a) = g_{\mu\nu}(x^\mu, x^a), \quad B_{\mu\nu}(x^\mu, -x^a) = -B_{\mu\nu}(x^\mu, x^a), \tag{8.7.5}$$

$$g_{\mu a}(x^\mu, -x^a) = -g_{\mu a}(x^\mu, x^a), \quad B_{\mu a}(x^\mu, -x^a) = B_{\mu a}(x^\mu, x^a), \tag{8.7.6}$$

$$g_{ab}(x^\mu, -x^a) = g_{ab}(x^\mu, x^a), \quad B_{ab}(x^\mu, -x^a) = -B_{ab}(x^\mu, x^a). \tag{8.7.7}$$

As in orbifolds, there are fixed planes at points left invariant by the reflection. These are called orientifold planes. At an orientifold plane, the projection on the wave functions imposes a well-defined boundary condition. However, away from the orientifold planes, the physics is that of oriented string theory since all modes propagate. This should be compared with the type-I string where the Ω projection does not act at all to the space-time coordinates. As a result, B_{MN} is totally projected out at every point in space-time. If we T-dualize a circle of radius R, then there are two orientifold planes, one at $x = 0$ and the other at $x = \pi\tilde{R} = \pi\ell_s^2/R$. They correspond to the fixed points of the geometrical action. This T-dual version is known as type-I' theory.

There is also an important difference from orbifolds. In the orbifold case, the $\tau \to -\frac{1}{\tau}$ invariance of the torus amplitude implies the presence of twisted states localized on the fixed (orbifold) planes. In the orientifold case the relevant surface generating the Ω-projection is the Klein bottle, and $t \to 1/t$ is not a symmetry. Thus, there are no fluctuating modes localized on the orientifold planes. In this sense, they are distinct from the D-branes

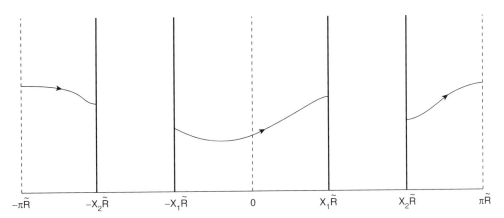

$$-\pi\tilde{R} \qquad -X_2\tilde{R} \qquad -X_1\tilde{R} \qquad 0 \qquad X_1\tilde{R} \qquad X_2\tilde{R} \qquad \pi\tilde{R}$$

Figure 8.5 D-brane configuration at the orientifold dual.

that can respond dynamically to closed string probes. All the orientifold plane does is to impose appropriate boundary conditions on closed string modes.

To conclude, the closed sector of the T-dual of the unoriented string is the orientifold, where the orientifold symmetry transformation is $\Omega\mathcal{I}$. There are more general orientifolds, and we will see some of them later.

We would like to analyze what happens in the open sector. As we have seen earlier, Ω symmetrizes or antisymmetrizes the CP factors, giving an Sp($2n$) or SO(n) gauge group. We will look at the SO(n) case for simplicity since Sp($2n$) is analogous.

We consider T-dualizing a circle of radius R. The dual manifold, as argued above, is the line segment $0 < x < \pi\tilde{R}$ with $\tilde{R} = \ell_s^2/R$. We first assume that the group is SO($2n$). A general SO($2n$) (exponential) Wilson line can be diagonalized to

$$e^{i\oint A} = \mathrm{diag}(e^{i\chi_1}, e^{-i\chi_1}, e^{i\chi_2}, e^{-i\chi_2}, \ldots, e^{i\chi_n}, e^{-i\chi_n}). \tag{8.7.8}$$

Upon T-duality, according to our discussion in section 8.2, we will obtain n D$_8$-branes located at $x_i = \tilde{R}\chi_i$ and their images under the inversion $\mathcal{I}$ at $\tilde{x}_i = -\tilde{R}\chi_i$. The possible open strings stretch between the n D$_8$-branes and their images. The generic gauge group is U(1)^n.

If m D-branes become coincident at a generic point, then the group is enhanced to U(m). To see this, consider the possible strings between these branes and their images. The m^2 strings among themselves are massless and generate U(m). On the other hand the strings between them and their images are stretched, not generating any more massless vectors. If, on the other hand, the m branes are coincident on one of the orientifold planes ($x = 0$ or $\pi\tilde{R}$) then also the second type of strings becomes also massless, enhancing the group to SO($2m$). The maximally enhanced group SO($2n$) occurs when all branes move on top of one of the orientifold planes.

When the group is SO($2n+1$), the general Wilson line can be brought to the form

$$e^{i\oint A} = \mathrm{diag}(e^{i\chi_1}, e^{-i\chi_1}, e^{i\chi_2}, e^{-i\chi_2}, \ldots, e^{i\chi_n}, e^{-i\chi_n}, \pm 1). \tag{8.7.9}$$

After T-duality, the D-brane corresponding to the last eigenvalue is fixed on one of the two orientifold planes. Since it has no image, it is a "half" D_8-brane. Its tension is half of the tension of a regular brane because of the orientifold projection. This a simple case of the general concept of "fractional" branes that abound in orientifolds.

Orientifold planes, like D-branes, couple to massless closed-string fields. The relevant calculation is similar to the one we described in section 8.4. The only difference is that now we have to consider other surfaces beyond the cylinder. In particular, the Klein-bottle calculation gives the interaction between two orientifold planes, while the Möbius strip calculation gives the interaction between a D-brane and an orientifold plane. This can be seen from figure 4.11 on page 113, where a boundary corresponds to a D-brane and a cross cap indicates the presence of an orientifold plane imposing orientation reversal.

We will reinterpret the tadpole cancellation of the type-I string in ten dimensions. The Klein bottle introduces tadpoles that are due to one nondynamical plane. This is the O_9 orientifold plane. It has negative tension equal to $-32T_9$ and a negative nine-brane charge density equal to $-32T_9$. Its energy and charge are canceled by D_9-branes. The vacuum therefore contains 32 D_9-branes that cancel the orientifold charge and maintain a flat space.

Unlike D-branes, O-planes are nondynamical in the sense that they cannot fluctuate and therefore cannot carry degrees of freedom. This is as well, since a fluctuating negative tension object has necessarily negative-norm states.

One-loop amplitudes are calculated in section 5.3 on page 517 for the bosonic string and in section 7.6 on page 170 for the superstring. In particular, in the superstring case we have found that the tadpoles cancel in the presence of $N = 2^5$ D_9-branes. T-dualizing on a circle, we will need 2^4 D_8-branes to cancel them since the other 2^4 are images. Thus, the tension of an O_8-plane is $2^4 T_8$. Continuing further, the T-dualization we conclude that the tension of an O_p-plane is $T_{O_p} = 2^{p-4}T_p$. The effective action of an O_p-plane is then

$$S_{O_p} = 2^{p-4}T_p \left[\int d^{p+1}\xi e^{-\Phi}\sqrt{-\det g} + i \int d^{p+1}\xi\, C_{p+1} \right]. \tag{8.7.10}$$

We should again stress that orientifold planes have no (intrinsic) dynamical degrees of freedom. They just carry charges (energy, R-R charge, etc.).

To summarize, a T-duality of type-I theory compactified on a circle of radius R produces an orientifold (termed type-I′ theory) with two O_8-orientifold planes, one at $x' = 0$ and the other at $x' = \pi \tilde{R}$. We now have 32 D_8-branes[11] and they are all on top of one of the orientifold planes. The gauge group is still SO(32). Turning on Wilson lines, the branes can start moving off the orientifold plane, reducing the gauge group via Higgsing as explained earlier.

The D_8 branes preserve 16 of the 32 supercharges of the original IIB theory. Thus, the open strings of the original type-I theory are now the ones associated with the D_8-branes. This process can be further continued by dualizing more dimensions.

---

[11] Sixteen of them are the images of the other 16 under the orientifold inversion.

8.8 D-branes as Supergravity Solitons

D-branes are minimally charged objects under the appropriate massless fields of the low-energy effective action. A D_p-brane is charged minimally under the C_{p+1}-form field and it also carries dilaton charge and energy. As we shall now see, it can also be viewed as a semiclassical solution of supergravity.

8.8.1 The supergravity solutions

To find the long-range fields of D-branes we may keep the relevant part of the IIA/IIB two-derivative effective action in the string frame, involving the metric, the dilaton, and C_{p+1},

$$S_p = \frac{1}{2\kappa_{10}^2} \int d^{10}x \sqrt{-g} \left[e^{-2\Phi}[R + 4(\nabla\Phi)^2] - \frac{1}{2(p+2)!} F_{p+2}^2 \right], \tag{8.8.1}$$

with $F_{p+2} = dC_{p+1}$.

We would like to find a solution that has the geometry of a flat p-brane, and is charged under C_{p+1}. Such a brane is pointlike in transverse space. Therefore, we expect its long-range fields to be spherically symmetric.

The equations stemming from varying the action (8.8.1) can be written in the form

$$R_{\mu\nu} + 2\nabla_\mu\nabla_\nu\Phi = \frac{e^{2\Phi}}{2(p+1)!} \left[F_{\mu\nu}^2 - \frac{g_{\mu\nu}}{2(p+2)} F^2 \right], \tag{8.8.2}$$

$$d^\star F_{p+2} = 0, \quad R = 4(\nabla\Phi)^2 - 4\Box\Phi. \tag{8.8.3}$$

An appropriate *Ansatz* for the string-frame metric is

$$ds_{10}^2 = \frac{-f(r)dt^2 + d\vec{x}\cdot d\vec{x}}{\sqrt{H_p(r)}} + \sqrt{H_p(r)} \left(\frac{dr^2}{f(r)} + r^2 d\Omega_{8-p}^2 \right), \tag{8.8.4}$$

where $d\Omega_{8-p}^2$ is the line element on the $(8-p)$-sphere of unit radius. The world-volume coordinates are t, x^i, $i = 1, 2, \ldots, p$, while r and the sphere coordinates are transverse. Solving the equations of motion (8.8.2),(8.8.3) we obtain that the dilaton is

$$e^{2\Phi} = g_s^2 \, H_p^{(3-p)/2}(r), \tag{8.8.5}$$

and H_p, f are harmonic functions in the transverse space. For the spherically symmetric solutions they are

$$H_p(r) = 1 + \frac{L^{7-p}}{r^{7-p}}, \quad f(r) = 1 - \frac{r_0^{7-p}}{r^{7-p}}, \tag{8.8.6}$$

where we have normalized them to 1 at infinity, $r \to \infty$, by rescaling the coordinates r and $\vec{x}$. This metric is asymptotically flat at spatial infinity, $r \to \infty$, in transverse space.

The R-R form is given by

$$C_{012\cdots p}(r) = \sqrt{1 + \frac{r_0^{7-p}}{L^{7-p}} \frac{H_p(r) - 1}{H_p(r)}}.$$

(8.8.7)

All other components vanish, except in the case of $p = 3$, where the self-duality condition

$$F_{\mu_1\cdots\mu_5} = \frac{1}{5!\sqrt{\det g}}\epsilon_{\mu_1\cdots\mu_5 \nu_1\cdots\nu_5} F^{\nu_1\cdots\nu_5}$$

(8.8.8)

requires nonzero p-form components in the transverse directions. The electric field is given by

$$F_{01\cdots pr} = -\sqrt{1 + \frac{r_0^{7-p}}{L^{7-p}}} \frac{H_p'(r)}{H_p^2(r)}.$$

(8.8.9)

Calculating the R-R charge of the solution by integrating $\frac{1}{2\kappa_{10}^2}{}^\star F_{p+2}$ over S^{8-p} and setting it equal to an integer N times the tension, we obtain

$$N = \frac{(7-p)\Omega_{8-p}}{2\kappa_{10}^2 T_p} L^{(7-p)/2}\sqrt{r_0^{7-p} + L^{7-p}},$$

(8.8.10)

where Ω_n is the volume of a unit n-dimensional sphere S^n,

$$\Omega_n = \frac{2\pi^{(n+1)/2}}{\Gamma((n+1)/2)},$$

(8.8.11)

and T_p and κ_{10} are given in (8.4.8),(8.4.13). N is equal to the number of D-branes. We can therefore write the electric field (8.8.9) as

$$F_{01\cdots pr} = \frac{g_s N}{\Omega_{8-p} H_p^2(r)} \frac{(2\pi\ell_s)^{7-p}}{r^{8-p}} \xrightarrow{r\to\infty} \frac{g_s N}{\Omega_{8-p}} \frac{(2\pi\ell_s)^{7-p}}{r^{8-p}} + \cdots .$$

(8.8.12)

This is the asymptotic electric field of a point charge in the transverse $(9-p)$-dimensional space. Note however, that in the neighborhood of $r \to 0$ the electric field behaves as

$$F_{01\cdots pr} = \frac{g_s N}{\Omega_{8-p}} \left(\frac{2\pi\ell_s}{L^2}\right)^{7-p} r^{6-p}\left[1 + \mathcal{O}(r^{7-p})\right].$$

(8.8.13)

It is therefore screened for $p < 6$, unlike the standard case of electric charges.

We may also calculate the asymptotic ADM mass of the solution from the g_{00} component to be

$$M = \frac{\Omega_{8-p} V_p}{2\kappa_{10}^2}\left[(8-p)r_0^{7-p} + (7-p)L^{7-p}\right],$$

(8.8.14)

where V_p is the common p-dimensional D-brane (flat) volume.

Solving (8.8.10) we obtain

$$L^{7-p} = \sqrt{\left(\frac{2\kappa_{10}^2 T_p N}{(7-p)\Omega_{8-p}}\right)^2 + \frac{1}{4}r_0^{2(7-p)}} - \frac{1}{2}r_0^{7-p}.$$

(8.8.15)

Some comments are in order here for special values of p. For $p = -1$, and upon an analytic continuation of time we have the D_{-1}-instanton. For $p = 7$, the transverse space is two dimensional, H_7, f are constants, and the metric is flat. If $f \neq 1$, the metric has a conical singularity in transverse space. The D_8-brane is not asymptotically flat, since the transverse space is one dimensional and the one-dimensional solution of the Laplacian grows linearly.

8.8.2 Horizons and singularities

To understand the meaning of the special points $r = 0$ and $r = r_0$, we must change coordinates to extend the metric[12] past $r = 0$:

$$\rho^{7-p} = L^{7-p} + r^{7-p}, \quad r_- = L, \quad r_+^{7-p} = r_0^{7-p} + L^{7-p}. \tag{8.8.16}$$

The metric (8.8.4) now becomes

$$ds^2 = -\frac{f_+(\rho)}{\sqrt{f_-(\rho)}}dt^2 + \sqrt{f_-(\rho)}d\vec{x}\cdot d\vec{x} + f_-(\rho)^{-1/2-(5-p)/(7-p)}\left[\frac{d\rho^2}{f_+(\rho)} + \rho^2 f_-(\rho)d\Omega_{8-p}^2\right], \tag{8.8.17}$$

with

$$f_\pm = 1 - \left(\frac{r_\pm}{\rho}\right)^{7-p}. \tag{8.8.18}$$

There is an inner Killing horizon at $\rho = r_-$ and an outer horizon at $\rho = r_+$. There is no curvature singularity at the horizon. There is however a singularity at the inner horizon. We may calculate the scalar curvature in the string metric to be

$$R_\sigma = \frac{(p+1)(3-p)(p-7)^2}{4\,r^{(p-3)/2}\,L^{(7-p)/2}}\left[1 + \mathcal{O}(r^{7-p})\right] \simeq \frac{(p+1)(3-p)(p-7)^2}{4\left(\rho^{7-p} - r_-^{7-p}\right)^{(p-3)/2(7-p)}L^{(7-p)/2}} + \cdots. \tag{8.8.19}$$

It therefore seems that there is no singularity for $p = 3$. For $7 > p > 3$ there is a curvature singularity at $r = r_-$. On the other hand, for $-1 \leq p < 3$ the curvature scalar is regular at $r = r_-$. However, a look at (8.8.5) shows that for $7 > p > 3$ the effective string coupling vanishes at the inner horizon, while for $-1 \leq p < 3$ it blows up. In fact, going to the Einstein frame metric

$$g_{\mu\nu}^E = e^{-\Phi/2}g_{\mu\nu}^\sigma, \tag{8.8.20}$$

the scalar curvature becomes

$$R_E = -\frac{e^{\Phi/2}}{8}R_\sigma \sim r^{-(p-3)^2/8} \tag{8.8.21}$$

and is always singular at the inner horizon except when $p = 3$.

Another interesting datum is the area of the inner horizon. This can be read off from the value of the radius of the unit $(8-p)$-sphere metric in (8.8.4). In the string frame it is given by

$$R_{\text{inner horizon}} = \lim_{r \to 0} rH^{1/4} \sim \lim_{r \to 0} r^{(p-3)/4}. \tag{8.8.22}$$

---

[12] The reader should also consult appendix F on page 519 where a similar metric in four dimensions, sharing many common features with the solutions here, is analyzed.

Thus, the horizon has zero area for $p > 3$, while it has finite area at $p = 3$. In the Einstein metric, the radius of the singular horizon is

$$R^E_{\text{inner horizon}} = \lim_{r \to 0} \; rH^{(p+1)/16} \sim \lim_{r \to 0} \; r^{(p-3)^2/16}, \tag{8.8.23}$$

and is always vanishing except when $p = 3$.

Thus, the $p = 3$ brane can be interpreted as a smooth solitonic object of the effective field theory. It has asymptotically flat long-range fields, a regular horizon, and is regular at the core. All other branes are singular solution resembling the Dirac monopole in four dimensions. Under normal circumstances, we would not take them seriously. However, as we will argue later on, duality symmetries will force us to include them in the quantum theory. Moreover their semiclassical singularities, will be resolved at least at weak string coupling.

8.8.3 The extremal branes and their near-horizon geometry

For an acceptable brane solution we must have $r_+ \geq r_-$ or equivalently $r_0 \geq 0$. Otherwise we will have a naked singularity. This and (8.8.10),(8.8.14) imply the Bogomolnyi bound

$$\frac{M}{V_p} \geq T_p \, N. \tag{8.8.24}$$

The extremal brane $r_0 = 0$ saturates the Bogomolnyi bound and is thus a BPS configuration, preserving half of the space-time supersymmetry (as you are asked to show in exercise 8.43 on page 218). The region around $r = 0$ is known as the "throat" of the solution or its near-horizon region. The throat size L is proportional to the charge

$$\left(\frac{L}{2\pi \ell_s}\right)^{7-p} = \frac{g_s N}{7 - p} \frac{\Gamma\left(\frac{9-p}{2}\right)}{2\pi^{(9-p)/2}}. \tag{8.8.25}$$

From (8.8.10),(8.8.14) we obtain

$$M = N V_p T_p, \tag{8.8.26}$$

that is, mass = charge as advertised. Since this is a BPS configuration, we can generalize the solution so that the charge is distributed at different points. Indeed the general solution of the Laplace equation in transverse space

$$\square \, H_p = 0 \tag{8.8.27}$$

is a superposition

$$H_p(\vec{r}) = 1 + \sum_{i=1}^{N} \frac{L_1^{7-p}}{|\vec{r} - \vec{r}_i|^{7-p}}, \tag{8.8.28}$$

where L_1 is given in (8.8.25) with N=1 and $\vec{r}_i$ are the positions of the individual branes in the transverse space.

As we have argued in the beginning of this chapter, there is a simple stringy description of D-branes, in terms of the fluctuations of open strings in trivial background fields (flat space-time metric, etc.). We have seen in this section, however, that a brane charged under the R-R fields has nontrivial long-range background fields. To study the compatibility of the two descriptions, we will need to understand the range of validity of each description.

Consider first Polchinski's description in terms of open strings. The perturbative description is set up in terms of diagrams of the open strings that have their end points attached on the D-branes. Each boundary on the Riemann surface carries a factor of the string coupling g_s as well as a factor of N from the CP trace, where N is the total number (charge) of D-branes. Thus, this perturbative description is valid when

$$\lambda \equiv 2\pi g_s\, N \ll 1. \tag{8.8.29}$$

Moreover, the distances that we can probe using the open strings are necessarily no smaller than the string length ℓ_s.

Let us investigate now the supergravity description of the extremal RR-charged solutions. To do this we will have to study the curvature invariants of the solutions. These will determine when the solution is trustworthy and when the solution can be approximated by flat space.

A straightforward calculation of the scalar curvature of the extremal D_p metric (8.8.4)

$$ds^2 = \frac{\eta_{\mu\nu}dx^\mu dx^\nu}{\sqrt{H_p(r)}} + \sqrt{H_p(r)}\, dx^i dx^i \tag{8.8.30}$$

with $r^2 = x^i x^i$ gives

$$R = -\frac{L^{2(7-p)}}{4\, r^{(p-3)/2}} \frac{(p+1)(p-3)(p-7)^2}{(r^{7-p}+L^{7-p})^{5/2}}. \tag{8.8.31}$$

As mentioned earlier this vanishes for $p = 3$. Because of this it is necessary to calculate a higher scalar invariant like the square of the Ricci or Riemann tensor. You are asked to do this in exercise 8.44 on page 218. We obtain the following asymptotics:

$$\ell_s^4\, R_{\mu\nu;\rho\sigma} R^{\mu\nu\,\rho\sigma} = \begin{cases} c_+(p)\left(\dfrac{r}{\ell_s}\right)^{-4}\left(\dfrac{L}{r}\right)^{2(7-p)} + \cdots, & r \gg L, \\[3mm] c_-(p)\left(\dfrac{r}{\ell_s}\right)^{-4}\left(\dfrac{r}{L}\right)^{7-p} + \cdots, & r \ll L, \end{cases} \tag{8.8.32}$$

with

$$c_+(p) = 2(7-p)^2(8-p)(9-p), \quad c_-(p) = \frac{(p+1)(7-p)^2(p^2-13p+50)}{16}, \tag{8.8.33}$$

which are characteristic of the behavior of curvature invariants.

From (8.8.25) we see that in the perturbative regime $\lambda \ll 1$, the throat size L is much smaller than the string length ℓ_s. The throat region is therefore "invisible" to the string probes. The supergravity description is also not reliable at such distances.

The regime $r \gtrsim \ell_s$ that we probe in perturbative string theory is captured by the $r \gg L$ form of the curvature in (8.8.32). We observe that in this region the curvature and other invariants[13] are very small in string units. This is a reflection of the fact that the Planck mass

---

[13] Other relevant invariants like $(\nabla\Phi)^2$ or $e^{2\Phi}F_{p+2}^2$ are of the same order of magnitude by the equations of motion.

scales as $1/g_s^2$, while the D-brane energy density as $1/g_s$. Therefore, the D-brane gravitational back-reaction can be neglected when $\lambda \ll 1$. A look at the effective string coupling in (8.8.5) indicates that it is very small when $r \gg L$, so the effective action results are reliable.

This explains why, when $\lambda \ll 1$, we can describe the D_p-branes to leading order in open-string perturbation theory by assuming the background to be trivial. Of course at higher orders, tadpoles will appear indicating that the D-brane has long-range fields that need to be taken into account.

It is interesting to investigate the situation in the opposite limit, namely, $\lambda \gg 1$. In this case the open string description is not very useful as we are in the nonperturbative regime. When $\lambda \gg 1$, the throat size $L \gg \ell_s$ is macroscopic (see (8.8.25)). We may therefore probe the interior structure of the p-brane solution, without having to look at distances smaller than the string scale.

We will investigate first the simplest case, namely, three-branes ($p = 3$). Here, the dilaton is a constant and we take $g_s \ll 1$ so that the loop corrections to the gravitational action are small. As shown in the previous subsection, the metric is also nonsingular. From (8.8.32), the curvature is weak outside the throat $r \gg L$ and thus the solution reliable. In the interior of the throat, the curvature invariant is essentially constant,

$$
\ell_s^4 \, R_{\mu\nu;\rho\sigma} R^{\mu\nu\rho\sigma} \simeq 80 \left(\frac{\ell_s}{L}\right)^4 \sim \frac{1}{\lambda}. \tag{8.8.34}
$$

When $\lambda \gg 1$ (possible when $N \gg 1$), the curvature is weak and the solution is reliable to arbitrarily small distance. As we shall see in chapters 12 and 13, this fact is at the heart of the counting of black-hole microstates in string theory and of the AdS/CFT correspondence.

The situation when $p \neq 3$ is similar, albeit a bit more complicated. Again from (8.8.32) the solution is weakly curved for $r \gg \ell_s$. However, this description may break down at shorter distances. You are invited to investigate it in exercise 8.46 on page 218.

The arguments above indicate that the throat (or near-horizon) geometry is accessible. We may study it by focusing at distances $r \ll L$. In this limit, we may neglect the 1 in the harmonic function H_p. The near-horizon metric is most interesting in the D_3 case,

$$
ds^2 = \frac{L^2}{r^2} dr^2 + \frac{r^2}{L^2} \left(-dt^2 + d\vec{x} \cdot d\vec{x}\right) + L^2 d\Omega_5^2. \tag{8.8.35}
$$

It factorizes into two constant-curvature metrics, as hinted in (8.8.34). The first is that of AdS$_5$, a constant negative curvature manifold with radius of curvature L. The geometry of AdS$_5$ is detailed in appendix K on page 541. The second is the metric of S^5 with radius L.

The dilaton is constant and there is a four-form flux through both S^5 and AdS$_5$

$$
C_{0123} = -\frac{r^4}{L^4}. \tag{8.8.36}
$$

There is an SO(2,4) isometry group, associated with the AdS$_5$ part of the metric as well as an SO(6) isometry associated with the S^5 part. All of this demonstrates explicitly that the near-horizon region of D$_3$-branes is not only regular but also highly symmetric. The full extremal solution can be thought of as an interpolation between the AdS$_5 \times$S^5 manifold at $r \simeq 0$ and flat ten-dimensional space at $r \to \infty$. These observations will be crucial when we will describe the AdS/CFT correspondence in chapter 13.

8.9 NS$_5$-branes

Apart from the R-R forms there is another massless form that is omnipresent in the string theory effective actions, the NS two-form $B_{\mu\nu}$. The fundamental string is charged under it. This coupling is minimal.[14] According to our discussion in section 8.1 on page 187, the magnetic dual of the fundamental string should be a five-brane. It will be (minimally) magnetically charged under B. This is known as the NS$_5$-brane.

The NS$_5$-brane is dual to the fundamental string that has tension $T = \frac{1}{2\pi\ell_s^2}$. The DNT quantization condition (8.4.12) on page 196 with $n = 1$ implies that its minimal tension $\tilde{T}_5$ is

$$T\tilde{T}_5 = \frac{2\pi}{2\kappa_{10}^2} \Rightarrow \tilde{T}_5 = \frac{1}{(2\pi)^5 \, \ell_s^6 \, g_s^2}. \tag{8.9.1}$$

The tension of the NS$_5$-brane scales as $1/g_s^2$. This is a characteristic of field-theory solitons. It is therefore "heavier" than the D-branes at weak coupling.

To find the respective solution to the low-energy effective action we start from the universal sector of the closed string theories that includes the metric, B-field and the dilaton

$$S = \frac{1}{2\kappa_{10}^2} \int d^{10}x \sqrt{-g} \, e^{-2\Phi} \left[R + 4(\nabla\Phi)^2 - \frac{1}{2 \cdot 3!}(dB)^2 \right]. \tag{8.9.2}$$

Following similar considerations as in section 8.8 the appropriate *Ansatz* is

$$ds_{10}^2 = -f(r)dt^2 + d\vec{x} \cdot d\vec{x} + H(r)\left[\frac{dr^2}{f(r)} + r^2 d\Omega_3^2 \right], \tag{8.9.3}$$

$$e^{2\Phi} = g_s^2 \, H(r), \quad dB = \frac{H'}{H} \, \omega, \tag{8.9.4}$$

where the three-form ω is the normalized volume form on the S^3,

$$\omega = \sin^2\psi \sin\theta \, d\psi \wedge d\theta \wedge d\phi, \quad d\Omega_3^2 = d\psi^2 + \sin^2\psi(d\theta^2 + \sin^2\theta d\phi^2). \tag{8.9.5}$$

The functions f, H are harmonic functions and can be normalized to

$$H = 1 + \frac{L^2}{r^2}, \quad f = 1 - \frac{r_0^2}{r^2}. \tag{8.9.6}$$

As in the case of D-branes, r_0 is the position of a regular outside horizon, while $r = 0$ is a regular inner horizon. At the extremal limit, $r_0 = 0$, the solution is BPS, preserving half of the space-time supersymmetry. By integrating dB over S^3, the total NS$_5$-brane charge N can be related to L via

$$L^2 = N \, \ell_s^2. \tag{8.9.7}$$

In the extremal limit the solution can be generalized to the multibrane solution

$$H = 1 + \sum_i \frac{N_i \ell_s^2}{|\vec{r} - \vec{r}_i|^2}, \tag{8.9.8}$$

that describes N$_i$ NS$_5$-branes at the point $\vec{r} = \vec{r}_i$ in the four-dimensional transverse space.

---

[14] This is valid for an infinite straight string. A closed string behaves like a dipole.

In the near-horizon region $r \ll L$, an infinite throat appears with an S^3 cross section. The metric and dilaton become

$$ds_{10}^2 = -dt^2 + d\vec{x} \cdot d\vec{x} + L^2 \left[\frac{dr^2}{r^2} + d\Omega_3^2 \right], \quad e^{2\Phi} = \frac{g_s^2 N \ell_s^2}{r^2}. \tag{8.9.9}$$

Introducing a new coordinate $y = -\sqrt{N} \, \ell_s \log \frac{r}{g_s \ell_s \sqrt{N}}$, the near-horizon geometry is

$$ds_{10}^2 = -dt^2 + d\vec{x} \cdot d\vec{x} + dy^2 + N \ell_s^2 \, d\Omega_3^2, \quad \Phi = \frac{y}{\sqrt{N} \ell_s}. \tag{8.9.10}$$

This background is described by an exact conformal field theory. The six coordinates $(t, \vec{x})$ associated with the world-volume are free. The extra coordinate y has a linear dilaton background. Finally the S^3 part is described by the $SU(2)_N$ (supersymmetric) WZW model.

Since we are considering superstring theories, the level N is the level of the super-affine $\widehat{SU}(2)$ algebra. This is the sum of the bosonic level k and a 2 coming from the three free fermionic partners of the currents, $N = k + 2$.

Another characteristic of the NS_5 solution is that, although the metric is regular in the near-horizon region, the effective string coupling e^{Φ} is unbounded as $r \to 0$. This indicates that this geometry is probably unreliable because of potential string quantum corrections.

As all string theories contain fundamental strings, all string theories contain NS_5-branes.[15] The massless world-volume excitations of the NS_5-branes, as well as the world-volume supersymmetry depend on the type of superstring theory in question. There are however, some universal features. In particular, the broken translation invariance implies the presence of at least four massless scalars on the world-volume corresponding to the transverse coordinates of the brane. Moreover, the NS_5-brane is a 1/2-BPS configuration, breaking half of the original supersymmetry.

In the IIB theory, the world-volume theory has nonchiral $\mathcal{N} = (1, 1)_6$ supersymmetry. The massless world-volume fields of a single NS_5-brane form a vector multiplet containing a vector, four scalars, and two spinors of opposite chirality.

In the IIA theory the world-volume theory has chiral $\mathcal{N} = (2, 0)_6$ supersymmetry. The massless world-volume fields of a single NS_5-brane form an $\mathcal{N} = (2, 0)_6$ tensor multiplet containing an anti-self-dual two-index antisymmetric tensor, five scalars, and two fermions of the same chirality. Four of the scalars describe the transverse coordinates of the NS_5 brane. The interpretation of the fifth will become clear in chapter 11.

In the heterotic theory, the NS_5-brane, has a world-volume theory with $\mathcal{N} = 1_6$ supersymmetry. It is more complicated to describe, compared to its type-II siblings. In particular, the associated supergravity solution involves instanton configurations of the O(32) or $E_8 \times E_8$ gauge group. At finite instanton size, the massless world-volume fields belong to $\mathcal{N} = 1_6$ hypermultiplets. Each hypermultiplet contains four real scalars and a Weyl spinor. The brane thickness becomes small when the instanton size is small. There are interesting effects at zero instanton size that will be uncovered in section 11.7.2 on page 341.

---

[15] The exception is the type-I string, in which the NS-NS two-index antisymmetric tensor is absent because of the orientifold projection.

Bibliography

The books by Polchinski [8] and Johnson [9] contain a detailed discussion of D-branes and their effects. Szabo's short book [11] has an extensive coverage of the anomalous couplings of D-branes. There are several reviews on the subject, ranging from extensive ones to short introductions [156, 157,158,159,160,161,162,163] The generalization of the Dirac quantization condition can be found in [164,165]. An extensive review on T-duality in closed strings, which is important in understanding simple features of D-branes as well as some more nontrivial issues in curved backgrounds can be found in [70]. A pedestrian introduction to T-duality and D-branes can be found in [12].

The introduction of D-branes and the calculation of their charge can be found in the original paper [166]. The arguments for the existence of D-brane effects based on the divergence of closed string perturbation theory can be found in [123].

The Born-Infeld action in string theory was first discussed in [167–169]. The interplay between T-duality and the D-brane action is discussed in [170,171]. A review of classical DBI dynamics can be found in [172]. The technique of superembeddings is very useful in order to construct world-volume actions for supersymmetric branes. It is reviewed in [173].

The dynamics of D-branes turns out to be very interesting. In particular, they behave differently from fundamental strings in sub-Planckian energies. They are interesting probes that can reach regimes not accessible by fundamental strings. D-brane dynamics and scattering has been discussed in [174–176] The role of the $\mathcal{CP}$-odd part of the D-brane effective action on anomaly inflow has been discussed in [177]. Higher-derivative gravitational corrections have been discussed in [178,179].

The nonabelian dynamics of D-branes was first discussed in [180]. Further work on the nonabelian Born-Infeld action as well as the problems at higher orders in α' can be found in [181–187]. The interplay of T-duality and the nonabelian dynamics, as well as the D-brane dielectric effect, are discussed in [188]. Bound states of D-branes have been discussed in [189,190,191,192,193,194,195]. The engineering of gauge theories using D-branes is reviewed in [196].

The boundary state formalism has not been treated here. Boundary states describe the coupling of D-branes to closed strings. The reader will find good reviews in [92,93,94,95,96]. Another aspect of D-branes is their relation to K-theory. D-branes with their world-volume gauge fields induce a physical picture for fiber bundles and sub-bundles and give a physical meaning to the K-theory notion of the subtraction of bundles. Interesting work in this direction can be found in [197,198,199,200]. Discussion of non-BPS branes can be found in [201] while reviews can be found in [202,203,204].

Good reviews on branes as solutions of the string theory effective action equations can be found in [205] (summarizing some prescient pieces of work), and [206,207,208,163,209]. The D-brane solutions to the effective field theory equations have been found in [210].

The NS$_5$-brane was described as a string soliton in [211,212,213]. The CFT of its near-horizon region was described earlier in [214,215]. It is an important ingredient of the brane construction of gauge theories as detailed in [196].

We do not treat a large and interesting subject: the unstable D-branes and the associated tachyon dynamics. Good places to start are [216,217,218,219].

Exercises

8.1. Find the generalization of the Dirac monopole solution for a p-form gauge field.

8.2. Derive the Hamiltonian (8.2.4) on page 189.

8.3. In section 2.2 on page 14 it was shown that for a coordinate satisfying Dirichlet boundary conditions, the associated momentum is not conserved. How do you interpret this fact?

8.4. Consider the action of U(N) Yang-Mills theory on $M_9 \times S^1$, with a Wilson line background (8.2.6) on page 189. Derive the spectrum and verify (8.2.8).

8.5. Calculate the four-point scattering amplitude, for four transverse scalars on a D_p-brane. Take the low-energy limit and compare with (8.5.9) on page 199.

8.6. Do the previous problem for two coincident D_p-branes, taking properly into account the CP factors. Study again the low-energy limit and compare with the effective action (8.6.1) on page 200.

8.7. Calculate the disk one-point functions of the graviton and dilaton in the presence of a D_p-brane. Does the NS antisymmetric tensor couple to the D-brane?

8.8. Calculate the disk one-point function of an R-R $(p+1)$-form in the presence of a D_p-brane. The appropriate vertex operator must be in the $(-3/2, -1/2)$ picture. Fix the normalization so that you agree with the calculation in section 8.4 on page 193.

8.9. Calculate the disk amplitude of two massless gravitons/antisymmetric tensors in the presence of a D_p-brane. Study its Regge limit and show that there are intermediate open string poles. What is its physical interpretation? Compare its low-energy limit with the effective action (8.6.1) on page 200.

8.10. Use T-duality on spinors as described in section (7.7) to show that the supercharges preserved by a D_p-brane are $Q + \delta^{\perp} \tilde{Q}$, where $\delta^{\perp}$ is defined in (8.3.2) on page 193.

8.11. Consider a D_p-brane and a parallel $D_{p'}$-brane inside the original one ($p' < p$). Find for various values of p, p' the type of supersymmetry preserved by the configuration. Use this to guess when the static force between the two branes vanishes.

8.12. Consider again a D_p-brane and a $D_{p'}$-brane with $p \neq p'$ which are parallel and a distance r apart in the transverse space. Calculate the force between them generalizing the calculation in section 8.4 on page 193. What massless fields mediate the force? You may need the results of section 4.22 on page 112.

8.13. Consider a D_2-brane with a constant magnetic field F_{12} on its world-volume. Do a T-duality transformation in the x^2 direction. Show that in the T-dual coordinates we have a D_1-brane tilted by an angle θ in the $x^1 - x^2$ plane, where $\tan\theta = -2\pi \ell_s^2 F_{12}$. Use this result to derive the Born-Infeld action from the world-line length action, and vice versa. Generalize this to a ten-dimensional constant magnetic field.

8.14. Consider two D_1-strings. The first is stretching along the x^1-axis. The second is rotated by an angle θ with respect to the first in the $x^1 - x^2$ plane, while it is at distance r apart in transverse space. Calculate the force between them.

8.15. Consider two D_4-branes. The first stretches along the x^1, x^2, x^3, x^4 directions. The second, compared to the first, is rotated by an angle θ_1 on the $x^1 - x^5$ plane, θ_2 on the $x^2 - x^6$ plane, θ_3 on the $x^3 - x^7$ plane, and θ_4 on the $x^4 - x^8$ plane. Find the force between them as a function of the distance in the transverse space.

8.16. Consider a D_p-brane with a world-volume electric field $\vec{E}$. Show that in the T-dual picture this describes a D-brane moving in transverse space with velocity $\vec{v} = \vec{E}\ell_s^2$.

8.17. Find the appropriate boundary conditions on the string coordinates in the presence of a world-volume electric field. Calculate the cylinder amplitude in the presence of the electric field and show that it has poles for discrete real values of the modulus. By deforming appropriately the integration contour calculate the imaginary part of the one-loop amplitude. What is its interpretation?

8.18. Use the result of the previous exercise to study the scattering of two parallel D_p-branes, moving with relative velocity $\vec{v}$ and impact parameter b. Investigate the region of small impact parameter and show that D-branes can probe distances down to $\sim \ell_s\sqrt{v}$. Combine this with the standard quantum mechanical uncertainty relation to show that D-branes can probe distances down to order $\sim g_s^{1/3}\ell_s$, that is, in perturbation theory, much smaller then the string scale.

8.19. Consider the (bosonic) action for N D_0-branes

$$S = \frac{1}{g_s\,\ell_s}\mathrm{Tr}\left[(\dot{X}^I + [A_t, X^I])^2 + \frac{1}{2(2\pi\ell_s^2)^2}[X^I, X^j]^2\right]. \tag{8.1E}$$

Calculate the Hamiltonian. By redefining variables appropriately, show that the length scale is set by $g_s^{1/3}\,\ell_s$ while the energy scale is $\frac{g_s^{1/3}}{\ell_s}$ in agreement with the previous exercise.

8.20. Consider the Born-Infeld action in (8.5.1) on page 198. Dimensionally reduce it to $p + 1$ dimensions (take the fields to be independent of the $9 - p$ coordinates) and show that the result agrees with the D_p-brane action (8.5.4) on page 198 in the static gauge.

8.21. The action of the $\mathcal{N} = 1_{10}$ SYM theory is given by

$$\mathcal{L}_{\mathrm{SYM}} = -\frac{1}{4}\mathrm{Tr}\,F_{\mu\nu}F^{\mu\nu} - \mathrm{Tr}\,\bar{\chi}\,\Gamma^\mu D_\mu\chi, \tag{8.2E}$$

$$F_{\mu\nu} = \partial_\mu A_\nu - \partial_\nu A_\mu + [A_\mu, A_\nu], \quad D_\mu\chi = \partial_\mu\chi + [A_\mu, \chi]. \tag{8.3E}$$

Dimensionally reduce $9-p$ dimensions to find the complete action in $(p + 1)$ dimensions.

8.22. Calculate the masses of the non-Cartan gauge bosons A_μ^{ij} with $i \neq j$ by starting from the action (8.6.1) on page 200. Show that they are proportional to the distance in transverse space between the ith and jth branes.

8.23. Consider a D_1-brane, tilted in the 1-2 plane. Write its world-volume coupling to the R-R two-form, and after a T-duality in direction 2, show that this now includes a coupling to a R-R three-form and a subleading coupling to a R-R one-form. This was first described in [188].

8.24. Consider the $\mathcal{CP}$-odd terms in the D_2-brane action (8.5.10) on page 199. Show that they are invariant under the gauge transformations of the type-IIA R-R three-form C and one-form A as well as the two-form B, in the frame described in exercise 7.26 on page 185. What are the $\mathcal{CP}$-odd terms in the frame of equation (H.14) on page 527?

8.25. Show that the D_1 effective action is invariant under gauge transformations of the axion C_0.

8.26. Consider the abelian DBI action for a D_p-brane in trivial background bulk fields. Derive the nonlinear equations of motion for the transverse scalars and the gauge field. Make a rotationally invariant *Ansatz* where X^9 is a function of the radial distance on the brane, the electric field satisfies the BPS condition $\vec{E} = \pm \vec{\nabla} X^9$ and the magnetic field is zero. Solve the equations, and interpret the solution.

8.27. Turning on a world-volume gauge field strength on a D-brane provides a source for $B_{\mu\nu}$. It is therefore expected that this type of configurations are bound-states of the D-brane and fundamental strings. Consider the D_1-brane action (8.5.12) in trivial background fields. Turn on a (quantized) electric field and, by evaluating the Hamiltonian, compute the tension on the string. Show that it is given by

$$T = \frac{1}{2\pi \ell_s^2 g_s} \sqrt{1 + m^2 g_s^2},$$ (8.4E)

where $m \in \mathbb{Z}$ is the quantized electric flux or equivalently the number of fundamental strings bound to the D_1.

8.28. As obvious from (8.5.10), turning on nontrivial $F_{\alpha\beta}$ on the world-volume of a D-brane generates an effective R-R charge for lower-dimensional forms. Consider a D_3-brane with a nontrivial $F_{\alpha\beta}$, namely, a four-dimensional instanton configuration with instanton number n. Show that this describes a bound state of one D_3- and n D_{-1}-branes. Describe the supersymmetry of the configuration and discuss the moduli space.

8.29. Consider the world-volume gauge theory of a D_5-brane. Turn on again a configuration of a (four-dimensional) instanton, with instanton number n along directions x^2, x^3, x^4, x^5. Show that this describes the bound state of a D_5-brane and n D_1-strings stretched along x^1. Show that the instanton action provides the correct D_1-brane tension. Generalize this to N D_5-branes. How is the moduli space of such bound states related to the moduli space of $U(N)$ Yang-Mills instantons?

8.30. Consider three D_5-branes extended along directions $(12345),(12367),(12389)$. How much supersymmetry is left unbroken by this configuration?

8.31. Add a D_1-string along direction 1 to the previous configuration. How many supersymmetries are now left unbroken? Answer the same question when the D_1-string is along direction 4.

8.32. T-dualize the type-I string with an $O(2)$ Wilson line $(1, -1)$ on the circle. What is the D-brane picture after dualization?

8.33. Consider the T-dualization of an open string theory, with an $Sp(2N)$ group. Go through the same analysis as in section 8.7 on page 201 to find the various gauge groups as a function of the position of the D-branes and the orientifold planes.

8.34. Consider a fluctuating negative-tension hyperplane. Show that it has negative norm fluctuations.

8.35. Consider the supersymmetric variation of the gravitini in pure $\mathcal{N} = 2_4$ supergravity

$$\delta\psi_\mu = \left(\nabla_\mu - \frac{1}{4}F^-_{\nu\rho}\gamma^\nu\gamma^\rho\gamma_\mu\right)\epsilon, \tag{8.5E}$$

where ∇_μ is the standard covariant derivative action on spinors, and $F^\pm_{\mu\nu} = \frac{1}{2}(F_{\mu\nu} \pm {}^\star F_{\mu\nu})$ are the self-dual and anti-self-dual parts of the graviphoton field strength. Use this to find how many supersymmetries are left unbroken by the extremal Reissner-Nordström black hole in appendix F on page 519.

8.36. Consider the metric and gauge field *Ansatz* in equation (F.8) of appendix F. Define the three standard cartesian coordinates x^i as

$$d\rho^2 + \rho^2 d\Omega_2^2 = dx^i dx^i. \tag{8.6E}$$

Show that equations (F.2) imply that H is harmonic:

$$\Box H \equiv \partial_i\partial_i H = 0. \tag{8.7E}$$

The multicenter Reissner-Nordström solution in equation (F.9) on page 520 is the most general solution to (8.7E). What is its physical interpretation?

8.37. Show that the Bertotti-Robinson solution (F.12) on page 521 has eight Killing spinors. Use the supersymmetry transformations of exercise 8.35.

8.38. By varying (8.8.1) derive the equations (8.8.2),(8.8.3) on page 205. You may wish to also look at exercise 6.2.

8.39. Show that (8.8.4), (8.8.5), and (8.8.7) solve the equations (8.8.2),(8.8.3).

8.40. Integrate $^{\star}F_{p+2}$ on a sphere at spatial infinity and verify (8.8.10).

8.41. Upon compactifying x^i on a p-dimensional torus, the D-brane solution in section 8.8 becomes pointlike in the transverse $(9-p)$-dimensional space, and is that of a Reissner-Nordström black hole in $(9-p)$ dimensions. Calculate its ADM mass and verify (8.8.14).

8.42. Show that the metric (8.8.4) on page 205 after the coordinate transformation (8.8.16) becomes (8.8.17).

8.43. Consider the D_p-brane solution, and use the supersymmetry variations of IIA/B theory given in appendix H on page 525 to show that when $r_0 = 0$ the solution preserves 16 supercharges.

8.44. Calculate the scalar curvature, the square of the Ricci tensor $R_{\mu\nu}R^{\mu\nu}$ as well as the square of the Riemann tensor $R_{\mu\nu\rho\sigma}R^{\mu\nu\rho\sigma}$ of the extremal p-brane metric (8.8.30) on page 209. Verify (8.8.32).

8.45. Consider the solution of extremal D_p-branes with metric (8.8.30) on page 209 satisfying (8.8.27) on page 208 where the transverse space is $S^1 \times T^{8-p}$. T-dualizing along S^1 and using the result of exercise 7.7 on page 184, show that the background describes a D_{p+1}-brane wrapping the S^1.

8.46. Consider the reliability of D_p-brane solutions when $\lambda \gg 1$. Show that they are all reliable when $r \gg \ell_s$. When $p > 3$ show that the curvature becomes large at shorter scales. Find the typical distance where this happens. When $p < 3$, the curvature remain small at short distances but the string coupling grows. Thus, beyond a certain distance, the tree-level effective action is no longer reliable. Find the typical distance where this happens.

8.47. Consider the NS_5-brane discussed in section 8.9 on page 211. Show that (8.9.3) and (8.9.4) on page 211 solve the equations stemming from the action (8.9.2).

8.48. Show that the NS_5-brane solution (8.9.3) and (8.9.4) on page 211 preserves half of the original supersymmetry in the type-II theory.

8.49. Find the solution to the low-energy string theory action (8.9.2) on page 211 that describes a macroscopic fundamental string.

8.50. Consider IIA/IIB superstring theory in ten dimensions. Consider orbifolding by the following symmetry: $\mathcal{I}_4 (-1)^{\mathsf{F}_L}$ where $\mathcal{I}_4$ is the standard reflection of four Euclidean coordinates, while F_L is the left world-sheet fermion number. The twisted sector is localized on a five-plane. Find its supersymmetry and massless spectrum. Show that it coincides with the supersymmetry and massless spectrum of a single NS_5-brane.

9 | Compactification and Supersymmetry Breaking

We have considered so far superstring theories in ten noncompact dimensions. However, our direct physical interest is in theories with four large dimensions. One way to obtain them is to make use of the Kaluza-Klein idea: consider some of the dimensions to be curled up into a compact manifold, leaving only four noncompact dimensions.

As we have seen in the case of the bosonic strings, exact solutions to the equations of motion correspond to a CFT. The classical geometric picture is only appropriate at large volume (α'-expansion). In the case of type-II string theory, vacua correspond to an $\mathcal{N} = (1,1)_2$ SCFT. In the heterotic case, vacua correspond to $\mathcal{N} = (1,0)_2$ SCFT.

We generalize the concept of compactification to four dimensions, by replacing the original flat noncompact CFT with another one, where four dimensions are still flat but the rest is described by an arbitrary unitary[1] CFT with the appropriate central charge. This type of description is more general than that of a geometrical compactification, since there are CFTs with no geometrical interpretation. In the following, we will examine both the geometric point of view and the CFT point of view, mainly via orbifold compactifications.

9.1 Narain Compactifications

The simplest possibility is the "internal compact" manifold to be a (flat) torus. This can be considered as a different background of the ten-dimensional theory, where we have given constant expectation values to the internal metric and other background fields.

Consider first the case of the heterotic string compactified to $D < 10$ dimensions. It is rather straightforward to construct the partition function of the compactified theory. There are now $D - 2$ transverse noncompact coordinates, each contributing $\sqrt{\tau_2}\eta\bar{\eta}$. There is no change in the contribution of the left-moving world-sheet fermions and 16 right-moving

---

[1] The spectrum of dimensions of this CFT should be discrete in order to correspond to a "compactification."

compact coordinates. Finally the contribution of the $10 - D$ compact coordinates is given by (4.18.40) on page 99. Putting everything together we obtain

$$Z_D^{\text{heterotic}} = \frac{\Gamma_{10-D,10-D}(G, B) \, \bar{\Gamma}_H}{\tau_2^{(D-2)/2} \eta^8 \bar{\eta}^8} \frac{1}{2} \sum_{a,b=0}^{1} (-1)^{a+b+ab} \frac{\vartheta^4[{}^a_b]}{\eta^4}, \tag{9.1.1}$$

where $\bar{\Gamma}_H$ stands for the partition function of either $\text{Spin}(32)/\mathbb{Z}_2$ or $E_8 \times E_8$ lattice; $G_{\alpha\beta}$, $B_{\alpha\beta}$ are the constant expectation values of the internal $(10 - D)$-dimensional metric and antisymmetric tensor.

We now analyze the massless spectrum. The original ten-dimensional metric gives rise to the D-dimensional metric $G_{\mu\nu}$, $(10 - D)$ U(1) gauge fields, $G_{\mu\alpha}$ and $\frac{1}{2}(10 - D)(11 - D)$ scalars, $G_{\alpha\beta}$. The antisymmetric tensor produces a D-dimensional antisymmetric tensor, $B_{\mu\nu}$, $(10 - D)$ U(1) gauge fields, $B_{\mu\alpha}$, and $\frac{1}{2}(10 - D)(9 - D)$ scalars, $B_{\alpha\beta}$. The ten-dimensional dilaton gives rise to another scalar. Finally the $\dim(H)$ ten-dimensional gauge fields give rise to $\dim(H)$ gauge fields, A_μ^a, and $(10 - D) \cdot \dim(H)$ scalars, $A_\alpha^a \equiv Y_\alpha^a$. Similar remarks apply to the fermions.

We will consider in more detail the scalars Y_α^a coming from the ten-dimensional vectors, where a is the adjoint index and α the internal index taking values $D + 1, \dots, 10$. The nonabelian field strength (8.3E) on page 215 contains a term without derivatives. This is the commutator of two gauge fields. Upon dimensional reduction this gives rise to a potential term for the (Higgs) scalars Y_α^a:

$$V_H \sim G^{\alpha\gamma} G^{\beta\delta} \text{Tr}[Y_\alpha, Y_\beta][Y_\gamma, Y_\delta] \sim f^a{}_{bc} f^a{}_{b'c'} \, G^{\alpha\gamma} \, G^{\beta\delta} \, Y_\alpha^b \, Y_\beta^c \, Y_\gamma^{b'} \, Y_\delta^{c'}, \tag{9.1.2}$$

where $Y_\alpha = Y_\alpha^a T^a$. This potential is minimized when the matrices Y_α are commuting. They then have arbitrary expectation values in the Cartan subalgebra. These expectations values are moduli (flat directions or continuous families of minima). We will label these values by Y_α^I, $I = 1, 2, \dots, 16$. This is a normal Brout-Englert-Higgs phenomenon and it generates a mass matrix for the gauge fields

$$[m^2]^{ab} \sim G^{\alpha\beta} f^{ca}{}_d f^{cb}{}_{d'} Y_\alpha^d Y_\beta^{d'}. \tag{9.1.3}$$

This mass matrix has $\text{rank}(H)$ generic zero eigenvalues. The gauge fields belonging to the Cartan subalgebra remain massless while all the other gauge fields get a nonzero mass. Consequently, the gauge group is broken to $\text{U}(1)^{\text{rank}(H)}$. If we turn on these expectation values, the heterotic compactified partition function becomes

$$Z_D^{\text{heterotic}} = \frac{\Gamma_{10-D,26-D}(G, B, Y)}{\tau_2^{(D-2)/2} \eta^8 \bar{\eta}^8} \frac{1}{2} \sum_{a,b=0}^{1} (-1)^{a+b+ab} \frac{\vartheta^4[{}^a_b]}{\eta^4}, \tag{9.1.4}$$

where the structure of the $\Gamma_{10-D,26-D}$ lattice sum is described in detail in Appendix D on page 513.

The $(10 - D)(26 - D)$ scalar fields G, B, Y are called moduli since they can have arbitrary expectation values. Thus, the heterotic string compactified down to D dimensions provides a continuous family of vacua parametrized by the expectation values of the moduli that describe the geometry of the internal manifold (G, B) and the (flat) gauge bundle (Y).

Consider now the tree-level effective action for the bosonic massless modes in the toroidally compactified theory. It can be obtained by direct dimensional reduction of the ten-dimensional heterotic effective action, which in the string frame is given by (6.1.10) on page 146 with the addition of the gauge fields[2]

$$S_{10}^{\text{heterotic}} = \frac{1}{2\kappa_{10}^2} \int d^{10}x \sqrt{-\det G_{10}} e^{-2\Phi} \left[R + 4(\nabla\Phi)^2 - \frac{1}{12}\hat{H}^2 - \frac{1}{4}\text{Tr}[F^2] \right] + \mathcal{O}(\alpha'). \tag{9.1.5}$$

The massless fields in D dimensions are obtained from those of the ten-dimensional theory by assuming that the latter do not depend on the internal coordinates x^α. Moreover we keep only the Cartan gauge fields since they are the only ones that will remain massless for generic values of the Wilson lines Y_α^I, $I = 1, 2, \ldots, 16$. So, the gauge kinetic terms become abelian $\text{Tr}[F^2] \to \sum_{I=1}^{16} F_{\mu\nu}^I F^{I,\mu\nu}$ with

$$F_{\mu\nu}^I = \partial_\mu A_\nu^I - \partial_\nu A_\mu^I. \tag{9.1.6}$$

Also

$$\hat{H}_{\mu\nu\rho} = \partial_\mu B_{\nu\rho} - \frac{1}{2}\sum_I A_\mu^I F_{\nu\rho}^I + \text{cyclic}, \tag{9.1.7}$$

where we have neglected the gravitational Chern-Simons contribution, since it is of higher order in α'.

There is a standard *Ansatz* to define the D-dimensional fields, such that the gauge invariance of the compactified theory is simple. This is given in Appendix E on page 516. In this way we obtain

$$S_D^{\text{heterotic}} = \int d^D x \sqrt{-\det G} e^{-2\phi} \left[R + 4\partial^\mu\phi\partial_\mu\phi - \frac{1}{12}\hat{H}^{\mu\nu\rho}\hat{H}_{\mu\nu\rho} \right.$$
$$\left. - \frac{1}{4}(\hat{M}^{-1})_{ij} F_{\mu\nu}^i F^{j\mu\nu} + \frac{1}{8}\text{Tr}(\partial_\mu\hat{M}\partial^\mu\hat{M}^{-1}) \right], \tag{9.1.8}$$

where $i = 1, 2, \ldots, 36 - 2D$. ϕ is the D-dimensional dilaton and

$$\hat{H}_{\mu\nu\rho} = \partial_\mu B_{\nu\rho} - \frac{1}{2}L_{ij}A_\mu^i F_{\nu\rho}^j + \text{cyclic}, \tag{9.1.9}$$

where L_{ij} is the invariant metric of $O(10 - D, 26 - D)$.

The moduli scalar matrix $\hat{M}$ is given in (D.4) on page 514. The action (9.1.8) has a continuous $O(10 - D, 26 - D)$ symmetry. If $\Omega \in O(10 - D, 26 - D)$ is a $(36 - 2D) \times (36 - 2D)$ matrix then

$$\hat{M} \to \Omega \, \hat{M} \, \Omega^T, \quad A_\mu \to \Omega \cdot A_\mu, \tag{9.1.10}$$

leaves the effective action invariant. The presence of the massive states originating from the lattice, breaks this symmetry to the discrete infinite subgroup $O(10 - D, 26 - D, \mathbb{Z})$. This is the group of T-duality symmetries. The action for the $(10 - D)(26 - D)$ scalars in (9.1.8) is the $O(10 - D, 26 - D)/(O(10 - D) \times O(26 - D)) \sigma$-model.

---

[2] We have rescaled the gauge fields $\ell_s^4 A_\mu/\sqrt{2} \to A_\mu$ so that now they are dimensionless; see (H.42) in appendix H.5 on page 526.

In the Einstein frame (using (6.1.11)), the action becomes

$$S_D^{\text{heterotic}} = \int d^D x \sqrt{-\det G_E} \left[R - \frac{4}{D-2} \partial^\mu \phi \partial_\mu \phi - \frac{e^{-8\phi/(D-2)}}{12} \hat{H}^{\mu\nu\rho} \hat{H}_{\mu\nu\rho} \right.$$

$$\left. - \frac{e^{-4\phi/(D-2)}}{4} (\hat{M}^{-1})_{ij} F^i_{\mu\nu} F^{j\mu\nu} + \frac{1}{8} \text{Tr}(\partial_\mu \hat{M} \partial^\mu \hat{M}^{-1}) \right]. \tag{9.1.11}$$

In section 4.18.6 on page 100 we have described the effect of gauge symmetry enhancement. This applies to the toroidal compactifications of the heterotic string. Whenever new currents appear on the nonsupersymmetric side, new massless gauge bosons appear in the effective theory. This is very much like the bosonic string. There is a difference, however, here: currents that appear on the supersymmetric side do not generate new massless gauge bosons. The reason is that the massless states on the supersymmetric sides come from the fermionic oscillators $\psi^i_{-1/2}$ which are not affected by changing the torus moduli. Therefore, there is no symmetry enhancement coming from the supersymmetric sector. The abelian gauge bosons originating from the supersymmetric side (i.e., $\psi^I_{-1/2} \bar{a}^\mu_{-1} |p\rangle$) are graviphotons.[3]

Whenever the lattice contains as a sublattice, the root lattice of a Lie algebra $\mathfrak{g}$, the gauge group contains G as a gauge group.[4] Moving away from that point is equivalent to the Brout-Englert-Higgs breaking (sometimes partially) of the G symmetry.

We will now pay special attention to the $D = 4$ compactifications. Here, the ten-dimensional gravitino produces four four-dimensional Majorana gravitini. Consequently, the four-dimensional compactified theory has $\mathcal{N} = 4_4$ local SUSY. The relevant massless $\mathcal{N} = 4_4$ supermultiplets are the supergravity multiplet and the vector multiplet. The supergravity multiplet contains the metric, six vectors (the graviphotons), a scalar and an antisymmetric tensor, as well as four Majorana gravitini and four Majorana spin-$\frac{1}{2}$ fermions. The vector multiplet contains a vector, four Majorana spin $\frac{1}{2}$ fermions and six scalars. In total we have, apart from the supergravity multiplet, 22 vector multiplets.

In $D = 4$, the antisymmetric tensor is equivalent (on shell) via a duality transformation, to a pseudoscalar a, the "axion." The relation (in the Einstein frame) is

$$e^{-4\phi} \hat{H}_{\mu\nu\rho} = E_{\mu\nu\rho}{}^\sigma \nabla_\sigma a \tag{9.1.12}$$

with the E tensor defined as in (B.12) on page 506. This relation is such that the $B_{\mu\nu}$ equations of motion $\nabla^\mu (e^{-4\phi} \hat{H}_{\mu\nu\rho}) = 0$ are automatically solved by substituting (9.1.12). The Bianchi identity for $\hat{H}$ from (9.1.9) is

$$E^{\mu\nu\rho\sigma} \partial_\mu \hat{H}_{\nu\rho\sigma} = -L_{ij} F^i_{\mu\nu} \tilde{F}^{j,\mu\nu}, \quad \tilde{F}^{\mu\nu} = \frac{1}{2} E^{\mu\nu\rho\sigma} F_{\rho\sigma}. \tag{9.1.13}$$

Using (9.1.12), it becomes the equation of motion for the axion:

$$\nabla^\mu (e^{4\phi} \nabla_\mu a) = -\frac{1}{4} L_{ij} F^i_{\mu\nu} \tilde{F}^{j,\mu\nu}. \tag{9.1.14}$$

---

[3] It is known that making some of the graviphotons part of a nonabelian symmetry is equivalent to gauging the associated supergravity. Gauged supergravities are very interesting and useful but they rarely have flat supersymmetric vacua. They correspond typically to compactifications with fluxes.

[4] Only simply laced (A-D-E) algebras with rank at most $26 - D$ can appear.

This equation can be obtained from the "dual" action

$$\tilde{S}_{D=4}^{\text{heterotic}} = \int d^4x \sqrt{-\det g_E} \left[R - 2\partial^\mu \phi \partial_\mu \phi - \frac{1}{2}e^{4\phi}\partial^\mu a \partial_\mu a \right.$$

$$\left. -\frac{1}{4}e^{-2\phi}(\hat{M}^{-1})_{ij}F^i_{\mu\nu}F^{j,\mu\nu} + \frac{1}{4}aL_{ij}F^i_{\mu\nu}\tilde{F}^{j,\mu\nu} + \frac{1}{8}\text{Tr}(\partial_\mu \hat{M}\partial^\mu \hat{M}^{-1}) \right]. \tag{9.1.15}$$

We define the complex axion-dilaton S field

$$S = S_1 + iS_2 = a + ie^{-2\phi}, \tag{9.1.16}$$

and write the action as

$$\tilde{S}_{D=4}^{\text{heterotic}} = \int d^4x \sqrt{-\det g_E} \left[R - \frac{1}{2}\frac{\partial^\mu S \partial_\mu \bar{S}}{S_2^2} - \frac{1}{4}S_2(\hat{M}^{-1})_{ij}F^i_{\mu\nu}F^{j,\mu\nu} \right.$$

$$\left. +\frac{1}{4}S_1 L_{ij}F^i_{\mu\nu}\tilde{F}^{j,\mu\nu} + \frac{1}{8}\text{Tr}(\partial_\mu \hat{M}\partial^\mu \hat{M}^{-1}) \right]. \tag{9.1.17}$$

From the definition (9.1.16), S_2 is the string loop expansion parameter (heterotic string coupling constant). The scalar field S takes values in the upper half plane $\mathcal{H}_2 = \text{SL}(2,\mathbb{R})/\text{U}(1)$. The scalars $\hat{M}$ parametrize the coset space $O(6,22)/O(6)\times O(22)$. As we will see later on, the four-dimensional heterotic string has a nonperturbative $\text{SL}(2,\mathbb{Z})$ action on S by fractional linear transformations. It entails electric-magnetic duality transformations on the abelian gauge fields as described in appendix G on page 522.

We will briefly describe here the toroidal compactification of type-II string theory to four dimensions. As discussed in section 7.7.1 on page 174, upon toroidal compactification the IIA and IIB theories are equivalent. Consequently, we need only examine the compactification of the type-IIA theory.

We compactify on a six-torus to four dimensions. The two Majorana-Weyl gravitini and fermions give rise to eight $D = 4$ Majorana gravitini and 48 spin-$\frac{1}{2}$ Majorana fermions. Therefore, the $D = 4$ theory has maximal $\mathcal{N} = 8_4$ supersymmetry. The ten-dimensional metric produces the four-dimensional metric, six U(1) vectors, and 21 scalars. The antisymmetric tensor produces (after four-dimensional dualization), six U(1) vectors and 16 scalars. The dilaton gives an extra scalar. The R-R U(1) gauge field gives one gauge field and six scalars. The R-R three-form gives a three-form (no physical degrees of freedom in four dimensions) 15 vectors and 26 scalars. All the degrees of freedom form the $\mathcal{N} = 8_4$ supergravity multiplet that contains the graviton, 28 vectors, 70 scalars, eight gravitini, and 56 fermions. We will see more on the symmetries of this theory in chapter 11. We note that there is no perturbative gauge symmetry enhancement in type-II string theory.

9.2 World-sheet versus Space-time Supersymmetry

There is an interesting relation between world-sheet and space-time supersymmetry. To uncover it, we consider first the case of the heterotic string compactified to $D = 4$. The four dimensions are described by a flat Minkowski space.

An N-extended supersymmetry algebra in four dimensions is generated by N Weyl supercharges Q_a^I and their Hermitian conjugates $\bar{Q}_{\dot{\alpha}}^I$ satisfying the algebra

$$\{Q_\alpha^I, Q_\beta^J\} = \epsilon_{\alpha\beta} Z^{IJ},$$
$$\{\bar{Q}_{\dot{\alpha}}^I, \bar{Q}_{\dot{\beta}}^J\} = \epsilon_{\dot{\alpha}\dot{\beta}} \bar{Z}^{IJ},$$
$$\{Q_\alpha^I, \bar{Q}_{\dot{\alpha}}^J\} = \delta^{IJ} \sigma_{\alpha\dot{\alpha}}^\mu P_\mu, \tag{9.2.1}$$

where Z^{IJ} is the antisymmetric matrix of central charges.

As we have seen in section 7.5 on page 168, the space-time supersymmetry charges can be constructed from the massless fermion vertex at zero momentum. In our case we have

$$Q_\alpha^I = \frac{1}{2\pi i} \oint dz\, e^{-\phi/2} S_\alpha \Sigma^I, \quad \bar{Q}_{\dot{\alpha}}^I = \frac{1}{2\pi i} \oint dz\, e^{-\phi/2} C_{\dot{\alpha}} \bar{\Sigma}^I, \tag{9.2.2}$$

where S, C are the spinor and conjugate spinor of O(4) and $\Sigma^I, \bar{\Sigma}^I$ are operators in the R sector of the internal CFT with conformal weight $\frac{3}{8}$. We will also need

$$: e^{q_1\phi(z)} :: e^{q_2\phi(w)} : \doteq (z-w)^{-q_1 q_2} : e^{(q_1+q_2)\phi(w)} : + \cdots, \tag{9.2.3}$$

$$S_\alpha(z) C_{\dot{\alpha}}(w) = \frac{1}{\sqrt{2}} \sigma_{\alpha\dot{\alpha}}^\mu \psi^\mu(w) + \mathcal{O}(z-w), \tag{9.2.4}$$

$$S_\alpha(z) S_\beta(w) = \frac{\epsilon_{\alpha\beta}}{\sqrt{z-w}} + \mathcal{O}(\sqrt{z-w}),$$

$$C_{\dot{\alpha}}(z) C_{\dot{\beta}}(w) = \frac{\epsilon_{\dot{\alpha}\dot{\beta}}}{\sqrt{z-w}} + \mathcal{O}(\sqrt{z-w}). \tag{9.2.5}$$

Imposing the anticommutation relations (9.2.1) we find that the internal operators must satisfy the following OPEs:

$$\Sigma^I(z) \bar{\Sigma}^J(w) = \frac{\delta^{IJ}}{(z-w)^{3/4}} + (z-w)^{1/4} J^{IJ}(w) + \cdots, \tag{9.2.6}$$

$$\Sigma^I(z) \Sigma^J(w) = (z-w)^{-1/4} \Psi^{IJ}(w) + \cdots,$$

$$\bar{\Sigma}^I(z) \bar{\Sigma}^J(w) = (z-w)^{-1/4} \bar{\Psi}^{IJ}(w) + \cdots, \tag{9.2.7}$$

where J^{IJ} are some weight-1 operators of the internal CFT and $\Psi^{IJ}, \bar{\Psi}^{IJ}$ have weight 1/2. The central charges are given by $Z^{IJ} = \oint \Psi^{IJ}$. The R fields $\Sigma, \bar{\Sigma}$ have square root branch cuts with respect to the internal supercurrent

$$G^{\text{int}}(z) \Sigma^I(w) \sim (z-w)^{-1/2}, \quad G^{\text{int}}(z) \bar{\Sigma}^I(w) \sim (z-w)^{-1/2}. \tag{9.2.8}$$

BRST invariance of the fermion vertex implies that the OPE $(e^{-\phi/2} S_\alpha \Sigma^I)(e^\phi G)$ has a single pole term. This in turn implies that there are no more singular terms in (9.2.8).

Consider an extra scalar X with two-point function

$$\langle X(z) X(w) \rangle = -\log(z-w). \tag{9.2.9}$$

Construct the dimension-$\frac{1}{2}$ operators

$$\lambda^I(z) = \Sigma^I(z) e^{iX/2}, \quad \bar{\lambda}^I(z) = \bar{\Sigma}^I(z) e^{-iX/2}. \tag{9.2.10}$$

Using (9.2.6) and (9.2.7) we can verify the following OPEs:

$$\lambda^I(z)\bar{\lambda}^J(w) = \frac{\delta^{IJ}}{z-w} + \hat{J}^{IJ} + \mathcal{O}(z-w), \quad \hat{J}^{IJ} = J^{IJ} + \tfrac{i}{2}\delta^{IJ}\partial X, \tag{9.2.11}$$

$$\lambda^I(z)\lambda^J(w) = e^{iX}\Psi^{IJ} + \mathcal{O}(z-w), \quad \bar{\lambda}^I(z)\bar{\lambda}^J(w) = e^{-iX}\bar{\Psi}^{IJ} + \mathcal{O}(z-w). \tag{9.2.12}$$

Thus, $\lambda^I, \bar{\lambda}^I$ are N complex free fermions and they generate an $O(2N)_1$ current algebra. Moreover, this immediately shows that $\Psi^{IJ} = -\Psi^{JI}$. Thus, the fields Ψ^{IJ} belong to the coset $O(2N)_1/U(1)$. It is not difficult to show that as current algebras, $O(2N)_1 \sim U(1) \times SU(N)_1$. The $U(1)$ is precisely the one generated by ∂X.

We may now compute the OPE of the Cartan currents $\hat{J}^{II}$,

$$\hat{J}^{II}(z)\hat{J}^{JJ}(w) = \frac{\delta^{IJ}}{(z-w)^2} + \text{regular}. \tag{9.2.13}$$

Using (9.2.11) we finally obtain

$$J^{II}(z)J^{JJ}(w) = \frac{\delta^{IJ} - 1/4}{(z-w)^2} + \text{regular}. \tag{9.2.14}$$

9.2.1 $\mathcal{N} = 1_4$ space-time supersymmetry

In this case, there is a single field Σ and a single current that we will call J

$$J = 2J^{11}, \quad J(z)J(w) = \frac{3}{(z-w)^2} + \text{regular}, \tag{9.2.15}$$

and no Ψ operator because of the antisymmetry. From (9.2.6) we compute the three-point function to find

$$\langle J(z_1)\Sigma(z_2)\bar{\Sigma}(z_3)\rangle = \frac{3}{2}\frac{z_{23}^{1/4}}{z_{12}z_{13}}. \tag{9.2.16}$$

We learn that $\Sigma, \bar{\Sigma}$ are affine primaries with $U(1)$ charges $3/2$ and $-3/2$ respectively. Bosonize the $U(1)$ current and separate the charge degrees of freedom

$$J = i\sqrt{3}\partial\Phi, \quad \Sigma = e^{i\sqrt{3}\Phi/2}W^+, \quad \bar{\Sigma} = e^{-i\sqrt{3}\Phi/2}W^-, \quad \langle\Phi(z)\Phi(w)\rangle = -\log(z-w), \tag{9.2.17}$$

where $W^\pm$ do not depend on Φ. If we write the internal Virasoro operator as $T^{\text{int}} = \hat{T} + T_\Phi$ with $T_\Phi = -(\partial\Phi)^2/2$, then $\hat{T}$ and T_Φ commute. The fact that the dimension of the Σ fields is equal to the $U(1)$ charge squared over 2 implies that $W^\pm$ have dimension zero and thus must be proportional to the identity. Consequently $\Sigma, \bar{\Sigma}$ are pure vertex operators of the field Φ.

Now consider the internal supercurrent and expand it in operators with well-defined $U(1)$ charge

$$G^{\text{int}} = \sum_{q\geq 0} e^{iq\Phi}T^{(q)} + e^{-iq\Phi}T^{(-q)}, \tag{9.2.18}$$

where the operators $T^{(\pm q)}$ do not depend on Φ. Then, (9.2.8) implies that q in (9.2.18) can only take the value $q = 1/\sqrt{3}$. We can write $G^{\text{int}} = G^+ + G^-$ with

$$J(z)G^{\pm}(w) = \pm \frac{G^{\pm}(w)}{(z-w)} + \cdots . \tag{9.2.19}$$

Finally the $\mathcal{N} = (1,0)_2$ superconformal algebra satisfied by G^{int} implies that, separately, $G^{\pm}$ are Virasoro primaries with weight 3/2. Moreover the fact that G^{int} satisfies (4.13.8) on page 78 implies that J, $G^{\pm}$, T^{int} satisfy the $\mathcal{N} = (2,0)_2$ superconformal algebra (4.13.16)–(4.13.21) with $c = 9$. The reverse argument is obvious: if the internal CFT has $\mathcal{N} = (2,0)_2$ invariance, then one can use the (chiral) operators of charge $\pm 3/2$ to construct the space-time supersymmetry charges. In section 4.13.2 on page 79 we have shown, using the spectral flow, that such R operators are always in the spectrum since they are the images of the NS ground state.

We now describe how the massless spectrum emerges from the general properties of the internal $\mathcal{N} = (2,0)_2$ superconformal algebra. As discussed in section 4.13.2, in the NS sector of the internal $\mathcal{N} = (2,0)_2$ CFT, there are two relevant ground states, the vacuum $|0\rangle$ and the chiral ground states $|\Delta, q\rangle = |1/2, \pm 1\rangle$. We have also the four-dimensional left-moving world-sheet fermion oscillators ψ_r^{μ} and the four-dimensional right-moving bosonic oscillators $\bar{\alpha}_n^{\mu}$. In the right-moving sector of the internal CFT, we have, apart from the vacuum state, a collection of $\bar{\Delta} = 1$ states. Combining the internal ground states, we obtain

$$|\Delta, q; \bar{\Delta}\rangle: \quad |0,0;0\rangle, \quad |0,0;1\rangle^I, \quad |1/2,\pm 1;1\rangle^i, \tag{9.2.20}$$

where the indices $I = 1, 2, \ldots, M$, $i = 1, 2 \ldots, \bar{M}$ count the various such states. The physical massless bosonic states are

- $\psi_{-1/2}^{\mu} \bar{a}_{-1}^{\nu} |0,0;0\rangle$, which provide the graviton, antisymmetric tensor, and dilaton,

- $\psi_{-1/2}^{\mu} |0,0;1\rangle^I$, which provide the massless vectors of the gauge group with dimension M,

- $|1/2, \pm 1; 1\rangle^i$, which provide $\bar{M}$ complex scalars.

Taking into account also the fermions from the R sector, we can organize the massless spectrum in multiplets of $\mathcal{N} = 1_4$ supersymmetry. Using the results of appendix D, we obtain the $\mathcal{N} = 1_4$ supergravity multiplet, one tensor multiplet (equivalent under a duality transformation to a chiral multiplet), M vector multiplets, and $\bar{M}$ chiral multiplets.

9.2.2 $\mathcal{N} = 2_4$ space-time supersymmetry

In this case there are two fields $\Sigma^{1,2}$ and four currents J^{IJ}. Define $J^s = J^{11} + J^{22}$, $J^3 = (J^{11} - J^{22})/2$ in order to diagonalize (9.2.14):

$$J^s(z)J^s(w) = \frac{1}{(z-w)^2} + \cdots,$$

$$J^3(z)J^3(w) = \frac{1/2}{(z-w)^2} + \cdots,$$

$$J^s(z)J^3(w) = \cdots . \tag{9.2.21}$$

As before we compute, using (9.2.6), (9.2.7) the three-point functions $\langle J\Sigma\Sigma\rangle$. From these we learn that under (J^s, J^3), Σ^1 has charges $(1/2, 1/2)$, Σ_2 has $(1/2, -1/2)$, $\bar{\Sigma}^1$ $(-1/2, -1/2)$, and $\bar{\Sigma}^2$ $(-1/2, 1/2)$. Moreover, their charges saturate their conformal weights so that if we bosonize the currents then the fields Σ, $\bar{\Sigma}$ are pure vertex operators

$$J^s = i\partial\phi, \quad J^3 = \frac{i}{\sqrt{2}}\partial\chi, \tag{9.2.22}$$

$$\Sigma^1 = \exp\left[\frac{i}{2}\phi + \frac{i}{\sqrt{2}}\chi\right], \quad \Sigma^2 = \exp\left[\frac{i}{2}\phi - \frac{i}{\sqrt{2}}\chi\right], \tag{9.2.23}$$

$$\bar{\Sigma}^1 = \exp\left[-\frac{i}{2}\phi - \frac{i}{\sqrt{2}}\chi\right], \quad \bar{\Sigma}^2 = \exp\left[-\frac{i}{2}\phi + \frac{i}{\sqrt{2}}\chi\right]. \tag{9.2.24}$$

Using these in (9.2.6) we obtain that $J^{12} = \exp[i\sqrt{2}\chi]$ and $J^{21} = \exp[-i\sqrt{2}\chi]$. Thus, J^3, J^{12}, J^{21} form the current algebra SU(2)$_1$. Moreover, $\Psi^{12} = \exp[i\phi]$, $\bar{\Psi}^{12} = \exp[-i\phi]$.

We again consider the internal supercurrent and expand it in charge eigenstates. Using (9.2.5) we can verify that the charges that can appear are $(\pm 1, 0)$ and $(0, \pm 1/2)$. We can split

$$G^{int} = G_{(2)} + G_{(4)}, \quad G_{(2)} = G_{(2)}^+ + G_{(2)}^-,$$

$$G_{(4)} = G_{(4)}^+ + G_{(4)}^-, \tag{9.2.25}$$

where $G_{(2)}^{\pm}$ have charges $(\pm 1, 0)$ and $G_{(4)}^{\pm}$ have charges $(0, \pm 1/2)$. This is attested by the following OPEs:

$$J^s(z)G_{(2)}^{\pm}(w) = \pm\frac{G_{(2)}^{\pm}(w)}{z-w} + \cdots, \quad J^3(z)G_{(4)}^{\pm}(w) = \pm\frac{1}{2}\frac{G_{(4)}^{\pm}(w)}{z-w} + \cdots, \tag{9.2.26}$$

$$J^s(z)G_{(4)}^{\pm}(w) = \text{finite}, \quad J^3(z)G_{(2)}^{\pm}(w) = \text{finite}, \quad G_{(2)}^{\pm} = e^{\pm i\phi}Z^{\pm}. \tag{9.2.27}$$

$Z^{\pm}$ are dimension-1 operators. They can be written in terms of scalars as $Z^{\pm} = i\partial X^{\pm}$. The vertex operators $e^{\pm i\phi}$ are those of a complex free fermion. Thus, the part of the internal theory corresponding to $G^{(2)}$ is a free two-dimensional CFT with $c = 3$. Finally it can be shown that the SU(2) algebra acting on $G_{(4)}^{\pm}$ supercurrents generates two more supercurrents that form the $\mathcal{N} = (4,0)_2$ superconformal algebra (4.13.29)–(4.13.31) on page 81 with $c = 6$.

Since there is a complex free fermion $\psi = e^{i\phi}$ in the $c = 3$ internal CFT we can construct two massless vector boson states $\psi_{-1/2}\bar{a}_{-1}^{\mu}|p\rangle$ and $\bar{\psi}_{-1/2}\bar{a}_{-1}^{\mu}|p\rangle$. One of them is the graviphoton belonging to the $\mathcal{N} = 2_4$ supergravity multiplet while the other is the vector belonging to the vector-tensor multiplet (to which the dilaton and $B_{\mu\nu}$ also belong). The vectors of massless vector multiplets correspond to states of the form $\psi_{-1/2}^{\mu}\bar{J}_{-1}^a|p\rangle$, where $\bar{J}^a$ is a right-moving affine current. The associated massless complex scalar of the vector multiplet corresponds to the state $\psi_{-1/2}\bar{J}_{-1}^a|p\rangle$. Massless hypermultiplet bosons arise from the $\mathcal{N} = (4,0)_2$ internal CFT. As already described in section 4.13.3 on page 81, an $\mathcal{N} = (4,0)_2$ superconformal CFT with $c = 6$ always contains states with $\Delta = \frac{1}{2}$ that transform as two conjugate doublets of the SU(2)$_1$ current algebra. Combining them with a right-moving operator with $\bar{\Delta} = 1$ gives the four massless scalars of a hypermultiplet.

In the maximal case, namely $\mathcal{N} = 4_4$ space-time supersymmetry, the internal CFT must be free (toroidal). You are invited to show this in exercise 9.4 on page 287. The six graviphotons participating in the $\mathcal{N} = 4_4$ supergravity multiplet are states of the form

$\bar{a}^\mu_{-1}\psi^I_{-1/2}|p\rangle$ where $I = 1, \ldots, 6$ and the ψ^I are the fermionic partners of the six left-moving currents of the toroidal CFT mentioned above.

In our previous discussion, there are no constraints due to space-time SUSY on the right-moving side of the heterotic string.

To summarize, in the $D = 4$ heterotic string, the internal CFT has at least $\mathcal{N} = (1,0)_2$ invariance. If it has $\mathcal{N} = (2,0)_2$ then we have $\mathcal{N} = 1_4$ space-time SUSY. If we have a $(c = 3)$ $\mathcal{N} = (2,0)_2 \oplus (c = 6)$ $\mathcal{N} = (4,0)_2$ CFT then we have $\mathcal{N} = 2_4$ in space-time. Finally, if we have six free left-moving coordinates then we have $\mathcal{N} = 4_4$ in four-dimensional space-time.

In the type-II theory, the situation is similar, but here the supersymmetries can come from either the right-moving and/or the left-moving side. For example, $\mathcal{N} = 1_4$ space-time supersymmetry needs a $\mathcal{N} = (2,1)_2$ or $\mathcal{N} = (1,2)_2$ world-sheet SUSY. For $\mathcal{N} = 2_4$ space-time supersymmetry there are two possibilities. Either we must have $\mathcal{N} = (2,2)_2$, in which one supersymmetry comes from the right-moving sector and the other from the left-moving sector, or $(c = 3)$ $\mathcal{N} = (2,1)_2 \oplus (c = 6)$ $\mathcal{N} = (4,1)_2$ CFT in which both space-time supersymmetries come from one side.

9.3 Orbifold Reduction of Supersymmetry

We are interested in vacua with a four-dimensional flat space-time times some compact internal manifold. In the most general case, such vacua are given by the tensor product of a four-dimensional noncompact flat CFT and an internal (compact) CFT. A CFT with appropriate central charge and world-sheet symmetries is an exact solution of the (tree-level) string equations of motion to all orders in α'. In the heterotic case, this internal CFT must have $\mathcal{N} = (1,0)_2$ invariance and $(c, \bar{c}) = (9, 22)$. In the type-II case it must have $\mathcal{N} = (1,1)_2$ superconformal invariance and $(c, \bar{c}) = (9, 9)$. If the CFT has a large volume limit, then an α'-expansion is possible and we can recover the leading σ-model (geometrical) results.

In this section we will consider orbifold CFTs which will provide compactification spaces that reduce the maximal supersymmetry in four dimensions. The advantage of orbifolds is that they are exactly soluble CFTs and yet they have the essential characteristics of nontrivial curved compactifications. In the next few sections we will give examples of orbifolds with $\mathcal{N} = 2_4$ and $\mathcal{N} = 1_4$ supersymmetry. We will focus first on the heterotic string.

We have already seen in section 9.1 that the toroidal compactification of the heterotic string down to four dimensions, gives a theory with $\mathcal{N} = 4_4$ supersymmetry. We have to find orbifold symmetries under which some of the four four-dimensional gravitini are not invariant. They will be projected out of the spectrum and we will be left with a theory that has less supersymmetry. To find such symmetries we have to look carefully at the vertex operators of the gravitini first. We will work in the light-cone gauge and it will be convenient to bosonize the eight transverse left-moving fermions ψ^i into four left-moving scalars. Pick a complex basis for the fermions

$$\Psi^0 = \frac{1}{\sqrt{2}}(\psi^3 + i\psi^4), \quad \Psi^1 = \frac{1}{\sqrt{2}}(\psi^5 + i\psi^6), \tag{9.3.1}$$

$$\Psi^2 = \frac{1}{\sqrt{2}}(\psi^7 + i\psi^8), \quad \Psi^3 = \frac{1}{\sqrt{2}}(\psi^9 + i\psi^{10}), \tag{9.3.2}$$

and similarly for $\bar{\Psi}^I$. They satisfy

$$\langle \Psi^I(z)\bar{\Psi}^J(w)\rangle = \frac{\delta^{IJ}}{z-w}, \quad \langle \Psi^I(z)\Psi^J(w)\rangle = \langle \bar{\Psi}^I(z)\bar{\Psi}^J(w)\rangle = 0. \tag{9.3.3}$$

The four Cartan currents of the left-moving $O(8)_1$ current algebra $J^I = \Psi^I \bar{\Psi}^I$ can be written in terms of four free bosons as

$$J^I(z) = i\partial_z \phi^I(z), \quad \langle \phi^I(z)\phi^J(w)\rangle = -\delta^{IJ} \log(z-w). \tag{9.3.4}$$

In terms of the bosons

$$\Psi^I =: e^{i\phi^I} :, \quad \bar{\Psi}^I =: e^{-i\phi^I} : . \tag{9.3.5}$$

The spinor primary states are given by

$$V(\epsilon_I) =: \exp\left[\frac{i}{2}\sum_{I=0}^{3} \epsilon_I \phi^I\right] :, \tag{9.3.6}$$

with $\epsilon_I = \pm 1$. This operator has $2^4 = 16$ components and contains both the S and the C $O(8)$ spinors.

The fermionic system has an $O(8)$ global symmetry (the zero-mode part of the $O(8)_1$ current algebra). Its $U(1)^4$ abelian subgroup acts as

$$\Psi^I \to e^{2\pi i\theta^I}\Psi^I, \quad \bar{\Psi}^I \to e^{-2\pi i\theta^I}\bar{\Psi}^I. \tag{9.3.7}$$

This acts equivalently on the bosons as

$$\phi^I \to \phi^I + 2\pi\,\theta^I. \tag{9.3.8}$$

A $\mathbb{Z}_2$ subgroup of the $U(1)^4$ symmetry, namely, $\theta^I = 1/2$ for all I, is the $(-1)^{F_R}$ symmetry. Under this transformation, the fermions are odd. The spinor vertex operator transforms with a phase $\exp[i\pi(\sum_I \epsilon^I)/2]$. Therefore,

- $\sum_I \epsilon^I = 4k$, $k \in \mathbb{Z}$ corresponds to the spinor S,

- $\sum_I \epsilon^I = 4k + 2$, $k \in \mathbb{Z}$ corresponds to the conjugate spinor C.

The standard GSO projection picks one of the two spinors, let us say the S. Consider the massless physical vertex operators given by

$$V^{\pm,\epsilon} = \bar{\partial}X^{\pm}\, V_S(\epsilon)e^{ip\cdot X}, \quad X^{\pm} = \frac{1}{\sqrt{2}}(X^3 \pm iX^4). \tag{9.3.9}$$

The boson ϕ^0 was constructed from the $D = 4$ light-cone space-time fermions and thus carries four-dimensional helicity. The $X^{\pm}$ bosons also carry four-dimensional helicity ± 1. The subset of the vertex operators in (9.3.9) that corresponds to the gravitini are $\bar{\partial}X^+ V(\epsilon^0 = 1)$, with helicity $3/2$, and $\bar{\partial}X^- V(\epsilon^0 = -1)$, with helicity $-3/2$. Taking also into account the GSO projection we find four helicity $(\pm 3/2)$ states, as we expect in an $\mathcal{N} = 4_4$ theory.

Consider the maximal subgroup $O(2) \times O(6) \subset O(8)$ where the $O(2)$ corresponds to the four-dimensional helicity. The $O(6)$ symmetry is an internal symmetry from the four-dimensional point of view. It is the so-called R-symmetry of $\mathcal{N} = 4_4$ supersymmetry, since the supercharges transform as the four-dimensional spinor of O(6). O(6) is an automorphism of the $\mathcal{N} = 4_4$ supersymmetry algebra. Since the supercharges are used to generate the states of an $\mathcal{N} = 4_4$ supermultiplet, the various states inside the multiplet have well-defined transformation properties under the O(6) R-symmetry. Here are some useful examples.

The $\mathcal{N} = 4_4$ SUGRA multiplet. It contains the graviton (singlet of O(6)) four Majorana gravitini (spinor of O(6)), six graviphotons (vector of O(6)), four Majorana fermions (conjugate spinor of O(6)), and two scalars (singlets).

The massless spin-3/2 multiplet. It contains a gravitino (singlet), four vectors (spinor), seven Majorana fermions (vector plus singlet), and eight scalars (spinor + conjugate spinor).

The massless vector multiplet. It contains a vector (singlet), four Majorana fermions (spinor), and six scalars (vector).

To break the $\mathcal{N} = 4_4$ symmetry, it is enough to break the O(6) R-symmetry.

We now search for symmetries of the CFT that will reduce, after orbifolding, the supersymmetry. In order to preserve Lorentz invariance, the symmetry should not act on the four-dimensional supercoordinates X^μ, ψ^μ. The rest are symmetries acting on the internal left-moving fermions and a simple class are the discrete subgroups of the $U(1)^3$ subgroup of O(6) acting on the fermions. There are also symmetries acting on the bosonic $(6, 22)$ compact CFT. An important constraint on such symmetries is that they leave the internal supercurrent

$$G^{\text{int}} = \sum_{i=5}^{10} \psi^i \, \partial \, X^i \tag{9.3.10}$$

invariant. The reason is that G^{int} along with $G^{D=4}$ (which is invariant since we are not acting on the $D = 4$ part) define the constraints responsible for the absence of ghosts. Messing them up will jeopardize the unitarity of the orbifold theory.

The generic symmetries of the internal toroidal theory are translations and SO(6) rotations of the (6,6) part as well as gauge transformations of the (0,16) part. Thus a generic orbifold group will be a combination of them all. Translations and gauge transformations do not affect the massless gravitini. Under the SO(6) rotations the gravitini transform as a four-dimensional spinor. We must therefore study the transformation of the spinor under an SO(6) rotation. Any rotation can be conjugated to the Cartan subalgebra, so it will be a combination of three O(2) rotations in the three planes of T^6. Let, θ^1 be the angle of rotation in the 5-6 plane, θ^2 in the 7-8 plane, and θ^3 in the 9-10 plane. Then, the respective fermions transform as in (9.3.7) and the transformation of the spinors can be obtained from (9.3.6). If we like to preserve a single gravitino, lets say the one corresponding to $(+ + + +), (- - - -)$, then the condition on the rotation angles is

$$\theta^1 + \theta^2 + \theta^3 = 0 \pmod{2\pi}. \tag{9.3.11}$$

The original four-dimensional SO(6) spinor decomposes as $4 \to (1 + 3)$ under the group $G \subset SO(6)$. Therefore, G can be at most SU(3). The final result is that orbifold rotations inside an SU(3) subgroup of SO(6) preserve at least $\mathcal{N} = 1_4$ space-time supersymmetry.

9.4 A Heterotic Orbifold with $\mathcal{N} = 2_4$ Supersymmetry

We will describe here a simple example of a $\mathbb{Z}_2$ orbifold that will produce $\mathcal{N} = 2_4$ supersymmetry.

Consider the toroidal compactification of the heterotic string. Set the Wilson lines to zero and pick appropriately the internal six-torus G, B so that the $(6, 22)$ lattice factorizes as $(2, 2) \otimes (4, 4) \otimes (0, 16)$. This lattice has a symmetry that changes the sign of all the $(4,4)$ bosonic coordinates. To keep the internal supercurrent invariant we must also change the sign of the fermions ψ^i, $i = 7, 8, 9, 10$. This corresponds to shifting the associated bosons

$$\phi^2 \to \phi^2 + \pi, \quad \phi^3 \to \phi^3 - \pi. \tag{9.4.1}$$

Under this transformation, two of the gravitini vertex operators are invariant while the other two transform with a minus sign. This is exactly what we need. It turns out, however, that this simple orbifold action does not give a modular-invariant partition function.

We must make a further action somewhere else. What remains is the $(0, 16)$ part. Consider the case in which it corresponds to the $E_8 \times E_8$ lattice. As we have mentioned already, $E_8 \ni [248] \to [120] \oplus [128] \in O(16)$. Decomposing further with respect to the $SU(2) \times SU(2) \times O(12)$ subgroup of $O(16)$, we obtain:

$$[120] \to [3, 1, 1] \oplus [1, 3, 1] \oplus [1, 1, 66] \oplus [2, 1, 12] \oplus [1, 2, 12], \tag{9.4.2}$$

$$[128] \to [2, 1, 32] \oplus [1, \bar{2}, 32]. \tag{9.4.3}$$

We choose the $\mathbb{Z}_2$ action on E_8 to take the spinors (the $[2]$'s) of the two $SU(2)$ subgroups to minus themselves, but keep the conjugate spinors (the $[\bar{2}]$'s) invariant. This projection keeps the $[3, 1, 1], [1, 3, 1], [1, 1, 66], [1, \bar{2}, 32]$ representations that combine to form the group $SU(2) \times E_7$. This can be seen by decomposing the adjoint of E_8 under its $SU(2) \times E_7$ subgroup.

$$E_8 \ni [248] \to [1, 133] \oplus [3, 1] \oplus [2, 56] \in SU(2) \times E_7, \tag{9.4.4}$$

where in this basis the above transformation corresponds to $[3] \to [3]$ and $[2] \to -[2]$. The reason why we considered a more complicated way in terms of orthogonal groups is that, in this language, the construction of the orbifold blocks is straightforward.

We will now construct the various orbifold blocks. The left-moving fermions contribute

$$\frac{1}{2} \sum_{a,b=0}^{1} (-1)^{a+b+ab} \frac{\vartheta^2[{}^a_b] \vartheta[{}^{a+h}_{b+g}] \vartheta[{}^{a-h}_{b-g}]}{\eta^4}. \tag{9.4.5}$$

The bosonic $(4,4)$ blocks can be constructed in a similar fashion to $(4.21.10)$ on page 110. We obtain

$$Z_{(4,4)}[{}^0_0] = \frac{\Gamma_{4,4}}{\eta^4 \bar{\eta}^4}, \quad Z_{(4,4)}[{}^h_g] = 2^4 \frac{\eta^2 \bar{\eta}^2}{\vartheta^2[{}^{1-h}_{1-g}] \bar{\vartheta}^2[{}^{1-h}_{1-g}]}, \quad (h, g) \neq (0, 0). \tag{9.4.6}$$

The blocks of the E_8 factor in which our projection acts are given by

$$\frac{1}{2} \sum_{\gamma,\delta=0}^{1} \frac{\bar{\vartheta}[^{\gamma+h}_{\delta+g}]\bar{\vartheta}[^{\gamma-h}_{\delta-g}]\bar{\vartheta}^6[^{\gamma}_{\delta}]}{\bar{\eta}^8}. \tag{9.4.7}$$

Finally there is a (2,2) toroidal and an E_8 part that are not touched by the projection. Putting all things together we obtain the heterotic partition function of the $\mathbb{Z}_2$ orbifold

$$Z_{N=2}^{\text{heterotic}} = \frac{1}{2} \sum_{h,g=0}^{1} \frac{\Gamma_{2,2}\,\bar{\Gamma}_{E_8}Z_{(4,4)}[^h_g]}{\tau_2\eta^4\bar{\eta}^{12}} \frac{1}{2} \sum_{\gamma,\delta=0}^{1} \frac{\bar{\vartheta}[^{\gamma+h}_{\delta+g}]\bar{\vartheta}[^{\gamma-h}_{\delta-g}]\bar{\vartheta}^6[^{\gamma}_{\delta}]}{\bar{\eta}^8}$$

$$\times \frac{1}{2} \sum_{a,b=0}^{1} (-1)^{a+b+ab} \frac{\vartheta^2[^a_b]\vartheta[^{a+h}_{b+g}]\vartheta[^{a-h}_{b-g}]}{\eta^4}. \tag{9.4.8}$$

This partition function is modular invariant. The massless spectrum is as follows: from the untwisted sector ($h = 0$) we obtain the graviton, an antisymmetric tensor, vectors in the adjoint of $G = U(1)^4 \times SU(2) \times E_7 \times E_8$, a complex scalar in the adjoint of the gauge group G, 16 more neutral scalars as well as scalars transforming as four copies of the $[2, 56]$ representation of $SU(2) \times E_7$. From the twisted sector ($h = 1$), we obtain scalars only, transforming as 32 copies of the $[1, 56]$ and 128 copies of the $[2, 1]$.

As mentioned before, this theory has $\mathcal{N} = 2_4$ local supersymmetry. The associated R-symmetry is $SU(2)$, which rotates the two supercharges. We will describe the relevant massless representations and their transformation properties under the R-symmetry.

The SUGRA multiplet contains the graviton (singlet), two Majorana gravitini (doublet), and a vector (singlet).

The vector multiplet contains a vector (singlet), two Majorana fermions (doublet), and a complex scalar (singlet).

The vector-tensor multiplet contains a vector (singlet), two Majorana fermions (doublet), a real scalar (singlet), and an antisymmetric tensor (singlet).

The hypermultiplet contains two Majorana fermions (singlets) and four scalars (two doublets).

The vector-tensor multiplet and the vector multiplet are related by a duality transformation of the two-form.

We can now arrange the massless states into $\mathcal{N} = 2_4$ multiplets. We have the SUGRA multiplet, a vector-tensor multiplet (containing the dilaton), a vector multiplet in the adjoint of $U(1)^2 \times SU(2) \times E_7 \times E_8$; the rest are hypermultiplets transforming under $SU(2) \times E_7$ as $4[1, 1] + [2, 56] + 8[1, 56] + 32[2, 1]$.

We will investigate further the origin of the $SU(2)$ R-symmetry. Consider the four real left-moving fermions $\psi^{7,\dots,10}$. Although they transform with a minus sign under the orbifold action, their $O(4) \sim SU(2) \times SU(2)$ currents, being bilinear in the fermions, are invariant. Relabel the four real fermions as ψ^0 and ψ^a, $a = 1, 2, 3$. Then, the $SU(2)_1 \times SU(2)_1$ current algebra is generated by

$$J^a = -\frac{i}{2}\left[\psi^0\psi^a + \frac{1}{2}\epsilon^{abc}\psi^b\psi^c\right], \quad \tilde{J}^a = -\frac{i}{2}\left[\psi^0\psi^a - \frac{1}{2}\epsilon^{abc}\psi^b\psi^c\right]. \tag{9.4.9}$$

Although both SU(2)'s are invariant in the untwisted sector, the situation in the twisted sector is different. The O(4) spinor ground state decomposes as $[\mathbf{4}] \to [\mathbf{2}, \mathbf{1}] + [\mathbf{1}, \mathbf{2}]$ under $SU(2) \times SU(2)$. The orbifold projection acts trivially on the spinor of the first $SU(2)$ and with a minus sign on the spinor of the second. The orbifold projection breaks the second $SU(2)$ invariance. The remaining $SU(2)_1$ invariance becomes the R-symmetry of the $\mathcal{N} = 2_4$ theory. Moreover, the only operators at the massless level that transform non-trivially under the $SU(2)$ are the linear combinations

$$V_{\alpha\beta}^{\pm} = \pm i(\delta_{\alpha\beta}\psi^0 \pm i\sigma_{\alpha\beta}^a \; \psi^a), \tag{9.4.10}$$

which transform as the $[\mathbf{2}]$ and $[\bar{\mathbf{2}}]$, respectively, as well as the $[\mathbf{2}]$ spinor in the R-sector. We obtain

$$V_{\alpha\gamma}^+(z) V_{\gamma\beta}^+(w) = V_{\alpha\gamma}^-(z) V_{\gamma\beta}^-(w) = \frac{\delta_{\alpha\beta}}{z - w} - 2\sigma_{\alpha\beta}^a (J^a(w) - \bar{J}^a(w)) + \mathcal{O}(z - w), \tag{9.4.11}$$

$$V_{\alpha\gamma}^+(z) V_{\gamma\beta}^-(w) = \frac{3\delta_{\alpha\beta}}{z - w} + 4\sigma_{\alpha\beta}^a \bar{J}^a(w) + \mathcal{O}(z - w), \tag{9.4.12}$$

$$V_{\alpha\gamma}^-(z) V_{\gamma\beta}^+(w) = \frac{3\delta_{\alpha\beta}}{z - w} - 4\sigma_{\alpha\beta}^a J^a(w) + \mathcal{O}(z - w), \tag{9.4.13}$$

where a summation over γ is implied.

This $SU(2)_1$ current algebra combines with four operators of conformal weight 3/2 to make the $\mathcal{N} = (4, 0)_2$ superconformal algebra in any theory with $\mathcal{N} = 2_4$ space-time supersymmetry. This agrees with the general discussion of section 9.2.

In an $\mathcal{N} = 2_4$ theory, the complex scalars that are partners of the gauge bosons, belonging to the Cartan of the gauge group, are moduli (they have no potential). If they acquire generic expectation values, they break the gauge group down to the Cartan. All charged hypermultiplets acquire masses in such a case.

A generalization of the above orbifold, where all Higgs expectation values are turned on, corresponds to splitting the original (6,22) lattice to (4,4)⊕(2,18). We perform a $\mathbb{Z}_2$ reversal in the (4, 4) part, which will break $\mathcal{N} = 4_4 \to \mathcal{N} = 2_4$. In the (2,18) lattice we can only perform a $\mathbb{Z}_2$ translation (otherwise the supersymmetry will be broken further). We perform a translation by $\epsilon/2$, where $\epsilon \in L_{2,18}$. Then the partition function is

$$Z_{N=2}^{\text{heterotic}} = \frac{1}{2} \sum_{h,g=0}^{1} \frac{\Gamma_{2,18}(\epsilon)[{}_g^h] Z_{(4,4)}[{}_g^h]}{\tau_2 \eta^4 \bar{\eta}^{20}} \frac{1}{2} \sum_{a,b=0}^{1} (-1)^{a+b+ab} \frac{\vartheta^2[{}_b^a]\vartheta[{}_{b+g}^{a+h}]\vartheta[{}_{b-g}^{a-h}]}{\eta^4}; \tag{9.4.14}$$

the shifted lattice sum $\Gamma_{2,18}(\epsilon)[{}_g^h]$ is described in Appendix B.

The theory depends on the 2×18 moduli of $\Gamma_{2,18}(\epsilon)[{}_g^h]$ and the 16 moduli in $Z_{4,4}[{}_0^0]$. There are, apart from the vector-tensor multiplet, another 18 massless vector multiplets. The 2×18 moduli are the scalars of these vector multiplets. There are also four neutral hypermultiplets whose scalars are the untwisted (4,4) orbifold moduli. At special submanifolds of the vector multiplet moduli space, extra massless vector multiplets and/or hypermultiplets can appear. We have seen such a symmetry enhancement already at the level of the CFT.

The local structure of the vector moduli space is $O(2, 18)/O(2) \times O(18)$. From the real moduli, $G_{\alpha\beta}, B_{\alpha\beta}, Y_\alpha^I$ we can construct the 18 complex moduli $T = T_1 + iT_2$, $U = U_1 + iU_2$, $W^I = W_1^I + iW_2^I$ as follows:

$$G = \frac{T_2 - \frac{W_2^I W_2^I}{2U_2}}{U_2} \begin{pmatrix} 1 & U_1 \\ U_1 & |U|^2 \end{pmatrix},$$

$$B = \left(T_1 - \frac{W_1^I W_2^I}{2U_2} \right) \begin{pmatrix} 0 & 1 \\ -1 & 0 \end{pmatrix}, \tag{9.4.15}$$

and $W^I = -Y_2^I + UY_1^I$. There is also one more complex scalar, the S field with $\mathrm{Im}\,S = S_2 = e^{-\phi}$, whose real part is the axion a, which comes from dualizing the antisymmetric tensor. The tree-level prepotential and Kähler potential are[5]

$$\mathcal{F} = S(TU - \tfrac{1}{2} W^I W^I), \quad K = -\log(S_2) - \log\left[U_2 T_2 - \tfrac{1}{2} W_2^I W_2^I \right]. \tag{9.4.16}$$

The hypermultiplets belong to the quaternionic manifold $O(4, 4)/O(4) \times O(4)$. $\mathcal{N} = 2_4$ supersymmetry does not permit neutral couplings between vector and hypermultiplets at the two-derivative level. The dilaton belongs to a vector multiplet. Therefore, the hypermultiplet moduli space does not receive perturbative or nonperturbative corrections.

In this class of $\mathcal{N} = 2_4$ vacua, we will consider the helicity supertrace B_2 which traces the presence of $\mathcal{N} = 2_4$ (short) BPS multiplets.[6] The computation is straightforward, using the definitions of appendix J on page 537 and is the subject of exercise 9.8 on page 288. We find

$$\tau_2 B_2 = \tau_2 \langle \lambda^2 \rangle = \Gamma_{2,18}{\textstyle\left[{0 \atop 1}\right]} \frac{\bar\vartheta_3^2 \bar\vartheta_4^2}{\bar\eta^{24}} - \Gamma_{2,18}{\textstyle\left[{1 \atop 0}\right]} \frac{\bar\vartheta_2^2 \bar\vartheta_3^2}{\bar\eta^{24}} - \Gamma_{2,18}{\textstyle\left[{1 \atop 1}\right]} \frac{\bar\vartheta_2^2 \bar\vartheta_4^2}{\bar\eta^{24}} = \frac{\Gamma_{2,18}{\textstyle\left[{0 \atop 0}\right]} + \Gamma_{2,18}{\textstyle\left[{0 \atop 1}\right]}}{2} \bar F_1$$

$$- \frac{\Gamma_{2,18}{\textstyle\left[{0 \atop 0}\right]} - \Gamma_{2,18}{\textstyle\left[{0 \atop 1}\right]}}{2} \bar F_1 - \frac{\Gamma_{2,18}{\textstyle\left[{1 \atop 1}\right]} + \Gamma_{2,18}{\textstyle\left[{1 \atop 0}\right]}}{2} \bar F_+ - \frac{\Gamma_{2,18}{\textstyle\left[{1 \atop 0}\right]} - \Gamma_{2,18}{\textstyle\left[{1 \atop 1}\right]}}{2} \bar F_- \tag{9.4.17}$$

with

$$\bar F_1 = \frac{\bar\vartheta_3^2 \bar\vartheta_4^2}{\bar\eta^{24}}, \quad \bar F_\pm = \frac{\bar\vartheta_2^2 (\bar\vartheta_3^2 \pm \bar\vartheta_4^2)}{\bar\eta^{24}}. \tag{9.4.18}$$

For all $\mathcal{N} = 2_4$ heterotic vacua, B_2 transforms as

$$\tau \to \tau + 1: \quad B_2 \to B_2, \quad \tau \to -\frac{1}{\tau}: \quad B_2 \to \tau^2 B_2. \tag{9.4.19}$$

All functions $\bar F_i$ have positive Fourier coefficients and have the expansions

$$F_1 = \frac{1}{q} + \sum_{n=0}^{\infty} d_1(n) q^n = \frac{1}{q} + 16 + 156q + \mathcal{O}(q^2), \tag{9.4.20}$$

$$F_+ = \frac{8}{q^{3/4}} + q^{1/4} \sum_{n=0}^{\infty} d_+(n) q^n = \frac{8}{q^{3/4}} + 8q^{1/4}(30 + 481q + \mathcal{O}(q^2)), \tag{9.4.21}$$

$$F_- = \frac{32}{q^{1/4}} + q^{3/4} \sum_{n=0}^{\infty} d_-(n) q^n = \frac{32}{q^{1/4}} + 32q^{3/4}(26 + 375q + \mathcal{O}(q^2)). \tag{9.4.22}$$

---

[5] The definitions of the prepotential and Kähler potential may be found in appendix I.2 on page 535. The expressions in (9.4.16) can be obtained from these definitions and the kinetic terms of the moduli scalars from (E.22 on page 518).

[6] You will find the definition of helicity supertraces and their relation to BPS multiplicities in appendix J on page 537.

The lattice sums $\frac{1}{2}(\Gamma_{2,18}[^h_0] \pm \Gamma_{2,18}[^h_1])$ also have positive multiplicities. An overall plus sign corresponds to vectorlike multiplets,[7] while a minus sign corresponds to hypermultiplets. A vectorlike multiplet contributes 1 to the supertrace, and a hypermultiplet -1.

The contribution of the generic massless multiplets is given by the constant coefficient of F_1; it agrees with what we expected: $16 = 20 - 4$ since we have the supergravity multiplet and 19 vector multiplets contributing 20 and 4 hypermultiplets contributing -4.

We will analyze the BPS mass formulas associated with (9.4.17). We will use the notation for the shift vector $\epsilon = (\vec{\epsilon}_L; \vec{\epsilon}_R, \vec{\zeta})$, where ϵ_L, ϵ_R are two-dimensional integer vectors and ζ is a vector in the Spin$(32)/\mathbb{Z}_2$ lattice. We also have the modular-invariance constraint $\epsilon^2/2 = \vec{\epsilon}_L \cdot \vec{\epsilon}_R - \vec{\zeta}^2/2 = 1 \pmod 4$.

Using the results of Appendix D, we can write the BPS mass formulas associated with the lattice sums above. In the untwisted sector ($h = 0$), the mass formula is

$$M^2 = \frac{|-m_1 U + m_2 + Tn_1 + (TU - \frac{1}{2}\vec{W}^2)n_2 + \vec{W} \cdot \vec{Q}|^2}{4S_2 \left(T_2 U_2 - \frac{1}{2}\text{Im}\,\vec{W}^2\right)}, \tag{9.4.23}$$

where $\vec{W}$ is the 16-dimensional complex vector of Wilson lines. When the integer

$$\rho = \vec{m} \cdot \vec{\epsilon}_R + \vec{n} \cdot \epsilon_L - \vec{Q} \cdot \vec{\zeta} \tag{9.4.24}$$

is even, these states are vectorlike multiplets with multiplicity function $d_1(s)$ of (9.4.20) and

$$s = \vec{m} \cdot \vec{n} - \frac{1}{2}\vec{Q} \cdot \vec{Q}; \tag{9.4.25}$$

when ρ is odd, these states are hyperlike multiplets with multiplicities $d_1(s)$.

In the twisted sector ($h = 1$), the mass formula is

$$M^2 = \left|(m_1 + \tfrac{1}{2}\epsilon_L^1)U - (m_2 + \tfrac{1}{2}\epsilon_L^2) - T(n_1 + \tfrac{1}{2}\epsilon_R^1)\right.$$
$$- (TU - \tfrac{1}{2}\vec{W}^2)(n_2 + \tfrac{1}{2}\epsilon_R^2)$$
$$\left.- \vec{W} \cdot (\vec{Q} + \tfrac{1}{2}\vec{\zeta})\right|^2 / 4S_2 \left(T_2 U_2 - \tfrac{1}{2}\text{Im}\,\vec{W}^2\right). \tag{9.4.26}$$

The states with ρ even are vector-multiplet-like with multiplicities $d_+(s')$, with

$$s' = \left(\vec{m} + \frac{\vec{\epsilon}_L}{2}\right) \cdot \left(\vec{n} + \frac{\vec{\epsilon}_R}{2}\right) - \frac{1}{2}\left(\vec{Q} + \frac{\vec{\zeta}}{2}\right) \cdot \left(\vec{Q} + \frac{\vec{\zeta}}{2}\right), \tag{9.4.27}$$

while the states with ρ odd are hypermultiplets with multiplicities $d_-(s')$.

9.5 Spontaneous Supersymmetry Breaking

We have seen in the previous section that we can break the maximal supersymmetry by the orbifolding procedure. The extra gravitini are projected out of the spectrum. This type of orbifold breaking of supersymmetry we will call explicit breaking.

It turns out that there is an important difference between freely acting and non-freely-acting orbifolds with respect to the restoration of the broken supersymmetry. The example

---

[7] Vectorlike multiplets are, the vector multiplets, vector tensor multiplets, and the supergravity multiplet.

of the previous section (explicit breaking) corresponded to a non-freely-acting orbifold action.

To make the difference transparent, consider the $\mathbb{Z}_2$ twist on T^4 described before, under which two of the gravitini transform with a minus sign and are thus projected out. Consider also performing at the same time a $\mathbb{Z}_2$ translation (by a half period) in one direction of the extra (2,2) torus. Take the two cycles to be orthogonal, with radii R, R', and do an $X \to X + \pi$ shift on the first cycle. The oscillator modes are invariant but the vertex operator states $|m, n\rangle$ transform with a phase $(-1)^m$.

This is a freely-acting orbifold, since the action on the circle is free. Although the states of the two gravitini, $\bar{a}^\mu_{-1}|S^I_a\rangle$ $I = 1, 2$ transform with a minus sign under the twist, the states $\bar{a}^\mu_{-1}|S^I_a\rangle \otimes |m = 1, n\rangle$ are invariant! They have the space-time quantum numbers of two gravitini, but they are no longer massless. In fact, in the absence of the state $|m = 1, n\rangle$ they would be massless, but now we have an extra contribution to the mass coming from that state:

$$m_L^2 = \frac{1}{4}\left(\frac{1}{R} + \frac{nR}{\ell_s^2}\right)^2, \quad m_R^2 = \frac{1}{4}\left(\frac{1}{R} - \frac{nR}{\ell_s^2}\right)^2. \tag{9.5.1}$$

The matching condition $m_L = m_R$ implies $n = 0$, so that the mass of these states is $m^2 = 1/4R^2$. These are massive (KK) gravitini and in this theory, the $\mathcal{N} = 4_4$ supersymmetry is broken spontaneously to $\mathcal{N} = 2_4$. In field theory language, the effective field theory is a gauged version of $\mathcal{N} = 4_4$ supergravity where the supersymmetry is spontaneously broken to $\mathcal{N} = 2_4$ at the minimum of the potential.

We will note here some important differences between explicit and spontaneous breaking of supersymmetry.

• In spontaneously broken supersymmetric vacua, the behavior at high energies is softer than the case of explicit breaking. If supersymmetry is spontaneously broken, there are still broken Ward identities that govern the short distance properties of the theory. In such theories, there is a characteristic energy scale, namely the gravitino mass $m_{3/2}$ above which supersymmetry is effectively restored. A scattering experiment at energies $E \gg m_{3/2}$ will reveal supersymmetric physics. This has important implications on effects such as the running of low-energy couplings. We will return to this later, towards the end of section 10.5 on page 309.

• There is also a technical difference. As we already argued, in the case of the freely acting orbifolds, the states coming from the twisted sector have moduli-dependent masses that are generically nonzero (although they can become zero at special points of the moduli space). This is unlike non-freely-acting orbifolds, where the twisted sector masses are independent of the original moduli and one obtains generically massless states from the twisted sector.

• In vacua with spontaneously broken supersymmetry, the supersymmetry-breaking scale $m_{3/2}$ is an expectation value since it depends on compactification radii. If at least one supersymmetry is left unbroken, then the radii are moduli with arbitrary expectation values. In particular, there are corners of the moduli space where $m_{3/2} \to 0$, and physics becomes supersymmetric at all scales. These points are an infinite distance away using the natural metric of the moduli scalars.

In our simple example above, $m_{3/2} \sim 1/R \to 0$ when $R \to \infty$. At this point, an extra dimension of space-time becomes non-compact and supersymmetry is restored in five dimensions. This behavior is generic in all vacua where the free action originates from translations.

If however there is no leftover supersymmetry, then generically there is a potential for the radii. In such a case $m_{3/2}$ is dynamically determined.

Consider the class of $\mathcal{N} = 2_4$ orbifold vacua we described in (9.4.14). If the (2,18) translation vector ϵ lies within the (0,16) part of the lattice, then the breaking of $\mathcal{N} = 4_4 \to \mathcal{N} = 2_4$ is "explicit." When, however, $(\vec{\epsilon}_L, \vec{\epsilon}_R) \neq (\vec{0}, \vec{0})$ then the breaking is spontaneous.

In the general case, there is no global identification of the massive gravitini inside the moduli space due to surviving duality symmetries. To illustrate this in the previous simple example, consider instead the $(-1)^{m+n}$ translation action. In this case there are two candidate states with the quantum numbers of the gravitini: $\bar{a}^{\mu}_{-1}|S^I_a\rangle \otimes |m = 1, n = 0\rangle$ with mass $m_{3/2} \sim 1/R$, and $\bar{a}^{\mu}_{-1}|S^I_a\rangle \otimes |m = 0, n = 1\rangle$ with mass $\tilde{m}_{3/2} \sim R$. In the region of large R, the first set of states behaves like light gravitini, while in the region of small R it is the second set that is light.

Freely acting orbifolds breaking supersymmetry are stringy versions of Scherk-Schwarz compactifications.

9.6 A Heterotic $\mathcal{N} = 1_4$ Orbifold and Chirality in Four Dimensions

So far, we have used orbifold techniques to remove two of the four gravitini, ending up with $\mathcal{N} = 2_4$ supersymmetry. We will carry this procedure one step further in order to reduce the supersymmetry to $\mathcal{N} = 1_4$.

For phenomenological purposes, $\mathcal{N} = 1_4$ supersymmetry is optimal, since it is the only supersymmetric case that admits chiral representations in four dimensions. Although the very low-energy world is not supersymmetric, we seem to need some supersymmetry beyond Standard-Model energies to explain the gauge hierarchy.

Consider splitting the (6,22) lattice in the $\mathcal{N} = 4_4$ heterotic string as

$$(6, 22) = \oplus^3_{i=1}(2, 2)_i \oplus (0, 16). \tag{9.6.1}$$

Label the coordinates of each two-torus as $X^{\pm}_i$, $i = 1, 2, 3$. Consider the following $\mathbb{Z}_2 \times \mathbb{Z}_2$ orbifolding action: The element g_1 of the first $\mathbb{Z}_2$ acts with a minus sign on the coordinates of the first and second two-torus, the element g_2 of the second $\mathbb{Z}_2$ acts with a minus sign on the coordinates of the first and third torus, and $g_1 g_2$ acts with a minus sign on the coordinates of the second and third torus. Only one of the four four-dimensional gravitini survives this orbifold action. You are invited to verify this in exercise 9.9 on page 288.

To ensure modular invariance we also have to act on the gauge sector. We will consider the $E_8 \times E_8$ string, with the E_8's fermionically realized. We will split the 16 real fermions realizing the first E_8 into groups of 10+2+2+2. The $\mathbb{Z}_2 \times \mathbb{Z}_2$ projection will act in a similar way in the three groups of two fermions each, while the other ten will be invariant.

The partition function for this $\mathbb{Z}_2 \times \mathbb{Z}_2$ orbifold is:

$$
Z_{\mathbb{Z}_2 \times \mathbb{Z}_2}^{N=1} = \frac{1}{\tau_2 \, \eta^2 \bar{\eta}^2} \frac{1}{4} \sum_{h_1,g_1=0,h_2,g_2=0}^{1} \frac{1}{2} \sum_{\alpha,\beta=0}^{1} (-)^{\alpha+\beta+\alpha\beta}
$$

$$
\times \frac{\vartheta[^\alpha_\beta]}{\eta} \frac{\vartheta[^{\alpha+h_1}_{\beta+g_1}]}{\eta} \frac{\vartheta[^{\alpha+h_2}_{\beta+g_2}]}{\eta} \frac{\vartheta[^{\alpha-h_1-h_2}_{\beta-g_1-g_2}]}{\eta} \frac{\bar{\Gamma}_8}{\bar{\eta}^8} Z_{2,2}^1[^{h_1}_{g_1}] Z_{2,2}^2[^{h_2}_{g_2}] Z_{2,2}^3[^{h_1+h_2}_{g_1+g_2}]
$$

$$
\times \frac{1}{2} \sum_{\bar{\alpha},\bar{\beta}=0}^{1} \frac{\bar{\vartheta}[^{\bar{\alpha}}_{\bar{\beta}}]^5}{\bar{\eta}^5} \frac{\bar{\vartheta}[^{\bar{\alpha}+h_1}_{\bar{\beta}+g_1}]}{\bar{\eta}} \frac{\bar{\vartheta}[^{\bar{\alpha}+h_2}_{\bar{\beta}+g_2}]}{\bar{\eta}} \frac{\bar{\vartheta}[^{\bar{\alpha}-h_1-h_2}_{\bar{\beta}-g_1-g_2}]}{\bar{\eta}}. \tag{9.6.2}
$$

We will classify the massless spectrum in multiplets of $\mathcal{N} = 1_4$ supersymmetry. We obtain de facto the $\mathcal{N} = 1_4$ supergravity multiplet. Next we consider the gauge group of this vacuum. It originates in the untwisted sector. The orbifold group here contains the $\mathbb{Z}_2$ of the orbifold as a subgroup. We can therefore obtain the gauge group by imposing the extra $\mathbb{Z}_2$ projection on the gauge group of the $\mathcal{N} = 2_4$ vacuum of section 9.4. The graviphoton, the vector partner of the dilaton, and the two U(1)'s coming from the T^2 are now projected out. The second E_8 survives. The extra $\mathbb{Z}_2$ projection on $E_7 \times SU(2)$ gives $E_6 \times U(1) \times U(1)'$. The adjoint of E_6 can be written as the adjoint of O(10) plus the O(10) spinor plus a U(1) (singlet).

Therefore, the gauge group of this vacuum is $E_8 \times E_6 \times U(1) \times U(1)'$ and we have the associated vector multiplets. There is also the linear multiplet containing the antisymmetric tensor and the dilaton.

We now consider the rest of the states that form $\mathcal{N} = 1_4$ chiral multiplets. Notice first that there are no massless multiplets charged under the E_8.

The charges of chiral multiplets under $E_6 \times U(1) \times U(1)'$ and their multiplicities are given in the tables 9.1 and 9.2 below. You are invited to verify them in exercise 9.10 on page 288.

Table 9.1 Nonchiral Massless States of the $\mathbb{Z}_2 \times \mathbb{Z}_2$ Orbifold.

E_6	U(1)	U(1)$'$	Sector	Multiplicity
27	1/2	1/2	Untwisted	1
27	−1/2	1/2	Untwisted	1
27	0	−1	Untwisted	1
1	−1/2	3/2	Untwisted	1
1	1/2	3/2	Untwisted	1
1	1	0	Untwisted	1
1	1/2	0	Twisted	32
1	1/4	3/4	Twisted	32
1	1/4	−3/4	Twisted	32

Table 9.2 Chiral Massless States of the $\mathbb{Z}_2 \times \mathbb{Z}_2$ Orbifold.

E_6	U(1)	U(1)$'$	Sector	Multiplicity
27	0	1/2	Twisted	16
27	1/4	−1/4	Twisted	16
27	−1/4	−1/4	Twisted	16
1	0	3/2	Twisted	16
1	3/4	−3/4	Twisted	16
1	−3/4	−3/4	Twisted	16

As we can see, the spectrum of the theory is chiral. For example, the number of **27**'s minus the number of $\overline{\mathbf{27}}$'s is 3×16. The theory is free of gauge anomalies as it can be checked using the formulas of section 7.9 on page 176.

More complicated orbifolds give rise to different gauge groups and spectra. A guide of such constructions is provided at the end of this chapter.

9.7 Calabi-Yau Manifolds

We provide in this section information about a special class of complex manifolds, Calabi-Yau manifolds. As will become evident shortly, such manifolds are an indispensable tool in string compactifications.

We will start by introducing briefly the idea of cohomology for a generic real manifold. The exterior derivative (defined in appendix B) is nilpotent, $d^2 = 0$. We may therefore introduce a cohomology similar to the definition of physical states using the nilpotent BRST operator in section 3.7 on page 40.

A p-form A_p is called *closed*, if it is annihilated by the exterior derivative: $dA_p = 0$. It is called *exact*, if it can be written as the exterior derivative of a $(p-1)$-form: $A_p = dA_{p-1}$. Any closed form is locally exact but not globally. The pth de Rham cohomology group $H^p(K)$ of a D-dimensional manifold K is the space of closed p-forms modulo the space of exact p-forms. This is a group that depends only on the topology of K. Its dimension is known as the pth Betti number b_p. The Euler number of the manifold is given by the alternating sum

$$\chi(K) = \sum_{p=0}^{D} (-1)^p b_p. \tag{9.7.1}$$

The Laplacian on p-forms can be written in terms of the exterior derivative and the Hodge star operator as

$$\Box \equiv \star d \star d + d \star d \star = (d + \star d\star)^2. \tag{9.7.2}$$

A harmonic p-form satisfies $\Box A_p = 0$. It can be shown that the harmonic p-forms are in one-to-one correspondence with the generators of $H^p(K)$. The $\star$ operator maps every harmonic p-form to a harmonic $(D-p)$-form, so that $b_P = b_{D-p}$.

An almost complex manifold has a (1,1) tensor $J^i{}_j$, known as the almost complex structure, that squares to minus one:

$$J^i{}_j J^j{}_k = -\delta^i{}_k. \tag{9.7.3}$$

It can be used to define complex coordinates at any given point, since it plays the role of the imaginary number i locally. An interesting question is whether the definition of complex coordinates at a point, extends to a local neighborhood. This happens when the Nijenhuis tensor

$$N^k{}_{ij} = J^l{}_i(\partial_l J^k{}_j - \partial_j J^k{}_l) - J^l{}_j(\partial_l J^k{}_i - \partial_i J^k{}_l) \tag{9.7.4}$$

vanishes. In that case the manifold is called a *complex manifold*. Such a manifold can be covered by patches of complex coordinates (defined via the complex structure) with holomorphic transition functions. In any given patch we can choose

$$J^a{}_b = i\delta^a{}_b, \quad J^{\bar a}{}_{\bar b} = -i\delta^{\bar a}{}_{\bar b}. \tag{9.7.5}$$

On complex manifolds the notion of holomorphic functions is independent of the coordinates.

A *Hermitian metric* on a complex manifold is one for which

$$g_{ab} = g_{\bar a \bar b} = 0. \tag{9.7.6}$$

We may have a finer definition of forms: a (p,q)-form is a $(p+q)$-form with p anti-symmetrized holomorphic indices and q antiholomorphic antisymmetrized indices. The exterior derivative can also be separated as

$$d = \partial + \bar\partial, \quad \partial = dz^a \partial_a, \quad \bar\partial = d\bar z^{\bar a} \bar\partial_{\bar a}. \tag{9.7.7}$$

∂ takes a $(p,q) \to (p+1,q)$ while $\bar\partial$ takes a $(p,q) \to (p,q+1)$. Moreover

$$\partial^2 = \bar\partial^2 = 0, \quad \partial\bar\partial + \bar\partial\partial = 0. \tag{9.7.8}$$

∂ and $\bar\partial$ can be used to define a refined cohomology on a complex manifold, the Dolbeault cohomology groups $H^{p,q}_{\bar\partial}(K)$ of dimension $h^{p,q}$, containing the (p,q) forms that are $\bar\partial$-closed but not $\bar\partial$-exact. Using the natural inner product for (p,q) forms we define adjoints $\partial^\dagger, \bar\partial^\dagger$ for $\partial, \bar\partial$ and construct the two Laplacians

$$\Delta_\partial = \partial\partial^\dagger + \partial^\dagger\partial, \quad \Delta_{\bar\partial} = \bar\partial\bar\partial^\dagger + \bar\partial^\dagger\bar\partial. \tag{9.7.9}$$

The $\Delta_{\bar\partial}$-harmonic (p,q)-forms are in one-to-one correspondence with the generators of $H^{p,q}_{\bar\partial}(K)$.

On a complex manifold we may impose a stronger condition, namely, that the complex structure $J^i{}_j$ is covariantly constant. In this case we obtain a *Kähler manifold*. From J we can also construct the Kähler two-form $k_{ij} = g_{ik}J^k{}_j$. It is a closed form, $dk = 0$. In holomorphic coordinates we have that

$$k_{a\bar b} = -ig_{a\bar b} = -k_{\bar b a}, \quad k_{ab} = k_{\bar a \bar b} = 0, \tag{9.7.10}$$

where g is the Hermitian metric. The Kähler form is a closed (1,1) form. This means that

$$\partial k = \bar\partial k = 0, \tag{9.7.11}$$

from which it follows that locally

$$k = -i\partial\bar{\partial}K. \tag{9.7.12}$$

K is a zero-form (function) known as the Kähler potential. It is not uniquely determined since any transformation $K \to K + F + \bar{F}$ where F is holomorphic does not change the Kähler form. From relation (9.7.10) we obtain a local expression for the metric

$$g_{a\bar{b}} = g_{\bar{b}a} = \frac{\partial^2 K}{\partial z^a \partial z^{\bar{b}}}, \tag{9.7.13}$$

from which the Christoffel connections can be calculated. The only nonzero ones are

$$\Gamma^a_{bc} = g^{a\bar{d}}\partial_b g_{c\bar{d}}, \quad \Gamma^{\bar{a}}_{\bar{b}\bar{c}} = g^{\bar{a}d}\partial_{\bar{b}}g_{\bar{c}d}. \tag{9.7.14}$$

The only nonzero components of the Riemann tensor are $R_{a\bar{b}c\bar{d}}$ and the cyclic identity gives

$$R_{a\bar{b}c\bar{d}} = R_{c\bar{b}a\bar{d}} = R_{a\bar{d}c\bar{b}}. \tag{9.7.15}$$

The Ricci tensor can be calculated to be

$$R_{a\bar{b}} = -\partial_a\partial_{\bar{b}}\log\det g. \tag{9.7.16}$$

Ricci flatness leads to the Monge-Ampère equation.

9.7.1 Holonomy

The notion of holonomy is central in geometry. A Riemannian manifold of dimension D has a spin connection ω that is generically an $SO(D)$ gauge field. This implies, in analogy with standard gauge fields, that a field ϕ transported around a path γ, transforms to $W\phi$ where

$$W = P\, e^{\int_\gamma \omega \cdot dx}. \tag{9.7.17}$$

ω above is taken in the same $SO(D)$ representation as ϕ and P stands for path ordering. The $SO(D)$ matrices W form a group $H \subset SO(D)$. It is called the holonomy group of the manifold. Generically $H = SO(D)$, but there are special cases where this is not so.

Consider the possibility that the manifold admits a covariantly constant spinor ζ: $\nabla_i\zeta = 0$. As in gauge theories, a covariantly constant field has trivial holonomy: by definition it does not change along a path. This means that for any group element $W \in H$, $W\zeta = \zeta$. We are interested in finding what subgroup H can have this property.

In the $D = 6$ case, $SO(6)$ is locally equivalent to $SU(4)$. The spinor and conjugate spinor representations of $SO(6)$ are the fundamental **4** and antifundamental $\bar{\mathbf{4}}$ representations of $SU(4)$. Without loss of generality we can assume that ζ transforms as the **4**.

This special spinor, can always be brought to the form $(0, 0, 0, \zeta_0)$ by a $SU(4)$ rotation. In this frame, it is obvious that the subgroup of $SU(4)$ that preserves the spinor is $SU(3)$ and acts on the first three components. Because $\mathbf{4} \to \mathbf{3} + \mathbf{1}$, a manifold that has an $SU(3)$ holonomy, has necessarily just one covariantly constant spinor. If there are more, the holonomy group must be smaller. For example, for two distinct covariantly constant spinors, the holonomy group must be $SU(2)$. This is the case for the manifold $K3 \times T^2$ that we will meet later.

It will be shown in the next section, that the existence of a single covariantly-constant spinor will be associated with the presence of $\mathcal{N} = 1_4$ supersymmetry in the heterotic string.

9.7.2 Consequences of SU(3) holonomy

Once we have a covariantly constant spinor, we can construct several closed forms as bilinears in this spinor. One is the two-form, $k_{ij} = \bar{\zeta}\Gamma_{ij}\zeta$ where Γ_{ij} are the antisymmetrized product of the $SO(6)$ Γ-matrices, and the associated complex structure $J^i{}_j = g^{ik}k_{kj}$. The second is a three-form $\Omega_{ijk} = \zeta^T\Gamma_{ijk}\zeta$. The one-form $\Omega_i = \zeta^T\Gamma_i\zeta$ vanishes because of the six-dimensional Fierz identities.

We first focus on $J^i{}_j$. It is an $SO(6)$ matrix acting on vectors. By construction it is real, traceless and $SU(3)$ invariant. There is a unique such matrix in $SO(6)$ up to normalization and group conjugation

$$
J = \begin{pmatrix}
0 & +1 & 0 & 0 & 0 & 0 \\
-1 & 0 & 0 & 0 & 0 & 0 \\
0 & 0 & 0 & +1 & 0 & 0 \\
0 & 0 & -1 & 0 & 0 & 0 \\
0 & 0 & 0 & 0 & 0 & +1 \\
0 & 0 & 0 & 0 & -1 & 0
\end{pmatrix}. \tag{9.7.18}
$$

Therefore, J satisfies $J^2 = -1$ and is an almost complex structure. It can be used to define complex coordinates over K. From its construction in terms of ζ, J is covariantly constant, and the Nijenhuis tensor thus vanishes: K is a complex manifold.

Since the two-form k_{ij} constructed in terms of the spinor, is closed, the manifold K is also a Kähler manifold according to our previous definition. A Kähler manifold does not admit a unique metric, but two different Kähler metrics g and g' are related as

$$
g'_{a\bar{b}} = g_{a\bar{b}} + \partial_a\partial_{\bar{b}}\phi, \tag{9.7.19}
$$

where ϕ is an arbitrary function on the manifold.

A generic six-dimensional Kähler manifold has $U(3)$ holonomy. To obtain the restricted $SU(3)$ holonomy, we must impose extra conditions on the manifold K. The spin connection of a Kähler manifold is a $U(3)$ gauge field. The $U(1)$ part is a gauge field that we will call A. $F = dA$ is a closed two-form, an element of $H^2(K)$. Its class is known as the first Chern class of the manifold $c_1(K)$. For $SU(3)$ holonomy such a gauge field must have a vanishing field strength, F. This can happen if this $U(1)$ bundle (the canonical bundle) is topologically trivial. In this case, it has a global section. Moreover, $c_1(K) = 0$.

We will now show that a vanishing first Chern class implies Ricci flatness of K. Remember that the $U(1) \subset SO(6)$ in question is (9.7.18), namely, the complex structure J. Given an antisymmetric matrix (in the Lie algebra of $SO(6)$), its $U(1)$ part is given by $\text{tr}(JM) = J^i{}_j M^j{}_i$. Consider now the Riemann form generated by the Riemann tensor. Its $U(1)$ part is

$$
F_{ij} = \text{tr}(JR_{ij}) = R_{ij;kl}J^{kl}. \tag{9.7.20}
$$

From the previous equation, the nonzero components of F are

$$F_{a\bar{b}} = -F_{\bar{b}a} = R_{a\bar{b}}{}^k{}_l J^l{}_k = iR_{a\bar{b}}{}^c{}_c - iR_{a\bar{b}}{}^{\bar{c}}{}_{\bar{c}}. \tag{9.7.21}$$

Since

$$R_{a\bar{b}}{}^c{}_c = R_{a\bar{b}\bar{d}c}g^{\bar{d}c} = -R_{a\bar{b}c\bar{d}}g^{c\bar{d}} = -R_{a\bar{b}}{}^{\bar{d}}{}_{\bar{d}}, \tag{9.7.22}$$

we obtain

$$F_{a\bar{b}} = -F_{\bar{b}a} = 2iR_{a\bar{b}}{}^c{}_c = -2iR_{a\bar{b}}. \tag{9.7.23}$$

What we have shown is, that the U(1) component of the holonomy, in U(1)×SU(3)⊂SO(6), is generated by the Ricci form. Thus, Ricci flatness implies SU(3) holonomy. The converse is also true by the Yau theorem. Such spaces, are known as Calabi-Yau (CY) manifolds.

We have also constructed a closed (covariantly constant) three-form Ω_{ijk} on K. This is expected for the following reason: the vector of SO(6) decomposes as $\mathbf{6} \to \mathbf{3} + \bar{\mathbf{3}}$ under SU(3). We construct an SU(3) singlet out of the antisymmetrized product of three $\mathbf{3}$s, which on the other hand would transform under U(3). The existence of SU(3) holonomy is equivalent to the existence of such a covariantly constant (3,0) form Ω_{ijk}. In fact, a manifold of SU(3) holonomy has a unique non-vanishing (3,0) form that is covariantly constant. It is a section of the (topologically trivial) canonical bundle on K.

We now proceed to discuss the Dolbeault cohomology of compact CY manifolds.

Reality implies that $h^{p,q} = h^{q,p}$ and Poincaré duality $h^{p,q} = h^{3-p,3-q}$. $h^{0,0} = 1$ corresponding to the constant solution of the Laplacian on any connected compact manifold. Since there are no harmonic one-forms, $h^{1,0} = h^{0,1} = 0$. The relation $h^{p,0} = h^{3-p,0}$ valid for CY manifolds[8] then implies that $h^{2,0} = h^{0,2} = 0$. Finally, the uniqueness of the (3,0) form implies $h^{3,0} = h^{0,3} = 1$. We arrive at the following Hodge diamond, characteristic of CY manifolds:

$$
\begin{array}{ccccccc}
 & & & 1 & & & \\
 & & 0 & & 0 & & \\
 & 0 & & h^{1,1} & & 0 & \\
1 & & h^{2,1} & & h^{2,1} & & 1 \\
 & 0 & & h^{1,1} & & 0 & \\
 & & 0 & & 0 & & \\
 & & & 1 & & &
\end{array}
\tag{9.7.24}
$$

The Euler number is

$$\chi = 2(h^{1,1} - h^{2,1}). \tag{9.7.25}$$

9.7.3 The CY moduli space

Once we have a CY manifold, there may be continuous deformations that preserve this property. This will give rise to a moduli space. Its structure is important in order to understand the effective field theory of CY compactifications.

We will start by describing the *deformations of the complex structure J* defined as

$$J^i{}_k J^k{}_j = -\delta^i{}_j, \quad N^i{}_{jk} = 0, \tag{9.7.26}$$

---

[8] This isomorphism is obtained by contracting with the (3,0)-form Ω.

where N was defined in (9.7.4). An infinitesimal deformation of the complex structure

$$\tilde{J}^i_j = J^i_j + \tau^i_{\ j} \tag{9.7.27}$$

must satisfy to leading order (9.7.26). The first of the equation sets (in complex coordinates) $\tau^a_{\ b} = \tau^{\bar{a}}_{\ \bar{b}} = 0$. Moreover, the only nonzero components of the Nijenhuis tensor are

$$N^a_{\ \bar{b}\bar{c}} = \bar{\partial}_{\bar{b}}\tau^a_{\ \bar{c}} - \bar{\partial}_{\bar{c}}\tau^a_{\ \bar{b}}, \tag{9.7.28}$$

and its complex conjugate. We may view $\tau^a_{\ \bar{b}}$ as a (0,1)-form with values in the holomorphic tangent bundle T. Then, the vanishing of N can be written as

$$\bar{\partial}\tau^a = 0. \tag{9.7.29}$$

This says that τ is an element of $H^1(T)$, the closed one-forms with values in the tangent bundle. We may now consider the (2,1) form

$$\eta_{ab\bar{c}} = \Omega_{abd}\tau^d_{\ \bar{c}}. \tag{9.7.30}$$

Since Ω is covariantly constant, η is a harmonic form if and only if τ is. Thus, $H^1(T) \sim H^{2,1}(K)$, and the non-trivial complex structure deformations are in one-to-one correspondence with the (2,1)-harmonic forms. They form a moduli space, called a *complex structure moduli space* $\mathcal{M}_C$ of complex dimension $h^{2,1}$. It can be shown that they are related to deformations of the metric δg_{ab} and $\delta g_{\bar{a}\bar{b}}$ which preserve the CY condition.

There is another perturbation of the metric, namely, $\delta g_{a\bar{b}}$. The condition for this to preserve Ricci flatness is that $\delta g_{a\bar{b}}dz^a \wedge d\bar{z}^{\bar{b}}$ is harmonic. Thus the number of independent such deformations is $h^{1,1}$. These are known as deformations of the Kähler structure and their moduli space, *Kähler moduli space* $\mathcal{M}_K$.

Thus the total moduli space of CY metrics is a direct product $\mathcal{M}_K \times \mathcal{M}_C$ of real dimension $h^{1,1} + 2h^{2,1}$.

In string theory (compactified on a CY manifold) the metric comes always together with the two-index antisymmetric tensor and the dilaton. The two-index antisymmetric tensor, B, being a two-form, will give another $h^{1,1}$ real moduli (scalars), as well as a four-dimensional two-tensor, that in four dimensions is equivalent to a pseudoscalar. The $h^{1,1}$ moduli of B combine with the Kähler moduli and complexify the Kähler moduli space. From now on by $\mathcal{M}_K$ we will denote the complexified Kähler moduli space of real dimension $2h^{1,1}$.

Both $\mathcal{M}_K$ and $\mathcal{M}_C$ are themselves Kähler manifolds. The Kähler potential for $\mathcal{M}_K$ is given by

$$\mathcal{K} = -\log \int_K J \wedge J \wedge J, \tag{9.7.31}$$

while for $\mathcal{M}_C$

$$\mathcal{K}_C = -\log\left(i\int_K \Omega \wedge \bar{\Omega}\right). \tag{9.7.32}$$

In fact, these manifolds are special Kähler manifolds, whose geometry is determined from a holomorphic function $\mathcal{F}$, the prepotential. As we will see later, this is related to

$\mathcal{N} = 2_4$ supersymmetry (see appendix I on page 533). In special geometry, the Kähler potential can be obtained from the holomorphic prepotential $\mathcal{F}$ as

$$K = -\log\left[i(\bar{z}^i \partial_{z_i}\mathcal{F} - z^i \partial_{\bar{z}_i}\bar{\mathcal{F}})\right]. \tag{9.7.33}$$

9.8 $\mathcal{N} = 1_4$ Heterotic Compactifications

We have seen how orbifolds provide solvable compactifications of string theory. The disadvantage of the orbifold approach is that it describes explicitly a small subspace of the relevant moduli space. To obtain a potentially wider view of the space of $\mathcal{N} = 1_4$ compactifications, we must work perturbatively in α' (σ-model approach).

In the effective field theory approach (to leading order in α'), we assume that some bosonic fields acquire expectation values that satisfy the equations of motion, while the expectation values of the fermions are zero (to preserve $D = 4$ Lorentz invariance). Generically, such a background breaks all the supersymmetries of flat ten-dimensional space. Some supersymmetry will be preserved, if the associated variation of the fermion fields vanish. This gives a set of first order equations. If they are satisfied for at least one supersymmetry, then the full equations of motion will also be satisfied to leading order in α'. Another way to state this is by saying that every compact manifold that preserves at least one SUSY, is a solution of the equations of motion.

We will consider here the case of the heterotic string on a space that is locally $M_4 \times K$ with M_4 the four-dimensional Minkowski space and K some six-dimensional compact manifold. We split indices into Greek indices for M_4 and Latin indices for K.

The ten-dimensional Γ-matrices can be constructed from the $D = 4$ matrices γ^μ, and the internal matrices γ^m, $m = 4, 5, \ldots, 9$ as

$$\Gamma^\mu = \gamma^\mu \otimes 1_6, \quad \Gamma^m = \gamma^5 \otimes \gamma^m, \tag{9.8.1}$$

$$\gamma^5 = \frac{i}{4!}\epsilon_{\mu\nu\rho\sigma}\gamma^{\mu\nu\rho\sigma}, \quad \gamma = \frac{i}{6!}\sqrt{\det g}\,\epsilon_{mnrpqs}\gamma^{mnrpqs}. \tag{9.8.2}$$

γ is the analog of γ^5 for the internal space.

The supersymmetry variations of fermions in the heterotic string were given (in the Einstein frame) in appendix H.5 on page 530. Using the decomposition above, they can be written as

$$\delta\psi_\mu \sim \nabla_\mu\epsilon + \frac{e^{-\Phi/2}}{96}\left(\gamma_\mu\gamma_5 \otimes H\right)\epsilon, \tag{9.8.3}$$

$$\delta\psi_m \sim \nabla_m\epsilon + \frac{e^{-\Phi/2}}{96}\left(\gamma_m H - 12 H_m\right)\epsilon, \tag{9.8.4}$$

$$\delta\lambda \sim -(\gamma^m\partial_m\Phi)\epsilon + \frac{1}{12}e^{-\Phi/2}H\epsilon, \tag{9.8.5}$$

$$\delta\chi^a \sim -\frac{1}{4}e^{-\Phi/4}F^a_{mn}\gamma^{mn}\epsilon, \tag{9.8.6}$$

where ψ is the gravitino, λ is the dilatino, and χ^a are the gaugini; ϵ is a spinor (the parameter of the supersymmetry transformation). Furthermore, we used

$$H = H_{mnr}\gamma^{mnr}, \quad H_m = H_{mnr}\gamma^{nr}. \tag{9.8.7}$$

If, for some value of the background fields, the equations $\delta(\text{fermions}) = 0$ admit a solution, namely, a nontrivial, globally defined spinor ϵ, then the background is $\mathcal{N} = 1_4$ supersymmetric. If more than one solution exists, then we will have extended supersymmetry. For simplicity, we will make the assumption here that $H_{mnr} = 0$.

We assume a factorized spinor *Ansatz* $\epsilon = \chi \otimes \xi$. Vanishing of (9.8.3) when $H_{mnr} = 0$ implies that the four-dimensional spinor χ is constant. The vanishing of (9.8.4) implies that the internal manifold K must admit a Killing spinor ξ,

$$\nabla_m \, \xi = 0. \tag{9.8.8}$$

The vanishing of the dilatino variation (9.8.5) implies that the dilaton must be constant.

Applying one more covariant derivative to (9.8.8) and antisymmetrizing we obtain

$$[\nabla_m, \nabla_n]\xi = \frac{1}{4} R_{rs;mn} \gamma^{rs} \, \xi = 0. \tag{9.8.9}$$

Since $R_{rs;mn}\gamma^{rs}$ is the generator of the holonomy of the manifold, (9.8.9) implies that the holonomy group is smaller than the generic one, O(6). By multiplying (9.8.9) by γ^n and using the properties of the Riemann tensor we also obtain Ricci flatness ($R_{mn} = 0$). The holonomy is thus reduced to SU(3)$\subset$ SU(4)$\sim$O(6) so that the spinor decomposes as $4 \to 3 + 1$. Moreover the manifold has to be a Kähler manifold. Finally the background (internal) gauge fields must satisfy

$$F_{mn}^a \gamma^{mn} \, \xi = 0, \tag{9.8.10}$$

which again implies that $F_{mn}^a \gamma^{mn}$ acts as an SU(3) matrix.

Equation (7.9.33) on page 182 becomes

$$R^{rs}_{\ [mn} R_{pq]rs} = \frac{1}{30} F_{[mn}^a F_{pq]}^a. \tag{9.8.11}$$

We now take into account the discussion of the previous section to conclude that a compactification of the heterotic string on a CY manifold (SU(3) holonomy) with a gauge bundle satisfying (9.8.10) and (9.8.11) gives $\mathcal{N} = 1_4$ supersymmetric vacua.

9.8.1 The low-energy $\mathcal{N} = 1_4$ heterotic spectrum

We may now proceed with analyzing the effective theory of the heterotic string compactified on a CY manifold.

We must choose a gauge bundle on the CY manifold. A simple way to solve (9.8.10) and (9.8.11) is to embed the spin connection $\omega \in$ SU(3) into the gauge connection $A \in$ O(32) or $E_8 \times E_8$. The only embedding of SU(3) in O(32) that satisfies (9.8.11) is the one in which O(32) $\ni$ 32 $\to$ 3 + $\bar{3}$ + singlets $\in$ SU(3). In this case O(32) is broken down to U(1) $\times$ O(26) (this is the subgroup that commutes with SU(3)).

The U(1) is "anomalous," namely, the sum of the U(1) charges $\rho = \sum_i q^i$ of the massless states is not zero. This anomaly is only apparent, since the underlying string theory is not anomalous. What happens is that the Green-Schwarz mechanism implies that there is a one-loop coupling of the form $B \wedge F$. This gives a mass to the U(1) gauge field. The associated gauge symmetry is therefore broken at low-energy. This is discussed further in section 10.4. The leftover gauge group O(26) has only nonchiral representations.

More interesting is the case of $E_8 \times E_8$. E_8 has a maximal $SU(3) \times E_6$ subgroup, under which the adjoint of E_8 decomposes as $E_8 \ni 248 \rightarrow (8, 1) \otimes (3, 27) \otimes (\bar{3}, \overline{27}) \otimes (1, 78) \in SU(3) \times E_6$. Embedding the spin connection in one of the E_8 in this fashion solves (9.8.11). The unbroken gauge group in this case is $E_6 \times E_8$. Let N_L be the number of massless left-handed Weyl fermions in four dimensions transforming in the **27** of E_6 and N_R the same number for the $\overline{27}$. The number of net chirality (number of "generations") is $|N_L\text{-}N_R|$; it can be obtained by applying the Atiyah-Singer index theorem on the CY manifold. The **27**'s transform as the **3** of $SU(3)$ and the $\overline{27}$ transform in the $\bar{3}$ of $SU(3)$. Thus, the number of generations is the index of the Dirac operator on K for the fermion field $\psi_{\alpha A}$, where α is a spinor index and A is a **3** index. It can be shown that the index of the Dirac operator, and thus the number of generations, is equal to $|\chi(K)/2|$, where $\chi(K)$ is the Euler number of the manifold K.

The compactification of the $E_8 \times E_8$ theory provides a low-energy theory involving the E_6 gauge group that is known to be phenomenologically attractive. Moreover, below the string scale, there are no particles charged under both E_8's. Therefore, the other E_8 forms the "hidden sector": it contains particles that interact to the observable ones only via gravity and other universal interactions. This sector seems very weakly coupled to normal particles to have observable consequences. However, it can trigger supersymmetry breaking. Its strong self-interactions may force gaugini to condense, breaking supersymmetry. The breaking of supersymmetry can then be transmitted to the observable sector by the gravitational interaction.

The considerations in this section are correct to leading order in α'. At higher orders we expect (generically) corrections. It turns out that most of the statements above survive these corrections.

9.9 K3 Compactification of the Type-II String

As another example, we will consider the compactification of type II theory on the K3 manifold down to six dimensions. K3 denotes the class of four-dimensional compact, Ricci-flat, Kähler manifolds without isometries. Such manifolds have $SU(2) \subset O(4)$ holonomy and are also hyper-Kähler. The hyper-Kähler condition is equivalent to the existence of three integrable complex structures that satisfy the $SU(2)$ algebra.[9]

It can be shown that a left-right symmetric $\mathcal{N} = (1, 1)_2$ supersymmetric σ-model on such manifolds is exactly conformally invariant and has extended $\mathcal{N} = (4, 4)_2$ superconformal symmetry (see section 4.13.3 on page 81). Moreover, K3 has two covariantly constant spinors, so that the type-II theory compactified on it, has $\mathcal{N} = 2_6$ supersymmetry in six dimensions (and $\mathcal{N} = 4_4$ if further compactified on a two-torus).

It is useful for later purposes to briefly describe the cohomology of K3. There is a harmonic zero-form that is constant (since the manifold is compact and connected). There

---

[9] See also the discussion in appendix I.2 on page 535.

are no harmonic one-forms or three-forms. There is one (2,0) and one (0,2) harmonic forms as well as 20 (1,1) forms. The (2,0), (0,2), and one of the (1,1) Kähler forms are self-dual, the other 19 (1,1) forms are anti-self-dual. There is a unique four-form (the volume form).

We will consider first the type-IIA theory and derive the massless bosonic spectrum in six dimensions. To find the massless states originating from the ten-dimensional metric G, we make the following decomposition

$$G_{MN} \sim h_{\mu\nu}(x) \otimes \phi(y) + A_\mu(x) \otimes f_m(y) + \Phi(x) \otimes h_{mn}(y), \qquad (9.9.1)$$

where x denotes the six-dimensional noncompact flat coordinates and y are the internal (K3) coordinates. Also $\mu = 0, 1, \ldots, 5$ and $m = 1, 2, 3, 4$ is a K3 index. Applying the ten-dimensional equations of motion to the metric G, we obtain that $h_{\mu\nu}$ (the six-dimensional graviton) is massless if

$$\Box_y \phi(y) = 0. \qquad (9.9.2)$$

The solutions to this equation are the harmonic zero-forms on K3, and there is only one of them. Thus, there is one massless graviton in six dimensions. $A_\mu(x)$ is massless if $f_m(y)$ is covariantly constant on K3. Thus, it must be a harmonic one-form and there are none on K3. Consequently, there are no massless vectors coming from the metric. $\Phi(x)$ is a massless scalar if $h_{mn}(y)$ satisfies the Lichnerowicz equation

$$-\Box h_{mn} + 2R_{mnrs}h^{rs} = 0, \quad \nabla^m h_{mn} = g^{mn}h_{mn} = 0. \qquad (9.9.3)$$

The solutions of this equation can be constructed out of the three self-dual harmonic two-forms S_{mn} and the 19 anti-self-dual two-forms A_{mn}. Being harmonic, they satisfy the following equations (R_{mnrs} is anti-self-dual)

$$\Box f_{mn} - R_{mnrs}f^{rs} = \Box f_{mn} + 2R_{mrsn}f^{rs} = 0, \qquad (9.9.4)$$

$$\nabla_m f_{np} + \nabla_p f_{mn} + \nabla_n f_{pm} = 0, \quad \nabla^m f_{mn} = 0. \qquad (9.9.5)$$

Using these equations and the self-duality properties, it can be verified that solutions to the Lichnerowicz equation are given by

$$h_{mn} = A_m^p S_{pn} + A_n^p S_{pm}. \qquad (9.9.6)$$

Thus, there are $3 \cdot 19 = 57$ massless scalars. There is an additional massless scalar (the volume of K3) corresponding to constant rescalings of the K3 metric, that obviously preserves the Ricci-flatness condition. We obtain in total 58 scalars. The ten-dimensional dilaton also gives an extra massless scalar in six dimensions.

There is a similar expansion for the two-index antisymmetric tensor:

$$B_{MN} \sim B_{\mu\nu}(x) \otimes \phi(y) + B_\mu(x) \otimes f_m(y) + \Phi(x) \otimes B_{mn}(y). \qquad (9.9.7)$$

The masslessness condition implies that the zero-, one-, and two-forms (ϕ, f_m, B_{mn}, respectively) must be harmonic. We therefore obtain one massless two-index antisymmetric tensor and 22 scalars in six dimensions.

From the R-R sector we have a one-form that gives a massless vector in six dimensions. We also have a three-form that gives a massless three-form, and 22 vectors in six dimensions. A massless three-form in six dimensions is equivalent to a massless vector via a Poincaré duality transformation.

In total we have a graviton, an antisymmetric tensor, 24 vectors, and 81 scalars. The two gravitini in ten dimensions give rise to two Weyl gravitini in six dimensions. Their internal wave-functions are proportional to the two covariantly constant spinors that exist on K3. The gravitini preserve their original chirality. They have therefore opposite chirality. The relevant representations of $\mathcal{N} = (1, 1)_6$ supersymmetry in six dimensions are

- The vector multiplet. It contains a vector, two Weyl spinors of opposite chirality and four scalars.

- The supergravity multiplet. It contains the graviton, two Weyl gravitini of opposite chirality, four vectors, an antisymmetric tensor, a scalar, and four Weyl fermions of opposite chirality.

We conclude that the six-dimensional massless content of type-IIA theory on K3 consists of the supergravity multiplet and 20 U(1) vector multiplets. $\mathcal{N} = (1, 1)_6$ supersymmetry is sufficient to fix the two-derivative low-energy couplings of the massless fields. The bosonic part is (in the string frame)

$$
S^{IIA}_{K3} = \int d^6x \sqrt{-\det G_6} e^{-2\Phi} \left[R + \nabla^\mu \Phi \nabla_\mu \Phi - \frac{1}{12} H^{\mu\nu\rho} H_{\mu\nu\rho} + \frac{1}{8} \mathrm{Tr}(\partial_\mu \hat{M} \partial^\mu \hat{M}^{-1}) \right]
$$
$$
- \frac{1}{4} \int d^6x \sqrt{-\det G} (\hat{M}^{-1})_{IJ} F^I_{\mu\nu} F^{J\mu\nu} + \frac{1}{16} \int d^6x \epsilon^{\mu\nu\rho\sigma\tau\upsilon} B_{\mu\nu} F^I_{\rho\sigma} \hat{L}_{IJ} F^J_{\tau\upsilon}, \tag{9.9.8}
$$

where $I = 1, 2, \ldots, 24$. Φ is the six-dimensional dilaton.

Supersymmetry and the fact that there are 20 vector multiplets restricts the $4 \cdot 20$ scalars to live on the coset space $O(4, 20)/O(4) \times O(20)$. The scalars are therefore parameterized by the matrix $\hat{M}$ as in (D.4) on page 514 with $p = 4$, where $\hat{L}$ is the invariant O(4,20) metric. The action (9.9.8) is invariant under the continuous O(4,20) global symmetry. Here $H_{\mu\nu\rho}$ does not contain any Chern-Simons term. Note also the absence of the dilaton-gauge field coupling. This is due to the fact that the gauge fields come from the R-R sector.

Observe that type-IIA theory on K3 gives exactly the same massless spectrum as the heterotic string theory compactified on T^4. The low-energy actions (9.1.8) and (9.9.8) are different, though. As we will see in chapter 11, there is a nontrivial and interesting relation between the two.

Now consider the type-IIB theory compactified on K3 down to six dimensions. The NS-NS sector bosonic fields (G, B, Φ) are the same as in the type-IIA theory and we obtain again a graviton, an antisymmetric tensor, and 81 scalars.

From the R-R sector we have another scalar, the axion, which gives a massless scalar in $D = 6$. There is another two-index antisymmetric tensor, which gives, in six dimensions, a two-index antisymmetric tensor and 22 scalars. Finally there is the self-dual four-index antisymmetric tensor, which gives three self-dual two-index antisymmetric tensors and 19 anti-self-dual two-index antisymmetric tensors and a scalar. Since we can split a two-index antisymmetric tensor into a self-dual and an anti-self-dual part we can summarize the bosonic spectrum in the following way: a graviton, five self-dual and 21 anti-self-dual antisymmetric tensors, and 105 scalars.

Here, unlike the type-IIA case we obtain two massless Weyl gravitini of the same chirality. They generate a chiral $\mathcal{N} = (2, 0)_6$ supersymmetry. The relevant massless representations are

- The SUGRA multiplet. It contains the graviton, five self-dual antisymmetric tensors, and two left-handed Weyl gravitini.

- The tensor multiplet. It contains an anti-self-dual antisymmetric tensor, five scalars, and two Weyl fermions of chirality opposite to that of the gravitini.

The total massless spectrum forms the supergravity multiplet and 21 tensor multiplets. The theory is chiral but anomaly-free. The scalars live on the coset space $O(5, 21)/O(5) \times O(21)$ and there is a global $O(5,21)$ symmetry. Since the theory involves self-dual tensors, there is no covariant action principle, but we can write covariant equations of motion.

9.10 $\mathcal{N} = 2_6$ Orbifolds of the Type-II String

In section 9.9 we considered the compactification of the ten-dimensional type II string on the four-dimensional manifold K3. This provided a six-dimensional theory with $\mathcal{N} = 2_6$ supersymmetry. Upon toroidal compactification on an extra T^2 we obtain a four-dimensional theory with $\mathcal{N} = 4_4$ supersymmetry.

We will now consider a $\mathbb{Z}_2$ orbifold compactification to six dimensions with $\mathcal{N} = 2_6$ supersymmetry. We will also argue that it provides an alternative description of the geometric compactification on K3, considered earlier.

The $\mathbb{Z}_2$ orbifold transformation will act on the T^4 by reversing the sign of all four coordinates (and similarly for the world-sheet fermions on both the left and the right). This projects out half of the original gravitini. The partition function is

$$
Z_{6-d}^{II-\lambda} = \frac{1}{2} \sum_{h,g=0}^{1} \frac{Z_{(4,4)}[{}_g^h]}{\tau_2^2 \eta^4 \bar\eta^4} \times \frac{1}{2} \sum_{a,b=0}^{1} (-1)^{a+b+ab} \frac{\vartheta^2[{}_b^a]\vartheta[{}_{b+g}^{a+h}]\vartheta[{}_{b-g}^{a-h}]}{\eta^4}
$$
$$
\times \frac{1}{2} \sum_{\bar a,\bar b=0}^{1} (-1)^{\bar a+\bar b+\lambda \bar a \bar b} \frac{\bar\vartheta^2[{}_{\bar b}^{\bar a}]\bar\vartheta[{}_{\bar b+g}^{\bar a+h}]\bar\vartheta[{}_{\bar b-g}^{\bar a-h}]}{\bar\eta^4}. \tag{9.10.1}
$$

$Z_{4,4}[{}_g^h]$ are the $T^4/\mathbb{Z}_2$ orbifold blocks in (9.4.6) and $\lambda = 0, 1$ corresponds to type-IIB and type-2A, respectively.

We now focus on the massless bosonic spectrum. In the untwisted NS-NS sector we obtain the graviton, antisymmetric tensor, the dilaton (in six dimensions) and 16 scalars (the moduli of the $T^4/\mathbb{Z}_2$). In the NS-NS twisted sector we obtain 4·16 scalars. The total number of scalars (apart from the dilaton) is 4·20. Thus, the massless spectrum of the NS-NS sector is the same as that of the K3 compactification in section 9.9.

In the R-R sector we will have to distinguish IIA from IIB. In the type-IIA theory, we obtain seven vectors and a three-form from the R-R untwisted sector and another 16 vectors from the R-R twisted sector. In type IIB we obtain four two-index antisymmetric tensors and eight scalars from the R-R untwisted sector and 16 anti-self-dual two-index antisymmetric tensors and 16 scalars from the R-R twisted sector. Again this agrees with the K3 compactification.

To further motivate the fact that we are describing a CFT realization of the string moving on the K3 manifold, let us look more closely at the cohomology of the orbifold. We will use the two complex coordinates that describe the T^4, $z^{1,2}$. The T^4 has one zero-form, the constant, two (1,0) one-forms (dz^1, dz^2), two (0,1) one-forms ($d\bar{z}^1, d\bar{z}^2$), one (2,0) form ($dz^1 \wedge dz^2$) one (0,2) form ($d\bar{z}^1 \wedge d\bar{z}^2$), and four (1,1) forms ($dz^i \wedge d\bar{z}^j$). Finally there are four three-forms and one four-form.

Under the orbifolding $\mathbb{Z}_2$, the one- and three-forms are projected out and we are left with a zero-form, a four-form, a (0,2), (2,0), and 4 (1,1) forms. However the $\mathbb{Z}_2$ action has 16 fixed points on T^4, which become singular in the orbifold. To make a regular manifold we excise a small neighborhood around each singular point. The boundary is $S^3/\mathbb{Z}_2$ and we can paste a Ricci-flat manifold with the same boundary. The relevant manifold with this property is the Eguchi-Hanson gravitational instanton. This is the simplest of a class of four-dimensional noncompact hyper-Kähler manifolds known as asymptotically locally Euclidean (ALE) manifolds. These manifolds asymptote at infinity to a cone over S^3/Γ, with Γ one of the simple finite subgroups of SU(2). The SU(2) action on S^3 is the usual group action (remember that S^3 is the group manifold of SU(2)). This action induces an action of the finite subgroup Γ. The finite simple SU(2) subgroups have an A-D-E classification. The A series corresponds to the $\mathbb{Z}_N$ subgroups. The Eguchi-Hanson space corresponds to N = 2. The D series corresponds to the dihedral D_N subgroups of SU(2), which are $\mathbb{Z}_N$ groups augmented by an extra $\mathbb{Z}_2$ element. Finally, the three exceptional cases correspond to the tetrahedral, octahedral, and icosahedral groups.

The Eguchi-Hanson space carries an anti-self-dual (1, 1) form. Thus, in total, we will obtain 16 of them. We have eventually obtained the cohomology of the K3 manifold, at a submanifold of the moduli space where the metric has conical singularities. We can also compute the Euler number. Suppose we have a manifold M that we divide by the action of an abelian group G of order g; we excise a set of fixed points F and we paste some regular manifold N back. Then the Euler number is given by

$$\chi = \frac{1}{g}[\chi(M) - \chi(F)] + \chi(N). \tag{9.10.2}$$

Here $\chi(T^4) = 0$, F is the set of 16 fixed points with $\chi = 1$ each, while $\chi = 2$ for each of the 16 Eguchi-Hanson instantons, so that in total $\chi(T^4/\mathbb{Z}_2) = 24$. This is indeed the Euler number of K3.

The orbifold can be desingularized by moving away from zero S^2 volumes. This procedure is called a "blow-up" of the orbifold singularities. In the orbifold CFT description, it corresponds to marginal perturbations by the orbifold twist operators. In string theory language this corresponds to changing the expectation values of the scalars that are generated by the 16 orbifold twist fields. Note that at the orbifold limit, although the K3 geometry is singular, the associated string theory is not. The reason is that the shrinking spheres that become singularities have an NS two-form flux trapped in. The string couples to the flux and this prevents the development of divergences. There are points in the K3 moduli space though, where string theory does become singular. We will return in sections 11.9.1 and 11.10 to the interpretation of such singularities.

Before we move on, we will briefly describe other T^4 orbifolds associated to K3. They are of the $\mathbb{Z}_3$, $\mathbb{Z}_4$, and $\mathbb{Z}_6$ type. They are equivalent to associated orbifold limits of K3.

We use complex coordinates for the internal T^4 as

$$z^1 = x^6 + ix^7, \quad z^2 = x^8 + ix^9, \tag{9.10.3}$$

with the torus identifications $z^i \sim z^i + 1 \sim z^i + i$ for N = 2, 4 and $z^i \sim z^i + 1 \sim z^i + e^{i\frac{\pi}{3}}$ for N = 3, 6. Such identifications specify submoduli spaces of the T^4 moduli space for which $\mathbb{Z}_N$ is a symmetry, that you may explore in exercise 9.21 on page 289. The $\mathbb{Z}_N$ then acts as

$$(z^1, z^2) \rightarrow (e^{2\pi i/N} z^1, e^{-2\pi i/N} z^2). \tag{9.10.4}$$

In exercise 9.22 you are invited to construct the one-loop partition functions of these orbifolds, and read off the associated massless spectra.

9.11 CY Compactifications of Type-II Strings

We will study in this section some simple aspects of the compactification of type-II string theory in four dimensions on a CY manifold. We have already seen in section 9.8 on page 245 that in the compactification of the heterotic string on a CY manifold, the $\mathcal{N} = 1_{10}$ supersymmetry was reduced to an $\mathcal{N} = 1_4$ supersymmetry. The ten-dimensional gravitino gave a single massless gravitino in four dimensions. The type-II string has two gravitini in ten dimensions. Consequently, upon compactification on a CY manifold we obtain two massless gravitini and $\mathcal{N} = 2_4$ supersymmetry. In such a compactification, one of the supersymmetries is originating in the left-moving sector and the other in the right-moving sector.

We will now derive the massless spectrum of such compactifications. An important ingredient is the number of various harmonic forms of a CY threefold as discussed in section 9.7 on page 239. There is a single zero-form and no one-forms. There are $h^{1,1}$ (1,1)-forms and no (2,0)- or (0,2)-forms. A characteristic of CY manifolds is that there are unique (3,0)- and (0,3)-forms Ω and $\bar{\Omega}$. Ω is used to define the period integrals of the manifold. There are also $h^{2,1}$ (2,1)- and (1,2)-forms. The rest of the forms are given by Poincaré duality.

Let us first describe the massless spectrum of type-IIA theory compactified on a CY manifold. In the NS-NS sector, the ten-dimensional metric gives rise to a four-dimensional metric and $(h^{1,1} + 2h^{1,2})$ scalars (see section 9.7.3 on page 243). The $h^{1,1} + 2h^{1,2}$ scalars are the moduli of the CY manifold.

The NS antisymmetric tensor gives rise to a four-dimensional antisymmetric tensor (equivalent to an axion) as well as $h^{1,1}$ scalars, while the dilaton gives an extra scalar. So far in the NS-NS sector we have a metric as well as $2h^{1,1} + 2h^{1,2} + 2$ scalars.

In the R-R sector, the three-form gives $h^{1,1}$ vectors and $(2h^{1,2} + 2)$ scalars (descending from the three-forms), while the vector gives a vector in four dimensions. In total, apart from the supergravity multiplet, we have $N_V = h^{1,1}$ vector multiplets and $N_H = h_{12} + 1$

hypermultiplets. An important observation here is that, in contrast to the heterotic string, the dilaton belongs to a hypermultiplet.

Since the scalars of the vector multiplets are associated with the (1,1)-forms, the classical vector moduli space is the same as the moduli space of complexified Kähler structures, $k + iB$. Moreover, $\mathcal{N} = 2_4$ supersymmetry forbids neutral couplings between vector multiplets and hypermultiplets. Since the dilaton (string coupling) is in a hypermultiplet, this means that the tree-level geometry of the vector-multiplet moduli space M_V is exact! Notice that all vectors come from the R-R sector and thus have no perturbative charged states. On the other hand, the hypermultiplets are $h_{21} + 1$ in number. One contains the dilaton, while the others come from the metric and antisymmetric tensor. Therefore, the classical hypermultiplet moduli space is a product of the moduli space of complex structures and the SU(2,1)/U(2) coset parametrizing the geometry of the dilaton hypermultiplet. This space is affected by quantum corrections both perturbative and non-perturbative.

Let us now focus at the type-IIB theory compactified on a CY manifold. The NS-NS sector obviously remains similar. However, the content of the R-R sector is different. The ten-dimensional axion gives a lower-dimensional axion while the two-index antisymmetric tensor gives $h^{1,1} + 1$ scalars, the last one coming from dualizing the four-dimensional antisymmetric tensor. The self-dual four-form gives $h^{1,1}$ scalars and $h^{2,1} + 1$ vectors. The last one comes from the unique (3,0)-form of a CY. In total we have $h^{1,2}$ vector multiplets and $h^{1,1} + 1$ hypermultiplets. Thus, in type-IIB compactifications the vector moduli space $\mathcal{M}_V$ parametrizes the space of complex structures of the CY manifold. The hypermultiplet moduli space parameterizes the complexified Kähler structures. As in the type-IIA case, the dilaton is part of a hypermultiplet.

9.12 Mirror Symmetry

We have seen in the previous section that type-IIA and type-IIB theory compactified on a CY manifold are related by exchanging the complex structure and Kähler moduli spaces. This is reminiscent of the action of toroidal T-duality described in the end of section 7.2 on page 157. We will see that this resemblance is more than a coincidence.

Before we delve into three-complex-dimensional CY manifolds we will warm up by looking at one-complex-dimensional CY manifolds. Here the holonomy should be by definition SU(1) and since this is trivial the manifold is flat. Thus, a compact CY_1 is a T^2. The CFT on the torus has four moduli: the metric and the antisymmetric tensor

$$G = \frac{T_2}{U_2} \begin{pmatrix} 1 & U_1 \\ U_1 & |U|^2 \end{pmatrix}, \quad B = \begin{pmatrix} 0 & T_1 \\ -T_1 & 0 \end{pmatrix}. \tag{9.12.1}$$

Since U defines the complex coordinates on the torus as $z = \sigma_1 + U\sigma_2$ it is the complex structure modulus. $T = T_1 + iT_2$ is the complexified Kähler modulus.[10] For a rectangular torus with radii R_1, R_2 and no B field,

$$T = iR_1 R_2, \quad U = i\frac{R_1}{R_2}. \tag{9.12.2}$$

[10] The Kähler form is the volume form on the two-torus and T_2 is the volume of T^2.

From this we can see that a single T-duality in the second direction implements $T \leftrightarrow U$ interchange. Moreover, if this torus forms part of the type-II string compactification manifold, then as we have argued in section 7.2 this T-duality interchanges IIA $\leftrightarrow$ IIB. We therefore observe a similar phenomenon, a IIA/IIB interchange accompanied by an interchange of Kähler and complex structure moduli.

We now return to CY_3. We start by describing more closely the world-sheet supercon-formal field theory that is relevant in type-II CY compactifications. Since the string theory background is $M_4 \times K$, the $\mathcal{N} = (2,2)_2$ world-sheet theory on the CY has $(c, \bar{c}) = (9, 9)$. This should be generated by the supersymmetric σ-model on the CY.

In section 4.13.2 on page 79 we have described in some detail the general structure of $\mathcal{N} = (2,2)_4$ superconformal theories. An important class of states are the chiral primary states with $\Delta = \frac{q}{2}$ and the antichiral primary states with $\Delta = -\frac{q}{2}$. We have shown that they both form a ring under OPE, the chiral (c) and the antichiral (a) rings.

An important ingredient is the spectral flow (4.13.24) on page 80, that maps NS states to R states and vice versa. In particular, the $(\Delta, q) = (0, 0)$ NS vacuum is mapped to the states $(\frac{3}{2}, \pm 3)$, carrying the maximal possible U(1) charge. An important constraint that is imposed by space-time supersymmetry is that the spectrum of the U(1) charge must be integral in the NS sector. This is required in order to guarantee locality with the space-time supercharges. Thus, the charge can take integer values in the range $-3, \ldots, 3$.

Therefore, in the NS sector the chiral primaries have

$$(\Delta, q) = (0, 0), \left(\frac{1}{2}, \pm 1\right), (1, \pm 2), \left(\frac{3}{2}, \pm 3\right). \tag{9.12.3}$$

Only $(\frac{1}{2}, \pm 1)$ will give massless states in the type-II compactification.

We now take into account also the right-moving part of the theory. Then we have four chiral rings: (c,c), (c,a), (a,c), (a,a). The two last ones are related by charge conjugation to the two first. The question we would like to answer is this: What is the relationship between the two independent chiral rings (c,c) and (a,c) and the geometry of the CY manifold?

The (c,c) ring contains (massless) states with charges $(q, \bar{q}) = (1, 1)$, while the (c,a) ring contains $(q, \bar{q}) = (1, -1)$ massless states. All of them have conformal weights $(\frac{1}{2}, \frac{1}{2})$ and generate massless states. We will now compare them with the cohomology of the related CY manifold.

The (c,c) ring contains the unique state $(q, \bar{q}) = (3, 0)$ with the maximal U(1) charge which should correspond to the $(3, 0)$ Ω form, as well as its conjugate $(0, 3)$ that should correspond to $\bar{\Omega}$. It also contains the $(3, 3)$ states that should correspond to $\Omega \wedge \bar{\Omega}$.

The $(1, 1)$ states of the (c,c) ring should correspond to the complex structure moduli. This can be seen as follows. Let $\psi^i, \bar{\psi}^i, i = 1, 2, 3$ be the left-moving world-sheet fermions, while $\lambda^i, \bar{\lambda}^i, i = 1, 2, 3$ are the right-moving world-sheet fermions. The left and right U(1) currents are $J_L = \psi^i \bar{\psi}^i$, $J_R = \lambda^i \bar{\lambda}^i$. The lowest dimension field corresponding to the $(1, 1)$ state is $g_{ij} \psi^i \lambda^j$. We obtain the top state in the superfield by acting with $G^+_{-1/2} \bar{G}^-_{-1/2}$ to obtain $g_{ij} \partial X^i \bar{\partial} X^j$. This is the complex structure deformation operator in the σ-model.

On the other hand the $(-1, 1)$ states of the (a,c) ring by spectral flow can be mapped to $(1, 1)$ moduli. They can be written as $g_{i\bar{j}} \psi^i \bar{\lambda}^{\bar{j}}$ whose top component is $g_{i\bar{j}} \partial X^i \bar{\partial} \bar{X}^{\bar{j}}$ and corresponds to Kähler deformations.

This correspondence between the chiral rings, massless states and the cohomology of the CY manifold can be made more precise by identifying $G_0^+ \sim \partial$, $\bar{G}_0^+ \sim \bar{\partial}$ in the R sector and $G_{-1/2}^+ \sim \partial$, $\bar{G}_{-1/2}^+ \sim \bar{\partial}$ in the NS sector. There is a Hilbert space decomposition which parallels the Hodge decomposition for forms.

We conclude that the (c,c) chiral ring is associated with the complex structure moduli and the (a,c) ring with the Kähler moduli.

The simple observation is that the relative sign of the right U(1) current is a matter of convention and can be changed at will. This is an obvious symmetry of the CFT. However, the implications for the geometry are far reaching. The change of sign, interchanges the roles of the Complex structure and Kähler moduli spaces. This is known as *mirror symmetry*.

Define a mirror CY manifold K^* as a CY space with cohomology

$$h_{K^*}^{p,q} = h_K^{3-p,q}. \tag{9.12.4}$$

K and K^* are said to form a mirror pair.

Mirror symmetry in CFT is the statement that the supersymmetric σ-models on K and K^* give rise to the same CFT.

Once this $\mathcal{N} = (2, 2)_2$ CFT is embedded in type-II string theory, the mirror symmetry transformation interchanges type IIA and type-IIB because it is similar to T-duality. This is in agreement with our observations in the previous section.

9.13 Absence of Continuous Global Symmetries

An important result in string theory is the absence of continuous global symmetries. Physicists for a long time had a prejudice against continuous global symmetries. The rough argument is that one needs to rotate fields all over space-time at once. This is at odds with the "spirit" of relativity. Moreover, it is plausible that gravity in the quantum regime involves baby-universe processes. This leads to the conclusion that such global symmetries will be spoiled by quantum gravity, since global charge will leak out to baby universes and will never be retrieved.

We will give here an argument which indicates that all internal symmetries must be local symmetries in string theory.

We start from bosonic strings and consider a continuous symmetry with a conserved charge which acts on the physical spectrum of the theory. This guarantees the existence of a local current,

$$Q = \frac{1}{2\pi i} \oint (dz J_z - d\bar{z} j_{\bar{z}}). \tag{9.13.1}$$

If such a symmetry is continuous and appears in the compact sector of the CFT then it is conformal. That is the current J_z is a $(1,0)$ operator while $\bar{J}_{\bar{z}}$ is a $(0,1)$ operator. Then the following states are massless gauge bosons in space-time

$$A_\mu \sim J_z \bar{\partial} X^\mu : e^{ip \cdot x} :, \quad B_\mu \sim \bar{J}_{\bar{z}} \partial X^\mu : e^{ip \cdot x} : . \tag{9.13.2}$$

Thus, the symmetry is also local. It is not necessary in general that there will be two gauge bosons. Sometimes the symmetry is purely left moving and it will be associated with a single gauge boson. This is also what happens in the open string case.

When the world-sheet is supersymmetric, one can write (9.13.1) in superspace

$$Q = \frac{1}{2\pi i} \oint (dz d\theta J - d\bar{z} d\bar{\theta} \bar{J}),$$

(9.13.3)

where by superconformal invariance J is a (1/2,0) superfield and $\bar{J}$ is a (0,1/2) superfield. We can again construct gauge bosons

$$a_\mu \sim J_z \bar{\psi}^\mu : e^{ip \cdot x} :, \quad b_\mu \sim \bar{J}_{\bar{z}} \psi^\mu : e^{ip \cdot x} :.$$

(9.13.4)

This also generalizes to the heterotic case.

The two assumptions made so far are important. The loophole consists in the existence of a conserved current

$$\partial_z \bar{J} + \partial_{\bar{z}} J = 0, \quad Q = \int_{t=\text{constant}} dx J^0,$$

(9.13.5)

whose charge is conserved and commutes with L_0, but $J, \bar{J}$ are not conformal operators. This can happen in noncompact CFTs and the prototype example is provided by the Lorentz symmetry of the string. The currents are

$$J_\tau^{\mu\nu} = X^\mu \dot{X}^\nu - X^\nu \dot{X}^\mu, \quad J_\sigma^{\mu\nu} = X^\mu X'^\nu - X^\nu X'^\mu.$$

(9.13.6)

These currents generate a symmetry, the associated charge is conserved, and it commutes with the Virasoro algebra. However, the local currents $J_z^{\mu\nu}, \bar{J}_{\bar{z}}^{\mu\nu}$ are not good conformal operators due to IR divergences. No gauge bosons are associated with these currents. Although no general proof exists, no such occurrence seems to exist in a compact CFT.

The other possibility is a "compact" CFT with a σ-model description and a continuous symmetry whose current is not conformal in the sense described above. This is the case of a large class of parafermionic CFTs. What happens in this case is that nonperturbative world-sheet effects break the continuous symmetry to a discrete one. You are invited to work out the simplest case in exercise (9.66) on page 293.

We will comment on another case that is worth mentioning: that of *approximate global symmetries*. It is typical in orientifold vacua of the type-II string (generalizations of the type-I string) for the gauge group to contain several anomalous U(1) factors. The anomaly is canceled via a lower-dimensional version of the Green-Schwarz mechanism involving a pseudoscalar (axion). This breaks the gauge symmetry and gives a mass to the gauge field. However, in some regions of the moduli space the global part of the gauge symmetry remains intact in perturbation theory. It is broken by instanton effects to a discrete symmetry but this breaking can be made arbitrarily small at sufficiently weak coupling.

In exercise 13.53 on page 469 you are invited to use holography in order to uncover another reason for the absence of continuous global symmetries in a large class of string theory vacua.

9.14 Orientifolds

In sections 7.3 and 7.6 on page 170 the construction of the unoriented (type-I) string theory was described. It was performed through quotiening the IIB theory by the orientation reversal transformation Ω. This is the simplest example of an *orientifold*. It is a generalization of an orbifold, where the symmetry group involves also orientation reversal, generically combined with other symmetry transformations. In this language, the type-I orientifold group is $G = \{1, \Omega\}$.

In this section we will construct more general orientifolds by hybridization of the orbifold concept and orientation reversal. They are important vacua of string theory containing both open and closed strings. D-branes and orientifold planes also enter in an essential manner.

We will consider orientifolds that break half of the supersymmetry of the original ten-dimensional theory. They may be viewed as compactifications of the type-I theory on orbifold limits of the K3 manifold. Although not phenomenologically relevant as such, they are simple enough to illustrate the issues involved.

9.14.1 K3 orientifolds

In section 9.10 we presented in detail the $\mathbb{Z}_2$ orbifold of type-II string theory. This described the string compactification to six dimensions on a $\mathbb{Z}_2$ orbifold limit of K3. We also briefly described the $\mathbb{Z}_3$, $\mathbb{Z}_4$, and $\mathbb{Z}_6$ orbifolds that are equivalent to other limits of K3.

Our present aim is to analyze unoriented strings moving in an orbifold K3 compactification. This will be implemented by adding the orientation reversal to the orbifold group. The construction of the closed part of the theory was described in section 9.10 on page 250. We will describe in detail here the construction of the open string sectors since this involves novel features.

The orbifold action on the T^4 coordinates was specified in (9.10.4). There are two distinct orientifold groups possible:

$$Y_N = \{1, \Omega, g_k, \Omega_k\}, \quad k = 1, 2, \dots, N, \quad g_k \equiv e^{2\pi i k/N}, \quad \Omega_k \equiv e^{2\pi i k/N}\Omega, \tag{9.14.1}$$

and

$$W_N = \{1, g_{2k-2}, \Omega_{2k-1}\}, \quad k = 1, 2, \dots, \frac{N}{2}, \quad N \text{ even.} \tag{9.14.2}$$

Both Y_N and W_N form groups since Ω commutes with the orbifold elements and $\Omega^2 = 1$.

Another point to stress is that for Ω to be a symmetry of the T^4 lattice sum, we must put restrictions on the moduli. We will take here the internal components $B_{ij} = 0$.[11]

We will now elaborate the action of the orientifold groups on the states in the open string sector, on D-branes. A generic state can be written as $\lambda_{ij}|X, ij\rangle$ where i, j label the end points of the open strings, λ is a CP matrix, and X collectively labels the world-sheet oscillators that are involved in that state.

The orientifold elements have two possible actions on a generic D-brane state. In addition to the obvious action on the oscillator states, they also act on the CP indices with a matrix representation of the orientifold group. It is generated via matrices γ_g

$$g_k : |X, ij\rangle \rightarrow \epsilon_k (\gamma_k)_{ii'} |g_k \cdot X, i'j'\rangle (\gamma_k^{-1})_{j'j}, \tag{9.14.3}$$

$$\Omega_k : |X, ij\rangle \rightarrow \epsilon_{\Omega_k} (\gamma_{\Omega_k})_{ii'} |\Omega_k \cdot X, j'i'\rangle (\gamma_{\Omega_k}^{-1})_{j'j}, \tag{9.14.4}$$

---

[11] In exercise 9.28 you are requested to find all values of the T^4 moduli so that Ω is a symmetry.

where $\epsilon_k, \epsilon_{\Omega_k}$ are signs. Note that the Ω_k elements interchange also the string end points. The group property $g_k = (g_1)^k$ and $g_N = 1$ implies

$$\gamma_k = \pm(\gamma_1)^k, \quad (\gamma_k)^N = \pm 1. \tag{9.14.5}$$

Furthermore, the condition that Ω^2

$$\Omega^2: \ |X, ij\rangle \to \epsilon_\Omega^2 \, (\gamma_\Omega(\gamma_\Omega^T)^{-1})_{ii'} \, |X, i'j'\rangle \, (\gamma_\Omega^T \gamma_\Omega^{-1})_{j'j}, \tag{9.14.6}$$

is equal to the identity requires that

$$\gamma_\Omega = \zeta \gamma_\Omega^T, \quad \zeta^2 = 1. \tag{9.14.7}$$

Note that the adjoint action on the CP indices implies that the representation of the orientifold group on the CP sector is defined up to a sign.

These transformations do not completely fix the orientifold group transformations. There can be several CP matrices γ up to basis change that satisfy the group algebra. As we have seen however in section 5.3 on page 133, at the one-loop level, extra constraints emerge from tadpole cancellation.

We will now consider the implementation of the orientifold action at one-loop order. We focus on the $\mathbb{Z}_2$ case for simplicity.

9.14.2 The Klein bottle amplitude

The Klein bottle amplitude arises from the orientation projection in the closed string sector. In the operator formulation, according to our discussion in sections 5.3 on page 133 and 7.6 on page 170, the amplitude can be written as

$$Z_K = \mathrm{Tr}_{\mathrm{NS\text{-}NS+R\text{-}R}}^{U+T} \left[\frac{\Omega}{2} \cdot \frac{1+g}{2} \cdot \frac{1+(-1)^{F_L}}{2} \, e^{-2\pi t(L_0 + \bar{L}_0 - c/12)} \right]. \tag{9.14.8}$$

The trace is taken both in the $\mathbb{Z}_2$ untwisted and twisted sector. As usual, because of the Ω insertion, only the left-right symmetric sectors (NS-NS and R-R) contribute to the trace. Only the left GSO projection was inserted for the same reason. g is the $\mathbb{Z}_2$ orbifold element. To evaluate these traces, we require the action of the orientation reversal on the bosonic oscillators, given in (3.4.2) on page 33 as well as on the fermionic ones,

$$\Omega \, \psi_r \, \Omega^{-1} = \bar{\psi}_r, \quad \Omega \, \bar{\psi}_r \, \Omega^{-1} = -\psi_r. \tag{9.14.9}$$

The extra minus sign is inserted in order for the product $\psi_r \bar{\psi}_r$ to be orientation invariant. This is mostly for convenience: this choice does not affect the GSO-invariant states.

We now compute the traces. We start from the T^4 lattice states. Since the orientation reversal acts on momenta and windings as

$$\Omega \, |m_i, n_i\rangle = |m_i, -n_i\rangle, \tag{9.14.10}$$

only momenta survive the Klein bottle trace when no $\mathbb{Z}_2$ element g is inserted

$$\langle -m_i, -n_i | \, \Omega \, |m_i, n_i\rangle = \prod_{i=1}^{4} \delta_{n_i,0}. \tag{9.14.11}$$

On the other hand, since $g\,|m_i, n_i\rangle = |-m_i, -n_i\rangle$ we obtain

$$\langle -m_i, -n_i|\,\Omega\, g\,|m_i, n_i\rangle = \prod_{i=1}^{4}\delta_{m_i,0}. \tag{9.14.12}$$

Concerning the action of Ω on the bosonic and fermionic oscillators, we obtain a nonzero contribution in the trace only if the state has the same left and right oscillators. This effectively sets $L_0 + \bar{L}_0 \to 2L_0$ for such symmetric states.

It is useful at this point to introduce the $\mathbb{Z}_2$-twisted GSO-projected partition functions

$$T[^h_g](\tau) \equiv \sum_{a,b=0}^{1}(-1)^{a+b}\,\vartheta^2[^a_b](\tau)\,\vartheta[^{a+h}_{b+g}](\tau)\,\vartheta[^{a-h}_{b-g}](\tau). \tag{9.14.13}$$

Putting everything together, we find in the untwisted sector

$$\Lambda_{K_1} = \frac{1}{4}\int_0^\infty \frac{dt}{t}\,\mathrm{Tr}^U\left[\Omega\cdot\frac{1+(-1)^{F_L}}{2}\,e^{-2\pi t(L_0+\bar{L}_0-c/12)}\right]$$

$$= i\,\frac{V_6\sqrt{G}}{2(2\pi\ell_s)^6}\int_0^\infty \frac{dt}{8t}\,\frac{T[^0_0](2it)}{t^5\,\eta^{12}(2it)}\sum_{m^i\in\mathbb{Z}}\exp\left[-\frac{\pi}{t}\,G_{ij}m^i m^j\right], \tag{9.14.14}$$

where $i = 1, 2, 3, 4$.

It is important in this computation, to start with the lattice sum in the Hamiltonian form. It is this form that is proper in the operator formalism and the windings and momenta in equations (9.14.11) and (9.14.12) are those of the Hamiltonian form. Once the projection to windings or momenta only is made, we may then Poisson-resum at will. In (9.14.14) we have in fact Poisson-resumed the lattice sum over all momenta.

Taking the decompactification limit for the T^4, $\sqrt{G}\to\infty$, we obtain the associated ten-dimensional type-I Klein bottle amplitude in (7.6.5) on page 171 up to a factor of two, originating from the $\mathbb{Z}_2$ projection.

To obtain the same trace with the $\mathbb{Z}_2$ element g inserted, we may note that for states that are left-right symmetric and therefore survive the Ω projection, the g action is trivial. Therefore, the only nontrivial consequence of the insertion of g is to keep the T^4 windings instead of the momenta as documented in (9.14.12)

$$\Lambda_{K_2} = \frac{1}{4}\int_0^\infty \frac{dt}{t}\,\mathrm{Tr}^U\left[\Omega\cdot g\cdot\frac{1+(-1)^{F_L}}{2}\,e^{-2\pi t(L_0+\bar{L}_0-c/12)}\right]$$

$$= i\,\frac{V_6}{2(2\pi\ell_s)^6}\int_0^\infty \frac{dt}{8t}\,\frac{T[^0_0](2it)}{t^3\,\eta^{12}(2it)}\sum_{n^i\in\mathbb{Z}}\exp\left[-\pi t\,G_{ij}n^i n^j\right]. \tag{9.14.15}$$

We now turn to the twisted sector. Here the partition functions, before Ω projection, can be found in section 9.10 on page 250. Note that there is no lattice sum here because twisted states are localized, and therefore carry no windings or momenta. Only symmetric states survive the Ω projection, so that $L_0 + \bar{L}_0 \to 2L_0$. The g insertion in the trace is trivial, since left and right pieces transform similarly under the g projection after the Ω projection.[12]

---

[12] For a general orbifold, the insertion of the Ωg element, implies that this sector is equivalent to the sector with $L_0 + \bar{L}_0 \to 2L_0$ and the insertion of g^2 in the trace.

This implies that we may take $g \to 1$ in the trace. We therefore find

$$
\Lambda_{K_3} = \frac{1}{4} \int_0^\infty \frac{dt}{t} \operatorname{Tr}^T \left[\Omega \cdot (1+g) \cdot \frac{1+(-1)^{F_L}}{2} \, e^{-2\pi t \left(L_0 + \bar{L}_0 - c/12 \right)} \right]
$$
$$
= i \frac{2^4 V_6}{(2\pi \ell_s)^6} \int_0^\infty \frac{dt}{8t} \frac{T\!\left[{}^1_0 \right](2it)}{t^3 \, \eta^6(2it) \vartheta_4^2(2it)}. \tag{9.14.16}
$$

We may now collect the full Klein bottle amplitude, and transform it to the transverse (closed string) channel along the lines of section 5.3.3 on page 138 in order to expose the tadpoles. We use

$$
\ell = \frac{\pi}{2t}, \quad \vartheta_2(2it) = \frac{1}{\sqrt{2t}} \vartheta_4\!\left(i\frac{\ell}{\pi} \right), \quad \vartheta_3(2it) = \frac{1}{\sqrt{2t}} \vartheta_3\!\left(i\frac{\ell}{\pi} \right), \quad \eta(2it) = \frac{1}{\sqrt{2t}} \eta\!\left(i\frac{\ell}{\pi} \right),
$$
$$
\tag{9.14.17}
$$

to find

$$
\Lambda_K = i \frac{2 V_6}{\pi (2\pi \ell_s)^6} \int_0^\infty d\ell \, \frac{T\!\left[{}^0_0 \right]\!\left(i\frac{\ell}{\pi} \right)}{\eta^{12}\!\left(i\frac{\ell}{\pi} \right)} \left[V_4 \sum_{m^i \in \mathbb{Z}} \exp\!\left[-2\ell \, G_{ij} m^i m^j \right] + \frac{1}{V_4} \sum_{n_i \in \mathbb{Z}} \exp\!\left[-2\ell \, G^{ij} n_i n_j \right] \right]
$$
$$
+ i \frac{2^4 V_6}{\pi (2\pi \ell_s)^6} \int_0^\infty d\ell \, \frac{T\!\left[{}^0_1 \right]\!\left(i\frac{\ell}{\pi} \right)}{\eta^6\!\left(i\frac{\ell}{\pi} \right) \vartheta_2^2\!\left(i\frac{\ell}{\pi} \right)}, \tag{9.14.18}
$$

where we Poisson-resummed the winding contribution and set $V_4 = \sqrt{G}$ (dimensionless).

We may now extract the diverged part of the Klein bottle, i.e., the tadpole

$$
\mathcal{T}_K = i \frac{2^{10} V_6}{32\pi (2\pi \ell_s)^6} \int_0^\infty d\ell \left[V_4 + \frac{1}{V_4} \right]. \tag{9.14.19}
$$

We note that the twisted sector contribution does not give rise to a tadpole. This occurs only in $\mathbb{Z}_2$ sectors of orientifolds.

Due to supersymmetry, the R tadpoles are opposite in sign to the NS ones. We have kept all contributions even though their sum formally vanishes, since it will not vanish in more complicated amplitudes.

The tadpole contribution linear in V_4 is the one that survives the decompactification to ten dimensions. It does indeed agree with the ten-dimensional result in (7.6.6) on page 172 once a factor of 2 coming from the $\mathbb{Z}_2$ projection is accounted for.

The tadpole will be canceled by the insertion of D_9-branes filling all ten dimensions. The term inversely proportional to V_4 is related to the previous one by inverting the volume of T^4. As this operation turns D_9-branes to D_5-branes, the tadpole must be canceled by the addition of D_5-branes.

We therefore conclude that the tadpoles are due to O_9- and O_5-planes.

9.14.3 D-branes on $T^4/\mathbb{Z}_2$

We now turn to the open sector. According to the previous section, we must include D_9- and D_5-branes. Although there are no options on D_9-branes, since they fill all ten dimensions, there are options for D_5-branes. They will be stretching in the six noncompact dimensions.

They are also pointlike on T^4. The orbifold now acts on the transverse positions of the branes. Therefore, there are two main options to consider.

We may consider a group of branes sitting at a fixed point of the orbifold action. In such a case there is no further restriction on the transverse position. We may also consider a group of branes at a generic position x^i on T^4. Orbifold invariance imposes that we also include a mirror brane group at the position $-x^i$.

Branes placed at an orbifold fixed point, are sometimes fixed to it. Such branes are also known as "fractional" branes. One reason for this is that to move off the fixed point they must split in mirror pairs and sometimes this is impossible. An equivalent reason is that the scalar fields, corresponding to the transverse brane coordinates are all projected out by the orbifold projection. Another reason is that their world-volume fields are charged under vectors localized on the orbifold planes.

Not every set of branes localized at an orbifold fixed point represents fractional branes. In the orientifold we are considering, the D_5-branes will have vanishing twisted tadpoles and therefore will not be fractional.

In order to accommodate the orbifold action on the CP factors of D_9- and D_5-branes we must introduce matrices $\gamma_{g,9}$ and $\gamma_{g,5}$. They satisfy the constraints, (9.14.5)–(9.14.7) coming from the orbifold group property.

It is important to determine the signs entering in the orientifold projections. According to the detailed discussion in section 7.3 on page 162, in the NS sector there is an ϵ phase for each of the 9-9, 5-5, and 9-5 strings as follows

$$\Omega \, |9-9,p\,;ij\rangle_{\mathrm{NS}} = \epsilon_{99} \, (\gamma_{\Omega,9})_{ii'} \, |9-9,p\,;,j'i'\rangle_{\mathrm{NS}} \, (\gamma_{\Omega,9})_{j'j}^{-1}, \tag{9.14.20}$$

$$\Omega \, |5-5,p\,;ij\rangle_{\mathrm{NS}} = \epsilon_{55} \, (\gamma_{\Omega,5})_{ii'} \, |5-5,p\,;,j'i'\rangle_{\mathrm{NS}} \, (\gamma_{\Omega,5})_{j'j}^{-1}. \tag{9.14.21}$$

Similar arguments as in section 7.3 fix

$$\epsilon_{99}^2 = \epsilon_{55}^2 = -1, \quad \gamma_{\Omega,5,9} = \zeta_{5,9}\gamma_{\Omega,5,9}^T, \quad \zeta_5^2 = \zeta_9^2 = 1. \tag{9.14.22}$$

In the 5-9, 9-5 sectors, however, we may write

$$\Omega \, |5-9,p\,;ij\rangle_{\mathrm{NS}} = \epsilon_{59}(\gamma_{\Omega,5})_{ii'} \, |9-5,p\,;,j'i'\rangle_{\mathrm{NS}} \, (\gamma_{\Omega,9})_{j'j}^{-1},$$

$$\Omega \, |9-5,p\,;ij\rangle_{\mathrm{NS}} = \epsilon_{59}(\gamma_{\Omega,9})_{ii'} \, |5-9,p\,;,j'i'\rangle_{\mathrm{NS}} \, (\gamma_{\Omega,5})_{j'j}^{-1}. \tag{9.14.23}$$

Imposing $\Omega^2 = 1$ we obtain

$$\epsilon_{59}^2 \zeta_5 \zeta_9 = 1. \tag{9.14.24}$$

The phase ϵ_{59} captures the transformation properties under Ω of the SO(4) twisted spinor as well of the NS open string vacuum. If two 9-5 states interact, they may produce a 5-5 or a 9-9 state. Therefore, a nontrivial coupling of two 9-5 states to the massless 9-9 or 5-5 states should be allowed. This implies that $\epsilon_{59}^2 = -1$. Therefore, from (9.14.24), the CP projection is opposite for five-branes compared to that of nine-branes,

$$\zeta_5 \zeta_9 = -1. \tag{9.14.25}$$

In particular, the type-I D_5-branes have symplectic gauge group, a fact supported by other considerations in section 11.7.2. Similar considerations apply in the R sector. You are asked in exercise 9.29 on page 289 to carefully work them out.

We will now describe the light open string spectrum. For the 9-9 strings we have the following bosonic states: The vectors

$$\psi^\mu_{-1/2}|p;ij\rangle\lambda_{ij}, \quad \lambda = \gamma_{g,9}\,\lambda\,\gamma_{g,9}^{-1}, \quad \lambda = -\gamma_{\Omega,9}\,\lambda^T\,\gamma_{\Omega,9}^{-1} \tag{9.14.26}$$

are singlets under the SO(4) R-symmetry that rotates the four transverse dimensions. The scalars

$$\psi^i_{-1/2}|p;ij\rangle\lambda_{ij}, \quad \lambda = -\gamma_{g,9}\,\lambda\,\gamma_{g,9}^{-1}, \quad \lambda = -\gamma_{\Omega,9}\,\lambda^T\,\gamma_{\Omega,9}^{-1} \tag{9.14.27}$$

transform in the vector of SO(4). The fermionic states originating in the R sector can be obtained from the fact that the theory has $\mathcal{N} = 1_6$ supersymmetry and will not be consider further in this section.

D_5 branes can be localized at a fixed point a, with associated CP matrices $\gamma_{g,5a}$ and $\gamma_{\Omega,5a}$ or at a generic point x^i, together with a copy at the image point $-x^i$ with a CP matrix $\gamma_{\Omega,5x}$. For the low-lying spectrum of the 5a-5b strings we obtain

$$\psi^\mu_{-1/2}|p;ij\rangle\lambda_{ij}, \quad \lambda = \gamma_{g,5a}\,\lambda\,\gamma_{g,5b}^{-1}, \quad \lambda = -\gamma_{\Omega,5a}\,\lambda^T\,\gamma_{\Omega,5b}^{-1}, \tag{9.14.28}$$

$$\psi^i_{-1/2}|p;ij\rangle\lambda_{ij}, \quad \lambda = -\gamma_{g,5a}\,\lambda\gamma_{g,5b}^{-1}, \quad \lambda = \gamma_{\Omega,5a}\,\lambda^T\gamma_{\Omega,5b}^{-1}. \tag{9.14.29}$$

A point to stress here is that the Ω action on the DD directions is the opposite from NN, as explained in section 7.3. If $a = b$, these states are massless. For $a \neq b$, they have a mass proportional to the distance between the fixed points. Consider now the massless spectrum of the 5x-5x strings

$$\psi^\mu_{-1/2}|p;ij\rangle\lambda_{ij}, \quad \lambda = -\gamma_{\Omega,5x}\,\lambda^T\,\gamma_{\Omega,5x}^{-1}, \tag{9.14.30}$$

$$\psi^i_{-1/2}|p;ij\rangle\lambda_{ij}, \quad \lambda = \gamma_{\Omega,5x}\,\lambda^T\gamma_{\Omega,5x}^{-1}. \tag{9.14.31}$$

Note that the $\mathbb{Z}_2$ transformation g, relates them to the 5(−x)-5(−x) strings and poses no other constraint. All such strings so far give pairs of a vector and a hypermultiplet of $\mathcal{N} = 1_6$ supersymmetry.

Consider now the 9-5a strings. These have DN boundary conditions along the T^4 directions. Therefore the massless (bosonic) state is a space-time scalar but an internal SO(4) spinor

$$|s,s';ij\rangle\lambda_{ij}, \quad \lambda = \gamma_{g,9}\,\lambda\,\gamma_{g,5a}^{-1}. \tag{9.14.32}$$

There are two such scalars. The Ω projection relates these states to the states of the 5a-9 strings and therefore provides no further constraints. We obtain hypermultiplets in this sector.

Similarly, for 9-5x strings we have

$$|s,s';ij\rangle\lambda_{ij}. \tag{9.14.33}$$

The $\mathbb{Z}_2$ projection relates them to the 9-5(−x) strings, and the Ω projection to the 5x-9 strings.

9.14.4 *The cylinder amplitude*

We may now proceed to evaluate the cylinder amplitude. We should remember the following general properties: NN directions have only momenta, DD only windings, and DN none of the above.

In operator form, the amplitude is

$$\Lambda_C = \int_0^\infty \frac{dt}{2t} \, \mathrm{Tr}_{NS,R}^{99+55+95+59} \left[\frac{1}{2} \cdot \frac{1+g}{2} \cdot \frac{1+(-1)^F}{2} \, e^{-2\pi t(L_0-c/24)} \right]. \tag{9.14.34}$$

We start with the untwisted contributions. The 9-9 strings contribute

$$\Lambda_{C_{99}^U} = i \frac{V_6 V_4}{2^6 (2\pi \ell_s)^6} \mathrm{Tr}(\gamma_{1,9})^2 \int_0^\infty \frac{dt}{8t} \frac{T_0^{[0]}(it)}{t^5 \, \eta^{12}(it)} \sum_{m^i \in Z} e^{-(\pi/t) G_{ij} m^i m^j}$$

$$= i \frac{V_6 V_4}{2^9 \pi (2\pi \ell_s)^6} \mathrm{Tr}(\gamma_{1,9})^2 \int_0^\infty d\ell \frac{T_0^{[0]}\left(i\frac{\ell}{\pi}\right)}{\eta^{12}\left(i\frac{\ell}{\pi}\right)} \sum_{m^i \in Z} e^{-\ell \, G_{ij} m^i m^j}, \tag{9.14.35}$$

where, as usual for the cylinder, $\ell = \pi/t$ and we included the 1/2 from the Ω projection and the 1/2 from the $\mathbb{Z}_2$ projection. $\gamma_{1,9}$ is the unit matrix in the nine-brane sector and therefore $\mathrm{Tr}(\gamma_{1,9}) = N_9$, the number of D9-branes. We also set $V_4 = \sqrt{G}$. This amplitude decompactifies properly to recover (7.6.7) on page 172 as expected, up to an extra factor of 1/2 coming from the $\mathbb{Z}_2$ orbifold projection.

We now consider the (untwisted) contribution of the 5-5 strings. We will label the D$_5$-branes by the index a. A subset will be localized at the (16) orbifold fixed points. We will label the fixed points with the letter I. The lattice sum is here a winding sum. We must sum over all paths connecting the D$_5$-branes. Let the brane coordinates on T^4 be X_a^i with $i = 1, 2, 3, 4$ and a labeling the particular set of D$_5$-branes. The compact coordinates are normalized so as to have integer periodicity. Then, on T^4 the distance between the two sets is $G_{ij}(X_a^i - X_b^i + n^i)(X_a^j - X_b^j + n^j)$ where n^i are arbitrary integers (windings). Using (2.3.37) on page 24 we may write the cylinder contribution of this configuration as

$$\Lambda_{C_{5_a5_b}^U} = i \frac{V_6}{2^6 (2\pi \ell_s)^6} \mathrm{Tr}(\gamma_{1,5_a}) \mathrm{Tr}(\gamma_{1,5_b}) \int_0^\infty \frac{dt}{8t} \frac{T_0^{[0]}(it)}{t^3 \, \eta^{12}(it)} \sum_{n^i \in Z} e^{-\pi t G_{ij}(X_a^i - X_b^i + n^i)(X_a^j - X_b^j + n^j)}$$

$$= i \frac{V_6}{2^9 \pi (2\pi \ell_s)^6 V_4} \mathrm{Tr}(\gamma_{1,5_a}) \mathrm{Tr}(\gamma_{1,5_b}) \int_0^\infty d\ell \frac{T_0^{[0]}\left(i\frac{\ell}{\pi}\right)}{\eta^{12}\left(i\frac{\ell}{\pi}\right)} \sum_{n_i \in Z} e^{-\ell \, G^{ij} n_i n_j - 2\pi i n_i (X_a^i - X_b^i)}, \tag{9.14.36}$$

where as before $\mathrm{Tr}(\gamma_{1,5_a}) = N_5^a$ is the number of D$_5$-branes located at X_a^i.

Lastly, we have the (untwisted) contributions of the 9-5$_a$ strings. Here the torus coordinates have DN boundary conditions and are therefore $\mathbb{Z}_2$ twisted. Therefore, the oscillator trace here can be obtained from the chiral $h = 1, g = 0$ part of the closed string orbifold in section 9.10 on page 250. The amplitude then is

$$\Lambda_{C_{9-5a}^U} = i \frac{V_6}{2^5 (2\pi \ell_s)^6} \mathrm{Tr}(\gamma_{1,9}) \mathrm{Tr}(\gamma_{1,5_a}) \int_0^\infty \frac{dt}{8t} \frac{T_0^{[1]}(it)}{t^3 \, \eta^6(it) \vartheta_4^2(it)}$$

$$= i \frac{V_6}{2^8 \pi (2\pi \ell_s)^6} \mathrm{Tr}(\gamma_{1,9}) \mathrm{Tr}(\gamma_{1,5_a}) \int_0^\infty d\ell \frac{T_1^{[0]}\left(i\frac{\ell}{\pi}\right)}{\eta^6\left(i\frac{\ell}{\pi}\right) \vartheta_2^2\left(i\frac{\ell}{\pi}\right)}. \tag{9.14.37}$$

We have included a factor of 2, due to the two orientations of the 9-5 strings. The transverse channel 9-5 contribution is zero in both the NS and the R sectors, because this is so for $T[^0_1]$. This is accidental for $\mathbb{Z}_2$ orientifold sectors.

We now move to the twisted contributions, which arise by inserting the $\mathbb{Z}_2$ element g in the cylinder trace. For the 9-9 strings we obtain

$$
\Lambda_{C^T_{99}} = i \frac{V_6}{2^6 (2\pi \ell_s)^6} \mathrm{Tr}(\gamma_{g,9})^2 \int_0^\infty \frac{dt}{8t} \frac{T[^0_1](it)}{t^3 \, \eta^8(it)} \left(\frac{2\eta(it)}{\vartheta_2(it)} \right)^2
$$
$$
= i \frac{V_6}{2^7 \pi (2\pi \ell_s)^6} \mathrm{Tr}(\gamma_{g,9})^2 \int_0^\infty d\ell \, \frac{T[^1_0] \left(i\frac{\ell}{\pi} \right)}{\eta^6 \left(i\frac{\ell}{\pi} \right) \vartheta_4^2 \left(i\frac{\ell}{\pi} \right)}, \tag{9.14.38}
$$

where the last contribution comes from the T^4 bosons. This effectively follows from the $h = 0, g = 1$ chiral part of the closed $T^4/\mathbb{Z}_2$ partition function in section 9.10.

Consider now 5_a-5_b strings. In order for the trace to be nonzero, $a = b$ and the associated D_5-branes should be located at the orbifold fixed points. The presence of the four DD directions does not otherwise affect the trace:

$$
\Lambda_{C^T_{5_I - 5_I}} = i \frac{V_6}{2^4 (2\pi \ell_s)^6} \mathrm{Tr}(\gamma_{g,5_I})^2 \int_0^\infty \frac{dt}{8t} \frac{T[^0_1](it)}{t^3 \, \eta^6(it) \vartheta_2^2(it)}
$$
$$
= i \frac{V_6}{2^7 \pi (2\pi \ell_s)^6} \mathrm{Tr}(\gamma_{g,5_I})^2 \int_0^\infty d\ell \, \frac{T[^1_0] \left(i\frac{\ell}{\pi} \right)}{\eta^6 \left(i\frac{\ell}{\pi} \right) \vartheta_4^2 \left(i\frac{\ell}{\pi} \right)}. \tag{9.14.39}
$$

Finally we consider the 9-5 strings. The presence of four DN directions effectively twists the four bosonic and fermionic coordinates. Therefore the oscillator trace here can be obtained from the chiral $h = 1, g = 1$ part of the closed string orbifold in 9.10. The presence of the $\mathbb{Z}_2$ element in the trace implies that only D_5-branes localized at the orbifold fixed points can contribute:

$$
\Lambda_{C^T_{9-5_I}} = i \frac{V_6}{2^5 (2\pi \ell_s)^6} \mathrm{Tr}(\gamma_{g,9}) \mathrm{Tr}(\gamma_{g,5_I}) \int_0^\infty \frac{dt}{8t} \frac{T[^1_1](it)}{t^3 \, \eta^6(it) \vartheta_3^2(it)}
$$
$$
= i \frac{V_6}{2^8 \pi (2\pi \ell_s)^6} \mathrm{Tr}(\gamma_{g,9}) \mathrm{Tr}(\gamma_{g,5_I}) \int_0^\infty d\ell \, \frac{T[^1_1] \left(i\frac{\ell}{\pi} \right)}{\eta^6 \left(i\frac{\ell}{\pi} \right) \vartheta_3^2 \left(i\frac{\ell}{\pi} \right)}. \tag{9.14.40}
$$

We have again multiplied by a factor of 2, to account for the two possible orientations.

The cylinder tadpoles extracted from (9.14.40) are

$$
\mathcal{T}_C = i \frac{V_6}{2^5 \pi (2\pi \ell_s)^6} \int_0^\infty d\ell \left[V_4 \, (\mathrm{Tr}[\gamma_{1,9}])^2 + \frac{\left(\sum_a \mathrm{Tr}[\gamma_{1,5_a}] \right)^2}{V_4} \right.
$$
$$
\left. + \frac{1}{16} \sum_{I=1}^{16} \left(\mathrm{Tr}[\gamma_{g,9}] - 4\mathrm{Tr}[\gamma_{g,5_I}] \right)^2 \right]. \tag{9.14.41}
$$

The minus sign in the 9-5 twisted contribution is due to the $\mathbb{Z}_2$ element g in the associated trace.

9.14.5 The Möbius strip amplitude

We now turn to the Möbius strip, which implements the Ω projection in the open sector. We must calculate the same traces as on the cylinder but with an extra insertion of Ω,

$$\Lambda_M = \int_0^\infty \frac{dt}{2t} \operatorname{Tr}_{NS,R}^{99+55} \left[\frac{\Omega}{2} \cdot \frac{1+g}{2} \cdot \frac{1+(-1)^F}{2} e^{2\pi t(L_0 - c/24)} \right]. \tag{9.14.42}$$

Since Ω changes the orientation of the string, 9-5 strings do not contribute to the trace. For the same reason, only strings starting and ending on the same D$_5$-brane contribute. For the CP factors, using (9.14.4) we may evaluate the trace as in (5.3.24) on page 139.

We start from the untwisted sector. The contribution of the 9-9 strings is

$$\Lambda_{M_{99}^U} = -i \frac{V_6 V_4}{2^6 (2\pi \ell_s)^6} \operatorname{Tr}(\gamma_{\Omega,9}^T \gamma_{\Omega,9}^{-1}) \int_0^\infty \frac{dt}{8t} \frac{\hat{T}[^0_0](it)}{t^5 \, \hat{\eta}^{12}(it)} \sum_{m^i \in Z} e^{-(\pi/t) G_{ij} m^i m^j}$$

$$= -i \frac{V_6 V_4}{2^3 \pi (2\pi \ell_s)^6} \operatorname{Tr}(\gamma_{\Omega,9}^T \gamma_{\Omega,9}^{-1}) \int_0^\infty d\ell \, \frac{\hat{T}[^0_0]\left(i\frac{\ell}{\pi}\right)}{\hat{\eta}^{12}\left(i\frac{\ell}{\pi}\right)} \sum_{m^i \in Z} e^{-4\ell \, G_{ij} m^i m^j}, \tag{9.14.43}$$

where for the Möbius strip, $\ell = \pi/(4t)$ and the overall sign is a convention. The various characters have been replaced with careted characters as is standard for the Möbius strip. They are defined in (7.6.10) on page 172 and some of their properties presented in (C.28) and (C.29) on page 510. We also used in the second line, the transformation properties of the fermionic characters from appendix C on page 507.

For the 5_a-5_a strings, according to (7.3.5) and (7.3.6) on page 163, the T^4 directions have an extra minus sign because they now carry DD boundary conditions. This is equivalent to an insertion of the $\mathbb{Z}_2$ element g in the trace. We obtain

$$\Lambda_{M_{5_a5_a}^U} = i \frac{V_6}{2^6 (2\pi \ell_s)^6} \operatorname{Tr}(\gamma_{\Omega,5_a}^T \gamma_{\Omega,5_a}^{-1}) \int_0^\infty \frac{dt}{8t} \frac{\hat{T}[^0_1](it)}{t^3 \, \hat{\eta}^8(it)} \left(\frac{2\hat{\eta}(it)}{\hat{\vartheta}_2(it)} \right)^2$$

$$= i \frac{V_6}{2^4 \pi (2\pi \ell_s)^6} \operatorname{Tr}(\gamma_{\Omega,5_a}^T \gamma_{\Omega,5_a}^{-1}) \int_0^\infty d\ell \, \frac{\hat{T}[^0_1]\left(i\frac{\ell}{\pi}\right)}{\eta^6\left(i\frac{\ell}{\pi}\right) \vartheta_2\left(2i\frac{\ell}{\pi}\right) \vartheta_4\left(2i\frac{\ell}{\pi}\right)}, \tag{9.14.44}$$

where we have used $\hat{\vartheta}_2^2(it) = 2\vartheta_2(2it)\vartheta_4(2it)$ and $\hat{T}[^0_1](it) = -\hat{T}[^0_1]\left(i\frac{\ell}{\pi}\right)$.

We now proceed to calculate the traces in the twisted sector,

$$\Lambda_{M_{99}^T} = i \frac{V_6}{2^6 (2\pi \ell_s)^6} \operatorname{Tr}(\gamma_{g\Omega,9}^T \gamma_{g\Omega,9}^{-1}) \int_0^\infty \frac{dt}{8t} \frac{\hat{T}[^0_1](it)}{t^3 \, \hat{\eta}^8(it)} \left(\frac{2\hat{\eta}(it)}{\hat{\vartheta}_2(it)} \right)^2$$

$$= i \frac{V_6}{2^4 \pi (2\pi \ell_s)^6} \operatorname{Tr}(\gamma_{g\Omega,9}^T \gamma_{g\Omega,9}^{-1}) \int_0^\infty d\ell \, \frac{\hat{T}[^0_1]\left(i\frac{\ell}{\pi}\right)}{\eta^6\left(i\frac{\ell}{\pi}\right) \vartheta_2\left(2i\frac{\ell}{\pi}\right) \vartheta_4\left(2i\frac{\ell}{\pi}\right)}. \tag{9.14.45}$$

Before computing the twisted trace for the 5-5 strings we first observe that not only the 5_I-5_I strings contribute, as on the cylinder but also the 5_x-5_{-x} strings for any $x \in T^4$. To see this, consider a 5-5 string stretched between points x and y on T^4. Since in these directions the boundary conditions are DD, the expansion (2.3.28) on page 23 is relevant with center-of-mass coordinate x and winding $w \sim y - x$. We have the following actions on the string ground state:

$$\Omega|x, w\rangle = |y, -w\rangle, \quad g|x, w\rangle = |-x, -w\rangle, \tag{9.14.46}$$

where, as usual, Ω interchanges the end points of the string. Therefore,

$$\langle x', w'|g \cdot \Omega|x, w\rangle = \langle x', w'|-y, w\rangle = \delta(y + x')\delta(w - w') = \delta(y + x')\delta(x + y'). \tag{9.14.47}$$

The trace vanishes unless $y = -x$, that is the string stretches from an arbitrary D$_5$-brane to its image under the $\mathbb{Z}_2$ transformation g. Note that this property ceases to be true for other $\mathbb{Z}_N$ orbifold actions.

We may now evaluate the trace as

$$\Lambda_{M^T_{5_a 5_a}} = i\frac{V_6}{2^6(2\pi \ell_s)^6}\mathrm{Tr}\big(\gamma^T_{g\Omega,5_a}\gamma^{-1}_{g\Omega,5_a}\big) \int_0^\infty \frac{dt}{8t}\frac{\hat{T}[^0_0](it)}{t^3\,\hat{\eta}^{12}(it)} \sum_{n^i \in \mathbb{Z}} e^{-\pi t G_{ij}(2X^i_a + n^i)(2X^j_a + n^j)}$$

$$= -i\frac{V_6}{2^3\pi(2\pi \ell_s)^6 V_4}\mathrm{Tr}\big(\gamma^T_{g\Omega,5_a}\gamma^{-1}_{g\Omega,5_a}\big) \int_0^\infty d\ell\,\frac{\hat{T}[^0_0]\big(i\frac{\ell}{\pi}\big)}{\hat{\eta}^{12}\big(i\frac{\ell}{\pi}\big)} \sum_{n_i \in \mathbb{Z}} e^{-4\ell\,G^{ij}n_i n_j - 4\pi n_i X^i_a}, \tag{9.14.48}$$

where the projection is reversed in the DD directions.

We collect the tadpoles as

$$\mathcal{T}_M = -i\frac{2V_6}{\pi(2\pi \ell_s)^6} \int_0^\infty d\ell \left[V_4\,\mathrm{Tr}(\gamma^T_{\Omega,9}\gamma^{-1}_{\Omega,9}) + \frac{\sum_a \mathrm{Tr}\big(\gamma^T_{g\Omega,5_a}\gamma^{-1}_{g\Omega,5_a}\big)}{V_4} \right], \tag{9.14.49}$$

where the contributions proportional to $T[^0_1]\big(i\frac{\ell}{\pi}\big)$ vanish identically for the $\mathbb{Z}_2$ orbifold.

9.14.6 Tadpole cancellation

We are now ready to discuss the cancellation of tadpoles. Due to the unbroken supersymmetry, the NS and R tadpoles are equal and opposite. Collecting the various contributions from (9.14.19), (9.14.41), and (9.14.49) we obtain

$$\mathcal{T} = \frac{iV_6}{32\pi(2\pi \ell_s)^6} \int_0^\infty d\ell \Big[\big(2^{10} + (\mathrm{Tr}[\gamma_{1,9}])^2 - 2^6\mathrm{Tr}[\gamma^T_{\Omega,9}\gamma^{-1}_{\Omega,9}]\big)\,V_4$$

$$+ \frac{\big(2^{10} + (\sum_a \mathrm{Tr}[\gamma_{1,5_a}])^2 - 2^6\sum_a \mathrm{Tr}[\gamma^T_{g\Omega,5_a}\gamma^{-1}_{g\Omega,5_a}]\big)}{V_4} + \frac{1}{16}\sum_{I=1}^{16} \big(\mathrm{Tr}[\gamma_{g,9}] - 4\mathrm{Tr}[\gamma_{g,5_I}]\big)^2 \Big]. \tag{9.14.50}$$

Tadpole cancellation conditions thus require the cancellation of the ten-form R-R charge

$$2^{10} + (\mathrm{Tr}[\gamma_{1,9}])^2 - 2^6\mathrm{Tr}[\gamma^T_{\Omega,9}\gamma^{-1}_{\Omega,9}] = 0, \tag{9.14.51}$$

six-form R-R charge

$$2^{10} + \Big(\sum_a \mathrm{Tr}[\gamma_{1,5_a}]\Big)^2 - 2^6\sum_a \mathrm{Tr}[\gamma^T_{g\Omega,5_I}\gamma^{-1}_{g\Omega,5_I}] = 0, \tag{9.14.52}$$

and the twisted-form R-R charges,

$$\mathrm{Tr}[\gamma_{g,9}] - 4\mathrm{Tr}[\gamma_{g,5_I}] = 0, \quad \forall\ I = 1, 2, \ldots, 16. \tag{9.14.53}$$

We will now try to find a simple solution to these conditions. We assume that all D_5-branes are located at a single fixed point, that we will take to be the origin. As shown in (9.14.7) we must have

$$\gamma_{\Omega,9} = \zeta_9 \, \gamma_{\Omega,9}^T, \quad \gamma_{g\Omega,9} = \tilde{\zeta}_9 \, \gamma_{g\Omega,9}^T, \quad \gamma_{\Omega,5} = \zeta_5 \, \gamma_{\Omega,5}^T, \quad \gamma_{g\Omega,5} = \tilde{\zeta}_5 \, \gamma_{g\Omega,5}^T. \tag{9.14.54}$$

Then (9.14.51), (9.14.52) become

$$(N_9 - 32\zeta_9)^2 = 0, \quad (N_5 - 32\tilde{\zeta}_5)^2 = 0, \tag{9.14.55}$$

with obvious solution

$$N_9 = N_5 = 32, \quad \zeta_9 = 1, \quad \tilde{\zeta}_5 = 1. \tag{9.14.56}$$

Moreover, from (9.14.25), $\zeta_5 = -\zeta_9 = -1$.

We may therefore take

$$\gamma_{\Omega,9} = \gamma_{g\Omega,5} = \mathbf{1}_{32}, \quad \gamma_{g,9} = \gamma_{g\Omega,9} = \gamma_{g,5} = \gamma_{\Omega,5} = \begin{pmatrix} 0 & i\mathbf{1}_{16} \\ -i\mathbf{1}_{16} & 0 \end{pmatrix}, \tag{9.14.57}$$

where the subscripts stand for the dimension of the matrix blocks. It can be directly verified that these matrices satisfy the group relations and also satisfy the remaining twisted tadpole conditions (9.14.53). In exercise 9.33 on page 289 you are asked to investigate other solutions to the tadpole conditions.

Note also, that for this solution to the tadpole conditions, the twisted tadpoles vanish. This implies that the D_5-branes are not fractional branes. They are expected to be allowed to move off the orbifold fixed points.

9.14.7 *The open string spectrum*

We have determined the consistent projection in the open spectrum, by asking for the absence of tadpoles. We may now solve the projection conditions of section 9.14.3 to obtain the open string (massless) spectrum. We will split the CP matrices λ into 16×16 blocks, to accommodate the structure of the projection matrices in (9.14.57).

In the 9-9 sector, solving (9.14.26) we find that the vectors have

$$\lambda_V = \begin{pmatrix} A & S \\ -S & A \end{pmatrix}, \tag{9.14.58}$$

where A stands for a Hermitian antisymmetric matrix and S for a Hermitian symmetric matrix. Such matrices form the Lie algebra of the U(16) group. Therefore, taking into account the fermions, we have a U(16) vector multiplet of $\mathcal{N} = 1_6$ supersymmetry.

For the 9-9 scalars, solving (9.14.27) we obtain

$$\lambda_S = \begin{pmatrix} A_1 & A_2 \\ A_2 & -A_1 \end{pmatrix}, \tag{9.14.59}$$

where again $A_{1,2}$ are Hermitian antisymmetric matrices. We therefore obtain two anti-symmetric representations of U(16): $\mathbf{120} + \overline{\mathbf{120}}$. The scalars come in multiples of four (transforming as the **4** of the *R*-symmetry SO(4)) We therefore obtain two hypermultiplets transforming in the **120** of U(16).[13]

For the spectrum of 5-5 strings emerging from the 32 D$_5$-branes all in one of the fixed points we must solve (9.14.28) and (9.14.29). The solution is the same as in the 9-9 sector and we obtain another U(16) vector multiplet as well as two hypermultiplets in the **120**.

In the 5-9 sector we must solve (9.14.32). The solution is

$$\lambda_{95} = \begin{pmatrix} H_1 & H_2 \\ -H_2 & H_1 \end{pmatrix}, \tag{9.14.60}$$

where $H_{1,2}$ are Hermitian matrices. We therefore obtain the $(\mathbf{16},\overline{\mathbf{16}})$ and $(\overline{\mathbf{16}},\mathbf{16})$ representations of U(16)×U(16). Taking into account the multiplicity of scalars, this is a single hypermultiplet transforming as a $(\mathbf{16},\overline{\mathbf{16}})$.

We have assumed a very special D$_5$ brane configuration where all of them are on a single fixed point. We expect to be able to move them away, to other fixed points or in pairs in the bulk of T^4. Consider $2n_a$ D$_5$-branes at the ath fixed point. This number must be even so that (9.14.57) makes sense. Consider also n_x branes at point x and the same number at its image $-x$. The solution to the tadpole conditions gives a gauge group

$$U(16) \times \prod_{a=1}^{16} U(n_a) \prod_x Sp(2n_x), \quad \sum_{a=1}^{16} n_a + \sum_x n_x = 16, \tag{9.14.61}$$

where the U(16) factor originates from the 9-9 strings. There are two 9-9 hypermultiplets transforming in the **120** of U(16). There are also two hypermultiplets transforming in the antisymmetric representation for each U(n_a) group. There is one hypermultiplet in the $(\mathbf{16}, \mathbf{\bar{n}}_a)$ for each U(n_a) factor. There is one hypermultiplet in the antisymmetric representation plus a singlet for each symplectic factor. Finally, there is one hypermultiplet in the $(\mathbf{16},m_x)$ for each symplectic factor.

You are invited to derive this spectrum in exercise 9.34 by solving the tadpole conditions and implementing the projections. In exercise 9.35 on page 290 you are asked to give a field theory derivation of the same massless spectrum by Higgsing the U(16)×U(16) gauge symmetry.

9.15 D-branes at Orbifold Singularities

An important ingredient of the Standard Model of the fundamental interactions is the chirality of the particle spectrum. As already discussed in the case of the heterotic string in section 9.6 on page 237, to obtain a four-dimensional chiral spectrum the supersymmetry of the string vacuum should be at most $\mathcal{N} = 1_4$.

In orientifolds, as we will argue in section 9.17, matter is expected to arise from the open string sector, localized on D-branes. An attractive way to produce a chiral spectrum, is to place D-branes at an orbifold singularity as we will now show.

---

[13] The hypermultiplet being nonchiral, we do not need to distinguish a representation from its conjugate. In fact if a complex scalar transforms in the representation R, the second complex scalar transforms in the representation $\bar{R}$. The same applies to the two Weyl fermions of the hypermultiplet.

When a D-brane is placed transverse to an orbifold singularity, the orbifold projection acts directly on its world-volume spectrum. By an appropriate choice of projection, the spectrum will be chiral. This is to be contrasted with D-branes placed in a generic bulk point. Such D-branes, in order to be invariant under the orbifold projections, must have mirror copies placed in related points. The orbifold projection in this case gives a spectrum that is identical to one of the original D-brane copies. Therefore the spectrum is not chiral in this case, due to the effective extended supersymmetry that remains.

We will therefore analyze branes transverse to orbifold singularities. The orbifold action being local, we may ignore global issues when we discuss the invariant spectrum. Global issues will become important when we wish to implement tadpole cancellation.

We will examine orbifold fixed points whose local structure is $\mathbb{R}^6/\mathbb{Z}_N$ for some integer N. We will therefore consider D-branes transverse to a $\mathbb{R}^6/\mathbb{Z}_N$ singularity.

As we have seen in the previous section, the branes we simply obtain during orientifold compactifications of the type-I string are D_9- and D_5-branes. These can be dualized to D_3- and D_7-branes and it is in this incarnation that we will describe our brane configuration.

We will first consider n D_3-branes transverse to the $\mathbb{R}^6/\mathbb{Z}_N$ singularity. We split the ten-dimensional indices into the four-dimensional Minkowski ones denoted by $\mu, \nu, \ldots$ and the six internal ones that we package in three complex pairs and label as $k, l, \ldots$. The $\mathbb{Z}_N$ rotation acts on the internal $\mathbb{R}^6$. It equivalently acts on the SO(6) R-symmetry quantum numbers of the massless D-brane fields. The vectors A_μ transform in the singlet, the fermions in the spinor and the scalars in the vector.

Complexifying the scalars[14] in pairs, the $\mathbb{Z}_N$ rotation acts on them as

$$R_\theta = \text{diag}\left(e^{2\pi i b_1/N}, e^{-2\pi i b_1/N}, e^{2\pi i b_2/N}, e^{-2\pi i b_2/N}, e^{2\pi i b_3/N}, e^{-2\pi i b_3/N} \right), \tag{9.15.1}$$

with $b_i \in \mathbb{Z}_N$. In exercise 9.40 on page 290 you are asked to show that on the four-dimensional spinor representation of SO(6), the rotation acts as

$$S_\theta = \text{diag}\left(e^{2\pi i a_1/N}, e^{2\pi i a_2/N}, e^{2\pi i a_3/N}, e^{2\pi i a_4/N} \right), \tag{9.15.2}$$

with

$$a_1 = \frac{b_2 + b_3 - b_1}{2}, \quad a_2 = \frac{b_1 - b_2 + b_3}{2}, \quad a_3 = \frac{b_1 + b_2 - b_3}{2}, \quad a_4 = -\frac{b_1 + b_2 + b_3}{2}. \tag{9.15.3}$$

We can parametrize the action of the rotation on the CP indices without loss of generality using the matrices

$$\gamma_{3,\theta} = \text{diag}\left(\mathbf{1}_{n_0}, \theta \, \mathbf{1}_{n_1}, \ldots, \theta^{N-1} \mathbf{1}_{n_{N-1}} \right), \tag{9.15.4}$$

where $\theta = e^{2\pi i/N}$ is the generating $\mathbb{Z}_N$ rotation, $n = \sum_{i=0}^{N-1} n_i$ and $\mathbf{1}_n$ is the unit $n \times n$ matrix.

The orbifold action on the gauge boson state is

$$A_\mu \sim \psi_{-1/2}^\mu |0; \lambda\rangle \to \psi_{-1/2}^\mu |0; \gamma_{3,\theta}\, \lambda\, \gamma_{3,\theta}^{-1}\rangle, \tag{9.15.5}$$

where the matrix λ keeps track of the CP indices: $|0; \lambda\rangle \equiv \lambda^{ij}|0; ij\rangle$.

---

[14] These are in one-to-one correspondence with the six transverse coordinates of the D_3-branes.

Therefore, the gauge bosons must satisfy $\lambda = \gamma_{3,\theta} \, \lambda \, \gamma_{3,\theta}^{-1}$. The solutions to this equation are $n_i \times n_i$ block diagonal matrices: the invariant gauge bosons are in the adjoint of $\prod_{i=0}^{N-1} U(n_i)$.

The three complex scalars Φ_k obtained from the complexification of the six real scalars transform as

$$\Phi_k \sim \psi_{-1/2}^k |0; \lambda\rangle \to e^{-2\pi i b_k/N} \, \psi_{-1/2}^k |0; \gamma_{3,\theta} \, \lambda \, \gamma_{3,\theta}^{-1}\rangle. \tag{9.15.6}$$

The invariant scalars must therefore satisfy $\lambda = e^{2\pi i b_k/N} \gamma_{3,\theta} \, \lambda \, \gamma_{3,\theta}^{-1}$. In exercise 9.41 you are asked to solve this condition explicitly and show that the invariant scalars transform in the following representation of the gauge group

$$\text{scalars} \to \oplus_{k=1}^{3} \oplus_{i=0}^{N-1} \left(n_i, \bar{n}_{i-b_k} \right). \tag{9.15.7}$$

Finally the fermions are labeled as

$$\psi_a \sim |\lambda; s_1, s_2, s_3, s_4\rangle, \tag{9.15.8}$$

where $s_i = \pm\frac{1}{2}$ are spinorial indices, with $\sum_{i=1}^{4} s_i = \text{odd}$ (due to the GSO projection). The states with $s_4 = -\frac{1}{2}$ correspond to left-handed, four-dimensional Weyl fermions while $s_4 = \frac{1}{2}$ corresponds to right-handed, four-dimensional Weyl fermions. The $s_{1,2,3}$ spinor quantum numbers are R-symmetry spinor quantum numbers. We can thus label the 8 on-shell fermion states as $|\lambda; \alpha, s_4\rangle$ where $\alpha = 1, 2, 3, 4$ is the R-spinor quantum number. The fermions then transform as

$$|\lambda; \alpha, s_4\rangle \to e^{2\pi i a_\alpha/N} \, |\gamma_{3,\theta} \, \lambda \, \gamma_{3,\theta}^{-1}; \alpha, s_4\rangle, \tag{9.15.9}$$

and the invariant fermions must satisfy $\lambda = e^{2\pi i a_\alpha/N} \gamma_{3,\theta} \, \lambda \, \gamma_{3,\theta}^{-1}$. The solution to this equation gives left-handed Weyl fermions in the following representation of the gauge group:

$$\text{left-handed fermions} \to \oplus_{\alpha=1}^{4} \oplus_{i=0}^{N-1} \left(n_i, \bar{n}_{i+a_\alpha} \right), \tag{9.15.10}$$

a representation that is generically chiral.

When $\sum_{i=1}^{3} b_i = 0$ so that $a_4 = 0$ we have an $\mathcal{N} = 1_4$ supersymmetric configuration (the rotation $\in SU(3) \subset SO(6)$). The associated fixed point is known as an $\mathcal{N} = 1_4$ orbifold singularity. The a_4 fermions become the gaugini, while the $a_{1,2,3}$ fermions are the $\mathcal{N} = 1_4$ supersymmetric partners of the scalars.

We now add D$_7$-branes. They are in general needed to cancel the twisted tadpoles. There are three generically distinct ways of adding the D$_7$-branes. They may be transverse to the third plane (and therefore wrap the 1 and 2 complex internal dimensions). They could also be transverse to the first or second plane. We will discuss only the first case explicitly, leaving the other two cases to the reader as an exercise (9.42). We therefore place m D$_{7_3}$-branes that we take to be transverse to the last complex coordinate (the third plane).

For the 7_3-7_3 strings, the story is similar, with a new CP matrix parametrized as

$$\gamma_{7_3,\theta} = \begin{cases} \text{diag}\left(\mathbf{1}_{m_0}, \theta \, \mathbf{1}_{m_1}, \ldots, \theta^{N-1} \, \mathbf{1}_{m_{N-1}} \right), & b_3 \text{ even}, \\ \text{diag}\left(\theta \, \mathbf{1}_{m_0}, \theta^3 \, \mathbf{1}_{m_1}, \ldots, \theta^{2N-1} \, \mathbf{1}_{m_{N-1}} \right), & b_3 \text{ odd}. \end{cases} \tag{9.15.11}$$

The extra fields that are localized on the D_3-brane world-volume come from the 3-7_3 and 7_3-3 strings. For such strings, there are four DN directions which provide four zero modes in the NS sector (directions 4,5,6,7), while from the NN and DD directions we have zero modes in the R sector (directions 2,3,8,9).

The invariant (complex) scalars (NS sector) must satisfy

$$\lambda_{3-7_3} = e^{-i\pi(b_1+b_2)/N} \; \gamma_{3,\theta} \, \lambda_{3-7_3} \, \gamma_{7_3,\theta}^{-1}, \tag{9.15.12}$$

$$\lambda_{7_3-3} = e^{-i\pi(b_1+b_2)/N} \; \gamma_{7_3,\theta} \, \lambda_{7_3-3} \, \gamma_{3,\theta}^{-1}, \tag{9.15.13}$$

with spectrum

$$\oplus_{i=0}^{N-1} \left[\left(n_i, \bar{m}_{i-(b_1+b_2)/2}\right) + \left(m_i, \bar{n}_{i-(b_1+b_2)/2}\right) \right], \quad b_3 \text{ even},$$

$$\oplus_{i=0}^{N-1} \left[\left(n_i, \bar{m}_{i-(b_1+b_2+1)/2}\right) + \left(m_i, \bar{n}_{i-(b_1+b_2-1)/2}\right) \right], \quad b_3 \text{ odd}. \tag{9.15.14}$$

The invariant fermions coming from the R sector must satisfy

$$\lambda_{3-7_3} = e^{i\pi b_3/N} \; \gamma_{3,\theta} \, \lambda_{3-7_3} \, \gamma_{7_3,\theta}^{-1}, \tag{9.15.15}$$

$$\lambda_{7_3-3} = e^{i\pi b_3/N} \; \gamma_{7_3,\theta} \, \lambda_{7_3-3} \, \gamma_{3,\theta}^{-1}, \tag{9.15.16}$$

with spectrum

$$\oplus_{i=0}^{N-1} \left[\left(n_i, \bar{m}_{i+b_3/2}\right) + \left(m_i, \bar{n}_{i+b_3/2}\right) \right], \quad b_3 \text{ even},$$

$$\oplus_{i=0}^{N-1} \left[\left(n_i, \bar{m}_{i+(b_3-1)/2}\right) + \left(m_i, \bar{n}_{i+(b_3+1)/2}\right) \right], \quad b_3 \text{ odd}. \tag{9.15.17}$$

We observe that such brane configurations provide a generically chiral spectrum of four-dimensional fermions. Model building involves putting together such sets of branes on a compact orbifold so that the tadpoles are canceled. It turns out that several of the U(1) factors of the gauge group have triangle anomalies. These are canceled by a variation of the Green-Schwarz mechanism, which at the same time renders the U(1)'s massive.

9.16 Magnetized Compactifications and Intersecting Branes

So far we have seen how compactification on tori, combined with orbifold projections reduce the space-time supersymmetry, in our quest for realistic vacua of string theory.

In this section, we will describe another method of breaking supersymmetry during compactification. It involves turning on constant internal magnetic fields. Considering the internal manifold to be a torus this provides with vacua with reduced supersymmetry, where calculations can be performed.

In the case of closed string theory, turning on a constant internal magnetic field must be accompanied by a nontrivial deformation of the metric, in order to satisfy the classical equations of motion. Although such exact solutions exist, model building is complicated.

If the internal magnetic field originates in the open sector, the gravitational back-reaction appears in the next order of perturbations theory (at one loop). It is therefore easier to tune the appropriate brane configurations.

It turns out that T-duality changes magnetized branes into intersecting branes and vice versa. This gives an alternative (geometric) view of some important effects, like chirality generation in such compactifications.

In the sequel, we will analyze magnetized and intersecting branes in simple contexts in order to illustrate the important effects.

9.16.1 Open strings in an internal magnetic field

We will consider open strings compactified on T^6. We take for simplicity values for the T^6 moduli in order for the torus to have the factorized form $T^2 \times T^2 \times T^2$. We consider a D$_p$ brane wrapping one of the tori, in the x^4-x^5 plane. We turn on a constant magnetic field H, in the Cartan of the D-brane gauge group U(n):

$$A_4 = 0, \quad A_5 = Hx^4. \tag{9.16.1}$$

This is the magnetic monopole solution on T^2. The flux quantization condition implies that

$$n\,(2\pi \ell_s R_4)(2\pi \ell_s R_5)\,qH = 2\pi m \;\to\; (2\pi \ell_s^2)qH = \frac{1}{R_4 R_5}\frac{m}{n}, \quad m, n \in \mathbb{Z}, \tag{9.16.2}$$

where q is the minimum charge and m, n are relatively prime. We have assumed that the brane wraps the T^2 n times. We have also assumed that the T^2 is orthogonal with the two radii being $R_{4,5}$.

It is obvious from (9.16.2), that an internal magnetic field is not a continuous modulus of the compactification. It is inversely proportional to the volume and is characterized by a rational number m/n.

We now consider an open string with one (or both) end points on the D$_p$-brane under consideration. One or both end points will in general carry electric charges $q_{L,R}$ under the magnetic field. The charge that couples to the magnetic field is $q = q_L + q_R$.

Before quantizing this open string exactly, we would like to look at the modifications to the massless spectrum due to the magnetic field. The first obvious modification affects the momenta on T^2. They no longer commute, rather their commutator is proportional to the gauge field as in the Landau problem,

$$[p_4, p_5] = iqH. \tag{9.16.3}$$

For various fields on T^2, the modification to the mass formula has the form

$$\delta M^2 = \left(N + \frac{1}{2}\right)|2qH| + 2qH\Sigma_{45}, \quad N = 0, 1, 2, \ldots, \tag{9.16.4}$$

where N labels the Landau levels and Σ_{45} is the projection of the angular operator on the 45 plane. For fermions,[15] $\Sigma_{45} = \frac{i}{4}[\Gamma^4, \Gamma^5]$. The lowest level is degenerate as we will show below.

Consider a spin-1/2 state. The $\Sigma_{45} = 1/2$ component, has, according to (9.16.4) a lowest mass of $\delta M^2 = 2|qH|$ ($N = 0$) while the $\Sigma_{45} = -1/2$ component is massless

---

[15] For example, for spin 1/2, $\delta M^2 = (\Gamma^4 p_4 + \Gamma^5 p_5)^2$.

at the lowest Landau level. Therefore, we have massless chiral fermions ($\Sigma_{45} = -1/2$, $N = 0$) and at the first massive level an equal number of massive Dirac fermions ($\Sigma_{45} = -1/2$, $N = 1 \oplus \Sigma_{45} = 1/2$, $N = 0$). The generation of chirality can be understood from the index theorem, since the Dirac index is proportional to the integral of the magnetic field on the two-torus,

$$\text{Index}(\partial) = \frac{q}{2\pi} \int dx^4 dx^5 F_{45}. \tag{9.16.5}$$

Consider now an internal massless vector. The state with helicity on T^2, $\Sigma_{45} = 1$ is massive, while the one with helicity $\Sigma_{45} = -1$ has a mass at the lowest Landau level, $\Delta M^2 = -|qH|$ and is tachyonic. This is the well-known Nielsen-Olesen instability of field theory due to constant chromomagnetic fields.

The presence of the magnetic field breaks supersymmetry. This is obvious from the fact that the masses depend nontrivially on the spin component. This breaking is spontaneous since $\text{Str}[\delta M^2] = 0$.

Consider now independent magnetic fields H_I, $I = 1, 2, 3$, on each of the three T^2's. Then, scalars are all massive with lowest masses

$$\delta M_0^2 = \sum_{I=1}^{3} |q_I H_I|. \tag{9.16.6}$$

Fermions have a single massless chiral mode, with $\Sigma_{45} = \Sigma_{67} = \Sigma_{89} = -1/2$. All others are massive with minimum masses

$$\delta M_{1/2}^2 = 2|q_I H_I|, \quad 2(|q_I H_I| + |q_J H_J|), \quad 2\sum_{I=1}^{3} |q_I H_I|. \tag{9.16.7}$$

Note that chirality in four dimensions requires that all $q_I H_I$ are nonzero.

Finally, the vectors have minimal masses

$$\delta M_1^2 = |q_1 H_1| + |q_2 H_2| - |q_3 H_3|, \quad |q_1 H_1| - |q_2 H_2| + |q_3 H_3|, \quad -|q_1 H_1| + |q_2 H_2| + |q_3 H_3|. \tag{9.16.8}$$

Depending on the values of the magnetic fields, the masses in (9.16.8) maybe positive or tachyonic. In the second case, that may be used to trigger spontaneous symmetry breaking in the open sector. If one of the masses in (9.16.8) vanishes, some supersymmetry remains unbroken.

After this field-theoretic description of the effect of the internal magnetic fields on the massless sector, we now turn to a stringy description. We will describe the string quantization of the coordinates of the 4-5 plane, the others being similar. We start with the action for the X^4 and X^5 and the partner fermions

$$S = \frac{1}{4\pi \ell_s^2} \int d\tau \int_0^\pi d\sigma \left[\partial_\alpha X^I \partial^\alpha X^I - \frac{i}{2} \psi^I (\partial_\tau + \partial_\sigma) \psi^I - \frac{i}{2} \bar{\psi}^I (\partial_\tau - \partial_\sigma) \bar{\psi}^I \right]$$

$$+ q_L H_L \int d\tau \left[X^4 \partial_\tau X^5 - \frac{i}{4} (\psi^4 \psi^5 + \bar{\psi}^4 \bar{\psi}^5) \right] \Big|_{\sigma=0}$$

$$+ q_R H_R \int d\tau \left[X^4 \partial_\tau X^5 - \frac{i}{4} (\psi^4 \psi^5 + \bar{\psi}^4 \bar{\psi}^5) \right] \Big|_{\sigma=\pi}. \tag{9.16.9}$$

The boundary terms incorporate the presence of the magnetic field. We allowed for different magnetic fields at the two endpoints, since strings can start and end at different magnetized branes. We vary the action, being careful to keep the boundary terms in the σ direction. After integrations by parts, we obtain the usual bulk equations

$$\Box X^I = 0, \quad (\partial_\tau + \partial_\sigma)\psi^I = (\partial_\tau - \partial_\sigma)\bar{\psi}^I = 0, \tag{9.16.10}$$

together with the boundary conditions

$$\partial_\sigma X^4 - \beta_L \, \partial_\tau X^5\big|_{\sigma=0} = 0, \quad \partial_\sigma X^5 + \beta_L \, \partial_\tau X^4\big|_{\sigma=0} = 0, \tag{9.16.11}$$

$$\psi^4 - \bar{\psi}^4 + \beta_L(\psi^5 + \bar{\psi}^5)\big|_{\sigma=0} = 0, \quad \psi^5 - \bar{\psi}^5 - \beta_L(\psi^4 + \bar{\psi}^4)\big|_{\sigma=0} = 0, \tag{9.16.12}$$

$$\partial_\sigma X^4 + \beta_R \, \partial_\tau X^5\big|_{\sigma=\pi} = 0, \quad \partial_\sigma X^5 - \beta_R \, \partial_\tau X^4\big|_{\sigma=\pi} = 0, \tag{9.16.13}$$

$$\psi^4 - \beta_R\psi^5 + (-1)^a(\bar{\psi}^4 + \beta_R\bar{\psi}^5)\big|_{\sigma=\pi} = 0, \quad \psi^5 - \beta_R\psi^4 + (-1)^a(\bar{\psi}^5 - \beta_R\bar{\psi}^4)\big|_{\sigma=\pi} = 0, \tag{9.16.14}$$

where $a = 0$ for the NS sector and $a = 1$ for the R sector, in accordance with section 4.16.2 on page 88. We also defined

$$\beta_{L,R} \equiv 2\pi q_{L,R} \, H_{L,R} \, \ell_s^2. \tag{9.16.15}$$

Note that magnetic fields interpolate between Neumann and Dirichlet boundary conditions. For example, in the limit $\beta_L \to 0$, the $\sigma = 0$ end point has Neumann boundary conditions. In the opposite limit $\beta_L \to \infty$, the boundary conditions can be satisfied only when $\partial_\tau X^{4,5} = 0$, i.e., for Dirichlet boundary conditions.

Defining the complex coordinates

$$X_\pm = (X^4 \pm iX^5)/\sqrt{2}, \quad \psi_\pm = (\psi^4 \pm i\psi^5)/\sqrt{2}, \quad \bar{\psi}_\pm = (\bar{\psi}^4 \pm i\bar{\psi}^5)/\sqrt{2}, \tag{9.16.16}$$

we may rewrite the boundary conditions as

$$\partial_\sigma X_\pm \pm i\beta_L \, \partial_\tau X_\pm\big|_{\sigma=0} = 0, \quad \partial_\sigma X_\pm \mp i\beta_R \, \partial_\tau X_\pm\big|_{\sigma=\pi} = 0, \tag{9.16.17}$$

$$\left(\psi - \frac{1 + i\beta_L}{1 - i\beta_L}\bar{\psi}\right)\bigg|_{\sigma=0} = 0, \quad \left(\psi + (-1)^a\frac{1 - i\beta_R}{1 - i\beta_R}\bar{\psi}\right)\bigg|_{\sigma=\pi} = 0. \tag{9.16.18}$$

The boundary conditions are linear and are easily solved,

$$X_\pm = x^\pm + i\sqrt{2}\ell_s \sum_{n \in \mathbb{Z}} \frac{a_{n\mp\epsilon}^\pm}{n \mp \epsilon} e^{-i(n\mp\epsilon)\tau} \cos[(n \mp \epsilon)\sigma \pm \theta_L], \tag{9.16.19}$$

$$\psi_\pm = \sum_{\mathbb{Z}+\frac{1-a}{2}} b_{n\mp\epsilon}^\pm e^{i(n\mp\epsilon)(\tau-\sigma)\pm i\theta_L}, \quad \bar{\psi}_\pm = \sum_{\mathbb{Z}+\frac{1-a}{2}} b_{n\mp\epsilon}^\pm e^{i(n\mp\epsilon)(\tau+\sigma)\pm i\theta_L}. \tag{9.16.20}$$

We have set

$$\theta_{L,R} = \arctan(\beta_{L,R}), \quad \epsilon = \frac{1}{\pi}[\theta_L + \theta_R]. \tag{9.16.21}$$

The Hermiticity relations are

$$(a_{n+\epsilon}^-)^\dagger = a_{-n-\epsilon}^+, \quad (b_{n+\epsilon}^-)^\dagger = b_{-n-\epsilon}^+. \tag{9.16.22}$$

Note that the oscillator expansions are identical to those of the twisted sector of an orbifold, with twist angle $2\pi\epsilon$.

As in the orbifold case, $X_\pm$ carry no momentum and the oscillator frequencies are shifted from integer ones. Unlike the orbifold case, the phase here is continuous, and there is no summation over orbifold sectors.

The oscillator expansions must be supplemented by canonical commutations relations that as usual read

$$[a_{n-\epsilon}^+, a_{m+\epsilon}^-] = (n-\epsilon)\delta_{m+n}, \quad \{b_{n-\epsilon}^+, b_{m+\epsilon}^-\} = \delta_{m+n}. \tag{9.16.23}$$

The commutator of the zero modes, however, is a bit unusual. We will evaluate the equal-time commutator of the coordinates using the commutation relations in (9.16.23),

$$[X_+(\tau,\sigma), X_-(\tau,\sigma')] = [x^+, x^-] + 2\ell_s^2 \, J(\sigma, \sigma'), \tag{9.16.24}$$

$$J(\sigma, \sigma') = \sum_{n\in\mathbb{Z}} \frac{\cos\left[(n-\epsilon)\sigma + \theta_L\right]\cos\left[(n-\epsilon)\sigma' + \theta_L\right]}{n-\epsilon}. \tag{9.16.25}$$

This function has the property of being piecewise constant. This can be ascertained by evaluating $\partial_\sigma J$ and showing that apart from jumps at $\sigma = \sigma' = 0, \pi$, it is constant. Using

$$\sum_{n\in\mathbb{Z}} \frac{1}{n-\epsilon} = -\pi\cot(\pi\epsilon), \quad \sum_{n\in\mathbb{Z}} \frac{(-1)^n}{n-\epsilon} = -\frac{\pi}{\sin(\pi\epsilon)}, \tag{9.16.26}$$

we may evaluate

$$J(0,0) = \frac{\pi(\beta_L\beta_R - 1)}{(1+\beta_L^2)(\beta_L+\beta_R)}, \quad J(\pi,\pi) = \frac{\pi(\beta_L\beta_R - 1)}{(1+\beta_R^2)(\beta_L+\beta_R)},$$

$$J(0,\pi) = -\frac{\pi}{(\beta_L+\beta_R)}. \tag{9.16.27}$$

We must now impose that the commutator (9.16.24) vanishes, except at the end points. This fixes uniquely the zero-mode commutator to

$$[x^+, x^-] = \frac{2\pi\ell_s^2}{\beta_L+\beta_R}. \tag{9.16.28}$$

Moreover, at the end points the commutator (9.16.24) does not vanish. Rather,

$$[X_+(\tau,0), X_-(\tau,0)] = \frac{2\pi\ell_s^2 \,\beta_L}{1+\beta_L^2}, \quad [X_+(\tau,\pi), X_-(\tau,\pi)] = \frac{2\pi\ell_s^2 \,\beta_R}{1+\beta_R^2}. \tag{9.16.29}$$

We therefore observe that the end points of the string in the magnetic field, do not commute anymore. The associated effective theory can in fact be rewritten in terms of a noncommutative field theory, but we will not explore this further.

We may now discuss the spectrum. The vacuum is defined in analogy with orbifolds. We take without loss of generality $0 < \epsilon < \frac{1}{2}$. In the NS sector

$$a_{n-\epsilon}^+|0\rangle = 0, \quad n > 0, \quad a_{n+\epsilon}^-|0\rangle = 0, \quad n \geq 0, \tag{9.16.30}$$

$$b_{r-\epsilon}^+|0\rangle = 0, \quad r > 0, \quad b_{r+\epsilon}^-|0\rangle = 0, \quad r > 0. \tag{9.16.31}$$

In the R sector we have instead

$$b_{n-\epsilon}^+|0\rangle = 0, \quad n > 0, \quad b_{n+\epsilon}^-|0\rangle = 0, \quad n \geq 0. \tag{9.16.32}$$

The L_0 eigenvalue on the vacuum in the NS sector (associated with the two dimensions in question) is $\epsilon(1 - \epsilon)/2$ from the bosonic part and $\epsilon^2/2$ from the fermionic part. Therefore,

$$L_0|0\rangle_{NS,\epsilon} = \frac{\epsilon}{2} |0\rangle_{NS,\epsilon}. \tag{9.16.33}$$

On the other hand, the magnetic contributions cancel in the R sector ground state.

We observe that $a_{-n-\epsilon}^+$ and $b_{-r-\epsilon}^+$ raise the helicity on the plane by 1, and shift the L_0 eigenvalue by ϵ, while $a_{-n+\epsilon}^-, b_{-r-\epsilon}^-$ lower the helicity eigenvalue by 1, and shift the L_0 eigenvalue by $-\epsilon$. The operator $a_{-\epsilon}^+$ in particular creates the Landau states upon multiple action on the ground state. We may therefore write the generic state as

$$|\psi\rangle = \prod_{i=0} (a_{-i-\epsilon}^+)^{N_i} \prod_{j=1} (a_{-j+\epsilon}^-)^{\bar{N}_i} \prod_{i=0} (b_{-i-1/2-\epsilon}^+)^{n_i} \prod_{j=0} (b_{-j-1/2+\epsilon}^-)^{\bar{n}_i}|0\rangle_\epsilon, \tag{9.16.34}$$

with L_0 eigenvalue

$$L_0|\psi\rangle = L_0|\psi\rangle_{\epsilon=0} + \left(N_0 + \frac{1}{2}\right)\epsilon + \Sigma_{45}\epsilon, \tag{9.16.35}$$

and

$$\Sigma_{45} = \sum_{i=1}^{\infty} (N_i - \bar{N}_i + n_i - \bar{n}_i), \tag{9.16.36}$$

as expected.

An important question is the multiplicity of the ground state. This is determined by the commutation relations of the coordinate zero modes in (9.16.28). We define normalized angular coordinates as $X^i = (2\pi \ell_s R_i)\theta^i$, so that when going once around the circles we have $\theta^i \to \theta^i + 1$. We now translate (9.16.28) to

$$[\theta^4, \theta^5] = -\frac{i}{2\pi} \frac{n_L n_R}{m_L n_R + m_R n_L}, \tag{9.16.37}$$

using (9.16.2). Due to the periodicity on T^2, the appropriate operators are $e^{2\pi i\theta^{4,5}}$ instead of $\theta^{4,5}$. We will have to treat one of them as momentum and the other as a coordinate. Then we can generate states, by acting with one set of operators on the vacuum. Since both $\theta^{4,5}$ commute with all other oscillators, they will provide with an overall degeneracy of the vacuum. Using (9.16.37) we may compute

$$e^{2\pi I_{LR}i\theta^4} e^{2\pi i\theta^5} e^{-2\pi i I_{LR}\theta^4} = e^{2\pi i\theta^5}, \quad e^{2\pi I_{LR}i\theta^5} e^{2\pi i\theta^4} e^{-2\pi i I_{LR}\theta^5} = e^{2\pi i\theta^4}, \tag{9.16.38}$$

with

$$I_{LR} = m_L n_R + m_R n_L. \tag{9.16.39}$$

It is therefore obvious that the operators $e^{2\pi ik\theta^4}$, $k = 0, 1, \ldots, I_{LR} - 1$, generate independent states characterized by distinct eigenvalues of the "momentum" operator $e^{2\pi i\theta^5}$. We conclude that we must have I_{LR} ground states.

Putting everything together we obtain the shift of the low-lying string energy levels

$$\delta \mathcal{M}^2_{\text{string}} \sim (2n+1)|\epsilon| + 2\epsilon \Sigma_{45}. \tag{9.16.40}$$

For weak magnetic fields this agrees with the field theory expectation (9.16.4).

It is useful to compute the string partition sum in the presence of a magnetic field on a plane. For the two bosonic coordinates, taking into account the frequency shifts,

$$\text{Tr}[e^{-2\pi t(L_0 - c/24)}] = \frac{q^{\epsilon(1-\epsilon)/2 - 1/12}}{(1 - q^\epsilon)\prod_{n=1}^{\infty}(1 - q^{n+\epsilon})(1 - q^{n-\epsilon})} = i\frac{q^{-\epsilon^2/2 + 1/24}\prod_{n=1}^{\infty}(1 - q^n)}{\vartheta_1(it\epsilon|it)}, \tag{9.16.41}$$

with $q = e^{-2\pi t}$. The analogous trace on the fermions in the NS sector is

$$q^{\epsilon^2/2 - 1/24}\prod_{n=0}^{\infty}(1 + q^{n+1/2+\epsilon})(1 + q^{n+1/2-\epsilon}) = q^{-\epsilon^2/2 + 1/24}\frac{\vartheta_3(it\epsilon|it)}{\prod_{n=1}^{\infty}(1 - q^n)}. \tag{9.16.42}$$

Putting together fermions and bosons we finally have

$$\text{Tr}[e^{-2\pi t(L_0 - c/24)}]_{\text{NS}} = iI_{LR}\frac{\vartheta_3(it\epsilon|it)}{\vartheta_1(it\epsilon|it)}, \tag{9.16.43}$$

where we also included the degeneracy of the ground state. It is a straightforward exercise to derive the other relevant magnetized partition functions

$$\text{Tr}[(-1)^F e^{-2\pi t(L_0 - c/24)}]_{\text{NS}} = iI_{LR}\frac{\vartheta_4(it\epsilon|it)}{\vartheta_1(it\epsilon|it)}, \tag{9.16.44}$$

$$\text{Tr}[e^{-2\pi t(L_0 - c/24)}]_{R} = iI_{LR}\frac{\vartheta_2(it\epsilon|it)}{\vartheta_1(it\epsilon|it)}, \quad \text{Tr}[(-1)^F e^{-2\pi t(L_0 - c/24)}]_{R} = I_{LR}. \tag{9.16.45}$$

In particular, the last trace is nothing else than the Witten index which counts the number of ground states of the system.

9.16.2 Intersecting branes

We may now proceed to apply a T-duality transformation to one of the two T^2 coordinates of the previous section. For concreteness we will T-dualize along the X^5 coordinate. Apart from $R_5 \to 1/R_5$, the boundary conditions (9.16.11)–(9.16.14) will change, via $\partial_\sigma X^5 \leftrightarrow \partial_\tau X^5$. The new boundary conditions on the coordinates are

$$\partial_\sigma(X^4 - \beta_L X^5)\big|_{\sigma=0} = 0, \quad \partial_\tau(X^5 + \beta_L X^4)\big|_{\sigma=0} = 0, \tag{9.16.46}$$

$$\partial_\sigma(X^4 + \beta_R X^5)\big|_{\sigma=\pi} = 0, \quad \partial_\tau(X^5 - \beta_R X^4)\big|_{\sigma=\pi} = 0. \tag{9.16.47}$$

We now define rotated coordinates

$$\begin{pmatrix} X^4_{L,R} \\ X^5_{L,R} \end{pmatrix} = \begin{pmatrix} \cos\theta_{L,R} & \mp\sin(\theta_{L,R}) \\ \pm\sin(\theta_{L,R}) & \cos(\theta_{L,R}) \end{pmatrix} \begin{pmatrix} X^4 \\ X^5 \end{pmatrix}, \tag{9.16.48}$$

where the angles $\theta_{L,R}$ were defined in (9.16.21). We may now reinterpret the boundary conditions (9.16.46), (9.16.47). Let us call the branes on which the L/R end points of the open string end, the L/R-branes. Both have now only one dimension wrapping the two-torus. This is expected from the standard action of T-duality on D-branes.

The L-brane has a Neumann boundary condition along X_L^4 and a Dirichlet boundary condition along X_L^5. It is therefore rotated at an angle $-\theta_L$ with respect to the X^4 axis. On the other hand the R-brane is rotated at an angle θ_R with respect to the X^4 axis.

We are therefore describing a string stretching between two intersecting branes at an angle $\theta_L + \theta_R = \pi\epsilon$. The branes intersect at a point[16] on the X^4-X^5 plane, but the intersection may also stretch in other dimensions.

The magnetic flux quantization condition (9.16.2) becomes in the T-dual version

$$(2\pi\ell_s^2)q_{L,R}H_{L,R} = \frac{R_5}{R_4}\frac{m_{L,R}}{n_{L,R}} \rightarrow \tan\theta_{L,R} = \frac{R_5}{R_4}\frac{m_{L,R}}{n_{L,R}}. \tag{9.16.49}$$

The interpretation of (9.16.49) is that the L brane is winding around the two-torus by wrapping m_L times the x^5 cycle and n_L times the x^4 cycle. The R brane is wrapping $-m_R$ times the x^5 cycle and n_R times the x^4 cycle. $m_{L,R}$ and $n_{L,R}$ are therefore wrapping numbers of the branes on T^2. Moreover,

$$I_{LR} = m_L n_R + m_R n_L \tag{9.16.50}$$

is the (oriented) intersection number of the two branes on the T^2. It is satisfying that the number of ground states of the generalized Landau problem of the last section, namely I_{LR}, is the same as the number of geometrical brane intersections. The reason is that in the T-dual picture of intersecting branes we expect precisely this number of ground states. A string stretched between two intersecting branes will classically minimize its length (and energy) by sitting at an intersection. Upon quantization, the number of ground states of the string coordinates is equal to the number of intersections.

9.16.3 Intersecting D_6-branes

The simplest configuration of this setup involves an original system of D_9-branes on T^6. This is the type-I string on T^6. For simplicity we may take a factorizable torus $T^6 = \prod_{i=1}^{3}(T^2)_i$. We may turn on different magnetic fields H_I^i on different D_9-branes, labeled by I on the three T^2 labeled by $i = 1, 2, 3$. If we T-dualize one coordinate from each T^2 we end up with intersecting D_6-branes on $(T^2)^3$. In the T-dual picture, each brane is now characterized by three angles $\theta_I^i = \arctan(H_I^i)$. They are rotated by θ_I^i in each T^2 with respect to the standard axes. The angles are related to the two winding numbers per torus (n_I^i, m_I^i), and to the complex structure U_2^i as $\tan\theta_I^i = \frac{m_I^i}{n_I^i U_2^i}$.

In the following we shall swing back and forth between the magnetized and intersecting picture. The reason is that some features are easier to discern in one picture and others in the T-dual one.

After T-duality on each of the three tori, the Ω projection transforms to an $\Omega\mathcal{I}^3$ projection according to (8.7.4) on page 201, where the inversions act on three of the six torus coordinates. They can be thought of as an antiholomorphic involution on the three complex torus

---

[16] Because of the torus periodicity, there can be several intersection points.

coordinates $z_i \rightarrow \bar{z}_i$. Thus, the orientifold of the IIA string by $\Omega \mathcal{I}^3$ has an open sector that contains intersecting D_6-branes with winding numbers (n_I^i, m_I^i). In particular, for each brane a, its image under $\Omega \mathcal{I}^3$ is another brane a' with winding numbers $(n_I^i, -m_I^i)$. As mentioned earlier and advocated in exercise 9.49 on page 291, such generic configurations break supersymmetry completely.

Let us denote the number of the Ith brane by N_I. In exercise 9.58 you are asked to derive the R tadpole conditions and show that they are given by

$$\sum_I N_I n_I^1 n_I^2 n_I^3 = 16, \quad \sum_I N_I n_I^i m_I^j m_I^k = 0, \quad i \neq j \neq k \neq i. \tag{9.16.51}$$

A compact way of writing the tadpole conditions above is

$$\sum_I N_I \, \Pi_I = \Pi_{O_6}, \tag{9.16.52}$$

where Π_I is the homology cycle of the Ith brane and Π_{O_6} are the homology cycles of the Orientifold planes. They are the T-duals of the well-known ten-dimensional O_9 plane. Another way to rephrase the tadpole conditions in (9.16.51) in the magnetized picture is as follows: the first condition is the usual cancellation of the D_9 charge. The other three are the cancelations of the induced D_5 charges (see exercise 9.57) transverse to the three possible T^2s.

The orientifold projection $\Omega \mathcal{I}^3$ maps a generic brane a to its image a' which is spatially distinct. Therefore, for a generic brane, the group is expected to be $U(N_I)$. There are however two special cases. The first is a brane aligned with the axes (no original magnetic fields). This is equivalent to an unmagnetized D_9-brane and the orientifold projection is expected to give an $SO(N_I)$ group. The other extreme is a brane, where two of the original magnetic fields are infinite. As discussed in section 9.16.1, an infinite magnetic field imposes a Dirichlet boundary condition. Therefore, such branes correspond to D_5-branes. And for D_5-branes we have argued already that they must have an opposite projection compared to that of the D_9-branes (see (9.16.4)). Therefore branes equivalent to D_5-branes will have an $Sp(N_I)$ gauge group.

We will now consider a generic configuration of intersecting branes giving rise to unitary groups only, and describe the massless spectrum. We will assume for simplicity that all branes intersect pairwise non-trivially. Some general properties of the low-lying spectrum were already detailed in an earlier section:

- Strings starting and ending on the same brane do not feel the magnetic fields since they are neutral ($q_L H_L = -q_R H_R$). Here we have the full $\mathcal{N} = 4_4$ supersymmetry.

- Strings stretching between intersecting branes will have generically massive scalars and vectors and some of the fermions as explained in section 9.16.1. However, there will be a number of massless chiral four-dimensional fermions. This number is equal to the oriented intersection number of the branes.

Let us first consider strings that start at a set of branes I and their image I'.

- $\Omega \mathcal{I}^3$ maps II strings to $I'I'$ strings. Therefore we can take one set as the independent one. The massless states generate the $\mathcal{N} = 4_4$ $U(N_I)$ Yang-Mills theory.

- II' strings are mapped by $\Omega\mathcal{I}^3$ to themselves. Therefore here we obtain symmetric and antisymmetric representations. The intersection number $I_{II'} = 8\prod_{i=1}^{3} m_I^i n_I^i$ is generically nonzero so we obtain only massless chiral fermions here. You are invited in exercise 9.59 on page 292 to show that we obtain $8m_I^1 m_I^2 m_I^3$ fermions in the $\square$ representation of U(N_I) as well as $4m_I^1 m_I^2 m_I^3 (n_I^1 n_I^2 n_I^3 - 1)$ fermions in the $\boxminus$ and $\square\square$ representations.

Consider now strings stretching between different stacks:

- The sector IJ is mapped by $\Omega\mathcal{I}^3$ to I'J'. We obtain fermions in the bifundamental $(N_I, \overline{N}_J)$ with multiplicity

$$I_{IJ} = \prod_{i=1}^{3} (m_I^i n_J^i - m_J^i n_I^i). \tag{9.16.53}$$

- The sector IJ' is mapped by $\Omega\mathcal{I}^3$ to I'J. We obtain fermions in the bifundamental (N_I, N_J) with multiplicity

$$I_{IJ} = -\prod_{i=1}^{3} (m_I^i n_J^i + m_J^i n_I^i), \tag{9.16.54}$$

where the minus sign as usual implies opposite chirality.

The spectra thus obtained can be engineered to reproduce the chiral SM spectrum. You are invited to explore this in exercise 9.60.

9.17 Where is the Standard Model?

Different classes of string vacua have distinct ways of realizing the gauge interactions that could be responsible for the SM forces. Ten-dimensional gravity is always an ingredient, coming from the closed string sector. The simplest way to convert it to four dimensional gravity is via compactification and this is what we will assume here. In section 13.13 we will describe another way of turning higher-dimensional gravity to four-dimensional, but the implementation of this idea in string theory is still in its infancy.

From (8.4.8) on page 195, upon compactification to four dimensions on a six-dimensional manifold of volume $(2\pi\ell_s)^6 V_6$, the four-dimensional Planck scale is given at tree level by

$$M_P^2 = \frac{V_6}{2\pi g_s^2} M_s^2 = \frac{M_s^2}{g_u^2}, \tag{9.17.1}$$

where g_s is the string coupling constant and the volume V_6 is by definition dimensionless. We have implicitly defined also g_u, the effective four-dimensional string coupling constant.

9.17.1 The heterotic string

The ten-dimensional theory, apart from the gravitational supermultiplet, contains also a (super-)Yang-Mills (SYM) sector with gauge group $E_8 \times E_8$ or SO(32).

Here, the four-dimensional gauge fields descend directly from ten dimensions. The gauge field states are

$$|A_\mu^a\rangle = b_{-1/2}^\mu \bar{J}_{-1}^a |p\rangle, \qquad (9.17.2)$$

where b_r^μ are the modes of the left-moving world-sheet fermions and their vertex operators are given in (10.1.2) on page 296. The four-dimensional action and gauge coupling constants are given by

$$S_4 = -\frac{1}{4g_I^2}\text{Tr}[F_I F_I], \qquad \frac{1}{g_I^2} = \frac{V_6}{4\pi g_s^2}k_I, \qquad (9.17.3)$$

where the trace is in the fundamental representation and k_I is the level[17] of the associated affine algebra (see section 4.11 on page 69). You are invited to derive this relation in exercise 9.64 on page 292. Although vectors can also come from the metric, they cannot provide chirality [220]. Therefore, the essential part of the SM must come from the vectors arising from the non-supersymmetric side.

Tree-level relations like (9.17.1) or (9.17.3) are corrected in perturbation theory and the couplings run with energy. We will see this phenomenon in more detail in the next section. The tree-level couplings correspond to their values at the string (unification) scale up to some threshold corrections coming from integrating out the stringy modes. In a stable and reliable perturbative expansion, such corrections are small. There may be also corrections from KK modes. These can become important only if the KK masses are very light compared to M_s. This is typically not the case in the heterotic string.

Therefore, the order of magnitude estimates of couplings at the string scale are expected to be reliable. In order to comply with experimental data, $g_{YM} \sim \mathcal{O}(1)$ and (9.17.1), (9.17.3) imply that

$$M_P^2 = \frac{M_s^2}{g_I^2}\frac{2}{k_I}. \qquad (9.17.4)$$

Typically $k_I = 1$ for almost all semirealistic heterotic vacua. Also the values of the observable coupling constants are in the 1–10^{-2} range. We deduce from (9.17.4) that the string scale and the Planck scale have the same order of magnitude. This is an interesting prediction, valid for all realistic perturbative heterotic string vacua.

The issue of supersymmetry breaking is of crucial importance in order to eventually make contact with the low-energy dynamics of the Standard Model.

There are two alternatives here, gaugino condensation (dynamical) and Scherk-Schwarz (geometrical) supersymmetry breaking described in section 9.5 on page 235.

The first possibility can be implemented in the heterotic string. However, it involves nonperturbative dynamics and consequently is not well controlled in perturbation theory. We do not know how to describe this dynamics at the string level.

If supersymmetry is broken à la Scherk-Schwarz, then the supersymmetry breaking scale is related to the size R of an internal compact dimension as

$$M_{\text{SUSY}} \sim \frac{1}{R}. \qquad (9.17.5)$$

---

[17] For abelian groups, one must first normalize the charges in order to determine the level.

A successful resolution of the hierarchy problem requires that $M_{\text{SUSY}} \sim$ a few TeV so that $M_{\text{SUSY}}/M_P \ll 1$. This implies, $R \gg M_s^{-1}$ and from (9.17.3) $g_s \gg 1$ in order to keep $g_I \sim \mathcal{O}(1)$. Thus, we are pushed in the non-perturbative regime. In chapter 11 we will find out how to handle such strong couplings regions, and therefore open new model-building possibilities.

9.17.2 Type-II string theory

The perturbative type-II string is very restrictive when it comes to nonabelian gauge groups combined with chirality.

Gauge fields may come both from the NS-NS and R-R sectors. R-R sector gauge fields generate abelian gauge groups in perturbation theory. The reason is that, as we have argued in section 7.2.1 on page 159, they cannot have minimal couplings to any perturbative state. Therefore, no perturbative string state is charged under them. To put it mildly, they are phenomenologically worthless.[18]

We will not prove in detail here why it is impossible to embed the SM spectrum in the perturbative type-II string. We will give instead the basic hints why this is so. The curious and enterprizing reader is guided to exercise 9.63 on page 292.

- To construct a compactification to four flat dimensions, the internal SCFT must have $(c_L, c_R) = (9, 9)$ and $\mathcal{N} = (1, 1)$ supersymmetry on the world-sheet.

- Gauge groups in space-time are in one-to-one correspondence with right-moving or left-moving (super)current algebras of $\text{SCFT}_{\text{internal}}$.

- If there is a nonabelian left-moving current algebra, then the R_L-NS_R sector contains only massive fermions. An immediate corollary is that all R_L-R_R bosons are also massive. Similarly, with $L \leftrightarrow R$.

- If there is an abelian left-moving current algebra, then R_L-NS_R fermions are neutral with respect to it. Worse, in such a case the R_L-NS_R fermions are nonchiral.

- The upshot of the previous statements is that all the SM gauge symmetry must come from one side of the type-II string, say the left. Moreover, the vectors will come from the NS-NS sector. All the massless fermions of the SM model will then arise from the NS_L-R_R sector.

  Together with the constraint on the central charge, this substantially limits the possible gauge groups that can appear. They are $SU(2)^6$, $SU(4) \times SU(2)$, $SO(5) \times SU(3)$, $SO(5) \times SU(2) \times SU(2)$, $SU(3) \times SU(3)$, G_2, and their subgroups. So far the gauge group of the standard model is possible.

- If we further require that massless fermionic states transform in the representations of the SM, then it turns out that it is not possible to fit them with the allowed internal central charge (9,9).

---

[18] This statement ceases to be true, beyond perturbation theory.

Therefore, to embed the standard model in the type-II string we must go beyond perturbation theory. This turns out to be possible [221]. However, it is difficult in this case to do detailed calculations.

9.17.3 The type-I string

In the type-I vacua, gauge symmetries can arise from D_p-branes that stretch along the four Minkowski directions and wrap their extra $p - 3$ dimensions in a submanifold of the compactification manifold. Let us denote by $V_{\parallel}$ the volume of such a submanifold in string units.

The relation of the four-dimensional Planck scale to the string scale is the same as in (9.17.1) since gravity originates in the closed string sector. However, the four-dimensional YM coupling of the D-brane gauge fields now become[19]

$$\frac{1}{g_{YM}^2} = \frac{V_{\parallel}}{2\sqrt{2}\pi g_s}, \quad \frac{M_P^2}{M_s^2} = \frac{V_6}{\sqrt{2}g_s} \frac{1}{V_{\parallel}} \frac{1}{g_{YM}^2}, \tag{9.17.6}$$

where M_P is the four-dimensional Planck scale. M_s can be much smaller than M_P while keeping the theory perturbative, $g_s < 1$, by having the volume of the space transverse to the D_p-branes $\frac{V_6}{V_{\parallel}} \gg 1$. Therefore, in this context, the string scale M_s can be anywhere between the four-dimensional Planck scale and a few TeV without obvious experimental contradictions. The possibility of perturbative string model building with a very low string scale is intriguing and interesting for several reasons:

- If M_s is a few TeV, string effects will be visible at TeV-scale experiments in the near future. If nature turns out to work that way, the experimental signals will be forthcoming. In the other extreme case $M_s \sim M_P$, there seems to be little chance to see telltale signals of the string in TeV-scale experiments.

- Supersymmetry can be broken directly at the string scale without the need for fancy supersymmetry-breaking mechanisms (for example by direct orbifolding). Past the string scale, there is no hierarchy problem since there is no field-theoretic running of couplings.

The possibility of having the string scale and the supersymmetry-breaking scale in the TeV range renders the gauge hierarchy problem nonexistent. However, the realization of such vacua is difficult, since, as we have seen earlier, they require the presence of large internal volumes. Once supersymmetry is broken, the volume moduli will acquire potentials. The novel hierarchy problem is that such minima for the volume will be required to give $V_6 \ggg 1$. Although there are ideas in this direction, no fully successful vacuum is known yet.

---

[19] The origin of the factor of $\sqrt{2}$ can be found in exercise 11.23 on page 366.

9.18 Unification

The first attempt to unify the fundamental interactions beyond the SM employed the embedding of the SM group into a simple unified group. This provided tree-level relations between the different SM coupling constants of the form

$$\frac{1}{g_I^2} = \frac{k_I}{g_U^2}, \quad I = SU(3), SU(2), U(1)_Y, \tag{9.18.1}$$

where g_U is the unique coupling constant of the unified gauge group. The k_I are group-theoretic rational numbers, that depend on the way the SM gauge group is embedded in the unified gauge group.[20] All couplings are evaluated at the scale M_{GUT} where the unified gauge group is expected to break to the SM gauge group.

In string theory, unification in its general sense is a fact: the theory has no free parameters, but expectation values that should be determined in a given ground state by minimizing the appropriate potential. This picture, often fails in string theory, when some scalars, the moduli, have no potential. However this is a characteristic of supersymmetric vacua. In nonsupersymmetric vacua, all moduli are expected to have a potential, and baring accidents, they determine, among other things, the gauge coupling constants.

Remarkably, the (measured) gauge couplings constants of the SM, when extrapolated to high energy using the (supersymmetric) renormalization group, they seem to satisfy the relation (9.18.1) at an energy $M_{GUT} \sim 10^{16.1 \pm 0.3}$ GeV, with $1/\alpha_{GUT} \equiv (4\pi)/g_U^2 \simeq 25$. This is pictured in figure 9.1. The matching is not as good if the nonsupersymmetric running is used.

Of course, some assumptions must be made, in order to make such an extrapolation. The first, is that the only particles that contribute are those of the minimal supersymmetric standard model.[21] The second is that no important thresholds are met, until $E \sim M_{GUT}$.

What should we conclude from such an observation? Certainly, it is not a proof for the existence of a relation of the type (9.18.1). It is however, an intriguing piece of evidence that we cannot immediately discard.

In this section we would like to investigate what kind of gauge coupling relations we obtain in the two most promising sets of string theory vacua: heterotic and type I.

In the heterotic string theory, as we have seen, the four-dimensional gauge groups descend from the right-moving nonsupersymmetric sector. They are associated with a respective right-moving current algebra.

As we have shown in (9.17.3) a formula similar to (9.18.1) holds for the gauge couplings in the heterotic string at tree level with $g_U^2 = 4\pi g_s^2/V_6$. This type of coupling unification occurs naturally in the heterotic string.

Let us now consider the gauge couplings in type-I string theory. As already explained in section 9.17.3, we may consider the Ith gauge group factor coming from a D_{p+3}-brane.

---

[20] For the SU(5)-like embeddings, $k_{SU(3)} = k_{SU(2)} = 1$, $k_{U(1)_Y} = 5/3$.
[21] Groups of particles whose presence does not affect the running of the ratio of couplings, could be allowed.

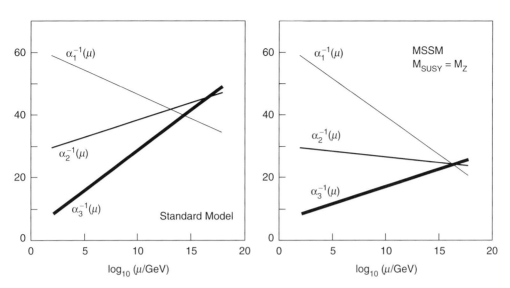

Figure 9.1 The running of gauge couplings $\alpha_i = g_i^2/(4\pi)$ in the SM and the supersymmetric standard model with supersymmetry breaking scale at M_Z. $\alpha_1 = \frac{5}{3}\alpha_Y$ where α_Y is the canonically normalized hypercharge coupling. The width of the lines is proportional to the respective experimental error.

It stretches along the four Minkowski dimensions. It also wraps $k_I \in \mathbb{Z}$ times a p-dimensional cycle of the compact six-dimensional manifold with volume $V_\parallel$ in string units. The associated tree-level gauge coupling is

$$\frac{1}{g_I^2} = k_I \frac{V_\parallel}{2\sqrt{2\pi}\,g_s}. \tag{9.18.2}$$

It is therefore obvious that, for gauge group factors originating from the same type of D_{p+3}-brane, we have a similar relation to (9.18.1) but the interpretation of k_I is different here. It is an (integer) wrapping number. When gauge group factors originate on different branes, then (9.18.1) ceases to hold.

The presence of other background fields, may change the relation (9.18.2) already at the tree level. For example, if twisted bulk moduli in orbifolds have nonzero expectation values, there are generic additive corrections to (9.18.2). Internal magnetic fields also alter (9.18.2). You are invited to investigate this in exercise 9.65.

We conclude that the "unification relation" (9.18.2) is not generic in type-I vacua. It will hold only if the SM gauge group originates from branes in the same stack.

An intermediate situation may arise in this case. It is known as "petite unification." It is the statement, that relation (9.18.2) is valid for a subset of the SM group factors. For example, $SU(3)$ and $U(1)_Y$ may originate from a D_9 stack of branes while $SU(2)$ from a D_5 stack.

We will revisit relation (9.18.2) and the associated one-loop corrections at the end of the next chapter.

Bibliography

A nice and comprehensive review of KK compactifications in supergravity and related issues can be found in [222].

Our discussion on the connection between space-time and world-sheet supersymmetry is based on [223]. In the same reference the absence of continuous global symmetries in string vacua was argued. Other general phenomenological issues in heterotic vacua are discussed in [224].

Basic references on the orbifold idea were given in the bibliography of chapter 4. Here we will provide a further guide towards the building of realistic string vacua. A large class of vacua has been constructed by utilizing free fermionic blocks in order to construct the CFT representing the internal six-dimensional compact part [225,226]. A similar construction using bosonic generalized lattices is reviewed in [73]. Using such algorithmic constructions, a partial computerized scanning gave several interesting heterotic models. The two most successful ones are [227,228]. A good review of heterotic orbifold model building in the eighties is [229]. [230] provides another review where more attention is paid to generic phenomenological properties of heterotic vacua and the structure of the relevant moduli interactions. A more extensive review for the late nineties is [231]. The PhD thesis [232] is also a detailed source of heterotic orbifold vacua.

In [233] a pedestrian description of low-energy theories relevant for string phenomenology is given. Another review that also includes higher (affine) level string model building is [234].

Coordinate-dependent compactifications were introduced in the context of field theory in [235]. They were implemented in closed string theory, to generate spontaneous supersymmetry breaking in [236,237]. The relationship between spontaneous supersymmetry breaking and freely acting orbifolds was detailed in [238]. For the open string version see [239].

Geometric compactifications of the heterotic string to four dimensions with $\mathcal{N} = 1_4$ supersymmetry are discussed in [240]. The complex geometry and CY manifold are described extensively in GSW [7] and the reviews [241,242]. A more extensive and higher level exposition is given in the book, [243]. Detailed information on the Eguchi-Hanson space and other hyper-Kähler manifolds can be found in [244]. A detailed discussion of the geometry and topology of the K3 manifold can be found in [245].

The review [68] contains a very good survey of both the geometry and the quantum geometry of CY manifolds. In particular it contains a nice description of $\mathcal{N} = (2, 2)_2$ CFTs and their diverse descriptions, mirror manifolds and mirror symmetry, examples of space-time topology change and the physics of conifold transitions, that we will also describe in section 11.10. Moreover, it contains a good introduction in toric geometry.

Mirror symmetry has been interpreted as T-duality in [246]. A more complete and rigorous discussion can be found in the AMS book [247] as well as in several good reviews [248,249,68].

Orientifolds were first described in [114–119]. D-branes in orbifolds and the related quiver theories were discussed in [250]. We follow here the Hilbert space notation of [251,252] on orientifolds. A very extensive and informative review can be found in [97]. This is also a good source for references in this direction. A comprehensive description of $\mathcal{N} = 1_4$ orientifolds of standard orbifolds can be found in [253]. Our general description of D-branes at singularities follows [254].

Magnetic fields in string theory have been discussed in [167,168] where the first derivation of the DBI action was given. Magnetic fields were used to break supersymmetry and generate chirality in the context of string theory in [255]. Further discussions of magnetized compactifications/intersecting branes and SM constructions can be found in [256–260].

A review that summarizes general features of D-brane model building in terms of branes at singularities and intersecting branes is [261]. Applications of D-branes to cosmology are also discussed. Concrete model building using intersecting branes is reviewed in [262,263]. A general overview can be found in [264].

Noncommutative aspects of magnetic fields in string theory and field theory can be found in [265]. A comprehensive review with a guide to the literature can be found in [266]. A detailed discussion of the commutators of the string coordinates in a magnetic field is in [267].

A general discussion of the gauge symmetries coming from the supersymmetric side of heterotic strings as well as type-II strings and the associated constraints on the perturbative spectrum are presented in [220].

Discussions of the large extra dimensions, the decompactification problem and suggestions on how it can be avoided, can be found in [268–278]. Reviews of the string theory related developments can be found in [279,280].

Unification in field theory is reviewed in [281]. The review of [282] provides an extensive discussion of gauge coupling unification in the heterotic string as well as several other phenomenological questions. The review of [230] contains among other things, a description of nonperturbative supersymmetry breaking due to gaugino condensation.

We have not addressed here the compactifications with nontrivial fluxes and the stabilisation of moduli. A review which is a good starting point is [283] that also contains a good guide to the literature. Warped compactifications with fluxes have been discussed in [284]. De Sitter spaces in fluxed compactifications have been described in [285,286]. The generalized geometry, capable of classifying supersymmetric compactifications with fluxes can be found in [287,288].

An emerging subject, not addressed in this book, involves cosmological applications of string theory. The reviews [289–291, 261] summarize our current knowledge on the subject.

Exercises

9.1. Consider the heterotic string compactified on a circle of radius R, with all sixteen Wilson lines Y_α turned on. Use the results of appendix D on page 513 to write the modular invariant partition function. Find how the Wilson lines transform under T-duality.

9.2. Apply the results of appendix E on page 516 to derive the heterotic effective action (9.1.8) on page 221. Show the invariance (9.1.10).

9.3. Start from the ten-dimensional type-IIA effective action in (H.16) on page 527. Use toroidal dimensional reduction (you will find relevant formulas in appendix E on page 516) and derive the four-dimensional effective action. Dualize all two-forms into axions.

9.4. In the $\mathcal{N} = 4_4$ space-time supersymmetric case of section 9.2 on page 223, bosonize the remaining three currents, write the $\Sigma, \bar{\Sigma}$ fields as vertex operators and show that in this case the left-moving internal CFT has to be a toroidal one.

9.5. Compute the partition function of the orbifold generated by the action (9.4.1) on page 231. Show that it is not modular invariant.

9.6. Show that the partition function (9.4.8) on page 232 is modular invariant. Verify that the massless bosonic spectrum is as claimed in the text.

9.7. Show that (9.4.14) is modular invariant if $\epsilon^2/2 = 1 \bmod(4)$.

9.8. Use the definition of the second helicity supertrace B_2 in appendix J on page 537 in order to derive (9.4.17) on page 234.

9.9. Show that only one of the four gravitini survives the $\mathbb{Z}_2 \times \mathbb{Z}_2$ projection described in section 9.6 on page 237.

9.10. Derive the massless spectrum of tables 9.1 and 9.2 from the partition function (9.6.2). Show that the spectrum is anomaly-free in four dimensions.

9.11. Consider a $\mathbb{Z}_3$ orbifold of the heterotic string, with generating rotation $\theta^1 = \theta^2 = \pi/3$, $\theta^3 = -2\pi/3$ in (9.3.8) on page 229. Show that this orbifold will give a vacuum with $\mathcal{N} = 1_4$ supersymmetry. Find the appropriate action on Γ_{16} so that a modular-invariant partition function is obtained. Derive the massless spectrum of this vacuum.

9.12. Show that the Nijenhuis tensor (9.7.4) on page 240 is indeed a tensor.

9.13. Consider the complex projective space CP^N: A space of $N+1$ complex variables, moded out by the scaling $\{Z_k\} \sim \lambda\{Z_k\}$ where λ is any nonzero complex number. Show that this space is compact, and that it is a complex manifold.

9.14. Start from equation (9.8.9) on page 246 and use the identity

$$\gamma^j \gamma^{kl} = \gamma^{jkl} + g^{jk}\gamma^l - g^{jl}\gamma^k \tag{9.1E}$$

and the properties of the Riemann tensor to show that the Ricci tensor vanishes.

9.15. Starting from the *Ansatz* (9.9.1) on page 248 for the ten-dimensional metric derive (9.9.2) and (9.9.3) on page 248.

9.16. Show that the *Ansatz* (9.9.6) on page 248 provides solutions to equations (9.9.3).

9.17. Consider compactifications of type-IIA,B theories to four dimensions. Greek indices describe the four-dimensional part, Latin ones the six-dimensional internal part. Repeat the analysis at the beginning of section 9.8 on page 245 and find the conditions for the internal fields g_{mn}, B_{mn}, Φ as well as A_m, C_{mnr} for type-IIA and $\chi, B^{\text{R-R}}_{mn}, F^+_{mnrst}$ for type-IIB so that the effective four-dimensional theory has $\mathcal{N} = 4_4$ supersymmetry in flat space.

9.18. Use the results of section 7.9 on page 176 to show that the $O(5,21)$, $\mathcal{N} = (2,0)_6$ supergravity obtained by compactifying the IIB string on K3 is anomaly-free.

9.19. Derive the massless spectrum of the $T^4/\mathbb{Z}_2$ orbifold described in section 9.10 on page 250.

9.20. Compute the elliptic genus $\text{Tr}_{R\text{-}R}[(-1)^{F_L+F_R}e^{izJ_0-i\bar{z}\bar{J}_0}]$ for the type-II K3 compactification. The trace is in the R-R sector. Hint: show that it is independent of the K3 moduli.

9.21. The $\mathbb{Z}_2$ orbifold transformation $x^i \to -x^i$ is a symmetry of the T^4 for all values of the moduli. This is not the case for $\mathbb{Z}_3$, $\mathbb{Z}_4$, and $\mathbb{Z}_6$ rotations. Find the submoduli space of $\mathbb{Z}_3$-, $\mathbb{Z}_4$-, and $\mathbb{Z}_6$-invariant T^4s.

9.22. Construct the torus partition function of the $\mathbb{Z}_3$, $\mathbb{Z}_4$, and $\mathbb{Z}_6$ orbifold compactifications with $\mathcal{N} = 1_6$ supersymmetry. Derive from this the massless spectrum and compare with the geometrical description in section 9.9 on page 247.

9.23. Describe the blowing up of the $\mathbb{Z}_3$, $\mathbb{Z}_4$, and $\mathbb{Z}_6$ orbifold points of K3.

9.24. Show that in type-II compactifications on CY manifolds, the dilaton belongs to a hypermultiplet.

9.25. Show that in type-IIA compactifications on CY manifolds, the gauge group is $U(1)^{h^{1,1}+1}$.

9.26. Consider the IIA/B theory compactified on the $T^6/\mathbb{Z}_3$ supersymmetry-preserving orbifold, with $\mathcal{N} = 2_4$ supersymmetry. Calculate the massless spectrum. Find the topological data of the CY threefold whose singular limit is the orbifold above.

9.27. Consider a collection of O_9 and O_5 planes on $T^4/\mathbb{Z}_2$. Consider their coupling to the metric and dilaton and by varying derive the tadpole conditions.

9.28. Consider the T^4 lattice sum. Find the values for the torus moduli so that this sum is invariant under the action of Ω.

9.29. Find from first principles the phases of the Ω action on the fermionic ground-states of the open strings in section 9.14.3 on page 260.

9.30. Derive from first principles (i.e., not relying on supersymmetry) the massless fermionic spectrum of open strings in the K3 orientifold of section 9.14.3 on page 260.

9.31. Derive the low-lying spectrum of the $5a - 5x$ and $5x - 5(-x)$ strings described in section 9.14.3 on page 260.

9.32. Show that the solution to the tadpole conditions (9.14.51)–(9.14.53) on page 266 and the group properties imply (9.14.57).

9.33. Consider the tadpole conditions of the $\mathbb{Z}_2$ orientifold in section 9.14.6 on page 266. Find the general solution, considering a general brane configuration.

9.34. Consider a general configuration of D_5-branes in the $T^4/\mathbb{Z}_2$ orientifold of section 9.14.3 on page 260. Solve the tadpole conditions and derive the massless spectrum that was presented in section 9.14.7 on page 267.

9.35. Consider the effective gauge theory of the $U(16) \times U(16)$ solution to the $T^4/\mathbb{Z}_2$ tadpole conditions. Giving expectation values to various scalars, show that you can obtain the more general spectrum, in (9.14.61) on page 268.

9.36. Turning on the T^4 Wilson lines, the 9-9 gauge group $U(16)$ of the $T^4/\mathbb{Z}_2$ orientifold is Higgsed. What is the most general remaining gauge group and the associated brane configuration?

9.37. Using the results of section 7.9 on page 176 and of exercise 7.33 on page 186 show that the $U(16) \times U(16)$ $T^4/\mathbb{Z}_2$ orientifold theory is free of gravitational and nonabelian gauge anomalies.

9.38. Show that the two $U(1)$ factors of the $U(16) \times U(16)$ gauge group of the $T^4/\mathbb{Z}_2$ orientifold have abelian as well as abelian/nonabelian mixed anomalies in six dimensions. Show how the Green-Schwarz mechanism can cancel the anomalies in this case. Verify that in the process, the two $U(1)$'s become massive.

9.39. Consider the open string sector of the supersymmetric $T^4/\mathbb{Z}_3$ orientifold. Derive the tadpole conditions and show that due to the absence of $\mathbb{Z}_2$ factors in the orbifold group, no D_5-branes are needed. Solve the tadpole conditions and show that the massless spectrum consists of a vector multiplet of the $U(8) \times SO(16)$ gauge group, with hypermultiplets transforming as $(8,16)$, $(\bar{8}, 16)$, $(28,1)$, $(\overline{28}, 1)$.

9.40. Using group theory show that when an $SO(6)$ $\mathbb{Z}_N$ rotation acts on the vector as in (9.15.1), it acts on the spinor as in (9.14.9).

9.41. Solve the invariance condition (9.14.14) on page 259 for the scalars explicitly in order to show that they transform in representation of the gauge group given in (9.15.7) on page 270. Do the same for the fermions to derive (9.15.10).

9.42. Consider in section 9.15) on page 268 D_{7_1}-branes transverse to the first plane and D_{7_2}-branes transverse to the second plane. Derive the massless spectrum of 3-7_1, 3-7_2, and 7_i-7_j, strings with $i, j = 1, 2, 3$.

9.43. Consider the massless spectrum of the D_3- and D_{7_i}-branes in section 9.15 on page 268. Calculate the four-dimensional nonabelian gauge anomalies. Impose the cancellation of the nonabelian anomalies to constraint the integers n_i, m_i.

9.44. Calculate the (mixed) gauge anomalies of the U(1) factors originating from the 3-3 strings in section 9.15. Determine the axion couplings responsible for their cancellation. Which linear combinations acquire masses in the process?

9.45. Consider the configuration of branes transverse to the orbifold singularity in section 9.15. Calculate the (massless) twisted tadpoles and show that the tadpole cancellation condition is

$$\text{Tr}\left(\gamma_{3,\theta}^k\right) \prod_{i=1}^{3}\left[2\sin\frac{\pi k b_i}{N}\right] + \sum_{i=1}^{3}\text{Tr}\left(\gamma_{7_i,\theta}^k\right)2\sin\frac{\pi k b_i}{N} = 0, \quad k = 1, 2, \ldots, N-1. \tag{9.2E}$$

What is its relation to the anomaly cancellation studied in exercise 9.43?

9.46. Consider the $T^6/\mathbb{Z}_3$ supersymmetric orbifold in four dimensions. Derive and solve the tadpole conditions, to find the massless spectrum. Is the spectrum chiral?

9.47. Consider D_3- and D_7-branes at a $\mathbb{Z}_N$ singularity. Try to construct a gauge group and a chiral spectrum of fermions as close to the Standard Model as possible.

9.48. Show that the mass formula (9.16.4) on page 272 implies that $\text{Str}[\delta M^2] = 0$.

9.49. Consider a D_9-brane wrapping a magnetized $(T^2)^3$. If H_i is the magnetic field through the ith torus show that $\mathcal{N} = 1_4$ supersymmetry is preserved if $|\theta_1| + |\theta_2| - |\theta_3| = 0$ (up to cyclic permutations). This corresponds to the statement that the SO(6) rotation, generated by $(\theta_1, \theta_2, \theta_3)$, is in fact in SU(3).

9.50. Solve equations (9.16.10) on page 274 together with the boundary conditions (9.16.17), (9.16.18), and verify the mode expansions (9.16.19), (9.16.20).

9.51. Verify explicitly, relations (9.16.24)–(9.16.29) on page 275.

9.52. Use the commutation relations (9.16.23) on page 275 to calculate the equal-time commutators $[X_I(\tau, \sigma), P^J(\tau, \sigma')]$ and $[P^I(\tau, \sigma), P^J(\tau, \sigma')]$. Observe that there are boundary contributions to the momentum operators.

9.53. The spectrum of strings starting and ending on the same magnetized brane, with $q_L = -q_R$, $H_L = H_R$, is not directly affected by the magnetic field. Quantize these strings carefully to find out the subtle effect of the magnetic field on the spectrum.

9.54. Derive from first principles the magnetized partition functions (9.16.41)–(9.16.45) on page 277.

9.55. Show geometrically, that (9.16.50) on page 278 is indeed the intersection number of the two branes on T^2.

9.56. Find the quantization of the magnetic flux threading a nonorthogonal T^2 carrying a constant antisymmetric tensor background. In type-I string theory, this background is discrete. Show that in this case, this is equivalent to the fact that the integer m in (9.16.2) on page 272 can take also half-integer values.

9.57. Consider a D_9-brane wrapping a magnetized $(T^2)^2$, with magnetic fields H_1 and H_2 through the two tori. Show that the D_9-brane acquires a D_5-brane charge. Discuss its quantization. Show that when the flux is infinite, the D_9 is equivalent to a D_5 brane stretching in the transverse directions.

9.58. Consider a set of magnetized D_9-branes wrapped on $(T^2)^3$ as those considered in section 9.16.3 on page 278. Derive the tadpole conditions using the magnetized partition functions and show that they are given by (9.16.51) on page 279.

9.59. Show that the massless states of strings stretched between a magnetized brane l and its orientifold image l' are $8m_l^1 m_l^2 m_l^3$ fermions in the $\boxminus$ representation of $U(N_l)$ as well as $4m_l^1 m_l^2 m_l^3 (n_l^1 n_l^2 n_l^3 - 1)$ fermions in the $\boxminus$ and $\Box$ representations.

9.60. Consider intersecting D_6-branes on T^6. Find a solution to the tadpole conditions, so that the gauge group is $U(3) \times U(2) \times U(1) \times U(1)$, and the chiral spectrum is that of the SM. You must identify the hypercharge as a linear combination of the $U(1)$ generators. Show that the other $U(1)$'s are anomalous and therefore massive.

9.61. Consider three D-brane stacks realizing the gauge group $U(3) \times U(2) \times U(1)$. Assume that the SM fermions and scalars (including one or more right-handed neutrino singlets) as arising from strings stretched between these three branes. Find all possible ways of doing this. An important ingredient is how hypercharge is realized as a linear combination of the three $U(1)$ symmetries present. Do the other two $U(1)$ gauge bosons remain massless?

9.62. Consider magnetized D_9-branes on T^6. The DBI action depends nontrivially on both the magnetic fields and the geometric moduli of the T^6. Show that this provides a potential for the torus moduli at the tree level. Minimize this potential and find which of the torus moduli can be stabilized. This is a special case of the more general stabilization mechanism of moduli by turning on fluxes of (generalized) gauge fields.

9.63. Before looking up reference [220], try to prove the key points, mentioned in section 9.17.2 on page 282 using basic properties of (super)conformal field theory and current algebra.

9.64. By considering the tree-level amplitudes of heterotic gauge bosons show the relation (9.17.3) on page 281.

9.65. Investigate how (9.18.2) on page 285 changes if the relevant brane wraps a flat internal magnetized cycle.

9.66. Consider the σ-model of the $SU(2)_k$ WZW model given in exercise 6.7 on page 152. We may gauge the $U(1)_L \times U(1)_R$ affine symmetry without including a standard kinetic term for the gauge fields. This preserves conformal invariance. Gauge fix and integrate out the gauge fields to find the resulting two-dimensional sigma-model. Describe its effective background fields and symmetry. Argue using CFT arguments that the continuous $U(1)$ remnant symmetry is broken to $\mathbb{Z}_k$.

10 | Loop Corrections to String Effective Couplings

So far, we have described several ways of obtaining four-dimensional string vacua with or without supersymmetry and with various particle contents. Some vacua have the correct structure at tree level to describe the supersymmetric Standard Model particles and interactions. However, to test further agreement with experimental data, loop corrections should be incorporated. In particular we know that, at low energy, coupling constants run with energy due to loop contributions of charged particles. So we need a computational framework to address similar issues in the context of string theory.

We have mentioned the relation between a "fundamental theory" (here ST) and its associated effective field theory (EFT), at least at tree level. Now we will have to take loop corrections into account. The heavy-particle loops must be incorporated in the EFT. Then we can use the EFT to calculate the quantum effects that are generated only by the light states.

To derive the loop-corrected EFT, we must apply the following procedure: For every given amplitude of light states we perform the computation in ST where both light and heavy states propagate in the loops. We subtract the same amplitude calculated in the EFT (with only light states propagating in the loops). The difference (known as the *threshold correction*) is essentially the contribution to a particular process of heavy states only. We will have to incorporate this into the effective action.

We address in detail here the essential issues of such computations. We will only deal with one-loop corrections. Higher-loop computations are in principle possible but in practice forbidding.

A one-loop amplitude in closed string theory will be obtained by integrating a modular invariant function on the fundamental domain of the torus. There are no UV divergences in closed string theory since the UV region of the Schwinger integration is missing. There is therefore, no need for an UV cutoff.

Such calculations are done in the first-quantized framework. This implies in particular that we are forced to work on-shell. In this context, IR divergences will be present, since massless on-shell particles can propagate in the loop. Formally, the amplitude will be infinite. This IR divergence is physical, and the way we deal with it in field theory is to allow the external momenta to be off-shell. The IR divergence will cancel when we subtract the EFT result from the ST result.

There are several methods to deal with the IR divergence in one-loop calculations, each with its merits and drawbacks.

• The original approach computes appropriate two-point functions of gauge fields on the torus, removes wave function factors that would make this amplitude vanish (such a two-point function on shell is required to vanish by gauge invariance), and regularizes the IR divergence by inserting a regularizing factor for the massless states. This procedure gave the gauge-group-dependent threshold corrections and the first concrete calculation of their moduli dependence. However, modular invariance is broken by such a regularization, and the prescription of removing vanishing wave-function factors does not *a priori* rest on a solid basis.

• Another approach is to calculate derivatives of threshold corrections with respect to moduli. Threshold corrections depend on moduli, since the masses of KK states do. This procedure is free of IR divergences (the massless states drop out) and modular-invariant. However, there are still vanishing wave-function factors that need to be removed by hand, and this approach cannot calculate moduli-independent constant contributions to the thresholds.

• An approach that provides a rigorous framework to calculate one-loop thresholds is based on the on-shell background field method. The idea is to curve four-dimensional space-time, in a way that provides a physical IR cutoff on the spectrum. This procedure is IR-finite, modular-invariant, free of ambiguities, and allows the calculation of thresholds. On the other hand, this IR regularization breaks maximal supersymmetry ($\mathcal{N} = 4_4$ in heterotic and $\mathcal{N} = 8_4$ in type II). It preserves, however, smaller fractions of supersymmetry.

The last method is the rigorous method of calculation. We will describe it, in a subsequent section without going into all the details. The results in several cases are not different from that obtained by the other methods. Therefore, for simplicity, we will do some of the calculations using also the other methods.

Loop computations can be made in any vacuum of the string. However, we will be mostly interested here in four-dimensional couplings, obtained after the compactification of six dimensions.

We will also typically consider string vacua that have $\mathcal{N} \geq 1_4$ supersymmetry. At one loop, this is mostly for convenience. Although we know that supersymmetry is broken in the low-energy world, for hierarchy reasons it should be probably broken at a low enough scale $\sim$ TeV. If we assume that the superpartners have masses that are not far away from the supersymmetry breaking scale, their contribution to thresholds can be estimated in the EFT. Therefore, without loss of generality, we will assume the presence of (at least) unbroken $\mathcal{N} = 1_4$ supersymmetry.

10.1 Calculation of Heterotic Gauge Thresholds

We will first consider $\mathcal{N} = 2_4$ heterotic vacua with an explicit T^2. This will provide a simple way to calculate derivatives of the correction with respect to the moduli. Afterwards, we will derive a general formula for the corrections. Such vacua have a geometrical interpretation as a compactification on K3$\times T^2$ with a gauge bundle of instanton number 24.

In $\mathcal{N} = 2_4$ vacua, the complex moduli that belong to vector multipletss are the S field ($1/S_2$ is the four-dimensional string coupling), the moduli T, U of the two-torus and several Wilson lines W^I, which we will set to zero, in order to have unbroken non-abelian groups. Because of $\mathcal{N} = 2_4$ supersymmetry, the gauge couplings can depend only on the vector moduli. We will focus here on the dependence on S, T, U. At tree level, the gauge coupling for the nonabelian factor G_i is given according to (9.17.3) on page 281 by

$$
\frac{1}{g_i^2}\bigg|_{\text{tree}} = \frac{k_i}{g_u^2} = k_i S_2, \quad g_u^2 = 4\pi \frac{g_s^2}{V_6},
\tag{10.1.1}
$$

where k_i is the central element of the right-moving affine G_i algebra, which generates the gauge group G_i. The gauge boson vertex operators are

$$
V_G^{\mu,a} = (\partial X^\mu + i(p \cdot \psi)\psi^\mu)\,\bar{J}^a\,e^{ip \cdot X}.
\tag{10.1.2}
$$

In the simplest vacua, all nonabelian factors have $k = 1$. We will keep however k arbitrary.

The term in the effective action we would like to calculate is

$$
\int d^4x \frac{1}{4g^2(T_i)} F_{\mu\nu}^a F^{a,\mu\nu},
\tag{10.1.3}
$$

where the coupling will depend in general on the vector moduli. To find the dependence on the moduli, we must calculate a three-point amplitude on the torus, with two gauge fields and one modulus, in the even spin structures. The odd spin structure gives a contribution proportional to the ϵ-tensor and is thus a contribution to the renormalization of the θ angle. The term that is quadratic in momenta will give us the derivative with respect to the appropriate modulus of the correction to the gauge coupling. The vertex operators for the torus moduli T, U are given by

$$
V_{\text{modulus}}^{IJ} = (\partial X^I + i(p \cdot \psi)\psi^I)\,\bar{\partial} X^J\,e^{ip \cdot X}.
\tag{10.1.4}
$$

So we must calculate[1]

$$
\begin{aligned}
I_{1-\text{loop}} &= \int \langle V^{a,\mu}(p_1, z) V^{b,\nu}(p_2, w) V_{\text{modulus}}^{IJ}(p_3, 0)\rangle \\
&\sim \delta^{ab}(p_1 \cdot p_2 \eta^{\mu\nu} - p_1^\nu p_2^\mu)f^{IJ}(T, U) + \mathcal{O}(p^4),
\end{aligned}
\tag{10.1.5}
$$

where $p_1 + p_2 + p_3 = 0$, $p_i^2 = 0$. Because of supersymmetry, in order to get a nonzero result we will have to contract the four fermions ψ^μ in (10.1.5), which gives us two powers

---

[1] As mentioned in the introduction, this amplitude is zero on shell. We may make it nonzero by not imposing momentum conservation in intermediate stages.

of momentum. Therefore, to quadratic order, we can set the vertex operators $e^{ip\cdot X}$ to 1. The only nonzero contribution to F^{IJ} is

$$f^{IJ} = \int_{\mathcal{F}} \frac{d^2\tau}{\tau_2^2} \int \frac{d^2 z}{\tau_2} \int d^2 w \; \langle \psi(z)\psi(w)\rangle^2 \; \langle \bar{J}^a(\bar{z})\bar{J}^b(\bar{w})\rangle \; \langle \partial X^I(0)\bar{\partial} X^J(0)\rangle. \tag{10.1.6}$$

The normalized fermionic two-point function on the torus for an even spin structure is given by the Szegö kernel[2]

$$\frac{2}{\ell_s^2} \langle \psi(z)\psi(0)\rangle|_b^a = S[^a_b](z) = \frac{\vartheta[^a_b](z)\vartheta'_1(0)}{\vartheta_1(z)\vartheta[^a_b](0)} = \frac{1}{z} + \cdots. \tag{10.1.7}$$

It satisfies the following identity:

$$S^2[^a_b](z) = \mathcal{P}(z) + 4\pi i \partial_\tau \log \frac{\theta[^a_b](\tau)}{\eta(\tau)}, \tag{10.1.8}$$

so that all the spin structure dependence is in the z-independent second term. We will have to weight this correlator with the partition function. Since the first term is spin-structure independent it will not contribute for vacua with $\mathcal{N} \geq 1_4$ supersymmetry, where the partition function vanishes. For the $\mathcal{N} = 2_4$ vacua described earlier, the spin-structure sum of the square of the fermion correlator can be evaluated directly:

$$\langle\langle \psi(z)\psi(0)\rangle\rangle = \frac{1}{2} \sum_{(a,b)\neq(1,1)} (-1)^{a+b+ab} \frac{\vartheta^2[^a_b]\vartheta[^{a+h}_{b+g}]\vartheta[^{a-h}_{b-g}]}{\eta^4} S^2[^a_b](z)$$

$$= 4\pi^2 \eta^2 \vartheta[^{1+h}_{1+g}]\vartheta[^{1-h}_{1-g}], \tag{10.1.9}$$

where we have used (C.11) and the Jacobi identity (C.21) on page 509.

The two-point function of the currents is

$$\langle \bar{J}^a(\bar{z})\bar{J}^b(0)\rangle = \frac{k\delta^{ab}}{4\pi^2}\bar{\partial}_{\bar{z}}^2 \log \bar{\vartheta}_1(\bar{z}) + \mathrm{Tr}[J_0^a J_0^b]$$

$$= \delta^{ab}\left(\frac{k}{4\pi^2}\bar{\partial}_{\bar{z}}^2 \log \bar{\vartheta}_1(\bar{z}) + \mathrm{Tr}[Q^2]\right), \tag{10.1.10}$$

where $\mathrm{Tr}[Q^2]$ stands for the conventionally normalized trace into the whole string spectrum of the quadratic Casimir of the group G. This can be easily computed by picking a single Cartan generator squared and performing the trace.

In terms of the affine characters $\chi_R(v_i)$ this trace is $\partial_{v_1}^2 \chi_R(v_i)/(2\pi i)^2|_{v_i=0}$. This is the normalization for the quadratic Casimir which is standard in field theory: for a representation R, $\mathrm{Tr}[T^a T^b] = I_2(R)\delta^{ab}$, where T^a are the matrices in the R representation. The field theory normalization corresponds to picking the squared length of the highest root to be 1. For the fundamental of $SU(N)$ this implies the value 1 for the Casimir. With this normalization, the spin j representation of $SU(2)$ gives $2j(j+1)(2j+1)/3$.

---

[2] We use the supersymmetric normalizations of section 4.13 on page 77.

Finally, $\langle \partial X^I(0) \bar{\partial} X^J(0) \rangle$ gets contributions from zero modes only, and it can be easily calculated, using the results of section 4.18, to be

$$\langle \partial X^I(0)\bar{\partial} X^J(0) \rangle = \frac{\sqrt{\det G}}{(\sqrt{\tau_2}\eta\bar{\eta})^2} \sum_{\vec{m},\vec{n}} (m^I + n^I\tau)(m^J + n^J\bar{\tau})$$

$$\times \exp\left[-\frac{\pi(G_{KL} + B_{KL})}{\tau_2}(m_K + n_K\tau)(m_L + n_L\bar{\tau})\right]. \tag{10.1.11}$$

A convenient basis for the T^2 moduli is given by (4.18.65) on page 103. In this basis we have

$$V_{T_i} = v_{IJ}(T_i)\partial X^I \bar{\partial} X^J, \tag{10.1.12}$$

with

$$v(T) = -\frac{i}{2U_2}\begin{pmatrix} 1 & U \\ \bar{U} & |U|^2 \end{pmatrix}, \quad v(U) = \frac{iT_2}{U_2^2}\begin{pmatrix} 1 & \bar{U} \\ U & \bar{U}^2 \end{pmatrix}, \tag{10.1.13}$$

$v(\bar{T}) = \overline{v(T)}, v(\bar{U}) = \overline{v(U)}$. Then

$$\langle V_{T_i} \rangle = -\frac{\tau_2}{2\pi}\partial_{T_i}\frac{\Gamma_{2,2}}{\eta^2\bar{\eta}^2}. \tag{10.1.14}$$

Using (C.34) we obtain for the one-loop correction to the gauge coupling in the $\mathcal{N} = 2_4$ vacuum

$$\frac{\partial}{\partial T_i}\frac{16\pi^2}{g_I^2}\bigg|_{\text{1-loop}} \sim \frac{\partial}{\partial T_i}\int_{\mathcal{F}}\frac{d^2\tau}{\tau_2^2}\frac{\tau_2\Gamma_{2,2}}{\bar{\eta}^4}\operatorname{Tr}_R^{\text{int}}\left[(-1)^F\left(Q_I^2 - \frac{k_i}{4\pi\tau_2}\right)\right] + \text{constant}. \tag{10.1.15}$$

The internal theory consists of the (4,20) part of the original theory, which carries $\mathcal{N} = (4,0)_2$ superconformal invariance on the left. The derivative with respect to the moduli kills the IR divergence due to the massless states.

For the remainder of this section, we will be cavalier about IR divergences and vanishing wave functions. This will be dealt with rigorously in the next section. For a general heterotic string vacuum (with or without supersymmetry), we can parametrize its partition function as

$$Z_{D=4}^{\text{heterotic}} = \frac{1}{\tau_2\eta^2\bar{\eta}^2}\sum_{a,b=0}^{1}\frac{(-1)^{a+b+ab}}{2}\frac{\vartheta[^a_b]}{\eta}\,C^{\text{int}}[^a_b], \tag{10.1.16}$$

where we have separated the (light-cone) bosonic and fermionic contributions of the four-dimensional part. $C[^a_b]$ is the partition function of the internal CFT with $(c, \bar{c}) = (9, 22)$ and at least $\mathcal{N} = (2, 0)_2$ superconformal symmetry. In particular, the ϑ-function carries the helicity-dependent contributions due to the fermions. What we are now computing is the two-point amplitude of two gauge bosons at one loop. When there is no supersymmetry, the ∂X factors of the vertex operators contribute $\langle \partial_z X(z)X(0) \rangle^2$, where we have to use the torus propagator for the noncompact bosons:

$$\frac{2}{\ell_s^2}\langle X(z, \bar{z})X(0) \rangle = -\log|\vartheta_1(z)|^2 + 2\pi\frac{\operatorname{Im}z^2}{\tau_2}, \quad \partial_z\partial_{\bar{z}}\langle X(z, \bar{z})X(0) \rangle = -\ell_s^2\pi\delta^{(2)}(z). \tag{10.1.17}$$

We obtain

$$\epsilon_1^\mu \epsilon_2^\nu \langle A_\mu(p_1) A_\nu(p_2) \rangle$$

$$= -\frac{\epsilon_1^\mu \epsilon_2^\nu}{2(2\pi \ell_s)^4} \int_{\mathcal{F}} \frac{d^2\tau}{\tau_2^2} \int \frac{d^2 z}{\tau_2} \langle (\partial_z X^\mu + ip_1 \cdot \psi \psi^\mu) \bar{J}^a \, e^{ip_1 \cdot X} (\partial_w X^\nu + ip_2 \cdot \psi \psi^\nu) \bar{J}^b \, e^{ip_2 \cdot X} \rangle$$

$$= \frac{(p_1 \cdot p_2)(\epsilon_1 \cdot \epsilon_2) - (\epsilon_1 \cdot p_2)(\epsilon_2 \cdot p_1)}{2(2\pi \ell_s)^4} \int_{\mathcal{F}} \frac{d^2\tau}{\tau_2^2} \int \frac{d^2 z}{\tau_2} \left(\langle X \partial_z X \rangle^2 - \langle \psi \psi \rangle^2 \right) \langle \bar{J}^a \bar{J}^b \rangle e^{-p_1 \cdot p_2 \langle XX \rangle}.$$

$$(10.1.18)$$

We must pick the terms quadratic in the momenta. In this case the z-integral we will have to perform is

$$\int \frac{d^2 z}{\tau_2} (S^2[{}^a_b](z) - \langle X \partial X \rangle^2) \left(\frac{k}{4\pi^2} \bar{\partial}^2 \log \bar{\vartheta}_1(\bar{z}) + \text{Tr}[Q^2] \right)$$

$$= 4\pi i \partial_\tau \log \frac{\vartheta[{}^a_b]}{\eta} \left(\text{Tr}[Q^2] - \frac{k}{4\pi \tau_2} \right),$$

$$(10.1.19)$$

where we have used (10.1.8), (C.33), and (C.35). The total threshold correction is, using (10.1.16),

$$Z_2^I = \frac{16\pi^2}{g_I^2}\bigg|_{\text{1-loop}} = \frac{1}{4\pi^2} \int_{\mathcal{F}} \frac{d^2\tau}{\tau_2} \frac{1}{\eta^2 \bar{\eta}^2} \sum_{\text{even}} 4\pi i \partial_\tau \left(\frac{\vartheta[{}^a_b]}{\eta} \right) \text{Tr}_{\text{int}} \left[Q_I^2 - \frac{k_I}{4\pi \tau_2} \right] [{}^a_b], \qquad (10.1.20)$$

where the trace is taken in the $[{}^a_b]$ sector of the internal CFT. Note that the integrand is modular invariant. This result is general. The measure $\int_{\mathcal{F}} \frac{d^2\tau}{\tau_2}$ will give an IR divergence as $\tau_2 \to \infty$ coming from constant parts of the integrand. The constant part precisely corresponds to the contributions of the massless states. The derivative on the ϑ-function gives a factor proportional to $s^2 - 1/12$, where s is the helicity of a massless state. The k/τ_2 factor accompanying the group trace gives an IR-finite part. Thus, the IR-divergent contribution to the one-loop result is

$$\frac{16\pi^2}{g_I^2}\bigg|_{\text{1-loop}}^{\text{IR}} = \int_{\mathcal{F}} \frac{d^2\tau}{\tau_2} \text{Str} \left[Q_I^2 \left(\frac{1}{12} - s^2 \right) \right], \qquad (10.1.21)$$

where Str stands for the supertrace.

Inserting by hand a regularizing factor $e^{-\ell_s^2 \mu^2 \tau_2}$ we obtain

$$\frac{16\pi^2}{g_I^2}\bigg|_{\text{1-loop}}^{\text{IR}} = b_I \log \left(\frac{\mu^2}{M_s^2} \right) + \text{finite}, \qquad (10.1.22)$$

where b_I is the conventional one-loop β-function coefficient

$$b_I = \text{Str} \left[Q_I^2 \left(\frac{1}{12} - s^2 \right) \right]\bigg|_{\text{massless}}. \qquad (10.1.23)$$

We will now try to better understand the origin of the various terms in (10.1.20). The term proportional to $\text{Tr}[Q^2]$ comes from conventional diagrams where the two external gauge bosons are coupled to a loop of charged particles (figure 10.1a).

The second term proportional to k seems unusual, since all particles contribute to it, whether charged or not. This is a correction to the gauge couplings due to the presence of

a)   b)

Figure 10.1 a) One-loop gauge coupling correction due to charged particles. b) Universal one-loop correction.

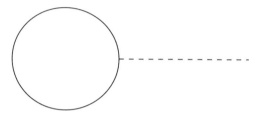

Figure 10.2 One-loop dilaton tadpole.

the gravitational sector. There is no analog of it in field theory. Roughly speaking this term would arise from diagrams like the one shown in figure 10.1b). Two external gauge bosons couple to the dilaton (remember that there is a tree-level universal coupling of the dilaton to all gauge bosons) and then the dilaton couples to a loop of any string state (the dilaton coupling is universal). One may object that this diagram, being one-particle reducible, should not be included as a correction to the coupling constants. Moreover, it seems to imply that there is a nonzero dilaton tadpole at one loop (figure 10.2). When at least $\mathcal{N} = 1_4$ supersymmetry is unbroken, we can show that the dilaton tadpole in Fig. 10.2 is zero. However, the diagram in figure 10.1b) still contributes owing to a delicate cancellation of the zero from the tadpole and the infinity coming from the dilaton propagator on shell. This type of term is due to a modular-invariant regularization, of the world-sheet short-distance singularity present when two vertex operators collide.

We will see in the next section that the universal terms arise in a background field calculation due to the gravitational back-reaction to background gauge fields. It is important to notice that such terms are truly universal, in the sense that they are independent of the gauge group in question. Their presence is essential for modular invariance.

We should also stress here that the identification of stringy contribution with field theory diagrams is not always unique. This is due to the underlying crossing symmetry of CFT correlators. For example the universal contribution also contains the "gravitational" corrections to the gauge couplings. The relevant diagrams are one-loop diagrams where a gauge bosons emits and reabsorbs a graviton, antisymmetric tensor, or dilaton.

There is an analogous diagram contributing to the one-loop renormalization of the θ-angles. At tree level, there is a (universal) coupling of the antisymmetric tensor to two gauge fields due to the presence of Chern-Simons terms. This gives rise to a parity-odd contribution like the one in figure 10.1b), where now the intermediate state is the antisymmetric tensor.

It was obvious from the previous calculation that the universal terms came as contact terms from the singular part of the correlator of affine currents. In open string theory, the gauge symmetry is not realized by a current algebra on the world-sheet, but by charges (CP factors) attached to the end points of the open string. Thus, one would think that such universal contributions are absent. However, even in the open string case, such terms appear in an indirect way, since the Planck scale has a nontrivial correction at one loop for $\mathcal{N} \leq 2_4$ supersymmetry, unlike the heterotic case.

10.2 On-shell Infrared Regularization

As mentioned in the previous section, the one-loop corrections to the effective coupling constants are calculated on shell and are IR divergent. Also, for comparison with low-energy data, the moduli-independent piece is also essential. In this section we will provide a framework for this calculation.

We will describe in detail the heterotic string, although the same method works in the type-II string without extra subtleties.

Any four-dimensional heterotic string vacuum is described by a world-sheet CFT, which is a product of a flat noncompact CFT describing Minkowski space with $(c, \bar{c}) = (6, 4)$ and an internal compact CFT with $(c, \bar{c}) = (9, 22)$. Both must have $\mathcal{N} = (1, 0)_2$ superconformal invariance on the left, necessary for the decoupling of ghosts.

To regulate the IR divergence on shell, we will modify the four-dimensional part. We will consider the theory in a background with nontrivial four-dimensional curvature and other fields $B_{\mu\nu}$, Φ so that the string spectrum acquires a mass gap. Thus, all states are massive on shell and there will be no IR divergences. The curved background must satisfy the exact string equations of motion. Consequently, it should correspond to an exact CFT. We will require the following properties:

- The string spectrum must have a mass gap μ^2. In particular, chiral fermions should be regularized consistently.

- We should be able to take the limit $\mu^2 \to 0$.

- It should have $(c, \bar{c}) = (6, 4)$ so that we will not have to modify the internal CFT.

- It should preserve as many space-time supersymmetries of the original theory as possible.

- We should be able to calculate the regularized quantities relevant for the effective field theory.

- The theory should be modular invariant (which guarantees the absence of anomalies).

- Such a regularization should be possible also at the effective field theory level. In this way, calculations in the fundamental theory can be matched without any ambiguity to those of the effective field theory.

There are several CFTs with the properties required above. It can be shown that the thresholds will not depend on which we choose. We will pick a simple one, which corresponds to the SO(3)$_N$ WZW model times a free boson with background charge. The background fields corresponding to this CFT are:

$$ds^2 = G_{\mu\nu}dx^\mu dx^\nu = (dX^0)^2 + \frac{N}{4}\ell_s^2\left(d\alpha^2 + d\beta^2 + d\gamma^2 + 2\sin(\beta)\,d\alpha d\gamma\right), \tag{10.2.1}$$

where the Euler angles take values $\beta \in [0, \pi]$, $\alpha, \gamma \in [0, 2\pi]$. Thus, the three-space is almost a sphere[3] of radius squared equal to N. For unitarity, N must be a positive even integer.

$$B_{\mu\nu}dX^\mu \wedge dX^\nu = \ell_s^2\frac{N}{4}\cos(\beta)\,d\alpha \wedge d\gamma, \quad \Phi = \frac{X^0}{\ell_s\sqrt{N+2}}. \tag{10.2.2}$$

We will work in Euclidean space. As described in section 4.14 on page 82, the spectrum of operators in the X^0 part of the CFT with background charge is given by $\Delta = \ell_s^2 E^2/4+$integers, where E is a continuous variable, the "energy." In the SO(3) theory the conformal weights are $j(j + 1)/(N + 2)+$ integers. The ratio $j(j + 1)/(N + 2)$ plays the role of $\vec{p}^2$ of flat space. We obtain

$$L_0 - \frac{1}{2} = \ell_s^2 E^2 + \frac{1}{4(N+2)} + \frac{j(j+1)}{(N+2)} + \cdots. \tag{10.2.3}$$

All the states now have masses shifted by a mass gap μ^2,

$$\mu^2 = \frac{M_s^2}{(N+2)}, \quad M_s = \frac{1}{\ell_s}. \tag{10.2.4}$$

Taking $N \to \infty$, $\mu \to 0$, we recover the flat space theory. Moreover, this CFT preserves the original supersymmetries.[4] The partition function of the new CFT is known, and after some manipulations we can write the partition function of the IR-regularized theory as

$$Z(\mu) = \Gamma(\mu/M_s)\,Z(0), \tag{10.2.5}$$

where $Z(0)$ is the original partition function and

$$\Gamma(\mu/M_s) = 4\sqrt{x}\frac{\partial}{\partial x}[\rho(x) - \rho(x/4)]\Big|_{x=N+2},$$
$$\rho(x) = \sqrt{x}\sum_{m,n\in\mathbb{Z}}\exp\left[-\frac{\pi x}{\tau_2}|m + n\tau|^2\right]. \tag{10.2.6}$$

Here, $\Gamma(\mu/M_s, \tau)$ is modular invariant and $\Gamma(0) = 1$.

The background we have employed has another interesting interpretation. It is related to the neutral heterotic five-brane of charge N and zero size, described in section 8.9 on page 211. As will become clear in the next chapter, this regularized computation can be interpreted as the calculation of the gauge thresholds in the background of a four-dimensional (zero-size) instanton.

We must turn on background gauge fields and compute the one-loop amplitude as a function of these background fields. This is essentially an on-shell background-field method.

---

[3] It really is $S^3/\mathbb{Z}_2$.

[4] Except when the original supersymmetry has sixteen supercharges. Then it is broken in half.

The quadratic part will provide the one-loop correction to the gauge coupling constants. The perturbation of the theory that turns on gauge fields is

$$\delta I = \int d^2 z \left(A^a_\mu(X) \partial X^\mu + F^a_{\mu\nu} \psi^\mu \psi^\nu \right) \bar{J}^a. \tag{10.2.7}$$

In this background, there is such a class of perturbations, which are exact solutions of the string equations of motion:

$$\delta I = \int d^2 z \, B^a \left(J^3 + i \psi^1 \psi^2 \right) \bar{J}^a, \tag{10.2.8}$$

where J^3 is the current belonging to the SO(3) current algebra of the WZW model and ψ^i, $i = 1, 2, 3$, are the associated free fermions. In fact, this is the analogue of a constant magnetic background on the four-dimensional manifold we have chosen.

For this choice, the one-loop free energy can be computed exactly as a function of B^a. The details of the calculation will not presented here. They form the content of exercise 10.3 on page 318.

The expression we obtain is (10.1.20), but with a factor of $\Gamma(\mu/M_s)$ inserted into the modular integral, which renders this expression IR finite. We will denote the one-loop regularized result as $Z_2^I(\mu/M_s)$. Therefore, to one-loop order, the gauge coupling can be written as the sum of the tree-level and one-loop results as

$$k_I \frac{16\pi^2}{g_u^2} + Z_2^I(\mu/M_s). \tag{10.2.9}$$

10.2.1 Evaluation of the threshold

In order to evaluate the thresholds, we must perform a similar calculation in the EFT and subtract the string from the EFT result. The EFT result (with the same IR regulator) can be obtained from the string result by the following operations:

- Do the trace on the massless sector only.

- Only the momentum modes contribute to the regularizing function $\Gamma(\mu/M_s)$. We will denote this piece by $\Gamma_{\text{EFT}}(\mu/M_s)$.

- The EFT result is UV-divergent. We will have to regularize separately this UV divergence. We will use dimensional regularization in the $\overline{DR}$ scheme. With these changes, the field theory result for the tree-level and one-loop contributions reads

$$\frac{16\pi^2}{g_{I\,\text{bare}}^2} + b_I (4\pi)^\epsilon \int_0^\infty \frac{dt}{t^{1-\epsilon}} \Gamma_{\text{EFT}} \left(\frac{\mu}{\sqrt{\pi} M_s}, t \right). \tag{10.2.10}$$

The extra factor of $\sqrt{\pi}$ comes in since $t = \pi \tau_2$ and we chose M_s as the EFT renormalization scale. In the $\overline{DR}$ scheme, the relation between the bare and running coupling constant is

$$\frac{16\pi^2}{g_{I\,\text{bare}}^2} = \frac{16\pi^2}{g_I^2(\mu)} - b_I (4\pi)^\epsilon \int_0^\infty \frac{dt}{t^{1-\epsilon}} e^{-t\mu^2/M^2}. \tag{10.2.11}$$

Putting (10.2.11) into (10.2.10) and identifying the result with (10.2.9), we obtain

$$\frac{16\pi^2}{g_I^2(\mu)}\bigg|_{\overline{DR}} = k_I \frac{16\pi^2}{g_u^2} + Z_2^I(\mu/M_s) - b_I(2\gamma + 2), \tag{10.2.12}$$

where $\gamma = 0.577\ldots$ is the Euler-Mascheroni constant. We can separate the IR piece from Z_2^I using

$$\int_{\mathcal{F}} \frac{d^2\tau}{\tau_2} \Gamma(\mu/M_s) = \log \frac{M_s^2}{\mu^2} + \log \frac{2e^{\gamma+3}}{\pi\sqrt{27}} + \mathcal{O}\left(\frac{\mu}{M_s}\right), \tag{10.2.13}$$

in order to rewrite the effective running coupling in the limit $\mu \to 0$ as

$$\frac{16\pi^2}{g_I^2(\mu)}\bigg|_{\overline{DR}} = k_I \frac{16\pi^2}{g_u^2} + b_I \log \frac{M_s^2}{\mu^2} + b_I \log \frac{2e^{1-\gamma}}{\pi\sqrt{27}} + \Delta_I, \tag{10.2.14}$$

$$\Delta_I = \int_{\mathcal{F}} \frac{d^2\tau}{\tau_2} \left[\frac{1}{|\eta|^4} \sum_{\text{even}} \frac{i}{\pi} \partial_\tau \left(\frac{\vartheta[^a_b]}{\eta} \right) \text{Tr}_{\text{int}} \left[Q_I^2 - \frac{k_I}{4\pi\tau_2} \right] [^a_b] - b_I \right]. \tag{10.2.15}$$

This is the desired result, which produces string corrections to the EFT running coupling in the $\overline{DR}$ scheme. We will call Δ_I the string threshold correction to the associated gauge coupling. It is IR finite for generic values of the moduli. However, as we will see below, at special values of the moduli, extra states can become massless. If such states are charged, then there will be an additional IR divergence in the string threshold, which will modify the β-function. They will show up as logarithmic singularities of the threshold.

10.3 Heterotic Gravitational Thresholds

Corrections to the gravitational couplings are also of interest. We will first show that there are no corrections, at one loop, to the Planck mass in heterotic vacua with $\mathcal{N} \geq 1_4$ supersymmetry. The vertex operator for the graviton is

$$V_{\text{grav}} = \epsilon_{\mu\nu}(\partial X^\mu + ip \cdot \psi \psi^\mu)\bar{\partial} X^\nu. \tag{10.3.1}$$

We must calculate the two-point function on the torus and keep the $\mathcal{O}(p^2)$ piece. On the left-moving side, only the fermions contribute (because of space-time supersymmetry) and produce a z-independent contribution. On the left, we obtain a correlator of scalars, which must be integrated over the torus.[5] The result is proportional to

$$\int \frac{d^2z}{\tau_2} \langle X \bar{\partial}_{\bar{z}}^2 X \rangle = \int \frac{d^2z}{\tau_2} \left(\bar{\partial}_{\bar{z}}^2 \log \bar{\vartheta}_1(\bar{z}) + \frac{\pi}{\tau_2} \right) = 0, \tag{10.3.2}$$

where we used (C.34) on page 510.

---

[5] Strictly speaking the amplitude is zero on shell but we can remove the wave-function factors. A rigorous way to calculate it, is by calculating the four-point amplitude of gravitons, and extract the $\mathcal{O}(p^2)$ piece, or use the background field method of section 10.2.

We conclude that in the presence of at least $\mathcal{N} = 1_4$ supersymmetry, there is no one-loop renormalization of the Planck mass in the heterotic string. Similarly it can be shown that there are no wave-function renormalizations for the other universal fields, namely the antisymmetric tensor and the dilaton.

Higher derivative, gravitational terms do obtain quantum corrections however. A specific example is the R^2 term[6] and its parity-odd counterpart $R \wedge R$ (four derivatives) whose one-loop β-function is the conformal anomaly in four dimensions. In theories without supersymmetry, the corrections to these terms are unrelated. In theories with at least $\mathcal{N} = 1_4$ supersymmetry, the two couplings are related.

The one-loop correction to R^2 can be obtained from the $\mathcal{O}(p^4)$ part of the one-loop two-graviton amplitude, summed over the even spin structures. The odd spin structure will give the renormalization of $R \wedge R$. Going through the same steps as above we obtain (assuming $\mathcal{N} \geq 1_4$ supersymmetry)

$$\Delta_{\text{grav}} = \int_{\mathcal{F}} \frac{d^2\tau}{\tau_2} \left[\frac{1}{|\eta|^4} \sum_{\text{even}} \frac{i}{\pi} \partial_\tau \left(\frac{\vartheta[^a_b]}{\eta} \right) \frac{\hat{\bar{E}}_2}{12} C^{\text{int}}[^a_b] - b_{\text{grav}} \right], \tag{10.3.3}$$

where the modular form $\hat{\bar{E}}_2$ is defined in (C.44) on page 511. The R^2 and $R \wedge R$ couplings run logarithmically in four dimensions.

The coefficient of the logarithmic term, equal to $90b_{\text{grav}}$, is the conformal anomaly. A scalar contributes 1 to the conformal anomaly, a Weyl fermion $\frac{7}{4}$, a vector -13, a gravitino $-\frac{233}{4}$, an antisymmetric tensor 91, and a graviton 212. Again Δ_{grav} is IR finite, since we have subtracted the contribution of the massless states b_{grav}.

10.4 One-loop Fayet-Iliopoulos Terms

Some $\mathcal{N} = 1_4$ vacua contain U(1) gauge fields that are "anomalous." We have seen this already in our earlier discussion on compactifications of the heterotic and type-I string in chapter 9.

The term "anomalous" indicates that the sum of U(1) charges of all massless states charged under the U(1) is not zero.[7] In a standard field theory, this would imply the existence of a mixed (gauge-gravitational) anomaly in the theory. However, in string theory things work a bit differently.

In the presence of an "anomalous" U(1), under a gauge transformation the effective action is not invariant. There is a one-loop term (gauge anomaly) proportional to $F \wedge F$. For the theory to be gauge invariant there should be some other term in the effective action that cancels the anomalous variation. Such a term is $B \wedge F$.

We have seen in section 7.9 on page 176 that B has an anomalous transformation law under gauge transformations, $\delta B = \epsilon F$. This gives precisely the term we need to cancel the

---

[6] There are several tensor structures *a priori* possible. It is however the Gauss-Bonnet combination that turns out to be relevant.

[7] Also the higher odd traces of the charge are nonzero.

one-loop gauge anomaly. There is another way to argue on the existence of this term. In ten dimensions there was an anomaly cancelling term of the form $B \wedge F^4$. Upon compactifying to four dimensions, this will give rise to a term $B \wedge F$ with proportionality factor $\int F \wedge F \wedge F$ computed in the internal theory.

The anomaly cancellation mechanism in four dimensions becomes transparent if we dualize the antisymmetric tensor to a pseudoscalar, the axion a.

The potentially nonzero triangle diagrams of an anomalous U(1) gauge boson are

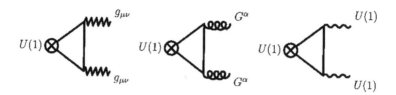

The first gives a mixed gauge-gravitational anomaly. It is nonzero if the trace of the U(1) charge $\text{Tr}[Q] \neq 0$. The second is nonzero if $\text{Tr}[QT^aT^a] \neq 0$ and contributes to the mixed U(1)-nonabelian anomaly. The third is nonzero if $\text{Tr}[Q^3] \neq 0$ and contributes to the standard U(1) gauge anomaly.

Under a gauge transformation of the anomalous U(1) gauge boson, $A_\mu \to A_\mu + \partial_\mu \epsilon$ the effective action varies as

$$\delta S_{\text{one-loop}} = C \int d^4x \, \epsilon \, \left(\text{Tr}[Q] \, R \wedge R + \text{Tr}[Q^3] \, F \wedge F + \text{Tr}[QT^aT^a]\text{Tr}[G \wedge G] \right), \qquad (10.4.1)$$

due to the triangle diagram contributions. In (10.4.1), R is the gravitational two-form, G the field strength of a nonabelian gauge group and $C = \frac{1}{24\pi^2}$.

This one-loop anomalous variation is canceled by an axion field a that couples appropriately to the action (in the Einstein frame):

$$S_{\text{eff}} = \int \sqrt{G} \left[-\frac{e^{-2\Phi}}{4g^2} F_{\mu\nu} F^{\mu\nu} - \frac{M_s^2}{2} e^{4\Phi} (\partial_\mu a + 2\zeta A_\mu)^2 \right.$$
$$\left. + \frac{C}{2} \frac{a}{\zeta} \left(\text{Tr}[Q] \, R \wedge R + \text{Tr}[Q^3] \, F \wedge F + \text{Tr}[QT^aT^a]\text{Tr}[G \wedge G] \right) \right]. \qquad (10.4.2)$$

Under the U(1) gauge transformation the axion transforms as $a \to a - 2\zeta\epsilon$, and since $\delta S_{\text{eff}} = -\delta S_{\text{triangle}}$, the full effective action, including the one-loop determinants is invariant. By dualizing the axion to an antisymmetric tensor, we obtain the (on-shell) equivalent effective action in the Einstein frame

$$S = \int \sqrt{G} \left[-\frac{e^{-2\Phi}}{4g^2} F_{\mu\nu} F^{\mu\nu} - \frac{e^{-4\Phi}}{12} \hat{H}_{\mu\nu\rho} \hat{H}^{\mu\nu\rho} \right] - \frac{\zeta}{2} M_s^2 \int \epsilon^{\mu\nu\rho\sigma} B_{\mu\nu} F_{\rho\sigma}, \qquad (10.4.3)$$

where $\hat{H}$ is the field strength of $B_{\mu\nu}$ up to gauge and gravitational Chern-Simons contributions. You are asked in exercise 10.7 to carefully derive the action (10.4.3) starting from the action (10.4.2).

In the heterotic string, the gauge-boson axion mixing term appears at one-loop order. The reason is that the crucial $B \wedge F$ mixing term cannot have a non-trivial dilaton dependence for reasons of gauge invariance. The coefficient of such a term at one loop can be computed directly. The coupling is parity violating so it will come from the odd spin structure.

To calculate ζ we will therefore consider the two-point function of an antisymmetric tensor and the "anomalous" U(1) gauge boson vertex operator in the odd spin structure of the torus. In this case, one of the vertex operators must be put in the zero picture and an insertion of the zero mode of the supercurrent is needed:

$$
\epsilon^1_{\mu\nu}\epsilon^2_\rho \langle B^{\mu\nu} A^\rho \rangle = \frac{\epsilon^1_{\mu\nu}\epsilon^2_\rho}{8(2\pi\ell_s)^4} \int_{\mathcal{F}} \frac{d^2\tau}{\tau_2^2} \int \frac{d^2z}{\tau_2} \left\langle (\partial X^\mu + ip_1 \cdot \psi \psi^\mu) \bar\partial X^\nu e^{ip_1 \cdot X} \right|_z
$$

$$
\times \left. \psi^\rho \bar{J} e^{ip_2 \cdot X}\right|_0 \oint \frac{dw}{2\pi i}\left(\frac{2i}{\ell_s^2}\psi^\sigma \partial X^\sigma + G^{\text{int}}\right)\bigg|_w \right\rangle_{\text{odd}},
\tag{10.4.4}
$$

where $\bar{J}$ is the right-moving U(1) current that generates the anomalous U(1) gauge boson.

If $\mathcal{N} \geq 2_4$ there are more than four fermion zero modes and the amplitude vanishes. For $\mathcal{N} = 1_4$ there are exactly four zero modes, and we will have to "use" the four fermions in order to obtain a nonzero answer:

$$
\epsilon^1_{\mu\nu}\epsilon^2_\rho \langle B^{\mu\nu} A^\rho\rangle = \frac{i\epsilon^1_{\mu\nu}\epsilon^2_\rho p_1^a}{8\pi(2\pi\ell_s)^4\ell_s^2}\int_{\mathcal{F}}\frac{d^2\tau}{\tau_2^2}\int\frac{d^2z}{\tau_2}\langle\psi^a\psi^\mu\psi^\rho\psi^\sigma\rangle\langle\partial X^\sigma\bar\partial X^\nu\rangle\langle\bar{J}\rangle + \mathcal{O}(p^2).
\tag{10.4.5}
$$

Note that only the expectation value of $\bar{J}$ is evaluated in the internal CFT. All other correlators are evaluated in the free CFT of four-dimensional Minkowski space.

The correlator $\langle\psi^a\psi^\mu\psi^\rho\psi^\sigma\rangle$ gets contributions from zero modes only. It is therefore position independent. $\langle\bar{J}\rangle$ is also position independent like all one-point functions on the torus. The only position-dependent correlator is $\langle\partial X^\sigma\bar\partial X^\nu\rangle$ We may calculate it, by integrating by parts and using (10.1.17)

$$
\int \frac{d^2z}{\tau_2}\langle\partial X^\sigma\bar\partial X^\nu\rangle = \pi\ell_s^2\,\delta^{\sigma\nu}.
\tag{10.4.6}
$$

The determinant with the insertion of two-fermionic zero modes can be computed as follows: Consider two fermions in the odd spin structure. The insertion of the two-fermion zero modes is equivalent to the insertion of the O(2) current zero mode $J_0^{ij} = \psi_0^i\psi_0^j + \cdots$. Therefore

$$
\langle\psi^i(z)\psi^j(w)\rangle_{\text{odd}} = \langle J_0^{ij}\rangle_{\text{odd}} = \frac{\epsilon^{ij}}{2\pi i}\frac{\partial}{\partial v}\text{Tr}[(-1)^F e^{2\pi ivJ_0^{ij}}]_{v=0} = \frac{\epsilon^{ij}}{2\pi i}\frac{\partial}{\partial v}\frac{-i\vartheta_1(v)}{\eta} = -\epsilon^{ij}\eta^2,
\tag{10.4.7}
$$

where we have used the affine characters in (4.12.27), (4.12.37), and (4.12.38) on page 76. Therefore, each pair of fermion zero modes produce a $-\eta^2$ and

$$
\langle\psi^a(z)\psi^\mu(z)\psi^\rho(0)\psi^\sigma(w)\rangle = \epsilon^{a\mu\rho\sigma}\,\eta^4.
\tag{10.4.8}
$$

There are also two (super)ghost zero modes here that contribute the inverse of two fermions, namely, $-1/\eta^2$. Moreover, we have a factor of $1/|\eta|^4$ originating from the four-dimensional bosonic coordinates. In total, we are left with a $-1/\bar\eta^2$ factor from the Minkowski part.

Using all this we may now extract ζ as

$$\zeta = -\frac{1}{8(2\pi)^4}\int_{\mathcal F}\frac{d^2\tau}{\tau_2^2}\frac{1}{\bar\eta^2}\mathrm{Tr}[(-1)^F\bar J]_R, \tag{10.4.9}$$

where the trace is taken in the R sector of the internal CFT. The odd spin-structure projects on the internal elliptic genus, which is antiholomorphic. Thus, the full integrand is anti-holomorphic, modular invariant, and has no pole at infinity (the tachyon is chargeless). Such a function is a constant and it can be found by considering the limit $\tau_2 \to \infty$, where only massless states contribute.

Using $\mathrm{Tr}[(-1)^F\bar J]_R = -2\pi\bar\eta^2\mathrm{Tr}[Q]_{\mathrm{massless}}$ and that the volume of the moduli space is $\pi/3$ we finally obtain

$$\zeta = \frac{\mathrm{Tr}[Q]_{\mathrm{massless}}}{48(2\pi)^2}. \tag{10.4.10}$$

An "anomalous" U(1) symmetry is spontaneously broken. From (10.4.2) by gauge fixing to a physical gauge $a = 0$ we observe that the anomalous U(1) gauge field has a (normalized) mass

$$M^2_{U(1)} = 4g^2\zeta^2 M_s^2. \tag{10.4.11}$$

The gauge symmetry is spontaneously broken. The fate of the associated global symmetry is of interest however. In supersymmetric $\mathcal N = 1_4$ vacua, there is a nontrivial D-term potential. At heterotic tree level, the associated potential contributions are

$$V_D = e^{-2\Phi}\left(\frac{1}{2}D^2 + D\left(\sum_i q_i|H_i|^2\right)\right), \tag{10.4.12}$$

where H_i are the complex scalars charged under the anomalous U(1), with charge q_i.

In the heterotic string, there is a one-loop contribution proportional to D. This can be seen directly, by noting that the vertex operator for the auxiliary D-term is $-\frac{i}{3}J\bar J$, where J is the world-sheet $\mathcal N = 2_2$ left-moving U(1) current, associated with the $\mathcal N = 1_4$ space-time supersymmetry. $\bar J$ is the right-moving U(1) current, generating the anomalous U(1) gauge symmetry. The total potential can be calculated either by using supersymmetry or by a direct loop-amplitude calculation. The result is

$$V_D = e^{-2\Phi}\left(\frac{1}{2}D^2 + D\left(\sum_i q_i|H_i|^2\right)\right) + \zeta D. \tag{10.4.13}$$

Eliminating the D-field by its equations of motion we obtain

$$V_D = -\frac{1}{2}e^{-2\Phi}\left(\sum_i q_i|H_i|^2 + \zeta e^{2\Phi}\right)^2. \tag{10.4.14}$$

The term linear in ζ is a (one-loop) mass term for the charged scalars. The term quadratic in ζ is a two-loop contact term.

When the string coupling is nonzero, $\langle e^{2\Phi} \rangle \neq 0$, this potential almost always triggers spontaneous breaking of the anomalous U(1) global symmetry. This implies the existence of a massless pseudoscalar (axion) in the spectrum. This particle is a linear combination of the universal axion, dual to the antisymmetric tensor, and the phases of the charged scalars that acquire expectation values. Such an axion can obtain a mass from instanton effects.

We may now move to the type-II string and discuss the possibility of anomalous U(1)'s. It is obvious that vectors emerging from the R-R sector cannot have that property, since they have no charged states in perturbation theory. In the NS-NS sector, in order to have chiral fermions in four dimensions, necessary for the existence of an anomalous U(1), the gauge symmetry must come from one sector of the string only (see section 9.17.2) on page 282. In exercise 4.19.3 on page 104 you are asked to show that in this case no $B \wedge F$ couplings can appear at one loop. There can be no anomalous U(1)'s in perturbative type-II vacua.

The situation in type-I vacua is richer. The axions, with which gauge fields from the open sector can mix are R-R states. Therefore the counting of couplings constants in the effective action (10.4.2) changes. Now, the mixing term comes from the disk (tree level). On the other hand, the term quadratic in the anomalous U(1) gauge field appears at one loop (cylinder). Moreover, since there are many axions in general in the R-R sector, there are typically several anomalous U(1) symmetries.

10.5 $\mathcal{N} = 1,2_4$ Examples of Threshold Corrections

We will examine here some sample evaluations of the one-loop threshold corrections described in the previous sections. Consider the $\mathcal{N} = 2_4$ heterotic vacuum described in section 9.4 on page 231. The partition function was given in (9.4.8). The gauge group is $E_8 \times E_7 \times SU(2) \times U(1)^2$ (apart from the graviphoton and the vector partner of the dilaton). From (10.1.23) we find that, up to the group trace, a vector multiplet contributes -1 and a hypermultiplet 1 to the β-function.

First we will compute the sum over the fermionic ϑ-functions appearing in (10.2.15).

$$\frac{i}{2\pi}\frac{1}{2}\sum_{even}(-1)^{a+b+ab}\,\partial_\tau\left(\frac{\vartheta[^a_b]}{\eta}\right)\frac{\vartheta[^a_b]\vartheta[^{a+h}_{b+g}]\vartheta[^{a-h}_{b-g}]}{\eta^3}\frac{Z_{4,4}[^h_g]}{|\eta|^4} = 4\frac{\eta^2}{\vartheta[^{1+h}_{1+g}]\vartheta[^{1-h}_{1-g}]}, \tag{10.5.1}$$

for $(h,g) \neq (0,0)$ and gives zero for $(h,g) = (0,0)$.

We will also compute the group trace for E_8. The level is $k = 1$ and the E_8 affine character is

$$\bar{\chi}_0^{E_8}(v_i) = \frac{1}{2}\sum_{a,b=0}^{1}\frac{\prod_{i=1}^{8}\bar{\vartheta}[^a_b](v_i)}{\bar{\eta}^8}. \tag{10.5.2}$$

Then

$$\left[\frac{1}{(2\pi i)^2}\partial_{v_1}^2 - \frac{1}{4\pi\tau_2}\right]\bar{\chi}_0^{E_8}(v_i)\bigg|_{v_i=0} = \frac{1}{12}(\hat{\bar{E}}_2\bar{E}_4 - \bar{E}_6), \tag{10.5.3}$$

which gives the correct value for the Casimir of the adjoint of E_8, namely, 60. Using also

$$\frac{1}{2} \sum_{(h,g)\neq(0,0)} \sum_{a,b=0}^{1} \frac{\bar{\vartheta}[^a_b]^6 \bar{\vartheta}[^{a+h}_{b+g}] \bar{\vartheta}[^{a-h}_{b-g}]}{\bar{\vartheta}[^{1+h}_{1+g}] \bar{\vartheta}[^{1-h}_{1-g}]} = -\frac{1}{4} \frac{\bar{E}_6}{\bar{\eta}^6}, \tag{10.5.4}$$

and putting everything together, we obtain $b_{E_8} = -60$ and

$$\Delta_{E_8} = \int_{\mathcal{F}} \frac{d^2\tau}{\tau_2} \left[-\frac{1}{12} \Gamma_{2,2} \frac{\hat{\bar{E}}_2 \bar{E}_4 \bar{E}_6 - \bar{E}_6^2}{\bar{\eta}^{24}} + 60 \right]. \tag{10.5.5}$$

In exercise 10.11 on page 318 you are invited to compute a similar threshold for the E_7 group,

$$\Delta_{E_7} = \int_{\mathcal{F}} \frac{d^2\tau}{\tau_2} \left[-\frac{1}{12} \Gamma_{2,2} \frac{\hat{\bar{E}}_2 \bar{E}_4 \bar{E}_6 - \bar{E}_4^3}{\bar{\eta}^{24}} - 84 \right]. \tag{10.5.6}$$

The difference between the two thresholds has a simpler form:

$$\Delta_{E_8} - \Delta_{E_7} = -144\Delta, \quad \Delta = \int_{\mathcal{F}} \frac{d^2\tau}{\tau_2} (\Gamma_{2,2} - 1). \tag{10.5.7}$$

The integral can be computed with the result

$$\Delta = -\log\left[4\pi^2 T_2 U_2 |\eta(T)\eta(U)|^4| \right]. \tag{10.5.8}$$

As we will show in the next section, (10.5.7) written as $\Delta_i - \Delta_j = (b_i - b_j)\Delta$ applies to all $K3 \times T^2$ vacua of the heterotic string. Taking the large volume limit $T_2 \to \infty$ in (10.5.8) we obtain

$$\lim_{T_2 \to \infty} \Delta = \frac{\pi}{3} T_2 + \mathcal{O}(\log T_2). \tag{10.5.9}$$

In the decompactification limit, the difference of gauge thresholds behaves as the volume of the two-torus. A similar result applies to the individual thresholds. This can be understood from the fact that a six-dimensional gauge coupling scales as [length].

It should be stressed, however, that the limits $\mu \to 0$ and $T_2 \to \infty$ in the IR regulated threshold do not commute. You are invited to investigate this question in exercise 10.14. This is important for the consistency of the formula (10.2.15) with decompactification. In particular, decompactifying to six dimensions we would expect the dimensionless six-dimensional gauge coupling to run in the IR as $g_6^2 \sim \mu^2$ consistent with scaling dimensions.

Consider now the $\mathcal{N} = 1_4$ $\mathbb{Z}_2 \times \mathbb{Z}_2$ orbifold vacuum described in section 9.6 with gauge group $E_8 \times E_6 \times U(1) \times U(1)'$. The partition function depends on the moduli (T_i, U_i) of the 3 two-tori ("planes"). In terms of the orbifold projection there are three types of sectors:

- $\mathcal{N} = 4_4$ sectors. They correspond to $(h_i, g_i) = (0, 0)$ and have $\mathcal{N} = 4_4$ supersymmetry structure. They give no correction to the gauge couplings.

- $\mathcal{N} = 2_4$ sectors. They correspond to one plane being untwisted while the other two are twisted. There are three of them and they have an $\mathcal{N} = 2_4$ structure. For this reason their contribution to the thresholds is similar to the ones we described above.

- $\mathcal{N} = 1_4$ sectors. They correspond to all planes being twisted. Such sectors do not depend on the untwisted moduli (T_i, U_i), but they may depend on twisted moduli.

The structure above is generic in $\mathcal{N} = 1_4$ orbifold vacua of the heterotic string.

In our example, the $\mathcal{N} = 1_4$ sectors do not contribute to the thresholds, so we can directly write down the the E_8 and E_6 threshold corrections as

$$\Delta_{E_8}^{N=1} = \int_{\mathcal{F}} \frac{d^2\tau}{\tau_2} \left[-\frac{1}{12} \sum_{i=1}^{3} \Gamma_{2,2}(T_i, U_i) \frac{\hat{\bar{E}}_2 \bar{E}_4 \bar{E}_6 - \bar{E}_6^2}{\bar{\eta}^{24}} + \frac{3}{2} 60 \right], \tag{10.5.10}$$

$$\Delta_{E_6}^{N=1} = \int_{\mathcal{F}} \frac{d^2\tau}{\tau_2} \left[-\frac{1}{12} \sum_{i=1}^{3} \Gamma_{2,2}(T_i, U_i) \frac{\hat{\bar{E}}_2 \bar{E}_4 \bar{E}_6 - \bar{E}_4^3}{\bar{\eta}^{24}} - \frac{3}{2} 84 \right]. \tag{10.5.11}$$

The extra factor $3/2$ in the β-function coefficient comes as follows. There is a $1/2$ because of the extra $\mathbb{Z}_2$ orbifold projection relative to the $\mathbb{Z}_2$ $\mathcal{N} = 2_4$ orbifold vacuum and a factor of 3 due to the three planes contributing. This is what we would expect from the massless spectrum. Remember that there are no scalar multiplets charged under the E_8. So the E_8 β-function comes solely from the $\mathcal{N} = 1_4$ vector multiplet and using (10.1.23) we can verify that it is $3/2$ times that of an $\mathcal{N} = 2_4$ vector multiplet.

The structure we have seen in the $\mathbb{Z}_2 \times \mathbb{Z}_2$ orbifold vacuum generalizes to more complicated $\mathcal{N} = 1_4$ orbifolds. It is always true that the untwisted moduli dependence of the threshold corrections comes only from the $\mathcal{N} = 2_4$ sectors.

We will also analyze here thresholds in $\mathcal{N} = 2_4$ vacua where $\mathcal{N} = 4_4$ supersymmetry is spontaneously broken to $\mathcal{N} = 2_4$. We will pick a simple vacuum described in section 9.4 on page 231. It is the usual $\mathbb{Z}_2$ orbifold acting on T^4 and one of the E_8 factors, but it is also accompanied by a $\mathbb{Z}_2$ lattice shift $X^1 \to X^1 + \pi$ in one of the coordinates of the left over two-torus. This is a freely acting orbifold and we have two massive gravitini in the spectrum. The geometrical interpretation is that of a compactification on a manifold that is locally of the form $K3 \times T^2$ but *not globally*. Its partition function is

$$Z_{\mathcal{N}=4_4 \to \mathcal{N}=2_4} = \frac{1}{2} \sum_{h,g=0}^{1} \frac{1}{\tau_2 |\eta|^4} \frac{\Gamma_{2,2}[^h_g]}{|\eta|^4} \frac{\bar{\Gamma}_{E_8}}{\bar{\eta}^8} Z_{(4,4)}[^h_g] \frac{1}{2} \sum_{\gamma,\delta=0}^{1} \frac{\bar{\vartheta}[^{\gamma+h}_{\delta+g}] \bar{\vartheta}[^{\gamma-h}_{\delta-g}] \bar{\vartheta}^6[^{\gamma}_{\delta}]}{\bar{\eta}^8}$$

$$\times \frac{1}{2} \sum_{a,b=0}^{1} (-1)^{a+b+ab} \frac{\vartheta^2[^a_b] \vartheta[^{a+h}_{b+g}] \vartheta[^{a-h}_{b-g}]}{\eta^4}, \tag{10.5.12}$$

where $\Gamma_{2,2}[^h_g]$ are the translated torus blocks described in Appendix B. In particular there is a $\mathbb{Z}_2$ phase $(-1)^{g\,m_1}$ in the lattice sum and n_1 is shifted to $n_1 + h/2$.

The gauge group of this vacuum is the same as the usual $\mathbb{Z}_2$ orbifold, namely $E_8 \times E_7 \times SU(2) \times U(1)^2$. The β-functions here are

$$b_{E_8} = -60, \quad b_{E_7} = -12, \quad b_{SU(2)} = 52. \tag{10.5.13}$$

After a straightforward evaluation we obtain that the thresholds can be written as

$$\Delta_I = b_I \Delta + \left(\frac{\tilde{b}_I}{3} - b_I \right) \delta - k_I Y, \tag{10.5.14}$$

where b_I are the β functions of this vacuum, while $\tilde{b}_I$ are those of the standard $\mathbb{Z}_2$ orbifold (without the torus translation). Moreover

$$\Delta = \int_{\mathcal{F}} \frac{d^2\tau}{\tau_2} \left[\sideset{}{'}\sum_{h,g} \Gamma_{2,2}[{}^h_g] - 1 \right] = -\log\left[\frac{\pi^2}{4} |\vartheta_4(T)|^4 |\vartheta_2(U)|^4 T_2 U_2 \right], \tag{10.5.15}$$

$$\delta = \int_{\mathcal{F}} \frac{d^2\tau}{\tau_2} \sideset{}{'}\sum_{h,g} \Gamma_{2,2}[{}^h_g] \bar{\sigma}[{}^h_g], \tag{10.5.16}$$

$$Y = \int_{\mathcal{F}} \frac{d^2\tau}{\tau_2} \sideset{}{'}\sum_{h,g} \Gamma_{2,2} \left[\frac{1}{12} \frac{\hat{\bar{E}}_2}{\bar{\eta}^{24}} \bar{\Omega}[{}^h_g] + \bar{\rho}[{}^h_g] + 40\bar{\sigma}[{}^h_g] \right], \tag{10.5.17}$$

where

$$\Omega[{}^0_1] = \frac{1}{2} E_4 \vartheta_3^4 \vartheta_4^4 (\vartheta_3^4 + \vartheta_4^4), \tag{10.5.18}$$

$$\sigma[{}^h_g] = -\frac{1}{4} \frac{\vartheta^{12}[{}^h_g]}{\eta^{12}}, \tag{10.5.19}$$

$$\rho[{}^0_1] = f(1-x), \quad \rho[{}^1_0] = f(x), \quad \rho[{}^1_1] = f(x/(x-1)), \tag{10.5.20}$$

with $x = \vartheta_2^4/\vartheta_3^4$ and

$$f(x) = \frac{4(8 - 49x + 66x^2 - 49x^3 + 8x^4)}{3x(1-x)^2}. \tag{10.5.21}$$

We will study the behavior of the above thresholds, in the limit in which $\mathcal{N} = 4_4$ supersymmetry is restored: $m_{3/2} \to 0$ or $T_2 \to \infty$. From (10.5.15), $\Delta \to -\log[T_2] + \cdots$ while the other contributions vanish in this limit.

This is different from the large volume behavior of the standard $\mathbb{Z}_2$ thresholds (10.5.5), which we have shown to diverge linearly with the volume T_2. The difference of behavior can be traced to the enhanced supersymmetry in the second example. There are two parts of the spectrum of the second vacuum: states with masses below $m_{3/2}$, which have effective $\mathcal{N} = 2_4$ supersymmetry and contribute logarithmically to the thresholds, and states with masses above $m_{3/2}$, which have effective $\mathcal{N} = 4_4$ supersymmetry and do not contribute. When we lower $m_{3/2}$, if there are always charged states below it, they will give a logarithmic divergence. This is precisely the case here. We could have turned on Wilson lines in such a way that there are no charged states below $m_{3/2}$ as $m_{3/2} \to 0$. In such a case the thresholds will vanish in the limit.

10.6 $\mathcal{N} = 2_4$ Universality of Thresholds

For vacua with $\mathcal{N} = 2_4$ supersymmetry, the threshold corrections have some universality properties. We will demonstrate this in vacua that come from $\mathcal{N} = 1_6$ theories compactified further to four dimensions on T^2. First, we observe that the derivative of the helicity ϑ-function that appears in the threshold formula essentially computes the supertrace of the helicity squared. In exercise 10.15 on page 319 you are invited to show that only short $\mathcal{N} = 2_4$ multiplets contribute to the supertrace and consequently to the thresholds. This projects on the elliptic genus of the internal CFT, which was defined in section 4.13.2 on page 79.

The gauge and gravitational thresholds can be written as

$$\Delta_I = \int_{\mathcal{F}} \frac{d^2\tau}{\tau_2} \left[\frac{\Gamma_{2,2}}{\bar{\eta}^{24}} \left(\mathrm{Tr}[Q_I^2] - \frac{k_I}{4\pi\tau_2} \right) \bar{\Omega} - b_I \right], \tag{10.6.1}$$

$$\Delta_{\mathrm{grav}} = \int_{\mathcal{F}} \frac{d^2\tau}{\tau_2} \left[\frac{\Gamma_{2,2}}{\bar{\eta}^{24}} \frac{\hat{\bar{E}}_2}{12} \bar{\Omega} - b_{\mathrm{grav}} \right]. \tag{10.6.2}$$

The function $\bar{\Omega}$ is constrained by modular invariance to be a weight-ten modular form, without singularities inside the fundamental domain. This is unique up to a constant

$$\bar{\Omega} = \xi \bar{E}_4 \bar{E}_6. \tag{10.6.3}$$

Consider further the integrand of $\Delta_I - k_I \Delta_{\mathrm{grav}}$ (without the b_I and b_{grav}). We find that it is anti-holomorphic with at most a single pole at $\tau = i\infty$; thus, it must be of the form $A_I \bar{j} + B_I$, where A_I, B_I are constants and j is the modular-invariant function defined in (C.45) on page 511.

Consequently, the thresholds can be written as

$$\Delta_I = \int_{\mathcal{F}} \frac{d^2\tau}{\tau_2} \left[\Gamma_{2,2} \left(\frac{\xi k_I}{12} \frac{\hat{\bar{E}}_2 \bar{E}_4 \bar{E}_6}{\bar{\eta}^{24}} + A_I \bar{j} + B_I \right) - b_I \right], \tag{10.6.4}$$

$$\Delta_{\mathrm{grav}} = \xi \int_{\mathcal{F}} \frac{d^2\tau}{\tau_2} \left[\Gamma_{2,2} \frac{\hat{\bar{E}}_2 \bar{E}_4 \bar{E}_6}{12\bar{\eta}^{24}} - b_{\mathrm{grav}} \right]. \tag{10.6.5}$$

We can now fix the constants as follows. In the gauge threshold, there should be no $1/\bar{q}$ pole (the tachyon is not charged), which gives

$$A_I = -\frac{\xi k_I}{12}. \tag{10.6.6}$$

Also the constant term is the β-function, which implies

$$744 A_I + B_I - b_I + k_I b_{\mathrm{grav}} = 0, \tag{10.6.7}$$

with $b_{\mathrm{grav}} = -22\xi$ from (10.6.5). Finally b_{grav} can be computed from the massless spectrum. Using the results of the previous section we find that

$$b_{\mathrm{grav}} = \frac{22 - N_V + N_H}{12}, \tag{10.6.8}$$

where N_V is the number of massless vector multiplets (excluding the graviphoton and the vector partner of the dilaton) and N_H is the number of massless hypermultiplets.

Since these vacua decompactify trivially to six dimensions we might use this to constrain further our data.

As advocated in exercise 7.35 on page 186, six-dimensional gravitational anomaly cancellation implies that

$$N_H^{d=6} - N_V^{d=6} - 29 N_T^{d=6} = 273 \tag{10.6.9}$$

where $N_{V,H}^{d=6}$ are the number of six-dimensional vectors and hypermultiplets while $N_T^{d=6}$ is the number of six-dimensional tensor multiplets. For perturbative heterotic vacua

$N_T^{d=6} = 1$ and we obtain $N_H^{d=6} - N_V^{d=6} = 244$. Upon toroidal compactification to four dimensions we obtain an extra two vector multiplets (from the supergravity multiplet). Thus, in four dimensions, $N_H - N_V = 242$ and from (10.6.8) we obtain $b_{\text{grav}} = 22, \xi = -1$ for all such vacua. The thresholds are now completely fixed in terms of the β-functions of massless states:

$$\Delta_I = b_I \, \Delta - k_I \, Y, \tag{10.6.10}$$

with

$$\Delta = \int_{\mathcal{F}} \frac{d^2\tau}{\tau_2} \left[\Gamma_{2,2}(T, U) - 1 \right]$$
$$= -\log \left(4\pi^2 |\eta(T)|^4 |\eta(U)|^4 \, \text{Im} T \, \text{Im} U \right), \tag{10.6.11}$$

$$Y = \frac{1}{12} \int_{\mathcal{F}} \frac{d^2\tau}{\tau_2} \, \Gamma_{2,2}(T, U) \left[\frac{\hat{\bar{E}}_2 \bar{E}_4 \bar{E}_6}{\bar{\eta}^{24}} - \bar{j} + 1008 \right], \tag{10.6.12}$$

and

$$\Delta_{\text{grav}} = -\int_{\mathcal{F}} \frac{d^2\tau}{\tau_2} \left[\Gamma_{2,2} \frac{\hat{\bar{E}}_2 \bar{E}_4 \bar{E}_6}{12 \, \bar{\eta}^{24}} - 22 \right]. \tag{10.6.13}$$

As can be seen from (10.2.14), the universal term Y can be absorbed into a redefinition of the tree-level string coupling. We can then write

$$\frac{16\pi^2}{g_I^2(\mu)} = k_I \frac{16\pi^2}{g_{\text{renorm}}^2} + b_I \log \frac{M_s^2}{\mu^2} + \hat{\Delta}_I, \tag{10.6.14}$$

where we have defined a "renormalized" string coupling by

$$g_{\text{renorm}}^2 = \frac{g_u^2}{1 - \frac{Y}{16\pi^2} g_u^2}. \tag{10.6.15}$$

Of course, such a coupling is meaningful, provided it appears as the natural expansion parameter in several amplitudes that are relevant for the low-energy string physics. In general, this might not be the case as a consequence of some arbitrariness in the decomposition (10.6.10), which is not valid in general. Examples of this kind arise in $\mathcal{N} = 1_4$ vacua as well as in certain more general $\mathcal{N} = 2_4$ vacua.

It is important to keep in mind that this "renormalized" string coupling is defined here in a *moduli-dependent* way. This moduli dependence can affect the string unification. Indeed, as we will see in the sequel, when proper unification of the couplings appears, namely when $\hat{\Delta}_I$ can be written as $b_I \Delta$, their common value at the unification scale is g_{renorm}, which therefore plays the role of a phenomenologically relevant parameter. Moreover, the unification scale turns out to be proportional to M_s. The latter can be expressed in terms of the "low-energy" parameters g_{renorm} and M_P, by using the fact that the Planck mass is not renormalized:

$$M_s = \frac{M_P g_{\text{renorm}}}{\sqrt{1 + \frac{Y}{16\pi^2} g_{\text{renorm}}^2}}. \tag{10.6.16}$$

How much Y, which is moduli dependent, can affect the running of the gauge couplings can be seen from its numerical evaluation. We take $T = iR_1 R_2$ and $U = iR_1/R_2$, which

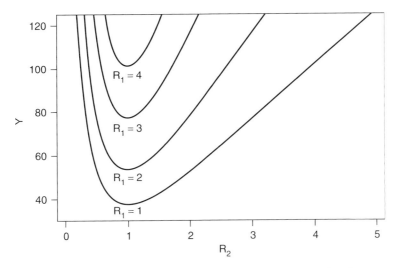

Figure 10.3 Plots of the universal threshold $Y(R_1, R_2)$ as a function of R_2 for $R_1 = 1, 2, 3, 4$ in string units.

corresponds to two orthogonal circles of radii $R_{1,2}$. The values of Y are plotted as functions of $R_{1,2}$ in Fig. 10.3.

10.7 Unification Revisited

In section 9.18 on page 284 we have discussed the apparent unification of observable gauge couplings and the realization of this setting in string theory. We have indicated in particular that this happens naturally in the heterotic string. It is time to look at this effect in detail, allowing for the corrections coming from one-loop effects.

Taking into account the one-loop running of couplings the unification relation in field theory becomes, in the $\overline{DR}$ scheme,

$$\frac{16\pi^2}{g_I^2(\mu)} = k_I \frac{16\pi^2}{g_U^2} + b_I \log \frac{M_U^2}{\mu^2}. \tag{10.7.1}$$

We may now use the results of the calculation of the running of gauge couplings in string theory, in the previous sections. This was essentially given in (10.2.14). We would like to express the various parameters in terms of a measurable mass scale such as the Planck mass, which is related to the string scale as in (9.17.1) on page 280. We will assume for simplicity the case of $\mathcal{N} = 2_4$ thresholds (10.6.10). We obtain a formula similar to (10.7.1) with

$$g_U^2 = \frac{g_u^2}{1 - \frac{g_u^2 Y}{16\pi^2}}, \quad M_{SU}^2 = \frac{2e^{1-\gamma}}{\pi\sqrt{27}} e^\Delta M_s^2 = \frac{2e^{1-\gamma}}{\pi\sqrt{27}} e^\Delta \frac{M_P^2 g_U^2}{1 + \frac{g_U^2 Y}{16\pi^2}}. \tag{10.7.2}$$

Both the "unified" coupling and "string unification mass," M_{SU}, are functions of the moduli. Moreover, they depend not only on the gauge-dependent threshold Δ but also on the gauge-independent correction Y.

To obtain an idea how far we are from the unification values, we evaluate M_{SU} from (10.7.2) using $M_P = 1.72 \times 10^{18}$ GeV, $g_U^2 = 4\pi/25$, and $\Delta = Y = 0$ to obtain

$$M_{SU} \simeq 5.27 \times 10^{17} \text{GeV}. \tag{10.7.3}$$

This is a factor of 40 larger than the unification scale as obtained from the meeting point of the SM coupling constants in section 9.18 on page 284, namely, $M_U = 10^{16.1}$ GeV. It is also a factor of 20 larger, than the larger value of $M_U = 10^{16.4}$ GeV allowed by the experimental error in the fit.

The discrepancy could be compensated by possible threshold corrections that give non-zero Δ and Y. A value of $\Delta \simeq -3.69$ would match the difference of 40. However, Δ is mostly positive, and attains its most negative value at the SU(3) point, $\Delta(T = U = e^{2\pi i/3}) \simeq -1.6$ which is not enough.

Similarly, a value of $Y \simeq 12 \times 10^3$ is needed to compensate the factor of 40. From figure 10.3 this requires a large volume for the torus. When the volume is large in string units, the KK energy scale $M_{KK} \simeq M_s/\sqrt{T_2}$ is well below the string scale: $M_{KK} \ll M_s$. In this case, the four-dimensional description of the physics is valid for energies $E \lesssim M_{KK}$. For energies above M_{KK}, and approaching M_s, the physics is effectively six-dimensional. The running of couplings in this range is different. The coupling constant has dimensions of length, and the dimensionless coupling runs proportional to the energy

$$g_6^2 \sim E^2. \tag{10.7.4}$$

If this is the case, the gauge couplings that start being of order 10^{-1}–10^{-2} at low energy quickly move in the nonperturbative region, above the KK scale. This is the *decompactification problem* for the gauge couplings. Therefore, generically, to avoid strong coupling problems, the KK scale should be very close to the string scale.

There are ways out of the decompactification problem. The idea behind them was indicated at the end of section 10.5. If the theory, upon compactification has a supersymmetry generated by 16 supercharges, then the higher-dimensional couplings do not run. This possibility has been developed in [270,271,292].

A third possibility is to make the gravitational coupling run faster [272]. This could be achieved if there is a new dimension whose KK states have gravitational couplings but not gauge couplings. In such a case, the gauge couplings will be insensitive to the presence of the new dimension, while the running of the gravitational coupling will be accelerated.

It is not known how to implement this option in the perturbative heterotic string. It is however possible to implement it in the strongly coupled heterotic string (to be discussed in section 11.8). Moreover, such a notion is natural in the type I string as was first pointed out in [272] and further analyzed in [276]. Consider a gauge group arising from D_3-branes. In this case, there are no KK states that are charged[8] under the gauge group. Therefore, a large dimension, although always felt by gravity, is invisible to the gauge group.

We should mention before ending, another , rather obvious possibility: contributions to the thresholds from particles at intermediate energies [293]. Such particles can arise because of gaugino condensation in the hidden sector.

---

[8] There are, however, winding states. Their masses are large, being proportional to the radius.

The analysis of the running of gauge couplings in more realistic heterotic string vacua and the associated problems are summarized in the review [282] to which we refer the reader for a more detailed account.

Bibliography

The subject of loop corrections in string theory is vast. A rather detailed discussion of various one-loop amplitudes, mostly in ten dimensions is given in [7]. Moreover various formalisms are used including the NSR formalism used here and the Green-Schwarz formalism.

The original approach to calculating gauge threshold in four-dimensional heterotic string compactifications can be found in [294]. The moduli dependence of differences of gauge couplings was calculated in [295]. Their incorporation in supergravity and its connection to σ-model anomalies is discussed in [296]. The approach of calculating the general moduli dependence was developed in [297,298]. Higher-genus corrections were evaluated in [299].

In [300] there is a lucid and rather complete discussion of the moduli dependent effective action of supersymmetric theories relevant to the superstring. Another excellent presentation of the moduli-dependent low-energy string effective action is given in [301].

The IR-regulated background field method for the gauge and gravitational thresholds in closed string theory was developed in [302]. The correlation of one-loop gauge thresholds with BPS states was developed in [303]. The universality of heterotic thresholds for standard or spontaneously broken supersymmetry was presented in [304,292]. A general review of one-loop threshold corrections in heterotic model building with an exhaustive list of references can be found in [282].

The one and two-loop evaluation of Fayet-Iliopoulos D-terms associated with the anomalous U(1) of the heterotic string is reviewed in [305] where references to previous work are provided.

Techniques to calculate higher-derivative threshold corrections associated with topological amplitudes were developed in [306,307].

Loop corrections to the ten-dimensional heterotic string for higher-derivative, anomaly related couplings were developed in [308,309]. A general overview of loop corrections in this context can be found in [310].

Higher orders in string perturbation theory are a difficult subject. There has been a lot of discussion on the proper formulation of the higher-genus vacuum diagrams. References [311,107] summarize what is known. Two-loop amplitudes in particular were analyzed more recently. A good starting point for the relevant literature is [312].

A good reference on open string threshold corrections to gauge couplings in four dimensions is [313], which also contains references to earlier work. Other one-loop open string threshold calculations can be found in [314–316]. A review of relevant issues in the open string sector can be found in [264].

Many other loop calculations, especially of higher derivative amplitudes, have been done, motivated by tests of nonperturbative symmetries. We will review some of them them in the next chapter.

Exercises

10.1. Derive, using CFT techniques the two-point function (10.1.10) on page 297 of currents on the torus.

10.2. Derive in detail (10.1.14) on page 298.

10.3. Consider the current-current perturbation (10.2.8) on page 303 that turns on a magnetic field. The one-loop correction to the gauge couplings can be evaluated, by computing the one-loop partition function in the presence of the magnetic field. It must be expanded to second order in the magnetic field. The coefficient is the one-loop gauge coupling threshold. Use the fact that current-current perturbations change the charge lattice of a CFT by $O(p, q)$ boosts. If this is too hard, you may find the answers in [302].

10.4. Calculate the one-loop two-graviton amplitude for the heterotic string in the even spin-structures. To avoid that it vanishes on shell, impose all physical state conditions except momentum conservation. Show that the correction to the Einstein term is indeed proportional to (10.3.2) on page 304. Keeping the $\mathcal{O}(p^4)$ terms derive the R^2 threshold (10.3.3).

10.5. Calculate the one-loop two-graviton amplitude in the heterotic string for the odd spin structure. Derive from this the $R \wedge R$ threshold. What is the relationship to the R^2 threshold if we have $\mathcal{N} = 1_4$ supersymmetry?

10.6. Consider the one-loop correction to the Einstein term in the type-II string. Unlike the heterotic string this is nonzero. Show that it is proportional to the $\text{Tr}[(-1)^{F_L+F_R}]_{RR}$. For compactifications on a CY threefold, this gives a correction proportional to the Euler number.

10.7. Dualize the four-dimensional axion field a into a two-index antisymmetric tensor and derive the action (10.4.3) dual to (10.4.2) on page 306.

10.8. Calculate the one-point function of $-\frac{i}{3}J\bar{J}$ on the torus and show that the result is again given by (10.4.9) on page 308. This justifies the result (10.4.13).

10.9. Consider the generation of a $B \wedge F$ one-loop correction in the type-II string. By considering all possibilities show that such a term cannot be generated.

10.10. Consider anomalous $U(1)$ gauge symmetries in six dimensions. Show that the structure of possible anomalous square diagrams is richer than four dimensions. What type of forms can be used for the cancellation of mixed anomalies? When is the gauge symmetry broken?

10.11. Calculate the threshold for the E_7 group in the $\mathcal{N} = 2_4$ vacuum constructed in section 9.4 on page 231, and analyzed in 10.5 on page 309. The E_7 group trace is given by

$$\left[\text{Tr}Q_{E_7}^2 - \frac{1}{4\pi\tau_2}\right] = \left[\frac{1}{(2\pi i)^2}\partial_v^2 - \frac{1}{4\pi\tau_2}\right]\frac{1}{2}\sum_{a,b}\frac{\bar{\vartheta}\begin{bmatrix}a\\b\end{bmatrix}(v)\bar{\vartheta}^5\begin{bmatrix}a\\b\end{bmatrix}\bar{\vartheta}\begin{bmatrix}a+h\\b+g\end{bmatrix}\bar{\vartheta}\begin{bmatrix}a-h\\b-g\end{bmatrix}}{\bar{\eta}^8}\Bigg|_{v=0}. \tag{10.1E}$$

Show that the β-function coefficient is 84 ($I_2(133) = 36, I_2(56) = 12$) and

$$\Delta_{E_7} = \int_{\mathcal{F}}\frac{d^2\tau}{\tau_2}\left[-\frac{1}{12}\Gamma_{2,2}\frac{\hat{\bar{E}}_2\bar{E}_4\bar{E}_6 - \bar{E}_4^3}{\bar{\eta}^{24}} - 84\right]. \tag{10.2E}$$

Show also that $b_{SU(2)} = 84$.

10.12. Show that the gauge group of the vacuum described by (10.5.12) on page 311 is the same as the usual $\mathbb{Z}_2$ orbifold, namely, $E_8 \times E_7 \times SU(2) \times U(1)^2$. Show that there are also four neutral massless hypermultiplets and one transforming as [2,56]. Confirm that there are no massless states coming from the twisted sector. Use (4.18.66) on page 103 to show that the mass of the two massive gravitini is given by

$$m_{3/2}^2 = \frac{|U|^2}{T_2 U_2}.$$ (10.3E)

Show that the β-function coefficients here are

$$b_{E_8} = -60, \quad b_{E_7} = -12, \quad b_{SU(2)} = 52.$$ (10.4E)

10.13. Calculate the leading decompactification limit to six dimensions of the universal threshold correction in (10.6.12) on page 314.

10.14. Consider the IR regulated gauge threshold in six noncompact dimensions. Match to the low-energy six-dimensional EFT along the lines of section 10.2 on page 301. Show that the six-dimensional gauge coupling now runs as $g^2 \sim \mu^2$. Deduce from this that the limits $\mu \to 0$ and $T_2 \to \infty$ in the four-dimensional one-loop correction do not commute.

10.15. Consider the various $\mathcal{N} = 2_4$ representations, described in appendix J on page 538. Show that only half-BPS representations give a non-zero contribution to the supertrace of the helicity squared, and therefore also to the gauge thresholds.

10.16. Consider the $T^4/\mathbb{Z}_2$ orientifold described in section 9.14.3 on page 260. Calculate the mass matrix of the two abelian gauge bosons of the gauge group, by calculating the cylinder diagram with the gauge bosons inserted at opposite boundaries. Show that they are both massive.

10.17. Consider the two anomalous $U(1)$'s of the $T^4/\mathbb{Z}_2$ orbifold. Turn on a magnetic field along one of them and show that the one-loop effective action has a non-zero linear piece in the magnetic field. Interpret its coefficient.

10.18. The threshold corrections to gauge couplings in open string theory can be calculated using the on-shell background field method: turn on a constant magnetic field, calculate the one-loop effective action and extract the coefficient of the quadratic term. Consider this calculation in the type-I string compactified to four-dimensions on T^6. Show that the gauge coupling does not renormalize. Why?

10.19. Consider the type I string on $T^4/\mathbb{Z}_2$ orientifold discussed in 9.14.3 on page 260. Compactify further on a T^2 torus to four dimensions. Calculate the threshold correction to the gauge couplings using the method of the previous exercise.

11 | Duality Connections and Nonperturbative Effects

In this chapter we will provide a guide to nonperturbative string dualities. Such dualities were uncovered in the mid-1990s and have improved our understanding of several nonperturbative features of string theory. They emerged as an answer to the question: what is the strong coupling limit of each of the five consistent ten-dimensional superstring theories. They indicated the presence of new degrees of freedom described by extended objects (branes) using supersymmetry in a crucial way.

Nonperturbative properties of string theory were developed in parallel with similar issues in the context of supersymmetric field theories. We will not discuss here the progress in the understanding of nonperturbative effects in supersymmetric field theories. We would like to point out, however, that nonperturbative field theory dynamics can be naturally understood in the context of string theory. This fact led to important cross-fertilization between the two disciplines.

We are now confident that the five different supersymmetric string theories in ten dimensions are connected. We also take seriously their (extended) solitonic excitations. One type of such excitations, D-branes, provide the appropriate microscopic degrees of freedom essential for reproducing the Bekenstein-Hawking (BH) entropy of black holes as it we will explain in chapter 12.

The credible picture today is that the underlying theory is unique and different perturbative string theories are distinct vacua of this theory. There also exist vacua without a perturbative string theory description. A notable example is the strong coupling limit of the type-IIA theory, which is a Lorentz-invariant eleven-dimensional theory whose low-energy effective action is that of $\mathcal{N} = 1_{11}$ supergravity. This is the vacuum with the highest symmetry. It is termed M-theory, although the name is also used to describe the underlying unified theory whose different vacua include the known string theories. To date, a definition/formulation of the underlying theory is still lacking.

We have already seen that in ten dimensions there are five distinct, consistent supersymmetric string theories: type IIA/B, heterotic (O(32), $E_8 \times E_8$) and the unoriented O(32) type-I theory that contains also open strings. The two type-II theories have $\mathcal{N} = 2_{10}$ supersymmetry while the others have only $\mathcal{N} = 1_{10}$.

We will find that the strong coupling behavior of each of these theories can be mapped to the weak-coupling behavior of another theory. This is the basic notion of duality, generically called S-duality. Sometimes, the strong-coupling behavior is equivalent to the weak-coupling behavior of the same theory. In this case we talk about "self-duality." There are also dualities that are perturbative in the string coupling. The basic example, T-duality, we have seen already in sections 4.18.7 on page 101 and 6.4 on page 151.

An important conceptual and practical question that arises is: how can we establish a duality conjecture, if we have no control over the strong coupling behavior of a given theory. It turns out that in certain supersymmetric theories, although we cannot solve the theory at strong coupling, there are certain data of the strong coupling physics that can still be under control. Such data include the spectra of special states, the BPS states, as well as special "BPS-saturated" effective couplings, whose quantum corrections can be controlled nonperturbatively because of supersymmetry. Using such data, one can directly compare strong coupling results of the original theory with perturbative (weak-coupling) results of the dual theory.

Once we are convinced of a given duality, we can put it to work: weak coupling calculations in the dual theory provide strong-coupling results in the original theory. Such results then can be interpreted in terms of nonperturbative objects, typically instantons.

The main duality connections can be summarized as follows: There is evidence that the type-IIB theory has an $SL(2,\mathbb{Z})$ symmetry that, among other things, includes the inversion of the coupling constant. Consequently, the strong coupling limit of type IIB is isomorphic to the perturbative type-IIB theory. Upon compactification on tori, this symmetry combines with the perturbative T-duality symmetries to produce a large discrete duality group known as the U-duality group, which is a discretization of the noncompact continuous symmetries of the effective supergravity theory.

It can be argued that the strong-coupling limit of type-IIA theory is described by an 11-dimensional theory named "M-theory." Its low-energy limit is 11-dimensional supergravity. Compactification of M-theory on a circle of very small radius gives the perturbative type-IIA theory. If instead we compactify M-theory on the $\mathbb{Z}_2$ orbifold of the circle $T^1/\mathbb{Z}_2$ then we obtain the heterotic $E_8 \times E_8$ theory. When the circle is large, the heterotic theory is strongly coupled, while for small radius it is weakly coupled.

Finally, the strong-coupling limit of the O(32) heterotic string theory is the type-I O(32) theory, and vice versa.

There is another nontrivial nonperturbative connection in six dimensions: the strong-coupling limit of the six-dimensional toroidally compactified heterotic string is given by the type-IIA theory compactified on K3 and vice versa.

We are thus led to suspect that there is an underlying "universal" theory whose various limits in its "moduli" space produce the weakly coupled ten-dimensional supersymmetric string theories. The correct description of this theory is unknown. Attempts to describe it using matrices have merits but such descriptions are background dependent. Matrix approaches to string theory are, however, very useful. We will study some of them in chapter 14.

In the following sections we will describe in some depth both the duality relations among different string theories as well as some quantitative tests of such dualities.

11.1 Perturbative Connections

By compactifying one dimension on a circle, we can connect the two heterotic theories as well as the two type-II theories. This is schematically represented with the broken arrows in figure 11.1.

We will first show how the heterotic O(32) and $E_8 \times E_8$ theories are connected in $D = 9$. Upon compactification on a circle of radius R we can turn on 16 Wilson lines according to our discussion in section 9.1 on page 219. The partition function of the O(32) heterotic theory can then be written as

$$Z_{D=9}^{O(32)} = \frac{1}{(\sqrt{\tau_2}\eta\bar{\eta})^7} \frac{\Gamma_{1,17}(R, Y^I)}{\eta\bar{\eta}^{17}} \frac{1}{2} \sum_{a,b=0}^{1} (-1)^{a+b+ab} \frac{\vartheta^4[^a_b]}{\eta^4}, \tag{11.1.1}$$

where the lattice sum $\Gamma_{1,17}$ was given explicitly in (D.3), (D.4). We will focus on some special values for the Wilson lines Y^I, namely, take eight of them to be zero and the other eight to be 1/2. Then the lattice sum (in Lagrangian representation (D.1)) can be rewritten as

$$\Gamma_{1,17}(R) = R \sum_{m,n\in\mathbb{Z}} \exp\left[-\frac{\pi R^2}{\tau_2}|m + \tau n|^2\right] \frac{1}{2} \sum_{a,b} \bar{\vartheta}^8[^a_b] \bar{\vartheta}^8[^{a+n}_{b+m}]$$

$$= \frac{1}{2} \sum_{h,g=0}^{1} \Gamma_{1,1}(2R)[^h_g] \frac{1}{2} \sum_{a,b} \bar{\vartheta}^8[^a_b] \bar{\vartheta}^8[^{a+h}_{b+g}], \tag{11.1.2}$$

where $\Gamma_{1,1}[^h_g]$ are the orbifold blocks under a $\mathbb{Z}_2$ translation of the circle partition function

$$\Gamma_{1,1}(R)[^h_g] = R \sum_{m,n\in\mathbb{Z}} \exp\left[-\frac{\pi R^2}{\tau_2}\left|\left(m + \frac{g}{2}\right) + \tau\left(n + \frac{h}{2}\right)\right|^2\right] \tag{11.1.3}$$

$$= \frac{1}{R} \sum_{m,n\in\mathbb{Z}} (-1)^{mh+ng} \exp\left[-\frac{\pi}{\tau_2 R^2}|m + \tau n|^2\right] \tag{11.1.4}$$

as calculated already in (4.21.25) and (4.21.26) on page 112.

In the $R \to \infty$ limit, (11.1.3) implies that only $(h, g) = (0, 0)$ contributes in the sum in (11.1.2) and we end up with the O(32) heterotic string in ten dimensions. In the $R \to 0$ limit the theory decompactifies again, but from (11.1.4) we deduce that all (h, g) sectors contribute equally in the limit. The sum on (a, b) and (h, g) factorizes and we end up with the $E_8 \times E_8$ theory in ten dimensions. Thus, both theories are different limiting points (boundaries) in the moduli space of toroidally compactified heterotic strings.

From the group theoretic point of view, the nine-dimensional theory constructed has a massless gauge group which is $SO(16) \times SO(16) \subset SO(32)$. There are two types of KK states. One set transforms as $(128,1) \otimes (1,128)$ under $SO(16) \times SO(16)$ with masses proportional to R. When $R \to 0$, they become massless. They combine with the adjoints of

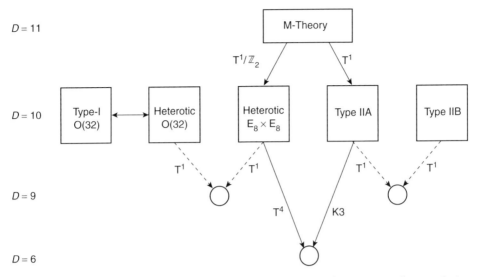

Figure 11.1 The web of duality symmetries between string theories. Broken lines correspond to perturbative duality connections. Type IIB in ten dimensions is supposed to be self-dual under SL(2,ℤ).

SO(16)×SO(16) to form the adjoint of $E_8 \times E_8$. The other set of KK states transforms as the (16,16) of SO(16)×SO(16), with masses proportional to $1/R$. They become massless as $R \to \infty$ completing the adjoint of SO(32).

In the type-II case the situation is similar. We compactify on a circle. Under an $R \to 1/R$ duality

$$\partial X^9 \to \partial X^9, \quad \psi^9 \to \psi^9, \quad \bar{\partial} X^9 \to -\bar{\partial} X^9, \quad \bar{\psi}^9 \to -\bar{\psi}^9. \tag{11.1.5}$$

As advocated in section 7.7 on page 174 T-duality here changes the GSO projection on the left-moving side of the string. Consequently the duality maps type IIA to type IIB, and vice versa. We can also phrase this in the following manner: the $R \to \infty$ limit of the toroidally compactified type-IIA string gives the type-IIA theory in ten dimensions. The $R \to 0$ limit gives the type-IIB theory in ten dimensions.

11.2 BPS States and BPS Bounds

The notion of BPS states is of capital importance in discussions of non-perturbative duality symmetries. Massive BPS particless appear in theories with extended supersymmetry. Even in the case of $\mathcal{N} = 1$ supersymmetry there is an analog of BPS states (shorter representations), namely the massless states.

It happens that supersymmetry representations are sometimes shorter than usual. This is due to some of the supersymmetry operators being "null," so that they cannot create new states. The vanishing of the action of some supercharges depends on the relation between the mass of a multiplet and some central charges appearing in the supersymmetry algebra. These central charges depend on electric and magnetic charges of the theory as well as on expectation values of scalars (moduli) or coupling constants in the rigid supersymmetric theories. In a sector with given charges, the BPS states are the lowest-lying states and they saturate the so-called BPS bound. For pointlike states, it is of the form

$$2M \geq \text{ maximal eigenvalue of } Z,$$ (11.2.1)

where Z is the central charge matrix. It appears in the anticommutator of supersymmetry charges (J.1) on page 537. All of the above are explicitly worked out for supersymmetry in four dimensions in Appendix J. There the representations of extended supersymmetry are described in detail.

BPS states behave in a very special way:

• At generic points in moduli space they are absolutely stable. The reason is the dependence of their mass on conserved charges. Charge and energy conservation prohibits their decay. Consider as an example, the BPS mass formula

$$M_{m,n}^2 = \frac{|m + n\tau|^2}{\tau_2},$$ (11.2.2)

where m, n are integer-valued conserved charges, and τ is a complex modulus. This BPS formula is relevant for $\mathcal{N} = 4_4$, SU(2) gauge theory, in a subspace of its moduli space. Consider a BPS state with charges (m_0, n_0), at rest, decaying into N states with charges (m_i, n_i) and masses M_i, $i = 1, 2, \ldots, N$. Charge conservation implies that $m_0 = \sum_{i=1}^{N} m_i$, $n_0 = \sum_{i=1}^{N} n_i$. The four-momenta of the produced particles are $(\sqrt{M_i^2 + \vec{p}_i^2}, \vec{p}_i)$ with $\sum_{i=1}^{N} \vec{p}_i = \vec{0}$. Conservation of energy implies

$$M_{m_0,n_0} = \sum_{i=1}^{N} \sqrt{M_i^2 + \vec{p}_i^2} \geq \sum_{i=1}^{N} M_i.$$ (11.2.3)

Also in a given charge sector (m,n) the BPS bound implies that any mass $M \geq M_{m,n}$, with $M_{m,n}$ given in (11.2.2). Thus, from (11.2.3) we obtain

$$M_{m_0,n_0} \geq \sum_{i=1}^{N} M_{m_i,n_i},$$ (11.2.4)

and the equality will hold if all particles are BPS and are produced at rest ($\vec{p}_i = \vec{0}$). Consider now the two-dimensional vectors $v_i = m_i + \tau n_i$ on the complex τ-plane, with length $||v_i||^2 = |m_i + n_i\tau|^2$. They satisfy $v_0 = \sum_{i=1}^{N} v_i$. Repeated application of the triangle inequality implies

$$||v_0|| \leq \sum_{i=1}^{N} ||v_i||.$$ (11.2.5)

This is incompatible with energy conservation (11.2.4) unless all vectors v_i are parallel. This may happen only if τ is real. For energy conservation it should also be a rational number. On the other hand, due to the SL(2,$\mathbb{Z}$) invariance of (11.2.2), the inequivalent choices for τ are in the SL(2,$\mathbb{Z}$) fundamental domain and τ is never real there. In fact, real rational values of τ are mapped by SL(2,$\mathbb{Z}$) to $\tau_2 = \infty$, and since τ_2 is the inverse of the coupling constant, this corresponds to the degenerate case of zero coupling. Consequently, for τ_2

finite, in the fundamental domain, the BPS states of this theory are absolutely stable. This is always true in theories with more than eight conserved supercharges (corresponding to $\mathcal{N} > 2_4$ supersymmetry). In cases corresponding to theories with 8 supercharges, there are regions in the moduli space, where BPS states, stable at weak coupling, can decay at strong coupling. However, there is always a large region around weak coupling where they are stable.

• Their mass formula is expected to be exact if one uses renormalized values for the mass and the moduli (couplings). The argument is that quantum corrections would spoil the relation of mass and charges, if we assume unbroken SUSY at the quantum level. This would give incompatibilities with the dimension of their representations. Of course this argument has a loophole: a specific set of BPS multiplets can combine into a long one. In such a case, the argument above does not prohibit corrections. Thus, we have to count BPS states modulo long supermultiplets. This is precisely what helicity supertrace formulae do for us. They are defined in Appendix J.

The BPS states described above can be realized as pointlike soliton solutions of the relevant effective supergravity theory. The BPS condition is equivalent to the statement that the soliton solution leaves part of the supersymmetry unbroken. The unbroken generators do not change the solution, while the broken ones generate the supermultiplet of the soliton, which is thus shorter than the generic supermultiplet.

So far we discussed pointlike BPS states. There are however BPS versions for extended objects (BPS p-branes). The D-branes described in chapter 8 are perfect examples of this. In the presence of extended objects, the supersymmetry algebra can acquire central charges that are not Lorentz scalars (as we assumed in Appendix J). Their general form can be obtained from group theory. They are antisymmetric tensors, $Z_{\mu_1 \ldots \mu_p}$. Such central charges have values proportional to the charges Q_p of p-branes. Then, the BPS condition would relate these charges with the energy densities (p-brane tensions) μ_p of the relevant p-branes. Such p-branes can be viewed as extended soliton solutions of the effective theory. The BPS condition is the statement that the soliton solution leaves some of the supersymmetries unbroken.

To conclude, the presence of BPS states can be determined at weak coupling and can be trusted in many cases at strong coupling. Their mass formula is valid beyond perturbation theory, although both sides can obtain nontrivial quantum corrections in $\mathcal{N} = 2_4$ supersymmetric theories. There are no such corrections in $\mathcal{N} \geq 3_4$ supersymmetric theories.

11.3 Nonrenormalization Theorems and BPS-saturated Couplings

There is another important issue in theories with supersymmetry, which is relevant in duality connections: the presence of special terms in the effective action, protected by supersymmetry. We will generically call such terms "BPS-saturated terms" for reasons that will become obvious in the sequel. We cannot at the moment give a rigorous definition of such terms for a generic supersymmetric theory, due to the lack of off-shell (superspace) formulation. However, for theories with an off-shell formulation the situation is better and all BPS-saturated terms are known.

Before we embark in a detailed discussion of various cases, we give some generic features of such terms.

• Supersymmetry constrains their (moduli-dependent) coefficients to have a special structure. The simplest situation is complex holomorphicity but there are several other cases where one would have special conditions associated with quaternions, as well as noncompact groups of the $O(p,q)$ type, etc. We will generically call such constraints "holomorphicity" constraints.

• In cases where there is a superfield formulation, such terms are (chiral) integrals over parts of superspace. This is at the root of the holomorphicity conditions of the previous item.

• Such terms satisfy special nonrenormalization theorems. It should be stressed here and it would be seen explicitly later that some of the nonrenormalization theorems depend crucially on the perturbation theory setup.

Before we enter further into a discussion of the nonrenormalization theorems we should stress in advance that they are generically only valid for the Wilsonian effective action. The reason is that many nonrenormalization theorems are violated due to IR divergences. There are several examples known, and we will mention here the case of $\mathcal{N} = 1_4$ supersymmetry: In the presence of massless contributions, a quantum correction to the Kähler potential (not protected by a nonrenormalization theorem) can be indistinguishable from a correction to the superpotential (not renormalized in perturbation theory). Because of such complications, from now on we will be always discussing the Wilsonian effective action.

To continue further, we will separate two cases.

Absolute nonrenormalization theorems. Such theorems state that a given term in the effective action of a supersymmetric theory does not get renormalized. This should be valid both for perturbative and nonperturbative corrections. A typical example of this is the case of two-derivative terms in the effective action of an $\mathcal{N} \geq 4_4$ supersymmetric theory (with or without gravity).

Partial (or perturbative) nonrenormalization theorems. Such theorems usually claim the absence of perturbative corrections for a given effective coupling, or that the only corrections appear in a few orders only in perturbation theory. Typically this happens at one-loop order but we also know of cases where renormalization can occur at a single, arbitrarily high, loop order. It is also common in the case of one-loop corrections only, that the appropriate effective couplings are related to an anomaly via supersymmetry. The appropriate Adler-Bardeen type theorem for anomalies guarantees the absence of higher-loop corrections, provided the perturbation theory is set up to respect supersymmetry. An example of such a situation is the case of two-derivative couplings in a theory with $\mathcal{N} = 2_4$ supersymmetry (global or local).

There are several examples that illustrate the general discussion above. We will mention some commonly known ones.

• Heterotic string theory on T^6 is dual to type-II string theory on K3 $\times T^2$. The R^2 effective coupling, has only a one-loop contribution on the type-II side. On the heterotic side apart from the tree-level contribution there are no other perturbative corrections. There are, however, nonperturbative corrections due to five-brane instantons.

• Heterotic string theory on K3$\times T^2$ is dual to type IIA on a Calabi-Yau (CY) manifold that is a K3 fibration. All two-derivative effective couplings for vector multiplets are tree level only on the type-II side. On the other hand they obtain tree-level, one-loop as well as nonperturbative corrections on the heterotic side. The situation is reversed for the hyper-multiple geometry. It has no loop corrections on the heterotic side but is loop corrected on the type-II side.

There is another general property that is shared by BPS-saturated terms: The quantum corrections to their effective couplings can be associated to BPS states. There are two concrete aspects of the statement above.

Their one-loop contributions are due to (perturbative) BPS multiplets only. The way this works out is that the appropriate one-loop diagrams come out proportional to helicity supertraces. In four dimensions, this is a supertrace of the helicity λ to an arbitrary even power (by $\mathcal{CPT}$ all odd powers vanish):

$B_{2n} = \text{Str}\,[\lambda^{2n}].$

The helicity supertraces are essentially indices to which only short BPS multiplets contribute. It immediately follows that such one-loop contributions are due to BPS states only. The appropriate helicity supertraces count essentially the numbers of "unpaired" BPS multiplets. It is only these that are protected from renormalization and can provide reliable information in strong coupling regions. Calling the helicity supertraces indices is more than an analogy. They provide the minimal robust information about IR-sensitive data. In particular, they do not depend on the moduli. Unpaired BPS states in lower dimensions are intimately connected with the chiral asymmetry (conventional index) of the ten-dimensional (or eleven-dimensional) theory.

It is well known that the ten-dimensional elliptic genus is the stringy generalization of the Dirac index. Projecting the elliptic genus on physical states in ten dimensions gives precisely the massless states, responsible for anomalies. In lower dimensions, BPS states are determined uniquely by the elliptic genus, as well as the compact manifold data.

We will describe here more detailed properties of helicity supertraces. Further information can be found in appendix J.

$\mathcal{N} = 2_4$ supersymmetry. Here we have only one kind of BPS multiplet the 1/2 multiplet (preserving half of the original $\mathcal{N} = 2_4$ supersymmetry). The trace B_2 is nonzero for the 1/2 multiplet but zero from the long multiplets. The long multiplets on the other hand contribute to B_4.

$\mathcal{N} = 4_4$ supersymmetry. Here we have two types of BPS multiplets, the 1/2 multiplets and the 1/4 multiplets. For all multiplets $B_2 = 0$. B_4 is nonzero for 1/2 multiplets only. B_6 is nonzero for 1/2 and 1/4 multiplets only. Long massive multiplets start contributing only to B_8.

A similar stratification appears also in the case of maximal $\mathcal{N} = 8_4$ supersymmetry.

There is a single "index" in the case of an $\mathcal{N} = 2_4$ supersymmetric theory, and it governs the one-loop corrections to the two-derivative action, in standard perturbation theory. In the $\mathcal{N} = 4_4$ case there are two distinct indices. B_4 controls one-loop corrections to several four-derivative effective couplings of the $\mathrm{tr} F^4$, $\mathrm{tr} F^2 \mathrm{tr} R^2$, and $\mathrm{tr} R^4$ types. The situation with B_6 is less clear.

BPS-saturated couplings may also receive instanton corrections. However, the instantons that contribute, parallel the BPS states that contribute in perturbation theory. They must preserve the same amount of supersymmetry. Since in string theory instantons are associated with Euclidean solitonic branes wrapped on compact manifolds, it is straightforward in most situations to classify possible instantons that contribute to BPS-saturated terms. We will see explicit examples in subsequent sections.

We summarize here the generic characteristics of BPS-saturated couplings.

- They obtain perturbative corrections from BPS states only.

- The perturbative corrections appear at a single order of perturbation theory, usually at one loop.

- They satisfy "holomorphicity constraints."

- They contain simple information about massless singularities.[1] This is because when one or more BPS states going around the loop becomes massless for some value of the moduli, an appropriate singularity is generated in the effective coupling.

- They obtain instanton corrections from "BPS instantons" (instanton configurations that preserve some fraction of the original supersymmetry).

- If there exists an off-shell formulation they can be easily constructed as chiral integrals over superspace.

11.4 Type-IIA Versus M-theory

The effective action of type-IIA supergravity is the dimensional reduction of that of 11-dimensional, $\mathcal{N} = 1_{11}$ supergravity.[2] As we shall see, this is not accidental.

We start by reviewing the spectrum of forms in type-IIA theory in ten dimensions.

- The NS-NS two-form B couples to a string (electrically) and a five-brane (magnetically). The string is the perturbative type-IIA string.

---

[1] A massless singularity in the effective field theory appears when the mass of a heavy particle that has been integrated out is made arbitrarily small. This can be achieved by tuning couplings and moduli appropriately.

[2] This is detailed in appendix H.

- The R-R U(1) gauge field A_μ can couple electrically to particles (zero-branes) and magnetically to six-branes. Since it comes from the R-R sector no perturbative state is charged under it.

- The R-R three-form $C_{\mu\nu\rho}$ can couple electrically to membranes ($p = 2$) and magnetically to four-branes.

- There is also the nonpropagating zero-form field strength and its ten-form dual field strength that would couple to eight-branes (see section 7.2.1 on page 159).

In appendix H.2 on page 527, we present the lowest-order type-IIA string-frame Lagrangian

$$\tilde{S}^{IIA} = \frac{1}{2\kappa_{10}^2} \left[\int d^{10}x \sqrt{g} \left[e^{-2\Phi} \left(R + 4(\nabla\Phi)^2 - \frac{1}{12} H_3^2 \right) - \frac{1}{2\cdot 4!} F_4^2 - \frac{1}{4} F_2^2 \right] \right.$$
$$\left. + \frac{1}{2} \int B_2 \wedge dC_3 \wedge dC_3 \right].$$
(11.4.1)

Note that the R-R kinetic terms do not couple to the dilaton as already argued in section 7.2.1.

In the type-IIA supersymmetry algebra there is a central charge proportional to the U(1) charge of the gauge field A:

$$\{Q^{1,\alpha}, Q_\beta^2\} = \delta^\alpha_{\ \beta} W.$$
(11.4.2)

A way to understand this is the following. The type-IIA supersymmetry algebra descends from $D = 11$ where in place of W there is the momentum operator of the eleventh dimension. Since the U(1) gauge field is the $G_{11,\mu}$ component of the metric, the momentum operator becomes the U(1) charge in the type-IIA theory. There is an associated BPS bound

$$M \geq T_0 |W|,$$
(11.4.3)

where $T_0 = (\ell_s g_s)^{-1}$. States that satisfy the equality are BPS saturated and form smaller supermultiplets. As mentioned earlier, all perturbative string states have $W = 0$. However, there is a soliton solution of type-IIA supergravity that is minimally charged under the one-form. This is none else but the D_0-brane described in chapter 8. We would expect that quantization of this solution would provide a (non-perturbative) particle state. Moreover, as we argued in a previous chapter, the D_0 charge is quantized. Therefore the spectrum of such D_0 states[3] is

$$M = \frac{|n|}{\ell_s g_s}, \quad n \in \mathbb{Z}.$$
(11.4.4)

At weak coupling these states are heavy. However, being BPS states, their mass can be reliably followed at strong coupling, where they become light, piling up at zero mass as the coupling becomes infinite. This is precisely the behavior of Kaluza-Klein (momentum) modes as a function of the radius. The effective type-IIA field theory is a dimensional reduction of the 11-dimensional supergravity, with $G_{11,11}$ becoming the string coupling.

---

[3] For $n < 0$ we have $\overline{D}_0$-branes.

We can then take this seriously and claim that as $g_s \to \infty$ type-IIA theory becomes an 11-dimensional theory whose low-energy limit is 11-dimensional supergravity.

We may calculate the relation between the radius of the 11th dimension and the string coupling. This was done essentially in appendix H, where we described the dimensional reduction of $\mathcal{N} = 1_{11}$ supergravity to ten dimensions. The radius of the eleventh dimension R follows from (11.4.4),

$$R = g_s \ell_s. \tag{11.4.5}$$

At strong type-IIA coupling, $R \to \infty$ and the theory decompactifies to 11 dimensions, while in the perturbative regime the radius is small and the theory is ten dimensional.

The gravitational couplings of the ten- and eleven-dimensional theories are related by

$$2\kappa_{11}^2 = (2\pi R)\, 2\kappa_{10}^2 = (2\pi)^8 \ell_s^9\, g_s^3, \tag{11.4.6}$$

where we have used (8.4.8) on page 195. Defining the 11-dimensional Planck mass as $2\kappa_{11}^2 = (2\pi)^8 (M_{11})^{-9}$ and its inverse, the 11-dimensional Planck length, $\ell_{11} = \frac{1}{M_{11}}$, we obtain

$$\ell_{11} = g_s^{1/3} \ell_s, \quad g_s = (M_{11} R)^{3/2}, \quad \ell_s^2 = \frac{\ell_{11}^3}{R}. \tag{11.4.7}$$

The 11-dimensional theory (which has been named M-theory) contains the three-form that can couple to a membrane M_2 and its magnetic dual, a five-brane M_5. An M_2 not winding around the circle is equivalent to the type-IIA D_2-brane, that couples to the type-IIA three-form. From this we obtain that the tensions of the M_2 and D_2 must be equal, $T_{M_2} = T_2$. On the other hand, the M_2-brane, wrapped around the circle, becomes the perturbative type-IIA string that couples to $B_{\mu\nu}$. In this case the relevant tension

$$(2\pi R)\, T_{M_2} = \frac{2\pi R}{(2\pi)^2 \ell_s^3 g_s} = \frac{1}{2\pi \ell_s^2} = T, \tag{11.4.8}$$

coincides with that of a fundamental string.

Consider now the M_5-brane. Its tension T_{M_5} can be determined from the DNT quantization condition

$$2\kappa_{11}^2 T_{M_2} T_{M_5} = 2\pi \to T_{M_5} = \frac{1}{(2\pi)^5 \ell_s^6 g_s^2} = \frac{1}{(2\pi)^5 \ell_{11}^6}, \tag{11.4.9}$$

where we have used (11.4.6). The M_5-brane descends to the type-IIA NS_5-brane. As expected, $T_{M_5} = \tilde{T}_5$ from (8.9.1) on page 211.

Wrapping one of the coordinates of the M_5-brane around the circle should produce the D_4-brane and we can confirm that

$$(2\pi R)\, T_{M_5} = T_4. \tag{11.4.10}$$

The 11-dimensional supergravity solutions representing the M_2- and M_5-branes are discussed in exercises 11.2 and 11.3.

All the known branes of the IIA theory, except the D_6, descend from the M-theory branes. To find the M-theory parent of the D_6-brane of IIA, we may argue as follows: the D_6 is the magnetic dual of the D_0-brane, who is charged under the U(1) R-R gauge field of IIA. Since the D_0 carries KK electric charge, the dual brane should be a KK monopole. Thus, it should be magnetically charged under the $g_{\mu,11}$ component of the 11-dimensional metric. You are invited to find this solution in exercise 11.5.

In section 8.9 on page 211 we have mentioned that the world-volume theory of the type-IIA NS_5-brane contains five massless scalars. Four of them are its transverse collective coordinates in ten dimensions. The origin of the fifth is now clear. It is the coordinate on the 11th-dimensional circle.

The world-volume spectrum of the M_5-brane is expected to be generated by open membranes ending on it. Their end points are strings, which are charged under the world-volume two-index form. Upon compactification the membranes become the open strings stretched between D_4-branes. The world-volume theory of N coincident M_5-branes is not well understood. Similar remarks apply to the IIA NS_5-brane.

11.5 Self-duality of the Type-IIB String

In appendix H.3 we point out that the low-energy effective action of the type-IIB theory in ten dimensions has an $SL(2,\mathbb{R})$ global symmetry. One of its transformations inverts the string coupling constant, $\tau \to -\frac{1}{\tau}$. We will argue here that this symmetry is instrumental in order to understand the strong coupling limit of this theory.

As described in section 7.2.1 on page 159, the type IIB theory in ten dimensions contains the following forms:

- The standard NS-NS two-form B_2 couples electrically to the perturbative type-IIB F_1-string and magnetically to the IIB NS_5-brane.

- The R-R scalar C_0 is a zero-form (there is a Peccei-Quinn symmetry associated with it) and couples electrically to the D_{-1}-brane. This is the D-instanton whose "world-volume" is a point in space-time. It also couples magnetically to D_7-branes.

- The R-R two-form C_2 couples electrically to the D_1-string and magnetically to the D_5-brane.

- The self-dual four-form C_4 couples to the self-dual D_3-brane.

The effective IIB action is presented in appendix H.3 on page 528. There we have shown that the effective theory is invariant under the $SL(2,\mathbb{R})$ symmetry, which acts by fractional linear transformations on the complex scalar S defined in (H.20) and linearly on the vector of two-forms (B_2, C_2), the four-form being invariant. The Einstein frame metric is also invariant. The part of $SL(2,\mathbb{R})$ that translates the R-R scalar is a symmetry of the full perturbative theory.

The charges/tensions of the strings charged under B_2 are quantized in units of the fundamental string tension T. Similarly we have seen that the charges of D-strings are quantized in units of T_1. Since $SL(2,\mathbb{R})$ rotates these charges, only its discrete $SL(2,\mathbb{Z})$ subgroup preserves the DNT quantization conditions.

One of the simplest transformations of $SL(2,\mathbb{Z})$ is the interchange of the fundamental string F_1 and the D-string, D_1. Both strings are 1/2-BPS states, preserving 16 supercharges.

According to the discussion in section 11.2, the charges and tensions/masses of 1/2-BPS states are protected quantities, receiving no corrections in theories with 32 supercharges. In the string frame, with $C_0 = 0$ for simplicity, this transformation acts as

$$g'_{\mu\nu} = e^{-\Phi} g_{\mu\nu}, \quad \Phi' = -\Phi, \quad B'_2 = C_2, \quad C'_2 = B_2, \quad C'_4 = C_4. \tag{11.5.1}$$

Since the SL(2,$\mathbb{Z}$) symmetry transforms the string metric but leaves the Einstein metric invariant, it is instructive to translate the string tensions in the Einstein frame. We obtain

$$T_{F_1} = \sqrt{g_s}\, T = \frac{\sqrt{g_s}}{2\pi \ell_s^2}, \quad T_{D_1} = \sqrt{g_s}\, T_1 = \frac{1}{2\pi \ell_s^2 \sqrt{g_s}}. \tag{11.5.2}$$

These relations are consequences of supersymmetry and thus exact, valid beyond perturbation theory.

At weak string coupling, the fundamental strings are much lighter than the D-strings. Thus, they are the relevant low-energy degrees of freedom. However at strong coupling $g_s \gg 1$, the light degrees of freedom are described by the D-string, while the fundamental string is heavy. It is tempting to believe that the strong-coupling region of the type-IIB string has a weak-coupling description, where now the fundamental string is the D-string. Supersymmetry and several consistency checks indicate that this is a reasonable definition of the strongly-coupled theory. A corollary is that there is no region in coupling constant space where the Planck scale is smaller than both string tensions.

We will describe here a few simple checks of this duality conjecture. In the way we described the F_1- and the D_1-strings they look very different. We will argue that this is an artifact of perturbation theory. To show this, we will match their collective coordinates, describing their low-energy dynamics.

We start from a straight D_1-string stretching along the x^1 direction. This configuration breaks the ten-dimensional Lorentz invariance as

$$SO(1,9) \rightarrow SO(8) \times SO(1,1), \tag{11.5.3}$$

where the SO(1,1) acts on time and the x^1 direction (the world-volume of the D-string), while the SO(8) acts on the transverse space. We also decompose the 16 Majorana-Weyl supercharges preserved by the solution under (11.5.3) as

$$\mathbf{16}_+ = \mathbf{8}_+ \otimes \mathbf{1}_+ + \mathbf{8}_- \otimes \mathbf{1}_-, \quad \mathbf{16}_- = \mathbf{8}_- \otimes \mathbf{1}_+ + \mathbf{8}_+ \otimes \mathbf{1}_-. \tag{11.5.4}$$

According to our standard convention, the subscripts indicate the chirality with respect to the appropriate Γ^{D+1}.

The D-string massless states coming from the NS sector are eight scalars and a two-dimensional gauge field that has no on-shell degrees of freedom. From the R sector we obtain a massless spinor obeying the two-dimensional Dirac equation. The GSO projection imposes $\Gamma^{11} = 1$ and selects $\mathbf{16}_+$. Since left-movers and right-movers of the Dirac equations have opposite chirality in two dimensions, we learn from (11.5.4) that this correlates their SO(8) quantum numbers.

We consider now the macroscopic fundamental string in static gauge. The world-sheet bosonic degrees of freedom are again 8 massless scalars (the transverse coordinates of the string). To match the world-volume fermions, we have to think of them as the Goldstinos of the supersymmetry broken by the fundamental string. The supersymmetry of the IIB string is generated by two supercharges transforming as the $\mathbf{16}_+$. The F_1 breaks one of them, so the fermionic degrees of freedom are the same as in the D_1-string. These can be mapped to the usual world-sheet fermions ψ^μ of the F_1 by SO(8) triality.

Starting from the F_1- and D_1-strings, and acting with SL(2,$\mathbb{Z}$) transformations, we can generate (p,q) strings, where p,q are relatively prime. In this notation, (1,0) is the F_1-string while (0,1) is the D_1-string. (p,q) strings can be thought of as bound states of p F_1-strings and q D_1-strings. Their tensions can be obtained by acting with SL(2,$\mathbb{Z}$) on the F_1 tension. We obtain in the Einstein frame

$$T_{p,q} = \frac{|p + qS|}{\sqrt{S_2}}\, T, \tag{11.5.5}$$

where S is defined in (H.20).

(11.5.5) is an example of a BPS energy-density formula, to which the remarks of section 11.2 apply. In particular, (p,q) strings are stable, for nonzero string coupling g_s.

Since the four-form does not transform under the SL(2,$\mathbb{Z}$) symmetry, the D_3-branes are "self-dual." This implies that the SL(2,$\mathbb{Z}$) symmetry acts inside the world-volume action. In particular, it acts as the Montonen-Olive strong-weak coupling duality on the low-energy $\mathcal{N} = 4_4$ gauge theory. You are invited to derive this in exercise 11.9.

The D_5-brane is magnetically charged under C_2. The NS$_5$-brane is magnetically charged under B_2. The SL(2,$\mathbb{Z}$) symmetry generates (p,q) five-branes that are the magnetic duals of (p,q) strings. We already know that the fluctuations of the D_5-brane are described by F_1-strings attached to it. From the SL(2,$\mathbb{Z}$) symmetry, the fluctuations of the NS$_5$-brane are expected to be due to D_1-strings attached to it. You are invited to explore this further in exercise 11.12.

We are now going to relate the SL(2,$\mathbb{Z}$) symmetry of the type-IIB string to the geometry of M-theory. By compactifying the type-IIB theory on a circle of radius R_B, it becomes equivalent to the IIA theory compactified on a circle of radius $R_A = \ell_s^2/R_B$. The nine-dimensional type-IIA theory is M-theory compactified on a two-torus.

From the type-IIB point of view, wrapping (p,q) strings around the tenth dimension provides a spectrum of particles in nine dimensions with masses

$$M_B^2 = \frac{m^2}{R_B^2} + (2\pi R_B n T_{p,q})^2 + 4\pi T_{p,q}(N_L + N_R), \tag{11.5.6}$$

where m is the Kaluza-Klein momentum integer, n the winding number, and $N_{L,R}$ the string oscillator numbers. The matching condition is $N_L - N_R = mn$, and BPS states are obtained for $N_L = 0$ or $N_R = 0$. We thus obtain the following BPS spectrum

$$M_B^2\big|_{\text{BPS}} = \left(\frac{m}{R_B} + 2\pi R_B n T_{p,q}\right)^2. \tag{11.5.7}$$

Since an arbitrary pair of integers (n_1, n_2) can be written as $n(p,q)$, where n is the greatest common divisor and p,q are relatively prime, we can rewrite the BPS mass formula above as

$$M_B^2\big|_{\text{BPS}} = \left(\frac{m}{R_B} + 2\pi R_B T\, \frac{|n_1 + n_2 S|}{\sqrt{S_2}}\right)^2. \tag{11.5.8}$$

In M-theory, we compactify to nine dimensions on a two-torus with area A and complex structure modulus[4] τ. We take the periods to be $2\pi R_{11}$ and $2\pi R_{11}\tau$. We have $A = (2\pi R_{11})^2\tau_2$.

We have two types of (pointlike) BPS states in nine dimensions. The first class consists of KK states on the torus with mass $(2\pi)^2|n_1 + n_2\tau|^2/(\tau_2 A_{11})$. There are also the states obtained by wrapping the M_2 brane m times around the two-torus, with mass mAT_{M_2}. Thus, the BPS spectrum is

$$M_{11}^2 = (m(2\pi R_{11})^2\tau_2 T_{11})^2 + \frac{|n_1 + n_2\tau|^2}{R_{11}^2\tau_2^2} + \cdots, \tag{11.5.9}$$

where the dots stand for contributions coming from M_2 brane/graviton bound states that are hard to calculate.

The two BPS mass spectra should be related by an overall scaling, $M_{11}^2 = \beta M_B^2$ since they are calculated in different metrics. In exercise 11.11 you are asked to carefully match the two effective actions in nine dimensions and thus derive the value of β from first principles. Here, we will obtain it by matching (11.5.8) and (11.5.9),

$$S = \tau, \quad \frac{1}{R_B^2} = TT_{M_2}A^{3/2}, \quad \beta = \sqrt{A}\,\frac{T_{M_2}}{T}. \tag{11.5.10}$$

An important outcome of this is that we obtain a geometric explanation of the SL(2,$\mathbb{Z}$) symmetry of the type-IIB string: the geometrical SL(2,$\mathbb{Z}$) symmetry of the M-theory two-torus becomes the SL(2,$\mathbb{Z}$) weak-strong coupling symmetry of the type-IIB theory.

11.6 U-duality of Type-II String Theory

So far we have seen two important links in what will eventually become a spider-web of relations in string theory. The first is the perturbative connection of type-IIA and type-IIB theories compactified on circle via T-duality. The other is the relation of IIA theory with the 11-dimensional M-theory. We already have seen that these two links put together, can explain the SL(2,$\mathbb{Z}$) symmetry of the ten-dimensional type-IIB theory in ten dimensions, by compactifying M-theory on a two-torus and then performing a T-duality.

It turns out that by further compactifying type-IIA and type-IIB theory on tori, perturbative symmetries like T-duality mix with the nonperturbative symmetries to generate larger and larger duality groups. These were termed U-duality groups. In table 11.1, the U-duality groups are given for various dimensions.

As argued earlier in 4.18.7 on page 101, upon compactification on an N-torus T^N, the (perturbative) T-duality group is O(N,N,$\mathbb{Z}$). An element with determinant 1 maps the type IIA to IIA and IIB to IIB. An element with determinant -1 maps IIA$\leftrightarrow$ IIB. However these are different descriptions of a unique underlying theory, and in this theory O(N,N,$\mathbb{Z}$) is a symmetry. These vacua can also be interpreted as M-theory compactified on T^{N+1}. Such compactifications have the obvious SL(N+1,$\mathbb{Z}$) symmetry of T^{N+1}. This symmetry

---

[4] See section 4.18.7 on page 101 for a parametrization of the two-torus metric.

Table 11.1 Cremmer-Julia Groups and U-Duality Symmetries of the Toroidally Compactified
Type II String.

Dimension	SUGRA symmetry	T-duality	U-duality
10A	$SO(1,1,\mathbb{R})/\mathbb{Z}_2$	1	1
10B	$SL(2,\mathbb{R})$	1	$SL(2,\mathbb{Z})$
9	$SL(2,\mathbb{R}) \times O(1,1,\mathbb{R})$	$\mathbb{Z}_2$	$SL(2,\mathbb{Z}) \times \mathbb{Z}_2$
8	$SL(3,\mathbb{R}) \times SL(2,\mathbb{R})$	$O(2,2,\mathbb{Z})$	$SL(3,\mathbb{Z}) \times SL(2,\mathbb{Z})$
7	$SL(5,\mathbb{R})$	$O(3,3,\mathbb{Z})$	$SL(5,\mathbb{Z})$
6	$O(5,5,\mathbb{R})$	$O(4,4,\mathbb{Z})$	$O(5,5,\mathbb{Z})$
5	$E_{6(6)}$	$O(5,5,\mathbb{Z})$	$E_6(\mathbb{Z})$
4	$E_{7(7)}$	$O(6,6,\mathbb{Z})$	$E_7(\mathbb{Z})$
3	$E_{8(8)}$	$O(7,7,\mathbb{Z})$	$E_8(\mathbb{Z})$
2	E_9	$O(8,8,\mathbb{Z})$	$E_9(\mathbb{Z})$
1	E_{10}	$O(9,9,\mathbb{Z})$	$E_{10}(\mathbb{Z})$

has nonperturbative pieces. If this symmetry is intertwined with the T-duality symmetry $O(N,N,\mathbb{Z})$, then it generates the U-duality group that we will label with $E_N(\mathbb{Z})$.

We will consider a simple example to show how it works: compactification of type-II string theory on T^2. We will look at it from the point of view of the IIB string. The T-duality here is $O(2,2,\mathbb{Z}) \simeq SL(2,\mathbb{Z})_T \times SL(2,\mathbb{Z})_U$ as explained in the end of section 4.18.7. $SL(2,\mathbb{Z})_U$ is one of the factors of the U-duality group $E_3(\mathbb{Z})=SL(3,\mathbb{Z}) \times SL(2,\mathbb{Z})$. The other factor arises when the nonperturbative $SL(2,\mathbb{Z})$ symmetry of the IIB string combines with $SL(2,\mathbb{Z})_T$.

We will describe how the spectrum of the eight-dimensional theory transforms under $E_3(\mathbb{Z})$. There is a single three-form descending from the self-dual four-form C_4. It is invariant under U-duality. There are three two-forms. Two of them descend directly from the two ten-dimensional two-forms B_2, C_2. The other comes from C_4. They transform as the **3** of $SL(3,\mathbb{Z})$. There are six vectors. Two come from the metric, and four from B_2, C_2. They transform in the **(3,2)** of $SL(3,\mathbb{Z}) \times SL(2,\mathbb{Z})$. Finally, there are seven scalars. Two of them, in the complex structure of T^2, belong to $SL(2,\mathbb{R})/U(1)$ manifold and transform with standard fractional transformations under $SL(2,\mathbb{Z})$ while they are neutral under $SL(3,\mathbb{Z})$. The other five belong to $SL(3,\mathbb{R})/O(3,\mathbb{R})$. One comes from the metric (the volume), two from the axion-dilaton S, and two from B_2, C_2. They transform with fractional linear transformations under $SL(3,\mathbb{R})$.

In general, the scalars belong to the coset manifold $E_N(\mathbb{R})/K$ where K is the maximal compact subgroup of $E_N(\mathbb{R})$. The maximality of K implies that the scalar kinetic terms are positive definite. The vectors transform linearly in the "vector" representation of $E_N(\mathbb{R})$. The associated charges also transform analogously. Since some of them come from NS-NS fields while others from R-R fields, they involve both perturbative charges like windings and momenta, as well as D-brane charges.

In two dimensions the group involved, E_9, is the affine algebra of E_8. In one dimension, E_{10} appears. It is a hyperbolic Kač-Moody algebra, with structure and implications that have still defied a controlled description. The structure of Dynkin diagrams of various Cremmer-Julia groups is detailed in table 11.2.

Exercises 11.14–11.17 ask you to derive in detail some of the statements above. They also invite you to explore the M-theory point of view.

11.6.1 U-duality and bound states

An important impact of U-duality transformations is that they mix perturbative and nonperturbative charges. We may thus map well-understood perturbative objects, to nonperturbative objects. A simple application of this is the prediction of nontrivial bound states of D-branes.

We will use a sequence that involves the nontrivial IIB S-duality transformation S and various T-duality transformations. We need to develop some notation so that the action of T-duality is transparent. We will take the space to be an orthogonal T^9 torus. An F^i stands for the winding charge under the vector coming from the NS antisymmetric tensor, in the x^i direction. A p_i is the momentum in the x^i direction. T-duality in direction x^i interchanges the two

$$T_i \colon F^i \leftrightarrow p_i. \tag{11.6.1}$$

A D-brane will be labeled by the spatial directions it wraps. Thus, D^{235} is a D_3-brane wrapping directions x^2, x^3, x^5. A D_0-brane will by denoted by D^0. $T_{ijk\ldots}$ denotes a simultaneous T-duality transformation in directions $x^i, x^j, x^k \ldots$.

The transformation S inverts the IIB coupling constant, and interchanges F^1 and D_1 while leaving a D_3 invariant.

The first example starts with a (p, q) string, a bound state of p fundamental and q D-strings. In our notation, it is (D^1, F^1) where we took the string to wrap the x^1 direction. We have the following chain of dualities:

$$(D^1, F^1) \xrightarrow{T_{23}} (D^{123}, F^1) \xrightarrow{S} (D^{123}, D^1) \xrightarrow{T_1} (D^{23}, D^0) \tag{11.6.2}$$

ending up with D_2-D_0 bound state.

A similar chain will produce a D_0-D_4 bound state out of a perturbative state with both winding and momentum

$$(F^1, p_1) \xrightarrow{S} (D^1, p_1) \xrightarrow{T_{1234}} (D^{234}, F^1) \xrightarrow{S} (D^{234}, D^1) \xrightarrow{T_1} (D^{1234}, D^0). \tag{11.6.3}$$

11.7 Heterotic/Type I Duality in Ten Dimensions

The O(32) type-I superstring theory and the O(32) heterotic theory have the same ten-dimensional effective action, namely, $\mathcal{N} = 1_{10}$ supergravity. As we will now argue, this is

Table 11.2 Dynkin Diagrams of the E_d Series. The Group Disintegration Proceeds by Omitting the Rightmost Node. The Integers Shown are the Coxeter Labels, that is, the Coordinates of the Highest Root on all Simple Roots.

$E_2 = A_1$

$E_3 = A_2 \oplus A_1$

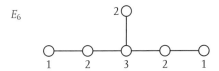

$E_4 = A_4$

$E_5 = D_5$

E_6

E_7

E_8

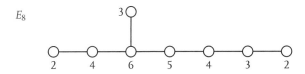

$E_9 = \hat{E}_8$

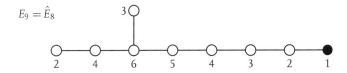

because each theory is the strong coupling dual of the other. We start our discussion by describing heterotic/type I duality in ten dimensions.

Consider first the O(32) heterotic string theory. At tree level (sphere) and up to two-derivative terms, the (bosonic) effective action in the string frame is

$$S^{het} = \int d^{10}x \sqrt{G} e^{-2\Phi} \left[R + 4(\nabla\Phi)^2 - \frac{1}{12}\hat{H}^2 - \frac{1}{4}F^2 \right]. \tag{11.7.1}$$

On the other hand, for the O(32) type-I string, the leading order two-derivative effective action is

$$S^{I} = \int d^{10}x \sqrt{G} \left[e^{-2\Phi} \left(R + 4(\nabla\Phi)^2 \right) - \frac{1}{4}e^{-\Phi}F^2 - \frac{1}{12}\hat{H}^2 \right]. \tag{11.7.2}$$

The different dilaton dependence here is as follows: the Einstein and dilaton terms originate from the closed sector on the sphere ($\chi = 2$). The gauge kinetic terms originate from the disk ($\chi = 1$). Since the antisymmetric tensor originates from the R-R sector of the closed superstring it does not have any dilaton dependence on the sphere.

We will now bring both actions to the Einstein frame, by redefining $G_{\mu\nu} = e^{\Phi/2}g_{\mu\nu}$:

$$S_E^{het} = \int d^{10}x \sqrt{g} \left[R - \frac{1}{2}(\nabla\Phi)^2 - \frac{1}{4}e^{-\Phi/2}F^2 - \frac{1}{12}e^{-\Phi}\hat{H}^2 \right], \tag{11.7.3}$$

$$S_E^{I} = \int d^{10}x \sqrt{g} \left[R - \frac{1}{2}(\nabla\Phi)^2 - \frac{1}{4}e^{\Phi/2}F^2 - \frac{1}{12}e^{\Phi}\hat{H}^2 \right]. \tag{11.7.4}$$

We observe that the two actions are related by $\Phi \to -\Phi$ while keeping the other fields invariant. This seems to suggest that the weak coupling of one is the strong coupling of the other, and vice versa. Of course, the fact that the two actions are related by a field redefinition is not a surprise. It is because $\mathcal{N} = 1_{10}$ ten-dimensional supergravity is completely fixed once the gauge group is chosen. It is interesting though, that the field redefinition here is just an inversion of the ten-dimensional coupling. The two theories have perturbative expansions that look very different.

In terms of the string frame fields, the duality maps is

$$g_{\mu\nu}^{I} = e^{-\Phi^h}g_{\mu\nu}^{h}, \quad \Phi^{I} = -\Phi^{h}, \quad \hat{H}^{I} = \hat{H}^{h}, \quad A_{\mu}^{I} = A_{\mu}^{h}. \tag{11.7.5}$$

In particular, the string scales of the two theories are related as

$$M_s^{I} = \frac{M_s^{het}}{\sqrt{g_s^{het}}}. \tag{11.7.6}$$

We would like to go further and check if there are nontrivial checks of this duality, suggested by the classical $\mathcal{N} = 1_{10}$ supergravity. Once we compactify one direction on a circle of radius R, we have an apparent problem. In the heterotic case, we have a spectrum that depends both on momenta m in the ninth direction as well as on windings n. The winding number is the charge that couples to the NS antisymmetric tensor. In particular, it is the electric charge of the gauge boson obtained from $B_{9\mu}$. On the other hand, in type

I theory, as we have shown earlier, we have momenta m but no windings: the open string Neumann boundary conditions forbid the string to wind around the circle. Another way is by noting that the NS-NS antisymmetric tensor that could couple to windings has been projected out by our orientifold projection.

We do have however the R-R antisymmetric tensor, but as we argued in section 7.2.1 on page 159, no perturbative states are charged under it. There are, however nonperturbative states that are charged under this antisymmetric tensor. The obvious guess is that these should come from the D_1-string of the type-I theory.

11.7.1 The type-I D_1-string

To investigate this, we study the D_1-string of the type-I string theory. We will localize it to the plane $X^2 = X^3 = \cdots = X^9 = 0$. Its world-sheet extends in the X^0, X^1 directions. Such an object is schematically shown in figure 11.2. Its fluctuations can be described by two kinds of open strings:

- D_1-D_1 strings starting and ending on the D_1 string.

- D_1-D_9 strings with one end point on the D_1 string and the other free (or equivalently ending on the 32 space-filling D_9 branes of type-I theory).

As we will see, this configuration breaks half of $\mathcal{N} = 1_{10}$ space-time supersymmetry possible in ten dimensions. It also breaks $SO(9, 1) \rightarrow SO(8) \times SO(1, 1)$. Moreover, we can put it anywhere in the transverse eight-dimensional space, so we expect eight bosonic zero modes associated with the broken translational symmetry. We will find the spectrum of massless fluctuations by quantizing the relevant open string fluctuations.

We start with the D_1-D_1 strings. Here $X^I, \psi^I, \bar{\psi}^I, I = 2, \ldots, 9$ have DD boundary conditions while $X^\mu, \psi^\mu, \bar{\psi}^\mu, \mu = 0, 1$ have NN boundary conditions.

The GSO projection removes the tachyon. The bosonic massless spectrum consists of a vector $A_\mu(x^0, x^1)$ corresponding to the state $\psi^\mu_{-1/2}|0\rangle$ and eight bosons $\phi^I(x^0, X^1)$ corresponding to the states $\psi^I_{-1/2}|0\rangle$. We now consider the action of the orientation reversal $\Omega: \sigma \rightarrow -\sigma, \psi \leftrightarrow \bar{\psi}$. Using (4.16.18) and (4.16.19) on page 89,

$$\Omega\, b^\mu_{-1/2}|0\rangle = \bar{b}^\mu_{-1/2}|0\rangle = -b^\mu_{-1/2}|0\rangle, \tag{11.7.7}$$

$$\Omega\, b^I_{-1/2}|0\rangle = \bar{b}^I_{-1/2}|0\rangle = b^I_{-1/2}|0\rangle. \tag{11.7.8}$$

Thus, the vector in the NS-NS sector is projected out, while the eight bosons survive the projection.

We now analyze the R sector, where fermionic degrees of freedom would come from. The massless ground state $|R\rangle$ is an $SO(9,1)$ spinor satisfying the usual GSO projection

$$\Gamma^{11}|R\rangle = |R\rangle. \tag{11.7.9}$$

Consider now the Ω projection on that spinor. In the usual D_9 case with only NN boundary

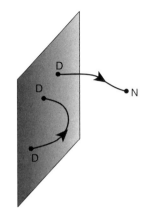

Figure 11.2 Open-string fluctuations of a D1-brane.

conditions, Ω can be taken to commute with $(-1)^F$ and acts on the spinor ground state as -1. This is important in order to produce the anomaly-free gauge group O(32).

Here, with eight DD coordinates, the action of Ω must be modified. It can be obtained by doing eight T-duality transformations as in (7.7.1) on page 174 to the D_9 case. Thus, on the spinor, this action is

$$\Omega|R\rangle = -\Gamma^2 \cdots \Gamma^9|R\rangle = |R\rangle. \tag{11.7.10}$$

From (11.7.9), (11.7.10) we also obtain

$$\Gamma^0\Gamma^1|R\rangle = -|R\rangle. \tag{11.7.11}$$

If we decompose the spinor under $SO(8)\times SO(1,1)$ as in (11.5.4) the surviving piece transforms as $\mathbf{8}_-$ under $SO(8)$ while from (11.7.11) it has negative $SO(1,1)$ chirality.

To recapitulate, in the D_1-D_1 sector we have found the following massless fluctuations moving on the world-sheet of the D1-string: eight bosons and eight chirality minus fermions.

Consider now the D_1-D_9 fluctuations. In this case we have 32 CP factors at the free string end. Moreover, the transverse directions have now DN boundary conditions. According to (2.3.31) on page 24, the bosonic oscillators are half-integrally moded as in the twisted sector of $\mathbb{Z}_2$ orbifolds. Thus, the ground-state conformal weight is 8/16=1/2. Also the modding for the fermions has been reversed between the NS and R sectors as shown in section 4.16.2 on page 88.

In the NS sector the fermionic ground state is also a spinor with ground-state conformal weight 1/2. The total ground state has conformal weight 1 and only massive excitations are obtained in this sector.

In the R sector there are massless states coming from the bosonic ground state combined with the O(1,1) spinor ground state from the longitudinal R fermions. The usual GSO projection here is $\Gamma^0\Gamma^1 = 1$. This gives a single ground state. We have also the CP multiplicity. Thus, the massless modes of the D_1-D_9 string are 32 chirality plus fermions.

In total, the world-sheet theory of the D-string contains exactly what we would expect from the O(32) heterotic string in the physical gauge! This is a nontrivial argument in favor of heterotic/type I duality.

11.7.2 The type-I D_5-brane

The R-R two-form couples to the D_1-brane and magnetically to a D_5-brane. We now describe in more detail the D_5-brane, since it involves some novel features.

To construct a five-brane, we have to impose Dirichlet boundary conditions in four transverse directions. We will again have DD and NN sectors, as in the D_1 case. Since half of the original supersymmetry is broken, we expect that the world-volume theory will have $\mathcal{N} = 1_6$ supersymmetry.

The massless fluctuations will form multiplets of this supersymmetry. The relevant multiplets are the vector multiplet, containing a vector and a gaugino, as well as the hyper-multiplet, containing four real scalars and a fermion. Supersymmetry implies that, just as in four dimensions, the manifold of the hypermultiplet scalars is a hyper-Kähler manifold. When the hypermultiplets are charged under the gauge group, the gauge transformations are isometries of the hyper-Kähler manifold, of a special type: they are compatible with the hyper-Kähler structure. You will find more on this geometry in appendix I.2 on page 535.

It will be important for our later purposes to describe the Brout-Englert-Higgs effect in this case. When a gauge theory is in the Higgs phase, the gauge bosons become massive by combining with some of the Higgs scalars. The low-energy theory (for energies well below the gauge boson mass) is described by the scalars that have not combined with the gauge bosons. In our case, supersymmetry implies that a massless vector multiplet, in order to become massive, must combine with a massless hypermultiplet.

The left-over low-energy theory of the scalars will be described by a smaller hyper-Kähler manifold. This is because supersymmetry is not broken during the Higgs phase transition. This manifold is constructed by a mathematical procedure known as the hyper-Kähler quotient. The procedure "factors out" the isometries of a hyper-Kähler manifold associated with the broken gauge symmetries, in order to produce a lower-dimensional manifold which is still hyper-Kähler. To conclude, the hyper-Kähler quotient construction describes the ordinary Brout-Englert-Higgs effect in $\mathcal{N} = 1_6$ gauge theory.

The D_5-brane we are about to construct is mapped via heterotic/type I duality to the NS_5-brane of the heterotic theory. The NS_5-brane has been constructed as a soliton of the effective low-energy heterotic action. The nontrivial fields, in the transverse space, are essentially configurations of axion-dilaton instantons, together with four-dimensional instantons embedded in the O(32) gauge group. Such instantons have a size that determines the "thickness" of the NS_5-brane.

The massless fluctuations are essentially the moduli of the instantons. There is a mathematical construction of this moduli space, as a hyper-Kähler quotient. This leads us to suspect that the interpretation of this construction is as a Brout-Englert-Higgs effect in the six-dimensional world-volume theory. In particular, the mathematical hyper-Kähler

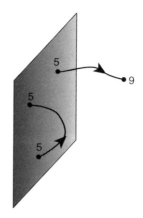

Figure 11.3 A type-I D_5 brane and its fluctuations.

quotient construction of the instanton moduli space implies that for N coincident NS_5-branes, an $Sp(2N)$ gauge group is completely Higgsed.

For a single five-brane, the gauge group should therefore be $Sp(2) \sim SU(2)$. If the size of the instanton is not zero, the massless fluctuations of the NS_5-brane form hypermultiplets only. When the size becomes zero, the moduli space has a singularity, which can be interpreted as the restoration of the gauge symmetry: at this point the gauge bosons become massless again. All of this indicates that the world-volume theory of a single five-brane should contain an $SU(2)$ gauge group, while in the case of N five-branes the gauge group is enhanced to $Sp(2N)$. In fact, in a previous section, 9.14.3, we showed from first principles that D_5-branes of the type-I theory have in fact symplectic rather than orthogonal groups. This is consistent with the discussion above.

We will now describe the massless fluctuations of a single D_5-brane. We will have two sources of world-volume fields, those coming from D_5-D_5 strings and those coming from D_5-D_9 strings (as shown in figure 11.3). As discussed earlier, to obtain the appropriate gauge fields, we must introduce CP factors for the D_5-brane. For a single D5-brane, this factor takes two values,[5] $i = 1, 2$. Thus, the massless bosonic states in the D_5-D_5 sector are of the form

$$b^{\mu}_{-1/2}|p; i, j\rangle, \quad b^I_{-1/2}|p; i, j\rangle, \quad I = 1, 2, 3, 4. \tag{11.7.12}$$

We have also seen in sections 7.3 on page 162 and 9.14.3 on page 260, that the orientifold projection Ω acts as

$$\Omega \, b^{\mu}_{-1/2}|p; i, j\rangle = -b^{\mu}_{-1/2}(\gamma_{\Omega})_{ii'}|p; j', i'\rangle(\gamma_{\Omega}^{-1})_{jj'},$$
$$\Omega \, b^I_{-1/2}|p; i, j\rangle = b^I_{-1/2}(\gamma_{\Omega})_{ii'}|p; j', i'\rangle(\gamma_{\Omega}^{-1})_{jj'}, \quad I = 1, 2, 3, 4, \tag{11.7.13}$$

with $\gamma_{\Omega} = -\gamma_{\Omega}^T$. Therefore, we obtain an $Sp(2)$ vector and four singlet scalars. They are the expected transverse coordinates of the D_5-brane.

---

[5] Another explanation for this can be found in exercise 11.23.

In the R sector, in a physical gauge, we decompose the spinor under $SO(8) \rightarrow SO(4) \times SO(4)$ as $\mathbf{8}_+ = \mathbf{2}_+ \otimes \mathbf{2}_+ \oplus \mathbf{2}_- \otimes \mathbf{2}_-$. In exercise 11.27 you are asked to impose the Ω projection and obtain three six-dimensional fermions of negative chirality completing the vector multiplets of Sp(2) and one fermion of positive chirality completing a hypermultiplet with the four transverse coordinates of the brane.

In the D_5-D_9 sector we have an $i = 1, 2$ CP factor on the D_5 end point and an $\alpha = 1, 2, \ldots, 32$ factor on the D_9 end point. We have six NN coordinates and four DN coordinates. In the NS sector the ground-state energy is $1/2$, $1/4$ coming from the four twisted transverse coordinates and $1/4$ from the transverse fermions with R boundary conditions. The ground state transforms as the $\mathbf{4}$ of transverse SO(4). It gives four massless scalars transforming in the $(\mathbf{2}, \mathbf{32})$ under Sp(2) $\times$ O(32) where Sp(2) is the world-volume gauge group and O(32) is the original (space-time) gauge group of the type-I theory. In the R sector, we obtain an SO(1,5) spinor in $(\mathbf{2}, \mathbf{32})$. Thus, in total we obtain a hypermultiplet in the $(\mathbf{2}, \mathbf{32})$ of Sp(2) $\times$ O(32).

For N parallel coinciding D_5-branes, the only difference is that the D_5 CP factor now takes 2N values. In exercise 11.28 you are invited to show that D_5-D_5 strings give a massless Sp(2N) vector multiplet, one singlet hypermultiplet, and another one transforming in the two-index anti-symmetric representation of Sp(2N) of dimension $2N^2 - N - 1$. The D_5-D_9 strings provide another hypermultiplet in the $(\mathbf{2N}, \mathbf{32})$ of Sp(2N) $\times$ O(32).

11.7.3 Further consistency checks

There are further checks of heterotic/type I duality in ten dimensions. BPS-saturated terms in the effective action, like F^4, $F^2 R^2$, and R^4, match appropriately between the two theories. This is not unexpected since they are anomaly related. Their matching is a consequence of supersymmetry and absence of anomaly in ten dimensions.

The comparison of BPS-saturated terms becomes more involved and nontrivial upon toroidal compactification. First, the spectrum of BPS states is richer. They are different, in perturbation theory, for the two theories. Second, by adjusting moduli both theories can be compared in the weak coupling limit.

In the heterotic string, the F^4, $F^2 R^2$, and R^4 terms, obtain perturbative corrections at one loop only. Their nonperturbative corrections are due to instantons that preserve half of the space-time supersymmetry. In the heterotic string, the only relevant nonperturbative configuration is the NS$_5$-brane. Taking its world-volume to be Euclidean and wrapping it supersymmetrically around a compact manifold (so that the classical action is finite), it provides the instanton configurations. We need at least a six-dimensional compact manifold to wrap it. We can therefore deduce that the BPS-saturated terms do not have nonperturbative corrections for toroidal compactifications with more than four noncompact directions. Thus, for $D > 4$, the full heterotic result emerges from tree level and one loop.

In the type-I string the situation is different. Here both the D_1-brane and the D_5-brane, after Euclideanization, can provide instanton configurations. As before, the D_5-brane will contribute in four or less noncompact dimensions.

The D_1-brane has a two-dimensional world-sheet and may contribute in eight or fewer noncompact dimensions. We conclude that, in nine dimensions, the two theories can be compared in perturbation theory. The type-I theory has very few BPS states that propagate in the loop and correct the effective action. They are the Kaluza-Klein states of the circle. On the other hand the heterotic theory has many more BPS states that contribute at one loop. On top of the usual Kaluza-Klein states there are the circle winding states and the whole tower of states charged under the gauge group which come from the $\text{Spin}(32)/\mathbb{Z}_2$ lattice. Such states are generated by the (nonperturbative) D_1-brane in the type-I side. We might think that in order for the two results to agree we would have include in the loop the D_1 solitonic states on the type-I side.

This turns out not to be the case. In fact, the type-I contribution is integrated over the field-theory-like Schwinger-parameter space (moduli space of the cylinder) while the heterotic one-loop result is integrated over the fundamental domain of the torus. Using a well-known unfolding trick, it can be shown that the two results agree. On the type-I side, however, duality also implies "contact" contributions for the factorizable terms $(\text{tr}R^2)^2$, $\text{tr}F^2\text{tr}R^2$, and $(\text{tr}F^2)^2$ coming from surfaces with Euler number $\chi = -1, -2$.

In eight dimensions, the perturbative heterotic result is mapped via duality to perturbative as well as nonperturbative type-I contributions coming from the D_1-instanton. The heterotic results can be expanded, and the type-I D-instanton terms can be identified. The calculation of the classical action is straightforward and agrees with the heterotic result. The determinants as well as the multi-instanton summation are more subtle but they can also be done in the type-I theory. This gives a strong test of heterotic/type I duality as well as it gives hints on the rules of D-instanton calculus.

To conclude, there is a single theory, with two weak-coupling descriptions. One is the heterotic O(32) string. The other is the O(32) type-I string. Each is the strong coupling extrapolation of the other.

11.8 M-theory and the $E_8 \times E_8$ Heterotic String

We have discovered that M-theory compactified on S^1 gives the type-IIA string theory. There is another possible manifold that upon compactification will provide a ten-dimensional theory. This is the one-dimensional orbifold $S^1/\mathbb{Z}_2$.

So far we have discovered the nature of strong-coupling limits of four out of the five ten-dimensional superstring theories. $E_8 \times E_8$ still remains to be investigated. It turns out that it is related to the $S^1/\mathbb{Z}_2$ compactification of M-theory.

M-theory has a parity symmetry, $x^{11} \to -x^{11}$. The three-form is odd under parity. Upon compactifying x^{11} and passing to the type-IIA description, this symmetry can be identified with $(-1)^{F_L}$ on the zero modes.[6] We will consider an orbifold of M-theory compactified on a circle of radius R, where the orbifolding symmetry is $x^{11} \to -x^{11}$.

---

[6] As you are required to show in exercise 11.29 orbifolding the IIA string by that symmetry gives the IIB string and vice versa.

The untwisted sector can be obtained by keeping the fields invariant under the projection. It is not difficult to verify that the ten-dimensional metric and dilaton survive the projection, while the R-R gauge boson is projected out. Also the three-form is projected out, while the two-form survives. Half of the fermions survive, a Majorana-Weyl gravitino and a Majorana-Weyl fermion of opposite chirality. Thus, in the massless spectrum, we are left with the $\mathcal{N} = 1_{10}$ supergravity multiplet.

We know that this theory is anomalous in ten dimensions. We must have some "twisted sector" that should arrange itself to cancel the anomalies. As we discussed in the section on orbifolds, $S^1/\mathbb{Z}_2$ is a line segment, with the fixed-points $x^{11} = 0, \pi R_{11}$ at the boundary. We obtain two fixed planes that are two copies of ten-dimensional flat space. This is sketched in figure 11.4. States coming from the twisted sector must be localized on these planes. There is also a symmetry exchanging the fixed planes (translation by πR_{11}). We therefore expect isomorphic massless content coming from the two fixed planes. It can also be shown that half of the anomalous variation is localized at one fixed plane and the other half at the other. The only $\mathcal{N} = 1_{10}$ multiplets that can cancel the anomaly symmetrically are vector multiplets, and we must have 248 of them at each fixed plane. The possible anomaly-free groups satisfying this constraint are $E_8 \times E_8$ and $U(1)^{496}$. As we will argue differently below, the natural choice is also the correct one: the gauge group is $E_8 \times E_8$.

We have thus obtained the $E_8 \times E_8$ heterotic string theory. A similar argument as in section 11.4 shows that $g_s \sim R_{11}^{3/2}$. In the perturbative heterotic string, the two ten-dimensional planes are on top of each other and they move further apart as the coupling grows. We observe that in this picture, the hidden[7] E_8 is displaced in the eleventh dimensional space-time from the observable E_8. They interact via gravitational interactions through the 11-dimensional bulk.

We will describe now a chain of dualities that brings us from the $E_8 \times E_8$ heterotic string theory to the $S^1/\mathbb{Z}_2$ orbifold of M-theory. We must first compactify the $E_8 \times E_8$ heterotic string on a circle of radius R_9^e in order to profit from the perturbative connection to the $O(32)$ theory. According to section 11.1, the theory is equivalent to the $O(32)$ heterotic string compactified at radius

$$R_9^o \sim \frac{1}{R_9^e}, \quad g_s^o \sim \frac{g_s^e}{R_9^e}, \tag{11.8.1}$$

where g_s^o is the ten-dimensional coupling of the $O(32)$ heterotic string and g_s^e the analogous coupling of the $E_8 \times E_8$ heterotic string.

From the $O(32)$ heterotic string we may now perform an S-duality to go to the type-I $O(32)$ string using (11.7.5):

$$R_9^I = \frac{R_9^o}{\sqrt{g_s^o}} \sim \frac{1}{\sqrt{g_s^e R_9^e}}, \quad g_s^I = \frac{1}{g_s^o} \sim \frac{R_9^e}{g_s^e}. \tag{11.8.2}$$

The strong-coupling limit $g_s^e \to \infty$ of the heterotic theory corresponds to the weak-coupling, small-radius limit of the type-I theory. We must T-dualize again to pass to the

---

[7] See the discussion at the end of section 9.8.1 on page 246.

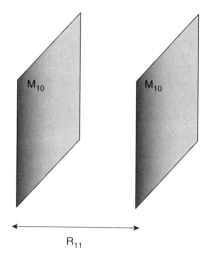

Figure 11.4 The brane picture of heterotic M-theory: An 11-dimensional bulk with two 9-branes as its boundary.

type-I$'$ theory (equivalent to type IIA on $S^1/\mathbb{Z}_2$):

$$R_9^{I'} = \frac{1}{R_9^I} \sim \sqrt{g_s^e R_9^e}, \quad g_s^{I'} = \frac{g_s^I}{R_9^I} \sim \frac{(R_9^e)^{3/2}}{\sqrt{g_s^e}} = \frac{(R_9^e)^2}{R_9^{I'}}. \tag{11.8.3}$$

Here, our compact space is the interval $[0, \pi R_9^{I'}]$ with eight D$_8$-branes localized at each end point. Taking $R_9^e \to \infty$ to recover the ten-dimensional $E_8 \times E_8$ heterotic string theory, while keeping $R_9^{I'}$ fixed, we obtain M-theory on $S^1/\mathbb{Z}_2$. Since $g_s^e \sim 1/R_9^e, g_s^e \to \infty$. With this chain of dualities we obtain M-theory on the finite interval $S^1/\mathbb{Z}_2$ with eight nine-planes on each boundary as the strong-coupling dual of $E_8 \times E_8$ heterotic string theory. The SO(16) symmetry localized at each boundary (fixed point) must enhance in the limit to the $E_8 \times E_8$ symmetry. In the original theory, as $R_9^e \to \infty$, the spinor of SO(16) becomes massless and combines with the adjoint to form the adjoint of E_8. This is what must happen also on the M-theory side.

The M-theory membrane survives in the orbifold only if one of its dimensions is wound around the $S^1/\mathbb{Z}_2$. It provides the perturbative heterotic string. This can be seen from the chain of dualities

$$F^8 \xrightarrow{T_9} F^8 \xrightarrow{S} D^8 \xrightarrow{T_9} D^{89} \xrightarrow{S} M^{8,10}, \tag{11.8.4}$$

using the notation of section 11.6.1.

On the other hand, the five-brane survives as such, and it cannot wind around the orbifold direction. It provides the heterotic NS$_5$-brane. This is in accord with what we would expect from the heterotic string. Upon compactification to four dimensions, the NS$_5$-brane will give rise to magnetically charged pointlike states (monopoles).

The picture is now clear: at strong heterotic coupling the walls are at a finite distance apart and the heterotic string is really an open membrane. At weak coupling, the two walls come on top of each other and the eleventh dimension becomes invisible.

11.8.1 Unification at strong heterotic coupling

We have seen in section 10.7 on page 315 that there is a factor of 20–40 discrepancy between the unification scale of the (extrapolated) SM gauge couplings and the string unification scale. Another way to present the same discrepancy is to say that once we fit the SM couplings in heterotic compactifications, the Planck scale is a factor of 20–40 lower than its measured value.

As advocated in section 10.7, in the strongly coupled heterotic string, we may accommodate this factor by using the fact that gravity propagates in the 11th dimension but not the SM particles. We will now develop this idea further.

The relevant parts of the 11-dimensional heterotic action are

$$S_{11} = \frac{1}{2\kappa_{11}^2} \int d^{11}x \sqrt{g}\, R + \sum_{i=1}^{2} \frac{1}{8\pi \left(4\pi \kappa_{11}^2\right)^{2/3}} \int d^{10}x \sqrt{\hat{g}}\, \mathrm{Tr}[F_i^2]. \tag{11.8.5}$$

$i = 1, 2$ labels the two ten-dimensional boundaries of the 11-dimensional bulk, where the two E_8 gauge groups are located. Upon compactification on $M_{10} \times S^1/\mathbb{Z}_2$, with $x^{11} \in [0, \pi R_{11}]$ we can match the eleven-dimensional coupling with the $E_8 \times E_8$ string parameters.

$$R_{11} = 2g_s \ell_s, \quad 2\kappa_{11}^2 = (2\pi)^8 \ell_s^9 g_s^3. \tag{11.8.6}$$

Consider now compactifying the theory further to four dimensions, on a CY manifold with volume V_6 in the 11-dimensional metric. Then the four-dimensional couplings are

$$M_P^2 = \frac{\pi R_{11} V_6}{2\kappa_{11}^2}, \quad \alpha_{\mathrm{GUT}} = \frac{\left(4\pi \kappa_{11}^2\right)^{2/3}}{2V_6}. \tag{11.8.7}$$

Picking the correct number for $\alpha_{\mathrm{GUT}} = 1/25$, implies that the heterotic nine-branes are weakly coupled to the 11-dimensional bulk, because α_{GUT} is small. Moreover, we would like the KK scale $V_6^{-1/6}$ to be at least equal to the GUT scale. We see from the relations above that we may make M_P as large as we wish by taking the radius of the 11th dimension large compared to the 11th-dimensional Planck length.

There is a however an upper bound on R_{11}. The equations of motion impose that the volume of the six-dimensional CY manifold must vary over the eleventh dimension. If V_6 is its value at the position of the "observable" 9-brane then the volume decreases as we approach the other brane. If R_{11} is large enough, the volume will turn negative before we reach the other brane. Despite this, it turns out that we gain some extra freedom in this context compared to the perturbative heterotic string.

We now take $M_{\mathrm{GUT}} = V_6^{-1/6}$. The eleven-dimensional Planck scale and R_{11} can be determined in terms of M_{GUT}, α_{GUT}, and M_P

$$2\kappa_{11}^2 = \frac{1}{2\pi} \frac{(2\alpha_{\mathrm{GUT}})^{3/2}}{M_{\mathrm{GUT}}^9}, \quad \frac{1}{R_{11}} = \frac{2\pi^2}{(2\alpha_{\mathrm{GUT}})^{3/2}} \frac{M_{\mathrm{GUT}}^3}{M_P^2}. \tag{11.8.8}$$

Setting $M_{\mathrm{GUT}} = 10^{16.1}$ GeV, $\alpha_{\mathrm{GUT}}^{-1} = 25$, and $M_P = 1.72 \times 10^{18}$ GeV we obtain

$$\frac{1}{R_{11}} \simeq 6 \times 10^{14} \text{ GeV}, \quad \frac{1}{2\kappa_{11}^2} \simeq (2.35 \times 10^{16} \text{ GeV})^9 \sim M_{\mathrm{GUT}}^9. \tag{11.8.9}$$

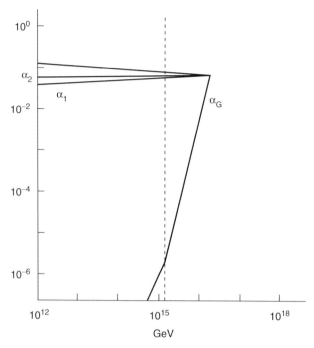

Figure 11.5 The running of gauge and gravitational couplings in compactified heterotic M-theory.

The physical picture is as follows. At energies below $R_{11}^{-1} \sim 10^{15}$ GeV, physics is four dimensional. For energies 10^{15} GeV $< E < 10^{16}$ GeV $\sim M_{\mathrm{GUT}}$, we are probing distances larger than the radius of the 11th dimension. The gravitational physics is five dimensional, and the gravitational coupling runs as $G_N \simeq E^{-3}$ in this region. However, physics is still four-dimensional for the matter on the observable brane, since it is localized at the boundary of the five-dimensional space. In particular, the gauge couplings run as in four dimensions. Above 10^{16} GeV, physics becomes 11 dimensional and at the same time the supergravity approximation breaks down. The behavior of the couplings is summarized in figure 11.5.

11.9 Heterotic/Type II Duality in Six Dimensions

There is another nontrivial duality that is present in six noncompact dimensions. It involves the heterotic string compactified on T^4 and the type-IIA string compactified on K3. Both theories have $\mathcal{N} = (1, 1)_6$ supersymmetry. They have the same massless spectrum, containing the $\mathcal{N} = (1, 1)_6$ supergravity multiplet and 20 vector multiplets.

The six-dimensional tree-level heterotic effective action in the string frame was derived in section 9.1 on page 219. We reproduce it here for convenience:

$$S_{T^4}^{\text{het}} = \int d^6x \sqrt{-\det G} e^{-2\Phi} \left[R + 4\partial^\mu \Phi \partial_\mu \Phi - \frac{1}{12} \hat{H}^{\mu\nu\rho} \hat{H}_{\mu\nu\rho} \right.$$
$$\left. - \frac{1}{4} (\hat{M}^{-1})_{ij} F^i_{\mu\nu} F^{j\mu\nu} + \frac{1}{8} \text{Tr}(\partial_\mu \hat{M} \partial^\mu \hat{M}^{-1}) \right], \tag{11.9.1}$$

where $i = 1, 2, \ldots, 24$ and

$$\hat{H}_{\mu\nu\rho} = \partial_\mu B_{\nu\rho} - \frac{1}{2} \hat{L}_{ij} A^i_\mu F^j_{\nu\rho} + \text{cyclic}. \tag{11.9.2}$$

The moduli scalar matrix $\hat{M}$ is the symmetric matrix

$$\hat{M} = \begin{pmatrix} G^{-1} & G^{-1}C & G^{-1}Y^t \\ C^t G^{-1} & G + C^t G^{-1}C + Y^t Y & C^t G^{-1}Y^t + Y^t \\ YG^{-1} & YG^{-1}C + Y & 1_{16} + YG^{-1}Y^t \end{pmatrix}, \tag{11.9.3}$$

where 1_{16} is the 16-dimensional unit matrix and

$$C_{\alpha\beta} = B_{\alpha\beta} - \frac{1}{2} Y^I_\alpha Y^I_\beta. \tag{11.9.4}$$

The scalar moduli are the metric of the internal T^4, $G_{\alpha\beta}$, $\alpha, \beta = 1, 2, 3, 4$, the associated antisymmetric tensor $B_{\alpha\beta}$, as well as the internal components of the ten-dimensional Cartan gauge fields (Higgs scalars) Y^I_α, $I = 1, 2, \ldots, 16$. The moduli matrix $\hat{M}$ satisfies

$$\hat{M}^T \hat{L} \hat{M} = \hat{M} \hat{L} \hat{M} = \hat{L}, \quad \hat{M}^{-1} = \hat{L} \hat{M} \hat{L}, \tag{11.9.5}$$

where $\hat{L}$ is the invariant O(4,20) metric. Thus, $\hat{M} \in$ O(4,20). The low-energy effective action (11.9.1) is invariant under continuous O(4,20;$\mathbb{R}$) transformations,

$$A^i_\mu \to U^{-1}_{ij} A^j_\mu, \quad \hat{M} \to U \hat{M} U^T \tag{11.9.6}$$

with $U \in$ O(4,20):

$$U \hat{L} U^T = \hat{L}. \tag{11.9.7}$$

The matrix $\hat{M}$ is not the most general O(4,20;$\mathbb{R}$) matrix. It lies in the coset O(4,20;$\mathbb{R}$)/O(4;$\mathbb{R}$)$\times$O(20;$\mathbb{R}$) of real dimension 4$\times$20.

The continuous global symmetry O(4,20;$\mathbb{R}$), is broken to the discrete T-duality symmetry O(4,20;$\mathbb{Z}$) by the massive states coming from the internal lattice (KK and winding states). This is described in more detail in appendix D on page 513. Here we should note that the lattice momentum and winding integers, $q^i \sim (m_\alpha, n_\alpha, Q^I)$ are the electric charges of the gauge fields A^i_μ. (m_α, n_α) are the integer momenta and windings on T^4 while Q^I are the 16 right-moving integer momenta on the Spin(32)/$\mathbb{Z}_2$ lattice.

Going to the Einstein frame by $G_{\mu\nu} \to e^\Phi G_{\mu\nu}$, we obtain

$$\tilde{S}_{T^4}^{\text{het}} = \int d^6x \sqrt{-G} \left[R - \partial^\mu \Phi \partial_\mu \Phi - \frac{e^{-2\Phi}}{12} \hat{H}^{\mu\nu\rho} \hat{H}_{\mu\nu\rho} \right.$$
$$\left. - \frac{e^{-\Phi}}{4} (\hat{M}^{-1})_{ij} F^i_{\mu\nu} F^{j\mu\nu} + \frac{1}{8} \text{Tr}(\partial_\mu \hat{M} \partial^\mu \hat{M}^{-1}) \right]. \tag{11.9.8}$$

We should remind the reader that, for generic values of the moduli, the gauge group is U(1)^{24}. As discussed in section 9.1, for special values of the moduli, nonabelian symmetry appears. Vector multiplets, charged under the Cartan gauge bosons, become massless. They combine with the abelian vector multiplets to various nonabelian ADE groups with maximal rank 20. This enhancement phenomenon was discussed in its most simple form in section 4.18.6 on page 100.

We will now consider the effective action of type-IIA theory compactified on K3. The tree-level type-IIA effective action in the string frame was given in (9.9.8) on page 249 and we reproduce it here:

$$S_{K3}^{IIA} = \int d^6x \sqrt{-\det G_6}\, e^{-2\Phi} \left[R + 4\partial^\mu \Phi \partial_\mu \Phi - \frac{1}{12} H^{\mu\nu\rho} H_{\mu\nu\rho} + \frac{1}{8}\text{Tr}(\partial_\mu \hat{M}\partial^\mu \hat{M}^{-1}) \right]$$
$$- \frac{1}{4} \int d^6x \sqrt{-\det G}(\hat{M}^{-1})_{ij} F^i_{\mu\nu} F^{j\mu\nu} + \frac{1}{16} \int d^6x \epsilon^{\mu\nu\rho\sigma\tau\upsilon} B_{\mu\nu} F^i_{\rho\sigma} \hat{L}_{ij} F^j_{\tau\upsilon}, \tag{11.9.9}$$

where $i, j = 1, 2, \ldots, 24$. Some comments about the form of the effective action are in order here. Although all terms of this effective action come from the sphere (tree level), the factors of the dilaton are different from the heterotic one. The terms associated with the supergravity multiplet $(g_{\mu\nu}, B_{\mu\nu}, \Phi)$ have the standard dilaton dependence since they come from the NS-NS sector. The gauge fields, however, are descendants of the R-R forms, and thus have the unusual dependence discussed in section 7.2.1 on page 159.

The $\mathcal{CP}$-odd term in the effective action is a descendant of the $\mathcal{CP}$-odd term of 11-dimensional supergravity cubic in the three-form (see appendix H.1 on page 526). Consequently, this term has also no dilaton dependence. Transforming to the Einstein frame we obtain

$$\tilde{S}_{K3}^{IIA} = \int d^6x \sqrt{-G} \left[R - \partial^\mu \Phi \partial_\mu \Phi - \frac{1}{12}e^{-2\Phi} H^{\mu\nu\rho} H_{\mu\nu\rho} - \frac{1}{4}e^{\Phi}(\hat{M}^{-1})_{ij} F^i_{\mu\nu} F^{j\mu\nu} \right.$$
$$\left. + \frac{1}{8}\text{Tr}(\partial_\mu \hat{M}\partial^\mu \hat{M}^{-1}) \right] + \frac{1}{16} \int d^6x \epsilon^{\mu\nu\rho\sigma\tau\varepsilon} B_{\mu\nu} F^i_{\rho\sigma} \hat{L}_{ij} F^j_{\tau\varepsilon}, \tag{11.9.10}$$

where $\hat{L}$ is the O(4,20)-invariant metric. Notice the following differences between the two Einstein actions (11.9.8),(11.9.10): the heterotic $\hat{H}_{\mu\nu\rho}$ contains the Chern-Simons term (11.9.2) while the type IIA one does not. The type-IIA action instead contains a parity-odd term coupling the gauge fields and $B_{\mu\nu}$. Both effective actions have a continuous O(4,20,$\mathbb{R}$) symmetry, which is broken in the string theory to the T-duality group O(4,20,$\mathbb{Z}$).

We will denote by a prime the Einstein-frame fields of the type-IIA theory and without a prime those of the heterotic theory.

You are invited in exercise 11.33 to show that the two sets of equations of motion stemming from the actions (11.9.8) and (11.9.10) are equivalent through the following (duality) transformations:

$$\Phi' = -\Phi, \quad G'_{\mu\nu} = G_{\mu\nu}, \quad \hat{M}' = \hat{M}, \quad A^{i}_\mu = A^i_\mu, \tag{11.9.11}$$

$$e^{-2\Phi} \hat{H}_{\mu\nu\rho} = \frac{1}{6}\frac{\epsilon_{\mu\nu\rho}{}^{\sigma\tau\varepsilon}}{\sqrt{-G}} H'_{\sigma\tau\varepsilon}, \tag{11.9.12}$$

where the data on the right-hand side are evaluated in the type-IIA theory. This is the statement of heterotic/IIA duality.

Since the string coupling g_s is related to the dilaton as $g_s = \langle e^\Phi \rangle$, the duality above gives

$$g_s^{\text{het}} \leftrightarrow \frac{1}{g_s^{\text{IIA}}}. \tag{11.9.13}$$

Therefore, heterotic/IIA duality is a nonperturbative correspondence. It may be viewed as the generalization to six dimensions of the Montonen-Olive duality. Here the role of the gauge fields is played by the antisymmetric tensor, whose field strength is dualized in (11.9.12). At the same time, the coupling constant is also inverted. The analogues of electric and magnetic charges are carried by strings rather than pointlike particles.

There are some further arguments for the duality described above. In the heterotic theory, the fundamental heterotic string is the string that couples electrically to $B_{\mu\nu}$. In order for the duality to be correct, there should be a solitonic string, which should couple magnetically to $B_{\mu\nu}$.[8] Moreover, this solitonic string should be the fundamental string of the type-IIA theory compactified on K3. This string can be obtained by wrapping the heterotic NS$_5$-brane on T^4. Since it descends from the NS$_5$-brane, it is indeed magnetically charged under $B_{\mu\nu}$. Similarly, the heterotic string, as viewed from the perturbative type-IIA string, is a solitonic string. It can be obtained by wrapping the IIA NS$_5$-brane around K3.

This duality can also be viewed from an 11-dimensional vantage point. Consider the compactification of M-theory on $S^1 \times K3$. According to our discussion in section 11.4 on page 328, this is equivalent to the type-IIA theory compactified on K3. All vectors descend from the R-R one- and three-forms of the ten-dimensional type-IIA theory. In turn, these descend from the three-form of M-theory, to which the M$_2$- and M$_5$-branes couple. The membrane wrapped around S^1 would give a string in six dimensions. As in ten dimensions, this is the perturbative type-IIA string. There is another string, however, obtained by wrapping the five-brane around the whole K3. It can be explicitly shown that this string has the world-volume dynamics of the heterotic string. Moreover, the 11-dimensional electric-magnetic duality between the M$_2$- and M$_5$-branes translates into the IIA/heterotic duality in six dimensions.

As we have seen in section 7.9 on page 176, at tree level, the heterotic string has a gravitational Chern-Simons correction to the field strength of $B_{\mu\nu}$. This survives compactification on T^4. This correction is of higher order in derivatives, but it is important for anomaly cancellation. In terms of ω, the spin-connection one-form, the curvature two-form can be written as $R = d\omega + \omega^2$. The gravitational Chern-Simons three-form is

$$CS_{\text{grav}} = \text{Tr}\left[\omega d\omega + \frac{2}{3}\omega^3\right], \quad d\, CS_{\text{grav}} = \text{Tr}\left[R \wedge R\right]. \tag{11.9.14}$$

The anomaly-related coupling of the heterotic string is obtained by substituting[9]

$$\hat{H} = dB + CS_{\text{grav}} + CS_{\text{gauge}} \tag{11.9.15}$$

in (11.9.2). Going through the heterotic/IIA duality map, this term implies the presence of the one-loop term $B \wedge R \wedge R$ term in the IIA theory compactified on K3. In exercise 11.35 you are invited to study the matching of these two terms.

---

[8] In six dimensions the magnetic dual of a string is still a string.

[9] In this chapter, for convenience, we have redefined fields to absorb the factor κ^2/g^2 in (7.9.13) on page 179.

11.9.1 Gauge symmetry enhancement and singular K3 surfaces

There is another effect associated with heterotic/IIA duality that has an interesting resolution.

It is well known, that on the heterotic side, at special values of the (4,20) moduli, we may have gauge symmetry enhancement. At such points, extra gauge bosons become massless, and since they are charged under the $U(1)^{24}$ generic gauge symmetry, they generate a non-abelian group. Four out of the 24 U(1) gauge bosons belong to the supergravity multiplet (those that come from the supersymmetric side of the string). These cannot participate in the gauge group enhancement.

If heterotic/IIA duality is correct, we would like to see a similar phenomenon on the type IIA side. In type IIA all gauge bosons come from the R-R sector. As we already know, there are no charged states under R-R forms in the perturbative type-IIA string. So if there are such charged gauge bosons that become massless for a special value of the moduli, then such states should be nonperturbative on the type-IIA side.

The gauge bosons in six dimensions descend from the R-R one- and three-forms in ten dimensions. The one-form in ten dimensions is paired with the unique closed zero-form (constant) on K3 to give a vector in six dimensions. The three-form can be expanded into a sum of 22 gauge bosons in six dimensions times the 22 harmonic two-forms on K3. Finally, the three-form in ten dimensions also gives a three-form in six dimensions paired with the zero-form of K3. A three-form in six dimensions is dual to a one-form, so this gives rise to an extra vector that can be associated with the volume form of K3. Thus, the 24 six-dimensional gauge bosons are in a one-to-one correspondence with the even cohomology of K3. The O(4,20) invariant metric $\hat{L}$ is the intersection form of the even cohomology of K3.

We can count the number of moduli as follows. The moduli space of Ricci-flat metrics for K3 is isomorphic to the coset $O(3,19)/(O(3) \times O(19))$. There are also moduli associated with the antisymmetric tensor on K3, in a one-to-one correspondence with the 22 harmonic two-forms on K3 (three of them are self-dual and 19 are anti-self-dual). Finally we have the constant mode of the dilaton. It can be shown that the full moduli space is now $O(4,20)/(O(4) \times O(20))$, the same as in the heterotic case. It takes a more delicate analysis to show that the discrete group $O(4,20,\mathbb{Z})$ is also a symmetry on this moduli space.

To come back to our problem, we have to find non-perturbative states charged under the six-dimensional U(1) gauge bosons. In ten dimensions, we know of nonperturbative states charged under the R-R forms: they are the D-branes. In the IIA theory, the D_0-brane is electrically charged under the one-form, the D_2-brane is electrically charged and the D_4-brane magnetically charged under the three-form.

Before we continue looking for candidate charged gauge bosons, we will have to ensure a specific property: that these states can become massless at some special points (or subvarieties) of the moduli space. Massless supermultiplets contain many less states than generic massive ones. Therefore, massless states can appear only if the massive states are in short (BPS) supermultiplets. This is good news, since BPS multiplets are protected from quantum corrections and provide a reliable window to strong-coupling physics.

The D_0-brane is a particle in both ten and six dimensions which, by its definition, carries the electric charge of the ten-dimensional vector. Moreover, it is a BPS state preserving half of the supersymmetry. Furthermore, it is known from ten dimensions that there are marginal bound states of m D_0-branes for any charge $m \in \mathbb{Z}$. The existence of such bound states is required by the type-IIA/M-theory connection as explained in section 11.4. Thus, these are some of the states we are looking for.

The D_4-brane can also give rise to BPS particles in six dimensions, when its world-volume is wrapped around the whole of K3, preserving half of the original supersymmetry. Thus, a wrapped D_4-brane is a BPS state breaking half of the original supersymmetry. Moreover, it was magnetically charged under the three-form in ten dimensions, which means that it is electrically charged under the vector in six dimensions obtained by dualizing the three-form. Multiple wrappings produce states with multiple charge.

So far we have found (nonperturbative) BPS states that are charged under the two six-dimensional gauge bosons dual to the zero- and four-form of K3. The rest correspond to D_2-branes wrapped supersymmetrically around nontrivial supersymmetric two-cycles of K3. They are in a one-to-one correspondence with the second cohomology of K3. Therefore, they provide the leftover 22 charges. The mass of such gauge bosons is proportional to the tension of the appropriate D_2-brane times the area of the compact surface it is wrapped around. By adjusting moduli, one or more two-cycles can be brought to have vanishing area. The associated charged particles will then become massless and provide an enhancement of the gauge group.

We introduce a basis in the cohomology of K3. The elements are the zero- and four-forms ω_0 and ω_4 as well as the 22 two-forms ω_2^i. They are normalized as

$$\int_{K3} \omega_0 \wedge \omega_4 = 1, \qquad \int_{K3} \omega_2^i \wedge \omega_2^j = \hat{L}_{ij}. \tag{11.9.16}$$

$\hat{L}_{ij}$ is the invariant metric of O(3,19). In total, we obtain the O(4,20) invariant intersection form. There is a dual basis of two-homology cycles C_i so that

$$\int_{C_i} \omega_2^j = \delta_{ij}. \tag{11.9.17}$$

A general two-cycle can be written as an integral linear combination of the generating cycles. By changing the moduli of the surface, some generating two-cycles may go to zero volume. The K3 surface is singular at such a limit. The $\mathbb{Z}_2$ orbifold singularity described in section 9.10 on page 250 is an example of such a singularity when a single two-cycle shrinks to zero. In that case, there is some B-flux trapped in the shrinking cycle that prevents the CFT from becoming singular. However, there are other points in the moduli space where cycles shrink to zero without any trapped flux. At such points the K3 CFT is singular. It is at such points that the gauge symmetry enhancement is expected to take place in IIA string theory.

The singularities of K3 admit an interesting ADE classification related to the ADE classification of finite simple subgroups of $SU(2)$. The geometry around the degenerating region can be shown to be isometric to a four-dimensional asymptotically locally Euclidean (ALE) space. ALE spaces are noncompact and hyper-Kähler, with asymptotic geometry isomorphic to S^3/Γ, where Γ is a finite (Kleinian) subgroup of $SU(2)$. They are in the same moduli space as the C^2/Γ orbifolds, where Γ acts via the SU(2) action on the two complex coordinates of C^2. To preserve the maximal supersymmetry, Γ must be in the SU(2) subgroup of the SO(4) symmetry of $\mathbb{R}^4$ that preserves the complex structure.

The case of $\Gamma = \mathbb{Z}_2$, corresponds to the Eguchi-Hanson space. It was discussed in association with the orbifold limit of the K3 surface in section 9.10. The $\Gamma = \mathbb{Z}_N$ generalization is described in exercise 11.34.

There is another description of the K3 surface as an algebraic variety that makes contact with singularity theory. If the K3 moduli are tuned appropriately, K3, defined as the vanishing locus of a polynomial may be written locally (near the singularity and in some suitable coordinate patch) as

$$W_{K3} = \epsilon[W_{ADE}] + \mathcal{O}(\epsilon^2) = 0, \tag{11.9.18}$$

where $\epsilon \to 0$ is a parameter controlling the degeneration. The surface $W_{ADE}(x_i) = 0$ is isomorphic to the ALE space described above. The appropriate polynomials are essentially given by the ADE simple singularities, up to adding extra quadratic pieces that do not change the structure of the singularity:

$$W_{A_{n-1}} = W_{A_{n-1}}(x_1) + x_2{}^2 + x_3{}^2, \quad W_{D,E} = W_{D,E}(x_1, x_2) + x_3{}^2, \tag{11.9.19}$$

with

$$W_{A_{n-1}}(x, u) = x^n, \quad W_{D_n}(x_1, x_2) = x_1{}^{n-1} + \frac{1}{2}x_1 x_2{}^2,$$

$$W_{E_6}(x_1, x_2) = x_1{}^3 + x_2{}^4, \quad W_{E_7}(x_1, x_2) = x_1{}^3 + x_1 x_2{}^3, \tag{11.9.20}$$

$$W_{E_8}(x_1, x_2) = x_1{}^3 + x_2{}^5.$$

At an ADE singularity, a collection of two-cycles of the surface have vanishing volume. They are in one-to-one correspondence with the nonzero roots $\vec{\alpha}$ of the associated Lie algebra. We may write for the vanishing cycles

$$C_{\vec{\alpha}} = \vec{\alpha} \cdot C \equiv \alpha^i\, C_i, \tag{11.9.21}$$

and we have one such vanishing cycle for every root of the appropriate ADE group. Consider now a D_2-brane wrapping once such a vanishing cycle. It will give rise to a massless vector multiplet in six dimensions, as the mass will be proportional to the volume of the wrapped cycle. We will calculate the charge of such a massless state with respect to the abelian gauge group. Consider the 22 massless U(1) gauge bosons A_μ^i obtained from the three-form by decomposing it as

$$C_{\mu ab} = A_\mu^i \omega_{ab}^i, \tag{11.9.22}$$

in the basis of the harmonic two-forms of the K3 surface. From the WZ term of the D_2-brane action we obtain

$$S_{WZ}^{D_2} = iT_2 \int \hat{C} = iT_2 \int d\tau\, \partial_\tau x^\mu A_\mu^i \int_{C_{\vec{\alpha}}} \omega_2^i = iT_2 \alpha^i \int d\tau\, \partial_\tau x^\mu A_\mu^i, \tag{11.9.23}$$

where we have used (11.9.17) and (11.9.21). Equation (11.9.23) indicates that the massless multiplets have charges under the abelian gauge bosons that are given by the components of the roots of the associated nonabelian simply laced group. They are therefore the nonperturbative states responsible for the enhancement of the gauge group, as advertised.

11.9.2 Heterotic/type II duality in four dimensions

We may now further compactify both theories on a two-torus to four dimensions and examine the consequences of the six-dimensional string/string duality. We will show that, among other things, it implies the Montonen-Olive duality for the four-dimensional gauge theory.

Both the type-IIA and heterotic theories, upon further compactification on T^2 to four dimensions, give $\mathcal{N} = 4_4$ supergravity coupled to 22 abelian vector multiplets. Compared to six dimensions, there are four extra gauge bosons, two of them graviphotons and another two in vector multiplets. The scalars now parametrize the moduli space

$$\frac{SL(2, \mathbb{R})}{U(1)} \times \frac{O(6, 22)}{O(6) \times O(22)}.$$

In addition to the $O(4,20)/(O(4) \times O(20))$ scalars that descend directly from six dimensions, there are the four scalars $G_{\alpha\beta}$, $B_{\alpha\beta}$, $\alpha, \beta = 1, 2$, parametrizing the metric and antisymmetric tensor on T^2, as well as 2×24 scalars corresponding to the components of the $(4,20)$ six-dimensional gauge fields along T^2.

The heterotic/IIA map in (11.9.11) implies that

$$e^{-2\phi} = \sqrt{\det G'_{\alpha\beta}}, \quad e^{-2\phi'} = \sqrt{\det G_{\alpha\beta}}, \tag{11.9.1}$$

$$\frac{G_{\alpha\beta}}{\sqrt{\det G_{\alpha\beta}}} = \frac{G'_{\alpha\beta}}{\sqrt{\det G'_{\alpha\beta}}}, \quad A'^\alpha_\mu = A^\alpha_\mu, \tag{11.9.2}$$

$$g_{\mu\nu} = g'_{\mu\nu}, \quad M' = M, \quad A^i_\mu = A'^i_\mu, \quad Y^i_\alpha = Y'^i_\alpha. \tag{11.9.3}$$

Here ϕ is the four-dimensional dilaton, A^α_μ are the two gauge fields coming from the metric, and the four-dimensional metrics $g_{\mu\nu}$ are in the Einstein frame.

The duality relation (11.9.12) implies

$$a = B', \quad a' = B, \tag{11.9.4}$$

and

$$\frac{1}{2} \frac{\epsilon_{\mu\nu}{}^{\rho\sigma}}{\sqrt{-g}} \epsilon^{\alpha\beta} F^{B'}_{\beta,\rho\sigma} = e^{-2\phi} G^{\alpha\beta} \left[F^B_{\beta,\mu\nu} - C_{\beta\gamma} F^{A,\gamma}_{\mu\nu} - \hat{L}_{ij} Y^i_\beta F^j_{\mu\nu} \right] - \frac{1}{2} a \frac{\epsilon_{\mu\nu}{}^{\rho\sigma}}{\sqrt{-g}} F^{A,\alpha}_{\rho\sigma}. \tag{11.9.5}$$

Here a is the four-dimensional axion obtained by dualizing the four-dimensional NS antisymmetric tensor $B_{\mu\nu}$ and B is defined as $B_{\alpha\beta} = B\epsilon_{\alpha\beta}$. Equation (11.9.5) is an electric-magnetic[10] duality transformation on the $B_{\alpha,\mu}$ gauge fields descending from the six-dimensional NS antisymmetric tensor.

In exercise 11.39 you are invited to derive the detailed KK reduction and calculation of the two effective actions. The interesting observation is that in the heterotic theory, the scalars that parametrize the $SL(2,\mathbb{R})/U(1)$ factor of the moduli space are the axion a and

---

[10] This is described in full generality in appendix G.

the dilaton ϕ combined in a complex scalar as

$$S = S_1 + iS_2 = a + i\,e^{-2\phi}. \tag{11.9.6}$$

All other scalars parametrize the $O(6,22)/(O(6) \times O(22))$ via the standard symmetric $O(6,22)$ matrix M_{IJ}. The relevant part of the heterotic effective action is

$$S_{D=4}^{\text{het}} = \int d^4x \sqrt{-g} \left[R - \frac{1}{2} \frac{\partial^\mu S \partial_\mu \bar{S}}{S_2^2} - \frac{1}{4} S_2 (M^{-1})_{IJ} F^I_{\mu\nu} F^{J,\mu\nu} \right.$$
$$\left. + \frac{1}{4} S_1\, L_{IJ} F^I_{\mu\nu} \tilde{F}^{J,\mu\nu} + \frac{1}{8} \text{Tr}(\partial_\mu M \partial^\mu M^{-1}) \right], \tag{11.9.7}$$

where the indices I,J take 28 values, $F^I_{\mu\nu}$ are the field strengths of the 28 gauge fields, and L_{IJ} is the invariant $O(6,22)$ metric. The low-energy theory has a continuous $SL(2;\mathbb{R}) \times O(6,22;\mathbb{R})$ invariance. The quantized momentum and winding states on T^6 break $O(6,22;\mathbb{R})$ to $O(6,22;\mathbb{Z})$, the perturbative T-duality of the toroidal heterotic string. The $SL(2;\mathbb{R})$ symmetry acts by fractional linear transformations on the field S. It also acts by electric-magnetic duality transformations on the gauge fields A^I_μ as in appendix G on page 522. The Dirac quantization condition breaks the continuous symmetry $SL(2;\mathbb{R})$ to $SL(2;\mathbb{Z})$, the Montonen-Olive duality of $\mathcal{N} = 4_4$ gauge theories, interchanging electric and magnetic charges.

In the IIA side, the complex scalar that parameterizes the $SL(2;\mathbb{R})/U(1)$ manifold is the complexified Kähler modulus of T^2:

$$T' \equiv T'_1 + iT'_2 = B' + i\det(G'_{\alpha\beta}), \tag{11.9.8}$$

while the axion and dilaton belong to the $O(6,22)$ part of the moduli space. Moreover from (11.9.1), (11.9.4) we obtain that heterotic/IIA duality implies

$$T' = S, \quad S' = T. \tag{11.9.9}$$

The heterotic string coupling is exchanged with the volume of the T^2. The perturbative T-duality, on the IIA side acting as $SL(2;\mathbb{Z})$ on the T-modulus, becomes the nonperturbative Montonen-Olive duality acting on the S-modulus on the heterotic side.

Heterotic-IIA duality thus provides a geometrical/stringy aspect to the strong-weak coupling duality of four-dimensional gauge theories.

11.10 Conifold Singularities and Conifold Transitions

We have seen that singularities in the moduli space of K3 compactifications of the type IIA string signal interesting low-energy physics, namely, the enhancement of the abelian gauge symmetry to a nonabelian gauge group. This enhancement phenomenon is nonperturbative on the IIA side while it is perturbative on the heterotic side. Moreover, the two-derivative effective field theory is smooth at the locus where the gauge symmetry is enhanced.[11]

---

[11] Quantum corrections cannot modify the two-derivative effective action in theories with 16 supercharges. They can modify the higher-derivative effective action though. In particular, F^4 terms have singular corrections at the point of enhanced symmetry.

The singularities of the K3 surface are a special case of more general singularities of Ricci-flat Kähler manifolds known as conifold singularities.

In this section we study further such singularities in CY threefolds and the physics they imply for the associated type-II string compactifications. Conifold singularities are essentially of two types. There are singularities in the complex structure moduli space where a collection of three-cycles shrinks to zero size. There are also singularities in the Kähler moduli space where two-cycles shrink to zero.

Before we discuss the associated physics, we will analyze the simplest conifold singularity. It is described by the complex algebraic equation on $\mathbb{C}^4$

$$y_1 y_2 - y_3 y_4 = 0. \tag{11.10.1}$$

(11.10.1) describes a three-complex dimensional cone. It is a cone since any multiple of a solution is a solution. The point $(y_1, y_2, y_3, y_4) = (0, 0, 0, 0)$ is the tip of the cone, and it is a singular point. Changing variables to

$$y_1 = z_1 + i z_2, \quad y_2 = z_1 - i z_2, \quad y_3 = z_3 + i z_4, \quad y_4 = -z_3 + i z_4, \tag{11.10.2}$$

with $z_i \in \mathbb{C}$, the conifold equation can be equivalently written as

$$z_1^2 + z_2^2 + z_3^2 + z_4^2 = 0. \tag{11.10.3}$$

Introducing real variables a_i, b_i by $z_i = a_i + i b_i$ the conifold equation may also be written as

$$|a|^2 - |b|^2 = 0, \quad a \cdot b = 0, \tag{11.10.4}$$

with $|a|^2 = \sum_{i=1}^{4} a_i^2$.

It is convenient to introduce slices of the conical geometry by using the overall scale of the variables as a coordinate

$$|a|^2 + |b|^2 = 2r^2. \tag{11.10.5}$$

On this slice, (11.10.4) becomes

$$|a|^2 = r^2, \quad |b|^2 = r^2, \quad a \cdot b = 0, \tag{11.10.6}$$

so that a_i and b_i describe two S^3's. For fixed a, the constraint $a \cdot b = 0$ implies that $b \in S^3$ is orthogonal to a given direction. It therefore lies on an S^2. We have thus shown that at fixed r, we have a space isomorphic to $S^3 \times S^2$. Therefore, the conifold (11.10.1) is a cone with cross section $S^3 \times S^2$, each sphere having radius r. Also $0 < r < \infty$, with the tip of the cone at $r = 0$. This is depicted schematically in figure 11.6b.

We would like now to smooth out the conifold singularity. There are two independent ways of doing so. The first is to replace the defining equation (11.10.1) by

$$y_1 y_2 - y_3 y_4 = \epsilon^2, \quad \epsilon \in \mathbb{R}. \tag{11.10.7}$$

Introducing real variables and the radial coordinate, instead of (11.10.6) we now obtain

$$|a|^2 = r^2, \quad |b|^2 = r^2 - \epsilon^2, \quad a \cdot b = 0. \tag{11.10.8}$$

Observe that now $r \geq \epsilon$. At the extreme value $r = \epsilon$, the S^2 shrinks to zero volume. The S^3 has now finite volume at this point. This is the deformed conifold described pictorially

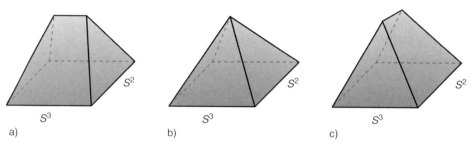

Figure 11.6 a) The deformed conifold. b) The conifold. c) The resolved conifold. In all cases the square at the base represents the sixes of S^3 and S^2. In a) the tip of the cone is singular. In b) it is replaced by a finite size S^3. In c) it is replaced by a finite size S^2.

in figure 11.6a. It is a smooth manifold and is topologically equivalent to the cotangent bundle of the three-sphere, T^*S^3.

There is a second way of resolving the singularity known as a "blow-up" in algebraic geometry. It amounts to replacing the defining equation (11.10.1) with a system of two equations

$$
\begin{pmatrix} \gamma_1 & \gamma_3 \\ \gamma_4 & \gamma_2 \end{pmatrix} \begin{pmatrix} W \\ Z \end{pmatrix} = 0.
\tag{11.10.9}
$$

Here W, Z are projective coordinates on $\mathbb{CP}^1 \sim S^2$

$$
(Z, W) \neq 0, \quad (W, Z) \sim \eta\,(W, Z), \quad \eta \in \mathbb{C} - \{0\}.
\tag{11.10.10}
$$

When any of the γ_i is nonzero, say γ_1, we can solve (11.10.9) for $W = -\frac{\gamma_3}{\gamma_1} Z$, and substitute in the other equation to obtain (11.10.1) since by assumption $(W, Z) \neq (0, 0)$. Therefore, in that region the geometry is that of the original conifold. However, at the putative tip where all $\gamma_i = 0$, the system in (11.10.9) is solved for arbitrary values of Z, W. Therefore the tip has now been replaced by an finite-size S^2. This is the geometry of the resolved conifold, and it is pictorially portrayed in figure 11.6c.

Singularities as the one we described here, arise in compact CY manifolds for special values of the moduli. The conifold and its resolutions that we described above are non-compact manifolds. In the compact case, we should think of them as local descriptions as we zoom in near the singularity. The neighborhood in moduli space around the conifold point is then described by one of the two resolutions we presented above.

As we have seen in section 9.11 on page 252, the type-II string compactified on a CY threefold has a $\mathcal{N} = 2_4$ supergravity theory[12] as its effective field theory. In the IIA case, the space of vector multiplets $\mathcal{M}_V$ is associated with the space of Kähler structures of complex dimension $h^{1,1}$, while the space of hypermultiplets $\mathcal{M}_H$ to the space of complex structures with quaternionic dimension $h^{1,2} + 1$. In the IIB case, the roles are reversed by mirror symmetry: $\mathcal{M}_V$ is associated with the space of complex structures of complex dimension $h^{1,2}$, and $\mathcal{M}_H$ with the space of Kähler structures with quaternionic dimension $h^{1,2} + 1$.

---

[12] Information on the couplings of $\mathcal{N} = 2_4$ supergravity can be found in appendix I.2 on page 535.

Moreover, in both cases the dilaton is in a hypermultiplet. Thus, string-loop corrections affect only $\mathcal{M}_H$ while $\mathcal{M}_V$ can be computed at tree level and receives no string-loop corrections.

We will concentrate here, for concreteness, on conifold singularities of the complex structure moduli space, associated with shrinking three-cycles in a type-IIB CY compactification.

Consider a basis for three-cycles of a CY manifold.[13] The basis is $2(1 + h_{12})$ dimensional with elements A_I, B^I satisfying

$$\langle A_I | B^J \rangle = -\langle B^J | A_I \rangle = \delta_I{}^J, \quad \langle A_I | A_J \rangle = \langle B^I | B^J \rangle = 0, \tag{11.10.11}$$

where $\langle \ | \ \rangle$ stands for oriented intersection. This basis is the three-dimensional analog of the a and b cycles of a two-torus. It is unique up to $\mathrm{Sp}(1+h_{12}, \mathbb{Z})$ transformations:

$$\begin{pmatrix} A_I \\ B^I \end{pmatrix} \to g \begin{pmatrix} A_I \\ B^I \end{pmatrix}, \quad g \begin{pmatrix} 0 & \mathbf{I} \\ -\mathbf{I} & 0 \end{pmatrix} g^T = \begin{pmatrix} 0 & \mathbf{I} \\ -\mathbf{I} & 0 \end{pmatrix}. \tag{11.10.12}$$

The choice of complex structure is determined by the periods of the unique holomorphic $(3,0)$ form Ω:

$$\mathcal{F}_I = \int_{A_I} \Omega, \quad Z^I = \int_{B^I} \Omega. \tag{11.10.13}$$

The Z^I, $I = 1, 2, \ldots, h_{12} + 1$, are projective coordinates in the moduli space of complex structures. The reason for projectivity is that multiplying Ω by a constant does not change the complex structure. The coordinates on the moduli space, are in one-to-one correspondence with the complex scalars (and vectors) of the $h^{1,2}$ vector multiplets. In an appropriate coordinate system, $\mathcal{F}_I \sim \frac{\partial \mathcal{F}(Z)}{\partial Z^I}$, and $\mathcal{F}(Z)$ is holomorphic. It is the prepotential of the effective $\mathcal{N} = 2_4$ supergravity describing the low-energy limit of the type-IIB compactification.

We will consider first a CY manifold with a single three-cycle shrinking to zero. This is the simplest kind of conifold singularity. Locally it is described by the deformed conifold geometry of figure 11.6a. We pick a basis for three-cycles such that the shrinking cycle is B^1. Then, (11.10.13) implies that the conifold singularity is at $Z^1 = 0$. This is a locus of complex codimension 1. The monodromy around it can be shown to be

$$\begin{pmatrix} Z^1 \\ \mathcal{F}_1 \end{pmatrix} \to \begin{pmatrix} Z^1 \\ \mathcal{F}_1 + Z^1 \end{pmatrix}. \tag{11.10.14}$$

It implies that near $Z^1 = 0$,

$$\mathcal{F}_1 \sim \frac{1}{2\pi i} Z^1 \log Z^1 + \text{regular}, \tag{11.10.15}$$

while the other coordinates have trivial monodromy. If we use the relation between the prepotential and the Kähler potential (I.13) we obtain for the Kähler metric

$$G_{1\bar{1}} = \partial_1 \bar{\partial}_1 K \sim \log |Z^1|^2. \tag{11.10.16}$$

---

[13] The complex geometry of CY manifolds is explained in section 9.7 on page 239.

Therefore, the kinetic term of the scalar Z^1 is singular. Moreover, since the logarithm is integrable, the singularity is at a finite distance in the moduli space. The curvature is singular at $Z^1 = 0$.

The effective action for the vector multiplets is thus singular at the conifold singularity. Moreover, we have argued above that the vector moduli space does not receive any quantum corrections. Therefore, it remains singular in perturbation theory.

The logarithmic divergence is reminiscent of the effect of integrating out a state with mass $\sim |Z^1|$. Consider a particle of mass m, minimally coupled to the appropriate vector multiplet. Its one-loop contribution to the wave-function renormalization of the vector multiplet is proportional to the logarithm $\log \frac{m^2}{\Lambda^2}$ where Λ is a dynamical scale. Therefore, integrating out a single charged hypermultiplet with mass proportional to $|Z^1|$ will give precisely such a singularity in the vector moduli space. Turning this around, the singularity indicates that a nonperturbative charged hypermultiplet has become massless at the conifold point. The validity of the effective theory will be restored if the light hypermultiplet is included in the effective action (i.e., "integrated in"). This phenomenon is already familiar from the Seiberg-Witten solution to $\mathcal{N} = 2_4$ SU(2) gauge theory.

Since Z^1 is the (complexified) volume of the shrinking three-cycle, a natural candidate for such a hypermultiplet is one generated from a D$_3$-brane wrapping this cycle. Indeed, the D$_3$-brane of type IIB can wrap around a nontrivial three-cycle and give a massive BPS hypermultiplet. Its mass is proportional to the volume of the three-cycle and hence vanishes on the conifold locus.

One may expect that integrating out a nonperturbative state would produce corrections to the effective action that have a nontrivial string-coupling dependence. In the situation at hand this is not the case. The reason is the peculiar (in)dependence of R-R gauge couplings on the dilaton.

The upshot of the previous discussion is that the simple conifold singularity of the effective theory indicates the presence of a nonperturbative, almost massless hypermultiplet. The singularity of the effective field theory is resolved if this extra hypermultiplet is explicitly included in the effective field theory.

Let us now consider the general case of conifold singularities. There is a number k of three-cycles γ^i, $i = 1, 2, \ldots, k$, which shrink to zero size simultaneously. The singularity is at $Z^1 = Z^2 = \cdots = Z^k = 0$ with $Z^i = \int_{\gamma^i} \Omega$. There may be homology relations between the degenerating cycles γ^i. In that case k is larger than the number N of independent degenerating cycles. In order to determine the monodromy, we need to understand the effect of going around the γ^i cycle on a generic three-cycle δ:

$$\delta \to \delta + \langle \delta | \gamma^i \rangle \, \gamma^i, \tag{11.10.17}$$

where $\langle \, | \, \rangle$ denotes oriented intersection.

We may consider a transition between CY$_3$ manifolds as follows. N independent three-cycles shrink to zero in the general conifold singularity. At the conifold point, we can replace the singular three-cycles by $(k-N)$ singular two-cycles and then blow them up to nonzero size, obtaining a smooth CY$_3$ manifold with different Betti numbers. This is a well-defined mathematical construction, known as a *conifold transition*. We will now interpret this construction in terms of the low-energy physics of the associated IIB compactification.

We expand the vanishing cycles in the basis of independent cycles (we neglect cycles that do not participate in the transition from now on)

$$\gamma^i = n^i_a B^a, \quad i = 1, 2, \ldots, k, \quad a = 1, 2, \ldots, N, \tag{11.10.18}$$

where $k \geq N$.

As shown in section 9.11, the independent cycles B^a participating in (11.10.18) are in one-to-one correspondence with the harmonic three-forms and therefore with the U(1) gauge fields of the vector multiplets A^a_μ. D_3-branes wrapped supersymmetrically around the γ^i cycle give BPS hypermultiplets with mass

$$M^i \sim |n^i_a Z^a|, \tag{11.10.19}$$

where n^i_a is the charge of the state with respect to the A^a_μ gauge boson. Thus, we obtain k hypermultiplets that become massless at the conifold point $Z^a = 0$. As argued in the case of the simple conifold singularity, in order to smooth out the effective field theory we must include the hypermultiplets H^i in the low-energy effective field theory.

Massless hypermultiplets in an $\mathcal{N} = 2_4$ field theory generically have a potential. Let us assemble the four scalars of the hypermultiplet H^i into two complex scalars $H^i_\alpha, \alpha = 1, 2$. $\mathcal{N} = 2_4$ supersymmetry implies that the potential has the form

$$V_H = \sum_{\alpha, \beta=1}^{2} \sum_{a,b=1}^{N} M^{ab} D^{\alpha\beta}_a D^{\alpha\beta}_b, \tag{11.10.20}$$

where the D-terms are given by

$$D^{\alpha\beta}_a = \sum_{i=1}^{k} n^i_a \left(\bar{H}^i_\alpha H^i_\beta + \bar{H}^i_\beta H^i_\alpha \right), \tag{11.10.21}$$

and M^{ab} is a positive-definite matrix. The index a labels the U(1) gauge fields.

This potential has flat directions given by $D^{\alpha\beta}_a = 0$. They provide $3N$ constraints on $4k$ real variables. We must also fix N gauge transformations so we are left in total with a $4(k-N)$-dimensional space of solutions. In this generic case, all N gauge bosons are Higgsed and disappear from the low-energy spectrum. They absorb $4N$ scalars (according to the $\mathcal{N} = 2_4$ super Brout-Englert-Higgs effect) and we are left with $k-N$ massless hypermultiplets (moduli). Their scalars are the parameters of the generic vanishing D-terms. Thus, $k-N$ hypermultiplets remain massless and must be retained in the low-energy spectrum.

This agrees precisely with the geometrical picture. The new manifold obtained by the conifold transitions has

$$\tilde{h}^{1,2} = h^{1,2} - N, \quad \tilde{h}^{1,1} = h^{1,1} + k - N, \tag{11.10.22}$$

in agreement with the fact that k three-cycles are lost, whereas $k-N$ new two-cycles are blown up. Thus, conifold transitions between topologically distinct CY manifolds correspond to Brout-Englert-Higgs transitions in the effective $\mathcal{N} = 2_4$ supergravity of the IIB string.

What is the physical picture of the same singularities in the IIA CY compactification? There, the moduli of the complex structures that control the vanishing of the area of three-cycles belong to the hypermultiplet moduli space. Thus, at such a conifold singularity, the vector moduli space is regular but the tree-level hypermultiplet moduli space is singular. The type-IIA string has no D_3-brane which can give rise to massless particles in the conifold limit.

However, in contrast to the vector moduli space, the hypermultiplet moduli space has nontrivial string-loop corrections. The reason is that the dilaton is in a hypermultiplet. Due to the quaternionic constraint and the fact that there are no R-R charged states in perturbation theory, no perturbative corrections are expected for the hyper-moduli space. On the other hand, instanton corrections are expected due to the IIA Euclidean D_2-brane, wrapped around the vanishing three-cycles. Near a conifold singularity, such corrections are expected to be important and smooth out the singularity.

Bibliography

We have touched the central ideas in nonperturbative string dualities. There is an extended literature on this subject and we will try to mention some starting points.

Good comprehensive reviews on branes as string solitons and their role in the nonperturbative duality conjectures can be found in [205,206,209]. General reviews on nonperturbative string dualities include [157,317,318,319,320,321,161]. A rather extensive and comprehensive review is [322]. BPS states, BPS-saturated terms in the effective action, supersymmetric non-renormalization theorems, instanton effects, and duality conjectures are reviewed in [310]. Moreover, various quantitative duality tests are also described. Nonrenormalization theorems in the context of $\mathcal{N} = 1_4$ supergravity are reviewed in [300].

In [323] and [324] the picture of most strong-coupling duality conjectures was presented and the concept of M-theory was coined. The missing item, heterotic M-theory was developed in [325,326]. The notion of D-branes [166] was also catalytic for these and further developments.

A short but dense review of M-theory can be found in [327]. A detailed review of M-theory, the U-dualities of its toroidal compactifications, and the associated BPS states can be found in [328] A set of lectures on heterotic M-theory may be found in [329] Special issues arising in six dimensions, in relation to nonperturbative string dualities are reviewed in [330,321]. In order to obtain nonabelian symmetries and chirality from direct compactifications of M-theory the compactification manifolds must be singular. Such efforts are extensively reviewed in [331]. Nonperturbative dualities are used in [332] to map phenomenologically interesting regions of the heterotic string to other string theories like type I and type II. The running of couplings in the context of the heterotic M-theory compactifications has been described in [272]. The phenomenology of M-theory is reviewed in [333].

Several tests in heterotic type-I duality have been performed in [334]. The issue of small instanton transitions relevant for the five-branes was described in [335]. The quaternionic geometry, and the moment maps are described in detail in appendix B of [336]. A self-duality of the heterotic theory compactified on K3 has been discussed in [337].

Instantons are an integral part of most discussions of non-perturbative string physics. Their emergence from wrapped Euclidean branes can be found in [338]. Rules for calculating their contributions to the superpotential are discussed in [339]. Their contributions play an important role in the structure of the hypermultiplet moduli space near the conifold singularity, [340]. They are responsible for non-trivial corrections to R^2 couplings in maximally supersymmetric heterotic compactifications

[341]. D$_{-1}$-instantons are responsible for the non-perturbative corrections to the R^4 threshold in maximally supersymmetric type-II string theory [342]. They are also crucial for the nonperturbative agreement of the F^4 thresholds in heterotic–type I duality below nine dimensions [343]. A review of their effects can be found in [310].

Heterotic–type II duality was proposed in [323]. A detailed review of the K3 geometry is in [245]. In the same reference, the interplay between geometry and effective field theory dynamics, related to gauge symmetry enhancement is analyzed. The most detailed direct test of heterotic–type II duality involves the comparison of the F^4 thresholds [344]. Indirect but powerful tests exploit its avatar: heterotic strings compactified on K3$\times T^2$ are dual to type-II strings on an appropriate elliptically fibered CY$_3$. This is reviewed in [345,346]. Information on the mathematics of the conifold singularities and conifold transitions can be found in [347] and [243]. The conifold transitions are interpreted in [348,349].

Another framework that captures nonperturbative information in the context of type II string theory is known as "F-theory" [350,351,352,353]. The idea is that in IIB string theory, compactifications with nontrivial axion-dilaton field S in (H.20) can be given a geometrical description. Consider the compactification of $2d$ dimensions in the IIB theory, with varying S. This can be described as a compactification of a 12-dimensional theory on an elliptically fibered CY$_{d+1}$ manifold. The elliptic fibration is defined as a T^2 fibered over a d-complex-dimensional manifold M_d. The modulus of T^2 is changing over M_d. This modulus can be identified with the IIB S field. Some aspects of F-theory are reviewed in [322].

The field of nonperturbative string dualities was triggered by associated advances in supersymmetric quantum field theories. It was conjectured rather early that $\mathcal{N} = 4_4$ SYM is invariant under the SL(2,$\mathbb{Z}$) S-duality [354]. Similar ideas led to advances in supersymmetric theories with reduced supersymmetry, [355,356,357]. There are many reviews of QFT dualities at varying depth and focus, [358,359,360,361,362,363]. The realization of field theory dualities in the stringy domain is reviewed in [345,346].

The engineering of brane configurations can be used to construct geometric realizations of gauge theories [364,365]. This line of thought in reviewed in [196].

Extended supersymmetry and the notion of BPS states are related to topological field and string theories. Their dynamics, tools and applications are reviewed in [366,367,368,242].

Exercises

11.1. Start from (11.1.1) on page 322 and use (D.3) and (D.4) on page 514 to obtain (11.1.2).

11.2. A natural *Ansatz* for the extremal (BPS) M$_2$ solution of 11-dimensional supergravity is

$$ds^2 = H(r)^{-2/3}\, d\vec{x} \cdot d\vec{x} + H(r)^{1/3}\, dy^a dy^a, \quad C_3 = \frac{dx^0 \wedge dx^1 \wedge dx^2}{H}, \tag{11.1E}$$

where x^i are the world-volume coordinates and $r^2 = y^a y^a$. Show that this solves the equations of motion when H is harmonic:

$$H = 1 + \frac{32\pi^2 N \ell_{11}^6}{r^6}. \tag{11.2E}$$

Calculate the M$_2$-brane charge. Is the curvature of this solution regular? What is the geometry of the near-horizon region?

11.3. For the extremal (BPS) M_5 solution of 11-dimensional supergravity you may try

$$ds^2 = H(r)^{-1/3} \, d\vec{x} \cdot d\vec{x} + H(r)^{2/3} \, dy^a dy^a, \quad (dC_3)_{abc} = \epsilon_{abcd} \partial_d H, \tag{11.3E}$$

where ϵ_{abcd} is the flat-space ϵ-tensor. Show that this solves the equations of motion when H is harmonic:

$$H = 1 + \frac{\pi N \ell_{11}^3}{r^3}. \tag{11.4E}$$

Calculate the M_5-brane charge. Is the curvature of this solution regular? What is the geometry of the near-horizon region?

11.4. Consider a uniform distribution of M_5-branes in one traverse direction. Show that the solution matches the NS_5-brane solution of IIA in ten dimensions, as expected.

11.5. To find the six-brane of M-theory, whose descendant is the D_6-brane of IIA we should search for a branelike solution of 11-dimensional supergravity with nontrivial metric only, of the form

$$ds^2 = dy^a dy^a + V d\vec{x} \cdot d\vec{x} + V^{-1}(dy + \vec{A} \cdot d\vec{x})^2, \tag{11.5E}$$

where the coordinates y^a parametrize the flat seven-dimensional world-volume, y is the 11th coordinate, $\vec{A}$ is the U(1) gauge field, and x^i parametrize the transverse three-dimensional space. V and $\vec{A}$ depend on x^i only and the metric must be rotationally invariant in the three-dimensional transverse space. Show that the 11-dimensional Einstein equations imply that

$$F_{ij} = \partial_i A_j - \partial_j A_i = \epsilon_{ijk} \partial_k V, \quad \partial_i \partial_i V = 0, \quad V = 1 + \frac{2N}{|\vec{x}|}, \tag{11.6E}$$

where ϵ_{ijk} is the flat space ϵ tensor. What is the D_6 charge of this brane?

11.6. Start from the DBI action for the D_2-brane in IIA theory. Dualize the three-dimensional world-volume gauge field to a scalar $\phi = (2\pi \ell_s^2) X^{10}$. Show that the new action is the world-volume action of the M_2-brane embedded in 11 dimensions.

11.7. Find the form of the S-duality transformation that replaces (11.5.1) on page 332 when $C_0 \neq 0$.

11.8. Show the relations in (11.5.2).

11.9. Under the type-IIB S-duality, the self-dual four-form is invariant. Show that this implies that a D_3 remains a D_3-brane, but the couplings of its world-volume $\mathcal{N} = 4_4$ SYM theory undergo a Montonen-Olive SL(2,$\mathbb{Z}$) transformation.

11.10. Consider the D_1-string solution described in section 8.8 on page 205. Apply on it the SL(2,$\mathbb{Z}$) symmetry to obtain the solutions for the (p, q) strings. What is their singularity structure?

11.11. Compactify the M-theory bosonic Lagrangian to nine dimensions. Do the same for the IIB theory. Note that in nine dimensions, unlike ten, there is a covariant Lagrangian for the IIB theory. T-dualize to match the two Lagrangians and derive this way the value of β in (11.5.10) on page 334.

11.12. A fundamental string stretched between two D_5-branes in IIB theory is S-dual to a D_1-string stretching between two NS_5-branes. Use the metric of several NS_5-branes in (11.6.3) on page 336 to calculate the energy of a D_1-string and compare it to that of an F_1-string stretched between two D_5-branes.

11.13. The magnetic dual of (p,q) strings in type-IIB theory are (p,q) five-branes. Find the supergravity solutions that describe such branes.

11.14. Start from the ten-dimensional type-IIB Lagrangian and derive the eight-dimensional scalar Lagrangian after compactification of the theory on T^2. Show that the metric of the scalar kinetic terms is that of the coset $SL(3,\mathbb{R})/O(3,\mathbb{R})\times SL(2,\mathbb{R})/U(1)$.

11.15. In the previous problem derive the kinetic terms of the three two-forms and show that they are invariant under $SL(3,\mathbb{Z})\times SL(2,\mathbb{Z})$.

11.16. Consider M-theory compactified on T^3. Derive the Lagrangian of the scalar sector and show that now $SL(3,\mathbb{Z})$ is manifest acting on the unit volume metric of T^3 while $SL(2,\mathbb{Z})$ acts on the volume of T^3 and the scalar coming from C_3. Show that this is none else but the T modulus of the torus, in the type-IIA theory, and the $SL(2,\mathbb{Z})$ is the perturbative T-duality acting on T. From this point of view rederive the results of section 11.6 on page 14.

11.17. In the previous problem, calculate the kinetic terms of the three-form in eight dimensions. Show that the $SL(2,\mathbb{Z})$ part of the U-duality group acting on the T modulus acts by an electric-magnetic duality transformation on the three-form. Show that it interchanges D_2-branes in eight dimensions and D_4-branes with two dimensions wrapped on the torus, as expected.

11.18. Consider the type-II theory compactified on T^5. Find out how many vectors there are and how they transform under the T-duality and the U-duality group. What objects carry the associated charges?

11.19. Use heterotic/type I duality to show relation (11.7.6) on page 338. This indicates that when the heterotic coupling is strong, type-I strings have a larger size than heterotic strings.

11.20. Consider the heterotic string compactified to four dimensions. Assume that five or six of the dimensions are much larger than the string scale. Show that if four-dimensional gauge couplings are $\mathcal{O}(1)$ then $g_s \gg 1$. Show that the theory is equivalent to the weakly coupled type-I string.

11.21. Show that under heterotic/type I duality the heterotic NS_5-brane is mapped to the type-I D_5-brane.

11.22. Show that heterotic/type I duality, together with T-duality can reproduce all known string dualities. If you have a hard time, look at [369].

11.23. Show that the tensions of a single D_1-brane and a single D_5-brane of the type I theory T_1^I are given by

$$T_1^I = \frac{1}{\sqrt{2}} T_1, \quad T_5^I = \sqrt{2} T_5, \tag{11.7E}$$

where T_p are the analogous tensions in the type-II string. Thus, they still satisfy the DNT quantization condition. This is another explanation why we need two CP indices for the type-I D_5-brane. Show also that

$$T_9^I = \frac{1}{\sqrt{2}} T_9. \tag{11.8E}$$

11.24. Consider the D_9 effective action in type-I string theory. Turn on a four-dimensional instanton background along directions 6789. Show that this describes a bound state of several D_5 with the D_9-branes. What is the relation between the instanton number and the number of D_5-branes.

11.25. Consider the general case of D_p-branes in type-I theory. Show that nontrivial configurations exist (compatible with GSO and Ω projections) preserving half of the supersymmetry, for $p = 1,5,9$. The case $p = 9$ corresponds to the usual open strings moving in ten-dimensional space.

11.26. Consider N parallel coincident D_1-branes in the type-I theory. Show that the massless excitations are a two-dimensional vector in the adjoint representation of $SO(N)$, eight scalars in the symmetric representation of $SO(N)$, eight left-moving fermions in the adjoint of $SO(N)$, and right-moving fermions transforming as $(N, 32)$ of $SO(N) \times SO(32)$.

11.27. Consider the R sector of the D_5-D_5 strings in type-I string theory. Impose the Ω projection and show that you obtain three six-dimensional fermions of negative chirality completing the vector multiplets of $Sp(2)$ and one fermion of positive chirality completing a hypermultiplet with the four transverse coordinates of the brane.

11.28. Consider N parallel coincident type-I D_5-branes. Show that the massless spectrum of the D_5-D_5 strings give an $Sp(2N)$ vector multiplet, one singlet hypermultiplet, and another one transforming in the two-index anti-symmetric representation of $Sp(2N)$ of dimension $2N^2 - N - 1$. The D_5-D_9 strings provide another massless hypermultiplet in the $(2N, 32)$ of

Sp(2N) × O(32). Investigate the massless spectrum if $m < N$ D$_5$-branes are pulled some distance away from the initial stack.

11.29. Consider the orbifold of the IIA string by the symmetry $(-1)^{F_L}$, where F_L, according to our conventions, is the left-moving space-time fermion number. It is 1 for the left NS sector and -1 for the left R sector. Show that you obtain the IIB string. What do you obtain if you orbifold by the same symmetry the IIB string? Explain in what sense the M-theory parity orbifold $x^{11} \rightarrow -x^{11}$ is different from the above.

11.30. Show that the heterotic momentum in the nineth direction becomes under the chain of dualities (11.8.4) on page 346 the momentum in the 11th dimension.

11.31. Consider 11-dimensional supergravity on the $S^1/\mathbb{Z}_2$ orbifold. Explore the bulk and boundary couplings needed in order to implement the Green-Schwarz mechanism of anomaly cancellation of the $E_8 \times E_8$ theory. You may check you answer against [326].

11.32. Show the $O(4,20)$ invariance of (11.9.1) on page 349.

11.33. Derive the equations of motion from the actions (11.9.8) and (11.9.10). Show that the two sets of equations of motion are equivalent through the following transformations

$$\Phi' = -\Phi, \quad G'_{\mu\nu} = G_{\mu\nu}, \quad \hat{M}' = \hat{M}, \quad A'^i_\mu = A^i_\mu, \tag{11.9E}$$

$$e^{-2\Phi}\hat{H}_{\mu\nu\rho} = \frac{1}{6} \frac{\epsilon_{\mu\nu\rho}{}^{\sigma\tau\varepsilon}}{\sqrt{-G}} H'_{\sigma\tau\varepsilon}. \tag{11.10E}$$

Data on the right-hand side refer to the type-IIA theory.

11.34. Consider the four-dimensional multicentered Eguchi-Hanson metric

$$ds^2 = V \, d\vec{x} \cdot d\vec{x} + V^{-1} \, (dy + \vec{A} \cdot d\vec{x})^2, \tag{11.11E}$$

with y an angular variable. $\vec{A}$ depends on x^i only, satisfying

$$\partial_i A_j - \partial_j A_i = \epsilon_{ijk}\partial_k V, \quad \partial_i\partial_i V = 0, \quad V = \sum_{i=1}^{N-1} \frac{1}{|\vec{x} - \vec{x}_i|}. \tag{11.12E}$$

Show that it is Ricci flat, and hyper-Kähler. Show also that at asymptotic infinity, $|\vec{x}| \rightarrow \infty$, the geometry is isomorphic to $S^3/\mathbb{Z}_N$.

11.35. Redo the heterotic/type IIA duality transformation by now keeping track of the extra gravitational Chern-Simons term in (11.9.15) on page 351. Show that it implies the presence of a $B \wedge \text{Tr}[R \wedge R]$ term in the type-IIA string effective action and calculate its coefficient. Such a term comes from a type-IIA one-loop diagram and its coefficient has been calculated. The two results agree, [370]. Agreement is expected from the cancellation of anomalies of the two theories. Which anomaly is related to the $B \wedge \text{Tr}[R \wedge R]$ term in the type-II theory?

11.36. Consider the K3 compactification of the IIB described in section 9.9 on page 247. Describe what special happens at the points in the moduli space where the K3 surface becomes singular due to vanishing two-cycles. By compactifying on an extra S^1 relate this behavior to that of the K3 compactification of the IIA theory.

11.37. Consider a Euclidean D_2-brane, wrapping a holomorphic three-cycle C of a CY manifold. Such a configuration is an instanton. Evaluate the instanton action and show that it is given by

$$S_{\text{instanton}} = \frac{T_2}{g_s} |n_i \, Z^i|, \tag{11.13E}$$

where the three-cycle is written in terms of the basis as $C = n_i B^i$. Observe that although the instanton action is suppressed by a power of the string coupling, it vanishes when the associated three-cycle has zero volume. In this case, the instanton effects are important. You will find an explicit example worked out in [340].

11.38. Consider the IIA and IIB theories compactified on a CY manifold. There are singularities in the Kähler moduli space generated by various two-cycles shrinking to zero size. Discuss the physical resolution of such singularities.

11.39. Derive carefully the dimensional reduction of the six-dimensional heterotic on T^6 and IIA on K3 effective actions to four dimensions, and verify (11.9.1)–(11.9.5) on page 355.

11.40. Show that in the IIB theory

$$S(-1)^{F_L} S^{-1} = \Omega, \tag{11.14E}$$

where S is the $SL(2,\mathbb{Z})$ transformation that inverts the coupling constant. Consider first the orbifolds $\text{IIB}/(-1)^{F_L}$. According to exercise 11.29 this is the IIA theory. On the other hand IIB/Ω is the type-I string. Conclude that orbifolding does not commute with S-duality.

11.41. Show that M-theory compactified on K3 is equivalent to the heterotic or type-I theory compactified on T^3.

11.42. Show that M-theory compactified on the orbifold $T^5/\mathbb{Z}_2$ is equivalent to IIB theory compactified on K3. The $\mathbb{Z}_2$ transformation inverts all five coordinates of T^5. The M-theory three-form is odd under such a transformation.

12 | Black Holes and Entropy in String Theory

12.1 A Brief History

One of the deepest mysteries of the gravitational force involves the physics of black holes.

At the beginning of the 1970s it was observed that classical black-hole dynamics obeys laws that are reminiscent of thermodynamics.

The area theorem[1] was the first hint in this direction. This stated that the area of the horizon of a black hole cannot decrease during its physical evolution. The work extraction processes and the area theorem prompted the idea that there is an entropy associated with a black hole that satisfies the analog of the second law of thermodynamics. In particular, one could associate a temperature (known today as the Hawking temperature) to a black hole and an entropy (known today as the Bekenstein-Hawking entropy). Moreover, for every other global conserved charge of the black hole, (like electric charge or angular momentum) there are associated thermodynamic-like potentials. Finally, the quantities above, were shown to satisfy laws analogous to the laws of thermodynamics. The presence of the effective thermodynamic laws in black-hole physics emerged as a major puzzle.

Shortly afterwards, it was shown that a quantum treatment of matter in the background of a classical black hole indicates that the black hole radiates as a black body with an effective temperature (the Hawking temperature) being essentially the same as the one guessed from the previous thermodynamics laws. This fixed also the normalization of the black-hole entropy to one-fourth the horizon area in Planck units and introduced explicitly $\hbar$ in black-hole thermodynamics. This observation tied the black-hole thermodynamic laws better together but created even bigger puzzles in black-hole physics, namely, the famous black-hole information paradox and its avatars.

The information paradox in its simplest form is as follows: imagine matter in a pure state, collapsing to form a black hole. After the horizon forms, the black hole radiates thermal radiation that carries no information. Eventually the black hole will evaporate

---

[1] Motivated by earlier studies of reversible and irreversible work extraction processes in black holes.

destroying all the information stored in the infalling matter. This state of affairs is in apparent contradiction with the unitarity of quantum mechanics. It prompted several physicists to entertain the radical idea that quantum mechanics must be substantially modified in order to fit together with gravitational phenomena.

There could be several alternatives to the scenario above, and they have been also studied extensively by physicists. The simplest would be that the information comes out in subtle correlations in the Hawking radiation. Although such a possibility was considered from the beginning, it is in the past few years that it gained momentum after the discovery of bulk-boundary correspondence (which was inspired by holographic ideas for the physics of black holes).

In this chapter we will present a simplified exposition of the progress that has been made in this direction in the context of string theory. In particular, we will see that string theory can explain microscopically the BH entropy of black holes, and that it gives suggestions for the resolution of the black-hole information paradox.

The reader is assumed to be familiar with the basic physics of black holes. Suggestion for reading in this direction are provided in the bibliography at the end of this chapter.

12.2 The Strategy

The central idea is the following. Certain extremal black-hole solutions in string theory have an interpretation as bound states of the solitons that we have discussed so far, namely, D-branes. In particular, the solution we are going to present in section 12.5 is a bound state of D_5- and D_1-branes wrapped on $T^4 \times S^1$ together with some KK momentum along S^1. Such a solution, apart from its gravitational description, has an alternative description in terms of the world-volume theory living on the bound state of branes. The relevant degrees of freedom are the 1-5 strings stretched between the D_5- and D_1-branes. Their world-volume theory is a two-dimensional gauge theory that is conformally invariant at low energy.

The black-hole solution has a BH entropy. Such entropy should be counting the number of possible microstates accessible to the black hole. The counting of such microstates can in principle be done in the associated world-volume theory.

The validity region of the two calculations is however different. The gravitational solution is reliable when it has weak curvature at the horizon and therefore represents a bona fide macroscopic black hole. This happens when the coupling of the open strings on the world-volume theory is strong. When the open string coupling is weak, then the bound state looks more like a pointlike elementary state, rather than a black hole.

It turns out that the presence of supersymmetry in the extremal case, provides a quick path to extrapolating the weak-coupling calculation to strong coupling. Indeed, the counting of the microstates is an index, that is protected by supersymmetry. Needless to say, the microscopic D-brane calculation matches the macroscopic BH entropy. Even the subleading piece, associated with the higher derivative terms in the effective action, turns out to match.

The extremal black hole is a supersymmetric configuration that is absolutely stable. Its Hawking temperature is zero, and therefore there is no Hawking radiation. To study the Hawking radiation microscopically, we must go away from extremality. This breaks supersymmetry, and according to our previous argument, we may no longer trust the weak-coupling calculations, when the coupling becomes strong.

The microscopic rate of Hawking evaporation was however calculated in the D-brane picture and in the regime of near-extremality[2] it agrees with the semiclassical gravitational calculations. This is interesting, not only because it completes and strengthens the microscopic description of the black holes, but also because it indicates that special circumstances are at work. These calculations indeed paved the way to the notion of bulk-boundary correspondence that will be described in more detail in chapter 13.

12.3 Black-hole Thermodynamics

We will consider here the representative example of the Reissner-Nordström (RN) four-dimensional black hole, and describe its thermodynamics. Its geometry is described in appendix F on page 519.

An important quantity is the area of the horizon associated to the BH entropy. This entropy is equal to one-quarter of the horizon area in Planck units. For the RN black hole it is given by[3]

$$
S(M, Q) \equiv \frac{A}{4G} = \frac{\pi r_+^2}{G} = \pi G \left[M + \sqrt{M^2 - \left(\frac{Q}{G} \right)^2} \right]^2 ,
\tag{12.3.1}
$$

where r_+ is the radius of the exterior horizon and is given in (F.5). Note that on reinstating $\hbar$, the entropy is inversely proportional to $\hbar$.

The mass of the black hole plays the role of the thermodynamic energy. The charge plays the role of the particle number. In this sense, equation (12.3.1) can be thought of as an equation of state.

The Hawking temperature, is equal to the surface gravity κ divided by 2π. The surface gravity is defined as the acceleration needed to keep a particle stationary at the horizon (in Killing coordinates). In our case it is

$$
\kappa = \frac{r_+ - r_-}{2r_+^2}.
\tag{12.3.2}
$$

The (Hawking) temperature can be calculated as

$$
T \equiv \left. \frac{\partial M}{\partial S} \right|_Q = \frac{r_+ - r_-}{4\pi r_+^2} = \frac{\kappa}{2\pi}.
\tag{12.3.3}
$$

It is proportional to $\hbar$ indicating that quantum mechanics is somehow hidden in this framework. For the extremal black hole, we obtain

$$
Q = GM, \quad T = 0, \quad S_{BPS} = \pi G M^2,
\tag{12.3.4}
$$

---

[2] Small departures from the BPS condition will be detailed in section 12.6 on page 379.

[3] In units $\hbar = c = k_B = 1$.

while for the Schwarzschild black hole

$$Q = 0, \quad T = \frac{1}{8\pi \, GM}, \quad S = 4\pi \, GM^2. \tag{12.3.5}$$

The vanishing of the temperature for the extremal case is compatible with the interpretation of the extremal black hole as a ground state of possible RN black holes. In exercise 8.35 on page 217 you have shown that the extremal RN black hole preserves half of the space-time supersymmetry of $\mathcal{N} = 2_4$ supergravity. It is therefore a 1/2-BPS configuration. Supersymmetry then implies that the BPS bound $GM \geq Q$ is equivalent to the unitarity of the representations of $\mathcal{N} = 2_4$ supersymmetry.

The chemical potential is given essentially by the Coulomb potential at the horizon

$$\mu \equiv -T \left. \frac{\partial S}{\partial Q} \right|_M = \frac{Q}{G \, r_+}. \tag{12.3.6}$$

We may now state the basic laws of black-hole thermodynamics:

- *The zeroth law* of black-hole thermodynamics states that the surface gravity (temperature) of stationary black holes is constant over the horizon.

- *The first law* states the balance of energy during physical processes

  $$dM = TdS + \mu dQ. \tag{12.3.7}$$

  In general there might be more "work" terms, associated with other conserved local charges (angular momentum for example).

- *The second law* states that in all processes the BH entropy is nondecreasing. When one involves also matter outside of black holes the second law is replaced by:

- *The generalized second law* states that the sum of the BH entropy and the entropy of the rest of the world is never decreasing.

- *The third law* states that an infinite number of steps are required to bring the temperature of a black hole to zero.

The BH entropy seems to suggest that a black hole with given M and Q has a large number of degenerate microstates Ω so that

$$S = \log \Omega. \tag{12.3.8}$$

The classical theory of gravitation gives no information on the nature of such putative microscopic degrees of freedom.

12.3.1 *The Euclidean continuation*

An interesting and quick way to calculate the Hawking temperature is in terms of the Euclidean continuation of the Minkowski black-hole solution. We will briefly present the algorithm here. We will consider for concreteness the Reissner-Nordström metric (F.3) on page 519. The general metric is treated in exercise 12.1.

We analytically continue the black-hole metric to Euclidean space by rotating time as $t \to i\tau$ to obtain

$$ds^2 = f(r)d\tau^2 + \frac{dr^2}{f(r)} + r^2 d\Omega_2^2. \tag{12.3.9}$$

Now the manifold extends from $r = \infty$ down to $r = r_+$ only. In a thermal background, the Euclidean time τ is compact with period $\beta \equiv \frac{1}{T}$. It is easy also to observe that at the ex-horizon, $r = r_+$ there is generically a conical singularity. Indeed, going close to the horizon by $r = r_+ + \epsilon$, and taking $d\Omega_2 = 0$ for simplicity we obtain

$$ds^2 \simeq \frac{r_+ - r_-}{r_+^2} \epsilon d\tau^2 + \frac{r_+^2}{r_+ - r_-} \frac{d\epsilon^2}{\epsilon}. \tag{12.3.10}$$

Defining a new radial coordinate by $u = \frac{2r_+ \sqrt{\epsilon}}{\sqrt{r_+ - r_-}}$ we obtain

$$ds^2 \simeq \frac{(r_+ - r_-)^2}{4r_+^2} u^2 d\tau^2 + du^2. \tag{12.3.11}$$

Therefore, the length of a circle $\tau \to \tau + \beta$ of radius u around the origin $u = 0$ is $\beta(r_+ - r_-)u/2r_+^2$. The thermodynamic equilibrium is associated with a smooth Euclidean geometry where τ can be viewed as an angle in polar coordinates. Therefore

$$\beta = \frac{1}{T} = \frac{4\pi r_+^2}{(r_+ - r_-)}. \tag{12.3.12}$$

We thus obtain the Hawking temperature in terms of the geometrical data.

To conclude, the Euclidean black-hole manifold, contains only the region outside the outer horizon and is a smooth manifold with a single boundary at $r = \infty$.

Like in any other thermodynamic system, we may define the standard thermodynamic potentials. The relevant potential for the Reissner-Nordström case is the Gibbs free energy

$$\mathcal{F} = M - TS - \mu Q = \frac{1}{2}(M - \mu Q). \tag{12.3.13}$$

This is in turn related to the Euclidean action of this background

$$S_g = -\frac{1}{16\pi G} \int_M \sqrt{g}\, R - \frac{1}{8\pi G} \int_{\partial M} \sqrt{h}\, K - \frac{1}{16\pi G} \int_M \sqrt{g}\, F^2. \tag{12.3.14}$$

The second term is the Gibbons-Hawking boundary term, necessary to have a well-posed variational problem for the Einstein-Hilbert action. h is the induced boundary metric and K is the trace of the extrinsic curvature at the boundary ∂M. It is defined as

$$K_{\mu\nu} \equiv \frac{1}{2} n_\rho G^{\rho\sigma} \partial_\sigma G_{\mu\nu}, \quad K = h^{\mu\nu} K_{\mu\nu}, \tag{12.3.15}$$

where n_μ is the unit normal to the boundary. Taking a sphere at fixed large $r = r_0$ as the boundary we may evaluate the extrinsic curvature of the RN solution, using

$$n^\mu = \frac{1}{\sqrt{g_{rr}}} \left(\frac{\partial}{\partial r}\right)^\mu \tag{12.3.16}$$

to be

$$K = \sqrt{f}\, \frac{rf' + 4f}{2rf}\Bigg|_{r=r_0}, \tag{12.3.17}$$

where f was defined in equation (12.3.9). We may thus compute

$$\frac{1}{8\pi G} \int_{\partial M} \sqrt{h} K = \beta r^2 \sqrt{f} \left. \frac{K}{2G} \right|_{r=r_0} = \beta \left. \frac{r}{2G} (rf' + 4f) \right|_{r=r_0}$$
$$= \frac{\beta}{2G} \left(2r_0 - 3GM + \frac{Q^2}{r_0} \right). \tag{12.3.18}$$

The first term in the action (12.3.14) gives zero since the Ricci scalar of the RN solution is zero. The last term gives

$$\frac{1}{16\pi G} \int_M \sqrt{g}\, F^2 = \frac{2}{16\pi G} \int d\Omega_2 \int_0^\beta d\tau \int_{r_+}^{r_0} r^2 dr \frac{Q^2}{r^4} = \frac{Q^2 \beta}{2G} \left(\frac{1}{r_+} - \frac{1}{r_0} \right). \tag{12.3.19}$$

We observe that the Euclidean action of the black hole is infinite due to the term linear in r in equation (12.3.18). However this term is insensitive to the the RN solution, and is also present for flat space. It is therefore reasonable to define a renormalized action by subtracting the action of (Euclidean) flat space. There is an ambiguity in this procedure that involves the determination of the periodicity of the τ coordinate in flat space (i.e., $M = Q = 0$) solution. It turns out that the correct prescription reproducing the thermodynamic potentials is to choose as the inverse temperature of flat space, the appropriately red-shifted RN temperature $\beta\sqrt{f}$. Therefore

$$S_{\text{flat}} = -\frac{\beta}{G} r_0 \sqrt{f(r_0)} = -\frac{\beta}{2G} \left[2r_0 - 2GM + \mathcal{O}\left(\frac{1}{r_0} \right) \right]. \tag{12.3.20}$$

Subtracting this contribution we obtain for the renormalized Euclidean black-hole action as the boundary approaches infinity $r_0 \to \infty$

$$I_{RN} \equiv S_{RN} - S_{\text{flat}} = \frac{\beta}{2} (M - \mu Q) = \beta \mathcal{F}, \tag{12.3.21}$$

where we have used the relations (12.3.3) and (12.3.6). We have therefore obtained the Gibbs free energy times the inverse temperature. This indicates that we can associate the Euclidean path integral with the thermodynamic partition function of the black hole

$$Z \equiv \sum_E e^{-\beta E} = \int \mathcal{D}[g, A]\, e^{-S_g}. \tag{12.3.22}$$

The gravitational Euclidean path integral is ill defined beyond the semiclassical approximation. String theory provides a perturbative definition of this path integral around a fixed background. A reliable nonperturbative definition of this path integral exists in cases of bulk-boundary correspondence.

12.3.2 Hawking evaporation and greybody factors

A black hole has an associated temperature. This implies that it may absorb or emit photons (or other massless particles). The absorption happens classically, and we may denote the absorption cross section by $\sigma_{\text{abs}}(\omega)$ where ω is the asymptotic energy of the absorbed photon. Such a cross section is also known as a greybody factor and can be calculated classically, by standard scattering theory. On the other hand, classically the emission rate of photons (or any other particles) is zero, due to the presence of the horizon.

However, particle creation in the background space-time of a black hole is responsible for an evaporation rate that is consistent with that of a black body with temperature equal to the Hawking temperature. The calculation proceeds by canonically quantizing the matter fields in the black-hole background. Specifying the in-vacuum in the asymptotic past to have no particles, it turns out via the standard Bogoliubov transform that the out-vacuum has a thermal density of particles, associated with the Hawking evaporation process. An intuitive (and approximate) picture of the effect is as follows: virtual particles and antiparticles are pair created outside the black-hole horizon. Before reannihilation the antiparticle crosses the horizon. The particle then is forced to materialize drawing energy away from the black hole.

Detailed balance gives the Hawking evaporation rate (in four dimensions) as

$$\Gamma_H = \frac{\sigma_{\text{abs}}(\omega)}{e^{\frac{\hbar\omega}{kT}} \mp 1} \frac{d^3 k}{(2\pi)^3}, \quad \omega = |\vec{k}|, \tag{12.3.23}$$

where the factor $\frac{d^3 k}{(2\pi)^3}$ is the density of states and the sign in the denominator changes between emitted bosons or fermions. The momentum k should not be confused with the Boltzmann constant k that was momentarily introduced here. The interpretation of (12.3.23) is as follows: massless particles are emitted from the horizon with the thermal rate $\frac{1}{e^{\frac{\hbar\omega}{kT}} \mp 1}$. The factor $\sigma_{\text{abs}}(\omega)$, takes into account the transmission of the particles from the horizon to the asymptotic area, while the last factor sums over possible final states. From now on we will set again $\hbar = k = 1$.

12.4 The Information Problem and the Holographic Hypothesis

The Hawking radiation is produced by quantum fluctuations of matter particles in the background of the gravitational field of the black hole. For very massive black holes, all curvature invariants at or outside the horizon are very small, and the gravitational back-reaction to the calculation is expected to be small. Moreover, the calculation of the evaporation rate does not really depend on the details of the matter dynamics.

If the black hole started as a pure state, and finally completely evaporates, unitarity (and information) will be lost, since the outgoing Hawking radiation is purely thermal, and therefore cannot carry any information about the original state of the black hole. This clashes with quantum mechanics that evolves pure states to pure states. For each virtual pair of particles created in the black-hole background, the outgoing particle of Hawking evaporation is entangled (correlated) with its antiparticle going down the black hole. Therefore, by the time the black hole evaporates, the system of outgoing radiation is in a complicated mixed state, and the proper description of the physics is in terms of a density matrix.

Accepting the above reasoning at face value we may imagine several alternatives. One is that the black hole does not evaporate completely but leaves a "remnant" of Planck size, which carries the leftover information and is entangled with the outgoing radiation. This alternative is problematic since it can be shown that this entanglement entropy must be larger than the black-hole entropy requiring an enormous degeneracy for the remnant. No reasonable model has been found that would reproduce similar behavior.

Hawking advocated that information indeed gets lost in quantum gravity, quantum mechanics must be modified, and a density matrix is the appropriate description of quantum gravitational phenomena. This proposal, apart from giving up a solid physical framework (quantum mechanics) seems to have also problems with the structure of virtual (quantum) processes.[4]

The conservative alternative is that Hawking radiation is not exactly thermal but carries out the information of the initial state in a subtle way. An analogous phenomenon is the radiation coming out of a burning piece of matter. The microscopic description of the microscopic degrees of freedom of the system (atoms) is perfectly unitary/quantum mechanical; however, the thermal nature of the outgoing radiation is due to tracing over irrelevant degrees of freedom.

This proposal also has some apparent conceptual problem. It is well known that there is no direct way of copying infalling quantum information to the outgoing Hawking radiation. Quantum duplication defined as

$$|\psi\rangle \to |\psi\rangle \otimes |\psi\rangle \tag{12.4.1}$$

is incompatible with the linearity of quantum mechanics. Starting with a large black hole, where the Hawking calculation is reliable until the very late stages, and the outgoing radiation is exactly thermal, it is too late to try to extract all the information at the last stages of the evaporation. It is widely accepted that a salvation of unitarity in the process of evaporation must involve nonlocal interactions in the gravitational theory that transfer the information efficiently from the interior of the horizon to the future asymptotic infinity.

String theory seems to have nonlocal interactions at distances of the order of the string scale. However, it is holography and the bulk-boundary correspondence that seem to be the appropriate framework for the nonlocality needed to salvage unitarity in black-hole evaporation.

Starting from the thesis that the black-hole evaporation process is unitary, we may developed a holographic hypothesis, generalizing Bekenstein's thesis: In a gravitational system confined in a given volume, the number of degrees of freedom scales with the surface area, rather than the volume. In particular, taking a microscopic view of the BH entropy, this implies that the number of gravitational (and other field-theoretic) degrees of freedom in a finite volume is finite. Moreover this finiteness is not just due to an underlying discrete lattice, since this would have given a logarithm of the number of states scaling as the volume rather that the surface area. Thus, it would seem that the degrees of freedom describe physics as in a hologram.

The general validity of the generalized second law of black-hole thermodynamics implies the holographic hypothesis: Consider a finite (spherical for simplicity) volume with an entropy larger than one quarter of its surface area. We add more matter so that a black hole forms with a horizon equal to the boundary of the original area. According to the gener-

---

[4] It seems that after supporting this view for many years, Hawking does not believe in this possibility anymore.

alized second law the entropy cannot diminish during this process and therefore remains larger than the associated BH entropy, contradicting the fact that a black hole has an entropy equal to one-quarter its horizon area. We deduce that in any finite region of space, the number of enclosed degrees of freedom is bounded above by the area in Planck units.

A potential way of making progress in the context of the information paradox, would be to try to understand the microscopic degrees of freedom responsible for the large degeneracy of black holes. If this is successful, one may derive the resulting thermodynamics by the usual corse-graining procedure as in the standard case of quantum statistical mechanics. As we will see in the sequel this is possible in string theory.

12.5 Five-dimensional Extremal Charged Black Holes

It turns out that the simplest type of black hole that has a nonzero horizon area and where its "constituents" are identifiable and tractable, is the RN black hole in five dimensions with three charges. This is a simple generalization of the four-dimensional RN solution. The interesting part of the story is that this black hole will be constructed as a bound state of three well known objects of type-IIB string theory: D_1- and D_5-branes along with momentum on the common intersection.

We will consider a $T^4 \times S^1$ compactification of type-IIB string theory to five noncompact dimensions. We will denote the T^4 coordinates by y^a, the S^1 coordinate by y and the four spatial noncompact coordinates by x^i. The D_5-brane is wrapping $T^4 \times S^1$ providing a pointlike particle in five dimensions. We will have Q_5 of these. The D_1 string wraps the S^1. We will have Q_1 of these. Finally, there will be Q_p units of momentum along the S^1.

As we saw earlier in section 8.8 on page 205, D_p-brane geometries generically have the tendency to shrink the longitudinal space and to expand the transverse space as one approaches the horizon. A combination of a D_1-brane inside its magnetic dual, a D_5-brane therefore will stabilize the volume longitudinal to the D_5 but transverse to D_1. This four-dimensional volume can therefore be used as part of the compactification volume. By adding momentum along the D_1-brane we also stabilize near the horizon the longitudinal dimension of the D_1-brane. This part can also be compactified, without danger of decompactification close to the horizon. Finally, the exponential of the dilaton will be proportional to the ratio of the two harmonic functions, as is standard in dyonic solutions, and it will also be bounded close to the horizon. Therefore, such a solution will have a controllable five-dimensional interpretation at all scales, larger than the compactification scale. To produce a four-dimensional solution more ingredients have to be added so that one more direction can become compact and stable. This produces a more complicated solution, and we will not discuss it here. You are invited to search for it in exercise 12.10.

The supersymmetry preserved by the solution plays an important role as we will see later. The compactified theory has maximal space-time supersymmetry generated by 32 supercharges. The presence of the D_5-branes leaves 16 unbroken supercharges, as described in section 8.8. The presence of parallel D_1-branes breaks another half of the supersymmetry leaving 8 supercharges. The addition of KK momentum along S^1 break supersymmetry by another half. We will be then considering a 1/8 BPS solution of the string equations of motion.

This configuration is related by dualities to other branes configurations. By T-dualizing S^1 and two of the T^4 directions we obtain a pointlike bound state of two D_2-branes and a fundamental string of type-IIA theory. By further going to strong coupling we obtain three different stacks of M_2-branes intersecting at a point in T^6.

The ten-dimensional extremal solution corresponding to this configuration is given by

$$ds_{10}^2 = \frac{-dudv + (H_p - 1)du^2}{\sqrt{H_1 H_5}} + \sqrt{H_1 H_5}\, dx^i dx^i + \sqrt{\frac{H_1}{H_5}}\, dy^a dy^a, \tag{12.5.1}$$

$$F_{05i} = -\frac{1}{2}\frac{\partial}{\partial x^i}\left(\frac{1}{H_1} - 1\right), \quad F_{ijk} = \epsilon_{ijkl}\frac{\partial}{\partial x^l}H_5, \tag{12.5.2}$$

$$e^{-2\Phi} = \frac{H_5}{H_1}, \quad H_n = 1 + \frac{r_n^2}{r^2}, \quad n = 1, 2, p, \quad r^2 = x^i x^i, \tag{12.5.3}$$

where $F_{\mu\nu\rho}$ is the field strength of the R-R two-form and $u = t + y, v = t - y$. The coordinates x^i, $i = 1, 2, 3, 4$, describe the common volume transverse to both the D_1- and D_5-branes. The coordinates y^a, $a = 1, 2, 3, 4$ are the T^4 coordinates, transverse to the D_1-branes and longitudinal for the D_5. The parameters r_1, r_5, and r_p are related to the three (quantized) charges of the solution

$$r_1^2 = c_1 Q_1, \quad c_1 = \frac{g_s}{V}\ell_s^2,$$

$$r_5^2 = c_5 Q_5, \quad c_5 = g_s \ell_s^2, \quad r_p^2 = c_p Q_p, \quad c_p = \frac{g_s^2}{VR^2}\ell_s^2. \tag{12.5.4}$$

$Q_{1,5,p}$ are non-negative integers. They are describing respectively the D_1 charge, the D_5 charge and the left-moving momentum on the S^1. V is the dimensionless volume of the T^4. The physical volume is $(2\pi\ell_s)^4 V$. R is the dimensionless radius of S^1. Its physical length is $(2\pi\ell_s)R$.

Special cases of this solution clarify better its interpretation. When $Q_1 = Q_p = 0$, this is a collection of Q_5 D_5-branes wrapped on $T^4 \times S^1$. This solution was already presented in section 8.8. It preserves 16 supercharges, namely $Q + \Gamma^{1234}\tilde{Q}$, as described in section 8.3 on page 191.

Turning on a nonzero Q_1 adds D_1-strings that wrap the S^1 inside the D_5-branes. The D_1 strings preserve again 16 supercharges but different ones, namely, $Q + \Gamma^{12346789}\tilde{Q} \sim \Gamma^{6789}Q + \Gamma^{1234}\tilde{Q}$. The common preserved supercharges are eight, since $(\Gamma^{6789})^2 = 1$. Turning on also Q_p adds left-moving KK momentum on the composite D_1-D_5 string (the KK momentum is Q_p/R). This breaks an extra half of the supersymmetry leaving finally four supercharges.

We would like now to reduce this solution to a five-dimensional solution, where the internal compact space will be $T^4 \times S^1$. Following the rules of appendix E, we obtain a five-dimensional metric, and a few scalars corresponding to the nontrivial diagonal metric factors as well as the dilaton. Finally there are three nontrivial gauge fields, whose charges will be associated to Q_1, Q_5, and Q_p. One of them will come from the off-diagonal component of the metric due to the left-moving wave while the other two descend from the R-R two-form. One of the two R-R vectors comes from the C_{05} component of the two form. The other is the five-dimensional dual of the C_{ij} components of the two-form.

The five-dimensional Einstein metric ds_E^2 can be obtained from (12.5.1) as

$$ds_{10}^2 = e^{2\chi} dy^a dy^a + e^{2\psi}(dy + A_\mu dx^\mu)^2 + e^{-(8\chi+2\psi+\phi)/3} ds_E^2, \tag{12.5.5}$$

with

$$ds_E^2 = -f^{-2/3} dt^2 + f^{1/3}(dr^2 + r^2 d\Omega_3^2), \quad f(r) = H_1(r) H_5(r) H_p(r). \tag{12.5.6}$$

This is the metric of an extremal five-dimensional RN black hole with three charges. The horizon is at $r = 0$. The radius of the three-sphere at $r = 0$ and the horizon area can be read from (12.5.6) to be

$$R_h = r f(r)^{1/6}\big|_{r=0} = (r_1 r_5 r_p)^{1/3}, \quad A = 2\pi^2 R_h^3 = 2\pi^2 r_1 r_5 r_p. \tag{12.5.7}$$

The five-dimensional Newton constant can be calculated from the ten-dimensional one as

$$\frac{1}{16\pi G_5} = \frac{(2\pi \ell_s R)(2\pi \ell_s)^4 V}{16\pi G_{10}} = \frac{(2\pi \ell_s R)(2\pi \ell_s)^4 V}{(2\pi)^7 \ell_s^8 g_s^2} = \frac{RV}{(2\pi)^2 \ell_s^3 g_s^2}. \tag{12.5.8}$$

We therefore obtain the BH entropy as

$$S = \frac{1}{4}\frac{A}{G_5} = 2\pi \sqrt{Q_1 Q_5 Q_p}. \tag{12.5.9}$$

A similar calculation of the Hawking temperature gives zero as expected for a supersymmetric BPS black hole.

12.6 Five-dimensional Nonextremal RN Black Holes

This BPS solution can be generalized beyond extremality. The departure from extremality is controlled by an extra parameter, that we will denote by r_0. The nonextremal solution is given in the string frame by

$$ds_{10}^2 = \frac{-dt^2 + dy^2 + \frac{r_0^2}{r^2}(\cosh a_p \, dt + \sinh a_p \, dy)^2}{\sqrt{H_1 H_5}} + \sqrt{H_1 H_5}\left[\frac{dr^2}{h} + r^2 d\Omega_3^2\right]$$

$$+ \sqrt{\frac{H_1}{H_5}}\, dy^a dy^a, \quad e^{-2\Phi} = \frac{H_5}{H_1}, \tag{12.6.1}$$

where

$$H_{1,5} = 1 + \frac{r_{1,5}^2}{r^2}, \quad r_{1,5} = r_0 \sinh a_{1,5}, \quad h = 1 - \frac{r_0^2}{r^2}, \tag{12.6.2}$$

$$F_{05i} = \frac{\coth a_1}{2}\frac{\partial}{\partial x^i} H_1, \quad F_{ijk} = \coth a_5 \epsilon_{ijkl}\frac{\partial}{\partial x^l} H_5. \tag{12.6.3}$$

The three charges can be calculated to be

$$Q_1 = \frac{V}{4\pi^2 g_s} \int_{S^3 \times T^4} e^{\Phi} \,{}^{\star}F = \frac{r_0^2 \sinh 2a_1}{2c_1}, \tag{12.6.4}$$

$$Q_5 = \frac{1}{4\pi^2 g_s} \int_{S^3} F = \frac{r_0^2 \sinh 2a_5}{2c_5}, \tag{12.6.5}$$

$$Q_p = \frac{r_0^2 \sinh 2a_p}{2c_p}, \tag{12.6.6}$$

where the constants c_i were defined in (12.5.4). There is an interesting interpretation of these charges in terms of branes and antibranes (as well as left-moving and right-moving momenta). If we define

$$N_{\pm 1} = \frac{r_0^2 e^{\pm 2a_1}}{4c_1}, \quad N_{\pm 5} = \frac{r_0^2 e^{\pm 2a_5}}{4c_5}, \quad N_{\pm p} = \frac{r_0^2 e^{\pm 2a_p}}{4c_p}, \tag{12.6.7}$$

then N_{+1} is the number of D_1-branes while N_{-1} is the number of D_1-antibranes and so on. Therefore,

$$Q_1 = N_{+1} - N_{-1}, \quad Q_5 = N_{+5} - N_{-5}, \quad Q_p = N_{+p} - N_{-p}. \tag{12.6.8}$$

The extremal limit is obtained by letting $r_0 \to 0$ taking at the same time $a_{1,5,p} \to \infty$ so that the total charges Q_i are finite. In this limit our solution reduces to the one presented in the previous section.

It should be noted that all curvatures are small in the limit

$$g_s Q_1 \gg 1, \quad g_s Q_5 \gg 1, \quad g_s^2 Q_p \gg 1. \tag{12.6.9}$$

Therefore, in this regime, the classical solution is a reliable description of the D-brane bound state.

Reducing this metric to five dimensions, we obtain the nonextremal RN black-hole metric with three charges

$$ds_E^2 = -f^{-2/3} h \, dt^2 + f^{1/3}\left(\frac{dr^2}{h} + r^2 d\Omega_3^2\right), \quad f(r) = H_1(r) H_5(r) H_p(r), \tag{12.6.10}$$

with

$$H_p = 1 + \frac{r_p^2}{r^2} = 1 + \frac{r_0^2 \sinh^2 a_p}{r^2}. \tag{12.6.11}$$

The exterior horizon is at $r = r_0$. $r = 0$ is the interior horizon. In the extremal limit, $r_0 \to 0$ the two coincide as in four dimensions. We are in a coordinate system[5] such that the singularity is hidden behind the inner horizon $r = 0$. Such a coordinate system has been discussed in the four-dimensional case in F. The horizon radius and area are now

---

[5] This is similar to the one described in appendix F on page 519.

$$R_h = r f(r)^{1/6}\big|_{r=r_0} = r_0 \left(\prod_{i=1,5,p} \cosh a_i \right)^{1/3},$$

$$A = 2\pi^2 R_h^3 = 2\pi^2 r_0^3 \prod_{i=1,5,p} \cosh a_i, \tag{12.6.12}$$

while the entropy is

$$S = \frac{A}{4G_5} = 2\pi (\sqrt{N_{+1}} + \sqrt{N_{-1}})(\sqrt{N_{+5}} + \sqrt{N_{-5}})(\sqrt{N_{+p}} + \sqrt{N_{-p}}). \tag{12.6.13}$$

The mass can be obtained from the $1/r^2$ term of g_{00} by comparing to (F.14) and (F.15). We obtain

$$M = \frac{\pi r_0^2}{8G_5}(\cosh 2a_1 + \cosh 2a_5 + \cosh 2a_p) = \frac{\pi}{2G_5}\left[\sum_{i=1,5,p} c_i(N_{+i} + N_{-i}) \right]. \tag{12.6.14}$$

Finally the Hawking temperature of the configuration can be calculated using the Euclidean rotation method, described in section 12.3.1 to be

$$\frac{1}{T_H} = 2\pi r_0 \cosh a_1 \cosh a_5 \cosh a_p. \tag{12.6.15}$$

It vanishes in the extremal limit $r_0 \to 0$, $a_i \to \infty$ keeping the charges fixed.

In our subsequent discussion we will leave extremality by adding opposite sign KK momentum to the system. The entropy (12.6.13) in this case will be

$$S = 2\pi \sqrt{N_{+1} N_{+5}}(\sqrt{N_{+p}} + \sqrt{N_{-p}}). \tag{12.6.16}$$

12.7 The Near-horizon Region

It is interesting, for future purposes, to study the region close to the inner horizon of the five-dimensional black hole we have presented. It will be shown to be relevant for the microscopic computation of the entropy and Hawking emission rate. Moreover, it will later lead us to the AdS/CFT correspondence.

In the near-horizon limit, $r \to 0$ and in the various harmonic functions H_i, appearing in the solution the r-dependent factor dominates over the 1. There is a controlled way of taking this limit by taking the string length to zero while approaching the horizon:

$$\ell_s \to 0, \quad r \to 0, \quad r_0 \to 0, \tag{12.7.1}$$

$$U = \frac{r}{\ell_s^2} = \text{fixed}, \quad U_0 = \frac{r_0}{\ell_s^2} = \text{fixed}, \tag{12.7.2}$$

as well as keeping V, R, and the six-dimensional string coupling $g_6 = g_s/\sqrt{V}$ fixed. Applying this limit to the ten-dimensional nonextremal metric (12.6.1) we obtain the product of $T^4 \times S^3$ as well as the three-dimensional BTZ black-hole metric:

$$ds^2 = \ell_s^2 \left(ds_{\text{BTZ}}^2 + ds_{S^3}^2 + ds_{T^4}^2 \right), \tag{12.7.3}$$

with

$$ds^2_{S^3} = \ell^2 d\Omega^2_3, \quad ds^2_{T^4} = \sqrt{\frac{Q_1}{VQ_5}} \, dy^a dy^a, \tag{12.7.4}$$

$$ds^2_{BTZ} = \frac{1}{\ell^2} U^2(-dt^2 + dy^2) + \frac{1}{\ell^2} U_0^2(\cosh a_p dt + \sinh a_p dy)^2 + \frac{\ell^2}{U^2 - U_0^2} dU^2, \tag{12.7.5}$$

and

$$\ell^2 = g_6 \sqrt{Q_1 Q_5}, \quad g_{\text{eff}} = e^\Phi \, |_{\text{horizon}} = g_6 \sqrt{\frac{Q_1}{Q_5}}. \tag{12.7.6}$$

The three-dimensional metric (12.7.5) can be brought to the standard BTZ form by the following coordinate transformation:

$$\phi = \frac{y}{R\ell_s}, \quad \tilde{t} = \frac{\ell}{R\,\ell_s} t, \quad \tilde{r}^2 = \frac{\ell^2}{\ell^2}(U^2 + U_0^2 \sinh^2 a_p) R^2. \tag{12.7.7}$$

Putting also back the multiplicative string scale from (12.7.3), we obtain

$$\frac{d\tilde{s}^2_{BTZ}}{\ell_s^2} = -\frac{(\tilde{r}^2 - r_+^2)(\tilde{r}^2 - r_-^2)}{\ell^2 \tilde{r}^2} d\tilde{t}^2 + \frac{\ell^2 \tilde{r}^2}{(\tilde{r}^2 - r_+^2)(\tilde{r}^2 - r_-^2)} d\tilde{r}^2 + \tilde{r}^2 \left(d\phi + \frac{r_+ r_-}{\ell \tilde{r}^2} d\tilde{t}\right)^2, \tag{12.7.8}$$

with

$$r_+ = \frac{\ell_s}{\ell} R U_0 \cosh a_p, \quad r_- = \frac{\ell_s}{\ell} R U_0 \sinh a_p. \tag{12.7.9}$$

All coordinates here are dimensionless. The mass and angular momentum of the BTZ black hole are given by

$$8G_3 M = 1 + \frac{r_+^2 + r_-^2}{\ell^2} = \frac{2(N_{+p} + N_{-p})}{Q_1 Q_5}, \quad 8G_3 J = 2\frac{\ell_s}{\ell} r_+ r_- = 2\ell_s \ell \frac{(N_{+p} - N_{-p})}{Q_1 Q_5}, \tag{12.7.10}$$

where G_3 is the effective three-dimensional Newton's constant. Note, that we adjusted an additive constant so that the mass is zero for global AdS$_3$, obtained when $r_- = 0, r_+^2 = -\ell^2$.

The extremal limit corresponds to $M\ell = J$ or $r_+ = r_-$, or $N_{-p} = 0$ or $U_0 \to 0$ and $a_p = 0$. In this limit, the metric (12.7.8) becomes, locally, that of three-dimensional anti de Sitter space AdS$_3$,

$$ds^2_{AdS_3} = \frac{U^2}{\ell^2}(-dt^2 + dy^2) + \frac{\ell^2}{U^2} dU^2. \tag{12.7.11}$$

Because the coordinate y is periodic with period 2π this space is a translation orbifold of standard AdS$_3$. Therefore the spectrum of M contains a point $M = 0$ corresponding to global AdS$_3$, a gap, and for $M \geq 1/8G_3$ a continuum.

The BTZ black hole is a solution of pure three-dimensional gravity with a cosmological constant. It is therefore asymptotically AdS$_3$, as can be verified by taking the $U \to \infty$ limit of (12.7.5).

An issue that will be important later on, is related to supersymmetry, and in particular the boundary conditions for the fermions. For fixed time, the topology of AdS$_3$ is that of a two-dimensional disk. Since this is contractible, there is only one possible boundary condition for the fermions namely antiperiodic (corresponding to periodic on the cylinder). In the case of the BTZ black hole, the fixed time slice has a singularity at zero radius. Therefore the circle around it is not contractible, and both antiperiodic and periodic boundary conditions are allowed for the fermions.[6] An analysis however of the Killing spinors in the zero mass BTZ black hole indicates that the fermions must be periodic.

12.8 Semiclassical Derivation of the Hawking Rate

So far we have described the D_1-D_5 black hole, as well as its thermodynamical data. We will now describe the semiclassical absorption and emission of Hawking radiation, which we would like to eventually understand microscopically from string theory. Classically the black hole can absorb, but it cannot emit radiation. At the semiclassical level the Hawking emission leads to the information puzzle. We would like to understand the microscopic mechanism, with the hope that it will give us interesting information about the structure of black holes and may shed light into the information paradox.

In this section we will present the semiclassical calculation of the greybody factors that are important ingredients of the emission and absorption rates, (12.3.23). To do this, we must derive the effective action of the various low-energy fluctuations of the theory, in the D_1-D_5 black-hole background, and then study their classical absorption cross section from the black hole. We will deal with the simplest such fluctuations: scalars. There are two types of scalar fluctuations. The so-called minimal scalars are massless and couple to the five-dimensional black-hole metric $g_{\mu\nu}$ as

$$S_{\min} = \frac{1}{16\pi G_5} \int d^5x \, \sqrt{g} \, g^{\mu\nu} \, \partial_\mu \phi \partial_\nu \phi, \tag{12.8.1}$$

where higher-derivative terms have been neglected at low enough energy. An example is presented in exercise 12.15.

The "fixed" scalars on the other hand also have non-derivative couplings to the background. Exercise 12.16 provides examples of fixed scalars.

We will study here the absorption of a minimal scalar in the s-wave mode for simplicity. Since we will be interested at low energies, the s-wave approximation is sufficient. We must solve the equation of motion $\Box \phi = 0$ in the black-hole metric (12.6.10). For the s-wave, the equation translates to

$$\left[\frac{h}{r^3} \frac{d}{dr} \left(h r^3 \frac{d}{dr} \right) + \omega^2 f \right] R_\omega(r) = 0, \tag{12.8.2}$$

where we have set $\phi = R_\omega(r) e^{-i\omega t}$. Redefining $\psi = r^{3/2} R(r)$ and introducing the tortoise coordinate

$$r_* = \int \frac{dr}{h} = r + \frac{r_0}{2} \log \left| \frac{r - r_0}{r + r_0} \right|, \tag{12.8.3}$$

---

[6] This is similar to the cylinder case described in section 4.12 on page 71. Around the non-contractible cycle we have the option of putting either periodic (Ramond) or antiperiodic (Neveu-Schwarz) boundary conditions.

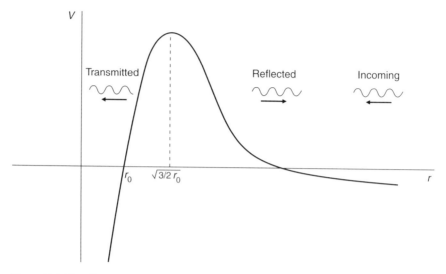

Figure 12.1 The effective potential for a minimal scalar.

we obtain

$$\left[-\frac{d^2}{dr_*^2} + V(r_*)\right]\psi = 0, \quad V(r_*) = -\omega^2 f(r) + \frac{3}{4r^2}\left(1 - \frac{r_0^2}{r^2}\right)\left(1 + 3\frac{r_0^2}{r^2}\right). \tag{12.8.4}$$

The effective potential is plotted in figure 12.1. Absorption is caused by tunneling to the central region of the potential.

The wave eqution above cannot be solved exactly. To calculate the cross section we must solve the wave equation very close to the horizon and asymptotically, and then match the two solutions.

We assume that we are near extremality and low energy so that

$$r_0, r_p \ll r_1, r_5, \quad \omega r_5 \ll 1, \tag{12.8.5}$$

where r_i were defined in (12.6.2) and (12.6.11). The length scales r_i determine the size of the throat due to the charge Q_i. The inequality above states that the departure from extremality is small (r_0 condition). It also implies that the travelling KK states are "dilute" as we will see later in the D-brane description. We will treat r_0/r_p and r_1/r_5 as order one quantities.

We also introduce two temperatures whose interpretation will become clear later:

$$T_L = \frac{1}{2\pi}\frac{r_0 e^{a_p}}{r_1 r_5}, \quad T_R = \frac{1}{2\pi}\frac{r_0 e^{-a_p}}{r_1 r_5}. \tag{12.8.6}$$

We assume that the energy is of order $\omega \sim T_{L,R}$.

To do the matching between the large- and short-distance behavior we must pick an intermediate point r_m at which the matching will be done. This can be chosen so that

$$r_0, r_p \ll r_m \ll r_1, r_5, \quad \omega r_1\frac{r_1}{r_p} \ll 1. \tag{12.8.7}$$

The first condition is guaranteed by (12.8.5) while the second condition is implied by the previous inequalities and $\omega \sim T_{L,R}$.

We are now ready to solve the equation in the "far region" $r > r_m$: Taking into account the previous inequalities (12.8.5),(12.8.7), the equation becomes a Bessel equation

$$\frac{d^2\psi}{d\rho^2} + \left(1 - \frac{3}{4\rho^2}\right)\psi = 0, \tag{12.8.8}$$

where $\rho = \omega r$. Note that in this region $r_* \simeq r$. The solution is

$$\psi = \sqrt{\frac{\pi\rho}{2}}\,[A\,J_1(\rho) + B\,N_1(\rho)]. \tag{12.8.9}$$

For $\omega r \gg 1$ we obtain for R

$$R \simeq \frac{1}{2r^{\frac{3}{2}}}\left[e^{i\omega r}\left(Ae^{-3i\pi/4} - Be^{-i\pi/4}\right) + e^{-i\omega r}\left(Ae^{3i\pi/4} - Be^{i\pi/4}\right)\right]. \tag{12.8.10}$$

In the near region $r < r_m$, the leading terms of equation (12.8.2) are

$$\left[\frac{h}{r^3}\frac{d}{dr}\left(hr^3\frac{d}{dr}\right) + \frac{(\omega r_1 r_5)^2}{r^4}\left(1 + \frac{r_p^2}{r^2}\right)\right]R(r) = 0, \tag{12.8.11}$$

with general solution given by hypergeometric functions

$$R = \left[\tilde{A}\,z^{-i(a+b)/2} + \tilde{B}\,z^{+i(a+b)/2}\right]\,{}_2F_1(-ia, -ib, 1 - ia - ib; z), \tag{12.8.12}$$

$$z = 1 - \frac{r_0^2}{r^2}, \quad a = \frac{\omega}{4\pi T_R}, \quad b = \frac{\omega}{4\pi T_L}. \tag{12.8.13}$$

Here we must impose that the solution contains only an incoming wave close to the horizon $z = 0$. This gives $\tilde{B} = 0$. We then match R and its first derivative R' at $r = r_m$. At $r = r_m$, $\omega r_m \ll 1$ due to (12.8.7) and we may approximate the far-region solution as

$$R \simeq \sqrt{\frac{\pi\omega^3}{2}}\left[\frac{A}{2} + \frac{B}{\pi}\left(-\frac{2}{\omega^2 r^2} + \log(\omega r) + \mathcal{O}(1)\right)\right]. \tag{12.8.14}$$

In exercise 12.17 you are asked to show that we can neglect B in the matching and in the computation of the cross section. There is already a hint of this in the previous equation: the term multiplying B is very large at the matching point, and therefore matching will require B to be small. Using

$${}_2F_1(-ia, -ib, 1 - ia - ib; 1) = \frac{\Gamma(1 - ia - ib)}{\Gamma(1 - ia)\Gamma(1 - ib)}, \tag{12.8.15}$$

we obtain

$$\sqrt{\frac{\pi\omega^3}{2}}\frac{A}{2} = \tilde{A}\,\frac{\Gamma(1 - ia - ib)}{\Gamma(1 - ia)\Gamma(1 - ib)}. \tag{12.8.16}$$

The conserved flux of the wave equation is

$$\mathcal{F} = \frac{1}{2i}\left[hr^3\,R^*\frac{dR}{dr} - cc\right], \quad \frac{d\mathcal{F}}{dr} = 0. \tag{12.8.17}$$

Using (12.8.10) we obtain for the incoming flux

$$\mathcal{F}_{in} \simeq -\omega\left|\frac{A}{2}\right|^2. \tag{12.8.18}$$

The absorbed flux is evaluated from the near-region solution at the horizon $z = 0$,

$$\mathcal{F}_{abs} \simeq -r_0^2(a+b)|\tilde{A}|^2, \tag{12.8.19}$$

and the ratio is

$$R_{abs} = \frac{\mathcal{F}_{abs}}{\mathcal{F}_{in}} = (\pi r_0 \omega)^2 \frac{ab \, e^{2\pi(a+b)}}{(e^{2\pi a} - 1)(e^{2\pi b} - 1)}, \tag{12.8.20}$$

where we have used $|\Gamma(1 - ia)|^2 = \frac{\pi a}{\sinh \pi a}$. Finally, to obtain the standard cross section we need to convert the incoming spherical wave into a plane wave using

$$e^{-i\omega z} = K \frac{e^{-i\omega r}}{r^{\frac{3}{2}}} Y_{000} + \text{higher partial waves} + \text{outgoing waves}, \tag{12.8.21}$$

where Z_{l,m_1,m_2} are the analogue of spherical harmonics for S^3, namely, the D-functions of SU(2). In particular $Y_{000} = 1/\sqrt{2\pi^2}$. You are asked in exercise 12.18 to compute K and show that, up to a phase, $K = \sqrt{4\pi/\omega^3}$. Therefore, the absorption cross section is

$$\sigma_{abs} = |K|^2 R_{abs} = \pi^3 (r_1 r_5)^2 \omega \frac{e^{\omega/T_H} - 1}{(e^{\omega/2T_L} - 1)(e^{\omega/2T_R} - 1)}, \tag{12.8.22}$$

where we have used

$$\frac{1}{T_H} = \frac{1}{2T_L} + \frac{1}{2T_R} \tag{12.8.23}$$

for the Hawking temperature (12.6.15). Note that the $\omega \to 0$ limit of the absorption cross section is equal to the horizon area as expected

$$\lim_{\omega \to 0} \sigma_{abs} = 4\pi^3 (r_1 r_5)^2 \frac{T_L T_R}{T_H} = A. \tag{12.8.24}$$

Finally we can calculate the Hawking evaporation rate by adapting (12.3.23) to five dimensions to obtain

$$\Gamma_H = \frac{\sigma_{abs}}{e^{\frac{\omega}{T_H}} - 1} \frac{d^4 k}{(2\pi)^4} = \pi^3 (r_1 r_5)^2 \omega \frac{1}{(e^{\frac{\omega}{2T_L}} - 1)} \frac{1}{(e^{\frac{\omega}{2T_R}} - 1)} \frac{d^4 k}{(2\pi)^4}. \tag{12.8.25}$$

There is an important comment concerning the semiclassical calculation of the absorption cross section. The near region $r < r_m$ is nothing else but the near-horizon region defined and explored in section 12.7. Moreover, in the far region, the wave function after neglecting the B coefficient (which is subleading) is

$$R \simeq \frac{A}{2} \frac{e^{-i\omega r}}{r^{3/2}}, \quad \mathcal{F}_{in} = |A|^2. \tag{12.8.26}$$

Therefore, there is no crucial information here that affects the Hawking rate. In particular, A cancels from the final rate. All that we used from this region is that it is asymptotically flat.

We arrive at the conclusion that *at sufficiently low energy, the absorption cross section as well as the Hawking evaporation rate depends only on the near-horizon geometry of the black hole.* This observation is a precursor of the AdS/CFT correspondence.

12.9 The Microscopic Realization

We have seen in section 12.5 that a five-dimensional black hole with regular horizon and finite entropy has three charges, Q_5 a D_5 brane charge, Q_1 a D_1 brane charge and Q_p, a momentum charge along S^1. This already suggests the way such a system could be realized microscopically in string theory. We could make a bound state of Q_5 D$_5$-branes, and Q_1 D$_1$-branes inside the D$_5$-branes. We could then add Q_p units of momentum on the D$_1$ world-volume as in the classical solution (12.5.1)–(12.5.3).

An important feature of this construction is that *a priori* it can be done by assembling the D-brane bound state out of branes in flat space. This is perfectly valid if $g_s Q_5, g_s Q_1 \ll 1$. In this case, the Polchinski description is reliable, and the dynamics of the bound state is described by the world-volume theory of the branes. Moreover, since the dynamics we are interested is at low energy, high-energy degrees of freedom, like open string excitations and KK excitations (with the exceptions of the KK states on S^1 that carry the Q_p units of momentum) can be neglected.

Therefore, the relevant world-volume theory will be a two-dimensional supersymmetric gauge theory and will be weakly coupled if $g_s Q_5, g_s Q_1 \ll 1$. In this case, we may count the states with the appropriate charges and determine their multiplicity. This should give us the entropy of the configuration and it will turn out that this calculation matches with the semiclassical result (12.6.16).

However, in the limit in which we work, namely $g_s Q_5, g_s Q_1 \ll 1$, the system does not look as a macroscopic black hole. As can be seen from the classical solution, close to the horizon the curvature is large. In the case of extremal black holes, the supersymmetry of the configuration will help us to extrapolate the weak coupling results at strong coupling, where the bound-state is a semiclassical black hole.

Supersymmetry is not enough to justify the agreement in the near-extremal case however. Moreover, as we will show, the emission rates for Hawking radiation will also turn out to match exactly between the semiclassical calculation presented in section 12.8 and the microscopic (weak coupling) calculation. It will turn out that the reason for this agreement is deeper and eventually leads to the bulk-boundary correspondence that will be treated in the next section.

12.9.1 The world-volume theory of the bound state

We will use the notation of section 12.5. We compactify IIB string theory on $S^1 \times T^4$ to five dimensions. We take Q_5 D$_5$-branes to wrap $S^1 \times T^4$ and Q_1 D$_1$-branes to wrap S^1 with coordinate y. We will assume that the volume of T^4 $V \sim \ell_s^4$ so that the KK particles in these directions have masses that are not smaller than the string scale. We will also take the radius R of S^1 to be $R \gg \ell_s$ so that the masses of the associated KK states (responsible for the Q_p charge) are much smaller than the string scale.

To summarize, if we focus on energies $E \ll M_s$ we can neglect the existence of T^4. We may also treat S^1 as a noncompact direction since the winding modes have energies larger than the string scale. Moreover, all stringy modes are invisible in the low-energy regime.

The D$_1$ branes can be generated on the D$_5$ world-volume gauge theory by turning on instanton configurations on T^4 as described in exercise 8.29 on page 216. The gauge theory has a Coulomb branch, where by giving appropriate expectation values to the world-volume scalars, we can freely move the branes apart. This is certainly not the bound state we are interested in, but a loose configuration of branes. There is, however, a Higgs branch, where the branes are bound together, the appropriate scalars have zero expectation values, and the moduli space is the $U(Q_5)$ instanton moduli space.

The D_1-D_5 bound state preserves 1/4 of the original supersymmetry, namely, eight supercharges (see exercise 8.11). Another way to see this is the following: the D_5 branes break 1/2 of the original supersymmetry. turning on an instanton background on the world-volume theory breaks an additional 1/2 of the supersymmetry.

We will now describe the basic ingredients of the World-volume theory. We keep modes with masses much smaller than M_s.

(1,1) Strings. This sector arises from fluctuations of open strings on the D_1-strings. It is a two-dimensional gauge theory along the t, y coordinates, isomorphic to the dimensional reduction of the ten-dimensional SYM.

Although this sector has invariance under 16 supercharges, we will arrange the fields in terms of multiplets of the final $\mathcal{N} = (4,4)_2$ supersymmetry in two dimensions. They are a $U(Q_1)$ vector multiplet, $(A_t^{(1)}, A_y^{(1)}, \Phi_i^{(1)})$ with $i = 1, 2, 3, 4$ corresponding to the noncompact x^i-directions, and a hypermultiplet $(\Phi_a^{(1)})$, with $a = 1, 2, 3, 4$ corresponding to the compact T^4 y^a-directions.

(5,5) Strings. Here the theory is again a dimensional reduction of the ten-dimensional SYM to 5+1 dimensions. Since the T^4 is along the world-volume and its volume is at the string scale, we can compactify further to two dimensions and forget about the KK states. The only difference from the (1,1) case is that the gauge group is $U(Q_5)$. We have again a vector multiplet $(A_t^{(5)}, A_y^{(5)}, \Phi_i^{(5)})$ and a hypermultiplet $(\Phi_a^{(5)})$.

(1,5)+(5,1) Strings. This is the sector that breaks an extra half of the supersymmetry. Since the string endpoints belong to different sets of branes the fields here transform as bifundamentals $(Q_5, \bar{Q}_1)$ or $(\bar{Q}_5, Q_1)$. Due to the DN boundary conditions on this sector, the conformal dimension of the NS and R ground states[7] is 1/2. This gives massless states without the action of any oscillators. We obtain therefore a hypermultiplet $Y_I^{(1,5)}$.

The dynamics of the low-energy configurations of the system is governed by a potential, containing three terms that we will denote by V_v, V_h, V_{vh}. The first part, V_v, comes from the maximally supersymmetric part of the gauge theory and involves the scalars of (1,1) and (5,5) vector multiplets. It can be obtained from the ten-dimensional kinetic terms by dimensional reduction. It is the trace of the square of the commutator of the adjoint scalars as in $\mathcal{N} = 4_4$ SYM.

The second term V_{vh} mixes the vector scalars with the hypermultiplet scalars. It is the square of a term that is linear in the vector scalars and the hypermultiplet ones. The third term V_h is the D-term potential, quadratic in the D terms. In the absence of FI terms this part is quartic in the hypermultiplet scalars.[8] Here, however, an important ingredient is the presence of FI terms that appear in the D-term potential.

At the supersymmetric minima, all three potential terms V_v, V_h, V_{vh} vanish. In the Coulomb branch, the Cartan scalars of the vector multiplets have nonzero vevs, breaking

---

[7] See also the discussion in section 11.7.1 on page 339.
[8] For a curved hypermultiplet space there are nontrivial vector fields involved implementing the moment map. This is not the case here.

the $U(Q_1) \times U(Q_5)$ gauge symmetry to a subgroup $(U(1)^{Q_1+Q_5}$ in the generic case). This is the moduli space of the unbound system.

In the Brout-Englert-Higgs phase, the vevs of the vector multiplets vanish (and the branes stay bound together). The vanishing of the V_{hv} and V_h terms imposes a set of non-trivial quadratic conditions on the hypermultiplets. Since their number is much larger than the dimension of the gauge group there is a large manifold of solutions (known as the hyper-Kähler quotient). These nontrivial hypermultiplet vevs generically Higgs the full gauge group (except the overall center of mass U(1)). A simple counting shows that the $Q_1^2 + Q_5^2 - 1$ vector multiplets will combine with an equal number of hypermultiplets to become massive. We are left with the center-of-mass vector multiplet and $Q_1 Q_5 + 1$ hypermultiplets constrained to live at the minimum of the D-term potential, implemented by the quadratic constraints. One of the hypermultiplets is the partner of the vector multiplet. It is free, and its scalars are the Wilson lines of the center-of-mass U(1) on T^4.

The kinetic terms give rise therefore to a nonlinear σ-model for the remaining multiplets. Mathematically, as described in appendix I.2 on page 535, the space of hypermultiplets is a hyperKähler manifold. The Brout-Englert-Higgs effect removes some of the dimensions, via the so called hyper-Kähler quotient[9] which produces a lower-dimensional hyper-Kähler manifold by removing the Higgsed dimensions.

At low energy, the theory therefore flows to a nontrivial $\mathcal{N} = (4, 4)_2$ superconformal field theory with target space, the (hyper-Kähler quotient) manifold specified by the minima of the potential. As argued earlier, apart from the center of mass vector and hypermultiplet, this is the moduli space $\mathcal{M}_{Q_1 Q_5}$ of Q_1 instantons of $U(Q_5)$ gauge theory. Its dimension is $4Q_1 Q_5$. The central charge of the SCFT on $\mathcal{M}_{Q_1 Q_5}$ can be obtained by a simple counting of the light degrees of freedom, and the fact that for (4,4) supersymmetry, although the σ-model manifold is curved, the value of the central charge is classical. We therefore obtain

$$c = 6Q_1 Q_5. \tag{12.9.1}$$

Our analysis above has been simplified. The details are more complicated, and the interested reader is referred to the bibliography.

12.9.2 *The low-energy SCFT of the D_1-D_5 bound state*

As mentioned in the previous section, the low-energy theory of the Higgs branch, describing the low-lying degrees of freedom of the D_1-D_5 bound state, is a $\mathcal{N} = (4, 4)_2$ SCFT represented by a supersymmetric σ-model on $R^4 \times T^4 \times \mathcal{M}_{Q_1 Q_5}$ where $\mathcal{M}_{Q_1 Q_5}$ is the moduli space of Q_1 instantons of a $U(Q_5)$ gauge theory. The extra T^4 comes from the Wilson lines of the overall (free) center-of-mass U(1) of the system. The R^4 are the scalars of the free center-of-mass vector multiplet.

The space $\mathcal{M}_{Q_1 Q_5}$ is known as the Hilbert scheme of $Q_1 Q_5$ points on an auxiliary four-torus $\tilde{T}^4$. It turns out that it is the resolution (blowup) of the orbifold $(\tilde{T}^4)^{Q_1 Q_5}/S(Q_1 Q_5)$

---

[9] A similar effect has been described in section 11.10 on page 356.

where $S(N)$ is the symmetric group of order N. This is a nontrivial mathematical result that we will try to motivate (but not prove) here.

What we will show is that $\mathcal{M}_{Q_1 Q_5}$ and $(\tilde{T}^4)^{Q_1 Q_5}/S(Q_1 Q_5)$ have the same cohomology. The classical cohomology corresponds to the short representations of the $\mathcal{N} = (4, 0)_2$ superconformal symmetry. They are the left-moving (supersymmetric) ground states of the SCFT.

To see this we will use nonperturbative dualities to map the D_1-D_5 bound state to a more tractable system. Let the common direction of the D_1-D_5 be x^5 and the rest of the world-volume directions of the D_5-brane be x^6, x^7, x^8, x^9. By a T_{56} T-duality we obtain a D_1-string along x^6 (it couples to $C_{\mu,6}$) and a D_3-brane along x^7, x^8, x^9 (it couples to $C^+_{\mu,789}$). Upon a sequence of dualities

$$ST_{6789}S: C_{\mu 6} \to G_{\mu 6}, \quad C^+_{\mu 789} \to B_{\mu 6}. \tag{12.9.2}$$

Therefore the D_1-D_5 brane charges (Q_1, Q_5) become winding and momentum along the x^6 direction. We may calculate the multiplicity of such a state in perturbative string theory.

Since these states are supersymmetric (they break half of the (4,4) supersymmetry) they are ground states on the left-moving part of the theory, and have arbitrary oscillator number $N_R \in \mathbb{Z}$ on the right-moving part (or vice versa). The level-matching condition for such a state is

$$N_R = Q_1 Q_5, \tag{12.9.3}$$

where now Q_1 is the momentum number along x^6 and Q_5 is the associated winding number. We will focus on the R sector. Because of supersymmetry the same counting applies also in the NS sector. The multiplicity $d(N_R)$ for such states can be obtained from the left-moving R partition function which for the type-IIB string is

$$\sum_{n=1}^{\infty} d(n) q^n = 256 \prod_{m=1}^{\infty} \left(\frac{1 + q^m}{1 - q^m} \right)^8, \tag{12.9.4}$$

where the $256 = (16)^2$ is the multiplicity of the massless states in the R sector. We conclude that the number of ground states of the D_1-D_5 bound state is given by $d(Q_1 Q_5)$ with $d(n)$ defined in (12.9.4).

It can be computed by direct methods that the dimension of the cohomology ring of $(\tilde{T}^4)^{Q_1 Q_5}/S(Q_1 Q_5)$ is given by $d(Q_1 Q_5)/256$.[10] Therefore the dimension of the cohomology of $R^4 \times T^4 \times (\tilde{T}^4)^{Q_1 Q_5}/S(Q_1 Q_5)$ is equal to

$$(16) \times (16) \times \frac{d(Q_1 Q_5)}{256} = d(Q_1 Q_5), \tag{12.9.5}$$

in agreement with our starting claim.

Therefore, the SCFT on $\mathcal{M}_{Q_1 Q_5}$, degenerates at a special point of its moduli space to the orbifold $(\tilde{T}^4)^{Q_1 Q_5}/S(Q_1 Q_5)$.

---

[10] This is a nontrivial result. You are invited to explore it in exercise 12.21 on page 402.

We will study here the free symmetric orbifold SCFT in order to present the notion of the "long strings."

The various twisted sectors are in one-to-one correspondence with the conjugacy classes of $S(Q_1 Q_5)$. The conjugacy classes are cyclic permutations of various lengths. The interesting elements are those which consist of a cycle of length n acting as

$$X_1^I \to X_2^I, \quad X_2^I \to X_3^I, \ldots, \quad X_n^I \to X_1^I, \quad X_{i>n}^I \to X_i^I. \tag{12.9.6}$$

The cyclic action of the n coordinates can be diagonalized as

$$Y_1^I \to \omega Y_1^I, \ldots, \quad Y_n^I \to \omega^n Y_n^I, \tag{12.9.7}$$

where Y_i^I are appropriate linear combinations of the X_i^I and $\omega = e^{2\pi i/n}$.

The ground state in this sector will be the product of bosonic and fermionic twist fields for $\mathbb{Z}_n$. The dimension of a bosonic $\mathbb{Z}_n$ twist field twisted by ω^k is $\Delta_k = \frac{k(n-k)}{2n^2}$ while the one for the corresponding fermionic twist field $\tilde{\Delta}_k = \frac{k(n+k)}{2n^2}$ so that the total dimension is

$$\Delta_n = \sum_{k=1}^n (\tilde{\Delta}_k + \tilde{\Delta}_k) = \frac{1}{2}(n-1). \tag{12.9.8}$$

In this sector, the modding of the oscillators is given by k/n where $k < N$ is an integer. Therefore, the states in this twisted sector have energies of the form

$$E_R = \frac{L_0}{\ell_s R} = \frac{N}{n \ell_s R}, \tag{12.9.9}$$

and they behave like strings wound around a circle of radius nR. The maximal twisted sector is obtained when $n = Q_1 Q_5$, corresponding to long strings of length $2\pi \ell_s Q_1 Q_5 R$.

There is a related observation here that is crucial for the microscopic thermodynamic calculations. As we will see in the next section, the number of states typically rises exponentially with the square root of oscillator number $\sim N$. Because of this and (12.9.9), at fixed energy, the multiplicity of states is maximized when n is maximal. Therefore, the largest multiplicity will be obtained by the maximally twisted sector, with $n = Q_1 Q_5$, giving rise to the longest string.

12.9.3 Microscopic calculation of the entropy

We have seen that the low-energy nontrivial dynamics of the D_1-D_5 bound system is described by the two-dimensional $(4,4)$ SCFT on $T^4 \times \mathcal{M}_{Q_1 Q_5}$ with central charge

$$c = 6(Q_1 Q_5 + 1), \tag{12.9.10}$$

where we have neglected the trivial R^4 part that describes the free motion of the bound state inside the four noncompact space dimensions. Different states of the SCFT correspond to different black holes in five dimensions. In particular, as argued in 12.7 and 12.8 the part of the bulk black-hole solutions that are relevant to the low-energy limit we are interested in, is the near-horizon region described by the BTZ black hole.

We have further seen in section 12.7 that the "ground state," the BTZ black hole of zero mass, has fermions with periodic boundary conditions. Therefore, it should correspond to the R ground state of the SCFT. The nontrivial black holes then correspond to excited states in the R sector. On the other hand, AdS$_3$ space has antiperiodic fermions and should correspond to the ground state of the NS sector.

Consider now a generic state in the R sector of this SCFT with

$$L_0 = N_R, \quad \bar{L}_0 = N_L. \tag{12.9.11}$$

Reinstating the fact that the space direction is a circle of radius R we obtain for the energy

$$E_R = \frac{N_R}{\ell_s R}, \quad E_L = \frac{N_L}{\ell_s R}. \tag{12.9.12}$$

Therefore $N_{L,R}$ is the left-moving/right-moving momentum on the circle.

When $N_L = 0$, half of the original $\mathcal{N} = (4, 4)_2$ supersymmetry is preserved. Such states, correspond to the extremal black hole of section 12.5 with charges Q_1, Q_5 and $Q_p = N_R$. The states giving rise to the entropy are BPS states, and we expect that they can be reliably counted at strong coupling.

The counting of BPS states at strong coupling can be subtle for various reasons. One immediate reason is that only states that cannot pair-up into long multiplets are protected by supersymmetry. The associated indices that count such states are known as helicity supertraces. More information on them can be found in appendix J. In the present case, all the states we are counting are hypermultiplets and therefore contribute with the same sign to the helicity supertrace. Therefore, their counting is robust.

A generic state with $N_L \neq 0$, $N_R \neq 0$ corresponds to the nonextremal black hole of section 12.6 with

$$N_{-5} = N_{-1} = 0, \quad N_{+1} = Q_1, \quad N_{+5} = Q_5, \quad N_{+p} = N_R, \quad N_{-p} = N_L, \tag{12.9.13}$$

and generically breaks all supersymmetry. The microscopic calculation of the entropy of such a state boils down to the calculation of its multiplicity in the SCFT.

For such configurations, we do not expect supersymmetry to be of much use. We will discover, however, that even in this case, our microscopic counting of states at weak coupling, will match (in the limit of large charges) the macroscopic entropy of the non-extremal black hole. The agreement at this stage is unexpected, and it will be elevated to a formalized correspondence in the next chapter.

The multiplicity of states, for large N_L, N_R, in a unitary CFT, is determined by the central charge c alone and is given by the asymptotic formula

$$\Omega(N_L, N_R) \sim e^{2\pi\sqrt{c/6}[\sqrt{N_L}+\sqrt{N_R}]}. \tag{12.9.14}$$

The entropy is then

$$\begin{aligned}
S = \log \Omega &= 2\pi\sqrt{\frac{c}{6}}\left[\sqrt{N_L} + \sqrt{N_R}\right] \\
&= 2\pi\left[\sqrt{Q_1 Q_5 N_L} + \sqrt{Q_1 Q_5 N_R}\right] + \mathcal{O}\left(\frac{1}{\sqrt{Q_1 Q_5}}\right),
\end{aligned} \tag{12.9.15}$$

in perfect agreement with the semiclassical result (12.6.16) which is valid to leading order as $Q_1 Q_5 \to \infty$.

So far, we have been working in a microcanonical ensemble where the left and right energies $E_{L,R}$ were fixed. It is well known that in the limit of large energies we can define an effective temperature and, under certain conditions, effectively work in the canonical ensemble.

To define the generalized canonical ensemble we introduce temperatures $T_{L,R}$ for the associated energies as

$$Z = \sum_{N_L, N_R = 1}^{\infty} \Omega(N_L, N_R) e^{-\beta_L E_L - \beta_R E_R}$$

$$\simeq \sum_{N_L, N_R = 1}^{\infty} \exp\left[2\pi \sqrt{Q_1 Q_5 N_L} - \beta_L \frac{N_L}{\ell_s R} + 2\pi \sqrt{Q_1 Q_5 N_R} - \beta_R \frac{N_R}{\ell_s R} \right]. \tag{12.9.16}$$

This is possible in two dimensions since left and right movers factorize.

The sums are dominated by the saddle points which define the effective temperatures

$$\sqrt{N_{L,R}} = \pi \ell_s R T_{L,R} \sqrt{Q_1 Q_5} \quad \Rightarrow \quad T_{L,R} = \frac{\sqrt{N_{L,R}}}{\pi \ell_s R \sqrt{Q_1 Q_5}}. \tag{12.9.17}$$

Therefore, we may treat the collection of states with energy $N_{L,R}/\ell_s R$ as a thermal ensemble with temperature $T_{L,R}$ given in (12.9.17). Note that they are equal to the analogous quantities (12.8.6) that emerged in the semiclassical calculation of the grey-body factors, using (12.5.4) and (12.6.6). We are probably on the right track! The Hawking temperature is then given by (12.8.23).

Although the result we obtained for the entropy is correct, the validity of the canonical thermodynamic description needs more discussion. It will be important for the microscopic calculation of the grey-body factors in the next section. In the end of the previous section we gave a heuristic argument, why most of the states that contribute to the entropy of the bound state, come for the maximally twisted sector (or the longer string). We will elaborate on this here and argue that it is in this sector that the two-dimensional thermodynamic ensemble we introduced above is reliable.

The degrees of freedom of interest are the (1,5) strings and they are $Q_1 Q_5$ in number. Each of these has eight transverse modes of fluctuation. Because, the D_1-D_5 system is bound (Higgs branch) only four of these modes are massless (corresponding to the D_1s' moving inside the D_5's). So in total we obtain $n_f = 4 Q_1 Q_4$ massless modes, that we will call "flavors." These flavors "see" the x^5 circle radius as R.

However, we have also seen in the previous section, that gauge symmetry arguments, and the mathematical structure of the appropriate instanton moduli space indicates that there are identifications of the (1,5) strings, implemented in the special case of the orbifold point $(\tilde{T}^4)^{Q_1 Q_5}/S(Q_1 Q_5)$ by the action of $S(Q_1 Q_5)$.

We have argued that the interpretation of the associated twisted sectors was that several (1,5) strings can join up to make a larger string of length $2\pi \ell_s n R$ with $n \leq Q_1 Q_5$. In the extreme case where they all make a single long string of length $2\pi \ell_s Q_1 Q_5 R$, the massless modes have only four flavors $n_f = 4$ but the effective radius of the circle is $Q_1 Q_5 R$.

Let us concentrate on the left-movers and rewrite for the various sectors the partition function as

$$Z_L(n) = \sum \exp\left[-\beta_L \sum_r \frac{n_r \, r}{\ell_s R_n}\right], \quad R_n = nR; \tag{12.9.18}$$

here n_r is the number of states that have left-moving energy $\frac{r}{R_n}$. You are invited, in exercise 12.23, to show that at high temperature $\beta/(\ell_s R_n) \ll 1$, the leading behavior of the partition function above is

$$\log Z_L(n) = \frac{3}{2} n_f \frac{(2\pi)^2 \ell_s R_n}{24\beta} + \mathcal{O}\left(\log \frac{\beta}{\ell_s R_n}\right). \tag{12.9.19}$$

This is the behavior of a superconformal theory with $c_{\text{eff}} = \frac{3}{2} n_f$ on a torus with modulus $\tau = i\frac{\beta}{2\pi \ell_s R_n}$. From this we may calculate the energy and entropy from the standard thermodynamic formulae

$$E_L = -\partial_\beta \log Z_L \simeq \frac{\pi^2 n_f \ell_s R_n}{4\beta^2}, \quad S_L = \log Z_L + \beta E \simeq \frac{\pi^2 n_f \ell_s R_n}{2\beta}, \tag{12.9.20}$$

and therefore

$$T_L = \sqrt{\frac{4E_L}{\pi^2 n_f \ell_s R_n}} = \frac{2S_L}{\pi^2 n_f \ell_s R_n}. \tag{12.9.21}$$

Matching this to the semiclassical result (12.6.12) we obtain

$$n_f R_n = 4Q_1 Q_5 R. \tag{12.9.22}$$

For the untwisted sector, $n = 1$ and $n_f = 4Q_1 Q_5$ satisfying (12.5.1). The same is true for the longest string sector with $n = Q_1 Q_5$, and $n_f = 4$.

We are now able to check the extensivity condition $T_L \ell_s R_n \gg 1$ which guarantees proper thermodynamics. In the long string sector

$$T_L \ell_s R_n \sim \sqrt{Q_1 Q_5 N_L} \gg 1. \tag{12.9.23}$$

The short $(n = 1)$ string sector on the other hand does not ensure extensivity since

$$T_L \ell_s R_n \sim \sqrt{\frac{N_L}{Q_1 Q_5}}. \tag{12.9.24}$$

Therefore, the dominant states thermodynamically are the longest open strings.

12.9.4 Microscopic derivation of Hawking evaporation rates

Here we will present the simplest stringy calculation of the absorption and Hawking emission rates, showing agreement with those calculated semiclassically in section 12.8.

The simplest process corresponds to a mode of the closed string hitting the bound state and exciting two open strings, one moving to the left the other to the right as in figure 12.2.

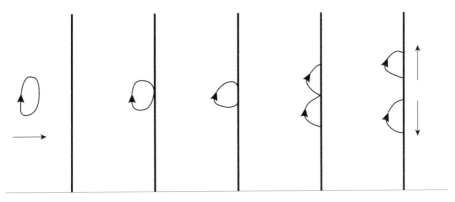

Figure 12.2 This figure, with time running to the right, describes the absorption of matter in a D-brane black hole. With time running to the left it describes the process of Hawking evaporation.

For the reasons mentioned in the end of the previous section, the most important (and reliable) contribution will come from the longest string sector corresponding to an effective radius $\tilde{R} = Q_1 Q_5 R$.

The simplest bulk mode is a minimal scalar corresponding to traceless fluctuations of the T^4 metric g^{ab}. It couples to the appropriate hypermultiplet scalars on the bound-state SCFT world-volume as

$$S = T \int d^2\xi \, g_{ab} \, \partial_\beta \phi^a \partial^\beta \phi^b, \tag{12.9.25}$$

where β is a world-sheet index and a, b are the T^4 indices. Expanding around the background metric we obtain

$$g_{ab} \simeq \delta_{ab} + \sqrt{2} \, \kappa_{10} \, h_{ab}(\xi^\alpha, y^a, x^i). \tag{12.9.26}$$

Since in the energy regime of interest we could drop the T^4 KK states, we can also ignore the dependence of h_{ab} on y^a. If we consider s-wave scattering, we can also neglect the x^i dependence. The normalization in front of h_{ab} in (12.9.26) is determined from the bulk supergravity action. The overall coefficient T in (9.8.10) on page 246 is irrelevant since it can be absorbed in the normalization of ϕ^a because the relevant interaction is quadratic in ϕ^a. The gravitational constant κ_{10} is given in (8.4.8) on page 195. The relevant interaction term is

$$S_{\text{interaction}} = \sqrt{2}\kappa_{10} \int d^2\xi \, h_{ab}(\xi) \, \partial_\beta \phi^a \partial^\beta \phi^b. \tag{12.9.27}$$

We go to momentum space and choose the bulk mode h_{ab} to have momentum ($k^0 = \omega, k^1, \ldots, k^4, k^5 = 0$) while the two open string modes corresponding to ϕ^I and ϕ^J have momenta p^0, p^5 and q^0, q^5. We may now write the expression for this decay rate, using standard QFT, as

$$\Gamma(p, q; k) = \frac{(2\pi)^2 \tilde{R} \, \delta(p^0 + q^0 - k^0)\delta(p^5 + q^5)}{(2p^0 \tilde{R})(2q^0 \tilde{R})(2k^0 (2\pi \ell_s)^4 V V_4 R)} \frac{2\kappa_{10}^2 (p \cdot q)^2}{(2\pi)^4} \frac{V_4 d^4 k}{(2\pi)^4}. \tag{12.9.28}$$

The origins of the factors are as follows: $(2\pi)^2 \tilde{R}$ accompanies the momentum-conserving δ-function and the interaction in (12.9.27) gives $\sqrt{2}\kappa_{10}\, p \cdot q$ for the vertex. $(2q^0 \tilde{R})$ and $(2p^0 \tilde{R})$ are the normalizations of the open string states while $(2k^0 (2\pi \ell_s)^4 VV_4 R)$ is the normalization of the closed string state, where here it is R that is relevant. V_4 is the volume of the noncompact x^i coordinates while $(2\pi \ell_s)^4 V$ is the physical volume of the T^4.

The modes are massless so that $(p^0)^2 = (p^5)^2$, $(q^0)^2 = (q^5)^2$ and from momentum conservation

$$q^5 = -p^5 = p^0 = q^0 = \frac{k^0}{2}, \quad p \cdot q = 2(p^0)^2. \tag{12.9.29}$$

The total rate for the emission of the closed string scalar is given by summing over initial states. As argued earlier, their distribution can be approximated by the thermal ensembles:

$$\Gamma(k) = \left(\frac{\tilde{R}}{2\pi}\right)^2 \int_{-\infty}^{\infty} dp^5 \int_{-\infty}^{\infty} dq^5 \, \rho(p^0, p^5)\, \rho(q^0, q^5)\, \Gamma(p, q; k). \tag{12.9.30}$$

Because the open string states have opposite momentum, one is L-moving while the other is R-moving. Therefore one of the thermal densities is ρ_L while the other is ρ_R.

Doing the integrals we finally obtain

$$\Gamma(k) = \pi^3 \omega (r_1 r_5)^2 \frac{1}{e^{\omega/2T_L} - 1} \frac{1}{e^{\omega/2T_R} - 1} \frac{d^4 k}{(2\pi)^4}, \tag{12.9.31}$$

where we have used

$$2\kappa_{10}^2 = (2\pi)^7 g_s^2 \ell_s^8, \quad (r_1 r_5)^2 = Q_1 Q_5 \frac{g_s^2 \ell_s^4}{V}. \tag{12.9.32}$$

This is in perfect agreement with the semiclassical Hawking emission rate in (12.8.25).

12.10 Epilogue

In the previous sections we gave a brief account of the microscopic determination of entropy in some string theory black holes as well as the associated Hawking evaporation rates. We have encountered a remarkable agreement between the semiclassical (gravity) and the microscopic (D-brane/gauge theory/SCFT) calculations. In this section, we will present some comments and hints on this matching. The next chapter will formalize these hints into the AdS/CFT and its generalization, the bulk-boundary correspondence.

One important issue in the microscopic calculation of absorption cross sections is the coupling of closed (bulk) and open string modes. We will write this coupling in general as

$$S_{\text{interaction}} = g \int d^2 z \, \phi_i(z, \bar{z}) O^i(z, \bar{z}). \tag{12.10.1}$$

Here ϕ_i is a closed string theory operator and O^i is an associated operator in the (open string) theory describing the bound D-branes. Both ϕ_i and O^i are assumed to have normalized two-point functions. g is the coupling between the bulk and the boundary operator.

This is the generalization of the interaction we used in the previous section to calculate microscopically the Hawking emission rate.

There are several problems with pinning down equation (12.10.1). The first is to find to which operator O_i of the D-brane theory the operator ϕ_i couples to. As we will see in the next chapter, there is a one-to-one correspondence between the bulk operators in the near-horizon geometry and the operators of the associated field theory on the branes. Next, normalizing the closed string operator is straightforward and can be obtained from the classical supergravity action. Normalizing the O_i operator is trickier than the reader might have guessed from our treatment in the previous section. The reason is that these are operators of a theory with many branes and cannot be obtained from the simple DBI action describing one-brane or from equivalence principle arguments.

A (general) way out is to use the bulk-boundary correspondence, to be developed in the next chapter, to obtain this normalization and the value of the coupling g (that turns out to be $g = 1$). This fixes the value of the couplings in the NS sector (corresponding to the AdS_3 space). It provides still a nontrivial test since the black-hole solution we are analyzing is in the R sector.

Another point of worry is that we have used the free orbifold CFT to perform several of our calculations, although we have argued that the true SCFT describing the D-brane bound state is a blowup of this one. The reason that this works is that several of the quantities that we have computed turn out to be moduli independent. This can be shown both from the point of view of the SCFT and that of the bulk supergravity.

The story we have presented can be generalized to four-dimensional near-extremal black holes with a similar success. Angular momentum may also be added.

Another recent development concerns the entropy counting of "small black holes." These are black holes carrying two electric charges, and which, according to our discussion have a singular horizon (zero area) and therefore zero entropy. This is unfortunate since the microscopic realization of such black holes involves perturbative (BPS) string states. Unlike the D-brane bound states discussed in this chapter, such perturbative states have multiplicities that are much easier to count. Although the classical horizon of such black holes is of zero area, one could expect that stringy effects would increase it to a size of the order of the string length. This expectation turns out to be correct, if one includes the higher-derivative corrections to the two-derivative string effective action. The gravitational entropy can again be calculated via a higher derivative generalization of the area formula. It turns out that the microscopic and macroscopic entropies again match properly.

It seems that we come close to understanding the microscopic nature of black holes. In some sense, they are described as elementary bound states in the theory, in a limit where their geometry is semiclassical. This picture is expected and in that sense it is satisfying. Moreover, it indicates that the black-hole evaporation process is described by an S-matrix and the thermal nature of the outgoing radiation is due to the large number of final states.

One question that still remains obscure, concerns the transition from the macroscopic to the microscopic description. In particular, what replaces the notion of the horizon in the microscopic description? Is there a phase transition (horizon formation) going from weak to strong coupling? It seems that understanding this will be a major step towards resolving the information paradox in black-hole physics.

Bibliography

There are many excellent books and lecture notes on the physics of classical black holes and the associated thermodynamics. Excellent introductory books are the ones by Schutz [371] and 't Hooft [372]. More advanced books are Carroll's [373] and Wald's [374]. Townsend's lectures [375] give a more rigorous introduction to the basic black-hole properties. The lectures of Jacobson [376] constitute an excellent introduction to the basics of black-hole thermodynamics. Hawking's calculation of the black-hole evaporation rate can be found in [377]. The calculation of the Hawking temperature from the Euclidean continuation can be found in [378]. The book [379] provides an introduction to the modern description of black holes, the information paradox and the holographic hypothesis.

Concise reviews of the information paradox are [380,381]. A review of the status of various singularities in string theory can be found in [382].

An extensive review of the microscopic calculation of black-hole entropy and Hawking radiation in string theory can be found in [383,384]. It also contains a comprehensive guide to the relevant literature. The thesis [385] is also a nice introduction into the subject. Other useful reviews include [207,208].

The relevance of the near-horizon region to the entropy counting is described in [386,387]. The detailed geometry of the BTZ black hole can be found in [388]. The appropriate boundary conditions on the fermions in this background are analyzed in [389].

The extremal five-dimensional black holes and the associated microstate counting can be found in [390,391]. The nonextremal five-dimensional black hole is described in [392]. The related nonextremal counting can be found on [393]. Four dimensional black holes have been constructed and the multiplicities were counted in [394,395]. Helicity supertraces, relevant for the correct counting of BPS states are described in appendix J. More detailed formulae can be found in appendix E of [19].

The higher-derivative terms in string theory, induce corrections to the Bekenstein-Hawking formula for the entropy. Such corrections were computed for the leading R^2 terms. They matched properly the finite-size corrections to the microscopic black-hole entropy [396].

The semiclassical emission rates for minimal scalars and the related ones from D-branes can be found in [397–399]. A review of comparisons in the case of fixed scalars can be found in [400].

Bound states of branes as due to world-volume instantons have been described in [401,402]. The IR CFT of the Higgs branch in two dimensions is analyzed in [403,404]. The Coulomb and Higgs branches of the D_1-D_5 system are further analyzed in [405] where the implications of the flat directions for the unbound system are analyzed for the IR CFT. The relation between the instanton moduli space and the Hilbert scheme is discussed in [406].

A discussion of the transition/matching between the microscopic and macroscopic regime can be found in [407]. There is shown that microscopic entropies of highly excited string states and black-hole entropies match in order of magnitude when the curvature is of the order of the string scale.

The recent developments on the entropy matching of small black holes can be accessed by starting from [408,409]. An alternative view of black holes, motivated from ideas of AdS/CFT has been proposed. This is reviewed in [410]. In turn this motivated the study of BPS solutions in AdS, and their associated dynamics [411].

We have not addressed in this book the thermodynamics of strings and the associated Hagedorn transition. A pedestrian introduction can be found in [12]. In [412] a review is presented as well as a guide to related literature.

Exercises

12.1. Consider the metric

$$ds^2 = -F(r)C(r)dt^2 + \frac{dr^2}{C(r)} + H(r)r^2 d\Omega_2^2. \tag{12.1E}$$

$C(r)$ vanishes at $r = r_0$, while all other functions are nonzero and positive for $r \geq r_0$. For asymptotically flat solutions they all asymptote to 1 as $r \to \infty$. Therefore, $r = r_0$ is a horizon. Use the Euclidean rotation method to calculate the Hawking temperature for this solution and show that

$$\frac{1}{T_H} = \beta = \frac{4\pi}{C'(r_0)\sqrt{F(r_0)}}. \tag{12.2E}$$

Is this formula valid for higher-dimensional black holes?

12.2. The area theorem places interesting constraints on classical black-hole processes:

- Two almost stationary Schwarzschild black holes, initially far apart, with masses $M_{1,2}$ form a single black hole radiating away some energy E. Show that the efficiency of energy extraction $E/(M_1 + M_2)$ is bounded above by $1 - \frac{1}{\sqrt{2}}$.

- Two extremal RN black holes each of mass M, collide and create a single charge zero black hole of mass M_f. Show that at most half of the original energy can be radiated away during this process.

- Show that a Schwarzschild black hole cannot decay into two Schwarzschild black holes plus radiation.

12.3. Derive equations (12.3.17), (12.3.18), and (12.3.21) on page 374.

12.4. Show that the specific heat of Schwarzschild black holes is negative. This implies that they are thermodynamically unstable. On the other hand show that RN black holes close to extremality are thermodynamically stable.

12.5. Show that the Kerr-Newman (KN) metric

$$ds^2 = -\frac{\Delta - a^2 \sin^2\theta}{\Sigma} dt^2 - \frac{2a \sin^2\theta (r^2 + a^2 - \Delta)}{\Sigma} dt d\phi$$

$$+ \frac{\Sigma}{\Delta} dr^2 + \Sigma d\theta^2 + \frac{(r^2 + a^2)^2 - \Delta a^2 \sin^2\theta}{\Sigma} \sin^2\theta d\phi^2,$$

$$\Sigma = r^2 + a^2 \cos^2\theta, \quad \Delta = r^2 + a^2 + Q^2 - 2GMr, \tag{12.3E}$$

is a solution of the Einstein-Maxwell theory (F.1) on page 519 with mass M, electric charge Q, and angular momentum $J = Ma$. Find the BH entropy and Hawking temperature as well as the other thermodynamic variables and write down the first law. When do we obtain an extremal black hole? When is the KN black hole thermodynamically stable?

12.6. Show how (12.3.23) on page 375 is modified if the particles emitted from the KN black hole have spin s and charge q.

12.7. Consider 11-dimensional supergravity compactified on T^6 to five dimensions. Find the solution describing Q_1 M_2 branes stretching along $x^9 - x^{10}$, Q_2 M_2 branes stretching along $x^7 - x^8$, and Q_3 M_2 branes stretching along $x^5 - x^6$. By descending the solution to IIA theory and T-dualizing, show that it is the same solution as the one described in section (12.5) on page 377.

12.8. Show that (12.5.1)–(12.5.3) on page 378 is a solution of the equations of motion coming from the IIB action

$$S = \int d^{10}x \sqrt{g}\left[e^{-2\Phi}(R + 4(\partial\Phi)^2) - \frac{1}{12}F^2\right],$$

(12.4E)

where $F_{\mu\nu\rho}$ is the field strength of the R-R two-form. The axion, NS-NS two-form, and self-dual four-form have been set to zero.

12.9. Consider the extremal solution (12.5.1)–(12.5.4) on page 378. Calculate the D_1 and D_5 charges and show that they are N_1 and N_5, respectively. Show also (12.6.4)–(12.6.6) on page 380.

12.10. Consider the possibility of constructing a controllable four-dimensional black hole from bound states of branes, in analogy with the five-dimensional case. You will need to stabilize one more compact dimension, by introducing an extra ingredient. You may want to check your guess against [385].

12.11. Derive the Hawking temperature (12.6.15) on page 381 of the non-extremal solution in section 12.6 on page 379.

12.12. Consider the nonextremal RN solution in section 12.6 on page 379. Show that the first law of thermodynamics

$$dM = TdS + \sum_{i=1,5,p} \mu_i dQ_i$$

(12.5E)

is valid with $\mu_i = \frac{\pi}{4G_5}c_i \tanh a_i$. Note that the variations do not involve arbitrary changes of the $N_{\pm i}$.

12.13. By choosing appropriate coordinates show that the BTZ black holes are orbifolds of AdS_3 by discrete subgroups of $SO(2,2)$.

12.14. Consider the IIB effective action in ten dimensions, and keep the fields that participate in the D_1-D_5 black hole, namely, the metric, dilaton, and R-R two-form. Dimensionally reduce to five dimensions on $T^4 \times S^1$ and show that after the standard KK redefinitions, the effective action in the Einstein frame is

$$(16\pi G_5) S_5 = \int d^5x \sqrt{-g} \left[R - \frac{4}{3}(\partial\phi)^2 - \frac{1}{4} G^{\alpha\beta} G^{\gamma\delta} (\partial G_{\alpha\gamma} \partial G_{\beta\delta} + e^{2\phi} \partial C_{\alpha\beta} \partial C_{\beta\delta}) \right.$$
$$- \frac{e^{-4\phi/3}}{4} G_{\alpha\beta} F^{\alpha}_{\mu\nu} F^{\beta,\mu\nu} - \frac{e^{2\phi/3}}{4} \sqrt{G} G^{\alpha\beta} H_{\mu\nu\alpha} H^{\mu\nu}_{\beta}$$
$$\left. - \frac{e^{4\phi/3}}{12} H_{\mu\nu\rho} H^{\mu\nu\rho} \right], \tag{12.6E}$$

where $\alpha, \beta, \gamma, \ldots$ refer to the compact space, $\mu, \nu, \ldots$ to the noncompact space, and H is the field strength of the R-R two-form. Derive the equations of motion and show that (12.6.10) on page 380 accompanied by the appropriate gauge fields is a solution.

12.15. Consider the fluctuation of $G_{\alpha\beta}$ around the black-hole metric with α, β in the T^4 directions, (6,7,8,9), $G_{\alpha\beta} = \sqrt{\frac{H_1}{H_5}} (\delta_{\alpha\beta} + h_{\alpha\beta})$ with h traceless. Use (12.6E) to find its effective action and show that $h_{\alpha\beta}$ is a collection of minimal scalars only.

12.16. Parameterize $G_{66} = \cdots = G_{99} = e^{2\nu}$, $G_{55} = e^{2\nu_5}$ and define the linear combinations, $\tilde{\phi} = \phi + \frac{1}{2}\nu_5$, $\lambda = \frac{3}{4}\nu_5 - \frac{1}{2}\phi$. Starting from (12.6E) show that their effective action is

$$S \sim \int d^5x \sqrt{-g} \left[R - (\partial\tilde{\phi})^2 - \frac{4}{3}(\partial\lambda)^2 - 4(\partial\nu)^2 \right.$$
$$\left. - \frac{e^{8\lambda/3}}{4} (F^5_{\mu\nu})^2 - \frac{e^{-(4/3)\lambda+4\nu}}{4} H^2_{5,\mu\nu} - \frac{e^{(4/3)\lambda+4\nu}}{12} H^2_{\mu\nu\rho} \right], \tag{12.7E}$$

where $F^5_{\mu\nu}$, $H_{5,\mu\nu}$, $H_{\mu\nu\rho}$ are the three nontrivial field strengths of the D_1-D_5 black-hole solution. Because of the potential terms, the values of these scalars at the horizon are fixed by the equations of motion. These scalars are therefore "fixed scalars."

12.17. Consider the matching of the solutions of (12.8.2) on page 383 without neglecting the outgoing wave, and show that they give $B/A \ll 1$.

12.18. Calculate the constant K in equation (12.8.21) on page 386 by integrating both sides on S^3 and keeping from the resulting Bessel function the ingoing part.

12.19. Consider a generic black hole in any dimension $d > 3$. By setting up the s-wave scattering problem for minimally coupled scalars prove that (12.8.24) on page 386 is always valid.

12.20. Consider the extremal D_3-brane solution studied in section 8.8 on page 205. Show that the dilaton fluctuation $\delta\Phi$ around that solution satisfies $\Box\delta\Phi = 0$. It is therefore a minimal scalar. Calculate the semiclassical absorption cross section for low energies $\omega L \ll 1$ where L is the throat size. Consider also s-waves and show that

$$\sigma_{\text{abs}} = \frac{\pi^4 L^8 \omega^3}{8}. \tag{12.8E}$$

Verify that the main characteristics of this cross section depend only on the near-horizon region. How does σ_{abs} scale with the energy for higher partial waves? What is the Hawking emission rate?

12.21. Consider the orbifold $(T^4)^N/S_N$. Compute directly the dimension of its cohomology ring for low values of N and verify that it is given by $d(N)/256$, where $d(N)$ is defined in (12.9.4) on page 390. A general proof can be found in section 4 of [413].

12.22. Consider a unitary CFT with central charge c and a discrete spectrum . By using saddle point methods, derive (12.9.14) on page 392. By calculating the determinant, derive also the next-to-leading correction.

12.23. Consider a (supersymmetric) gas of n_f free massless two-dimensional left-moving particles on a circle of radius R and at inverse temperature β. Show that the leading contribution to their partition function is given by

$$Z = \sum \exp\left[-\beta \sum_r \frac{n_r r}{R}\right] = \prod_{n=1}^{\infty} \frac{1}{(1 - e^{-(\beta/R)n})^{3n_f/2}}. \tag{12.9E}$$

Relating this to the η-function and using its modular properties show equation (12.9.19) on page 394. Show that the fermionic contributions that will appear in the numerator of (12.9E) correct the subleading logarithmic term in (12.9.19).

13 | The Bulk/Boundary Correspondence

There is a long suspected connection between confining gauge theories and string theory. In its earliest incarnation, string theory was invented to explain the properties of the strong interaction. It turned out shortly afterwards that $SU(3)$ gauge theory (QCD) seemed to explain well the properties of the strong interactions, and by now this evidence is overwhelming.

The perturbative gauge theory description is very well suited for the high-energy regime of the theory. On the other hand, the theory becomes strongly coupled, and therefore difficult to handle at low energies. In physical processes involving the strong interactions, there is always a part of the process (typically it is hadronization) which involves low-energy strongly coupled physics. There is a successful factorization of the hard and soft parts of the process. QCD can be used to calculate the hard parts of the amplitudes. The soft parts however have remained beyond the reach of perturbative techniques and are typically computed via a combination of resummation techniques and phenomenological parametrizations. The most useful phenomenological parametrizations used to date are variants of the simple string theory picture.

That a sort of string theory may describe the low-energy regime of a confining gauge theory was suspected early on. This emerged out of the fact that there is always flux connecting a quark-antiquark pair, and that confinement suggests that this flux will be squeezed into a tube with a typical thickness given by the dimensionfull scale of QCD. Formally, the flux tube is described by an associated Wilson line in the underlying gauge theory. The dynamics of a meson (quark-antiquark pair) is expected to be described, at sufficiently low energy by the dynamics of a string with two massive quarks at the end. On the other hand, a glueball, would look more like a closed tube of flux, namely, like a closed string.

A kind of duality is therefore expected to describe the physics of a confining gauge theory. At high enough energy perturbative gauge theory is a weakly coupled description of the dynamics, while at low energy where this description fails, a dual string theory description

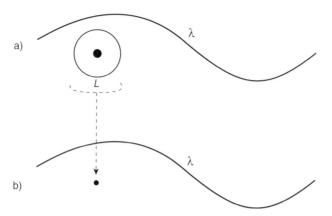

Figure 13.1 (A) Scattering of a large wavelength λ probe from a D-brane. When $\lambda \gg L$ we may approximate the brane as point-like in transverse space as in (B). The interaction with the brane is described in this regime by the world-volume gauge theory. The symbol for the wavelength λ here should not be confused with the symbol λ for the gauge theory 't Hooft coupling used in the text.

is suitable and weakly coupled. Finding this weakly coupled low-energy description for QCD has been a "holy grail" for modern theoretical physics.

In the previous sections of this book, we have seen another relationship emerge between gauge theories and gravity/string theory. The appropriate medium consists of D-branes. On the one hand, these are solitonlike objects of string theory, with long range gravitational interactions. On the other hand, we have a perturbative description of their fluctuations and dynamics in terms of open strings that at low energy reduce to gauge theories. Moreover, as we argued in section 5.3.6 on page 140 the dynamics of the associated closed and open strings is tightly intertwined together and open string amplitudes contain information on closed string interactions.

In section 8.8.3 on page 208 we have analyzed in detail the various regimes associated with p-branes. We will focus here on extremal D$_3$-branes to make the point.

We have seen, that the supergravity solution has a characteristic length scale, the throat size L. For distances larger than the throat size the fields of the D-brane are those of a pointlike object. Therefore, if we are considering probes with wavelengths that are much larger than L, to them the D$_3$-brane appears as a point-like object. Its internal structure can be described (at low energy) by the world-volume $\mathcal{N} = 4$ four-dimensional U(N) gauge theory. When the 't Hooft coupling of the world-volume theory is small, $\lambda \ll 1$, then $L \ll \ell_s$ and all stringy probes cannot penetrate and sample the throat region.

The conclusion is that when $\lambda \ll 1$, the world-volume theory is weakly coupled, all string probes feel the regions where curvatures are very weak and therefore all bulk fields (gravitational and others) can be neglected. This is the regime where the Polchinski description of branes in flat space is reliable. Figure 13.1 summarizes this regime.

On the other hand, when $\lambda \gg 1$, the world-volume theory is strongly coupled, and therefore difficult to use. In this case $L \gg \ell_s$, and the throat region is macroscopic. Moreover, the curvature in this region is of order $1/L^2$ and this is much smaller than $1/\ell_s^2$. It is therefore a region well described by the two-derivative effective supergravity action. Moreover, if we also have $g_s \ll 1$ then we can also neglect closed string loops.

The upshot of this discussion is that in the $\lambda \gg 1$ regime, although the gauge theory description of the branes is difficult to analyze, the gravitational description is both well behaved and covers the essential features close to the branes. We can therefore consider it as an equivalent (dual) description of the branes. From the gauge theory point of view, this is a weakly coupled dual description of the strongly coupled $\lambda \gg 1$ gauge theory.

The purpose of this chapter is to make this observation precise and eventually generalize it.

13.1 Large-N Gauge Theories and String Theory

The appropriate limit, where the low-energy effective string description of a gauge theory may be weakly coupled was first indicated by 't Hooft.

A pure four-dimensional SU(N) gauge theory has, apart from the standard dimensionless coupling constant g_{YM} (and its associated dimensionful scale Λ_{QCD}), another dimensionless parameter, namely the number of colors, N. An interesting limit is when the number of colors becomes large, $N \to \infty$.

It is important to know how to scale the gauge coupling constant when we take the large-N limit. The natural choice is that g_{YM} should be scaled so that the QCD mass scale Λ_{QCD} should remain finite. From the one-loop β-function

$$\mu \frac{dg_{YM}}{d\mu} = -\frac{11}{3} N \frac{g_{YM}^3}{(4\pi)^2} + \mathcal{O}(g_{YM}^5), \tag{13.1.1}$$

we may find the appropriate scaling by demanding that the leading terms are of the same order. This indicates that the combination

$$\lambda = g_{YM}^2 N \tag{13.1.2}$$

(known as the 't Hooft coupling) should be kept fixed as we send $N \to \infty$. This large-N limit is known as the 't Hooft limit. Adding fermions and scalars in the adjoint does not change the scaling above.

In an asymptotically-free gauge theory, keeping λ constant assures that the dimensionfull scale Λ_{QCD} is kept constant. When the theory is asymptotically conformal, then this is not necessary anymore, as we will see in the sequel.

The structure of large-N perturbation theory and its connection to gauge theory is a characteristic of any theory with fields transforming as adjoints of SU(N).[1] To make this

---

[1] They could also be symmetric or antisymmetric representations. One may also add up to an $\mathcal{O}(N)$ number of fundamentals. The structure of large-N perturbation theory is similar, with minor changes that reflect these new features.

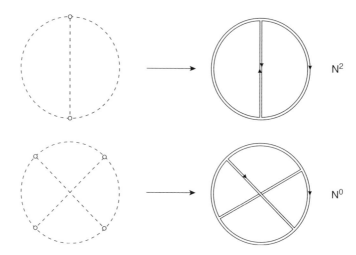

Figure 13.2 On the left are diagrams of adjoint fields (dashed lines) with interactions (circles). On the right the same diagrams in the 't Hooft double-line notation.

structure explicit, we will denote the adjoint fields by X_i and we will suppress their Lorentz indices. We will write schematically their action as

$$S = \text{Tr}[dX_i dX_i + g_{YM} c^{ijk} X_i X_j X_k + g_{YM}^2 d^{ijkl} X_i X_j X_k X_l + \cdots]. \tag{13.1.3}$$

Rescaling the fields as $X_i \to X_i/g_{YM}$ we obtain the action in the form

$$S = \frac{1}{g_{YM}^2} \text{Tr}[dX_i dX_i + c^{ijk} X_i X_j X_k + d^{ijkl} X_i X_j X_k X_l + \cdots], \tag{13.1.4}$$

where the full dependence on g_{YM} is solely in front of the action.

We may now estimate the behavior of correlation functions when $N \to \infty$ and λ is kept fixed. Since $\frac{1}{g_{YM}^2} = \frac{N}{\lambda}$ this seems equivalent to the classical limit. That would be the case were it not for the fact that the number of components of the fields X_i is becoming infinite in that limit.

A very convenient way to rewrite the Feynman diagrams involving the adjoint fields is the 't Hooft double-line notation (see figure 13.2). This substitutes for each line representing an adjoint field, a double line with opposite orientations, corresponding to the fundamental and antifundamental indices hidden in the adjoint. The interaction vertices are consistent with this notation in a U(N) theory. In an SU(N) theory there is a mixing term that is subleading in N, because

$$\langle X_i^{a\bar{b}} X_j^{c\bar{d}} \rangle \sim \delta^{a\bar{d}} \delta^{b\bar{c}} - \frac{1}{N} \delta^{a\bar{b}} \delta^{c\bar{d}}. \tag{13.1.5}$$

Therefore, any diagram of the adjoint fields, can be rewritten in terms of double lines. Such a double line notation turns each diagram into a closed oriented two-dimensional surface. Any closed loop is the face of a simplicial decomposition of the surface. A propagator (double line) is an edge in the simplicial decomposition, and an interaction vertex is a simplicial vertex on the surface. The surface becomes compact if we add to our space the point at infinity.

To estimate the factors of N and λ that are associated with a diagram, we must recall that according to (13.1.4) each interaction vertex brings a factor of N/λ, each propagator a factor of λ/N and each closed loop, a factor of N, because of the summation over N colors.

We therefore find that a vacuum diagram with V vertices, E propagators (edges), and F loops (faces) will have a coefficient proportional to

$$\left(\frac{\lambda}{N}\right)^E \left(\frac{N}{\lambda}\right)^V N^F = N^\chi \lambda^{E-V}, \tag{13.1.6}$$

where $\chi = V - E + F$ is the Euler number of the closed surface, computed from its simplicial decomposition. For compact closed surfaces, $\chi = 2 - 2g$ where g is the number of handles. Therefore, the standard perturbative expansion for any correlator can be written at large N (and small λ) as

$$\sum_{g=0}^{\infty} N^{2-2g} \sum_{i=0}^{\infty} c_{i,g} \lambda^i \equiv \sum_{g=0}^{\infty} N^{2-2g} Z_g(\lambda). \tag{13.1.7}$$

In the large-N limit the result is dominated by the surface with the minimal number of handles, typically a sphere. Such diagrams are called planar diagrams since they can be drawn on a plane (adding also the point at infinity this becomes topologically a sphere). Nonplanar diagrams correspond to surfaces with handles and are suppressed by additional factors of $1/N^2$.

The form of the expansion in (13.1.7) is very reminiscent of the topological expansion of string theory if the role of the string coupling constant is played by $1/N$. In particular this shows that the string theory will be weakly coupled at large N.

You are asked to analyze the large-N scaling of nontrivial correlators in exercises 13.3 and 13.5. There you will find that the three-point interaction of closed string states (glueballs) vanishes as $1/N$, while that of mesons as $1/\sqrt{N}$. This agrees well with the identification $g_s = 1/N$. Moreover, it shows that at $N = \infty$, the theory contains an infinite tower of absolutely stable single-particle states. They are expected to coincide with the various Regge trajectories of string theory.

This map is not yet rigorous. Perturbative expansions typically do not converge and therefore the identification is formal. Moreover, we obtain very little information on the nature of the string theory that, at least naively, should live in four dimensions. However we now know that this string theory should be subcritical and therefore the Liouville mode does not decouple. It could be therefore expected that the effective string theory is five dimensionals if we interpret the Liouville scalar as an extra coordinate in the spirit of section 6 on page 144. As we will see later in this chapter this expectation will turn out to be essentially correct.

As stated earlier, all nontrivial connected correlators vanish in the $N = \infty$ limit. This in particular implies that for local single-trace operators defined in (13.1E) on page 464

$$\left\langle \prod_{a=1}^{n} \Phi_a(x_a) \right\rangle = \prod_{a=1}^{n} \langle \Phi_a(x_a) \rangle + \mathcal{O}(1/N^2), \tag{13.1.8}$$

where we have displayed explicitly the space-time dependence of operators. Moreover, since the theory is Poincaré invariant, the one-point functions are independent of the points x_a. Therefore, to leading order the full result is factorized and independent of the insertion points. This indicates, that in the path integral formalism, the large-N limit is dominated by a single configuration (saddle point) that is moreover translationally invariant. This is known in the literature as the *master field*. We will see further below, in the context of AdS/CFT, that the notion of the master field takes an interesting (and unexpected) form.

13.2 The Decoupling Principle

In this section we will motivate and describe the decoupling limit, that is essential in AdS/CFT correspondence.

We begin by considering a number N of parallel coincident D_3-branes inside the ten-dimensional space-time. They are extended along a (3+1)-dimensional hyperplane and they are located at the same point of the transverse six-dimensional space. In this background, the theory contains two types of excitations: open strings and closed strings. The open strings are the excitations of the D_3-branes. The closed strings are the excitations of the ten-dimensional bulk space.

The low-energy description of the closed string excitations at energies below the string scale M_s is given by the IIB supergravity. Similarly, the low-energy description of the open degrees of freedom is given by $\mathcal{N} = 4$, U(N) super Yang-Mills theory. Both low-energy theories have corrections due to the exchange of the stringy modes. These are the higher-derivative corrections that we have discussed earlier. In the closed-string sector they include the quartic curvature terms while in the open string sector the higher-order terms of the DBI action. However, at energies small compared to M_s the effect of such terms is small and they can be treated perturbatively.

We may write the low-energy Lagrangian of the theory schematically as

$$S = S_{\text{bulk}} + S_{\text{brane}} + S_{\text{interaction}}, \qquad (13.2.1)$$

where the first and second terms contain the interactions of the closed and open string modes among themselves, while the third term contains the open-closed interactions. The leading such interaction terms have been obtained in section 8.5 on page 197 by covariantizing the DBI action.

We may now expand the bulk action around the free point, in powers of the gravitational constant κ. For example for the metric we expand as $g_{\mu\nu} = \eta_{\mu\nu} + \kappa h_{\mu\nu}$ to obtain

$$S_{\text{bulk}} \sim \frac{1}{2\kappa^2} \int d^{10}x \, \sqrt{g} \, R + \cdots \sim \int d^{10}x \, \left[(\partial h)^2 + \kappa \, h \, (\partial h)^2 + \cdots \right], \qquad (13.2.2)$$

where we have not explicitly indicated all the massless bulk fields for simplicity. The crucial observation here is that all interactions of closed string modes are proportional to positive powers of the gravitational constant κ. Consequently, at low energies, these interactions become weaker. This is equivalent to the fact that gravity is IR-free.

Doing the same expansion in the interaction Lagrangian we again obtain positive powers of κ only, as it befits gravitational interactions with sources.

$$S_{\text{interactions}} \sim \int d^4x \; \sqrt{g} \; \text{Tr}[F^2] + \cdots \sim \kappa \int d^4x \; h_{\mu\nu} \; \text{Tr}\left[F_{\mu\nu}^2 - \frac{\delta_{\mu\nu}}{4}F^2\right] + \cdots, \qquad (13.2.3)$$

where we have indicated the kinetic terms for the gauge bosons only for simplicity.

To obtain the low-energy limit we may take all energies to be small. We might also equivalently keep the energy fixed but take the characteristic scale of the theory $\ell_s \to 0$ keeping all other dimensionless parameters (like N or g_s) finite. This second version is more convenient for book-keeping purposes and this is what we will do from now on.

In this limit $\kappa \sim g_s \, \ell_s^4 \to 0$ as advertised, and all interaction terms in S_{bulk} as well as $S_{\text{interaction}}$ go to zero. Moreover, the higher derivative terms in S_{brane}, being proportional to positive powers of ℓ_s also vanish in this same limit. We end up with free bulk supergravity, and $\mathcal{N} = 4$, U(N) super Yang-Mills, not interacting with each other. $\mathcal{N} = 4$, U(N) super Yang-Mills is a very special four-dimensional theory: it is a scale invariant theory, and in fact a four-dimensional CFT.

13.3 The Near-horizon Limit

We would like now to consider the same system from a different point of view. D$_3$-branes act as sources to the bulk fields. We will therefore start from the background fields generated by the presence of the N D$_3$-branes in the previous section. These were presented in section 8.8 on page 205 and we reproduce here (8.8.4) for the special case of N D$_3$-branes,

$$ds^2 = H^{-1/2}(-dt^2 + d\vec{x} \cdot d\vec{x}) + H^{1/2}(dr^2 + r^2 d\Omega_5^2). \qquad (13.3.1)$$

The dilaton is constant while the self-dual four-form is given by

$$C_{0123} = 1 - \frac{1}{H}, \quad H = 1 + \frac{L^4}{r^4}, \quad L^4 = 4\pi g_s \ell_s^4 N, \qquad (13.3.2)$$

where we have also used (8.8.25).

Since the metric component g_{tt} is not constant, the energy measured is r dependent. In particular if we measure energy E_r at some position with radius r, then the energy that an observer at infinity would measure is

$$E_\infty = H^{-1/4} \, E_r, \qquad (13.3.3)$$

due to the red-shift. An object moving towards $r = 0$ would appear to have lower and lower energy to an observer at $r = \infty$.

We may now investigate the low-energy limit in this context. From the point of view of an observer at infinity there are two types of low-energy excitations in this background. The first type is the standard massless large-wavelength excitations that propagate in the whole bulk. The second, is any type of excitation that approaches $r = 0$ (the near-horizon region). As we argued above, any such excitations will have low energy due to the red-shift.

At low energies, these two types of excitations decouple from each other. The massless bulk excitations interact weakly with the near-horizon region. You were asked to compute this in exercise 12.20 and show that the coupling behaves as $\sim L^8 \omega^3$. It therefore vanishes at low energy as the third power of the energy. Moreover, the near-horizon modes, must climb a potential barrier very similar to the one in figure 12.1. At low energy, the modes on the two sides of the barrier cannot interact.

We conclude that in this description, at sufficiently low energy, we end up with two sets of non-interacting modes: Low-energy modes propagate far away from the branes, where the space is essentially flat. This is again described by the free limit of the IIB supergravity. There is also another set of modes confined in the near-horizon region. This region is obtained by going close to $r = 0$. This limit was already derived in the end of section 8.8.3 on page 208 and we reproduce the metric here for convenience

$$ds^2 = \frac{L^2}{r^2} dr^2 + \frac{r^2}{L^2}(-dt^2 + d\vec{x} \cdot d\vec{x}) + L^2 \, d\Omega_5^2. \tag{13.3.4}$$

We may simplify it further by changing radial coordinate $u = L^2/r$ to obtain

$$ds^2 = \frac{L^2}{u^2} \left[du^2 - dt^2 + d\vec{x} \cdot d\vec{x} \right] + L^2 \, d\Omega_5^2. \tag{13.3.5}$$

Its first part describes the highly symmetric space, anti–de Sitter space AdS$_5$ in Poincaré coordinates while the second the five-sphere S^5. The boundary of AdS$_5$ in these coordinates is at $u = 0$. The geometry of AdS$_5$ and its relation to four-dimensional conformal symmetry is detailed in appendix K on page 541.

In both descriptions, we found that in the decoupling (low-energy) limit the same system was a sum of two noninteracting systems. In both descriptions, one of the two systems was free IIB supergravity. It is natural therefore to expect, that the two remaining systems are equivalent. This way of thinking leads to *the conjecture that $\mathcal{N} = 4 \, U(N)$ super Yang-Mills theory is equivalent to IIB supergravity in AdS$_5 \times S^5$*. This may seem surprising at first. However, as we will see in the sequel, this seems to fit with the general ideas of the correspondence of large-N gauge theories and string theory.

13.4 Elements of the Correspondence

The correspondence we have just described comes under the name of AdS$_5$/CFT$_4$ correspondence, or AdS/CFT for short. As we will argue later, it is a special case of the more general bulk/boundary correspondence.

We would like now to describe the AdS/CFT correspondence in more detail. Consider first the limit on the energy. We are taking $\ell_s \to 0$ in the brane theory. We must keep fixed in the near-horizon region the energy in string units $\ell_s E_r$. In this way we may consider arbitrary string excitations there. Since

$$E_\infty \sim E_r \frac{r}{\ell_s} = (E_r \ell_s) \frac{r}{\ell_s^2}, \tag{13.4.1}$$

we must also keep fixed E_∞ since it is the energy as measured in the field theory. Moreover, by keeping it fixed as we go close to the horizon, we achieve the original goal of having small local energy for all such modes. Therefore we must take $r \to 0$ keeping $U = \frac{r}{\ell_s^2}$ fixed. In this coordinate, that has dimensions of energy, the near-horizon metric is

$$ds^2 = \ell_s^2 \left[\frac{U^2}{\sqrt{4\pi g_s N}} (-dt^2 + d\vec{x} \cdot d\vec{x}) + \sqrt{4\pi g_s N} \frac{dU^2}{U^2} + \sqrt{4\pi g_s N} d\Omega_5^2 \right], \quad U = \frac{r}{\ell_s^2}. \tag{13.4.2}$$

If we put an extra D_3-brane at distance r from the original stack of branes, then the ground-state energy of the string stretched between the two sets of branes would be $U = r/\ell_s^2$. We are therefore keeping this energy finite as we go in the near-horizon area.

A first step is to compare the global symmetries of the two descriptions. $\mathcal{N} = 4$ SU(N) SYM theory is conformal. Therefore, its global bosonic symmetries are generated by the conformal group SO(2,4) and the R-symmetry group SO(6). These are exactly the Killing symmetries of the $AdS_5 \times S^5$ background of the dual string theory.

The SYM theory also has fermionic symmetries. The usual $\mathcal{N} = 4$ supersymmetry implies the conservation of 16 supercharges, the supersymmetry preserved by the N D_3-branes. However, conformal invariance makes the symmetry larger. The superconformal group has an extra sixteen supercharges to complete a total of 32. This enhanced supersymmetry appears in the near-horizon limit. Indeed, string theory in the $AdS_5 \times S^5$ background is invariant under 32 supercharges. Therefore, the gauge theory and the string theory have the same (super)symmetries.

There is another point that we should clarify. A U(N) gauge theory is equivalent to a free supersymmetric U(1) theory and an SU(N) theory. There is only a correlation between the U(1) charges and U(N) representations. The U(1) factor represents the free motion of the position of the D_3-branes in the ten-dimensional space. When we consider the gravitational theory on $AdS_5 \times S^5$ all modes are coupled to gravity. Therefore the U(1) part of the gauge theory is not described by the bulk string theory on $AdS_5 \times S^5$. It is part of the free degrees of freedom that have been thrown away together with the bulk free supergravity during the decoupling limit.

The U(1) degrees of freedom can however be included in the $AdS_5 \times S^5$ theory as degrees of freedom living on th boundary of AdS_5. Such fields are known as "singletons" in the AdS_5 theory and correspond to very special (short) representations of the conformal group SO(2,4).

In the previous discussion the field theory was defined on $\mathbb{R}^{1,3}$. However we may also consider it as defined on $\mathbb{R} \times S^3$ by doing a conformal transformation and redefining its Hamiltonian. We have seen that the isometries of AdS_5 are in one-to-one correspondence with the generators of the conformal group. Therefore the AdS_5 generator of global time translations can be implemented via the appropriate conformal generator. This new definition is useful since there is no horizon in conformal coordinates. Also note that, on the field theory side, there is no Coulomb branch. On $\mathbb{R}^{1,3}$ the six massless scalars of $\mathcal{N} = 4$ SYM have flat directions that parameterize the Coulomb branch. However on S^3, since the scalars are conformally coupled via the $\int d^4x \, R \, \text{Tr}[\phi^2]$ coupling, they acquire a mass proportional to the constant curvature of S^3. This avoids IR divergences in the CFT.

When the gauge theory is defined over $\mathbb{R} \times S^3$, there is a discrete spectrum and a gap. The appropriate AdS$_5$ coordinates are the global coordinates. When the gauge theory is defined on $\mathbb{R}^{1,3}$ then the spectrum is continuous without gap. The appropriate AdS$_5$ coordinates are the Poincaré coordinates. They have a horizon, in agreement with the fact that the energy spectrum goes down to zero.

The number of colors N of the gauge theory appears as the four-form flux in the string theory

$$\frac{1}{2\kappa_{10}^2} \int_{S^5} F_5 = T_3 N. \tag{13.4.3}$$

Also we know already from the analysis of the DBI action that the Yang-Mills coupling is related to the string coupling. Putting also together the θ-angle, this can be written as

$$\tau \equiv \frac{4\pi i}{g_{YM}^2} + \frac{\theta}{2\pi} = \frac{i}{g_s} + \frac{\chi}{2\pi}, \tag{13.4.4}$$

where χ is the expectation value of the R-R axion scalar of IIB string theory.

A note on the normalization of the YM coupling constant is in order here. In (8.6.3) on page 200 we have derived the YM action from the DBI action as

$$\mathcal{L}_{YM} = \frac{1}{4(2\pi g_s)} \text{Tr}[F_{\mu\nu} F^{\mu\nu}] = \frac{c}{4(2\pi g_s)} F_{\mu\nu}^a F^{a,\mu\nu}, \quad F_{\mu\nu} = F_{\mu\nu}^a T^a, \quad \text{Tr}[T^a T^b] = c\, \delta^{ab}. \tag{13.4.5}$$

The definition of the gauge coupling constant depends therefore on the normalization of the nonabelian generators

$$g_{YM}^2 = \frac{2\pi g_s}{c}. \tag{13.4.6}$$

A popular choice in gauge field theory is $c = 1/2$ and it is this that we are using in this chapter (but not in the rest of this book).

We saw in section 11.5 on page 331 that the type-IIB theory has an SL(2,$\mathbb{Z}$) symmetry that acts with fractional transformations on the parameter τ as defined in (13.4.4) in terms of the string coupling and R-R axion expectation value. This symmetry implies a similar SL(2,$\mathbb{Z}$) symmetry acting on the complex coupling constant of $\mathcal{N} = 4$ SYM. This is nothing else but the Montonen-Olive duality.

The characteristic parameters on the gauge theory side are all dimensionless, g_{YM}, θ, and N as befits a CFT. We described above their interpretation on the string theory side. We do however have one extra dimensionfull parameter on the string theory side, the string length ℓ_s. This sets all the scales in string theory. The fact that it does not appear directly on the gauge theory side is because the relevant quantities are dimensionless ratios, which are independent of ℓ_s. There are three basic scales in string theory on AdS$_5$, all of them dimensionfull. They are the string length ℓ_s, the (common) radius of AdS$_5$, L, and the ten-dimensional Newton's constant G_{10}. The two independent dimensionless ratios are related to the two dimensionless parameters of the SYM theory as

$$\frac{L^4}{\ell_s^4} = 4\pi g_s N = g_{YM}^2 N = \lambda, \quad \frac{2\kappa_{10}^2}{\ell_s^8} = 16\pi \frac{G_{10}}{\ell_s^8} = (2\pi)^7 g_s^2 = 2^3 \pi^5 \frac{\lambda^2}{N^2}. \tag{13.4.7}$$

It is convenient on the string theory side to scale out the radius of curvature of AdS$_5$ × S^5 and use the radius-1 metric. In exercise 13.7 you are asked to show that with this convention the Newton constant scales as N^{-2} and the string scale as $\ell_s \sim (g_s N)^{-1/4}$. This implies that string theory quantities that can be calculated at the two-derivative supergravity level, should depend only on N and not on the 't Hooft coupling. Stringy effects will be proportional to (positive) powers of $(g_s N)^{-1/4}$.

We will now investigate the (perturbative) validity of the two descriptions. On the string theory side, since $G_N \sim 1/N^2$, quantum effects are suppressed when N $\gg$ 1. Higher stringy corrections are also small if the curvature of the background is much smaller than the string scale, $L^4 \gg \ell_s^4$, which happens if $\lambda \gg 1$ where we used (13.4.7). We conclude that the strong 't Hooft coupling limit of the large-N gauge theory is described well by the two-derivative action of classical IIB supergravity on AdS$_5$ × S^5.

On the other hand, the large-N gauge theory is weakly coupled in perturbation theory, when $\lambda \ll 1$, as explained in section 13.1. We therefore conclude that this correspondence is a duality between the two different descriptions. When the gauge theory description is strongly coupled the string theory is weakly coupled and vice versa. Like other duality symmetries described earlier, this duality is very useful, giving information on strong-coupling dynamics. The flip side is that it is very difficult to establish, and typical tests will have to use supersymmetry in essential way in order to compare protected (BPS) quantities. It should be stressed however, that our previous arguments for the correspondence suggest that supersymmetry is not essential, although it simplifies comparison of the two sides.

It might seem that by taking g_s large we may access the gravitational regime $g_s N \gg 1$ while N $\sim \mathcal{O}(1)$. This is not true since if $g_s > 1$ we may use the SL(2,$\mathbb{Z}$) symmetry of IIB string theory $g_s \to 1/g_s$ to bring it to a value that is smaller than 1. This is necessary since in this regime the D-string is light and the standard description invalid. It is therefore necessary that N is large so that the gravitational regime is accessible. This can also be seen from $L^4/\ell_p^4 \sim$ N, and the fact that the validity of the gravitational description requires N $\gg$ 1.

A weak form of the conjectured AdS/CFT correspondence requires it to be valid for large N only. In its strong form the correspondence should be valid at any N. There is no contradiction to date with the strong form of the correspondence.

13.5 Bulk Fields and Boundary Operators

The natural objects in a CFT are local operators, as explained in detail in chapter 4. For example, the $\mathcal{N} = 4$ gauge theory has a marginal operator $\mathcal{O}(x)$, (the Lagrangian density operator), which changes continuously the Yang-Mills coupling constant. From (13.4.4), changing the coupling constant corresponds to changing the dilaton expectation value in the AdS string theory. This expectation value is set as a boundary condition at the boundary of AdS$_5$.

We may consider the following additional term in the gauge theory action

$$\int d^4x \, \phi_0(x) \, \mathcal{O}(x). \tag{13.5.1}$$

According to our discussion above, the effect of such a term in the field theory is to change, in the dual description the boundary value of the dilaton by $\phi_0(x)$. In Poincaré coordinates (see appendix K) we may therefore write

$$\phi(x,r)|_{u=0} = \phi_0(x), \tag{13.5.2}$$

where x are the $\mathbb{R}^{1,3}$ Minkowski coordinates of the gauge theory and $u = 0$ is the boundary of AdS_5 .

This map between the gauge theory operator $\mathcal{O}$ and the bulk field ϕ (the dilaton) emerges directly from the usual DBI action of D_3-branes, which includes direct couplings of the bulk fields to the massless brane fields.

In view of this a quantitative form of the correspondence may be written symbolically as

$$\langle e^{\int d^4x \, \phi_0(x) \, \mathcal{O}(x)} \rangle_{\mathrm{CFT}_4} = \mathcal{Z}_{\mathrm{string}}\left[\phi(x,u)|_{u=0} = u^{4-\Delta}\phi_0(x)\right]. \tag{13.5.3}$$

On the left-hand side, the expectation value is taken in the gauge theory. $\phi_0(x)$ is an arbitrary source for the operator $\mathcal{O}$. The expectation value may be computed by expanding the exponential and evaluating the (off-shell) correlation functions of $\mathcal{O}$ in the gauge theory.

The right-hand side is calculated in the string theory on $AdS_5 \times S^5$. $\mathcal{Z}_{\mathrm{string}}$ is the generating functional of the (on-shell) string amplitudes in $AdS_5 \times S^5$ with the specific boundary condition for the bulk fields (the dilaton here) at the boundary of $AdS_5 \times S^5$. In a theory including gravity, like string theory, the only invariant data are boundary data. Therefore the theory calculates correlations/amplitudes of a set of boundary data. In asymptotically AdS spaces this corresponds to giving boundary conditions to the massless string fields at the boundary and then calculating the vacuum amplitude as a function of these boundary conditions. This is precisely what $\mathcal{Z}_{\mathrm{string}}$ is in (13.5.3).

In asymptotically flat spaces the boundary includes the conformal past and future boundaries and the boundary conditions correspond to the usual scattering data. Therefore $\mathcal{Z}_{\mathrm{string}}$ is the generating functional of the standard S-matrix elements. This is why sometimes we may use the name S-matrix elements also for the AdS amplitudes although they are not directly related to scattering processes.

Note also that while the operators and correlators in the gauge theory are off-shell, the data on the string theory side are on-shell.

The dilaton described above is representative of the general correspondence of any on-shell bulk field in the string theory on $AdS_5 \times S^5$ to any off-shell *gauge-invariant* operator of the gauge theory. Again, this follows from the coupling of the closed and open string modes in the original string theory before taking the low-energy limit. Therefore, equation (13.5.3) can be generalized to all such fields and operators as

$$\langle e^{\int d^4x \, \phi_i(x) \, \mathcal{O}^i(x)} \rangle_{CFT_4} = \mathcal{Z}_{\mathrm{string}}\left[\phi_i(x,u)|_{u=0} = u^{4-\Delta}\phi_i(x)\right]. \tag{13.5.4}$$

There is a direct relation between the scaling dimension Δ of a gauge theory operator $\mathcal{O}$ and the mass of the corresponding field in the $AdS_5 \times S^5$ theory. As shown in

Table 13.1 Correspondence Between the Spin of Bulk Fields in
AdS $_{d+1}$ and the Scaling Dimension of the Corresponding Gauge
Theory Operators. The Last Line Applies to a Massless Spin-two
Particle.

Spin	Dimension
0	$(d \pm \sqrt{d^2 + 4m^2 L^2})/2$
1/2	$(d + 2|m|L)/2$
p-form	$(d \pm \sqrt{(d - 2p)^2 + 4M^2 L^2})/2$
3/2	$(d + 2|m|L)/2$
2	d

appendix K 4.1, the wave equation in Euclidean space for a scalar of mass m in AdS$_5$ has
two solutions that near the boundary $r = 0$ behave as

$$\phi_{\pm}(u, x) \sim u^{\Delta_{\pm}} \phi_{\pm}(x), \quad \Delta_{\pm} = 2 \pm \sqrt{4 + m^2 L^2}. \tag{13.5.5}$$

Since $\Delta_- < \Delta_+$, the boundary condition at $u = 0$ is therefore imposed on the Δ_- solution.
This solution would generically vanish at $u = 0$, if Δ_- is positive. It is therefore appropriate
to put the boundary condition at $u = \epsilon \to 0$:

$$\phi(x, \epsilon) = \epsilon^{\Delta_-} \phi_0(x), \quad \Delta_- = 4 - \Delta_+, \tag{13.5.6}$$

and then take the limit $\epsilon \to 0$. As the bulk field ϕ is dimensionless we find that the
boundary value $\phi_0(x)$ has dimension $[\text{length}]^{-\Delta_-}$. From (13.5.1) we then find that the
scaling dimension of $\mathcal{O}$ is $[\text{mass}]^{4-\Delta_-}$ and therefore $\Delta = 4 - \Delta_- = \Delta_+$.

For general fields in AdS$_{d+1}$, the associated dimensions are tabulated in table 13.1.

From (13.5.6) we deduce that irrelevant scalar operators correspond to scalar fields in
AdS$_5$ with positive mass squared. Marginal perturbations correspond to massless fields,
while relevant perturbations corresponds to "tachyons." Unlike the case of flat space,
tachyonic scalars respecting the BF bound (see appendix K.4) are perfectly acceptable and
do not lead to instabilities.

The correspondence we have described, namely between $\mathcal{N} = 4$ SU(N) SYM in four
dimensions and string theory on AdS$_5 \times S^5$ can be generalized. The correspondence formulae (13.5.4) could be valid between a conformal field theory in d dimensions and a bulk
theory on AdS$_{d+1} \times K$ where K is a compact manifold. Since any CFT has a stress-energy
tensor, the dual bulk theory will always have a spin-2 particle. The conservation of the
stress tensor of CFT$_d$ implies diffeomorphism invariance associated with the spin-2 bulk
particle. It is therefore a graviton, and the bulk theory is a theory of gravity (see exercises 13.10 and 13.11). A CFT$_d$ is an UV-complete field theory. The same must be true
also for the bulk theory, so it is (most probably) a string theory.

The general arguments described above would apply also to non-conformal field theories
with the difference that the bulk geometry will not have the conformal Killing symmetry
anymore. Therefore the general context of the bulk-boundary correspondence would

involve a field theory in d dimensions, dual to string theory on a higher-dimensional manifold with a d-dimensional boundary. We will see more examples later on in this chapter.

13.6 Holography

We will now describe how AdS$_5$/CFT$_4$ realizes the idea of holography in AdS.

There is a general argument in theories of gravity due to Bekenstein, which states, that the maximal entropy contained in a closed region with surface (boundary) area A is given by $S_{\mathrm{max}} = A/4G_N$. This argument was presented in section 12.4 on page 375.

The holographic bound has been controversial since it seems to defy the prediction of quantum field theory, that the number of degrees of freedom inside a region scales with the volume rather than the area. It is indeed one aspect of the paradoxes that plague the marriage of general relativity and quantum mechanics. It has been argued that a successful theory of quantum gravity must satisfy the "holographic principle." In its simplest form the principle stipulates that the theory in a region of space-time is described by degrees of freedom that live on the boundary of that region.

In the example we are studying, the physics of gravity and string theory in AdS is equivalently described by a field theory living on the boundary. It seems therefore to be a concrete realization of the holographic idea.

To do a precise counting of degrees of freedom though we need to introduce cutoffs. For example, the area of the boundary is infinite, and the volume of AdS$_5$ is infinite.

We have seen in section 13.4 that the radial direction behaves like an energy scale. The argument involved in detail the decoupling limit. There are other arguments that indicate also that the radial direction is like an energy scale for the boundary field theory with the UV being at the boundary. The following is quite clear in Poincaré coordinates: (K.11) describes the scale transformation on the boundary coordinates and the radial distance. We learn that if we scale up the boundary coordinates (going therefore down in energy) we must scale up u (going away from the boundary that is at $u = 0$ in these coordinates). Therefore this suggests that the extreme UV is at $u = 0$ (the boundary) while the extreme IR is at $u = \infty$ (the horizon of the Poincaré coordinates).

This can be a source of confusion if one remembers the original position of the D$_3$-branes before we took the decoupling limit. Indeed, keeping carefully track of the coordinates, the position of the branes would be at $u = \frac{1}{r} = \infty$ of the Poincaré coordinates. However, this should not be taken to mean that when we discuss the AdS physics, we must take into account that the branes are anywhere (in particular the horizon, where they were originally, or the boundary as our words seem to imply). The AdS (bulk) theory does not contain branes. It involves only boundary conditions at the boundary of AdS. The associated physics is equivalent to the physics of the CFT$_4$ with the insertion of the associated sources.

In exercise 13.12 you are asked to show that the AdS$_5$ metric can be written as

$$ds^2 = L^2 \left[-\left(\frac{1 + \tilde{u}^2}{1 - \tilde{u}^2} \right)^2 d\tau^2 + \frac{4}{(1 - \tilde{u}^2)^2} \left(d\tilde{u}^2 + \tilde{u}^2 d\Omega_3^2 \right) \right]. \tag{13.6.1}$$

The boundary is at $\tilde{u} = 1$ and the interior at $0 \leq \tilde{u} < 1$.

Let us now consider an IR cutoff in AdS$_5$ where we cut off the space at $\tilde{u}^2 = 1 - \epsilon$ close to the boundary, where ϵ is a dimensionless small number. This would correspond, according to our previous description, to a UV cutoff in the gauge theory. This can be seen by computing the geodesic distance between two points $x_{1,2}$ on the cutoff-boundary sphere at $\tilde{u}^2 = 1 - \epsilon$ and showing that it scales as $\log(|x_1 - x_2|/\epsilon)$. We will see this in more detail when we compute the boundary two-point function from the bulk. Therefore we may view $\epsilon \ll 1$ as a small distance cutoff in the gauge theory. Putting back units the short-distance cutoff is $L\epsilon$.

We may now estimate the entropy (number of degrees of freedom). In the boundary gauge theory living on S^3, with a short distance cutoff $L\epsilon$, there are $1/\epsilon^3$ fundamental cells in the S^3 with radius L. Since the gauge theory has order N^2 degrees of freedom we expect that the entropy scales as the regulated volume of the boundary,

$$S_{\text{FT}} \sim \frac{N^2}{\epsilon^3}. \tag{13.6.2}$$

In the bulk AdS$_5$ theory, the area of the sphere at the regulated boundary is given by

$$A \sim L^3 \left. \frac{\tilde{u}^3}{(1 - \tilde{u}^2)^3} \right|_{\tilde{u}^2 = 1 - \epsilon} \sim \frac{L^3}{\epsilon^3}, \tag{13.6.3}$$

and the gravitational entropy as

$$S_{\text{AdS}} \sim \frac{A}{G_N} \sim \frac{N^2}{\epsilon^3}, \tag{13.6.4}$$

where we have used (13.6.3) and (13.4E). we find therefore that the AdS/CFT correspondence saturates the Bekenstein bound. Moreover, it provides a concrete example of the UV/IR correspondence, described first in section 5.3.6 on page 140.

It does indeed seem that gravity is holographic in AdS$_5$. However, as you will be convinced by doing exercise 13.14, the volume and the area in AdS$_5$ scale in the same way with ϵ close to the boundary. Holography, from that point of view, seems trivial! The nontriviality of our previous argument stems from the fact that we have an extra free parameter L, and the volume and surface scale with different powers of L.

13.7 Testing the AdS$_5$/CFT$_4$ Correspondence

We have argued earlier that the AdS/CFT correspondence is a form of duality. On the gauge theory side we can calculate using perturbation theory in the double expansion, large N, small λ. On the string theory side we calculate in perturbation theory, for large N (suppressed string loops), and large λ (large string tension, suppressed α' corrections.). There is therefore no regime where we can calculate reliably in both theories.

This state of affairs is not new. The nonperturbative dualities of string theory, analyzed in chapter 11, have exactly the same problem. The techniques used there, involved supersymmetry extensively, and compare quantities on the two sides whose coupling dependence is protected. These quantities can therefore be computed on the two sides, beyond perturbation theory, and eventually compared to each other.

Below we itemize several quantities in AdS$_5$/CFT$_4$ that can be reliably calculated and compared beyond perturbation theory.

- The global symmetries. They typically are independent of the coupling, except in the presence of phase transitions. This however is known not to happen in $\mathcal{N} = 4$ SYM. We have already indicated that the bosonic symmetries SO(2,4)×SO(6) match on the two sides. It can also be verified that the full supergroup, PSU(2,2|4) is the same on both sides. Both theories have a (conjectured) nonperturbative SL(2,$\mathbb{Z}$) symmetry that acts on the coupling constant. These are all symmetries for the gauge theory on $\mathbb{R}^{1,3}$.

- Chiral operators are protected due to the BPS property. In particular their spectrum of dimensions is independent of couplings and can be matched in perturbation theory.

- The moduli space of the theory is also independent of the coupling. On the gauge theory side it is $\mathbb{R}^{6(N-1)}/S_N$ corresponding to the expectation values of the six Cartan scalars modded out by the remaining gauge symmetry. This moduli space was also realized in the string theory before taking the near-horizon limit. It is however obscured once the near-horizon limit is taken since several regions would have large curvatures. You are invited to investigate this further in exercise 13.16.

- Correlation functions that are typically related to anomalies are protected from corrections. They can be evaluated in perturbation theory and the result is valid non-perturbatively. Also extremal correlators are protected.

- The qualitative behavior of the theory upon relevant or marginal chiral perturbations.

We will investigate below the successes of several of these comparisons.

13.7.1 The chiral spectrum of $\mathcal{N} = 4$ gauge theory

We will describe here the spectrum of the shortest chiral operators in the gauge theory. We will start by writing down the $\mathcal{N} = 4$ supersymmetry algebra as in (J.1) on page 537

$$\{Q^i_\alpha, \bar{Q}_{\dot\alpha j}\} = 2(\sigma^\mu)_{\alpha\dot\alpha}\, P_\mu, \quad \{Q^i_\alpha, Q^j_\beta\} = \{\bar{Q}_{\dot\alpha j}, \bar{Q}_{\dot\beta j}\} = 0, \tag{13.7.1}$$

where the index $i = 1, 2, 3, 4$ is the spinor index of the R-symmetry group SO(6), or the fundamental index if we take as R-symmetry group SU(4)∼SO(6).

The basic fields of the gauge theory belong to the vector multiplet. It includes the vector A_μ, the four Majorana fermions, $\lambda_{\alpha,i}$, and the six scalars ϕ^I transforming in the antisymmetric representation of SU(4), or the vector of SO(6). In total we have eight fermionic and eight bosonic degrees of freedom. They all transform in the adjoint of the gauge group. They also transform as follows under the supersymmetry:

$$[Q^i_\alpha, \phi^I] \sim \lambda_{\alpha j}, \quad \{Q^i_\alpha, \lambda_{\beta j}\} \sim (\sigma^{\mu\nu})_{\alpha\beta} F_{\mu\nu} + \epsilon_{\alpha\beta}[\phi^I, \phi^J],$$

$$\{Q^i_\alpha, \bar{\lambda}^j_{\dot\beta}\} \sim (\sigma^\mu)_{\alpha\dot\beta}\, D_\mu\phi^I, \quad [Q^i_\alpha, A_\mu] \sim (\sigma_\mu)_\alpha{}^{\dot\beta}\, \bar{\lambda}^i_{\dot\beta}. \tag{13.7.2}$$

There are similar expressions for the $\bar{Q}$ commutators. D_μ is the gauge-covariant derivative $[D_\mu, D_\nu] = F_{\mu\nu}$. There are also SU(4) Clebsch-Gordan coefficients in (13.7.2) that have been suppressed.

The spectrum of local gauge-invariant operators can be constructed by taking products of the vector-multiplet fields and their derivatives, and then taking a trace in the gauge indices. One may further take products of such traces. Taking a trace in the gauge indices corresponds to considering a closed string fluctuation on the string theory side. Therefore, a gauge theory operator that contains products of such traces (known as a *multitrace operator*) should correspond to multiparticle states in the string theory side. We will focus on the remainder on single-trace operators.[2] They form a generating set that can be compared to the string theory spectrum.

All possible single-trace operators can be organized into representations of the superconformal algebra. We would like to find the chiral ones, namely, the shortest representations of the superconformal algebra.

The representation theory of the $\mathcal{N} = 4$ supersymmetry in four dimensions is analyzed in appendix J. The difference here is that the group is larger, namely, the superconformal group. The number of fermionic raising operators is therefore double that of $\mathcal{N} = 4$ super-Poincaré algebra. The generic representation has 2^{16} states and the 1/2 BPS ones $2^8 = 128 + 128$. Moreover, for such BPS representations, the maximum spin is two, starting from a scalar ground state.

There are of course ultrashort representations. For a U(1) gauge theory, the (spin-1) vector multiplet operators are gauge-invariant and they form the so-called doubleton representation. This exists only in the abelian theory.

There is no known systematic way of constructing all BPS operators of the $\mathcal{N} = 4$ SYM theory. We will provide here a list that will eventually match analogous states in the dual string theory.

All such states fall in representations of the superconformal algebra. They are generated from a chiral primary state by the action of the raising operators.

A glance at the way the vector multiplet transforms under supersymmetry in (13.7.2) shows that single-trace operators containing fermions or gauge fields must be descendants. We should therefore look at operators involving the scalars alone. Such operators without derivatives can be written as

$$\mathcal{O}^{I_1 I_2 \cdots I_n} = \frac{1}{N}\text{Tr}[\phi^{I_1}\phi^{I_2} \cdots \phi^{I_n}], \quad I_i = 1, 2, \ldots, 6. \tag{13.7.3}$$

Since the commutator of two scalars can appear in the commutation relations (13.7.1), such operators are descendants when two indices are antisymmetric. For the fully symmetric operators only the traceless ones are chiral primaries. This can be inferred from the relation between the R-symmetry representations and their conformal dimension.

---

[2] Note that the concept of a singe-trace operator is well defined only at large N. At finite N some singe-trace operators are linear combinations of polynomials of traces.

For a chiral scalar operator transforming in an SU(4) representation characterized by the Dynkin indices (r_1, r_2, r_3), the relation is

$$\Delta = \sum_{i=1}^{3} r_i. \tag{13.7.4}$$

A symmetric traceless operator has dimension $\Delta = n$ and transforms in the $(0,n,0)$ representation. Taking a single trace, leaves the dimension the same, but reduces the representation to $(0, n - 2, 0)$, which does not satisfy the condition above.

The order n of the operators $\mathcal{O}^{l_1 l_2 \cdots l_n}$ cannot be arbitrarily large, since the ϕ fields are N × N matrices and for $n > $ N the higher traces can be written as polynomials of the smaller ones. Therefore, for the SU(N) theory, the independent operators have $n = 2, 3, \ldots, $ N. In the U(N) theory we also have $n = 1$ corresponding to the doubleton representation mentioned above.

The full representation can be built by acting on the operators in (13.7.3) with the lowering operators of the superconformal algebra. The dimension of the full representation is $256 \times \frac{n^2(n^2-1)}{12}$, where the 256 comes from the action of the supercharges and the rest is the dimension of the SU(4) representation. Since the basic fields are scalars, the spins contained in the representation go up to spin two. The spectrum of bosonic conformal dimensions inside the representation is $\Delta = n + k$, $k = 0, 1, \ldots, 4$.

The $n = 2$ chiral representation is special. It is the lowest dimension chiral representation and contains operators of dimension 2,3,4 (all other potential operators are null). The conserved currents of the superconformal algebra belong to this representation. Among others, it contains a complex scalar of dimension 2 in the **20** of SU(4) a complex scalar of dimension 3 in the **10**, a singlet scalar of dimension 4 (the Lagrangian density), the R-symmetry currents of dimension 3, and the stress-tensor of dimension 4.

13.7.2 Matching to the string theory spectrum

We expect the fields of string theory on $AdS_5 \times S^5$ to be in one-to-one correspondence with the single-trace operators. Multiple-trace operators should correspond to multiparticle states in the string theory. The quadratic Casimir of the conformal group $C_2 = \Delta(\Delta - 4)$ of the CFT_4 should be equated to that of AdS, $C_2 = m^2 L^2$, giving the relation (13.5.5).

It is not known how to compute the full string theory spectrum on $AdS_5 \times S^5$. It is only the field theory (supergravity) states that can be handled. One starts from the (two-derivative) IIB supergravity in ten dimensions and expands the supergravity fields around the $AdS_5 \times S^5$ background. This produces fields with spins up to 2, so they must belong to short multiplets of the superconformal symmetry. Therefore, they must be compared with the chiral operators in the SYM.

There should be other stringy states that have masses proportional to M_s. According to the mass-dimension relation (13.5.5) the dimensions of the corresponding gauge theory operators should behave as $\Delta \sim (g_s N)^{1/4}$. AdS_5/CFT_4 predicts that these states must be in long multiplets.

The IIB supergravity spectrum around AdS$_5 \times S^5$ can be computed as follows. The ten-dimensional fields are expanded into S^5 spherical harmonics. They are then expanded around the AdS$_5$ background and the equations for their fluctuations are diagonalized. At the quadratic level, we can recognize their quantum numbers (four-dimensional spin, and R-symmetry quantum numbers) as well as their mass.

For example, consider the ten-dimensional complex dilaton scalar τ. We may expand it as

$$\tau(x,y) = \sum_{k=0}^{\infty} \tau_k(x) \, Y^k(y), \quad x \in \text{AdS}_5, \quad \sum_{I=1}^{6} (y^I)^2 = 1. \tag{13.7.5}$$

Y^k are the scalar spherical harmonics on S^5. They are in representations of $SU(4)$ corresponding to traceless symmetric products of the **6**: $Y^k \sim y^{I_1} y^{I_2} \cdots y^{I_k}$. We therefore find a five-dimensional field τ_k on AdS$_5$ for every $(0,k,0)$ representation of $SU(4)$. These are the KK descendants of τ on S^5. You are asked in exercise 13.17 to calculate the masses of τ_k and show that

$$m_k^2 = \frac{k(k+4)}{L^2}. \tag{13.7.6}$$

According to the AdS/CFT dictionary in table 13.1, these scalar states would correspond to SYM operators with $\Delta = k + 4$. Performing the reduction of the full IIB supergravity it turns out that all fluctuations can be assembled in chiral representations with primaries transforming as $(0,n,0)$, with $n = 2, 3, \ldots$. The chiral primaries arise from linear combinations of the metric and the self-dual four-form, both of them with indices in the S^5 directions (so that they are scalars on AdS$_5$). Scalar fields with dimension $n + 1$ come from fluctuations of the two two-forms with indices in the S^5 directions. Finally the dimension $n + 2$ scalars come from the τ field described above.

The $n = 2$ multiplet is the supergraviton multiplet of $\mathcal{N} = 8 \, d = 5$ gauged supergravity. According to the AdS$_5$/CFT$_4$ correspondence this representation is matched with the analogous representation of the gauge theory that contains the conserved currents of the various symmetries. In particular, it incudes a massless graviton that corresponds to the stress-energy tensor and massless $SU(4)$ gauge fields that correspond to the conserved $SU(4)$ R-currents.

The $n = 1$ doubleton representation also appears, but it can be gauged away in the AdS bulk. As we argued earlier, dropping it, is equivalent to considering the $SU(N)$ gauge theory.

We have found a spectrum of chiral primaries with $n = 2, 3, \ldots$. In the gauge theory, n runs up to N. On the supergravity side, n, being a KK mode number, can be arbitrarily large. However, the supergravity results cannot be trusted once the KK masses become of the order of the string scale. Therefore we must have $\frac{n}{L} \lesssim \frac{1}{\ell_s}$ so that $n \lesssim (g_s N)^{1/4}$. For $g_s < 1$, always $(g_s N)^{1/4} < $ N. Therefore, the string theory and the gauge theory agree, in the common region of validity, $n = 2, 3, \ldots, (g_s N)^{1/4}$.

It therefore seems, that in the string theory on AdS$_5 \times S^5$ there are no nonchiral fields with masses below the string scale. This predicts on the gauge theory side, that at large $\lambda \sim g_s N$, all nonchiral operators acquire large anomalous dimensions, $\Delta \sim \lambda^{1/4}$.

13.7.3 $\mathcal{N} = 8$ five-dimensional gauged supergravity

We should give here a few more pieces of information about the IIB theory compactified on S^5 to five dimensions. As we have mentioned earlier, this is the $\mathcal{N} = 8$ gauged supergravity in five dimensions. A good way to describe it, is as follows. We start from the IIB theory compactified on T^5. As described in section 11.6 on page 334, this gives the standard $\mathcal{N} = 8$ supergravity with a bosonic spectrum that contains the graviton, 27 vectors and 42 scalars. The continuous symmetry of this theory is a maximally noncompact version of E_6 known as $E_{6(6)}$. In particular, the vectors transform in the **27** of this noncompact symmetry while the 42 scalars parameterize the coset $E_{6(6)}/Sp(8)$. The scalars have no potential.

There is a procedure in supergravity known as gauging. It takes some of the $U(1)$ gauge fields, and promotes them to a nonabelian group, by judiciously upgrading covariant derivatives to include also the nonabelian fields. Then, the Noether procedure is used to complete the action and the modified supersymmetry transformations. In general the nonabelian group will contain vectors that emerge both from the supergravity and matter multiplets. This is obviously not the case in the $\mathcal{N} = 8$ theory where there are no matter multiplets.

In the particular case of $\mathcal{N} = 8_5$ gauged supergravity, 15 of the 27 $U(1)$ gauge bosons make the nonabelian group $SO(6) \sim SU(4)$. This turns out to be the largest subgroup that can be gauged in this theory. The new effective action looks like

$$S \sim \frac{1}{16\pi\, G_5} \int d^5x \sqrt{-g} \left[R + \frac{12}{L^2} - G_{ij} \partial_\mu \phi^i \partial^\mu \phi^j - V(\phi^i) + \cdots \right], \tag{13.7.7}$$

where we have neglected the vectors and the fermions. We have also shown explicitly the constant part of the potential. An obvious highly symmetric solution is AdS_5, where scalars and vectors have zero expectation values. This is the vacuum that we are interested in.

In table 13.2 we show the gauge theory operators dual to the scalars of the gauged supergravity multiplet.

In conclusion, IIB supergravity compactified on S^5 to five dimensions reduces to $\mathcal{N} = 8$ gauged supergravity with $SU(4)$ gauge symmetry. In this language, the IIB theory expanded around $AdS_5 \times S^5$ gives the $\mathcal{N} = 8$ gauged supergravity expanded around AdS_5 plus an infinite tower of KK multiplets. This is a consistent truncation of the ten-dimensional supergravity. In particular, any solution of the five-dimensional theory can be lifted to a solution of the ten-dimensional theory.

13.7.4 Protected correlation functions and anomalies

There are special correlators in the gauge theory that are protected from quantum corrections. Such correlations functions are very useful in order to test the correspondence.

A prime example are correlations of global symmetry currents. We have mentioned already that the global symmetry of $\mathcal{N} = 4$ SYM is the $SU(4)$ R-symmetry. It is also known that in the presence of external gauge fields (i.e., sources) for the global currents, the global symmetry has an (Adler-Bardeen-Jackiw type) anomaly. The massless fermions of the theory transform in a chiral representation of the R-symmetry group: $\lambda_{\alpha,i} \in \mathbf{4}$,

Table 13.2 The 42 Scalar Fields of Gauged Supergravity and the Associated Gauge Theory (chiral) Operators.

Dimension	$m^2 L^2$	Operator	SO(6) representation
2	-4	$\mathrm{Tr}[\phi^{(I}\phi^{J)}] - \frac{1}{6}\delta^{IJ}\mathrm{Tr}[\phi^K\phi^K]$	20
3	-3	$\mathrm{Tr}[\lambda_{(i}\lambda_{j)}]$	10
3	-3	$\mathrm{Tr}[\bar{\lambda}^{(i}\bar{\lambda}^{j)}]$	$\overline{10}$
4	0	$\mathrm{Tr}[F_{\mu\nu}F^{\mu\nu} + \cdots]$	1
4	0	$\mathrm{Tr}[F \wedge F + \cdots]$	1

$\bar{\lambda}^i_{\dot{\alpha}} \in \bar{4}$. A direct computation of the triangle diagram gives the non-conservation of the R-currents as

$$(\mathcal{D}^\mu J_\mu)^a = \frac{N^2 - 1}{96(2\pi)^2}\, id^{abc} \epsilon^{\mu\nu\rho\sigma}\, F^b_{\mu\nu}\, F^c_{\rho\sigma}, \quad d^{abc} = \mathrm{Tr}[T^a(T^b T^c + T^c T^b)], \tag{13.7.8}$$

where $F^a_{\mu\nu}$ is the (nonabelian) field strength of the external gauge field sources and $\mathcal{D}$ is the covariant derivative using the external gauge connections. The standard Adler-Bardeen theorem moreover guarantees that upon proper definition of the R-currents, the anomalous variation in (13.7.8) does not receive further perturbative corrections.

The obvious question is, how is the global anomaly reproduced in the bulk supergravity. The answer is interesting: IIB supergravity has Chern-Simons-like corrections to the field strengths of various forms. These, reduced upon compactification on S^5, give a five-dimensional Chern-Simons term for bulk SU(4) gauge fields dual to the R-symmetry currents

$$S_{CS} = i\frac{N^2}{24(2\pi)^2} \int_{AdS_5} d^5 x\, (d^{abc}\, \epsilon^{\mu\nu\rho\sigma\tau} A^a_\mu \partial_\nu A^b_\rho \partial_\sigma A^c_\tau + \cdots), \tag{13.7.9}$$

where the ellipsis stands for the completion of the CS term with one and no derivatives. A defining relation of the CS term is that its gauge variation is a total derivative of a gauge-invariant form. Therefore, upon $A^a_\mu \to A^a_\mu + (\mathcal{D}_\mu E)^a$,

$$\delta S_{CS} = -i\frac{N^2}{96(2\pi)^2} \int_{\partial(AdS_5)} d^4 x\, d^{abc}\, \epsilon^{\mu\nu\rho\sigma} E^a\, F^b_{\mu\nu}\, F^c_{\rho\sigma} = -\int d^4 x\, E^a (\mathcal{D}^\mu J_\mu)^a. \tag{13.7.10}$$

This matches the field theory anomaly to leading order in $1/N$.

This state of affairs is generic. An anomaly in a global symmetry, in field theory, is related via the bulk/boundary correspondence, to CS terms for the associated bulk gauge symmetry. The noninvariance of the bulk gauge symmetry is due to the fact, that (i) the CS terms in the bulk theory shift by a total derivative under gauge transformations and (ii) the bulk space has a boundary.

The case of symmetries related to space-time, like the conformal symmetry of a CFT is more subtle. The anomaly in the conformal symmetry will be discussed in section 13.8.3.

There is another class of correlation functions that seem to be protected : the three-point functions of chiral operators. Although no rigorous nonrenormalization theorem is known, there are arguments that such correlators are protected for all values of the 't Hooft coupling.

13.8 Correlation Functions

The quantitative form of the AdS$_5$/CFT$_4$ correspondence is summarized in equation (13.5.4) which we reproduce here in the form

$$e^{-W_{\text{gauge}}} \equiv Z_{\text{gauge}}(\phi_0^i) = Z_{\text{string,M}}(\phi_0^i), \tag{13.8.1}$$

where

$$Z_{\text{gauge}}(\phi_0^i) \equiv \langle e^{-\int d^4x \, \phi_0^i(x)\mathcal{O}_i(x)} \rangle_{\partial M} \tag{13.8.2}$$

is the (quantum) generating functional of the correlation functions of *gauge-invariant, single-trace* gauge theory operators $\mathcal{O}_i$. It is a functional of the sources $\phi_0^i(x)$. W_{gauge} is the associated generating functional for connected correlation functions. $Z_{\text{string,M}}(\phi_0^i)$ is the generating functional of on-shell "S-matrix" amplitudes[3] for string theory in $M = \text{AdS}_5 \times S^5$. $Z_{\text{string,M}}$ is a functional of the boundary conditions $\phi_0^i(x)$ of the massless string theory fields at the boundary ∂M of AdS$_5 \times S^5$. The gauge theory is defined on a manifold conformally equivalent to the AdS$_5 \times S^5$ boundary.

Since string theory on AdS$_5 \times S^5$ is not exactly solvable so far we need some approximation. This has been described earlier: At large N we can neglect loops in the string theory while at large λ we can ignore α' corrections. Therefore we may substitute, to leading order, string theory with the two-derivative action of IIB supergravity on AdS$_5 \times S^5$. In this limit, the correspondence (13.8.1) becomes

$$Z_{\text{gauge}}(\phi_0^i) = e^{-I_{\text{SUGRA}}(\phi_0^i)}, \tag{13.8.3}$$

where I_{SUGRA} is the on-shell action of IIB supergravity. This means that it is evaluated on solutions to the classical supergravity equations satisfying the appropriate boundary conditions at the AdS boundary

$$\phi^i(u, x)\Big|_{u=0} = u^{4-\Delta}\phi_0^i(x). \tag{13.8.4}$$

We should mention here that the right-hand side of (13.8.3) in general must be replaced by a sum over semiclassical extrema with the same asymptotics. In the case of AdS$_5$/CFT$_4$ there is only one vacuum, namely, AdS$_5 \times S^5$, but as we will see further, in other situations there can be more than one competing extrema.

Also, the supergravity extremum realizes precisely the idea of the "master field" expected on general grounds from large-N arguments in the gauge theory. For $\mathcal{N} = 4$ SYM the analog of the master field is AdS$_5 \times S^5$. The extra twist, is that it is a configuration that is translationally invariant in four dimensions as expected but it is *not* a four-dimensional configuration.

The equations (13.8.1),(13.8.3) contain as an important input the specific one-to-one correspondence between SYM operators and string theory fields. Some examples were already

---

[3] See also the discussion after equation (13.5.3).

mentioned: the stress-tensor corresponds to the string theory metric, the R-currents to the $SU(4)$ vectors descending from the reduction of the metric and four-form on S^5, the axion-dilaton scalar to the CP-even and -odd parts of the SYM action etc. This follows from the linearized brane-bulk couplings of D_3 branes in flat space (although subtleties may arise at strong gauge coupling).

It now seems that (13.8.3) is a precise prescription for calculating on both sides. There are still subtleties with the usual renormalization issues. On the field theory side, this is well known. Even in a conformally invariant theory, composite operators must be carefully defined by subtracting UV divergences. In the SU(N) theory, any gauge-invariant operator is composite. Composite operators can be defined, for example, by point splitting: take the positions apart, subtract the divergences (here scheme dependence can enter) and then take the distance to zero.

There is a similar set of divergences on the string theory side. A preview of this appeared already in equation (13.5.6). The on-shell supergravity action (from which the S-matrix amplitudes can be extracted) is divergent.

Unlike the gauge theory side, these divergences come from the neighborhood of the boundary and they are long distance (IR) divergences. Both the sources and the string theory (supergravity action) must be renormalized. The procedure has already been sketched in section 13.6.

We define the boundary conditions at a distance ϵ from the boundary, calculate the contributions that diverge as $\epsilon \to 0$, subtract them by renormalizing operators (the analogue of "wave-function" renormalization) or renormalizing the action. As the IR infinities originate from the boundary, the counter-terms will be localized on the boundary and will therefore be four dimensional. We expect a one-to-one correspondence between the UV counterterms of the SYM theory and the IR counterterms of the string theory. This is yet another manifestation of the UV-IR correspondence at play here.

13.8.1 Two-point functions

We are now ready to apply the procedure outlined above in order to compute a two-point function in the "bulk" (supergravity) theory. We assume that we have compactified the theory over S^5 and therefore are dealing with the $\mathcal{N} = 8$ five-dimensional gauged supergravity coupled with an infinite number of KK multiplets and expanded around the AdS$_5$ "vacuum." We will choose Poincaré coordinates (u, x^i) in AdS$_5$ as in (K.10) on page 543 with the boundary at $u = 0$. As is usually the case, for the calculations to be well defined we Wick-rotate to Euclidean signature.

To keep things simple we will consider a massive scalar field. For the purpose of computing the two-point function the interactions can be neglected. We therefore start with the canonically normalized (Euclidean) action

$$S = \frac{1}{2} \int d^5x \sqrt{g} \left[(\partial \phi)^2 + m^2 \phi^2 \right] = \frac{1}{2} \int d^5x \sqrt{g} \left[-\phi (\Box - m^2)\phi + \partial_\mu (\phi \partial^\mu \phi) \right]. \tag{13.8.5}$$

To proceed with the calculation, we must evaluate the action on shell. To do so we need to write the solution of the equations of motion with appropriate (Dirichlet) boundary conditions at the boundary of AdS_5. In appendix K.4.2 it is explained how this can be done using the bulk-boundary propagator

$$K_{\Delta_+}(u, x; x') = C_3^{-1} \frac{u^{\Delta_+}}{(u^2 + |x - x'|^2)^{\Delta_+}}, \quad \Delta_\pm = 2 \pm \sqrt{4 + m^2 L^2}, \tag{13.8.6}$$

satisfying

$$(\Box_x - m^2) K_\Delta(u, x; x') = 0, \tag{13.8.7}$$

$$\lim_{u \to 0} K_\Delta(u, x; x') \to u^{4-\Delta} \left[\delta^{(4)}(x - x') + \mathcal{O}(u^2) \right] + u^\Delta \left[\frac{C_3^{-1}}{|x - x'|^{2\Delta}} + \mathcal{O}(u^2) \right]. \tag{13.8.8}$$

The constant $C_3 = \pi^2 \Gamma[\Delta - 2] / \Gamma[\Delta]$ has been calculated in (K.47). The part of the solution proportional to $u^{4-\Delta}$ is proportional to the source. The other part, proportional to u^Δ corresponds to the normalizable fluctuation and is determined by the source.

In terms of the bulk-boundary propagator, the solution with appropriate Dirichlet boundary conditions can be written as

$$\phi(u, x) = \int d^4x' \, K_{\Delta_+}(u, x; x') \, \phi_0(x'), \quad \lim_{u \to 0} \phi(u, x) \simeq u^{\Delta_-} \phi_0(x) + \cdots. \tag{13.8.9}$$

The boundary condition is implemented in accordance with (13.5.6) and involves a multiplicative renormalization of the field near the boundary.

We may now compute the on-shell action by inserting the solution (13.8.9) into (13.8.5). It is obvious that only the last, boundary term contributes and we obtain

$$\begin{aligned}
S_{\text{on-shell}} &= -\frac{1}{2} \int d^4 x \, \sqrt{g} g^{uu} \phi(u, x) \partial_u \phi(u, x) \Big|_{u=0} \\
&= -\frac{1}{2} \int d^4x_1 \, d^4x_2 \, \phi_0(x_1) \, \phi_0(x_2) \int d^4x \frac{K(u, x; x_1) \partial_u K(u, x; x_2)}{u^3} \Bigg|_{u=0}.
\end{aligned} \tag{13.8.10}$$

Substituting the near-boundary expansion (13.8.8) we obtain

$$\int d^4x \frac{K(u, x; x_1) \partial_u K(u, x; x_2)}{u^3} \simeq \Delta_- u^{2\Delta - 4} \delta(x_1 - x_2)$$

$$+ \frac{4}{C_3} \frac{1}{|x_1 - x_2|^{2\Delta_+}} + \frac{\Delta_+}{C_3^2} u^{2\Delta - 4} \int d^4x \frac{1}{|x - x_1|^{2\Delta_+} |x - x_2|^{2\Delta_+}} + \cdots. \tag{13.8.11}$$

The first term on the right-hand side is a contact term that diverges as we approach the boundary. It must be removed by renormalizing the bulk action. The way to do it is to move the boundary infinitesimally at $u^2 = \epsilon \ll 1$, add a counterterm to remove it and then take the limit $\epsilon \to 0$. The appropriate counterterm is a boundary term:

$$S_{\text{counter}} = -\frac{\Delta_-}{2} \epsilon^{\Delta_- - 2} \int d^4x \, \phi_0(x)^2 = -\frac{\Delta_-}{2} \int d^4x \, \sqrt{h^\epsilon} \, \phi(\epsilon, x)^2, \tag{13.8.12}$$

where $h_{ij}^\epsilon = \frac{1}{\epsilon} \delta_{ij}$ is the metric on the renormalization surface $u^2 = \epsilon$.

The relevant renormalized bulk action (to this order) is

$$S_{ren} = \frac{1}{2} \int_{M_5} d^5x \sqrt{g} \left[(\partial\phi)^2 + m^2\phi^2 \right] + \frac{\Delta_-}{2} \int_{\partial M_5} d^4x \sqrt{h}\, \phi^2. \tag{13.8.13}$$

With this subtraction, the second term in (13.8.11) will give a finite contribution and all the rest will yield vanishing contributions. We may therefore write, to quadratic order, after a simple rescaling of the sources $\phi_0 \to \sqrt{C_3}\,\phi_0/2$.

$$S_{ren}^{\text{on-shell}} = -\frac{1}{2} \int d^4x_1\, d^4x_2\, \frac{\phi_0(x_1)\phi_0(x_2)}{|x_1 - x_2|^{2\Delta_+}}. \tag{13.8.14}$$

According to the correspondence, (13.8.14) indicates the dual operator must have as two-point function

$$\langle \mathcal{O}(x_1)\mathcal{O}(x_2) \rangle = \frac{1}{|x_1 - x_2|^{2\Delta_+}}. \tag{13.8.15}$$

The sign in (13.8.14) is the correct one to guarantee that in Euclidean space $e^{-S(\phi_0)} = \langle e^{\int \phi_0 \mathcal{O}} \rangle$ gives a positive definite two-point function. Equation (13.8.15) is compatible with O being a CFT operator with scaling dimension[4]

$$\Delta = \Delta_+ = 2 + \sqrt{4 + m^2 L^2}. \tag{13.8.16}$$

We should stress that the relation (13.8.16) is valid beyond the supergravity states. It is this relation that indicates that stringy states correspond to gauge theory operators with $\Delta \sim \lambda^{1/4}$ at large λ.

Another remark concerns renormalization. We have seen two sources of renormalization. The first is a multiplicative wave-function renormalization $\phi(u = \epsilon, x) = \epsilon^{\Delta_-}\phi_0$ of the source. The second is an additive renormalization of the bulk action in (13.8.12). This situation is quite generic and reflects similar features of operator renormalization on the gauge theory side.

13.8.2 Three-point functions

We will now include the simplest three-point interaction in the bulk theory. Consider three distinct minimally coupled massive bulk scalars ϕ_i with a cubic interaction

$$S = \frac{1}{2} \int d^5x \sqrt{g} \left[\sum_{i=1}^{3} (\partial\phi_i)^2 + m_i^2\phi_i^2 + \xi\phi_1\phi_2\phi_3 \right]. \tag{13.8.17}$$

The equations of motion now read

$$(\Box - m_i^2)\phi_i = \xi\phi_j\phi_k. \tag{13.8.18}$$

They can be solved perturbatively in the coupling ξ. To leading order, the solution is the same as in the previous section. To obtain the first-order correction, we need also the

---

[4] When $1 < \Delta < 2$, Δ can be identified with either Δ_+ or Δ_-. The two choices correspond to different boundary theories. You will find a discussion of this in [414].

bulk-to-bulk propagator G_ϵ described in appendix K.4.3. Up to first order in the coupling the solution to (13.8.18) is

$$\phi_i(u, x) = \int d^4x' \, K_i(u, x; x') \, \phi_0^i(x')$$

$$+ \xi \int \frac{d^4x' du'}{u'^5} G_\epsilon^i(u, x; u', x') \int d^4x_1 \, d^4x_2 \, K_j(u', x'; x_1) \, K_k(u', x'; x_2) \, \phi_0^j(x_1) \phi_0^k(x_2).$$

$$(13.8.19)$$

It satisfies the same boundary condition as in (13.12.2) but differs in the subleading terms.

We rearrange also the on-shell action as

$$S = \frac{1}{2} \sum_{i=1}^{3} \int d^5x \, \partial_\mu(\sqrt{g} \, \phi_i \partial^\mu \phi_i) - \frac{\xi}{2} \int d^5x \, \sqrt{g} \, \phi_1 \phi_2 \phi_3. \tag{13.8.20}$$

We must also add the counterterms derived in the previous section, to cancel the infinities from the quadratic part.

There are two potential terms that can contribute to the three-point function. One is a bulk contribution from the cubic term. To obtain the cubic part in the external sources, we can keep the $\mathcal{O}(\xi^0)$ solution. We find the following contribution to the three-point function:

$$\langle \mathcal{O}_{\Delta_1}(x_1) \mathcal{O}_{\Delta_2}(x_2) \mathcal{O}_{\Delta_3}(x_3) \rangle = -\frac{\xi}{2} \int d^5x \sqrt{g} \, K_1(u, x; x_1) K_2(u, x; x_2) K_3(u, x; x_3)$$

$$= \frac{C}{|x_{12}|^{\Delta_{12}} |x_{13}|^{\Delta_{13}} |x_{23}|^{\Delta_{23}}},$$

$$C = -\frac{\xi}{2} \frac{\Gamma\left[\frac{\Delta_{12}}{2}\right] \Gamma\left[\frac{\Delta_{13}}{2}\right] \Gamma\left[\frac{\Delta_{23}}{2}\right] \Gamma\left[\frac{\Delta_1 + \Delta_2 + \Delta_3 - 4}{2}\right]}{2\pi^4 \Gamma[\Delta_1 - 2] \Gamma[\Delta_2 - 2] \Gamma[\Delta_3 - 2]}. \tag{13.8.21}$$

The definition of Δ_{ij} is the same as in exercise 4.5 in page 119.

The second potential contribution comes from the quadratic boundary terms in (13.8.20) keeping the $\mathcal{O}(\xi)$ piece of the solution. Here, the definition of the bulk-to-boundary propagator in appendix K.4.3 gives an advantage. In exercise 13.29 you are required to show that conditions (K.55) imply that the potential contribution of the cubic terms in the boundary terms of the on-shell action vanish.

Therefore, (13.8.21) is the full answer. It has the correct form implied by conformal invariance.

The structure of the computation of various n-point terms in AdS is summarized into diagrams like those in figure 13.3.

13.8.3 The gravitational action and the conformal anomaly

We will describe here the simplest aspects of the IR renormalization in AdS_5 involving the metric.

The effective action on AdS_5, after we compactify the five-dimensional theory on S^5 was sketched in section 13.7.3:

$$I_{\text{SUGRA}} = -\frac{1}{16\pi G_5} \int_{\text{AdS}} d^5x \sqrt{g} \left[R + \frac{12}{L^2} \right] - \frac{1}{8\pi G_5} \int_{\partial(\text{AdS})} d^4x \sqrt{h} \, K + \cdots, \tag{13.8.22}$$

where G_5 is given in (13.4E) and the second term is the usual GH term discussed in detail in section 12.3.1 on page 372. It is always necessary in the presence of boundaries. We have

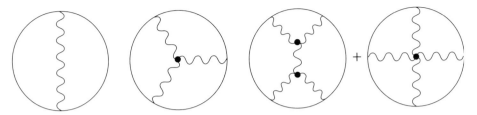

Figure 13.3 Diagrams representing contributions to the AdS "S-matrix". The boundary of a diagram represents the AdS boundary. Internal lines represent either bulk-to-boundary or bulk-to-bulk propagators. The vertices are provided by the interaction terms in the bulk action.

neglected the vectors, scalars and fermions that have vanishing expectation values in the AdS$_5$ ground state. The cosmological constant term descends from the F_5 flux through S^5.

The equations of motion stemming from this action are

$$R_{\mu\nu} - \frac{1}{2}g_{\mu\nu}R = \frac{6}{L^2}g_{\mu\nu} \Rightarrow R = -\frac{20}{L^2}, \quad R_{\mu\nu} = -\frac{4}{L^2}g_{\mu\nu}, \tag{13.8.23}$$

where we have ignored other fluctuations. We may substitute back to simplify the on-shell action as

$$I_{\text{sugra}} = \frac{1}{2\pi G_5 L^2} \int_{AdS} d^5x \sqrt{g} - \frac{1}{8\pi G_5} \int_{\partial(AdS)} d^4x \sqrt{h}\, K + \cdots. \tag{13.8.24}$$

It is already obvious from the previous on-shell action that there are divergences associated with both terms. The first term gives the infinite volume of AdS$_5$, while the second is proportional to the infinite volume of the boundary, as K is constant for AdS (see exercise 13.19).

Cutting off the Poincaré AdS$_5$ metric at $u = \epsilon$, we may calculate the leading singularity in the AdS$_5$ volume to be

$$\int_{AdS} d^5x \sqrt{g} = L^5 \int_\epsilon u^{-5}du \int d^4x = \frac{L^5}{4\epsilon^4} \int d^4x + \mathcal{O}\left(\frac{1}{\epsilon^3}\right)$$
$$= \frac{L^5}{4\epsilon^4}\left(\frac{\epsilon}{L}\right)^4 \int d^4x\, \sqrt{h^\epsilon} + \cdots = \frac{L}{4} \int d^4x\sqrt{h^\epsilon} + \cdots, \tag{13.8.25}$$

where h^ϵ is the induced metric at the $r = \epsilon$ surface (regulated boundary). Using also (13.19) we finally find

$$I_{\text{SUGRA}} = -\frac{3}{8\pi G_5 L} \int_{AdS} d^4x\sqrt{h^\epsilon} + \cdots. \tag{13.8.26}$$

The contributions calculated are singular because the induced metric on the $u = \epsilon$ surface is singular as $\epsilon \to 0$, as can be seen from (13.6.1). Therefore we must add a counterterm to the supergravity action that cancels this leading divergence

$$I_c = \frac{1}{8\pi G_5} \int_{AdS} d^4x\,\sqrt{h^\epsilon}\,\frac{3}{L}. \tag{13.8.27}$$

This is the first of the few counterterms necessary to remove all divergences from the five-dimensional gravitational action. The others involve the boundary sources.

We will now proceed to determine them systematically. We may parametrize an asymptotically AdS metric generally as

$$ds^2 = L^2 \left[\frac{du^2}{u^2} + \frac{1}{u^2} g_{ij}(u^2, x) dx^i dx^j \right], \tag{13.8.28}$$

$$g_{ij}(u^2, x) = g^{(0)}(x) + u^2 g^{(2)}(x) + u^4 [g^{(4)}(x) + \log u^2 \, h^{(4)}(x)] + \cdots. \tag{13.8.29}$$

This is known as the Fefferman-Graham form. $g^{(0)}(x)$ corresponds to the boundary condition for the metric. The equations of motion determine recursively the functions $g^{(2n)}$ and $h^{(4)}$ in terms of $g^{(0)}(x)$ except for $g^{(4)}$. This is in accordance with the fact that the second order equations of motion have two independent bulk solutions. The two independent functions are $g^{(4)}$ and $g^{(0)}$. $g^{(4)}$ will turn out to be related to the expectation value of the stress tensor in the dual field theory. After solving the equations of motion recursively in powers of u, $g^{(2n)}$ will be a functional of $g^{(0)}(x)$ involving $2n$ derivatives.

The logarithmic term proportional to $h^{(4)}(x)$ is determined by $g^{(0)}$ and turns out to be the metric variation of the conformal anomaly of the dual field theory.

Renormalization proceeds now as follows: The equations of motion are solved recursively in powers of u^2. The solution is afterwards substituted into the effective action (13.8.24). The boundary of the metric (13.8.28) is at $u = 0$. The regulated on-shell effective action is then evaluated with the boundary shifted to $u = \epsilon$. The singularities in ϵ are identified and then subtracted.

We first solve the equations of motion (13.8.23) to determine the functions $g^{(2n)}$. We obtain

$$g_{ij}^{(2)} = \frac{1}{2} R_{ij} - \frac{1}{12} R \, g_{ij}^{(0)}, \tag{13.8.30}$$

where the curvatures are constructed for the boundary metric $g^{(0)}$;

$$g_{ij}^{(4)} = \frac{1}{8} g_{ij}^{(0)} \left[(\mathrm{Tr} g^{(2)})^2 - \mathrm{Tr}[(g^{(2)})^2] \right] + \frac{1}{2} (g^{(2)})_{ij}^2 - \frac{1}{4} g_{ij}^{(2)} (\mathrm{Tr} g^{(2)}) + T_{ij}, \tag{13.8.31}$$

where $T_{ij}(x)$ is an "integration constant" satisfying

$$\nabla^i T_{ij} = 0, \quad T_i{}^i = -\frac{1}{4} \left[(\mathrm{Tr} g^{(2)})^2 - \mathrm{Tr}[(g^{(2)})^2] \right], \tag{13.8.32}$$

where the covariant derivative is taken with respect to $g^{(0)}$. It is the second independent function (apart from $g^{(0)}$) that appears in the solution, as expected from second order equations. As we will see below, it can be interpreted as the expectation value of the stress tensor in the dual field theory. For future convenience we introduce

$$\mathcal{A} = \frac{1}{2} \left[(\mathrm{Tr} g^{(2)})^2 - \mathrm{Tr}[(g^{(2)})^2] \right] = -\frac{1}{8} \left[R_{ij} R^{ij} - \frac{1}{3} R^2 \right], \tag{13.8.33}$$

where we have used (13.8.30) in the second step. Finally,

$$h_{ij}^{(4)} = -\frac{1}{2} T_{ij}^{\mathcal{A}} = -\frac{1}{\sqrt{g^{(0)}}} \frac{\delta \mathcal{A}}{\delta g^{(0), ij}}. \tag{13.8.34}$$

We are now in a position to evaluate the on-shell action (13.8.24). We obtain

$$I_{\text{sugra}} = \frac{L^3}{2\pi G_5} \int d^4x \left[\int_\epsilon \frac{du}{u^5} \sqrt{\det g(u, x)} - \frac{1}{u^4} \left(1 - \frac{1}{4} u \partial_u \right) \sqrt{\det g(u, x)} \Big|_{u=\epsilon} \right]. \tag{13.8.35}$$

The singularity structure will be of the following form:

$$I_{\text{SUGRA}} = \frac{L^3}{16\pi G_5} \int d^4x \sqrt{g^{(0)}} \left[\frac{1}{\epsilon^4} A_0 + \frac{1}{\epsilon^2} A_2 - \log \epsilon^2 \, A_4 + \text{finite} \right]. \tag{13.8.36}$$

The densities A_{2n} are local functionals of $g^{(0)}(x)$ and $h^{(4)}(x)$.

Our earlier computation in the beginning of this section corresponds to setting $g^{(0)} = $ constant and $h^{(4)} = 0$. From this we have obtained that $A_0 = -6$. Similarly we can compute by expanding (13.8.35)

$$A_2 = 0, \quad A_4 = \frac{1}{2} \left[(\text{Tr} g^{(2)})^2 - \text{Tr}[(g^{(2)})^2] \right] = \mathcal{A}. \tag{13.8.37}$$

We may now define the renormalized action by minimally subtracting these divergences. This is accomplished by adding the following counterterm action to (13.8.35)

$$\tilde{I}_c = -\frac{L^3}{16\pi G_5} \int d^4x \sqrt{g^{(0)}} \left[\frac{1}{\epsilon^4} A_0 + \frac{1}{\epsilon^2} A_2 - \log \epsilon^2 \, A_4 \right]. \tag{13.8.38}$$

This counterterm action can be rearranged by rewriting it in terms of the induced metric $h_{ij} = \frac{L^2}{u^2} g_{ij}(u, x)$ on the renormalization surface. Using the equations of motion and after a lengthy calculation the action (13.8.38) simplifies to

$$I_c = \frac{1}{16\pi G_5} \int d^4x \sqrt{h} \left[\frac{6}{L} + \frac{L}{2} R[h] - \frac{L^3}{8} \log \epsilon^2 \left(R_{ij}[h] R^{ij}[h] - \frac{1}{3} R[h]^2 \right) \right], \tag{13.8.39}$$

up to finite terms. The renormalized action

$$I_{\text{ren}} = I_{\text{SUGRA}} + I_c \tag{13.8.40}$$

is free of divergences and can be used to calculate the on-shell S-matrix elements by differentiating with respect to the source $g^{(0)}$. According to the correspondence, the first variation should be the expectation value of the gauge theory stress tensor.

$$\langle T_{ij} \rangle = \frac{2}{\sqrt{g^{(0)}}} \frac{\delta I_{\text{ren}}}{\delta g^{(0), ij}} = \lim_{\epsilon \to 0} \frac{2}{\sqrt{g(\epsilon, x)}} \frac{\delta I_{\text{ren}}}{\delta g^{ij}(\epsilon, x)} = \frac{L^2}{\epsilon^2} T_{ij}[h], \tag{13.8.41}$$

where in the last equality, it is the stress tensor of the theory with respect to the induced metric on the renormalization surface. We now proceed to compute it.

The stress-tensor contains two contributions

$$T_{ij}[h] = T_{ij}^{\text{SUGRA}} + T_{ij}^c. \tag{13.8.42}$$

The first piece comes from the variation of the regulated supergravity action with respect to the induced metric on the renormalization surface. It originates entirely from the GH term

$$\begin{aligned}
T_{ij}^{\text{SUGRA}} &= -\frac{1}{8\pi G_5} (K_{ij} - K h_{ij}) \\
&= \frac{L^3}{8\pi G_5} \left[\partial_\epsilon g_{ij}(\epsilon, x) - g_{ij}(\epsilon, x) \text{Tr}[g^{-1}(\epsilon, x) \partial_\epsilon g(\epsilon, x)] + \frac{3}{\epsilon} g_{ij}(\epsilon, x) \right].
\end{aligned} \tag{13.8.43}$$

The second piece comes from the variation of the counterterm action(13.8.39)

$$T_{ij}^c = -\frac{1}{8\pi G_5}\left[\frac{3}{L}h_{ij} - \frac{L}{2}\left(R_{ij} - \frac{1}{2}R\,h_{ij}\right) - \frac{L^3}{2}\log\epsilon^2\,T_{ij}^A\right], \tag{13.8.44}$$

where T^A was defined in (13.8.34).

These expressions can be further simplified to obtain the final result

$$\langle T_{ij}\rangle = \frac{L^3}{8\pi G_5}[2T_{ij} + 3h_{ij}^{(4)}]. \tag{13.8.45}$$

Note that the $h^{(4)}$ term in (13.8.45) is scheme-dependent since its coefficient can be changed by adding a multiple of $\mathcal{A}$ in the action. We choose the new counterterm action as

$$I_c' = I_c + \frac{3L^3}{16\pi G_5}\mathcal{A}, \tag{13.8.46}$$

so that

$$\langle T_{ij}\rangle = \frac{L^3}{4\pi G_5}\,T_{ij}. \tag{13.8.47}$$

Then, as advocated above, the integration constant T_{ij}, corresponding to the subleading solution near the boundary, is indeed proportional to the expectation value of the stress tensor. This is a general fact. Consider for instance a scalar bulk field behaving near the boundary as

$$\phi(u,x) = u^{4-\Delta}\left[\phi_0(x) + \mathcal{O}(u^2)\right] + u^{\Delta}\left[E(x) + \mathcal{O}(u^2)\right], \tag{13.8.48}$$

then

$$E(x) = \frac{1}{2\Delta - 4}\langle\mathcal{O}_\Delta(x)\rangle \tag{13.8.49}$$

gives the associated expectation value.

We may also compute the trace of the stress tensor using (13.8.32) to obtain

$$\langle T_i{}^i\rangle = -\frac{L^3}{8\pi G_5}\mathcal{A}_4 = \frac{N^2}{32\pi^2}\left(R_{ij}R^{ij} - \frac{1}{3}R^2\right), \tag{13.8.50}$$

where we have also used (13.4E). This result should coincide with the conformal anomaly of SYM. The (scheme-independent part of the) conformal anomaly of a CFT in four dimensions is given by

$$\langle T_i{}^i\rangle = a\,W + c\,G, \tag{13.8.51}$$

where W is the square of the Weyl tensor and G the Gauss-Bonnet term, that is also the Euler density in four dimensions,

$$W = R^{\mu\nu\rho\sigma}R_{\mu\nu\rho\sigma} - 2R^{\mu\nu}R_{\mu\nu} + \frac{1}{3}R^2, \quad G = R_{\mu\nu\rho\sigma}R^{\mu\nu\rho\sigma} - 4R_{\mu\nu}R^{\mu\nu} + R^2, \tag{13.8.52}$$

and the coefficients can be given in terms of the number of scalars, Dirac fermions and vectors as follows:

$$a = \frac{N_{scalar} + 6N_{fermion} + 12N_{vector}}{120(4\pi)^2}, \quad c = -\frac{N_{scalar} + 11N_{fermion} + 62N_{vector}}{360(4\pi)^2}. \tag{13.8.53}$$

For $\mathcal{N} = 4$ SYM we put $N_{scalar} = 6(N^2 - 1)$, $N_{fermion} = 2(N^2 - 1)$ (appropriate for four Majorana fermions), and $N_{vector} = N^2 - 1$, we find agreement with (13.8.50) to leading order in the large-N expansion.

The form of the conformal anomaly (13.8.50) obtained on the AdS side is of general validity (up to an overall coefficient). This implies that only CFTs with $a + c = 0$ to leading order in the large-N expansion have an AdS description that is weakly curved.

13.9 Wilson Loops

An important gauge-invariant variable in Yang-Mills theories is the Wilson loop

$$W(\mathcal{C}, R) = \text{Tr}_R \left[P\, e^{i \int_{\mathcal{C}} A} \right]. \tag{13.9.1}$$

It is given by a path-ordered exponential over the gauge field, around a closed path $\mathcal{C}$. The trace can be taken in any representation of the gauge group, although here we will discuss only the fundamental representation.

This observable contains important dynamical information about the gauge theory. For a large rectangular loop, we can extract from it the static quark potential. Consider a rectangular loop in the t-x plane, with length T along time and l along x. The Wilson loop describes the process of creating a heavy (nondynamical) quark-antiquark pair out of the vacuum, pulling them a distance l apart, letting them interact during time T, and then bringing them together and annihilating them. This gives an alternative formula for the expectation value

$$\langle W(\mathcal{C}) \rangle \simeq e^{-T\, E} \simeq e^{-TV(l)}. \tag{13.9.2}$$

We will describe here the computation of the expectation values of Wilson loops in the dual string theory in $AdS_5 \times S^5$. In fact, already from gauge theory intuition, we understand that the Wilson loop is related to the gauge theory string. We must however find the precise prescription in our case that will allow a quantitative evaluation of the expectation value.

We now present a rather simple argument that gives the right answer. We have argued earlier that we may define the Wilson loop by using heavy (nondynamical) quarks (fields transforming in the fundamental of the gauge group). We may achieve this in the $\mathcal{N} = 4$ theory by starting with a U(N+1) gauge group that is Higgsed to U(N)×U(1). This is equivalent to putting a probe D_3-brane in the transverse space, at some point U in the radial direction of AdS_5 and some point on S^5.

The strings stretched between the probe brane and the U(N) stack of branes, transform in the fundamental as far as the U(N) theory is concerned. We must make their ground-state mass infinite so that they become nondynamical quarks. This is achieved by pulling the probe brane, and therefore its strings to $U = \infty$, that is, to the boundary of AdS$_5$. We conclude that the Wilson loop $W(\mathcal{C})$ in AdS$_5$ should be at the boundary.

In the $\mathcal{N} = 4$ theory, the strings also couple to scalars, not only the gauge fields. It is natural to modify the Wilson loop definition in order to also include the scalars:

$$W(\mathcal{C}) = \mathrm{Tr}_R \left[\mathrm{P}\, e^{\int_{\mathcal{C}} \left[iA_\mu \dot{x}^\mu + n\cdot\phi\sqrt{\dot{x}^2} \right] d\tau} \right].$$
(13.9.3)

We work here in Euclidean space. n^I is a unit vector in $\mathbb{R}^6$ equivalent to a point in S^5, corresponding to the fact that the string is localized in the transverse space. The fermions also should be included, but they are expected to lead to subleading contributions and we will neglect them in the sequel.

The expectation value of this Wilson loop in the string theory side should be given by the string partition function with a string world-sheet ending at the loop $\mathcal{C}$ at the boundary of AdS$_5$ as indicated in figure 13.5. When $\lambda \gg 1$ this can be calculated reliably in the supergravity approximation. It amounts to computing the area of the string world-sheet with a given boundary loop $\mathcal{C}$.

It is not difficult to guess that the area, computed that way will turn out to be infinite. This is again due to the fact that it stretches up to the boundary of AdS$_5$. One way to interpret this divergences is due to the constant ground-state energy of the strings (alias W-boson mass) stretching for an infinite distance. This energy should be subtracted.[5] This divergence is equal to the length of the Wilson loop times the W-boson mass that is infinite.

We will compute now the simplest Wilson loop, namely the rectangular one in figure 13.4. We will take $T \to \infty$ since in this limit the problem is translationally invariant in the t-direction and therefore simpler, although we will still be able to recover the quark potential. We put the two quark lines at $x = \pm l/2$.

We mentioned above that we can ignore the world-sheet fermions. We also have the extra coordinates of the string along S^5. Since however, we have located the string at a point in S^5, they are constant and do not contribute in the dynamics. Finally, the four-form flux, to leading order, can be neglected as a fundamental string does not couple to it.

We will pick a static gauge where the world-sheet coordinates are $\tau = t$ and $\sigma = x$. We are also interested in a static configuration and take u to be a function of x only. The Nambu-Goto action in the AdS$_5$ metric becomes

$$S = \frac{1}{2\pi\ell_s^2} \int d^2\xi \sqrt{\det \hat{g}} = \frac{T\,L^2}{2\pi\ell_s^2} \int dx\, \frac{\sqrt{1 + u'^2}}{u^2},$$
(13.9.4)

where we have chosen Poincaré coordinates and the prime stands for differentiation with respect to x. The equations of motion can be integrated once to

$$u^2 \sqrt{1 + u'^2} = C^2.$$
(13.9.5)

---

[5] A careful consideration of the Neumann boundary conditions of the six coordinates of the string transverse to the boundary indicates that the proper amplitude is a certain Legendre transform of the Nambu-Goto action. We refer the reader to [415] for a detailed explanation. Here for simplicity reasons we will just subtract the divergence.

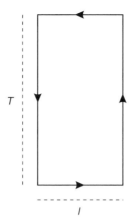

Figure 13.4 The Wilson loop in terms of the world-lines of quarks.

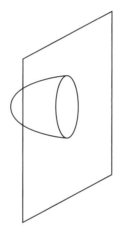

Figure 13.5 The Wilson loop in AdS.

Solving we find $x(u)$ as

$$x = C \int_{u/C}^{1} \frac{v^2 \, dv}{\sqrt{1 - v^4}}. \tag{13.9.6}$$

As $u \to 0$ x asymptotes to its maximum value $l/2$ so we obtain

$$\frac{l}{2} = C \int_{0}^{1} \frac{v^2 \, dv}{\sqrt{1 - v^4}} \to C = \frac{\Gamma\left[\frac{1}{4}\right]^2}{\sqrt{2}\pi^{3/2}} \frac{l}{2}. \tag{13.9.7}$$

We may now compute the action (which for static configurations coincides with the energy). We obtain

$$S = \frac{TL^2}{2\pi \ell_s^2} \int_{-l/2}^{l/2} dx \, \frac{\sqrt{1 + u'^2}}{u^2} = \frac{2TL^2}{2\pi \ell_s^2 C} \int_{0}^{1} \frac{dv}{v^2 \sqrt{1 - v^4}}. \tag{13.9.8}$$

It is divergent close to the boundary as expected. Regularizing by cutting the integral off at $r = \epsilon$ we obtain

$$S(\epsilon) = \frac{2TL^2}{2\pi\ell_s^2 C} \int_{\epsilon/C}^1 \frac{dv}{v^2\sqrt{1-v^4}} = \frac{2TL^2}{2\pi\ell_s^2 C}\left[\frac{C}{\epsilon} + \text{finite}\right]. \tag{13.9.9}$$

We therefore subtract the $1/\epsilon$ pole that is due to the W mass to obtain the renormalized energy as

$$S_{\text{ren}} = \frac{TL^2}{\pi\ell_s^2 C}\left[-1 + \int_0^1 \frac{dv}{v^2}\left(\frac{1}{\sqrt{1-v^4}} - 1\right)\right] = -\frac{TL^2}{\pi\ell_s^2 C}\frac{\sqrt{\pi}\,\Gamma\left[\frac{3}{4}\right]}{\Gamma\left[\frac{1}{4}\right]}. \tag{13.9.10}$$

Using (13.9.7) we finally obtain for the quark potential

$$V(l) = -\frac{4\pi^2\sqrt{2g_{\text{YM}}^2 N}}{\Gamma\left[\frac{1}{4}\right]^4}\frac{1}{l}. \tag{13.9.11}$$

The $1/l$ dependence is expected because of conformal invariance and dimensional analysis. We find that at large λ the potential scales as $\sqrt{\lambda}$. At weak coupling it scales rather as λ.

13.10 AdS$_5$/CFT$_4$ Correspondence at Finite Temperature

We have seen how the dynamics of conformally invariant strongly coupled $\mathcal{N} = 4$ SYM is encoded into the supergravity dynamics in AdS$_5 \times$ S^5. We would like to go further and study this bulk-boundary correspondence in other theories where in particular conformal invariance is not present. This is a large subject, actively investigated to this day. A modest step in that direction is to consider the CFT$_4$ at finite temperature T. This will turn out to be simple enough that we can handle it computationally, but still contains interesting effects.

A CFT at finite temperature is not conformally invariant anymore, but the breaking is soft. It is in fact analogous in this respect to turning on the scalar expectation values, that break the gauge group and break conformal invariance spontaneously. At the same time supersymmetry is broken, and we will have the example of a non-supersymmetric gauge theory. We will be getting closer to a theory interesting on physical grounds: quantum chromodynamics (QCD).

We will have to find the bulk background that will replace AdS$_5 \times$ S^5 in the finite temperature theory. Here the guess is not terribly difficult since we already know gravitational backgrounds that have a temperature associated to them: black holes.

13.10.1 $\mathcal{N} = 4$ super Yang-Mills theory at finite temperature

We will consider the Euclidean finite-temperature gauge theory in this section and discuss its expected behavior. The presence of finite temperature breaks supersymmetry softly.

The gauge theory lives now on S^1 (with radius β) times S^3 (with radius R). Taking $R \to \infty$ we recover the theory on $S^1 \times \mathbb{R}^3$. Because of the conformal invariance of the $T = 0$ theory, the partition function depends only on the dimensionless ratio $z = R/\beta$ that we will call the dimensionless temperature.

The theory on $S^1 \times S^3$ could have a phase transition at a particular value of z. Theories defined on finite volume cannot have a phase transitions. However, at large N, new phase transitions may appear that survive perturbatively in the 1/N expansion. We will see that in this case, there is a phase transition at a special value $z = z_*$. As we will argue below, the high-temperature phase is in a sense a "deconfined" phase, while the low temperature phase is a "confined" phase.

If we take $R \to \infty$ to recover the theory on $S^1 \times \mathbb{R}^3$ this corresponds to $z = \infty$, and therefore this theory cannot have a phase transition. It is in the high temperature phase of the finite volume theory.

There is a simple order parameter that can distinguish the confining from the deconfining phase. The key is the center C of the gauge group, that for SU(N) is isomorphic to $\mathbb{Z}_N$. Gauge transformations in the center act trivially on adjoint fields by definition, since for any $h \in C$ and any N×N SU(N) matrix, $hMh^{-1} = M$ by the definition of C. However, if there are fields that transform in the fundamental, then they transform nontrivially under the center.

Consider now an order parameter that transforms nontrivially under the center. If it has an expectation value, the symmetry under transformations of the center is spontaneously broken. This can be interpreted as equivalent to the fact that operators transforming in the fundamental condense, and this is a signal for deconfinement. In a phase where the center is unbroken, we expect to have confinement of flux in the fundamental representation.

A convenient operator, transforming nontrivially under the center of the gauge group is the temporal Wilson loop in the fundamental representation. It is defined as

$$W_T(\mathcal{C}) = \mathrm{Tr}_\square \left[P\, e^{i \oint_\mathcal{C} A_\mu \dot{x}^\mu \, d\tau} \right], \tag{13.10.1}$$

where the loop is wrapping once around the temporal S^1. Since the basic fields of the theory are invariant under transformations of the center, h, we have the right to consider transformations in the temporal direction that are periodic up to an element of the center

$$g(t + \beta) = g(t)\, h. \tag{13.10.2}$$

You are invited in exercise 13.35 to show that under such gauge transformations the temporal loop transforms as

$$W_T(\mathcal{C}) \to h\, W_T(\mathcal{C}). \tag{13.10.3}$$

A nonzero expectation value for W_T signals the breaking of the symmetry under the center. W_T transforms as a static quark, and a nonzero expectation value indicates that such a configuration has finite energy. Therefore fundamental charges are not confined. As we will see further on, $\langle W_T \rangle \neq 0$ in the high-temperature phase.

If we consider the finite-temperature theory in $\mathbb{R}^3$ (infinite volume), then we expect that at low energy it behaves as a pure SU(N) Yang-Mills theory in three dimensions. First, we have already argued that for any temperature, $z = \infty$, and therefore we are are effectively at $T = \infty$. Since $\beta = 0$ the S^1 shrinks to nothing and the theory lives in three dimensions. Moreover, because of the antiperiodic boundary conditions the fermions of the theory acquire already at the classical level (infinite) temperature induced masses and therefore decouple. The scalars remain massless at tree level, but obtain temperature dependent masses at one loop. They are expected to also decouple. thus, we are left with the field content of $d = 3$, $\mathcal{N} = 0$, SU(N) YM. This theory is expected to have confinement and a mass gap. In particular confinement shows up as an area-law behavior for a Wilson loop.

At large N, there is another criterion for confinement at finite volume: the dependence of the free energy on N. In a confined phase, where contributions to the free energy come from color singlets (mesons, glueballs, etc.), the free energy is $\mathcal{O}(1)$. In a deconfined phase, the free energy is $\mathcal{O}(N^2)$ reflecting the N^2 adjoint degrees of freedom contributing.

We will now argue that at low energies the $\mathcal{N} = 4$ SYM is expected to have an $\mathcal{O}(1)$ free energy and is therefore in a confining phase according to the previous criterion. As we mentioned earlier, on a round S^3 the SYM scalars have curvature induced masses and there is a unique (trivial) minimum of the potential. The vacuum state is unique, and all basic fields of the theory are trivial in the vacuum. Moreover, at finite volume the Gauss law stipulates that all physical states must be invariant under the global SU(N) symmetry. They are therefore color singlets, given by traces of the SU(N) matrix values fields. We have "kinematic" confinement. The multiplicities of color singlet states are N independent and the free energy is expected to scale as $\mathcal{O}(1)$.

At high temperatures z is large. We may have the same effect by taking R to be large. So we are considering a three-dimensional gauge theory at large volume. Due to the broken supersymmetry the one-loop result for the free energy is nonzero and scales with N^2 and the volume R^3: $E \sim N^2 R^3$. On dimensional grounds then $E \sim N^2 R^3 T^4$, which indicates deconfinement and we will verify this in the sequel.

13.10.2 The near-horizon limit of black D_3-branes

We have given in section 8.8 the supergravity solution describing a nonextremal collection of N D_3-branes. This has the structure of brane black hole, and it is reasonable to expect that its near-horizon region will be describing thermal SYM by generalizing our previous arguments in the $T = 0$ case. Taking the near-horizon limit we will have to arrange that energy above extremality remains finite to guarantee a finite temperature for the dual fields theory.

We reproduce here the metric of the nonextremal (black) D_3-branes as well as its most important characteristics, the ADM mass and D_3-brane charge:

$$ds^2 = \frac{-f(r)dt^2 + d\vec{x} \cdot d\vec{x}}{\sqrt{H(r)}} + \sqrt{H(r)}\left(\frac{dr^2}{f(r)} + r^2 d\Omega_5^2\right), \tag{13.10.4}$$

$$H(r) = 1 + \frac{\tilde{L}^4}{r^4}, \quad f(r) = 1 - \frac{r_0^4}{r^4}, \tag{13.10.5}$$

$$N = \frac{\tilde{L}^2\sqrt{r_0^4 + \tilde{L}^4}}{4\pi g_s \ell_s^4}, \quad M = \frac{V_3\left[5r_0^4 + 4\tilde{L}^4\right]}{2^7 \pi^4 g_s^2 \ell_s^8}, \quad T_H = \frac{r_0}{\pi\sqrt{r_0^4 + \tilde{L}^4}}, \tag{13.10.6}$$

$$S = \frac{V_3 \, r_0^3\sqrt{r_0^4 + \tilde{L}^4}}{2^5 \pi^3 g_s^2 \ell_s^8}, \quad \Phi = \frac{V_3}{(2\pi)^3 g_s \ell_s^4}\frac{\tilde{L}^2}{\sqrt{r_0^4 + \tilde{L}^4}}, \quad dM = TdS + \Phi dN, \tag{13.10.7}$$

where V_3 is the (regularized) volume of the three world-volume directions. We have put a tilde over L to distinguish it from the AdS length in (13.4.7), a definition that we will still use. You are invited in exercise 13.33 to derive the entropy, temperature and chemical potential of black D$_3$-branes.

In the near-horizon limit $r \ll \tilde{L}$. In order to remain outside the horizon $r > r_0$ we must take $r_0 \ll \tilde{L}$, implying that $T_H \ell_s \ll 1$. Expanding in the ratio $r_0/\tilde{L}$ we obtain

$$\tilde{L}^4 = L^4\left[1 - \frac{L^4}{2}(\pi T_H)^4 + \mathcal{O}((T_H\ell_s)^8)\right], \quad r_0^4 = L^8(\pi T_H)^4 + \mathcal{O}((T_H\ell_s)^8), \tag{13.10.8}$$

where L is the AdS length defined in (13.4.7). The near-horizon metric in the limit is

$$ds^2 = \frac{L^2}{u^2}\left[\tilde{f}\,dt^2 + d\vec{x}\cdot d\vec{x} + \frac{du^2}{\tilde{f}}\right] + L^2 d\Omega_5^2, \quad \tilde{f} = 1 - (\pi T_H)^4\,u^4, \tag{13.10.9}$$

where we have rotated to the Euclidean signature and we have also changed $r = L^2/u$. The Euclidean time t has radius $\beta = 1/T_H$. For the other thermodynamic variables we obtain in the near-horizon limit

$$M = NT_3 V_3 + \frac{3}{4}E_{\mathrm{YM}} + \mathcal{O}(\ell_s^4), \quad E_{\mathrm{YM}} = \frac{1}{2}\pi^2 N^2 V_3 T^4, \tag{13.10.10}$$

$$\Phi = T_3 V_3 - \frac{1}{4}\pi^2 N^2 V_3 T^4 + \mathcal{O}(\ell_s^4), \quad S = \frac{3}{4}S_{\mathrm{YM}} + \mathcal{O}(\ell_s^4), \quad S_{\mathrm{YM}} = \frac{2}{3}\pi^2 N^2 V_3 T^3, \tag{13.10.11}$$

where E_{YM} and S_{YM} are the analogous quantities in thermal SYM to leading order in λ ($\lambda \ll 1$), and we have dropped the subscript from the temperature.

If we identify the mass above extremality with the energy of the field theory and the Hawking temperature with the field theory temperature, then $E \sim N^2 V_3 T^4$ which is the correct scaling for the thermal energy E_{YM} of a CFT$_4$. The same is true for the entropy. This gives credence to the fact that the nonextremal black-brane geometry will be dual to thermal $\mathcal{N} = 4$ SYM. To leading order in $1/N$, we expect the energy of SYM to be of the form

$$E_{\mathrm{YM}} = \frac{1}{2}f(\lambda)\pi^2 N^2 V_3 T^4, \tag{13.10.12}$$

where $f(\lambda)$ is a function for the coupling constant. For weak λ it can be calculated in SYM perturbation theory. At large λ it can be calculated from the AdS$_5$/CFT$_4$ correspondence:

$$\lambda \ll 1 \quad \Rightarrow \quad f(\lambda) = 1 + \mathcal{O}(\lambda), \quad \lambda \gg 1 \quad \Rightarrow \quad f(\lambda) = \frac{3}{4} + \mathcal{O}\left(\lambda^{-3/2}\right). \tag{13.10.13}$$

The next-to-leading order at weak coupling comes from two loops in the field theory while the next-to-leading order at strong coupling comes from the leading α' corrections to the effective supergravity action, proportional to the (curvature)^4 terms.

The conclusion of the exercise presented above is that indeed the string theory dual background for thermal $\mathcal{N} = 4$ SYM on $S^1 \times \mathbb{R}^3$ (infinite volume) should be, as expected, the black hole in (13.10.9). A main reason is that we have been working in Poincaré coordinates and the AdS black hole's boundary in (13.10.9) is isomorphic to $\mathbb{R}^3$. This indicates that the dual thermal theory is at infinite volume. Similarly, the black-hole horizon is also isomorphic to $\mathbb{R}^3$.

We have also seen that the gravitational calculation of the energy indicates that the theory is in the high temperature deconfined phase as argued in section 13.10.1.

13.10.3 Finite-volume and large-N phase transitions

We will now describe the thermal correspondence at finite volume. Once we reduce the theory on the S^5 as we usually do, the background is an uncharged black hole in AdS$_5$.

We will now rotate to Euclidean space which is convenient both on the gravitational side and on the field theory side. As discussed in section 12.3.1, the Euclidean time will become an angle with range $\beta = 1/T$. The boundary is again at $u = 0$ and has the topology of $S^1 \times \mathbb{R}^3$ where S^1 is the Euclideanized compactified time.

It will be useful to go to global coordinates and rewrite the Euclidean black-hole metric as

$$ds^2 = \left(1 + \frac{r^2}{L^2} - \frac{wM}{r^2}\right) dt^2 + \left(1 + \frac{r^2}{L^2} - \frac{wM}{r^2}\right)^{-1} dr^2 + r^2 d\Omega_3^2, \quad w = \frac{16 G_5}{3\pi^2}, \tag{13.10.14}$$

where M is the ADM mass of the black hole, t is an angle with range β, and as usual L is the AdS length. For future convenience, we will call this manifold X. The radial coordinate is larger than r_+, where r_+ is the larger root of the red-shift factor

$$1 + \frac{r^2}{L^2} - \frac{wM}{r^2} = 0 \quad \Rightarrow \quad \frac{r_+^2}{L^2} = -\frac{1}{2} + \frac{1}{2}\sqrt{1 + \frac{4wM}{L^2}}. \tag{13.10.15}$$

The radius of the t coordinate can be computed from the requirement that there is no conical singularity. It is

$$\beta = \frac{2\pi L^2 r_+}{2 r_+^2 + L^2}. \tag{13.10.16}$$

Close to the boundary the radius of S^1 is $\beta_\infty = \frac{r}{L}\beta$ while the radius of S^3, $R_\infty = r$. Therefore,

$$z = \frac{R_\infty}{\beta_\infty} = \frac{L}{\beta} = \frac{2 r_+^2 + L^2}{2\pi L r_+}. \tag{13.10.17}$$

This should be identified with the effective "dimensionless" temperature of thermal SYM, as argued in section 13.10.1. It is an interesting feature of (13.10.16) that for a given value of the temperature, there are two corresponding values of r_+ and therefore two black holes, namely,

$$r_+ = \frac{\pi L^2}{2\beta}\left[1 \pm \sqrt{1 - \frac{2\beta^2}{\pi^2 L^2}}\right]. \tag{13.10.18}$$

Note also that β has a maximum, so that the temperature has a minimum. For the invariant temperature, $z \geq \sqrt{2}/\pi$.

We will call the black holes of the minus branch, small black holes. Their horizon size $R_+ < L$, therefore these black holes do not feel much the presence of the cosmological constant. They are expected to behave qualitatively as Schwarzschild black holes in flat space. In particular, they have negative specific heat and are therefore unstable due to Hawking evaporation.

We will call the black holes in the plus branch, large black holes. Their horizon size is larger than the AdS scale L, and they are affected by the AdS cosmological constant. Their specific heat is positive, and are therefore thermodynamically stable.

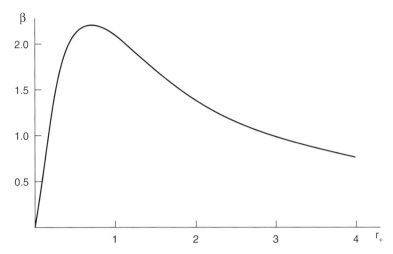

Figure 13.6 The inverse temperature as a function of the horizon radius for AdS$_5$ black holes. The AdS$_5$ length has been set to 1.

When $M = 0$, the metric is locally that of Euclidean AdS. It is related to the AdS$_5$ metric in (13.6.1) by the coordinate change $r \to 2rL/(1 - r^2)$. The only difference is that the coordinate t is now compact. We will call this manifold thermal AdS and denote it by X'. Unlike X, here the radius of the t coordinate can be anything we like. The boundary of the black-hole metric is at $r = \infty$ and has the topology of $S^1 \times S^3$. The same is true for X. There is, however, an important difference. For the black hole, the S^1 is the circle at infinity of the r, t plane. It is therefore a contractible S^1. In X', (thermal AdS) on the other hand the S^1 is not contractible.

When we specify a field theory living on a manifold M_4, its string dual will involve a five- (or higher-dimensional) manifold X whose boundary is M_4. The manifold X is interpreted as the "master field" or vacuum of the four-dimensional theory on M_4 and the precise correspondence is

$$Z_{\text{FT}}(M_4) = Z_{\text{string}}(X) \simeq e^{-I(X)} + \cdots, \tag{13.10.19}$$

where $I(X)$ is the supergravity action and the second inequality holds to leading order in the string tension. This is exactly what happened in AdS$_5 \times S^5$.

We may anticipate the more general situation where there are more than one competing ground-states. This will appear as different bulk manifolds X_i that all have the same boundary M_4. Then, the natural generalization of (13.10.19) is a sum over the semiclassical minima

$$Z_{\text{FT}}(M_4) = \sum_i Z_{\text{string}}(X_i) \simeq \sum_i e^{-I(X_i)} + \cdots. \tag{13.10.20}$$

In general, the supergravity action on X will scale with a positive power of N: $I(X_i) = N^\alpha F(X_i)$. For $\mathcal{N} = 4$ SYM, $\alpha = 2$.

In the large-N limit, the sum will be strongly dominated by the manifold X_* for which $F(X_*)$ is the least. Therefore,

$$\lim_{N \to \infty} -\frac{\log Z_{FT}(M_4)}{N^\alpha} = F(X_*). \tag{13.10.21}$$

It may happen, however, that by varying a parameter, two different manifolds can have the same free energy $F(X_i) = F(X_j)$. At that point the field theory free energy is expected to have a singularity, signaling the phase transition between two difference phases.

After this general discussion we return to our problem. We have argued that X=AdS black hole is a manifold which asymptotes correctly to $S^1 \times S^3$ and can give a bulk description of the thermal Yang-Mills. However, X'=thermal AdS$_5$, given by the $M = 0$ metric in (13.10.14) is also a good candidate.

We should therefore write[6]

$$Z_{FT}(S^1 \times S^3) = e^{-I(X)} + e^{-I(X')}. \tag{13.10.22}$$

For both manifolds we should match their invariant temperature z to that of the thermal SYM. According to our earlier discussion, this fixes two values for the mass, for X, and a single value for the radius of t, for X'. Moreover, for $z < \sqrt{2}/\pi$ the manifold X does not exist.

We are ready to study the actions for the two manifolds, X, X' as a function of the invariant temperature, z, in order to decide which dominates the partition function.

The relevant supergravity action in five dimensions, (13.8.22), reduces to essentially the volume on shell, as in (13.8.24). Moreover, the contribution of the GH boundary term is the same for X and X'. The reason is that the difference of their metrics vanishes fast enough close to the boundary where the GH term is evaluated. Therefore, the difference of the Euclidean actions is proportional to the difference of the volumes,

$$\Delta I = I(X) - I(X') = \frac{1}{2\pi G_5 L^2}[V(X) - V(X')]. \tag{13.10.23}$$

Both volumes are infinite, but the difference is finite. As usual we will cutoff close to the boundary $r \leq \frac{1}{\epsilon}$, evaluate the difference, and then take $\epsilon \to 0$. We obtain

$$V_\epsilon(X) = \int_0^\beta dt \int_{r=r_+}^{1/\epsilon} r^3 dr \int d\Omega_3 = \frac{\pi^3}{4}\beta\left[\frac{1}{\epsilon^4} - r_+^4\right], \tag{13.10.24}$$

$$V_\epsilon(X') = \int_0^{\beta'} dt \int_{r=0}^{1/\epsilon} r^3 dr \int d\Omega_3 = \frac{\pi^3}{4}\frac{\beta'}{\epsilon^4}. \tag{13.10.25}$$

We have to match properly though, the radius β' in X' so that the two asymptotic geometries are the same at the (regulated) boundary, so that they both correspond to the SYM temperature. This imposes

$$\beta'\sqrt{1 + \frac{1}{\epsilon^2 L^2}} = \beta\sqrt{1 + \frac{1}{\epsilon^2 L^2} - wM\epsilon^2}. \tag{13.10.26}$$

---

[6] This is the usual thermal partition function for the SYM, $\text{Tr}[e^{-\beta H}]$. One can write also a correspondence for the twisted partition function $\text{Tr}[(-1)^F e^{-\beta H}]$. However for this, only X' can contribute, because the appropriate spin structure (periodic fermions around S^1) cannot extend to X (see the discussion in [416]).

Finally

$$\Delta I = \lim_{\epsilon \to 0} \frac{1}{2\pi G_5 L^2} [V_\epsilon(X) - V_\epsilon(X')] = \frac{\pi^3}{8G_5} \frac{r_+^3(L^2 - r_+^2)}{2r_+^2 + L^2}. \tag{13.10.27}$$

It is therefore obvious that in the region $r_+ < L$, X' has smaller action. For, $r_+ > L$, it is X that dominates. In terms of the dimensionless SYM temperature (13.10.17) the transition happens at $z_* = 3/(2\pi)$. It is known as the Hawking-Page transition. At lower temperatures it is thermal AdS$_5$ that dominates the physics. At higher temperatures it is the large AdS$_5$ black hole that dominates. In particular, the thermal theory on $\mathbb{R}^3$ coincides with the $z = \infty$ theory, and we just found that it is consistent to describe it with the (flat) AdS$_5$ black hole as we did in section 13.10.2.

We may compute the other thermodynamic quantities

$$E = \frac{\partial \Delta I}{\partial \beta} = \frac{3\pi^3}{16\pi G_5} \left(\frac{r_+^4}{L^2} + r_+^2 \right) = M, \tag{13.10.28}$$

$$S = \beta E - \Delta I = \frac{\pi^3 r_+^3}{4G_5} = \frac{A}{4G_5}. \tag{13.10.29}$$

The entropy at large temperature, scales indeed as $S \sim \beta^{-3}$ which is the correct field theory behavior as we argued in section 13.10.2.

13.10.4 Thermal holographic physics

We will now discuss some of the questions raised in section 13.10.1. The first issue will be the calculation of the temporal Wilson loop $W_T(\mathcal{C})$, which will give us indications on confinement.

In section 13.9 we explained how to compute this in the supergravity approximation. A key ingredient was the existence of two-dimensional surface in the bulk geometry, $\mathcal{D}$ such that the loop is its boundary: $\mathcal{C} = \partial \mathcal{D}$. $\mathcal{D}$ is the appropriate string world-sheet. If $\mathcal{D}$ does not exist for topological reasons, then $\langle W(\mathcal{C}) \rangle = 0$.

In the low temperature phase, the manifold X' dominates the partition function. It has topology $\mathbb{R}^4 \times S^1$. The temporal Wilson loop is the noncontractible S^1 at the boundary. Therefore there is no $\mathcal{D}$ that asymptotes to $\mathcal{C}$, because otherwise $\mathcal{C}$ would have been contractible. Therefore $\langle W(\mathcal{C}) \rangle = 0$ is zero in the low temperature phase. According to the discussion in section in section 13.10.1 we are in a "confined" phase at low temperature.

In the high-temperature phase it is X that dominates the ensemble. At finite volume, the topology of X is $\mathbb{R}^2 \times S^3$ and the temporal S^1 at the boundary is the boundary of $\mathbb{R}^2$. Therefore the temporal loop now is contractible, and there is an obvious string world-sheet $\mathcal{D}$ that bounds $\mathcal{C}$. However, there is another degree of freedom that enters in the string classical action: the expectation value of the antisymmetric tensor B, via $\psi = \int_{\mathcal{D}} B$. Although $dB = 0$ in the background, a constant B is still allowed. In a manifold without boundary a constant B is trivial, but not when open strings are involved. In particular, the Wilson loop is proportional to $e^{i\psi}$. As discussed in earlier sections, B is defined modulo 2π, which indicates that $\psi \in [0, 2\pi]$. Integrating over this extra variable, sets the Wilson loop expectation value to zero.

Therefore, in the high temperature phase at finite volume, the $\mathbb{Z}_N$ symmetry is unbroken, consistent with the fact that it cannot break at finite volume.[7] However, when we consider the theory on $\mathbb{R}^3$ (infinite volume), the zero mode of ψ is not normalizable, and we must not integrate over it. It is a parameter, and labels the expectation value which now is nonzero:

$$\langle W(\mathcal{C}) \rangle \sim e^{i\psi}\, e^{-A(\mathcal{D})}, \tag{13.10.30}$$

where $A(\mathcal{D})$ is the (renormalized) minimum area world-sheet bounding $\mathcal{C}$. In standard examples of symmetry breaking, the degenerate vacua of the broken symmetry should be labeled by group elements, here $\mathbb{Z}_N$. What we find is a U(1) phase instead, presumably because we work in the large-N limit and in a sense $\lim_{N\to\infty}\mathbb{Z}_N = U(1)$. We conclude that in the high temperature phase at infinite volume we have deconfinement of flux.

13.10.5 Spatial Wilson loops in (a version of) QCD₃

It has been argued in section 13.10.1 that in the low-energy high-temperature regime, thermal $\mathcal{N} = 4$ SYM should have similar dynamics with pure QCD in three dimensions. Roughly, the radius of the temporal dimension $R = 1/T_H$ becomes very small at high temperature. We expect confinement and a mass gap in QCD_3. Of course the theory is not QCD_3 at all scales. Above $E \sim 1/R$ it becomes again the four-dimensional gauge theory.

We will calculate the spatial Wilson loop in the high-temperature phase at infinite volume. As $T \to \infty$, the temporal circle shrinks to zero and we expect to recover Euclidean QCD_3 along the three spatial dimensions x^i. The loop then will be a rectangle in the $x^1 = x^2$ plane. We will take $x^1 \gg x^2$ so that we can take the solution to be almost independent of x^1. If we have confinement then the exponential of the Wilson loop should scale as the area of the loop, signaling a linear potential between sources and therefore confinement.

The setup is similar to the analogous calculation in section 13.9 with a different metric now, namely, (13.10.9). We will denote the length of the loop in the x^1 direction by l_1 and in x^2 direction l_2. The string action now reads

$$S = \frac{1}{2\pi\ell_s^2}\int d^2\xi\sqrt{\det\hat{g}} = \frac{l_1\,L^2}{2\pi\ell_s^2}\int\frac{dx^2}{u^2}\sqrt{1+\frac{u'^2}{\tilde{f}}}, \tag{13.10.31}$$

where the prime stands for differentiation with respect to x^2.

The equations of motion can be integrated once to

$$u^2\sqrt{1+\frac{u'^2}{\tilde{f}}} = \frac{1}{\pi^2 T_0^2}, \quad \tilde{f} = 1 - (\pi T_H)^4 u^4, \tag{13.10.32}$$

where T_0 is an integration constant. Solving we find $x^2(u)$ as

$$x^2 = \frac{1}{\pi T_0} \int_{\pi T_0 u}^{1} \frac{v^2 \, dv}{\sqrt{(v^4 - 1)\left(\frac{T_H^4}{T_0^4} v^4 - 1\right)}}. \tag{13.10.33}$$

As u approaches the boundary, $u \to 0$, x^2 asymptotes to its maximum value $l_2/2$ so we obtain

$$\frac{l_2}{2} = \frac{1}{\pi T_0} \int_0^1 \frac{v^2 \, dv}{\sqrt{(v^4 - 1)\left(\frac{T_H^4}{T_0^4} v^4 - 1\right)}} = \frac{\sqrt{2\pi}}{\Gamma\left[\frac{1}{4}\right]^2 T_0} \, {}_2F_1\left[\frac{1}{2}, \frac{3}{4}; \frac{5}{4}; \frac{T_H^4}{T_0^4}\right]. \tag{13.10.34}$$

This determines T_0 in terms of the distance l_2 between the quarks.

We may now compute the action

$$S = \frac{l_1 L^2}{2\pi \ell_s^2} \int_{-l_2/2}^{l_2/2} \frac{dx^2}{u^2} \sqrt{1 + \frac{u'^2}{\tilde{f}}} = \frac{2 l_1 T_0 L^2}{2\ell_s^2} \int_0^1 \frac{dv}{v^2 \sqrt{1 - v^4} \sqrt{1 - \frac{T_H^4}{T_0^4} v^4}}. \tag{13.10.35}$$

It is divergent close to the boundary as expected. Regularizing by cutting the integral off at $u = \epsilon$ we obtain

$$S(\epsilon) = \frac{l_1 T_0 L^2}{\ell_s^2} \int_{\pi T_0 \epsilon}^1 \frac{dv}{v^2 \sqrt{1 - v^4} \sqrt{1 - \frac{T_H^4}{T_0^4} v^4}} = \frac{l_1 T_0 L^2}{\ell_s^2} \left[\frac{1}{\pi T_0 \epsilon} + \text{finite}\right]. \tag{13.10.36}$$

We therefore subtract the $1/\epsilon$ pole that is due to the W-boson mass to obtain the renormalized energy as

$$S_{\text{ren}} = \frac{l_1 T_0 L^2}{\ell_s^2} \left[-1 + \int_0^1 \frac{dv}{v^2} \left(\frac{1}{\sqrt{1 - v^4} \sqrt{1 - \frac{T_H^4}{T_0^4} v^4}} - 1\right)\right]$$

$$= -\frac{\sqrt{2\pi^3} \, l_1 T_0 L^2}{\Gamma\left[\frac{1}{4}\right]^2 \ell_s^2} \, {}_2F_1\left[\frac{1}{2}, -\frac{1}{4}; \frac{1}{4}; \frac{T_H^4}{T_0^4}\right]. \tag{13.10.37}$$

Finally, the quark potential can be obtained along similar lines with (13.9.5) as

$$V(l_2) = -\frac{\sqrt{2\pi^3 \lambda} \, T_0}{\Gamma\left[\frac{1}{4}\right]^2} \, {}_2F_1\left[\frac{1}{2}, -\frac{1}{4}; \frac{1}{4}; \frac{T_H^4}{T_0^4}\right], \tag{13.10.38}$$

and depends on l_2 and the temperature T_H via the relation (13.10.34).

We may now analyze the long-distance behavior of the potential. From (13.10.34) we obtain large l_2 when $T_0 \to T_H$

$$l_2 \simeq \frac{2\sqrt{2}}{\Gamma\left[\frac{1}{4}\right]\Gamma\left[-\frac{1}{4}\right] T_H} \log\left|1 - \frac{T_H^4}{T_0^4}\right| + \mathcal{O}(1), \quad V(l_2) \simeq -\frac{\sqrt{2\pi^2 \lambda} \, T_H}{\Gamma\left[\frac{1}{4}\right]\Gamma\left[-\frac{1}{4}\right]} \log\left|1 - \frac{T_H^4}{T_0^4}\right| + \mathcal{O}(1). \tag{13.10.39}$$

Eliminating T_0 we obtain the potential at large distances to be

$$V(l_2) = -T_s \, l_2 + \mathcal{O}(1), \quad T_s = \frac{\pi}{2} T_H^2 \sqrt{\lambda}. \tag{13.10.40}$$

The potential is linear as expected in a confining theory and T_s is the string tension.

We should pause here to understand the qualitative difference between this calculation showing area law behavior for the Wilson loop and that in section 13.9 which gave perimeter-law behavior. The difference is that, in the previous case, the string world-sheet stretches deep in the interior of AdS_5 to minimize its area, while here, it cannot go past the horizon. It therefore, deeps quickly straight to the horizon and then follows it approximately parallel with it. In fact, any background which is asymptotically AdS, but where the radial coordinate is bounded, will lead to a Wilson loop exhibiting the area law. For this, it is important that g_{00} and g_{xx} are nonvanishing at the "bottom." You are invited to explore this further in exercise 13.40.

In QCD_3 the coupling has dimensions of mass. It is the characteristic mass scale of the theory. Here we have a theory with cutoff $T_H = 1/R$ and a coupling $\lambda_3 = g_3^2 N = g_4^2 N/(2\pi R) = \lambda/(2\pi R)$. Therefore, the characteristic scale set by the coupling is much larger than the cutoff. This theory is not exactly QCD_3, but it confines nonetheless. We could control QCD_3, if we were able to send $\lambda \to 0$ so that the gauge theory scale is much smaller than the cutoff. This seems very difficult at present.

We may also define the four-dimensional analog of the same theory. This will describe a nonsupersymmetric theory in four dimensions, with a cutoff. You are invited to analyze this theory in exercise 13.41.

13.10.6 The glueball mass spectrum

We will now attempt to compute the low-energy spectrum in the cutoff QCD_3 defined in the previous section and verify our expectation that it has a mass gap. In a pure gauge theory, the particle states (glueballs) are generated from the vacuum by the action of gauge-invariant local operators. Their masses can be obtained from the behavior of their correlation functions. For example, if $\mathcal{O}$ is a gauge-invariant operator, then at large $|x - y|$,

$$\langle \mathcal{O}(x)\mathcal{O}(y) \rangle \simeq \sum_i C_i \, e^{-M_i |x-y|}. \tag{13.10.41}$$

Glueballs are classified by their quantum numbers, J^{PC}, where J is their spin, P their parity eigenvalue, and C their eigenvalue under charge conjugation $C : A_\mu^a \to -A_\mu^a$. C corresponds to the orientation reversal in the D-brane picture.

The simplest state is the 0^{++}. It is generated by the scalar operator of lowest dimension, namely $\text{Tr}[F^2]$. We must therefore compute its two-point function. We have argued already that in the holographic-dual, this operator corresponds to the dilaton. According to the holographic map, we must therefore solve, in the supergravity approximation and to linear order, the dilaton equation in the AdS_5 black-hole background (13.10.9):

$$\Box \Phi = 0. \tag{13.10.42}$$

In order to consider the lightest modes (as opposed to KK descendants on the S^1), we must take the solution to be independent of the S^1 coordinate t. We therefore write $\Phi = \phi(u)e^{ip\cdot x}$, for which the equation becomes

$$u^3 \partial_u \left(\frac{\tilde{f}}{u^3} \partial_u \phi \right) + M^2 \phi = 0, \quad M^2 = -p^2. \tag{13.10.43}$$

Upon continuation back to Minkowski space, M is the glueball mass. We must therefore solve (13.10.43) and find the spectrum of M^2. The coordinate u runs from the boundary $u = 0$ to the horizon $u = 1/(\pi T_H)$, which in the Euclidean picture is not a boundary but the center of the coordinate system. It is the analog of the center in spherical coordinates. Therefore in order to solve (13.10.43), we must impose normalizability of our solution at $u = 0$ and regularity at the horizon. This translates to $\phi' = 0$ at the horizon. At the boundary, there are two asymptotic solutions, behaving as $\sim u^4$ or constant. For normalizability we must impose that the solution asymptotes to $\phi \sim u^4$. These conditions determine completely the solution up to a multiplicative constant. For generic M^2 the solution will asymptote near $u = 0$ to a nonzero constant. Asking for a normalizable solution quantizes M^2 as in analogous quantum mechanical problems.

You are asked in exercise 13.43 to show using simple general arguments, that the spectrum of M^2 is discrete and strictly positive. This proves that the theory has a mass gap as expected. A more detailed analysis shows that

$$M_{0^{++}}^2 \simeq 1.44 \, n(n+1) \, T_H^2, \quad n = 1, 2, 3, \cdots. \tag{13.10.44}$$

Note that the glueball masses are quantized in units of the cutoff $T_H = 1/R$, which is much smaller than the effective string scale $\sqrt{T_s} \sim \lambda^{1/4} \, T_H$ in (13.10.40) since $\lambda \gg 1$. In general we expect stringy glueballs to have masses $\sim \sqrt{T_s}$. Such masses are much larger than the supergravity glueballs discussed here.

Although the theory is not exactly QCD$_3$, it turns out that the glueball spectrum is qualitatively similar to that calculated by lattice techniques.

13.11 AdS$_3$/CFT$_2$ Correspondence

We now return to the subject of the previous chapter, in order to view the results in a new light.

The black holes we have considered, were (generically nonextremal) bound states of D$_1$- and D$_5$-branes. It was also argued in section 12.9, that when the associated world-volume 't Hooft couplings $g_s Q_1$ and $g_s Q_5$ are small, the world-volume gauge theory is reliable in perturbation theory. This is the regime where the counting of microstates is made in order to reproduce the entropy.

However, the D$_1$-D$_5$ bound state has weak curvatures and a microscopic horizon in the opposite limit, $g_s Q_1, g_s Q_5 \gg 1$. It is in this limit that it resembles a macroscopic black hole.

The arguments put forward at the beginning of this chapter, can be carried, verbatim to the case of the D$_1$-D$_5$ bound state. We consider the extremal case first. The correspondence should relate the near-horizon geometry on one hand, and the IR limit of the world-volume theory on the branes, on the other.

The near-horizon geometry was already derived in section 12.7 on page 381, and it was shown to be AdS$_3 \times S^3 \times T^4$. The radii of the AdS$_3$ and S^3 are the same, and we will call them L. It should not be confused with the AdS$_5$ length defined in (13.3.2):

$$L^2 = \ell_s^2 \ell^2 = g_6 \sqrt{Q_1 Q_5} \, \ell_s^2 = \frac{g_s}{\sqrt{V}} \sqrt{Q_1 Q_5} \ell_s^2, \quad R_4^2 = \sqrt{\frac{Q_1}{V Q_5}}, \tag{13.11.1}$$

where g_6 is the six-dimensional coupling constant, after compactification on T^4, and R_4^2 is the overall factor in front of the T^4 metric of volume V in string units. After compactifying on $S^3 \times T^4$, we end up with a gauged supergravity on AdS$_3$, involving an infinite number of KK modes. The effective three-dimensional Newton constant is

$$G_N = \frac{g_6^2 \ell_s^4}{4L^3} = \frac{g_s^2 \, \ell_s^4}{4VL^3}. \tag{13.11.2}$$

As discussed in detail in sections 12.9.1 on page 387 and 12.9.2 on page 389, the IR limit of the world-volume gauge theory after reduction on T^4 is a two-dimensional $\mathcal{N} = (4, 4)$ SCFT, with central charge $c = 6Q_1 Q_5$. As in AdS$_5$/CFT$_4$, there is an analog for the number of colors, namely, $Q_1 Q_5$. The large-N limit $Q_1 Q_5 \to \infty$, suppresses quantum gravitational effects in AdS$_3$.

The global symmetries of the SCFT include the SO(1,2)$\times$ SO(1,2) global conformal symmetry as well the SU(2)$_L \times$SU(2)$_R$ R-symmetries associated with the left- and right-moving extended superconformal algebras. These bosonic symmetries are encoded in the near-horizon geometry of AdS$_3 \times S^3$: the SO(1,2) $\times$ SO(1,2) is the Killing symmetry of AdS$_3$ while the SU(2)$_L \times$SU(2)$_R$ is the Killing symmetry of S^3.

The boundary of AdS$_3$ is a two-dimensional cylinder, and therefore the dual SCFT lives on the cylinder. The fermions on the string theory side must have periodic boundary conditions on the circle (that translate to NS boundary conditions on the sphere) because the circle is contractible in AdS$_3$. Therefore, the dual SCFT contains only the NS-NS sector.

The reader is by now familiar with the fact that the conformal group is infinite-dimensional in two dimensions. However, we have seen here only its finite-dimensional subgroup, SO(1,2) $\times$ SO(1,2) realized explicitly. The reason is that, as explained in section 4.6 on page 59, it is only the SO(1,2) $\times$ SO(1,2) symmetry that is unbroken by the standard vacuum of a two-dimensional CFT. All other raising Virasoro generators create nontrivial states. This corresponds to the statement that the nontrivial conformal transformations will generate new string theory backgrounds that are asymptotically AdS$_3$.

The AdS$_3$ metric in global coordinates is as in (K.4)

$$\frac{ds_{\text{AdS}_3}^2}{L^2} = -\cosh^2 \rho d\tau^2 + d\rho^2 + \sinh^2 \rho d\phi^2 \tag{13.11.3}$$

and near the boundary $\rho = \infty$ it takes the asymptotic form

$$\frac{ds_{\text{AdS}_3}^2}{L^2} \sim -e^{2\rho} dx^+ dx^- + d\rho^2, \tag{13.11.4}$$

with $x^{\pm} = \tau \pm \phi$. The set of infinitesimal diffeomorphisms ϵ^i that preserve the asymptotic metric (13.11.4) are given by

$$\epsilon^+ = f(x^+) + \frac{e^{-2\rho}}{2} g''(x^-) + \mathcal{O}\left(e^{-4\rho}\right),$$

$$\epsilon^- = g(x^-) + \frac{e^{-2\rho}}{2} f''(x^+) + \mathcal{O}\left(e^{-4\rho}\right),$$

$$\epsilon^\rho = -\frac{f'(x^+)}{2} - \frac{g'(x^-)}{2} + \mathcal{O}\left(e^{-2\rho}\right). \tag{13.11.5}$$

The functions $f(x^+)$, $g(x^-)$ are arbitrary. They can be expanded as

$$f(x^+) = \sum_{n \in \mathbb{Z}} L_n\, e^{inx^+}, \quad g(x^-) = \sum_{n \in \mathbb{Z}} \bar{L}_n\, e^{inx^-}. \tag{13.11.6}$$

The coefficients L_n, $\bar{L}_n$ are in one-to-one correspondence with the Virasoro operators. You are asked in exercise 13.48 to verify that the diffeomorphisms in (13.11.5) satisfy an algebra that is isomorphic to the Virasoro algebra, with central charge

$$c = \frac{3L}{2G_N} = 6Q_1 Q_5, \tag{13.11.7}$$

where G_N is the three-dimensional Newton's constant and the SCFT central charge is obtained after substituting the expressions (13.11.1) and (13.11.2).

It can be verified that the infinitesimal diffeomorphisms for $n = 0, \pm 1$ can be extended to global diffeomorphisms, that are exact symmetries of the AdS_3 metric realizing its Killing isometries. The other infinitesimal diffeomorphisms also extend to global transformations that lead to new asymptotically AdS_3 metrics.

The quantitative form of the AdS_3/CFT_2 correspondence is similar to that of AdS_5/CFT_4. The master formula is again (13.5.4) with the obvious substitutions.

We will now discuss the nonextremal case. This implies that we consider the thermal SCFT on one hand and BTZ (AdS_3) black holes on the other. The geometry and thermodynamics of the BTZ black holes have been exposed in section 12.7 on page 381. The geometry of all BTZ black holes is locally that of AdS_3. They differ, however, in their global properties.

The (Euclidean) minimum mass black hole corresponds to $r_{\pm} = 0$ in (12.7.8). It is Euclidean AdS_3 with time compactified. It corresponds to the thermal AdS_5 ground state in AdS_5/CFT_4 correspondence. Since the thermal circle is not contractible here, we are allowed to put (supersymmetry preserving) R boundary conditions on the fermions. This background therefore should be dual to the ground-state of the R-R sector of the boundary SCFT.[8] It is already obvious from (12.7.10) that the correspondence with the SCFT dictates

$$L_0 + \bar{L}_0 = ML = \frac{c}{12} + N_{+p} + N_{-p}, \quad L_0 - \bar{L}_0 = J = N_{+p} - N_{-p}. \tag{13.11.8}$$

Therefore, thermal AdS_3 has a mass compatible with the minimum dimension of the R-R ground state namely, $L_0 = \bar{L}_0 = c/24$.

The nonzero-mass BTZ black hole may now be directly interpreted as a thermal (excited) state in the R-R sector of SCFT. The matching of the entropy, discussed in the previous chapter, between the microscopic (boundary) description and the gravitational (bulk) description, reflects once more the bulk-boundary duality, in this case the AdS_3/CFT_2 correspondence.

---

[8] We are also allowed to put supersymmetry-breaking NS boundary conditions. Then the dual will be an excited state of the NS-NS sector of the SCFT.

13.11.1 The greybody factors revisited

The string theory setup, described here, can be compared to an earlier calculation that we performed in the previous chapter. This is the calculation of the greybody factors for black-hole absorption in sections 12.8 and 12.9.4 on page 394. In a sense it is a calculation on one of the simplest S-matrix elements in the background of D_1-D_5 branes.

The problem involved also the asymptotically flat region of the branes. However we have seen in section 12.8 on page 383 that for the low-energy cross section, the calculation factorized in a trivial piece involving the asymptotically flat region that was independent of any dynamics, and a nontrivial part in the near-horizon region. We may restate this as follows: the asymptotically flat part of the wave fixes the boundary condition for the relevant field, in the boundary of the near-horizon region. This involves a source, that in the supergravity calculation of section 12.8 was a constant proportional to the coefficient A in (12.8.9). Then the linearized equations were solved in the near-horizon region, with the boundary conditions set by the source. The flux functional is proportional to the on-shell quadratic action. Finally the cross section is obtained by dividing by the square of the source. This is the prescription explained in detail in section 13.8.1 for the bulk calculation of the two-point function.

On the other hand, via AdS_3/CFT_2 correspondence this maps to the evaluation of the two-point function of the SCFT. Indeed, a bulk low-energy particle, when it is absorbed by the brane bound-state produces, to leading order, a pair of world-volume particles with opposite momenta. The cross section can be written as

$$
\begin{aligned}
\sigma_{\text{abs}} &\sim \frac{1}{N_i} \sum_{i,f} \left| \langle f | \int dt dx \, \mathcal{O}(x,t) \, e^{i(\omega t + qx)} |i\rangle \right|^2 \\
&\sim \frac{1}{N_i} \sum_i \int dt dx \, e^{i(\omega t + qx)} \langle i | \mathcal{O}(x,t) \mathcal{O}^\dagger(0,0) |i\rangle \\
&\sim \int dt dx \, e^{i(\omega t + qx)} \, \langle \mathcal{O}(x,t) \mathcal{O}^\dagger(0,0) \rangle_{\text{thermal}}.
\end{aligned}
\tag{13.11.9}
$$

In the second line we first summed over a complete set of intermediate states, and then expressed the sum over initial states as a thermal average of the two-point function.

Therefore, agreement of the greybody factors between gravity and world-volume SCFT, at low energies, is a corollary of the AdS_3/CFT_2 correspondence for the two-point functions.

13.12 The Holographic Renormalization Group

So far we have studied the bulk/boundary correspondence, in cases where the boundary theory is conformally invariant. We may contemplate however the case of non-conformal field theories.

In non-conformal theories, the couplings and the whole effective field theory are non-trivial functions of the energy. The energy dependence of the dynamics is controlled by the renormalization group (RG). We will study here the holographic picture of the RG, as it appears in the bulk supergravity theory. We will call it the holographic RG (HRG).

An important ingredient of the AdS/CFT correspondence, is that the radial direction in the bulk corresponds to the energy scale of the boundary field theory. Therefore, the bulk contains all possible energy slices of the field theory. We may even setup the quantization of the bulk theory, in a (quasi) Hamiltonian formalism, where instead of Hamiltonian time evolution, we use Hamiltonian radial evolution. This is more natural from the point of view of the correspondence. In this context, an operator perturbation and its expectation value are canonical conjugates. Most interestingly, the renormalization group flow towards the interior of the bulk space can be identified with the Hamiltonian evolution. Although such a formulation of the HRG is intuitively appealing, for simplicity of presentation, in the examples below we will be solving (when we can) the supergravity equations of motion.

We will specialize to $d = 4$, but our description easily generalizes to other dimensions.

13.12.1 Perturbations of the CFT$_4$

We may perturb away from the conformal point by adding an integrated Lorentz invariant operator to the fixed-point action

$$S = S_* + \mu \int d^4x \, \mathcal{O}(x). \tag{13.12.1}$$

When the dimension of $\mathcal{O}$ is $\Delta > 4$ this is an *irrelevant* perturbation.[9] Its effect is very strong in the UV. It is typically not interesting to analyze, since one needs a different UV description of the theory, which in the IR will reduce to (13.12.1).

When $\Delta < 4$ this is a *relevant* perturbation. Its effects are important in the IR. The CFT in the UV is driven by this perturbation to non-trivial RG trajectory. This trajectory may end up with another non-trivial fixed-point theory (CFT) in the IR, or it may end up in an IR-trivial theory. The later is a theory, that below a given energy scale, has no degrees of freedom. A theory with a mass gap is in this class.

Finally, when $\Delta = 4$, this is a *marginal* perturbation to leading order. If it is exactly marginal, then conformal invariance is preserved for all values of the coupling μ. If not then logs appear, and may turn the perturbation either *marginally relevant*, or *marginally irrelevant*. This implies that its fate will be the same as relevant or irrelevant perturbations, albeit with a slower (logarithmic) running of the coupling.

On the bulk theory side, we have seen in (13.8.9) that near the boundary, the bulk fields asymptote to the sources as

$$\phi(u, x) \simeq u^{4-\Delta} \, \phi_0(x) + u^{\Delta} \langle \mathcal{O} \rangle + \cdots, \tag{13.12.2}$$

where we have also used (13.8.16). ϕ_0 is the boundary source conjugate to the field theory operator $\mathcal{O}$. The Poincaré coordinate u in (13.12.2) is related to the energy scale U defined in section 13.4, as $u = L^2/(\ell_s^2 \, U)$. The AdS$_5$ boundary $u = 0$, $U = \infty$ is at the UV.

---

[9] Here it is assumed that the operator $\mathcal{O}$ is a single-trace operator. Multiple trace operators can also be used to perturb the field theory. The prescription for doing so can be found in [417]. The associated perturbations in the string theory are nonlocal on the world-sheet [418,419].

Consider first the non-normalizable solutions, proportional to the sources ϕ_0. The bulk fields corresponding to irrelevant ($\Delta > 4$) operators of the CFT, blow up at the boundary. The space will not be asymptotically AdS$_5$ anymore, and the UV of the theory changes. In order for the generating functionals (13.8.1),(13.8.2) to remain well defined, the associated sources $\phi_0(x)$ must be taken infinitesimal. This indicates that the generating functionals are good for generating the correlation functions of such operators, but we cannot deform the theory by adding a finite amount of such operators to the action. This is of course, what we already know about irrelevant operators in the field theory. We are now seeing the reflection of this fact in the dual bulk gravitational theory.

On the other hand, marginal and relevant operators asymptote smoothly to the boundary. For such operators, the sources, can be finite, and we have the possibility of perturbing the theory by a finite amount. The non-normalizable solutions correspond to changes of the theory. We have also the option of turning on the normalizable solutions. All of them, according to (13.12.2) are smooth at the boundary. They correspond to specifying different expectation values for the operators in the same theory.

The non-normalizable solution corresponding to a relevant operator changes the theory in the IR. This implies that the geometry in the interior will not be that of AdS$_5$ anymore. There are several possibilities. If the RG flow reaches an IR fixed point then we expect to find another AdS$_5$ region in the IR, the coordinates and the curvature will be in general different. It may also happen that during the flow the theory develops strong curvature. This will invalidate the effective supergravity description and may end up in a singularity. A horizon and/or a singularity may also appear when the theory flows to a trivial theory.

13.12.2 Domain walls and flow equations

We will now present the calculational setup for the discussion of the RG flows, from the supergravity point of view.

We perturb the field theory with scalar operators in order to preserve Lorentz invariance. Therefore, the dual field will be a supergravity scalar. As we discussed previously, the $\mathcal{N} = 8$ gauged supergravity in five dimensions, has 42 scalars, corresponding to the 42 chiral scalar operators of the $\mathcal{N} = 4$ SYM theory shown in table 13.2 on page 423.

The bulk equations of motion can be thought of as evolution equations in the radial variable. In that respect they resemble RG equations that specify the change of field theory operators/correlations functions with energy. The field theoretic RG equations are first order in the energy scale and therefore irreversible. The supergravity equations are second order, and therefore in principle reversible. The way to match the two contexts is to remember that the correspondence is asymmetric in the two independent solutions to the supergravity equations. When we perturb the field theory in the UV, we choose one of the two solutions, and this effectively amounts to a first order system.

Consider now the low-energy effective action of the $\mathcal{N} = 8$ gauged supergravity in (13.7.7). The gauge fields can be consistently be set to zero. The equations of motion therefore imply flows in the manifold of scalars, $E_{6(6)}/Sp(8)$, determined by the potential V. This has several minima. A domain wall solution (kink) starting from an AdS minimum

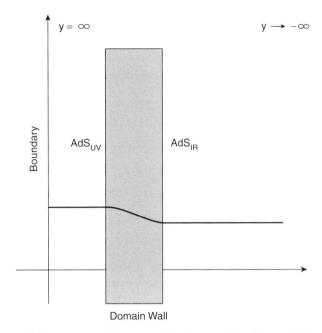

Figure 13.7 The geometry of the renormalization group flow from an UV to an IR fixed point.

and ending in another one, will be interpreted as a RG flow connecting an UV and an IR CFT.

After this general exposition we are ready to attack concrete flows. We will suppress other scalars in the bulk supergravity and focus on a single one, corresponding to the perturbing operator, that we will label as φ. The relevant five-dimensional action is

$$S = \frac{1}{16\pi G_5} \int d^5x \sqrt{-g}\left[R - 2\partial_\mu \varphi \partial^\mu \varphi - V(\varphi)\right]. \tag{13.12.3}$$

We are interested in solutions that preserve four-dimensional Poincaré symmetry. The appropriate *Ansatz* for the metric respecting this symmetry is

$$ds^2 = dy^2 + e^{2A(y)}(-dt^2 + dx \cdot dx), \tag{13.12.4}$$

where we opted for the use of exponential coordinates $-\infty < y < \infty$. It is related to our previous Poincaré coordinates by $u \to e^{-y/L}$. In these coordinates the AdS$_5$ solution is given by $A(y) = y/L$. The boundary is at $y = \infty$.

The fact that the theory at the UV is $\mathcal{N} = 4$ SYM implies that the solution must be asymptotically AdS$_5$ near the boundary: $\lim_{y \to \infty} A(y) \sim \frac{y}{L}$. Inserting the *Ansatz* into the equations of motion stemming from the bulk action (13.12.3) we obtain

$$6A'' + 12(A')^2 + 2(\varphi')^2 + V = 0, \tag{13.12.5}$$

$$12(A')^2 - 2(\varphi')^2 + V = 0, \quad \varphi'' + 4A'\varphi' - \frac{1}{4}\frac{\partial V}{\partial \varphi} = 0. \tag{13.12.6}$$

Two out of the three equations above are independent (as usual). This set of equations can be obtained from a (reduced) one-dimensional action[10]

$$S = \frac{1}{16\pi G_5} \int_{-\infty}^{+\infty} dy\, e^{4A} \left[2(\varphi')^2 - 12(A')^2 + V(\varphi) \right]. \tag{13.12.7}$$

It is often the case in supersymmetric theories that the potential is related to a superpotential W via

$$V(\varphi) = \frac{4}{L^2} \left[\frac{1}{2} \left(\frac{\partial W}{\partial \varphi} \right)^2 - \frac{4}{3} W^2 \right], \tag{13.12.8}$$

where the overall factor in front is conventional. Using this, the reduced action can be written as a quadratic form , up to a total derivative

$$S = \frac{1}{16\pi G_5} \int_{-\infty}^{+\infty} dy\, e^{4A} \left[2 \left(\varphi' \pm \frac{1}{L} \frac{\partial W}{\partial \varphi} \right)^2 - 12 \left(A' \mp \frac{2}{3L} W \right)^2 \right] \mp \frac{1}{4\pi G_5 L} e^{4A} W \Big|_{y=-\infty}^{y=\infty}. \tag{13.12.9}$$

Therefore, the first-order equations

$$A' = -\frac{2}{3L} W, \quad \varphi' = \frac{1}{L} \frac{\partial W}{\partial \varphi} \tag{13.12.10}$$

extremize the reduced action. They provide special solutions to the second order equations of motion (13.12.5),(13.12.6). They are in fact BPS equations, and the solutions preserve a fraction of the supersymmetry.[11] The particular fraction depends on W, and we will not discuss it further here.

The solutions of such a first order system can be naturally interpreted as the RG equations associated with the particular perturbation, in accordance with our general discussion in the beginning of this section.

13.12.3 A RG flow preserving $\mathcal{N} = 1$ supersymmetry

The spectrum of $\mathcal{N} = 4$ SYM can be reorganized into an $\mathcal{N} = 1$ language. It consists of a vector multiplet and three chiral multiplets, that we will label as Φ_i, $i = 1, 2, 3$, containing a complex scalar and a Weyl fermion. The $\mathcal{N} = 4$ potential descends from the superpotential

$$W = \mu\, \mathrm{Tr}\, (\Phi_3[\Phi_1, \Phi_2]) + \mathrm{H.c.}, \tag{13.12.11}$$

where μ is related by the $\mathcal{N} = 4$ supersymmetry to g_{YM}.

We will consider the $\mathcal{N}=1$ preserving deformation, giving a mass to one of the chiral multiplets

$$\delta S = \frac{m}{2} \int d^2\theta\, \Phi_3^2 + \mathrm{H.c.} \tag{13.12.12}$$

---

[10] Substituting an *Ansatz* inside the action, and then deriving the equations gives generically incorrect results. The correct procedure always is to substitute the *Ansatz* into the general equations of motion and then find an action that reproduces the reduced equations. Some times, especially when the *Ansatz* is dictated by symmetry the two procedures may give the same result.

[11] BPS solutions in general imply first-order equations. The opposite is not true in general.

This perturbations is obviously relevant, and breaks the SO(6)≃SU(4) symmetry to SU(2)×U(1) The SU(2) rotates Φ_1, Φ_2, while the U(1) is the $\mathcal{N}=1$ R-symmetry.

In order to find this RG flow in the supergravity side, we must identify the gauge-invariant chiral operators that give a mass to the scalars and fermions in Φ_3. One representative choice is

$$\varphi_1 \rightarrow \sum_{I=1}^{4} \text{Tr}[\phi^I \phi^I] - 2 \sum_{I=5}^{6} \text{Tr}[\phi^I \phi^I], \quad \varphi_2 \rightarrow \lambda_3 \lambda_3 + \text{Tr}[\phi^1, \phi^2] \phi^3, \tag{13.12.13}$$

where φ_1 belongs to the **20** and φ_2 to the **10**. The part of the superpotential of $\mathcal{N}=8$ supergravity describing the dual scalars is

$$W = \frac{1}{4} e^{-2\varphi_1} \left[\cosh(2\varphi_2)(e^{6\varphi_1} - 2) - (3e^{6\varphi_1} + 2) \right]. \tag{13.12.14}$$

The flow equations therefore become

$$\frac{\partial \varphi_1}{\partial y} = \frac{1}{6L} \frac{\partial W}{\partial \varphi_1} = \frac{e^{-2\varphi_1}}{6L} \left[e^{6\varphi_1}(\cosh(2\varphi_2) - 3) + 2\cosh^2 \varphi_2 \right], \tag{13.12.15}$$

$$\frac{\partial \varphi_2}{\partial y} = \frac{1}{L} \frac{\partial W}{\partial \varphi_2} = \frac{e^{-2\varphi_1}}{2L} (e^{6\varphi_1} - 2) \sinh(2\varphi_2), \tag{13.12.16}$$

$$\frac{\partial A}{\partial y} = -\frac{2}{3L} W, \tag{13.12.17}$$

while the potential is

$$V = \frac{1}{3L^2} \left(\frac{\partial W}{\partial \varphi_1} \right)^2 + \frac{1}{2L^2} \left(\frac{\partial W}{\partial \varphi_2} \right)^2 - \frac{16}{3L^2} W^2. \tag{13.12.18}$$

There are two fixed points generated by this potential. The first is the (trivial) UV fixed point $\varphi_1 = \varphi_2 = 0$. The other is at

$$\varphi_1 = \frac{1}{6} \log 2, \quad \varphi_2 = \pm \frac{1}{2} \log 3. \tag{13.12.19}$$

Integrating (13.12.17) around the fixed points we obtain the asymptotic forms of the warp factor A

$$A_{UV} \simeq \frac{y}{L}, \quad A_{IR} \simeq \frac{2^{5/3}}{3} \frac{y}{L}, \quad L_{IR} = \frac{3}{2^{5/3}} L. \tag{13.12.20}$$

Therefore, the IR fixed point corresponds to a CFT as expected.

The flow equations in (13.12.15)–(13.12.17) cannot be solved exactly. We can solve them either numerically, or asymptotically. In particular, near the UV fixed point we obtain

$$\varphi_1 \simeq A_0^2 \frac{2y}{3L} e^{-2y/L} + \frac{A_1}{\sqrt{6}} e^{-2y/L} + \cdots, \quad \varphi_2 \simeq A_0 e^{-y/L} + \cdots. \tag{13.12.21}$$

The operator φ_2 behaves as an operator of dimension three, (a fermion mass) and A_0 corresponds to the UV value of the mass. The operator φ_1 is a mixture of an operator of dimension two (A_0^2) and a vev of an operator of dimension 2.

In the IR the asymptotic behavior is

$$\varphi_1 \simeq \frac{1}{6}\log 2 - \frac{\sqrt{7}-1}{6}B_0\,e^{(\sqrt{7}-1)/L_{IR}\gamma} + \cdots, \quad \varphi_2 \simeq \frac{1}{2}\log 3 - B_0\,e^{(\sqrt{7}-1)/L_{IR}\gamma} + \cdots. \quad (13.12.22)$$

The IR CFT can be obtained by integrating out the Φ_3 chiral multiplet. A naive treatment using (13.12.11) and (13.12.12) gives a superpotential of the form

$$W_{IR} \sim \frac{\mu^2}{m}\mathrm{Tr}([\Phi_1,\Phi_2]^2). \quad (13.12.23)$$

In order for this to preserve conformal invariance, the operators $\Phi_{1,2}$ must have IR dimension 3/4. Then, the operator Φ_3 must have IR dimension 3/2 so that the perturbations are marginal. In exercise 13.52 you are asked to show that the two-loop β-function of this theory is zero.

13.13 The Randall-Sundrum Geometry

An interesting question is whether cutting off the UV in the field theory context has a counterpart in the bulk supergravity. We will start from AdS$_5$ in the coordinates of the last section (with $\gamma \to -\gamma$)

$$ds^2 = dy^2 + e^{-2\gamma/L}dx_i dx^i, \quad (13.13.1)$$

in which the boundary is now at $y = -\infty$. We will confine ourselves to five dimensions and we will neglect other compact dimensions in the rest of this section.

To remove the neighborhood near the boundary, we restrict the range of the radial coordinate to $y \geq 0$.[12] This restriction is not enough. The space with $y > 0$ is not complete. Waves can easily cross into the region $y < 0$. Appropriate boundary conditions are needed on the $y = 0$ hypersurface. The simplest way to achieve this is to put reflective boundary conditions. This can be formulated as the requirement of the invariance of the theory under the $\mathbb{Z}_2$ orbifold $y \to -y$.

The point $y = 0$ is a fixed point of the orbifold action, and a δ-function singularity appears into the Einstein equations that stem from the five-dimensional action

$$S_5 = \frac{1}{16\pi G_5}\int d^5 x\sqrt{g}\left[R + \frac{12}{L^2}\right]. \quad (13.13.2)$$

To see this, we will implement the $\mathbb{Z}_2$ orbifold action by keeping the original range in y, $-\infty < y < \infty$ but replace the metric in the $y < 0$ part by the mirror of the $y > 0$ part

$$ds^2 = dy^2 + e^{-2\frac{|y|}{L}}dx^i dx^i, \quad y \in \mathbb{R}. \quad (13.13.3)$$

A direct computation of the five-dimensional Einstein tensor components gives

$$E_{\mu\nu} \equiv R_{\mu\nu} - \frac{1}{2}g_{\mu\nu}R, \quad E_{\gamma\gamma} = \frac{6}{L^2}g_{\gamma\gamma}, \quad E_{ij} = \left[\frac{6}{L^2} - \frac{6}{L}\delta(\gamma)\right]g_{ij}, \quad (13.13.4)$$

all others being zero. As expected, we have the Einstein tensor of AdS$_5$ but for a δ-function singularity at $y = 0$, due to the orbifolding procedure. To make this compatible with

---

[12] Any other choice $y > a$ is related via a conformal transformation to the canonical one we chose.

equations of motion we are therefore required to insert a localized energy density at $y = 0$ of the appropriate size. This is essentially the (constant) tension of a three-brane embedded in five dimensions.

The effective five-dimensional action we must consider is

$$S = S_5 - \delta(y) \int d^4x \sqrt{\hat{g}} \, \Lambda_4, \tag{13.13.5}$$

where Λ_4 is the tension of the three-brane and $\hat{g}$ is the induced metric. According to (13.13.4) the tension must be

$$\Lambda_4 = 24 \frac{M^3}{L}, \quad M^3 \equiv \frac{1}{32\pi G_5} \tag{13.13.6}$$

in order for the equations of motion to be satisfied.

We have ended up with a cutoff AdS_5 (the boundary is missing), with a three-brane localized in the position of the $\mathbb{Z}_2$ orbifold singularity. The three-brane is therefore a UV brane that we will henceforth call the RS brane.

We now investigate the spectrum of fluctuations of the graviton in this background. An interesting set of fluctuations will be along the four Poincaré-invariant directions, parallel to the three-brane. This can be achieved by parametrizing the metric as

$$ds^2 = dy^2 + e^{-2|y|/L}(\eta_{ij} + h_{ij})dx^i dx^j, \quad \partial^i h_{ij} = h^i{}_i = 0, \tag{13.13.7}$$

and fixing a transverse-traceless gauge. Then, the fluctuation equations stemming from the action (13.13.5) are simply

$$\Box_5 \, h_{ij} = 0 \rightarrow \left[\partial_y^2 - e^{2|y|/L}\partial_i\partial^i - \frac{4}{L}H(y)\partial_y\right] h_{ij} = 0, \tag{13.13.8}$$

where $H(y)$ is the step function, $H(y) = \pm 1$ for $y \gtrless 0$. From now on we drop the ij indices as they are blind to the subsequent calculations. We may look for solutions with well defined four-dimensional momenta and masses $h = h_m(y) \, e^{ip\cdot x}$, $p_i p^i = m^2$. They satisfy

$$\left[\partial_y^2 + m^2 \, e^{2\frac{|y|}{L}} - \frac{4}{L}H(y)\partial_y\right] h_m = 0. \tag{13.13.9}$$

Defining a new function by $\phi_m = e^{-3|y|/2L}h_m$ and a new variable $z = L \, H(y)(e^{\frac{|y|}{L}} - 1)$ we obtain a Schrödinger-like problem

$$\left[-\frac{1}{2}\partial_z^2 + V(z)\right]\phi_m = m^2\phi_m, \quad V(z) = \frac{15}{8L^2\left(1 + \frac{|z|}{L}\right)^2} - \frac{3}{2L}\delta(z). \tag{13.13.10}$$

As is obvious from the potential plotted in figure 13.8, the spectrum of KK masses, m is continuous and extends down to $m = 0$. The massless end point corresponds to a bound state, due to the δ-function in the potential,

$$\phi_0 = \frac{1}{\left(1 + \frac{|z|}{L}\right)^{3/2}}. \tag{13.13.11}$$

It is normalizable, with the inherited norm

$$\|h\|^2 = \int d^5x \sqrt{g} \, g^{ij} g^{kl} \, h_{ik}h_{jl}. \tag{13.13.12}$$

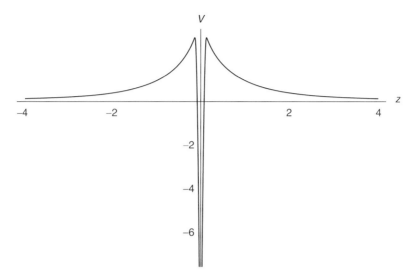

Figure 13.8 The effective potential ($L = 1$).

We have found that there is a massless four-dimensional graviton, localized on the three-brane at $z = 0$. This is unlike the AdS$_5$ case, where a similar procedure would not produce such a state. It would have been non-normalizable on the full AdS$_5$.

In a dual gauge-theory description we expect that the excision of the AdS$_5$ boundary, corresponds to the addition of an UV cutoff to the gauge theory. The fact that the non-normalizable four-dimensional graviton mode of AdS$_5$ has become normalizable, implies that the cutoff gauge theory must be supplemented with an explicit four-dimensional graviton field. The effective four-dimensional Planck mass, which determines its interactions with the four-dimensional matter, can be computed by calculating the coefficient of the quadratic fluctuation terms $h \Box h$ starting from (13.13.5). You are asked in exercise 13.55 to verify that

$$M_p^2 = M^3 L. \tag{13.13.13}$$

It is not, however, obvious at this stage that the full gravitational interaction will have a four-dimensional character. The reason is that there is no mass gap above the graviton zero mode. The continuum of KK modes touches $m = 0$.

You are invited in exercise 13.56 to derive the solutions for the continuum spectrum

$$\phi_m(w) = N(\tilde{m}) w^{1/2} \left[Y_2(\tilde{m}w) + F(\tilde{m}) J_2(\tilde{m}w) \right], \quad w \equiv 1 + \frac{|z|}{L}, \quad \tilde{m} = mL, \tag{13.13.14}$$

where J_2, Y_2 are the standard Bessel functions. The constant $F(\tilde{m})$ is fixed by the discontinuity in $\partial_y \phi_m(0)$ due to the presence of the δ-function

$$F(\tilde{m}) = -\frac{Y_1(\tilde{m})}{J_1(\tilde{m})}. \tag{13.13.15}$$

For $|z| \to \infty$, the KK modes become approximately plane waves

$$\phi_m(w) \simeq N(\tilde{m}) \sqrt{\frac{2}{\pi \tilde{m}}} \left[\sin\left(\tilde{m}w - \frac{5}{4}\pi \right) + F(\tilde{m}) \cos\left(\tilde{m}w - \frac{5}{4}\pi \right) \right]$$

$$\simeq N(\tilde{m}) \sqrt{\frac{2\left(1 + F^2(\tilde{m})\right)}{\pi \tilde{m}}} \sin\left(\tilde{m}w - \frac{5}{4}\pi + \beta(\tilde{m}) \right), \tag{13.13.16}$$

with $\beta = \arctan F$. As a result, for a noncompact fifth dimension, the KK modes have a continuous spectrum and their normalization is approximately that of plane waves

$$N(\tilde{m}) \sim \sqrt{\frac{\tilde{m}}{1 + F^2(\tilde{m})}}, \tag{13.13.17}$$

where we have neglected factors of order 1. The strength of the interaction of the KK graviton modes with the other fields on the brane is determined by the square of their wave functions at the position $z = 0$ of the brane. We find

$$\phi_m(y = 0) \sim \sqrt{\frac{\tilde{m}\left[Y_2(\tilde{m}) + F(\tilde{m})J_2(\tilde{m})\right]^2}{1 + F^2(\tilde{m})}} = -\frac{2}{\pi \sqrt{\tilde{m}\left(J_1^2(mt) + Y_1^2(\tilde{m})\right)}}. \tag{13.13.18}$$

For $\tilde{m} \ll 1$ equation (13.13.18) gives

$$\phi_m(z = 0) \sim \sqrt{\tilde{m}}. \tag{13.13.19}$$

We find that the coupling of low-lying KK modes to fields localized on the three-brane (at $z = 0$) is suppressed. On the other hand, for $\tilde{m} \gg 1$, $\phi_m(z = 0)$ is a constant of order one. The emission of high-energy KK modes is therefore unsuppressed.

13.13.1 An alternative to compactification

It is instructive to compare the behavior of the previous section, with a traditional compactification from five to four dimensions, on a circle of radius R.

There, the zero mode is constant on the circle, and mediates the four-dimensional interaction with effective Planck scale $M_p^2 = M^3 R$. The massive KK modes have a discretum of masses $m_n = |n|/R$, the usual KK tower. Due to the gap between the zero mode and the first massive mode, the gravitational interaction remains four-dimensional for distances $|x| \gg R$. At short distances (or high-energy) however, $|x| \ll R$, all KK modes couple with equal strength, and the gravitational interaction is five-dimensional.

All of the above can be summarized by the study of the static propagator in $\mathbb{R}^{1,3} \times S^1$ for the graviton fluctuations along $\mathbb{R}^3$. As in the previous section, it satisfies the equation

$$M^3 \Box_4\, G = \frac{1}{R}\delta^{(3)}(x)\delta(\theta), \tag{13.13.20}$$

where $\theta \to \theta + 2\pi$ parametrizes the fifth dimension, M is the five-dimensional Planck scale and we have also kept the three space coordinates x^i. The appropriate periodic solution is

$$G(x, \theta; x', \theta') = \frac{1}{(2\pi)^2 M^3} \sum_{n \in \mathbb{Z}} \frac{1}{(x^i - x'^i)^2 + R^2(\theta - \theta' + 2\pi n)^2}. \tag{13.13.21}$$

The summation over an infinite number of image charges implements the periodicity in the coordinate θ.

At short distances, $|x - x'| \ll R$, $|\theta - \theta'| \ll 1$, the dominant contribution to the propagator comes from the $n = 0$ image. In this limit we have

$$G(x, \theta; x', \theta') = \frac{1}{(2\pi)^2 M^3} \frac{1}{(x^i - x'^i)^2 + R^2(\theta - \theta')^2}, \tag{13.13.22}$$

which is the proper five-dimensional behavior for the interaction.

At large distances, $|x - x'| \gg R$, all images contribute importantly. To estimate the sum we must Poisson resum the integer n (see (C.54)) to obtain

$$G(x, \theta; x', \theta') = \frac{\sum_{n \in \mathbb{Z}} e^{-\frac{|n| \, |x-x'|}{R}}}{2(2\pi)^2 M^3 R |x - x'|} \simeq \frac{1}{4\pi \, M^3 (2\pi R) \, |x - x'|}, \tag{13.13.23}$$

exhibiting the correct four-dimensional behavior with Planck scale

$$M_p^2 = M^3 (2\pi R). \tag{13.13.24}$$

Therefore, compactification dominates the interactions in the IR (long distance). It is irrelevant in the UV, where interactions behave as higher dimensional.

It is instructive to compare this with the RS geometry. We will again compute the static propagator

$$M^3 \left(e^{2\frac{|y|}{L}} \Box_3 - \partial_y^2 + 4\frac{H(y)}{L} \partial_y \right) G(\vec{x}, y) = \delta^{(3)}(x) \delta(y), \tag{13.13.25}$$

where $\Box_3$ is the flat Laplacian in the three-spatial dimensions x^i. Going to momentum space for the three-dimensional part, we obtain the equivalent equation

$$M^3 \left(e^{2\frac{|y|}{L}} p^2 - \partial_y^2 + 4\frac{H(y)}{L} \partial_y \right) G(\vec{p}, r) = \delta(y), \tag{13.13.26}$$

with $p^2 = p_1^2 + p_2^2 + p_3^2$.

The symmetry of the problem implies that the propagator is a symmetric function of y, i.e., $G(\vec{p}, y) = G(\vec{p}, -y)$. The solution for $y > 0$ is given by

$$G(\vec{p}, y) = B w^2 K_2(wpL), \quad B = \frac{1}{2M^3 p \, K_1(pL)}, \tag{13.13.27}$$

with $w \equiv e^{\frac{y}{L}}$ and $K_{1,2}$ are the modified Bessel functions. We have also used the identity $K_2'(z) + 2K_2(z)/z = -K_1(z)$ to simplify B.

We will now focus on the static potential at the position of the 3-brane, $y = 0$. This can be obtained from

$$G(\vec{p}, y = 0) = \frac{1}{2M^3 p} \frac{K_2(pL)}{K_1(pL)}. \tag{13.13.28}$$

The corresponding potential due to a unit mass at $(\vec{x} = 0, z = 0)$ as viewed by an observer on the three-brane, at a distance $r = |\vec{x}|$ from the source, is

$$V(|\vec{x}|) = \int \frac{d^3p}{(2\pi)^3} e^{i\vec{p}\cdot\vec{x}} G(\vec{p}, y = 0) = \frac{1}{2\pi^2 |\vec{x}|} \int_0^\infty p\,dp \sin p|\vec{x}| G(\vec{p}, y = 0). \tag{13.13.29}$$

For the case at hand this leads to

$$V(|\vec{x}|) = \frac{1}{(2\pi)^2} \frac{1}{M^3 L |\vec{x}|} \int_0^\infty d\tilde{p} \, \frac{K_2(\tilde{p})}{K_1(\tilde{p})} \sin\left(\frac{\tilde{p}|\vec{x}|}{L}\right) = \frac{1}{4\pi} \frac{1}{M^3 L} \frac{1}{|\vec{x}|} + \delta V, \tag{13.13.30}$$

with

$$\delta V = \frac{1}{(2\pi)^2} \frac{1}{M^3 L |\vec{x}|} \int_0^\infty d\tilde{p} \, \frac{K_0(\tilde{p})}{K_1(\tilde{p})} \sin\left(\frac{\tilde{p}|\vec{x}|}{L}\right), \tag{13.13.31}$$

where, to obtain the second equality we used the identity $K_2(z) = 2K_1(z)/z + K_0(z)$.

For $\tilde{p} \to \infty$, the ratio $K_0(\tilde{p})/K_1(\tilde{p}) \to 1$ and the integral in (13.13.31) reduces to the ill-defined integral of $\sin \tilde{p}$ over the positive real axis. In order to evaluate (13.13.31) we multiply the integrand by $e^{-q\tilde{p}}$, perform the integration and then take the limit $q \to 0$.[13]

At large distances $|\vec{x}| \gg L$ the strongly oscillatory behavior of $\sin\left(\frac{\tilde{p}|\vec{x}|}{L}\right)$ results in a negligible contribution to the integral from large $\tilde{p}$. This means that we can employ the expansion of the Bessel functions for small $\tilde{p}$: $K_0(\tilde{p})/K_1(\tilde{p}) = -\tilde{p} \log \tilde{p}$. We obtain

$$\delta V \simeq \frac{1}{8\pi} \frac{L}{M^3} \frac{1}{|\vec{x}|^3}. \tag{13.13.32}$$

We conclude from (13.13.30),(13.13.32) that at distances on the brane much larger than the AdS length L, the leading gravitational interaction is four-dimensional with effective Planck scale $M_p^2 = M^3 L$, while the next to leading contribution, behaves as $1/|\vec{x}|^3$. This is an indication that although there is no gap in the spectrum of KK gravitons, their density is too weak to dominate the four-dimensional contribution of the zero-mode graviton.

In the short distance limit $|\vec{x}| \ll L$ the main contribution to the integral comes from large $\tilde{p}$, for which $K_0(\tilde{p})/K_1(\tilde{p}) = 1$. We find

$$\delta V \simeq \frac{1}{4\pi^2} \frac{1}{M^3} \frac{1}{|\vec{x}|^2}. \tag{13.13.33}$$

Note that in this regime, the contribution in (13.13.30) is subleading to the one in (13.13.33) that has five-dimensional behavior. Therefore, the gravitational interaction on the brane, at distances shorter than the AdS length is five-dimensional.

We are now ready to tie up the remaining loose ends. Standard compactification produces a gravitational interaction that is four dimensional at long distance scales (compared to the radius of compactification, R). It is, however, five dimensional at short distances.

The RS geometry has a fifth dimension that is noncompact but with finite volume. This gives a continuous KK spectrum in place of a discrete one. However, the gravitational interaction on the three-brane has a similar qualitative behavior as in the case of standard compactification if we substitute the AdS length L for the compactification radius R. It is therefore an alternative form of compactification.

---

[13] The integral depends on the regulator. However, the result remains the same if instead of $e^{-q\tilde{p}}$ one multiplies by $e^{-\alpha\tilde{p}^2}$ and then takes the limit $\alpha \to 0$, while a simple ultraviolet cutoff Λ on the $\tilde{p}$ leads to an ambiguous answer. A correct regulator should reproduce the well-known answer $V(|\vec{x}|) \sim 1/|\vec{x}|^2$ for the case $G(p, 0) \sim 1/p$, corresponding to five-dimensional behavior. The one used satisfies this criterion. Equivalently, one could just use $\int_{-\infty}^\infty dx \, e^{i\alpha x} = 2\pi\delta(\alpha)$ to define the ambiguous integrals mentioned above.

The behavior of the gravitational interaction is encoded into the couplings of KK gravitons to sources on the brane. We found that the low-mass states ($mL \ll 1$), are weakly coupled to the brane sources. At large distances or low energy, it is these states that mediate the interaction. Their weak coupling is subleading to the zero mode interaction. The leading behavior is therefore four dimensional.

On the other hand, at short distances or high energy, it is the high-mass KK states that mediate the interaction. We have found that their couplings to the brane sources are unsuppressed. They therefore dominate over the zero mode, and render the interaction five-dimensional.

An equivalent way of describing the RS gravity is the following. Gravitons are localized on the three-brane at low energy. At high energies, there is important leakage towards the five-dimensional bulk, making gravity five-dimensional.

An interesting question is whether we can embed the RS geometry into string theory. So far, no stringy solution is known that reproduces exactly the RS geometry.

It is expected however that geometries which have the qualitative characteristics of the RS geometry, can be parts of string theory compactification manifolds. In particular, in the presence of large fluxes, manifolds may develop throats that resemble the IR limit of the RS geometry. The UV region, with the three-brane and its appropriate boundary conditions should be replaced with the rest of the compactification manifold. An example of this is provided by the Klebanov-Strassler solution [420].

Bibliography

Perturbative gauge theories are reviewed in [421]. There are several good references on the large-N expansion of gauge theories. The original 't Hooft articles [422] are very interesting to read. Very good and extensive reviews include [423,424,425]. A more recent and extensive review is in [426]. The paper [427] also provides (among other things) a shorter introduction. Polyakov's book [428] provides several complementary views on the subject.

Discussions on conformal symmetry and its representations in dimensions $D > 2$ can be found in [429–431]. Early ideas on holography, motivated by black-hole issues were formulated in [432] and elaborated on in [433]. A review on holographic bounds can be found in [434].

The AdS/CFT correspondence has been conjectured in [435]. Its concrete quantitative interpretation has been elaborated in [414, 416, 436]. The thermal aspects as well as the possibility of phase transitions are explained in [437].

An excellent and comprehensive review covering developments up to early 1999 is [438]. It gives a complete guide to relevant literature and covers also aspects that are not covered in this chapter. Other reviews that are shorter or more specialized include [439–445]. A hitch-hiker's guide on higher-dimensional supergravities [446] is also very useful.

The calculation of Wilson loops in various contexts is reviewed in [447]. Early glueball calculations are reviewed in [448]. Later progress can be found in [449]. Detailed information on the AdS_3/CFT_2 correspondence is contained in [438], which describes also its relation to black-hole state counting. This is also addressed in [384]. The action of the Virasoro algebra in AdS_3 was described in [450]. AdS solutions corresponding to half-BPS states are described in [411].

The holographic renormalization group flow is treated nicely in the book [9] and is also reviewed in [451]. We have followed [452] in our exposition of the renormalization of the gravitational action in section 13.8.3. A complete treatment of renormalization that highlights relevant issues can be found in [453]. A nice exposition of the Hamilton-Jacobi method can be found in [454]. Renormalization group flows are associated to supergravity domain walls. A good review is [455].

The addition of flavor (quarks transforming in the fundamental representation) and calculations of associated meson spectra has a shorter history. Good places to start in order to follow the developments are [456,457,458].

Nonconformal holographic theories are reviewed in [459] and [460]. So far, there are two nonconformal supergravity solutions that are smooth and that describe confining theories in the IR. The first is the Maldacena-Nuñez solution [461,462]. It describes a confining four-dimensional gauge theory in the IR that becomes a higher-dimensional theory in the UV. The other is the Klebanov-Strassler solution [420]. It describes a duality cascade of $SU(N+M)\times SU(N)$ theories. The associated physics is reviewed in [463]. Useful modern reviews of supersymmetric gauge theories can be found in [358,360,464].

In the interest of removing the presence of internal compact manifolds, consideration of noncritical string theories has been revived, in the context of AdS/CFT correspondence. There are novel issues here and this line of investigation is little developed. A good starting point is [465].

Another class of backgrounds with a conjectured holographic description is related to little string theory [466] and linear dilaton backgrounds [467]. Their boundary theory is harder to analyze, being a non-gravitational string theory, with no tunable coupling. The bulk theory can on the other hand be described exactly using standard CFTs.

There are modified large-N limits of gauge theories, that correlate a large number of colors with other large quantum numbers (like R-symmetry charges for example). Via holography, they are described by backgrounds obtained via Penrose limits from AdS backgrounds [468]. Useful reviews of the correspondence can be found in [469,470]. Some properties of such limits can be obtained from semiclassical rotating string configurations, reviewed in [471]. They give rise to relations between SYM theory and integrable spin chains [472,473] raising the possibility that $\mathcal{N}=4$ SYM may be integrable.

An interesting case for holography is the presence of multiple AdS boundaries in the bulk spacetime. In the Lorentzian case they are necessarily separated by horizons. Holography then describes an entangled state of the the various field theories living on the boundaries [474]. The simplest example is the eternal AdS black hole. It has two AdS boundaries in its extended Penrose-Carter diagram. In this context, it seems that for eternal black holes information is lost [474]. However, for real black holes, the holographic correspondence suggests that a sum over relevant asymptotic geometries is appropriate. In this case, the sum includes the thermal AdS geometry. This is enough to provide the small correlations needed so that the information is transmitted to the outside world. The associated Euclidean case with multiple boundaries [475,476] is much less understood.

The Randall-Sundrum geometry has been described in [477]. Its particle physics and cosmological aspects are reviewed in [261,478,479,480].

Exercises

13.1. Show, by an explicit construction of the surface that the first of the diagrams in figure 13.2 corresponds to a sphere while the second one to a torus.

13.2. Consider a large-N theory that on top of the adjoint fields contains also fields transforming as fundamentals. Redo the counting of diagram factors as in (13.1.6) and show that the diagrams can now be mapped to open two-dimensional Riemann surfaces.

13.3. Consider single-trace gauge-invariant operators of the form

$$\Phi_a = \frac{1}{N} \text{Tr}\left[\prod_i X_i^{n_i}\right]. \tag{13.1E}$$

Their correlators can be studied by adding them in the action (13.1.4) on page 406 multiplied by a coupling constant $N\, g_a$, and further differentiating the vacuum diagrams with respect to g_a. This way, our previous estimates still hold. Use them to show that to leading order, and for connected correlators, $\langle \prod_{a=1}^m \Phi_a \rangle_c \sim N^{2-n}$. Such operators are closed string (glueball) operators. This estimate indicates that all such states are free at $N = \infty$, while $1/N$ controls directly their interactions.

13.4. Consider double-trace operators of the form

$$\Psi_a = \frac{1}{N^2} \text{Tr}\left[\prod_i X_i^{n_i}\right] \text{Tr}\left[\prod_i X_i^{m_i}\right]. \tag{13.2E}$$

Normalize their two-point functions and then compute their three-point functions. Compare with the result of the previous exercise.

13.5. Consider the large-N theory that also contains fundamentals $q_i, \bar{q}_{\bar{i}}$. Form the meson operators $q\bar{q}$, normalize their two-point functions and subsequently estimate the leading behavior of their multipoint correlation functions. Compare to open string theory states.

13.6. Consider pure SO(N) and Sp(2N) gauge theory. Generalize the discussion of the text, to develop the large-N expansion of the diagrams. What type of string theory would these correspond to?

13.7. Show that if you use the unit radius metric for AdS$_5 \times$ S^5 and therefore put all the radius dependence into the Newton constant G_N and the string scale ℓ_s, then they scale as

$$G_N \sim \frac{1}{N^2}, \quad \ell_s \sim (g_s N)^{-1/4}. \tag{13.3E}$$

13.8. Consider IIB string theory on AdS$_5 \times$ S^5. We may reduce the theory on S^5 to obtain the effective five-dimensional theory on AdS$_5$. Start from the ten-dimensional Newton constant $16\pi\, G_N = (2\pi)^7 \ell_s^8\, g_s^2$ and obtain the effective five-dimensional one as

$$G_5 = \frac{\pi L^3}{2N^2}. \tag{13.4E}$$

13.9. Consider a vector operator in a CFT$_d$ dual to AdS$_{d+1}$. Solve for the asymptotic behavior of the dual bulk massive vector close to the boundary, and from this derive the scaling dimension of the vector operators as function of mass quoted in the text:

$$\Delta_{\text{spin-1}} = \frac{1}{2}\left[d \pm \sqrt{(d-2)^2 + 4m^2 L^2}\right]. \tag{13.5E}$$

13.10. Consider a field theory in d dimensions with an abelian conserved global current, J_μ: $\partial_\mu J^\mu = 0$. Consider coupling this to a "source" A_μ as

$$W(A_\mu) = \langle e^{\int d^d x \, J^\mu A_\mu} \rangle. \tag{13.6E}$$

Show that the global invariance of the field theory associated to the current J^μ implies a (gauge) invariance of the W functional under $A_\mu \to A_\mu + \partial_\mu \epsilon$. Use this to argue that if this field theory has a $(d+1)$-dimensional bulk dual theory, then the bulk vector field dual to the conserved global current J^μ is a gauge field. A global U(1) symmetry of the boundary theory therefore translates to a U(1) gauge symmetry for the bulk theory.

13.11. Redo the argument of the previous exercise with a source coupled to the (conserved) stress-tensor of the boundary field theory. Show that this implies a diffeomorphism invariance for the spin-2 source. Argue that translation invariance of the boundary field theory translates into the presence of gravity in the dual bulk theory.

13.12. Start from the metric of AdS$_5$ in global coordinates and find the coordinate transformation that brings it to the form (13.6.1) on page 416.

13.13. Calculate the geodesic distance between two points $x_{1,2}$ on the cutoff-boundary sphere at $\tilde{u}^2 = 1 - \epsilon$ of the AdS metric given in (13.6.1) on page 416 and show that it scales as $\log(|x_1 - x_2|/\epsilon)$.

13.14. Calculate the regulated AdS$_5$ spatial volume up to $\tilde{u}^2 = 1 - \epsilon$ and show that it scales like the area in (13.6.3) on page 417 for $\epsilon \ll 1$.

13.15. Explain how the presence of a horizon in AdS$_5$ is related to the absence of a mass gap in the boundary theory.

13.16. Consider the general multi-center solution describing N D$_3$-branes at generically distinct positions in transverse space. This realizes the moduli space $\mathbb{R}^{6N}/S_N$ of $\mathcal{N} = 4$ U(N) gauge theory. Consider various distributions of the branes and investigate when they have weak curvature. Use this to ascertain whether this moduli space is visible at the near-horizon limit.

13.17. Consider the fluctuations of the axion-dilaton scalar τ, in IIB supergravity around AdS$_5 \times S^5$. Calculate the masses of the full KK tower of field τ_k and show that they are given by

$$m_k^2 = \frac{k(k+4)}{L^2}. \tag{13.7E}$$

13.18. Show that (K.40) satisfies that massive Laplace equation in AdS$_{p+2}$. Use a scaling argument to verify (K.47) on page 551 and calculate the constant C_p. Finally verify (K.49).

13.19. Use the definition of the extrinsic curvature in section 12.3.1 to calculate its trace at the boundary of AdS_{p+2}. Show that $K = (p+1)/L$.

13.20. Redo the two-point function calculation of the massive scalar in momentum space using the propagator (K.44) on page 550. Show that divergent contact terms appear as polynomials in momenta. Show that the two-point function comes from the first nonanalytic contribution in momenta. Compare with the result of in (13.8.15) on page 427.

13.21. Consider a massless vector in AdS_5. Go through the procedure described in section 13.8.1 on page 425 and derive the two-point function of the dual current.

13.22. Map carefully the IR renormalizations in AdS_5 described in section 13.8.1 on page 425 to UV operator renormalization in the gauge theory.

13.23. By substituting the form of the asymptotically AdS metric (13.8.28) on page 430 into the gravitational equations (13.8.23), derive the expressions for the coefficient functions (13.8.30), (13.8.31), and (13.8.34).

13.24. Expand (13.8.35) on page 431 and derive the expressions for A_0, A_2, and A_4.

13.25. Show that the difference between (13.8.38) and (13.8.39 on page 431) are finite terms as we approach the boundary.

13.26. Derive (13.8.45) on page 432 from (13.8.43) and (13.8.44).

13.27. Verify the (K.51) on page 551 provides the properly normalized bulk to bulk propagator, in AdS_{p+2}.

13.28. Use (K.53) on page 551 to show the relation (13.8.49) on page 432.

13.29. Calculate the contribution to the three-point function in section 13.8.2 on page 427 coming from the boundary term. Show that it vanishes.

13.30. Consider a bulk scalar with a cubic and a quartic vertex. Calculate the four-point function by calculating and summing the two relevant diagrams in figure 13.3 on page 429.

13.31. Use the AdS_5 string theory to compute a circular Wilson loop on a circle of radius R. Can you guess the qualitative form of the result using the symmetries?

13.32. Using similar techniques as in section 13.9 calculate the monopole-monopole static potential and show that it is compatible with S-duality.

13.33. Derive the Hawking temperature (13.10.6) on page 438 as well as the entropy and chemical potential in (13.10.7) for the black D_3-brane in (13.10.4) and show that they satisfy the first law of black-hole thermodynamics $dM = TdS + \Phi dN$.

13.34. Calculate the energy E_{YM} and entropy S_{YM} of thermal $\mathcal{N} = 4$ SYM at large N and to leading order at weak 't Hooft coupling and verify the expressions in (13.10.10) and (13.10.11) on page 439.

13.35. Show that under the gauge transformation (13.10.2), the temporal Wilson loop transforms as in (13.10.3) on page 437.

13.36. Calculate at one loop the scalar masses at finite temperature in $\mathcal{N} = 4$ SYM.

13.37. Calculate the specific heat of large and small AdS_5 black holes. Comment on their stability.

13.38. Consider the quantum mechanics in one dimension of the double well potential. Although, the quantum field theory intuition would suggest that the $\mathbb{Z}_2$ symmetry of the potential must spontaneously break, argue that this is not the case. Generalize to arbitrary (local) field theories in finite volume.

13.39. Calculate the rectangular Wilson loop in the x-t plane in thermal $\mathcal{N} = 4$ SYM and extract the interquark potential at finite temperature. Show that there are two regimes in the calculation: For distances $l \ll 1/T$, one obtains small modifications of the nonthermal potential. For distances $l \ll 1/T$ the quarks are free. How does this fit with the discussion of section 13.10.4 on page 443?

13.40. Consider a gauge theory dual to the following general bulk metric:

$$ds^2 = -g_{00}(r)dt^2 + g(r)dx^i dx^i + g_{rr}(r)dr^2 + \cdots, \tag{13.8E}$$

and compute the purely spatial Wilson loop along the lines of section 13.10.5 on page 444. Define $f^2 = g_{00}g$, $g^2 = g_{00}g_{rr}$, and let r_0 a point that corresponds to a minimum of f or a singularity of g. Show that if $f(r_0) \neq 0$ we obtain the area law at large distances for the Wilson loop.

13.41. Derive the near-horizon metric of a D_4-brane. In analogy with thermal SYM, derive the metric describing a compactification of a longitudinal direction on circle with supersymmetry-breaking conditions for the fermions. This is the analog of QCD_3 in four dimensions. Calculate the Wilson loop and derive the string tension. Is this theory closer to QCD compared to its three-dimensional analog, described in section 13.10.5 on page 444?

13.42. Derive the AdS-Reissner-Nordström black-hole solution, by minimizing the action $\int d^5x \sqrt{g} \left[R - \frac{1}{4}F^2 + \frac{12}{L^2} \right]$. Argue, that it is holographically dual to thermal $\mathcal{N} = 4$ SYM with a chemical potential for an abelian R-charge. By studying this solution determine the phases and possible phase transitions of thermal $\mathcal{N} = 4$ SYM with a chemical potential generalizing the discussion in section 13.10 on page 436.

13.43. Consider the equation (13.10.43) for the 0^{++} glueball spectrum in cutoff QCD_3. Using as an auxiliary tool the quadratic on-shell Euclidean action, show that the spectrum of M^2 is strictly positive and discrete. Analyze the same question in AdS_5. What are the qualitative differences?

13.44. Consider the M-theory solution that describes N M_5-branes in exercise 11.3 on page 364. Take the near-horizon limit and show that the geometry obtained is that of $AdS_7 \times S^4$. Is this background conformally flat? Formulate the bulk-boundary correspondence between M-theory on this background and an appropriate limit of the world-volume theory.

13.45. Consider the M-theory solution that describes N M_2-branes in exercise 11.2 on page 363. Take the near-horizon limit and show that the geometry obtained is that of $AdS_4 \times S^7$. What is the boundary theory in this case? Formulate the bulk-boundary correspondence.

13.46. Consider the appropriate near-horizon limit of the supergravity extremal D_2-brane solution. Analyze the validity of various descriptions as a function of the 't Hooft coupling and N by considering the effective string coupling. Show that as the holographic energy scale U decreases we pass from the perturbative SYM description to the D_2-brane supergravity description , to the wrapped M_2 brane supergravity description to the $\mathcal{N} = 8$ SCFT description. At what values of U are the transition regions?

13.47. Derive from first principles the effective three-dimensional Newton constant in (13.11.2) on page 448.

13.48. Derive the commutator of two Brown-Henneaux diffeomorphisms (13.11.5) on page 448 and show that it generates the Virasoro algebra.

13.49. Consider the leading correction to the AdS_3 beyond the asymptotic form (13.11.4) on page 448, due to a Virasoro diffeomorphism in (13.11.5). Show that this modifies the expectation value of the stress tensor along the lines discussed in section 13.8.3 on page 428. Compute it, and from this derive the Virasoro central charge verifying (13.11.7) on page 449.

13.50. Consider the holographic RG flows described in section 13.12.2 on page 452. Use the weak-energy condition $\rho + p_i \geq 0$, $\forall\ i$ for the diagonal supergravity stress tensor to argue that $A'' \leq 0$. Use this to show that the bulk space cannot develop a new boundary, except for the original UV boundary.

13.51. Derive the asymptotic expansions (13.12.21) on page 455 and (13.12.22).

13.52. Use the two-loop $\mathcal{N} = 1$ gauge beta-function to show that the IR theory with superpotential (13.12.23) on page 456 is conformally invariant.

13.53. In section 9.13 we have seen that string theory in the absence of R-R backgrounds does not have any continuous global (internal) symmetries. For string theories that have a (boundary) dual field theory, we may also make a similar statement: show that bulk/boundary duality prohibits global symmetries in the bulk theory unless they are also local.

13.54. Consider the product of two large-N conformal field theories. Consider a relevant interaction coupling them together in the UV. Describe the space-time picture of the interacting theory. Note that the dual bulk theory contains a massless and a massive graviton.

13.55. Obtain the kinetic terms of the massless graviton fluctuations in the RS geometry, and derive the effective four-dimensional Planck mass (13.13.13) on page 458.

13.56. Derive the expressions for the continuum of KK modes in the RS geometry, and verify expressions (13.13.14)–(13.13.17) on page 459.

13.57. Use Poisson resummation to derive the long-distance limit (13.13.23) on page 460 of the five-dimensional propagator (13.13.22).

13.58. Sum the Yukawa potentials e^{-mr}/r due to the continuum of KK modes in the RS geometry, to find the leading and subleading behavior of the static potential on the three-brane in (13.13.30) and (13.13.32) on page 461. When does this summation break down?

13.59. You are asked to use AdS/CFT ideas in order to recalculate the next-to-leading correction to the (static) gravitational interaction (13.13.32) on page 461. The $4d$ graviton couples to the CFT stress tensor. Show that the next-to-leading correction to the graviton propagator originates from the the stress-tensor two-point function. Use it to reproduce (13.13.32).

13.60. Consider the static gravitational potential on the hyperplane (three-brane) $y = y_0$ in the RS background. What is the effective Planck scale compared to that at $y_0 = 0$? How can this idea be used to generate a large hierarchy of scales?

13.61. Consider a RS brane that separates two slices of AdS_5 with different curvatures. Derive the analog of the RS solution and study the gravitational interactions on the RS brane.

13.62. Another alternative to compactification can be provided by a noncompact finite-volume manifold, whose Laplacian has a finite gap above zero in its spectrum. Find an example, and study the associated gravitational interaction on various distance scales.

14 | String Theory and Matrix Models

We have seen in the rest of this book that string theory does not contain only strings. Extended objects like branes seem to form an integral part of it. Moreover, in the previous chapter we have witnessed a deep connection between (large-N) gauge theory dynamics and string theory in specific gravitational backgrounds.

Such a connection has its roots in the IR-UV duality between open and closed strings. On one side, we have D-branes and their associated world-volume gauge theory. On the other, closed strings and gravity that reflect the back-reaction to the open string dynamics.

The discussion of the large-N expansion in section 13.1 on page 405 indicates that any theory with $N \times N$ matrix fields, and representations whose dimensions do not scale faster than N^2, has at large N the structure of a string theory.[1] There have been several proposals to give alternative descriptions of string theory using simple large-N gauge theories. We will present in this chapter some of these proposals. Although such proposals were motivated in different ways during their advent, the idea of open-closed string duality seems to lie behind each of them and unifies their validity.

Matrix descriptions of string theory typically have the advantage that they can access non-perturbative physics in a quantitative way. We also had indications on the way that matrix descriptions of string theory necessarily are background dependent. Therefore, such an attempt of defining string theory seems to be contrary in spirit to the picture that emerged after the advent of non-perturbative string theory dualities.

In the next sections we will describe two interesting matrix descriptions of specific string theory vacua. The first comes under the name of M(atrix) theory or the BFSS model, and describes flat 11-dimensional M-theory in the discrete light-cone quantization (DLCQ) formalism.

The second is today known as (old) matrix models. It provides a description of $D \leq 2$ noncritical string theory. It emerges via a direct discretization of the path integral over two-dimensional metrics.

---

[1] This string theory may, however, not be any of the ones we study in this book.

14.1 M(atrix) Theory

As we saw in chapter 11, the strong-coupling limit of the ten-dimensional IIA theory is an 11-dimensional theory, called M-theory, whose low-energy limit is $\mathcal{N} = 1_{11}$ supergravity. This is the vacuum of string theory with the highest symmetry, namely, SO(10,1) Lorentz invariance. It is suspected that it should be at the root of a nonperturbative, all-encompassing definition of string theory, also called M-theory.

A possible choice of degrees of freedom involves the pointlike field-theoretic states of the supergravity multiplet. However, as in string theory, we do not expect these to be the fundamental degrees of freedom. Instead they are expected to be the fluctuations of another fundamental object.

The other two classical excitations of the theory involve the M_2-membrane and its magnetic dual, the M_5-brane. The M_2-brane upon an S^1 compactification gives the fundamental IIA string. It has long been suspected that the 11-dimensional theory would arise from the quantization of the M_2-supermembrane, just as string theory arises from the quantization of the fundamental string.

14.1.1 Membrane quantization

Quantizing the supermembrane is a harder task than the analogous string quantization. The reason is that in the Polyakov formulation, the associated three-dimensional dynamical gravity has now nontrivial dynamics.

In this section we will sketch the problems associated with this approach, in order to put it in perspective. We will study the bosonic membrane for simplicity, as it is enough to indicate the issues.

The Nambu-Goto action for the membrane is

$$S_{M_2} = -T_2 \int d^3\xi \left[\sqrt{-\det \hat{g}} + \hat{C}_{\alpha\beta\gamma} \epsilon^{\alpha\beta\gamma} \right], \quad T_2 = \frac{1}{(2\pi)^2 g_s \ell_s^3} = \frac{1}{(2\pi)^2 \ell_{11}^3}, \tag{14.1.1}$$

with

$$\hat{g}_{\alpha\beta} = G_{\mu\nu} \partial_\alpha X^\mu \partial_\beta X^\nu, \quad \hat{C}_{\alpha\beta\gamma} = C_{\mu\nu\rho} \partial_\alpha X^\mu \partial_\beta X^\nu \partial_\gamma X^\rho. \tag{14.1.2}$$

$C_{\mu\nu\rho}$ is the three-form of the $\mathcal{N} = 1_{11}$ supergravity. $\xi^\alpha \sim (\tau, \sigma_1, \sigma_2)$ are the world-volume coordinates.

We obtain the analogous Polyakov action by introducing a world-volume metric $\gamma_{\alpha\beta}$

$$\tilde{S}_{M_2} = -T_2 \int d^3\xi \left[\frac{\sqrt{\gamma}}{2} \left(\gamma^{\alpha\beta} G_{\mu\nu} \partial_\alpha X^\mu \partial_\beta X^\nu - 1 \right) + \hat{C}_{\alpha\beta\gamma} \epsilon^{\alpha\beta\gamma} \right]. \tag{14.1.3}$$

The classical equations of (14.1.3) imply that

$$\gamma_{\alpha\beta} = \hat{g}_{\alpha\beta}. \tag{14.1.4}$$

The world-volume theory in (14.1.3) is a three-dimensional gravity theory, with a cosmological constant coupled to 11 scalar fields X^μ. Unlike the case of the string, in three dimensions gravity has nontrivial degrees of freedom, that cannot be gauged away.

A convenient gauge is

$$\gamma_{0i} = 0, \quad \gamma_{00} = -\det \hat{g}_{ij}, \tag{14.1.5}$$

where $i, j = 1, 2$ are labeling the space coordinates on the membrane. We may now use this choice of gauge and (14.1.4) to rewrite the action as

$$\tilde{S}_{M_2} = \frac{T_2}{2} \int d^3\xi \left(\dot{X}^\mu \dot{X}_\mu - \det \hat{g}_{ij} \right) = \frac{T_2}{2} \int d^3\xi \left(\dot{X}^\mu \dot{X}_\mu - \frac{1}{2} \{X^\mu, X^\nu\}\{X_\mu, X_\nu\} \right). \tag{14.1.6}$$

We have set $G_{\mu\nu} = \eta_{\mu\nu}$, $C_{\mu\nu\rho} = 0$ in (14.1.6) for simplicity. We have also introduced the Poisson bracket on the spacial slices, $\xi^0 = \text{constant}$, of the world-volume

$$\{F, G\} \equiv \epsilon^{ij} \partial_i F \partial_j G = \partial_1 F \partial_2 G - \partial_2 F \partial_1 G. \tag{14.1.7}$$

The associated equations of motion are

$$\ddot{X}^\mu = \{\{X^\mu, X^\nu\}, X_\nu\}. \tag{14.1.8}$$

The constraints following from the metric variations

$$\dot{X}^\mu \dot{X}_\mu + \frac{1}{2} \{X^\mu, X^\nu\}\{X_\mu, X_\nu\} = 0, \quad \dot{X}^\mu \partial_i X_\mu = 0, \tag{14.1.9}$$

must also be imposed. The last constraint implies

$$\{\dot{X}^\mu, X_\mu\} = \partial_1 (\dot{X}^\mu \partial_2 X_\mu) - \partial_2 (\dot{X}^\mu \partial_1 X_\mu) = 0. \tag{14.1.10}$$

As is usually the case, the constraints can be solved in a light-cone gauge. We introduce the light-cone coordinates $X^\pm = (X^0 \pm X^{10})/\sqrt{2}$ and pick the light-cone gauge condition

$$X^+(\tau, \sigma_1, \sigma_2) = \tau. \tag{14.1.11}$$

We may now solve the constraints determining X^-

$$\dot{X}^- = \frac{1}{2} \dot{X}^i \dot{X}_i + \frac{1}{4} \{X^i, X^j\}\{X_i, X_j\}, \quad \partial_i X^- = \dot{X}^j \partial_i X_j. \tag{14.1.12}$$

Translating to the Hamiltonian formalism we obtain

$$p^+ \equiv \int d^2\sigma \, P^+ = T_2 \int d^2\sigma \, \dot{X}^+ = V_2 T_2, \tag{14.1.13}$$

with V_2 the spatial volume of the membrane. The light-cone Hamiltonian is

$$H = \frac{T_2}{2} \int d^2\sigma \left(\dot{X}^i \dot{X}^i + \frac{1}{2} \{X^i, X^j\}\{X_i, X_j\} \right). \tag{14.1.14}$$

The only constraint that remains to be imposed is (14.1.10).

An attempt to quantize this system involves a discretization of the two-dimensional space. Take each interval of the $\sigma_{1,2}$ coordinates to have N lattice points (or bits). This turns the coordinates X^μ into Hermitian N×N matrices

$$X_\mu(\tau, \sigma_1, \sigma_2) \to X_\mu(\tau, I, J) \to X_\mu^{IJ}(\tau). \tag{14.1.15}$$

The Poisson brackets turn into matrix commutators and the integral over space becomes a trace:

$$\{\cdot,\cdot\} \to -\frac{i}{2}[\cdot,\cdot], \quad \frac{1}{V_2}\int d^2\sigma \to \frac{1}{N}\text{Tr}. \tag{14.1.16}$$

The matrix Hamiltonian can be therefore written as

$$H = \frac{T_2}{4}\text{Tr}\left(\frac{1}{2}\dot{X}^i\dot{X}^i - \frac{1}{4}[X^i,X^j][X^i,X^j]\right), \tag{14.1.17}$$

where we have appropriately rescaled the coordinates X^i and the time τ. It must be supplemented by the constraint

$$[\dot{X}^i, X^i] = 0. \tag{14.1.18}$$

We have ended up with a quantum mechanical system with $9N^2$ degrees of freedom.

The supermembrane action also has some fermionic partners θ of the bosonic coordinates. In the light-cone gauge, $\Gamma^+\theta = 0$, and the remaining fermionic coordinates form a 16-dimensional Majorana spinor of the SO(9) Lorentz symmetry, transverse to the light cone. The Hamiltonian is now

$$H = \frac{T_2}{4}\text{Tr}\left(\frac{1}{2}\dot{X}^i\dot{X}^i - \frac{1}{4}[X^i,X^j][X^i,X^j] + \frac{1}{2}\theta^T\Gamma_i[X^i,\theta]\right). \tag{14.1.19}$$

The Hamiltonians (14.1.17) and (14.1.19) are plagued by a continuous spectrum reflecting the classical instabilities of the membrane and its supersymmetric generalization.

Indeed the classical membrane action indicates that the membrane is unstable to the formation of spikes. Imagine a simple cylindrical spike of cross section ϵ and length L. It has a volume ϵL. Its energy is $E \sim \epsilon L T_2$. We observe that we may take $L \to \infty$, and $\epsilon \to 0$, keeping the energy finite and small. Therefore such very thin and long spikes are not suppressed by the action. You are required to show in exercise 14.4 that a similar instability does not exist for a string.

In the discrete formulation, the classical instability indicates that the membrane is unstable towards dissolving into the discrete "bits." Quantum effects could cure the classical instability. They indeed do so in the bosonic case. However, they do not in the supersymmetric case that is of interest here, due to the cancellation of the zero point energies between the bosonic and fermionic degrees of freedom.

14.1.2 Type-IIA D_0-branes and DLCQ

We may now compactify the eleven-dimensional theory on a circle, to return to the IIA description. Here we have point-particle states, the D_0-branes. The DLCQ is defined as a light-cone quantization with one compact light-cone direction X^- of radius (or period) R. Therefore, the light-cone momentum is quantized $P^+ = N/R$.

We will now argue, that in DLCQ, the D_0-brane dynamics is similar to that of the discrete supermembrane. Moreover, the instabilities indicated in the previous section are perfectly acceptable in this context: The D_0-bits are the proper degrees of freedom, unlike the membranes.

We start from a compact eleventh dimension, a circle of radius R_s. This implies that we identify

$$\begin{pmatrix} x^{10} \\ t \end{pmatrix} \sim \begin{pmatrix} x^{10} - R_s \\ t \end{pmatrix}. \tag{14.1.20}$$

According to our discussion in section 11.4, this is IIA string theory with

$$R_s = g_s \ell_s, \quad \ell_{11}^3 = g_s \ell_s^3. \tag{14.1.21}$$

Consider now boosting the theory to a new frame moving with velocity

$$v = \frac{1}{\sqrt{1 + 2\frac{R^2}{R_s^2}}} \simeq 1 - \frac{R_s^2}{R^2} + \mathcal{O}(R_s^4). \tag{14.1.22}$$

R is an arbitrary length scale that will eventually become the radius of the lightlike circle in the $R_s \to 0$ limit.

After the boost, the identification becomes

$$\begin{pmatrix} x^{10} \\ t \end{pmatrix} \sim \begin{pmatrix} x^{10} - \sqrt{\frac{R^2}{2} + R_s^2} \\ t + \frac{R}{\sqrt{2}} \end{pmatrix}. \tag{14.1.23}$$

If we take $R \gg R_s$, to leading order the theory is defined on a compactified lightlike circle of radius R,

$$\begin{pmatrix} x^{10} \\ t \end{pmatrix} \sim \begin{pmatrix} x^{10} - \frac{R}{\sqrt{2}} \\ t + \frac{R}{\sqrt{2}} \end{pmatrix} + \mathcal{O}\left(\frac{R_s^2}{R^2}\right). \tag{14.1.24}$$

Therefore, x^- is compact with radius R, and the associated momentum p^+ is quantized $p^+ = \frac{N}{R}$. As $R/R_s \to \infty$, $v \to 1$. Therefore we boost at the speed of light. The boosted frame is known as the *infinite-momentum frame*.

Consider now a collection of N D_0-branes with energy and momentum in the 11th direction

$$E = \frac{N}{R_s} + \Delta E, \quad P = \frac{N}{R_s}, \tag{14.1.25}$$

where ΔE are the fluctuations in energy. The light-cone energy and momentum in the infinite momentum frame are

$$P^- = \frac{E - P}{\sqrt{2}} = \frac{\Delta E}{\sqrt{2}} \sqrt{\frac{1+v}{1-v}} \simeq \frac{R}{R_s} \Delta E, \quad P^+ = \frac{E + P}{\sqrt{2}} = \frac{\sqrt{2}N}{R_s} \sqrt{\frac{1-v}{1+v}} \simeq \frac{N}{R}. \tag{14.1.26}$$

We are now ready to take the lightlike limit,

$$R_s \to 0, \quad \ell_{11} \to \text{fixed}, \tag{14.1.27}$$

which implies that $g_s \to 0$ and $\ell_s \to \infty$. This is the tensionless limit of IIA string theory, a rather difficult regime. However, the energies of interest to us,

$$\frac{\Delta E}{M_s} = \frac{R_s}{R} P^- \ell_s = \frac{P^-}{R} \sqrt{R_s \ell_{11}^3} \to 0, \tag{14.1.28}$$

are well below the string scale in this limit. Because of this, the world-volume action for the D_0-branes reduces to its nonrelativistic limit. This is just as well since we do not know most of the higher-order corrections to the nonabelian D-brane action.

To summarize, the DLCQ of M-theory with momentum $P^+ = N/R$ is described by N D_0-branes, in the limit $g_s \to 0$, $\ell_s \to \infty$ keeping ℓ_{11} fixed, in the infinite momentum frame. In that limit, the ten-dimensional Planck scale is given by $M_P^8 \sim \ell_{11}^9/R_s \to \infty$. Therefore, we can neglect the gravitational/closed string back-reaction to the D_0-branes even if their number N is large.

The relevant Lagrangian for the N D_0-branes according to our previous discussion is

$$\mathcal{L} = \frac{1}{2g_s \ell_s} \text{Tr} \left[D_t X^i D_t X^i + \frac{1}{2}[X^i, X^j]^2 + \theta^T (iD_t \theta - \Gamma_i[X^i, \theta]) \right], \tag{14.1.29}$$

with

$$D_t X^i = \dot{X}^i - i[A_t, X^i]. \tag{14.1.30}$$

We have chosen the static gauge and rescaled variables appropriately. X^i are nine N×N Hermitian matrices. Similarly θ is an N×N set of 16-dimensional Majorana spinors of SO(9). Any one-dimensional gauge field on the real line is pure gauge and can therefore be gauged away. We will therefore set $A_t = 0$. As usual, the associated Gauss's law constraint imposes

$$[X^i, \dot{X}^i] = 0. \tag{14.1.31}$$

We may now compute the Hamiltonian (up to rescaling)

$$H = \frac{g_s \ell_s}{2} \text{Tr} \left[P^i P^i - \frac{1}{2}[X^i, X^j]^2 + \theta^T \Gamma_i[X^i, \theta] \right]. \tag{14.1.32}$$

It is obvious that (14.1.32) and (14.1.31) coincide with the discretized membrane Hamiltonian (14.1.19) and the associated constraint (14.1.18), up to a scaling.

In this language we may understand the membrane instabilities. Consider the equations of motion stemming from (14.1.29)

$$\ddot{X}^i = -[[X^i, X^j], X^j], \tag{14.1.33}$$

where we set the gauge field and the fermions to zero. We choose X^i to be matrices that are block diagonal, containing two blocks,

$$X^i = \begin{pmatrix} X_1^i & 0 \\ 0 & X_2^i \end{pmatrix}, \tag{14.1.34}$$

where X_1^i are $N_1 \times N_1$ matrices, X_2^i are $N_2 \times N_2$ matrices, and $N = N_1 + N_2$. It is obvious that the equation (14.1.33) factorizes into two noninteracting systems. This can be interpreted as the possibility for the system to describe multiple independent entities. These entities are collections of D_0-branes. Their center-of-mass positions are

$$x_1^i = \frac{1}{N_1}\text{Tr}[X_1^i], \quad x_2^i = \frac{1}{N_2}\text{Tr}[X_2^i]. \tag{14.1.35}$$

From this point of view, the theory for $N \to \infty$ can be thought of as a second quantized theory of D_0-bits.

14.1.3 Gravitons and branes in M(atrix) theory

A simple class of time-dependent solutions to the classical equations (14.1.33) are given by

$$X^i = \begin{pmatrix} x_1^i + v_1^i t & 0 & 0 & \ddots \\ 0 & x_2^i + v_2^i t & \ddots & 0 \\ 0 & \ddots & \ddots & 0 \\ \ddots & 0 & 0 & x_N^i + v_N^i t \end{pmatrix}. \tag{14.1.36}$$

They correspond to a classical N-graviton solution, where each graviton has

$$p_a^+ = 1/R, \quad p_a^i = v_a^i/R, \quad E_a = v_a^2/(2R) = (p_a^i)^2/2p^+. \tag{14.1.37}$$

In the special case

$$x_1^i = \cdots = x_N^i, \quad v_1^i = \cdots = v_N^i, \tag{14.1.38}$$

the trajectories of all the components are identical. We interpret this as a single graviton with $p^+ = N/R$.

In exercise 14.7 you are invited to derive the integrated space-time stress tensor of a graviton as

$$T^{IJ} = \frac{p^I p^J}{p^+}, \tag{14.1.39}$$

where

$$p^+ = N/R, \quad p^i = p^+ \dot{x}^i, \quad p^- = p_\perp^2/2p^+. \tag{14.1.40}$$

Extended objects include the membrane and the five-brane. The best way to identify their charges is via the supersymmetry algebra in (14.3E) on page 499, where they appear as central charges. The membrane charge $M^{\mu\nu\rho}$ is dual to the three-form. Its dual, $M^{\mu_1\mu_2\mu_3\mu_4\mu_5\mu_6}$ is the five-brane charge. We will give here the associated charges without derivation. The interested reader may consult the bibliography at the end of this chapter for their derivation.

$$M^{+-i} \sim i\text{Tr}\left(P^j[X^i, X^j] + [[X^i, \theta^\beta], \theta^\beta]\right), \quad M^{+ij} \sim -i\text{Tr}[X^i, X^j], \tag{14.1.41}$$

$$M^{+-ijkl} \sim \text{Tr}[X^{[i}X^jX^kX^{l]}]. \tag{14.1.42}$$

M^{+-i} corresponds to a membrane wrapped around the 11-dimensional circle to give a fundamental IIA string. Similarly, M^{+-ijkl} corresponds to a wrapped M_5.

14.1.4 The two-graviton interaction from M(atrix) theory

The case $N = 1$ describes a single D_0-brane, or from the M-theory point of view, a single supergraviton with $P^+ = 1/R$ with $R = g_s \ell_s (2\pi \ell_s^2)$. If we consider two noninteracting supergravitons, according to the discussion of the previous section, we should consider 2×2 diagonal matrices. Interactions will be generated by integrating out the off-diagonal elements. To set up the calculation, we start from the action in (14.1.29) and expand it around a general background

$$X^i = B^i + Y^i, \quad A_t = B_t + Q_t, \tag{14.1.43}$$

where B^i are the background values for X^i and B_t for the gauge field A_t. Y^i, Q_t describe the fluctuations. The background values are assumed to satisfy the classical equations of motion stemming from the action (14.1.29). We also pick the background-field gauge

$$D_B Q \equiv \dot{Q}_t - i[B_t, Q_t] = 0. \tag{14.1.44}$$

To implement it we add $(D_B Q)^2$ to the action together with the appropriate ghosts, C and $\bar{C}$. We may write the Euclidean gauge-fixed action in the background field gauge as

$$S = S_0 + S_2 + S_Q + S_f + S_{gh} + S_{\text{int}}, \tag{14.1.45}$$

$$S_0 = \frac{1}{2R} \int dt \, \text{Tr} \left[\dot{B}^i \dot{B}^i + \frac{1}{2} [B^i, B^j]^2 \right], \tag{14.1.46}$$

$$S_2 = \frac{1}{2R} \int dt \, \text{Tr} \left[\dot{Y}^i \dot{Y}^i - [B^i, Y^j][B^i, Y^j] - [B^i, B^j][X^i, X^j] \right], \tag{14.1.47}$$

$$S_Q = \frac{1}{2R} \int dt \, \text{Tr} \left[\dot{Q}_t \dot{Q}_t - [Q_t, B^i][Q_t, B^i] - 2i \dot{B}^i [Q_t, Y^i] \right], \tag{14.1.48}$$

$$S_f = \frac{1}{2R} \int dt \, \text{Tr} \left[\theta^T (\dot{\theta} - \Gamma_i [B^i, \theta]) \right], \quad S_{gh} = \frac{1}{2R} \int dt \, \text{Tr} \left[\dot{\bar{C}} \dot{C} - [B^i, \bar{C}][B^i, C] \right], \tag{14.1.49}$$

where under the Wick rotation, the gauge fields transform as $A \to -iA$. Here we will focus on a one-loop computation. S_{int} contains cubic and higher-order interactions that are therefore not relevant for the one-loop computation.

We must first choose the background corresponding to two supergravitons moving past each other. In view of the interpretation of X^i as the coordinates of D_0-branes, we choose

$$B^1 = -\frac{i}{2} \begin{pmatrix} vt & 0 \\ 0 & -vt \end{pmatrix}, \quad B^2 = \frac{1}{2} \begin{pmatrix} b & 0 \\ 0 & b \end{pmatrix}, \quad B^{i>2} = 0. \tag{14.1.50}$$

This background describes two supergravitons moving past each other with relative velocity v and impact parameter b in the x^2 direction.

In this background, the off-diagonal degrees of freedom are massive and we will integrate them out.

We now substitute (14.1.50) into the quadratic action. The ten bosonic fields Q_t, Y^i at quadratic level are harmonic oscillators with a mass matrix

$$(\Omega_B)^2 = \begin{pmatrix} r^2 & -2iv & 0 & \cdots & 0 \\ 2iv & r^2 & 0 & \ddots & 0 \\ 0 & 0 & r^2 & \ddots & \vdots \\ \vdots & \ddots & \ddots & \ddots & 0 \\ 0 & 0 & \cdots & 0 & r^2 \end{pmatrix}. \tag{14.1.51}$$

$r^2 = b^2 + (vt)^2$ is the separation between the supergravitons. We also have two complex off-diagonal ghosts with $\Omega^2 = r^2$. Moreover, there are 16 fermionic oscillators with a mass-squared matrix

$$(\Omega_F)^2 = r^2 \mathbf{1}_{16} + v\gamma_1. \tag{14.1.52}$$

To calculate the potential we also need the propagators. However, for the leading contribution at large distance, we may use the quasistatic approximation in which the effective potential is approximated by the vacuum energy

$$V_{\text{eff}} = \sum \omega_B - \frac{1}{2}\omega_F - \omega_{gh}. \tag{14.1.53}$$

The 1/2 in the fermion contribution is due to the fact that they are real oscillators (rather than complex). This approximation is valid when the frequencies $\omega \gg \frac{v}{r}$.

We may diagonalize the frequency matrices in (14.1.51),(14.1.52). We find for the bosons $\omega_b = r$ with multiplicity 8, $\omega_b = \sqrt{r^2 \pm 2v}$ with multiplicity one each. For the two ghosts, $\omega_{gh} = r$, while for the fermions, $\omega_b = \sqrt{r^2 \pm v}$ with multiplicity 8 each. Altogether,

$$V_{\text{eff}} = 6r + \sqrt{r^2 + 2v} + \sqrt{r^2 - 2v} - 4\sqrt{r^2 + v} - 4\sqrt{r^2 - v} \simeq -\frac{15}{16}\frac{v^4}{r^7} + \mathcal{O}\left(\frac{v^6}{r^{11}}\right). \tag{14.1.54}$$

We will now indicate that this is the correct behavior expected from classical linearized DLCQ supergravity in eleven dimensions.

The massless scalar propagator in DLCQ is

$$D(x) = \frac{1}{2\pi R}\sum_{n\in\mathbb{Z}}\int \frac{dp^- d^9 p_\perp}{(2\pi)^{10}}\frac{e^{-i\frac{n}{R}x^- - ip^- x^+ + ip_\perp \cdot x_\perp}}{2\frac{n}{R}p^- - p_\perp^2}. \tag{14.1.55}$$

The contribution of interest to us is the $n = 0$ term in (14.1.55) since it corresponds to the exchange of supergraviton with zero longitudinal momentum. It is given by

$$D_0(x - y) = -\frac{1}{2\pi R}\delta(x^+ - y^+)\frac{15}{32\pi^4}\frac{1}{|x_\perp - y_\perp|^7} = -\frac{15}{2(2\pi)^6 g_s \ell_s^3}\frac{\delta(x^+ - y^+)}{|x_\perp - y_\perp|^7}. \tag{14.1.56}$$

Both the graviton as well as the three-form exchange involve, apart from polarization-dependent pieces, the propagator (14.1.55).

The interaction of two gravitons is proportional to the square of their energy,

$$E_{1,2} = \frac{v^2}{2R} = \frac{v^2}{4\pi g_s \ell_s^3}.$$
(14.1.57)

Their interaction potential is then

$$V = -\kappa_{11}^2 E_1 E_2 D_0(r) = -\frac{15v^4}{16r^7},$$
(14.1.58)

where κ_{11} is given in (11.4.6) on page 330. This is in agreement with the matrix model calculation (14.1.54).

14.2 Matrix Models and $D = 1$ Bosonic String Theory

Another class of matrix models has been used in directly defining the Polyakov path integral over two-dimensional metrics. We will describe this for the simplest possible bosonic string background: that corresponding to a CFT with zero central charge. In other words, there is neither space, nor time. According to our discussion in sections 6.2 and 6.3 on page 149, there will still be a nontrivial Liouville mode dependence, that will dress the trivial CFT into a $c = 26$ CFT. This CFT will be given solely in terms of the Liouville mode. The theory has no physical propagating degrees of freedom. However, it will be instructive to analyze it in order to demonstrate the main issues before we pass to a non-trivial example.

For the $c = 0$ CFT, the (perturbative, unrenormalized) Polyakov partition function takes the form

$$\mathcal{Z}_p = \sum_h \int \mathcal{D}g \, e^{-bA + d\chi}.$$
(14.2.1)

The sum is over closed Riemann surfaces of genus h. A is the area of the surface, multiplied by b, the unrenormalized world-sheet cosmological constant. d controls the sum over genera, by coupling to the Euler number $\chi = 2 - 2h$.

The idea here is to discretize appropriately the closed Riemann surfaces, in order to render the sum over metrics computable. One will then need to take the continuum limit.

For example, we could randomly triangulate Riemann surfaces using equilateral triangles as shown in figure 14.1. Curvature is introduced if the number of triangles N_i incident a vertex i is different from six. The local form of the curvature is proportional to the deficit angle at each vertex $1 - \frac{N_i}{6}$. The Euler number is given by $\chi = \sum_i \left(1 - \frac{N_i}{6}\right)$. On the other hand, the discretized form of the area is essentially given by $A = \frac{1}{3} \sum_i N_i$. We have taken without loss of generality the area of a single triangle to be one.

A proper definition of the discretized theory should not depend on the particular form of the discretization. In particular if we use other polygons, we should recover the same result in the continuum limit. It will be verified later that this is indeed the case, unless the discretized theory is fine-tuned.

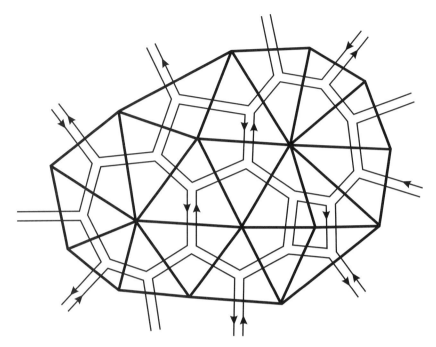

Figure 14.1 A random triangulation of a Riemann surface. Each triangle is dual to a three-vertex insertion of the matrix model.

In summary, the idea is to replace

$$\sum_h \int \mathcal{D}g \rightarrow \sum_{\text{triangulations}} .$$ (14.2.2)

One way in which this can be achieved quantitatively has been indicated in section 13.1 on page 405. There, it was shown that the diagrams of any U(N)-invariant matrix theory generate discretizations of Riemann surfaces. The strategy will be to pick a simple enough matrix theory, and use it to generate the combinatorics of the sum over surfaces, required in the Polyakov path integral.

The simplest U(N) invariant model which has a cubic vertex in order to generate the triangulation is given by[2]

$$e^{\mathcal{Z}_m} = \int dM \, e^{N[-\frac{1}{2}\text{tr}M^2 + g\text{tr}M^3]}.$$ (14.2.3)

M is an N×N hermitian matrix. According to its definition, $\mathcal{Z}_m$ generates the connected graphs. At large N, $\mathcal{Z}_m$ has a double power-series expansion in 1/N and g. As shown in section 13.1 on page 405, the 1/N expansion is a genus expansion

$$\mathcal{Z}_m = \sum_{h=0}^{\infty} N^{2-2h} Z_h(g).$$ (14.2.4)

---

[2] Due to the cubic term, the path integral is ill defined. For the moment we can define it in perturbation theory, where we expand in a power series in g. It can be rendered well defined if we add higher terms without affecting the continuum limit.

At fixed genus, $Z_h(g)$ is given by a sum over triangulations[3] weighted by g^n, where n is the number of cubic vertices of the diagram. This is equal to the number of triangles of the dual diagram and therefore to the total area A.

$$Z_h(g) = \sum_{\text{triangulations of genus } h} g^A. \tag{14.2.5}$$

Thus, $\mathcal{Z}_m$ is a discretization of $\mathcal{Z}_p$ with $g = e^{-b}$ and $N = e^d$.

The simplest operators of the matrix model are the single traces

$$O_n = \text{Tr}[M^n]. \tag{14.2.6}$$

The most general U(N)-invariant operators are products of the O_n. These operators are also known as loop operators. In exercise 14.13 you are invited to show that the Feynman diagrams for the one-point function of O_n correspond to a sum over discretized surfaces with a loop of length n cut-out.

14.2.1 The continuum limit

We are eventually interested in reproducing continuum Riemann surfaces. This requires taking the continuum limit. In our convention where the area of a single triangle is kept fixed, this amounts to arrange that the most important contributions to the path integral originate from triangulated surfaces with an infinite number of triangles.

To see how this may appear, consider a possible critical behavior of the sphere (leading order in $1/N$) matrix partition function $Z_0(g)$

$$Z_0(g) \sim (g - g_c)^{2-\gamma}. \tag{14.2.7}$$

In this case, the average area of the surfaces scales as

$$A = \langle n \rangle = g \frac{\partial \log Z_0}{\partial g} \sim g \frac{2 - \gamma}{g - g_c} \to \infty. \tag{14.2.8}$$

Therefore, as $g \to g_c$, we reach a continuum limit, where surfaces of infinite area dominate the path integral. This is the same limit as keeping the total area fixed, but taking the area of individual triangles to go to zero.

The exponent γ turns out to correspond to the observable that was discussed in the continuum formulation of the theory: the string critical exponent γ. To see this, remember that according to the definition in (6.3.7) on page 150, on the sphere

$$Z_0(A) \sim A^{\gamma-3}. \tag{14.2.9}$$

On the other hand, expanding (14.2.8) we obtain

$$Z_0 \sim (g_c - g)^{\gamma-2} \sim \sum_n n^{\gamma-3} \left(\frac{g}{g_c} \right)^n. \tag{14.2.10}$$

From (14.2.10) it follows that γ in (14.2.8) is the string critical exponent. It was computed in the continuum approach in (6.3.12).

---

[3] The triangles are dual to the matrix theory cubic vertices as shown in figure 14.1.

For pure two-dimensional gravity, $D = 0$ and we obtain

$$\gamma_{D=0} = -\frac{1}{2}. \tag{14.2.11}$$

In the next section we will indeed show that matrix models have such a critical behavior. Moreover, we will calculate the string exponent and show that it agrees with the continuum formulation.

14.2.2 Solving the matrix model

At large N, the matrix path integral in (14.2.3) can be approximated by saddle-point methods. Here however, we will introduce another method of computation, namely that of orthogonal polynomials. It is more powerful than other methods if we need the subleading corrections in $1/N$.

Let us consider the general connected matrix-model partition function

$$e^{\mathcal{Z}} \equiv \int dM \, e^{-\operatorname{tr} V(M)}. \tag{14.2.12}$$

We can diagonalize a Hermitian matrix by a U(N) transformation. We may therefore change variables, to an integral over the U(N) group and one over the N real eigenvalues λ_i. We normalize the U(N) volume to 1. We then obtain

$$e^{\mathcal{Z}} = \int \prod_{i=1}^{N} d\lambda_i \Delta^2(\lambda) e^{-\sum_i V(\lambda_i)}, \quad \Delta(\lambda) \equiv \prod_{i<j} (\lambda_i - \lambda_j). \tag{14.2.13}$$

The Vandermonde determinant $\Delta(\lambda)$ is the Jacobian for the change of variables from the matrix M to its eigenvalues. In exercise 14.15 you are invited to show that the Vandermonde determinant can be written also as

$$\Delta(\lambda) = \det \lambda_i^{j-1}. \tag{14.2.14}$$

In this language, the loop operator O_n amounts to introducing

$$O_n \to \sum_{i=1}^{N} \lambda_i^n \tag{14.2.15}$$

in the path integral.

We now introduce the infinite set of polynomials $P_n(\lambda)$, $n = 0, 1, 2, \ldots$, which are orthogonal under the inner product

$$\int_{-\infty}^{\infty} d\lambda \, e^{-V(\lambda)} P_n(\lambda) P_m(\lambda) = h_n \delta_{m,n}. \tag{14.2.16}$$

P_n is a polynomial in λ of degree n. They are normalized by choosing the coefficient of the leading term to be one, $P_n(\lambda) = \lambda^n + \cdots$. Using the properties of determinants we may now rewrite (14.2.14) as

$$\Delta(\lambda) = \det P_{j-1}(\lambda_i). \tag{14.2.17}$$

Using this relation, the integral in (14.2.13) is over products of orthogonal polynomials. In exercise 14.17 you are invited to show that expanding the determinant, and using the orthogonality properties of the polynomials we obtain

$$e^{\mathcal{Z}} = N! \prod_{i=0}^{N-1} h_i = N! \, h_0^N \prod_{k=1}^{N-1} f_k^{N-k}, \quad f_k \equiv \frac{h_k}{h_{k-1}}. \tag{14.2.18}$$

At large N, the variable $\zeta = \frac{k}{N}$ becomes a continuous real number $\zeta \in [0,1]$. Moreover, the ratio f_k/N becomes a continuous function $f(\zeta)$. In the large-N limit we may therefore rewrite (14.1.4) as

$$\frac{\mathcal{Z}}{N^2} \sim \frac{1}{N} \sum_k \left(1 - \frac{k}{N}\right) \log f_k \sim \int_0^1 d\zeta (1-\zeta) \log f(\zeta). \tag{14.2.19}$$

We will now proceed to determine $f(\zeta)$. For simplicity purposes only we will assume that the potential $V(M)$ is an even function[4] of M.

We first derive the recursion relation

$$\lambda P_n(\lambda) = P_{n+1}(\lambda) + r_n P_{n-1}(\lambda). \tag{14.2.20}$$

This can be justified as follows. From the definition of P_n, the highest power of λ on the left-hand side is λ^{n+1} with coefficient 1. This explains the first term on the right-hand side. The evenness of $V(\lambda)$ implies that parity $\lambda \to -\lambda$ is a symmetry, and $P_n(-\lambda) = (-1)^n P_n(\lambda)$. Therefore, only polynomials with the the same parity as P_{n+1} can appear on the right-hand side. Finally, P_m with $m < n-1$ cannot appear since $\int \lambda P_n P_m e^{-V}$ vanishes in that case, as seen by taking λ act on P_m. We may now compute in two different ways

$$h_n = \int d\lambda \, P_n(\lambda P_{n-1}) \, e^{-V} = \int (\lambda P_n) P_{n-1} \, e^{-V} = r_n h_{n-1}, \tag{14.2.21}$$

where in the second step we used (14.2.20). From (14.2.21) we deduce that $f_n = r_n$. We also find

$$n h_n = \int e^{-V} (\lambda P_n') P_n = \int e^{-V} P_n'(\lambda P_n) = r_n \int e^{-V} P_n' P_{n-1} = r_n \int e^{-V} V' P_n P_{n-1}, \tag{14.2.22}$$

where we have used again (14.2.20) and in the last step an integration by parts.

Equation (14.2.22) allows us to compute r_n recursively. This determines f_n, which is central in computing the partition function in (14.2.18), (14.2.19).

To proceed further, we pick a simple concrete potential function

$$V(\lambda) = \frac{1}{2g}\left(\lambda^2 + \frac{\lambda^4}{N}\right), \quad gV' = \lambda + 2\frac{\lambda^3}{N}, \tag{14.2.23}$$

and calculate

$$\int e^{-V} V' P_n P_{n-1} = \frac{1}{g} \int e^{-V}\left(\lambda + 2\frac{\lambda^3}{N}\right) P_n P_{n-1} = \frac{h_n}{g} + \frac{2}{gN} \int e^{-V} \lambda^3 P_n P_{n-1}. \tag{14.2.24}$$

---

[4] The procedure we describe is also valid in the general case. The formulas are just a bit more complicated.

We use (14.2.20) repeatedly to obtain

$$\lambda^3 P_{n-1} = P_{n+2} + (r_n + r_{n-1} + r_{n+1})P_n + (r_n r_{n-1} + r_{n-1}^2 + r_{n-1} r_{n-2})P_{n-2}$$
$$+ r_{n-1} r_{n-2} r_{n-3} P_{n-4} \tag{14.2.25}$$

which in turn implies that

$$\int e^{-V} V' P_n P_{n-1} = \frac{h_n}{g} + \frac{2h_n}{gN}(r_n + r_{n-1} + r_{n+1}). \tag{14.2.26}$$

Finally, the recursion relation (14.2.22) translates to

$$g\, n = r_n \left[1 + \frac{2}{N}(r_n + r_{n-1} + r_{n+1})\right]. \tag{14.2.27}$$

At large N, $r_n/N \to f(\zeta)$ and $r_{n\pm 1}/N \to f(\zeta \pm \epsilon)$ with $\epsilon = 1/N$. (14.2.27) becomes a finite difference equation,

$$g\zeta = f(\zeta)[1 + 2(f(\zeta) + f(\zeta + \epsilon) + f(\zeta - \epsilon))]. \tag{14.2.28}$$

To leading order in ϵ, (14.2.28) gives

$$g\zeta = f + 6f^2 \equiv W(f) = -\frac{1}{24} + 6\left(f + \frac{1}{12}\right)^2. \tag{14.2.29}$$

As we will soon see, at $f = f_c = -\frac{1}{12}$ and with $g_c = W(f_c) = -\frac{1}{24}$, we will obtain critical behavior. For this, we invert (14.2.29)

$$f(\zeta) - f_c = \sqrt{\frac{g\zeta - g_c}{6}}. \tag{14.2.30}$$

In exercise 14.18 you are invited to investigate the universality of the critical behavior of the continuum limit. In particular, for a generic polynomial potential,[5] without any tuning of parameters, $f(\zeta) \sim \sqrt{g\zeta - g_c}$. Continuing in exercise 14.19, you are invited to investigate the case of general polynomial potential and show that by fine-tuning the parameters of the potential, (14.2.30) generalizes to

$$f(\zeta) - f_c \sim (g\zeta - g_c)^{-\gamma}, \quad \gamma = -\frac{1}{m}, \quad m = 2, 3, \dots . \tag{14.2.31}$$

It is now straightforward to substitute (14.2.31) into (14.2.19), keep the most singular terms, and integrate by parts to obtain

$$\frac{\mathcal{Z}}{N^2} \sim \int_0^1 d\zeta\,(1 - \zeta)(g_c - g\zeta)^{-\gamma} = -\frac{1}{g(1-\gamma)}\,(1-\zeta)(g_c - g\zeta)^{-\gamma+1}\Big|_0^1$$
$$+ \frac{1}{g^2(1-\gamma)(2-\gamma)}\,(g_c - g\zeta)^{-\gamma+2}\Big|_0^1 \sim \frac{(g - g_c)^{-\gamma+2}}{g_c^2(1-\gamma)(2-\gamma)}$$
$$\sim \sum_n \frac{1}{g_c^\gamma \Gamma(\gamma)}\, n^{\gamma-3} \left(\frac{g}{g_c}\right)^n . \tag{14.2.32}$$

---

[5] This corresponds to a general discretization using various polygons.

This is precisely the critical behavior already advertised in (14.2.7). As explained in section 14.2.1, as $g \to g_c$, continuous surfaces dominate the discretized sum over surfaces.

Moreover, from (14.2.30),(14.2.31) the string critical exponent is $\gamma = -\frac{1}{2}$, compatible with the continuum Polyakov calculation (14.2.11).

Expanding f in a power series in N, we may use the equation (14.2.28) to calculate f, and therefore $\mathcal{Z}$ to arbitrary order in $1/N$.

14.2.3 The double-scaling limit

We have so far argued that to leading order in $1/N$ (sphere), the matrix model partition function, near the continuum limit scales as

$$\mathcal{Z}_{h=0} = N^2 \sum_n n^{\gamma-3} \left(\frac{g}{g_c}\right)^n \sim (g_c - g)^{2-\gamma}, \tag{14.2.33}$$

where as usual h is the number of handles, related to the Euler number as $\chi = 2 - 2h$ and n counts the number of triangles (or polygons) and therefore the area. The higher-order large-N contributions in the matrix model also reproduce the continuum scaling (6.3.11) on page 150

$$\mathcal{Z}_h \sim N^\chi \sum_n n^{(\gamma-2)\chi/2-1} \left(\frac{g}{g_c}\right)^n \sim (g_c - g)^{(2-\gamma)\chi/2}. \tag{14.2.34}$$

In exercise 14.20 you are invited to derive this for the case of the torus.

Equation (14.2.34) therefore implies that near the continuum limit the higher-genus contributions are enhanced. This suggests the possibility of a combined limit $N \to \infty$ and $g \to g_c$, so that all genus contributions survive.

To implement this, we transcribe (14.2.34) near the critical point as

$$\mathcal{Z}_m = \sum_{h=0}^{\infty} \Xi_h N^{2-2h} (g - g_c)^{(2-\gamma)(1-h)}. \tag{14.2.35}$$

We define the renormalized string coupling g_s as

$$\frac{1}{g_s} \equiv N(g - g_c)^{(2-\gamma)/2} \tag{14.2.36}$$

and rewrite (14.2.35) as

$$\mathcal{Z}_m = \sum_{h=0}^{\infty} \Xi_h g_s^{2h-2}. \tag{14.2.37}$$

The double-scaling limit is therefore defined as taking $N \to \infty$ and $g \to g_c$, while keeping their combination g_s in (14.2.36) fixed. It is the limit that should produce the continuum (perturbative) formulation, from the matrix model formulation. An extra advantage of the matrix model formulation is that it is not limited to the perturbative regime.

To proceed further, we will implement this limit in our simple matrix model of pure $2d$ gravity, described by (14.2.23), (14.2.27). We define the scaling variables a and z as

$$g - g_c = g_s^{-4/5} a^2, \quad \epsilon = \frac{1}{N} = a^{5/2}, \quad g_c - g\zeta = a^2 z. \tag{14.2.38}$$

The double-scaling limit is obtained by taking $a \to 0$. (14.2.30) implies that we may also define the scaling variable for the function $f(\zeta)$ as

$$f(\zeta) \equiv f_c + a \, u(z) = -\frac{1}{12} + a u(z). \tag{14.2.39}$$

We may rewrite (14.2.28) as

$$g\zeta = f(\zeta)[1 + 6f(\zeta) + 2(f(\zeta + \epsilon) + f(\zeta - \epsilon) - 2f(\zeta))] \tag{14.2.40}$$

and then use that as $\epsilon \to 0$

$$f(\zeta + \epsilon) + f(\zeta - \epsilon) - 2f(\zeta) \simeq \epsilon^2 f''(\zeta) = g_c^2 a^2 \frac{\partial^2 u}{\partial z^2}, \tag{14.2.41}$$

to finally derive in the limit $a \to 0$

$$z = -6u^2(z) + \frac{g_c^2}{6} u''(z). \tag{14.2.42}$$

By rescaling z and u, (14.2.42) is equivalent to the Painlevé I equation

$$z = u^2 - \frac{u''}{3}. \tag{14.2.43}$$

This equation determines the scaling function $u(z)$. In the scaling region, $\zeta \sim 1$ and from (14.2.38),

$$z = \lim \frac{g - g_c}{a^2} = g_s^{-4/5}. \tag{14.2.44}$$

The function $u(z)$ is essentially the second derivative of the matrix model partition function with respect to the string coupling,

$$u(z) \sim \frac{\partial^2 \mathcal{Z}}{\partial^2 z}. \tag{14.2.45}$$

You are invited to show this in exercise 14.23. It is therefore called the "specific heat" of the matrix model.

The Painlevé equation in (14.2.43) is well studied. Its solution can be expanded in a series

$$u(z) = z^{\frac{1}{2}} \left[1 - \sum_{k=1}^{\infty} u_k \, z^{-5k/2} \right], \tag{14.2.46}$$

where the coefficients u_k can be recursively computed from (14.2.43) and are all positive.

The leading term is $u \sim \sqrt{z} \sim g_s^{-2/5}$ and should correspond to the sphere contribution. (14.2.45) indeed gives

$$\mathcal{Z} \sim z^{5/2} \sim \frac{1}{g_s^2} \tag{14.2.47}$$

as required.

The generating set of observables of this theory are the single-trace operators O_n in (4.12.30). A convenient way of calculating their correlators is to start from the generalized partition function

$$e^{\mathcal{Z}(g_2,g_3,\cdots)} \equiv \int dM \exp\left[-N\sum_{n=2}^{\infty} g_n O_n\right]. \tag{14.2.48}$$

This action is of the general form that can be solved by the orthogonal polynomial method. The correlators can be obtained by appropriately differentiating $\mathcal{Z}$ in (14.2.48) with respect to the couplings.

The example that we have pursued so far is string theory in $D = 0$ dimensions. Although, it is ideal for exposing the matrix model ideas and techniques, it falls short of providing many physically interesting observables. There are $(m-1)$-matrix generalizations, as described in exercise 14.16, that in the scaling limit describe the coupling of two-dimensional gravity to the Virasoro minimal models with central charge $c = 1 - \frac{6}{m(m+1)}$. In these generalizations, the $(m-1)$ matrices are coupled together like in a finite chain with nearest-neighbor interactions.

We may think of the $(m-1)$ species of matrices as providing a new $\mathbb{Z}_{m-1}$ degree of freedom on the triangulations. It is known that the minimal models describe critical behavior in $\mathbb{Z}_{m-1}$ spin systems. Therefore, this seems to be the correct framework describing their coupling to two-dimensional gravity.

14.2.4 The free-fermion picture

We will now show that the matrix model formalism is related to free fermions. This reformulation is of practical interest, as it will be obvious in the sequel. This is already suggested in (14.2.13). The Vandermonde determinant can on one hand be exponentiated to form part of the action. This gives a logarithmic pairwise repulsion of the eigenvalues. However, it is instead possible to include a single power of the Vandermonde determinant in the wave functions. This has the effect that the wave functions become now completely antisymmetric under the interchange of two eigenvalues. In this case, there is no direct interaction between the eigenvalues in the action. In this picture, each eigenvalue feels the other only via the Pauli exclusion principle. It is for this fact that the formalism of free fermions is natural.

We start by defining the second-quantized free-fermion operators associated with a given one-matrix model as

$$\Psi(\lambda) \equiv \sum_{n=0}^{\infty} a_n \psi_n(\lambda), \quad \psi_n(\lambda) = \frac{1}{\sqrt{h_n}} P_n(\lambda) e^{-V(\lambda)/2}. \tag{14.2.49}$$

a_n are fermionic operators satisfying

$$\{a_n, a_m^{\dagger}\} = \delta_{n,m}. \tag{14.2.50}$$

(14.2.49) implies that in the Fock space, $a^{\dagger}$ creates a fermion in the nth state of the system.

An important ingredient is the fact that the system contain N fermions, associated with the N eigenvalues of the matrix model. The natural ground state is then the Fermi sea, $|N\rangle$. It is defined by filling the first N levels,

$$a_n|N\rangle = 0, \quad n \geq N, \quad a_n^\dagger|N\rangle = 0, \quad n < N. \tag{14.2.51}$$

As discussed earlier, the important operators in the matrix model are the loop operators O_n. Expressed in terms of eigenvalues, as in (14.2.15), they involve averages of λ^n in the path integral,

$$\left\langle \prod_i O_{n_i} \right\rangle = \frac{1}{\mathcal{Z}} \int dM \prod_i \left(\text{Tr} M^{n_i} \right) e^{-N\text{Tr}[V(M)]}. \tag{14.2.52}$$

The analogous operator in the free-fermion language is

$$O_n \to \Psi^\dagger \hat{\lambda}^n \Psi \equiv \int d\lambda \Psi^\dagger(\lambda) \lambda^n \Psi(\lambda). \tag{14.2.53}$$

(14.2.52) may be rewritten as

$$\left\langle \prod_i O_{n_i} \right\rangle = \langle N| \prod_i \left(\Psi^\dagger \hat{\lambda}^{n_i} \Psi \right) |N\rangle. \tag{14.2.54}$$

We can prove (14.2.54) for the one-point function as follows:

$$
\begin{aligned}
\langle O_n \rangle &= \frac{1}{\mathcal{Z}} \int \prod_i d\lambda_i \, \Delta^2(\lambda) \left(\sum_i \lambda_i^n \right) \prod_i e^{-V(\lambda_i)} \\
&= \frac{N}{N! \prod_i h_i} \int \prod_i d\lambda_i \left(\det P_{j-1}(\lambda_i) \right)^2 \lambda_1^n \prod_i e^{-V(\lambda_i)} \\
&= \frac{N(N-1)!}{N! \prod_i h_i} \sum_{j=0}^{N-1} \frac{\prod_i h_i}{h_j} \int d\lambda \left(P_j(\lambda) \right)^2 \lambda^n e^{-V(\lambda)} \\
&= \sum_{j=0}^{N-1} \langle \psi_j | \lambda^n | \psi_j \rangle = \langle N | \Psi^\dagger \hat{\lambda}^n \Psi | N \rangle,
\end{aligned} \tag{14.2.55}
$$

where we have used (14.2.17), (14.2.18), and (14.2.49).

We will see the fermionic formalism at work in the next section.

14.3 Matrix Description of $D = 2$ String Theory

We would like to describe the matrix model description of strings moving in 1+1 space-time dimensions. As we have argued in section 6.3 on page 149, this is equivalent to two-dimensional gravity coupled to a one-dimensional timelike $c = 1$ CFT of a scalar X. The extra space coordinate is provided by the conformal factor of the two-dimensional metric.

To discretize this theory, we must triangulate the two-dimensional surfaces, and add a new degree of freedom on every triangle i: the value of the X coordinate, X_i.

We must therefore consider the discretized path integral

$$\mathcal{Z}_{c=1} = \sum_{h=0}^{\infty} \sum_{\text{triangulations}} g_s^{2h-2} e^{-\lambda_0 A} \int \prod_{i=1}^{A} dX_i \; e^{-\sum_{\langle ij \rangle} [(X_i - X_j)^2 / 2\ell_s^2]}, \tag{14.3.1}$$

where $\langle ij \rangle$ stands for nearest neighbors, λ_0 is the bare world-sheet cosmological constant, and A is the number of triangles, or equivalently the area of the surface.

The goal is to find the appropriate set of matrices and interactions that generate the sum in (14.3.1). A hint in this direction is given by the fact that the $c = 1$ theory can be obtained as the large-m limit of the minimal models.[6] As $m \to \infty$, we obtain an infinite chain of $N \times N$ Hermitian matrices, that can be viewed as discretization of a time-dependent matrix $M(x)$. We are therefore lead to consider the quantum mechanics of a Hermitian $N \times N$ matrix $M(x)$ where x is the time.

We write the natural (Euclidean) matrix quantum mechanics (MQM) partition function as

$$e^{\mathcal{Z}} = \int \mathcal{D}M(x) \exp \left\{ -\beta \int_{-T/2}^{T/2} dx \; \text{Tr} \left[\frac{1}{2} \left(\frac{\partial M}{\partial x} \right)^2 + \frac{M^2}{2\ell_s^2} - \frac{M^3}{3!} \right] \right\}$$

$$= \int \mathcal{D}M(x) \exp \left\{ -N \int_{-T/2}^{T/2} dx \; \text{Tr} \left[\frac{1}{2} \left(\frac{\partial M}{\partial x} \right)^2 + \frac{M^2}{2\ell_s^2} - \frac{g \, M^3}{3!} \right] \right\}, \quad g^2 = \frac{N}{\beta}, \tag{14.3.2}$$

where in the second line we have rescaled appropriately the matrix M. T is the length of the time direction. In the Euclidean case it can also be interpreted as the inverse temperature. We will be interested here in the limit $T \to \infty$. Expanding in Feynman diagrams at large N, we obtain

$$\lim_{T \to \infty} \mathcal{Z} = \sum_{h=0}^{\infty} N^{2-2h} \sum_{\text{triangulations}} g^A \int \prod_{i=1}^{A} dX_i e^{-\sum_{\langle ij \rangle} (|X_i - X_j| / \ell_s^2)}, \tag{14.3.3}$$

where $|X_i - X_j|$ is the one-dimensional propagator. A is the number of vertices of the Feynman graph, or the number of triangles of the dual surface.

(14.3.3) is almost identical to (14.3.1), with $g = e^{-\lambda_0}$, but for the "kinetic" terms for the field X. However, the two propagators differ by irrelevant operators only. It is therefore conceivable that in the continuum limit we will still recover the continuum (1+1)-dimensional string theory à la Polyakov. This expectation is indeed correct.

To proceed further, we rewrite the Euclidean partition function as an amplitude using the time-evolution operator (or transfer matrix)

$$e^{\mathcal{Z}} = \langle f | e^{-\beta HT} | i \rangle. \tag{14.3.4}$$

If the initial and final states $|i\rangle, |f\rangle$ have a nonzero overlap with the ground state then

$$\lim_{T \to \infty} \frac{\mathcal{Z}}{T} = -\beta E_0. \tag{14.3.5}$$

We must therefore calculate the ground-state energy of MQM.

---

[6] There are subtleties in that limit, but they do not affect us here.

14.3.1 Matrix quantum mechanics and free fermions on the line

We return to Minkowski space and start from the Lagrangian

$$\mathcal{L} = \mathrm{Tr}\left[\frac{1}{2}\dot{M}^2 - V(M)\right]. \tag{14.3.6}$$

Using the U(N) symmetry we may parametrize the matrix M as

$$M = U^\dagger \Lambda U, \quad U \in \mathrm{U(N)}, \quad \Lambda \quad \text{diagonal.} \tag{14.3.7}$$

We may then compute

$$\mathrm{Tr}(\dot{M}^2) = \mathrm{Tr}(\dot{\Lambda}^2) + \mathrm{Tr}[\Lambda, \dot{U}U^\dagger]^2. \tag{14.3.8}$$

The anti-hermitian traceless matrix $\dot{U}U^\dagger$ can be decomposed in terms of the SU(N) generators

$$\dot{U}U^\dagger = \frac{i}{\sqrt{2}}\sum_{i<j}\left[\dot{a}_{ij}T_{ij} + \dot{b}_{ij}\hat{T}_{ij}\right] + \sum_{i=1}^{N-1}\dot{a}_i H_i. \tag{14.3.9}$$

Here T_{ij} is a matrix with zero entries except for the ij and ji elements which are 1. $\hat{T}_{ij}$ is a matrix with zero entries except for the ij element that is $-i$ and the ji element that is i. Finally H_i is a basis of the Cartan generators of SU(N), involving diagonal, real traceless matrices.

We may therefore write the Lagrangian (14.3.6) in the new basis as

$$\mathcal{L} = \sum_{i=1}^{N}\left[\frac{1}{2}\dot{\lambda}_i^2 - V(\lambda_i)\right] + \frac{1}{2}\sum_{i<j}(\lambda_i - \lambda_j)^2\left(\dot{a}_{ij}^2 + \dot{b}_{ij}^2\right). \tag{14.3.10}$$

The path-integral measure, on the other hand, transforms as

$$\mathcal{D}M(t) = \mathcal{D}U(t)\prod_i \mathcal{D}\lambda_i(t)\Delta^2(\lambda(t)), \tag{14.3.11}$$

with $\Delta(\lambda(t))$ the standard Vandermonde determinant and $\mathcal{D}U$ the SU(N) Haar measure.

We may now pass to the quantum Hamiltonian

$$H = -\frac{1}{2\beta^2\Delta(\lambda)}\sum_i \frac{d^2}{d\lambda_i^2}\Delta(\lambda) + \sum_i V(\lambda_i) + \sum_{i<j}\frac{\Pi_{ij}^2 + \hat{\Pi}_{ij}^2}{(\lambda_i - \lambda_j)^2}. \tag{14.3.12}$$

Π_{ij} and $\hat{\Pi}_{ij}$ are the momenta conjugate to the SU(N) variables a_{ij} and b_{ij} in (14.3.9). The appearance of the Vandermonde determinant in the kinetic part of the λ_i variables is due to the nontrivial path integral measure for λ_i. It is necessary for the self-adjointness of the Hamiltonian. The momentum operators Π_i conjugate to the Cartan variables a_i, are the Cartan generators of the SU(N) symmetry. They do not appear in the Hamiltonian. Therefore, they generate the constraints $\Pi_i = 0$ on the states.

The states transform into representations of the SU(N) symmetry. The constraints $\Pi_i = 0$ imply that only representations that contain states with weight zero are allowed. The ground state is necessarily an SU(N) singlet. This can be deduced from the fact,

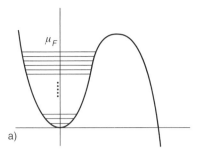 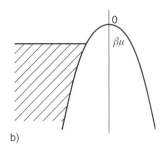

Figure 14.2 a) Filling the Fermi sea in the M^3 potential. b) The critical region near the top of the potential barrier.

that for nonsinglet states, the last part of the Hamiltonian in (14.3.11) is nonzero and positive.

Since we are eventually interested in the ground state we will drop the SU(N) generators from the Hamiltonian. Moreover, we will redefine the wave functions by including a factor of the Vandermonde determinant, $\psi \to \psi/\Delta$. The reduced Hamiltonian now is a sum of N one-particle Hamiltonians

$$H = \sum_{i=1}^{N} h_i, \quad h_i = -\frac{1}{2\beta^2} \frac{d^2}{d\lambda_i^2} + V(\lambda_i). \tag{14.3.13}$$

Due to the Vandermonde factor, the ground-state wave function must now be completely antisymmetric. We have ended up with the quantum mechanics of a system of N noninteracting fermions moving in the potential V. It is now straightforward to construct the ground state. It will be obtained by filling the first N energy levels of the potential V up to the Fermi energy μ_F.

In (14.3.2) we took a cubic potential in order to generate triangulations of Riemann surfaces. This is unbounded from below. We can however fill the states as indicated in figure 14.2a. Even this configuration, however, is unstable due to tunneling to the right side of the potential. If however we confine ourselves to perturbation theory ($\beta \to \infty$) and small fluctuations around this ground state, then the tunneling effects can be neglected. They are exponentially suppressed.

Another option is to add a stabilizing quartic term in the potential. Then we are lead to a picture similar to figure 14.3a. This is absolutely stable. We will see later on that this is the correct picture for two-dimensional vacua of the superstring.

We now proceed to determine the energy levels ϵ_n of the single-fermion Hamiltonians h_i semiclassically, using the Bohr-Sommerfeld quantization rules

$$\oint p_n d\lambda = \frac{2\pi}{\beta} n, \quad p_n = \sqrt{2(\epsilon_n - V)}. \tag{14.3.14}$$

The Fermi level $\mu_F = \epsilon_N$ in particular is given by

$$\int_{\lambda_-}^{\lambda^+} d\lambda \sqrt{2(\mu_F - V)} = \pi \frac{N}{\beta}, \tag{14.3.15}$$

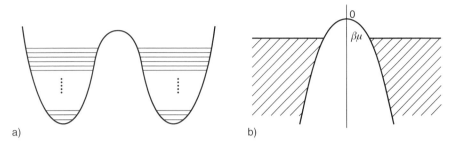

Figure 14.3 a) Filling the Fermi sea in the M^4 potential. b) The critical region near the top of the potential barrier.

with $\lambda_\pm$ the turning points of the classical path. The ground-state energy of the system is given by

$$E_0 = \sum_{i=1}^{N} \epsilon_i. \tag{14.3.16}$$

As in the pure gravity case, we must go to the continuum limit by approaching an appropriate critical point of the system. It is expected that this will occur in this system when the Fermi level reaches μ_c, the top of the potential barrier in figure 14.2. By appropriately tuning the coupling $g = \sqrt{\frac{N}{\beta}}$ of the model we will make $\mu_F \to \mu_c$ and reach the continuum limit. We will confirm this in the following.

14.3.2 The continuum limit

The matrix model we are studying has two dimensionless parameters: the number of colors, N, and the coupling constant $g^2 = \frac{N}{\beta}$ that controls the total area of the triangulations. $\beta = \frac{N}{g^2}$ on the other hand is the analogue of the inverse of Planck's constant $\hbar$ for the MQM.

We are taking the limit of large N in order to generate the triangulated surfaces We will also need to tune $g \to g_c$ in order to reach large surfaces, and therefore the continuum limit. The overall double scaling limit involves $N \to \infty$ and $g \to g_c$. This implies that also $\beta \to \infty$. Moreover, as we will see in the next section, we must keep one combination of parameters fixed. This will be equal to the renormalized string coupling. We are therefore both at the large N and the semiclassical ($\beta \to \infty$) regimes.

We are ready to tune the matrix model coupling $g \to g_c$ in order to obtain critical behavior and approach the continuum limit. In section 14.3 we have seen that g is related to the bare world-sheet cosmological constant as $g = e^{-\lambda_0}$. We may therefore define a renormalized cosmological constant, α,

$$\alpha = \pi \left(g_c^2 - g^2 \right) = \pi \left(g_c^2 - \frac{N}{\beta} \right). \tag{14.3.17}$$

The factor of π in front is for later convenience. $\alpha \to 0$ will parametrize our approach to the continuum limit. α is a function of the two independent parameters of the theory.

When tuned to zero, it will leave us with a continuum theory with only one parameter, the renormalized string coupling constant.

We are interested in the dependence of the matrix-model partition function on α. This is equivalently to knowing the vacuum energy $E(\alpha)$, where we drop the subscript 0 from now on. The analytic dependence is not universal and we will ignore it. As usual it is irrelevant for the critical behavior. The reason is developed in exercise 14.25.

We introduce the density of eigenvalues as

$$\rho(\epsilon) = \frac{1}{\beta} \sum_n \delta(\epsilon - \epsilon_n). \tag{14.3.18}$$

We further have

$$g^2 = \frac{N}{\beta} = \int_0^{\mu_F} \rho(\epsilon) d\epsilon, \quad \lim_{T \to \infty} -\frac{\log \mathcal{Z}}{T} = \beta E_0 = \beta^2 \int_0^{\mu_F} \rho(\epsilon) \epsilon \, d\epsilon. \tag{14.3.19}$$

We define as μ the distance of the Fermi level from the top of the potential barrier, μ_c

$$\mu \equiv \mu_c - \mu_F > 0. \tag{14.3.20}$$

Combining (14.3.17) and (14.3.19) we obtain

$$\frac{\partial \alpha}{\partial \mu} = -\frac{\partial \alpha}{\partial \mu_F} = \pi \rho(\mu_F), \tag{14.3.21}$$

$$\frac{\partial E}{\partial \alpha} = \frac{\partial E}{\partial \mu} \frac{\partial \mu}{\partial \alpha} = -\frac{\frac{\partial E}{\partial \mu_F}}{\pi \rho(\mu_F)} = -\frac{\beta \rho(\mu_F) \mu_F}{\pi \rho(\mu_F)} = \frac{\beta(\mu - \mu_c)}{\pi}. \tag{14.3.22}$$

By adding to E an irrelevant analytic term $E \to E + \frac{\beta \mu_c \alpha}{\pi}$ we finally obtain from (14.3.22)

$$\frac{\partial E}{\partial \alpha} = \frac{\beta \mu}{\pi}. \tag{14.3.23}$$

The previous equations indicate that the approach to the critical point is controlled by the density of states near the top of the potential barrier.

We must therefore compute $\rho(\mu_F)$. Since $\beta \to \infty$ we may use WKB theory to estimate

$$\rho(\mu_F) = \frac{1}{\pi} \int_{\lambda_-}^{\lambda_+} \frac{d\lambda}{\sqrt{2(\mu_F - V(\lambda))}} \simeq -\frac{1}{\pi} \log \mu + \mathcal{O}\left(\frac{1}{\beta^2}\right). \tag{14.3.24}$$

In exercise 14.26 you are invited to calculate the integral above, close the top of the potential barrier and verify the leading nonanalytic behavior.

Using (14.3.24) to integrate (14.3.21) and (14.3.23) we obtain in the scaling region $\mu \to 0$

$$\alpha = -\mu \log \mu + \cdots, \tag{14.3.25}$$

$$E = -\frac{\beta \mu^2 \log \mu}{2\pi} + \cdots \simeq -\frac{\beta \alpha^2}{2\pi \log \alpha} + \cdots. \tag{14.3.26}$$

It is important to note that the nonanalytic behavior above depends only on the quadratic maximum around the top of the potential barrier and not on other details of the potential (see figures 14.2b and 14.3b). For this it is convenient to introduce appropriate scaling variables

$$y = \sqrt{\beta}(\lambda - \lambda_c), \quad e = \mu_c - \epsilon. \tag{14.3.27}$$

λ_c is the position of the potential maximum. y is the local coordinate around the potential maximum. e is the energy as measured from the potential maximum.

The one-particle eigenvalue problem from (14.3.13) becomes in the new variables

$$\left[\frac{1}{2} \frac{d^2}{dy^2} + \frac{1}{2} y^2 + \mathcal{O}\left(\frac{y^3}{\sqrt{\beta}} \right) \right] \psi(y) = \beta e \psi(y). \tag{14.3.28}$$

This is the Schrödinger equation for an inverted harmonic oscillator. For finite values of βe all other terms except the quadratic ones are irrelevant. This explicitly shows that only the universal quadratic maximum is relevant for the double scaling limit. In particular, the continuum limit of the potentials in figures 14.2 and 14.3 should be equivalent. The only difference is that there is a doubling of the spectrum and the fermions fill both sides. We can therefore work with the nonperturbatively stable potential in figure 14.3. We must however divide by a factor of 2, to account for the duplication of the number of fermions.

We should also stress that this universality will be valid in the (renormalized) perturbation theory. Nonperturbative effects are not universal and in general depend on the details of the potential.

14.3.3 The double-scaling limit

The leading singular behavior in (14.3.24) emerges from surfaces with the topology of the sphere (i.e., planar graphs). We must, however, systematically expand in order to obtain also the rest of the surfaces.

The form of the scaled equation (14.3.28) indicates that the relevant limit is to take $\mu \to 0$ in order to reach the continuum critical point, and $\beta \to \infty$, by keeping $\beta \mu$ fixed. This combination will turn out to control the renormalized string coupling constant as we will see later on.

As is standard, we will write the density of states in terms of the resolvent of the Hamiltonian operator

$$\rho(\mu) = \text{Tr}\delta(h + \beta\mu) = \frac{1}{\pi} \text{ImTr}\left[\frac{1}{h + \beta\mu - i\epsilon} \right], \quad h = -\frac{1}{2} \frac{d^2}{dy^2} - \frac{1}{2} y^2. \tag{14.3.29}$$

The ϵ above is related to the usual $i\epsilon$ prescription for going around the poles. This can be evaluated in position space, using the known resolvent for the standard harmonic oscillator,

$$\left\langle y_f \left| \frac{1}{-\frac{1}{2} \frac{d^2}{dy^2} + \frac{\omega^2}{2} y^2 + \beta\mu - i\epsilon} \right| y_i \right\rangle = \int_0^\infty dT e^{-\beta\mu T} \frac{\sqrt{\omega} \exp\left[-\omega \frac{(y_i^2 + y_f^2) \cosh(\omega T) - 2y_i y_f}{2 \sinh(\omega T)} \right]}{\sqrt{2\pi \sinh(\omega T)}}. \tag{14.3.30}$$

We must rotate the frequency $\omega \to i\omega$, and $T \to iT$ to obtain that of the inverted oscillator,[7] relevant for our case,

$$\left\langle \gamma_f \left| \frac{1}{h + \beta\mu - i\epsilon} \right| \gamma_i \right\rangle = i \int_0^\infty dT e^{-i\beta\mu T} \sqrt{\frac{-i}{2\pi \sinh(T)}} \exp\left[i \frac{(\gamma_i^2 + \gamma_f^2)\cosh(T) - 2\gamma_i\gamma_f}{2\sinh(T)} \right].$$

(14.3.31)

Using (4.18.14) and (4.18.16) we may proceed to evaluate $\rho(\mu)$. The trace involved in (4.18.14) is formally divergent. The reliable way to compute it is to differentiate once with respect to $\beta\mu$, calculate the finite trace, and then integrate back adjusting appropriately the integration constant. We obtain

$$\frac{\partial \rho(\mu)}{\partial(\beta\mu)} = \frac{1}{\pi}\frac{\partial}{\partial(\beta\mu)} \text{Im} \int_{-\infty}^\infty d\gamma \left\langle \gamma \left| \frac{1}{h + \beta\mu - i\epsilon} \right| \gamma \right\rangle = \frac{1}{\pi}\text{Im} \int_0^\infty dT e^{-i\beta\mu T} \frac{T}{2\sinh\frac{T}{2}}.$$

(14.3.32)

Expanding (14.3.32) and integrating back we obtain

$$\rho(\mu) = \frac{1}{\pi}\left[-\log\mu + \sum_{k=1}^\infty \frac{|B_{2k}|}{k}(2\beta\mu)^{-2k} \right],$$

(14.3.33)

where B_{2k} are the Bernoulli numbers. We have fixed the integration constant so that the solution agrees with the WKB estimate (14.3.24).

To solve (14.3.21) it is convenient to parameterize the cosmological constant α as

$$\alpha = -\nu\log\nu.$$

(14.3.34)

Then (14.3.21) implies the implicit relation for α

$$\mu = \nu\left[1 - \frac{1}{\log\nu}\sum_{k=1}^\infty \left(2^{2k-1} - 1\right) \frac{|B_{2k}|}{k(2k-1)}(2\beta\nu)^{-2k} + \mathcal{O}\left(\frac{1}{\log^2\nu}\right) \right].$$

(14.3.35)

Finally, by integrating (14.3.23) we obtain the final matrix partition function per unit time, in the infinite-time-volume limit as

$$\mathcal{Z} = -\beta E = \frac{1}{8\pi}\left[(2\beta\nu)^2\log\nu - \frac{1}{3}\log\nu + \sum_{k=1}^\infty \frac{(2^{2k-1} - 1)|B_{2k+2}|}{k(k+1)(2k+1)}(2\beta\nu)^{-2k} \right].$$

(14.3.36)

We are now in the position to take the double-scaling limit

$$\nu \to 0, \quad \beta \to \infty, \quad \beta\nu = \frac{1}{g_s} \quad \text{fixed.}$$

(14.3.37)

We observe that (14.3.36) reproduces the sum over the Riemann surfaces. The coupling β is multiplicatively renormalized to give the string coupling g_s as in the case of pure gravity. However, unlike that case, the renormalization factor is not a power of the cosmological constant α, it is rather by $\nu \simeq \frac{\alpha}{|\log\alpha|}$.

As can be verified from (14.3.36), the sphere and torus contributions are divergent in the double-scaling limit. This infinity is expected. It is present in the continuum formulation

---

[7] It can be shown that this simple analytic continuation reproduces the correct result, properly done using the scattering wave functions of the inverted potential.

of the theory. It is due to integration over the zero mode of the Liouville field. You are invited to derive this in exercise 14.27.

There is another way to parametrize the double-scaled solution (14.3.34)–(14.3.37). It amounts to scale $\beta \to 1$, and use the (now finite) variable ν as the only remnant parameter related to the renormalized string coupling as $g_s = \frac{1}{\nu}$. In this redefinition, the distance of the Fermi sea from the top of the potential $\mu \sim \nu \sim \frac{1}{g_s}$ remains finite. This parametrization is used sometimes in the modern literature.

The calculation of the matrix model partition function (14.3.36) in the double-scaled continuum limit is the starting point for the computation of observables, namely, correlators and S-matrix elements. The natural operators of the MQM are the single-trace (loop) operators $O_n(x) = \mathrm{Tr} M^n(x)$. Their insertion in the path integral created a whole of length n. In the double-scaling limit, as the area becomes infinite, the hole shrinks to a puncture. Fourier transforming in x makes them eigenfunctions of space-time energy.[8] More precisely we define the finite-length loop operators

$$O(l, p) = \int dx\, e^{ipx} \mathrm{Tr}\, e^{-lM(x)} = \sum_{n=0}^{\infty} \frac{l^n}{n!} \int dx\, e^{ipx} \mathrm{Tr} O_n(x). \tag{14.3.38}$$

The tachyon vertex operators correspond to the leading asymptotics of $O(l, p)$

$$e^{ipX + (2 - |p|)\phi} \leftrightarrow \lim_{l \to 0} O(l, p). \tag{14.3.39}$$

The correlators of the loop operators can be calculated using the free-fermion formalism. We will not pursue this here. Suffice it to say that where they can be compared to the continuum correlators, they do agree. There are subtleties on the way associated to the fact that two-dimensional string theory, apart from the massless propagating "tachyon" field has an infinity of discrete, physical non-propagating modes. These are special to two dimensions, affect tachyon scattering and are closely connected with the infinite $W_{1+\infty}$ symmetry of $D = 2$ string theory (see the bibliography).

14.3.4 D-particles, tachyons, and holography

So far we have seen several examples of a correspondence between string theories on one hand and (generalized) gauge theories on the other. Those include, the $\mathrm{AdS}_{d+1}/\mathrm{CFT}_d$ correspondence and its avatars, the M(atrix) theory description of M-theory and the matrix model descriptions of $D \leq 1$ string theories.

The common ingredient in the examples of the AdS/CFT and M(atrix) theory, is that it is a limit of an open string/closed string duality. On the open string side are D-branes and the associated gauge theories are their world-volume theories. On the closed string theory side are the gravitational effects of the D-branes.

This correspondence seems *a priori* lacking in the case of the matrix models describing $D \leq 1$ string theories. The purpose of this section is to argue that even in this case, such

---

[8] Remember that x is the time coordinate of space-time.

matrix models can be thought to arise from the dynamics of D-branes of the associated string theories.

We will consider for concreteness the two-dimensional string theory, and its associated matrix model description. This is a bosonic string theory. It has D_0-branes, defined by Dirichlet boundary conditions in the Liouville direction and Neumann in time. Such D-branes do not couple to any massless gauge field and are in fact unstable. Their physical spectrum contains a "tachyon" field M, but no other degrees of freedom.[9]

The relevant DBI-like action describing the low-energy dynamics of the M field of a single D_0 is

$$S = -\int dt V(M)\sqrt{1 - \dot{M}^2}, \quad V(M) = \frac{1}{g_s \ell_s \cosh(M/2)}. \tag{14.3.40}$$

This potential is valid for $M > 0$. The potential is different for $M < 0$ and indeed unbounded from below in the bosonic theory we are studying. The properties relevant here is that the potential decays to zero exponentially as $M \to \infty$ while for small[10] T

$$V(M) \simeq \frac{1}{g_s \ell_s}\left[1 - \frac{M^2}{8} + \cdots\right]. \tag{14.3.41}$$

We now proceed to consider the dynamics of N D_0 particles. The associated tachyon field M is now a Hermitian $N \times N$ matrix and the associated DBI action is the nonabelian generalization of (14.3.40)

$$S = -\int dt \operatorname{Tr} V(M)\sqrt{1 - (D_t M)^2}, \quad D_t M = \dot{M} + [A_t, M], \tag{14.3.42}$$

where A_t is the usual one-dimensional U(N) world-volume gauge field of D_0 particles. It can be gauged away completely, and imposed the usual Gauss law: all physical states must be U(N) singlets. The overall U(1) degree of freedom is trivial and we will drop it from now on. We are therefore left with the quantum mechanics of Hermitian matrix field together with the SU(N) singlet constraint.

The rest of the procedure is identical to that described already in section 14.3 on page 488. In particular as we have seen, the details of the potential do not enter the double-scaling limit. The large-N double-scaled D_0 matrix theory is therefore equivalent to the two-dimensional string theory. Moreover it explains our previous "approximation": only SU(N) singlet states describe physical string configurations.

The coupling β of the $c = 1$ matrix model in section 14.3 is essentially $1/g_o$, where g_o is the open string coupling of the D_0 branes. The effective coupling of the closed strings g_c, obtained during the double-scaling limit, was given in (14.3.37) as $g_c = \frac{g_o}{v}$ where v is an explicit, albeit complicated function of the cosmological μ in (14.3.35).

---

[9] As in the closed string case, there are discrete nonpropagating modes but they are irrelevant for our discussion. They are, however, relevant for the full description of the rolling tachyon.

[10] There is a puzzle here between the fact that perturbative open string theory indicates a zero mass for the tachyon in two dimensions and the potential we are using here. The resolution is that the perturbative description seems not applicable near the top of the potential due to the Liouville wall. On the other hand, the action in (14.3.40) is compatible with known exact solutions.

Another interesting example is two-dimensional 0B superstring theory. This has been defined in exercise 7.5. It contains no space-time fermions. Its massless spectrum contains the usual scalar tachyon from the NS-NS sector and an axion scalar from the R-R sector. It also has unstable D_0-branes described by an action of the form (14.3.42), but now the potential (and the whole theory) is invariant under the $\mathbb{Z}_2$ symmetry $M \to -M$. This implies that the potential is of the type displayed in figure 14.3. This gives a nonperturbatively stable matrix model description. $\mathbb{Z}_2$-symmetric fluctuations correspond to the NS-NS tachyon, while $\mathbb{Z}_2$-antisymmetric ones to the R-R scalar.

Bibliography

The maximally supersymmetric matrix model in (14.1.19) has been first studied in [481]. Its connection to discretized membranes is discussed in [482]. In the BFSS proposal it is conjectured to describe the DLCQ of M-theory, [483,484]. The background field gauge expansion of the M(atrix) theory is discussed in [485].

There are several good reviews on the subject of M(atrix) theory with varying focus. We should mention [486,487,488,489,490]. They provide an overview plus a detailed guide to the literature. In particular, toroidal compactifications to lower dimensions, and the dynamics of branes are described.

There are matrix model proposals that are related to the one we have described here. The proposal in [491], described IIB theory in terms of the matrix model of D_{-1} instantons. This approach can be also viewed as a discretization of the D_1 string action. Another proposal is to use (1+1)-dimensional SYM in order to describe the DLCQ of IIA string theory. This is related by T-duality to the BFSS proposal. Its dynamics has been extensively analyzed in [492,493,494].

The matrix model approach to describe the Polyakov integrals on the sphere for $c \leq 1$ CFTs was developed in [496,497]. The double-scaling limit for pure gravity is described in [498,499,500]. The generalizations of the nonlinear string equations to other minimal models and the associated KdV formalism was developed in [501]. We have followed the $D = 1$ matrix model exposition of [502].

There are three extensive reviews of matrix models for $c \leq 1$ string theory. The reviews [137, 138] have a substantial overlap and they treat the subject comprehensively. Apart from the few points developed here, they have a complete discussion on states, operators and loop equations, the large-order behavior of perturbation theory, the KdV and integral hierarchies formalism, S-matrix calculations, and comparisons with the continuum approach, an analysis of the discrete states in $c = 1$ string theory, and their role in scattering amplitudes and the associated $W_{1+\infty}$ symmetry. The collective field theory of the $c = 1$ string theory is also reviewed. The third review [503] specializes in $c = 1$ string theory, and gives a complete description of the issues involved.

To explore the interpretation of matrix modes in terms of unstable D_0-branes the reader may start from [504,505,506].

Exercises

14.1. Derive the equations of motion of the Polyakov action (14.1.3) on page 471 for the membrane. Solve the equation for the metric and show that the theory is classically equivalent to the Nambu-Goto formulation.

14.2. Derive the Hamiltonian in (14.1.14) on page 472 from the Lagrangian.

14.3. Rescale coordinates and time appropriate to arrive at the matrix Hamiltonian (14.1.17) on page 473.

14.4. Repeat the argument of section 14.1.1 on page 471, indicating the classical instability of the membrane, for a string. Show that there is no instability in this case. Therefore, of all extended objects, only point particles and strings are generically stable.

14.5. Show that the Nambu-Goto membrane action (14.1.1) on page 471 can be written in terms of the classical Nambu bracket

$$\{F, G, H\} \equiv \epsilon^{\alpha\beta\gamma} \partial_\alpha F \partial_\beta G \partial_\gamma H \tag{14.1E}$$

as

$$S_{M_2} = -T_2 \int d^3\xi \left[\sqrt{-\frac{1}{6}\{X^\mu, X^\nu, X^\rho\}\{X_\mu, X_\nu, X_\rho\}} + \hat{C}_{\mu\nu\rho}\{X^\mu, X^\nu, X^\rho\} \right]. \tag{14.2E}$$

Is this bracket associative? No quantum analog of the Nambu bracket is currently known.

14.6. Derive the Hamiltonian (14.1.32) and the constraint (14.1.31) from the Lagrangian (14.1.29) on page 475.

14.7. Derive the integrated stress tensor (14.1.39) on page 476 for classical gravitons.

14.8. Show that the action obtained from the Lagrangian in (14.1.29) on page 475 is invariant under the supersymmetry

$$\delta X^i = -2\epsilon^T \Gamma^i \theta, \quad \delta\theta = \frac{1}{2}\left[\Gamma_i D_t X^i + \Gamma_- + \frac{1}{2}\Gamma_{ij}[X^i, X^j]\right]\epsilon + \epsilon', \quad \delta A = -2\epsilon^T \theta, \tag{14.3E}$$

where ϵ, ϵ' are two independent SO(9) spinors. Show that the supersymmetry algebra closes up to gauge transformations.

14.9. Gauge fix (14.1.29) on page 475 and derive the action (14.1.46)–(14.1.49).

14.10. Show that the M(atrix) theory proposal implies that DLCQ of M-theory with one transverse dimension compactified is described by (1+1)-dimensional maximally super-symmetric SYM.

14.11. Consider a discretization of Riemann surfaces using squares. Write the analog of (14.1.40) on page 476 and show that it indeed generates the discretized Polyakov sum.

14.12. Consider a discretization of Riemann surfaces using both triangles and squares. Write the relevant generating matrix model and interpret its parameters.

14.13. Consider the Feynman diagrams for the insertion of a single-trace loop operator O_n in (14.2.6) on page 481, into the matrix path integral. Show that they correspond to summing over triangulations of a Riemann surface with a loop of length n cut out. Correlate this with the picture of adding O_n with a given coupling constant to the matrix model action.

14.14. Change variables in (14.2.12) $M = U\Lambda U^\dagger$ with U unitary and Λ diagonal. Use the Faddeev-Popov method to derive (14.2.13) on page 482.

14.15. Show that (14.2.14) and (14.2.17) agree with the definition of the Vandermonde determinant in (14.2.13) on page 482.

14.16. Consider the generalized matrix model with $m - 1$ N×N Hermitian matrices M_i and partition function

$$\mathcal{Z} = \log \int \prod_{i=1}^{m-1} dM_i \exp\left[-\mathrm{Tr} \sum_{i=1}^{m-1} \left(V(M_i) + c_i M_i M_{i+1}\right)\right]. \tag{14.4E}$$

By integrating over the U(N) degrees of freedom show that it can be written as an integral over the eigenvalues

$$\mathcal{Z} = \log \int \prod_{i=1}^{m-1} \prod_{a=1}^{N} d\lambda_i^{(a)} \Delta(\lambda_1) e^{-\sum_{i,a}\left[V(\lambda_i^{(a)}) + c_i \lambda_i^{(a)} \lambda_{i+1}^{(a)}\right]} \Delta(\lambda_{m-1}). \tag{14.5E}$$

The continuum limits of such matrix models describe two-dimensional gravity coupled to the mth minimal CFT of the Virasoro algebra, with $c = 1 - \frac{6}{m(m+1)}$, mentioned in section 4.10. The case, studied in section 14.2 on page 479, is $m = 2$ and corresponds to $2d$ gravity coupled to the trivial CFT.

14.17. Expand the determinants in (14.2.13) and use the orthogonality properties of polynomials to derive (14.2.18) on page 483.

14.18. Consider a generic polynomial potential, and derive the analog of (14.2.27) on page 484. Show that in the large-N limit, without any tuning of parameters, we still have $f(\zeta) \sim \sqrt{g\zeta - g_c}$.

14.19. Consider a general polynomial potential and show that for appropriate choice of parameters the critical behavior (14.2.30) generalizes to (14.2.31) on page 484.

14.20. Expand (14.2.27) on page 484 to next order in 1/N to compute the next-to-leading (torus) contribution to the partition function. Show that at the critical point it scales as implied by (6.3.11) on page 150, namely, $Z \sim A^{-1}$.

14.21. Generalize the scaling *Ansatz* (14.2.39) on page 486 to the case of general γ.

14.22. Derive carefully (14.2.42) on page 486 from (14.2.28) in the double-scaling limit.

14.23. Derive the relation between the scaling function $u(z)$ and the second derivative of the matrix model partition function in the double-scaling limit.

14.24. Add a λ^6 term in (14.2.23) on page 483, and adjust its coefficient, to obtain a critical exponent $\gamma = -\frac{1}{3}$. Double-scale the theory and derive the following equation for the specific heat

$$z = u^3 - uu'' - \frac{1}{2}(u')^2 + \frac{1}{10}u''''. \tag{14.6E}$$

Show that it has a series solution of the form

$$u = z^{\frac{1}{3}}\left[1 + \sum_{k=1}^{\infty} u_k z^{-7k/3}\right], \tag{14.7E}$$

where not all u_k are positive. Although the critical exponent $\gamma = -\frac{1}{3}$ is that corresponding to the Ising model, this theory is nonunitary and corresponds to the CFT of the Lee-Yang edge singularity (see [495]).

14.25. Show that the matrix model partition function at fixed area is the Laplace transform of $E_0(\Delta)$. Deduce from this that analytic terms in Δ produce (derivatives of) δ-functions of the area, A. They are therefore irrelevant for the continuum limit, that in our conventions arises as $A \to \infty$.

14.26. Evaluate the WKB integral in (14.3.24) on page 493 and near the top of the potential barrier and verify the leading non-analytic behavior.

14.27. Consider the continuum $D = 1$ CFT coupled to gravity, described in section 6.3. Physical "tachyon" vertex operators are of the form $T = e^{iqX/\ell_s + (2-|q|)\phi}$. Consider the correlator of N such vertex operators. Split the Liouville field as $\phi = \phi_0 + \delta\phi$, where ϕ_0 is the constant zero mode, and first integrate over ϕ_0. Show that the correlator is proportional to $\Gamma(-s)$ with

$$s = 2(1 - h) - N + \frac{1}{2}\sum_{i=1}^{N}|q_i|. \tag{14.8E}$$

From this deduce that the partition function on the sphere and the torus are divergent.

Appendix A | Two-dimensional Complex Geometry

In this appendix we set up some conventions and describe the basics of two-dimensional complex geometry. All compact closed Euclidean two-manifolds are complex and even Kähler.

We use the following conventions:

$$z = \sigma_1 + i\sigma_2, \quad \bar{z} = \sigma_1 - i\sigma_2, \quad \frac{\partial}{\partial z} = \frac{1}{2}(\partial_1 - i\partial_2), \quad \frac{\partial}{\partial \bar{z}} = \frac{1}{2}(\partial_1 + i\partial_2). \tag{A.1}$$

A general metric can be always brought to the Kähler frame

$$ds^2 = 2g_{z\bar{z}}dzd\bar{z}, \quad g^{z\bar{z}} = \frac{1}{g_{z\bar{z}}}, \quad \sqrt{g} = g_{z\bar{z}}, \tag{A.2}$$

so that the volume measure is

$$\int \sqrt{\det(g_{ab})}d^2\sigma = \int \sqrt{g}d^2z, \quad d^2z = idz \wedge d\bar{z} \equiv 2d\sigma_1 d\sigma_2. \tag{A.3}$$

For the standard flat metric $ds^2 = d\sigma_1^2 + d\sigma_2^2$, $g_{z\bar{z}} = \frac{1}{2}$. The appropriate δ-function in complex coordinates is defined as

$$\delta^{(2)}(z, \bar{z}) = \frac{1}{2\sqrt{g}}\delta(\sigma_1)\delta(\sigma_2), \quad \int \sqrt{g}d^2z\, \delta^{(2)}(z, \bar{z}) = 1. \tag{A.4}$$

An important relation is

$$\partial_z \frac{1}{\bar{z}} = \partial_{\bar{z}} \frac{1}{z} = 2\pi\delta^{(2)}(z). \tag{A.5}$$

General coordinates change the complex structure in (A.2) generating g_{zz} and $g_{\bar{z}\bar{z}}$ components. Holomorphic transformations

$$z' = f(z), \quad \bar{z}' \to \bar{f}(\bar{z}), \quad g_{z'\bar{z}'} = \left(\frac{\partial f}{\partial z}\right)^{-1} \left(\frac{\partial \bar{f}}{\partial \bar{z}}\right)^{-1} g_{z\bar{z}} \tag{A.6}$$

preserve the complex structure.

Tensors can have upper or lower z and $\bar{z}$ indices. Due to the form of the metric in (A.2) an upper $\bar{z}$ index can be traded to a lower z index and vice versa,

$$t^{\cdots\bar{z}}{}_{\cdots} = g^{z\bar{z}}t^{\cdots}{}_{z\cdots}, \quad t^{\cdots}{}_{\bar{z}\cdots} = g_{z\bar{z}}t^{\cdots z}{}_{\cdots}. \tag{A.7}$$

Thus, without loss of generality we can take tensors to have n_u upper z indices and n_d lower z indices. Such a tensor, under the holomorphic coordinate transformations (A.6) transforms as

$$t^{z'\cdots z'}{}_{z'\cdots z'} = \left(\frac{\partial f}{\partial z}\right)^{n_u - n_d} t^{z\cdots z}{}_{z\cdots z}. \tag{A.8}$$

We call this tensor a tensor of (covariant) rank $(n_d - n_u, 0)$. A general tensor having n_u upper z indices, n_d lower z indices, $\bar{n}_u$ upper $\bar{z}$ indices, and $\bar{n}_d$ lower $\bar{z}$ indices transforms as

$$t^{z'\cdots\bar{z}'\cdots}{}_{z'\cdots\bar{z}'\cdots} = \left(\frac{\partial f}{\partial z}\right)^{n_u - n_d} \left(\frac{\partial \bar{f}}{\partial \bar{z}}\right)^{\bar{n}_u - \bar{n}_d} t^{z\cdots\bar{z}\cdots}{}_{z\cdots\bar{z}\cdots}. \tag{A.9}$$

This a tensor of rank $(n_d - n_u, \bar{n}_d - \bar{n}_u)$.

The covariant derivative on a tensor $t_{n,\bar{n}}$ of rank $(n, \bar{n})$ is

$$\nabla_z t_{n,\bar{n}} = (g_{z\bar{z}})^n \partial_z \left((g^{z\bar{z}})^n t_{n,\bar{n}}\right), \quad \nabla_{\bar{z}} t_{n,\bar{n}} = (g_{z\bar{z}})^{\bar{n}} \partial_{\bar{z}} \left((g^{z\bar{z}})^{\bar{n}} t_{n,\bar{n}}\right). \tag{A.10}$$

The Ricci tensor and the scalar curvature of the surface are

$$R_{z\bar{z}} = -\partial_z \left(g^{z\bar{z}}\partial_{\bar{z}}\right) g_{z\bar{z}}, \quad R = -2g^{z\bar{z}}\partial_z \left(g^{z\bar{z}}\partial_{\bar{z}} g_{z\bar{z}}\right). \tag{A.11}$$

Appendix B | Differential Forms

Antisymmetric tensors give rise to the concept of a differential form.

A p-form is a completely antisymmetric p-index tensor $A_{\mu_1\mu_2\cdots\mu_p}$. When we talk about forms we will omit the indices and we will write A_p instead. We can also introduce anti-commuting differentials dx^m to write

$$A_p = \frac{1}{p!}A_{\nu_1\nu_2\cdots\nu_p}dx^{\mu_1}dx^{\nu_2}\cdots dx^{\nu_p}. \tag{B.1}$$

We introduce the symbol for antisymmetrization [ ]. This stands for summing over all permutations of indices with the appropriate signs that antisymmetrize and dividing by the number of permutations. For example

$$A_{[\mu\nu]} \equiv \frac{1}{2}(A_{\mu\nu} - A_{\nu\mu}), \quad A_{[\mu\nu\rho]} = \frac{1}{3!}(A_{\mu\nu\rho} - A_{\nu\mu\rho} + A_{\rho\mu\nu} - A_{\mu\rho\nu} + A_{\nu\rho\mu} - A_{\rho\nu\mu}). \tag{B.2}$$

The exterior product of two forms is

$$\begin{aligned}(A_p \wedge B_q)_{\mu_1\mu_2\cdots\mu_{(p+q)}} &= (-1)^{pq}(B_q \wedge A_p)_{\mu_1\mu_2\cdots\mu_{(p+q)}} \\ &= \frac{(p+q)!}{p!q!}A_{[\mu_1\mu_2\cdots\mu_p}B_{\mu_{p+1}\mu_{p+2}\cdots\mu_{(p+q)}]}.\end{aligned} \tag{B.3}$$

Of particular use is the exponential of a form

$$e^A \equiv 1 + A + \frac{1}{2!}A \wedge A + \frac{1}{3!}A \wedge A \wedge A + \cdots. \tag{B.4}$$

The exterior derivative is defined as a differential map between p and $(p+1)$-forms

$$(dA_p)_{\mu_1\mu_2\cdots\mu_{(p+1)}} = (p+1)\partial_{[\mu_1}A_{\mu_2\mu_3\cdots\mu_{(p+1)}]}. \tag{B.5}$$

It can be shown to be nilpotent,

$$d^2 = 0, \tag{B.6}$$

and satisfies the graded distributive property

$$d(A_p \wedge B_q) = dA_p \wedge B_q + (-1)^p A_p \wedge dB_q. \tag{B.7}$$

We define the inner product of two p-forms as

$$A_p \cdot B_p \equiv \frac{1}{p!}(A_p)_{\mu_1 \cdots \mu_p}(B_p)^{\mu_1 \cdots \mu_p}. \tag{B.8}$$

A natural operation on a p-form is to integrate it over a p-dimensional manifold

$$\int A_d \equiv \int A_{\mu_1, \mu_2, \ldots, \mu_p} dx^{\mu_1} \wedge \cdots \wedge dx^{\mu_p}. \tag{B.9}$$

The reason is that such an integral does not depend on the metric of the manifold. Since a p-form transforms under reparametrizations as the volume form of the p-dimensional manifold, the integral (B.9) is well defined.

Another important ingredient is Stokes' theorem

$$\int_{M_p} dA_{p-1} = \int_{\partial M_p} A_{p-1}, \tag{B.10}$$

where ∂M is the boundary of M.

It is useful here to introduce the Levi-Cività symbol ϵ, a tensor density. It is defined as being completely antisymmetric with

$$\epsilon^{1,2,\ldots,D} = 1. \tag{B.11}$$

It becomes a (covariantly constant) tensor when divided by $\sqrt{-\det g}$ in the Minkowski case (and $\sqrt{\det g}$ in the Euclidean case)

$$E^{\mu_1 \cdots \mu_D} \equiv \frac{\epsilon^{\mu_1 \cdots \mu_D}}{\sqrt{-\det g}}. \tag{B.12}$$

Independent of signature it satisfies

$$\epsilon_{1,2,\ldots,D} = \det g \, \epsilon^{1,2,\ldots,D}. \tag{B.13}$$

We also obtain

$$E^{\mu_1 \cdots \mu_D} E_{\mu_1 \cdots \mu_D} = \pm D!, \tag{B.14}$$

where the upper sign is for Euclidean space and the lower for Minkowski space.

We may now introduce the Hodge star operation that involves the metric on the manifold. In a manifold of dimension D it maps $p \to (D - p)$ forms

$${}^\star A_{\mu_1 \mu_2 \cdots \mu_p} = \frac{1}{p!} E_{\mu_1 \mu_2 \cdots \mu_{D-p}}{}^{\nu_1 \nu_2 \cdots \nu_p} A_{\nu_1 \nu_2 \cdots \nu_p}. \tag{B.15}$$

The Hodge star verifies

$${}^{\star\star} = (-1)^{p(D-p)+1} \tag{B.16}$$

in Minkowski signature and picks up an extra minus sign in Euclidean signature.

Appendix C | Theta and Other Elliptic Functions

In this appendix we will give definitions and various useful formulae pertaining to elliptic functions.

C.1 ϑ and Related Functions

Definition

$$\vartheta[^a_b](v|\tau) = \sum_{n\in\mathbb{Z}} q^{(1/2)(n-a/2)^2} e^{2\pi i(v-b/2)(n-a/2)}, \tag{C.1}$$

where a, b are real and $q = e^{2\pi i\tau}$.

Periodicity properties

$$\vartheta[^{a+2}_b](v|\tau) = \vartheta[^a_b](v|\tau), \quad \vartheta[^{\ \ a}_{b+2}](v|\tau) = e^{i\pi a}\vartheta[^a_b](v|\tau), \tag{C.2}$$

$$\vartheta[^{-a}_{-b}](v|\tau) = \vartheta[^a_b](-v|\tau), \quad \vartheta[^a_b](-v|\tau) = e^{i\pi ab}\vartheta[^a_b](v|\tau) \quad (a, b \in \mathbb{Z}). \tag{C.3}$$

In the usual Jacobi/Erderlyi notation we have $\vartheta_1 = \vartheta[^1_1]$, $\vartheta_2 = \vartheta[^1_0]$, $\vartheta_3 = \vartheta[^0_0]$, $\vartheta_4 = \vartheta[^0_1]$.

Behavior under modular transformations

$$\vartheta[^a_b](v|\tau+1) = e^{-(i\pi/4)a(a-2)}\vartheta[^{\ \ a}_{a+b-1}](v|\tau), \tag{C.4}$$

$$\vartheta[^a_b]\left(\frac{v}{\tau} \Big| -\frac{1}{\tau}\right) = \sqrt{-i\tau}\, e^{(i\pi/2)ab+i\pi v^2/\tau}\vartheta[^{\ b}_{-a}](v|\tau). \tag{C.5}$$

Product formulas

$$\vartheta_1(v|\tau) = 2q^{1/8} \sin[\pi v] \prod_{n=1}^{\infty} (1 - q^n)(1 - q^n e^{2\pi i v})(1 - q^n e^{-2\pi i v}), \tag{C.6}$$

$$\vartheta_2(v|\tau) = 2q^{1/8} \cos[\pi v] \prod_{n=1}^{\infty} (1 - q^n)(1 + q^n e^{2\pi i v})(1 + q^n e^{-2\pi i v}), \tag{C.7}$$

$$\vartheta_3(v|\tau) = \prod_{n=1}^{\infty} (1 - q^n)(1 + q^{n-1/2} e^{2\pi i v})(1 + q^{n-1/2} e^{-2\pi i v}), \tag{C.8}$$

$$\vartheta_4(v|\tau) = \prod_{n=1}^{\infty} (1 - q^n)(1 - q^{n-1/2} e^{2\pi i v})(1 - q^{n-1/2} e^{-2\pi i v}). \tag{C.9}$$

We define the Dedekind η-function:

$$\eta(\tau) = q^{1/24} \prod_{n=1}^{\infty} (1 - q^n). \tag{C.10}$$

It is related to the v derivative of ϑ_1:

$$\frac{\partial}{\partial v} \vartheta_1(v)|_{v=0} \equiv \vartheta_1' = 2\pi \eta^3(\tau) \tag{C.11}$$

and satisfies

$$\eta\left(-\frac{1}{\tau}\right) = \sqrt{-i\tau}\, \eta(\tau). \tag{C.12}$$

v-periodicity formula

$$\vartheta[{}^a_b]\left(v + \frac{\epsilon_1}{2}\tau + \frac{\epsilon_2}{2}|\tau\right) = e^{-(i\pi\tau/4)\epsilon_1^2 - (i\pi\epsilon_1/2)(2v-b) - (i\pi/2)\epsilon_1\epsilon_2} \vartheta[{}^{a-\epsilon_1}_{b-\epsilon_2}](v|\tau). \tag{C.13}$$

Useful identities

$$\vartheta_2(0|\tau)\vartheta_3(0|\tau)\vartheta_4(0|\tau) = 2\eta^3, \tag{C.14}$$

$$\vartheta_2^4(v|\tau) - \vartheta_1^4(v|\tau) = \vartheta_3^4(v|\tau) - \vartheta_4^4(v|\tau). \tag{C.15}$$

Duplication formulas

$$\vartheta_2(2\tau) = \frac{1}{\sqrt{2}}\sqrt{\vartheta_3^2(\tau) - \vartheta_4^2(\tau)}, \quad \vartheta_3(2\tau) = \frac{1}{\sqrt{2}}\sqrt{\vartheta_3^2(\tau) + \vartheta_4^2(\tau)}, \tag{C.16}$$

$$\vartheta_4(2\tau) = \sqrt{\vartheta_3(\tau)\vartheta_4(\tau)}, \quad \eta(2\tau) = \sqrt{\frac{\vartheta_2(\tau)\eta(\tau)}{2}}. \tag{C.17}$$

Jacobi identity

$$\frac{1}{2} \sum_{a,b=0}^{1} (-1)^{a+b+ab} \prod_{i=1}^{4} \vartheta[{}^a_b](v_i) = -\prod_{i=1}^{4} \vartheta_1(v_i'), \tag{C.18}$$

where

$$v_1' = \frac{1}{2}(-v_1 + v_2 + v_3 + v_4), \quad v_2' = \frac{1}{2}(v_1 - v_2 + v_3 + v_4), \tag{C.19}$$

$$v_3' = \frac{1}{2}(v_1 + v_2 - v_3 + v_4), \quad v_4' = \frac{1}{2}(v_1 + v_2 + v_3 - v_4). \tag{C.20}$$

Using (C.18) and (C.13) we can show that

$$\frac{1}{2}\sum_{a,b=0}^{1}(-1)^{a+b+ab}\prod_{i=1}^{4}\vartheta[{a+h_i \atop b+g_i}](v_i) = -\prod_{i=1}^{4}\vartheta[{1-h_i \atop 1-g_i}](v_i'). \tag{C.21}$$

The Jacobi identity (C.21) is valid only when $\sum_i h_i = \sum_i g_i = 0$. There is also a similar (IIA) identity

$$\frac{1}{2}\sum_{a,b=0}^{1}(-1)^{a+b}\prod_{i=1}^{4}\vartheta[{a \atop b}](v_i) = -\prod_{i=1}^{4}\vartheta_1(v_i') + \prod_{i=1}^{4}\vartheta_1(v_i) \tag{C.22}$$

and

$$\frac{1}{2}\sum_{a,b=0}^{1}(-1)^{a+b}\prod_{i=1}^{4}\vartheta[{a+h_i \atop b+g_i}](v_i) = -\prod_{i=1}^{4}\vartheta[{1-h_i \atop 1-g_i}](v_i') + \prod_{i=1}^{4}\vartheta[{1+h_i \atop 1+g_i}](v_i). \tag{C.23}$$

The ϑ-functions satisfy the following heat equation:

$$\left[\frac{1}{(2\pi i)^2}\frac{\partial^2}{\partial v^2} - \frac{1}{i\pi}\frac{\partial}{\partial \tau}\right]\vartheta[{a \atop b}](v|\tau) = 0 \tag{C.24}$$

as well as

$$\frac{1}{4\pi i}\frac{\vartheta_2''}{\vartheta_2} = \partial_\tau \log \vartheta_2 = \frac{i\pi}{12}\left(E_2 + \vartheta_3^4 + \vartheta_4^4\right), \tag{C.25}$$

$$\frac{1}{4\pi i}\frac{\vartheta_3''}{\vartheta_3} = \partial_\tau \log \vartheta_3 = \frac{i\pi}{12}\left(E_2 + \vartheta_2^4 - \vartheta_4^4\right), \tag{C.26}$$

$$\frac{1}{4\pi i}\frac{\vartheta_4''}{\vartheta_4} = \partial_\tau \log \vartheta_4 = \frac{i\pi}{12}\left(E_2 - \vartheta_2^4 - \vartheta_3^4\right), \tag{C.27}$$

where the function E_2 is defined in (C.37).

We will also include here the transformation properties of the hatted ϑ-functions that appear in traces over the Möbius strip. The general definition of hatted characters is in (7.6.10) on page 172. Note that we are using a slightly different notation than [97]. For the η-function we obtain

$$\hat{\eta}(it) = \sqrt{\eta(2it)\vartheta_3(2it)}, \quad \hat{\eta}(it) = \frac{1}{\sqrt{2t}}\hat{\eta}\left(\frac{i}{4t}\right). \tag{C.28}$$

For the fermionic SO($2n$) characters of section 4.12 we have

$$
\begin{pmatrix} \hat{\chi}_O(it) \\ \hat{\chi}_V(it) \\ \hat{\chi}_S(it) \\ \hat{\chi}_C(it) \end{pmatrix} = \begin{pmatrix} c & s & 0 & 0 \\ s & c & 0 & 0 \\ 0 & 0 & e^{-in\pi/4}c & ie^{-in\pi/4}s \\ 0 & 0 & ie^{-in\pi/4}s & e^{-in\pi/4}c \end{pmatrix} \cdot \begin{pmatrix} \hat{\chi}_O(\frac{i}{4t}) \\ \hat{\chi}_V(\frac{i}{4t}) \\ \hat{\chi}_S(\frac{i}{4t}) \\ \hat{\chi}_C(\frac{i}{4t}) \end{pmatrix}
\tag{C.29}
$$

with $c = \cos(n\pi/4)$, $s = \sin(n\pi/4)$.

C.2 The Weierstrass Function

$$
\mathcal{P}(z) = 4\pi i \partial_\tau \log \eta(\tau) - \partial_z^2 \log \vartheta_1(z) = \frac{1}{z^2} + \mathcal{O}(z^2)
\tag{C.30}
$$

is even and is the unique analytic function on the torus with a double pole at zero.

$$
\mathcal{P}(-z) = \mathcal{P}(z), \quad \mathcal{P}(z+1) = \mathcal{P}(z+\tau) = \mathcal{P}(z),
\tag{C.31}
$$

$$
\mathcal{P}(z, \tau+1) = \mathcal{P}(z, \tau), \quad \mathcal{P}\left(\frac{z}{\tau}, -\frac{1}{\tau}\right) = \tau^2 \mathcal{P}(z, \tau).
\tag{C.32}
$$

The following torus integrals are useful for the one-loop calculations:

$$
\int \frac{d^2 z}{\tau_2} \left[\partial_z \log \vartheta_1(z) + 2\pi i \frac{\mathrm{Im}z}{\tau_2} \right]^2 = \int \frac{d^2 z}{\tau_2} \mathcal{P}(z, \tau) = 4\pi i \partial_\tau \log(\sqrt{\tau_2}\eta),
\tag{C.33}
$$

From this and (C.30) it follows that

$$
\int \frac{d^2 z}{\tau_2} \partial_z^2 \log \vartheta_1(z) = -\frac{\pi}{\tau_2}.
\tag{C.34}
$$

Finally the integral

$$
\int \frac{d^2 z}{\tau_2} \overline{\mathcal{P}(\bar{z}, \bar{\tau})} \left(\left[\partial_z \log \vartheta_1(z) + 2\pi i \frac{\mathrm{Im}z}{\tau_2} \right]^2 - \mathcal{P}(z) \right) = 0
\tag{C.35}
$$

is defined by analytic continuation. We must insert $e^{-\epsilon \langle X(z)X(0) \rangle}$ in the integrand, evaluate the integral for sufficiently large and positive ϵ and take $\epsilon \to 0$.

C.3 Modular Forms

In this section we collect some formulae on modular forms. They can be useful in analyzing the spectra of BPS states and appear in BPS-generated one-loop corrections to effective supergravity theories.

A (holomorphic) modular form $F_d(\tau)$ of weight d behaves as follows under modular transformations:

$$F_d(-1/\tau) = \tau^d F_d(\tau), \quad F_d(\tau + 1) = F_d(\tau). \tag{C.36}$$

We first list the Eisenstein series:

$$E_2 = \frac{12}{i\pi} \partial_\tau \log \eta = 1 - 24 \sum_{n=1}^{\infty} \frac{nq^n}{1 - q^n}, \tag{C.37}$$

$$E_4 = \frac{1}{2} \left(\vartheta_2^8 + \vartheta_3^8 + \vartheta_4^8 \right) = 1 + 240 \sum_{n=1}^{\infty} \frac{n^3 q^n}{1 - q^n}, \tag{C.38}$$

$$E_6 = \frac{1}{2} \left(\vartheta_2^4 + \vartheta_3^4 \right) \left(\vartheta_3^4 + \vartheta_4^4 \right) \left(\vartheta_4^4 - \vartheta_2^4 \right) = 1 - 504 \sum_{n=1}^{\infty} \frac{n^5 q^n}{1 - q^n}. \tag{C.39}$$

In counting BPS states in string theory the following combinations arise:

$$H_2 \equiv \frac{1 - E_2}{24} = \sum_{n=1}^{\infty} \frac{nq^n}{1 - q^n} \equiv \sum_{n=1}^{\infty} d_2(n) q^n, \tag{C.40}$$

$$H_4 \equiv \frac{E_4 - 1}{240} = \sum_{n=1}^{\infty} \frac{n^3 q^n}{1 - q^n} \equiv \sum_{n=1}^{\infty} d_4(n) q^n, \tag{C.41}$$

$$H_6 \equiv \frac{1 - E_6}{504} = \sum_{n=1}^{\infty} \frac{n^5 q^n}{1 - q^n} \equiv \sum_{n=1}^{\infty} d_6(n) q^n. \tag{C.42}$$

We have the following arithmetic formulas for d_{2k}:

$$d_{2k}(N) = \sum_{n|N} n^{2k-1}, \quad k = 1, 2, 3. \tag{C.43}$$

The E_4 and E_6 modular forms have weights four and six, respectively. They generate the ring of modular forms. However, E_2 is not exactly a modular form, but

$$\hat{E}_2 = E_2 - \frac{3}{\pi \tau_2} \tag{C.44}$$

is a modular form of weight 2 but is not holomorphic any more. The (modular-invariant) j-function and η^{24} can be written as

$$j = \frac{E_4^3}{\eta^{24}} = \frac{1}{q} + 744 + \cdots, \quad \eta^{24} = \frac{1}{2^6 \cdot 3^3} \left[E_4^3 - E_6^2 \right]. \tag{C.45}$$

We will also introduce the covariant derivative on modular forms

$$F_{d+2} = \left(\frac{i}{\pi} \partial_\tau + \frac{d/2}{\pi \tau_2} \right) F_d \equiv D_d F_d. \tag{C.46}$$

F_{d+2} is a modular form of weight $d + 2$ if F_d has weight d. The covariant derivative introduced above has the following distributive property:

$$D_{d_1+d_2}(F_{d_1} F_{d_2}) = F_{d_2}(D_{d_1} F_{d_1}) + F_{d_1}(D_{d_2} F_{d_2}). \tag{C.47}$$

The following relations and (C.47) allow the computation of any covariant derivative:

$$D_2\hat{E}_2 = \frac{1}{6}E_4 - \frac{1}{6}\hat{E}_2^2, \quad D_4E_4 = \frac{2}{3}E_6 - \frac{2}{3}\hat{E}_2E_4, \quad D_6E_6 = E_4^2 - \hat{E}_2E_6. \tag{C.48}$$

Here we will give some identities between derivatives of ϑ-functions and modular forms. They are useful for (one-loop) trace computations in string theory:

$$\frac{\vartheta_1'''}{\vartheta_1'} = -\pi^2 E_2, \quad \frac{\vartheta_1^{(5)}}{\vartheta_1'} = -\pi^2 E_2 \left(4\pi i\partial_\tau \log E_2 - \pi^2 E_2\right), \tag{C.49}$$

$$-3\frac{\vartheta_1^{(5)}}{\vartheta_1'} + 5\left(\frac{\vartheta_1'''}{\vartheta_1'}\right)^2 = 2\pi^4 E_4, \tag{C.50}$$

$$-15\frac{\vartheta_1^{(7)}}{\vartheta_1'} - \frac{350}{3}\left(\frac{\vartheta_1'''}{\vartheta_1'}\right)^3 + 105\frac{\vartheta_1^{(5)}\vartheta_1'''}{\vartheta_1'^2} = \frac{80\pi^6}{3}E_6, \tag{C.51}$$

$$\frac{1}{2}\sum_{i=2}^{4}\frac{\vartheta_i''\vartheta_i^7}{(2\pi i)^2} = \frac{1}{12}(E_2E_4 - E_6). \tag{C.52}$$

C.4 Poisson Resummation

A very useful tool to handle modular series on the torus is Poisson resummation. Consider a function $f(x)$ and its Fourier transform $\tilde{f}$ defined as

$$\tilde{f}(k) \equiv \frac{1}{2\pi}\int_{-\infty}^{+\infty} f(x)e^{ikx}\,dx. \tag{C.53}$$

Then, the Poisson resummation formula states that

$$\sum_{n\in\mathbb{Z}} f(2\pi n) = \sum_{n\in\mathbb{Z}}\tilde{f}(n). \tag{C.54}$$

Choosing as f an appropriate Gaussian function we obtain

$$\sum_{n\in\mathbb{Z}} e^{-\pi an^2 + \pi bn} = \frac{1}{\sqrt{a}}\sum_{n\in\mathbb{Z}} e^{-(\pi/a)(n+ib/2)^2}, \tag{C.55}$$

$$\sum_{n\in\mathbb{Z}} n\,e^{-\pi an^2 + \pi bn} = -\frac{i}{\sqrt{a}}\sum_{n\in\mathbb{Z}}\frac{\left(n+i\frac{b}{2}\right)}{a}e^{-(\pi/a)(n+ib/2)^2}, \tag{C.56}$$

$$\sum_{n\in\mathbb{Z}} n^2\,e^{-\pi an^2 + \pi bn} = \frac{1}{\sqrt{a}}\sum_{n\in\mathbb{Z}}\left[\frac{1}{2\pi a} - \frac{\left(n+i\frac{b}{2}\right)^2}{a^2}\right]e^{-(\pi/a)(n+ib/2)^2}. \tag{C.57}$$

The multidimensional generalization is (repeated indices are summed over)

$$\sum_{m_i\in\mathbb{Z}} e^{-\pi m_i m_j A_{ij} + \pi B_i m_i} = (\det A)^{-1/2}\sum_{m_i\in\mathbb{Z}} e^{-\pi(m_k + iB_k/2)(A^{-1})_{kl}(m_l + iB_l/2)}. \tag{C.58}$$

Appendix D | Toroidal Lattice Sums

We will consider here asymmetric lattice sums corresponding to p left-moving bosons and q right-moving ones. To have good modular properties $p - q$ should be a multiple of eight. We will consider here the case $q - p = 16$ relevant for the heterotic string. Other cases can be easily worked out using the same methods.

We will write the genus-1 action using p bosons and 16 complex right-moving fermions, $\psi^I(\bar{z})$, $\bar{\psi}^I(\bar{z})$:

$$
S_{p,q} = \frac{1}{4\pi} \int d^2\sigma \sqrt{\det g}\, g^{ab} G_{\alpha\beta} \partial_a X^\alpha \partial_b X^\beta + \frac{1}{4\pi} \int d^2\sigma\, \epsilon^{ab} B_{\alpha\beta} \partial_a X^\alpha \partial_b X^\beta
$$

$$
+ \frac{1}{4\pi} \int d^2\sigma \sqrt{\det g} \sum_I \psi^I (\bar{\nabla} + Y_\alpha^I (\bar{\nabla} X^\alpha)) \bar{\psi}^I, \tag{D.1}
$$

where the torus metric is given in (4.17.1) on page 90. We will take the fermions to be all periodic or antiperiodic. We have also set $\ell_s = 1$ in this appendix. A direct evaluation of the torus path integral along the line of section 4.18 on page 93 gives

$$
Z_{p,p+16}(G, B, Y) = \frac{\sqrt{\det G}}{\tau_2^{p/2} \eta^p \bar{\eta}^{p+16}}
$$

$$
\times \sum_{m^\alpha, n^\alpha \in \mathbb{Z}} \exp\left[-\frac{\pi}{\tau_2} (G + B)_{\alpha\beta} (m^\alpha + \tau n^\alpha)(m^\beta + \bar{\tau} n^\beta) \right]
$$

$$
\times \frac{1}{2} \sum_{a,b=0}^{1} \prod_{I=1}^{16} e^{i\pi (m^\alpha Y_\alpha^I Y_\beta^I n^\beta - b\, n^\alpha Y_\alpha^I)} \bar{\vartheta} \begin{bmatrix} a - 2n^\alpha Y_\alpha^I \\ b - 2m^\beta Y_\beta^I \end{bmatrix}
$$

$$
= \frac{\sqrt{\det G}}{\tau_2^{p/2} \eta^p \bar{\eta}^{p+16}} \sum_{m^\alpha, n^\alpha \in \mathbb{Z}} \exp\left[-\frac{\pi}{\tau_2} (G + B)_{\alpha\beta} (m^\alpha + \tau n^\alpha)(m^\beta + \bar{\tau} n^\beta) \right]
$$

$$
\times \exp\left[-i\pi \sum_I n^\alpha (m^\beta + \bar{\tau} n^\beta) Y_\alpha^I Y_\beta^I \right] \frac{1}{2} \sum_{a,b=0}^{1} \prod_{I=1}^{16} \bar{\vartheta}[{}^a_b](Y_\gamma^I (m^\gamma + \bar{\tau} n^\gamma)|\bar{\tau}).
$$

Under modular transformations

$$\tau \to \tau + 1, \quad Z_{p,p+16} \to e^{4\pi i/3} Z_{p,p+16}, \tag{D.2}$$

while it is invariant under $\tau \to -1/\tau$.

Performing a Poisson resummation in m^α we can cast it in Hamiltonian form $Z_{p,p+16} = \Gamma_{p,p+16}/\eta^p \bar\eta^{p+16}$ with

$$\Gamma_{p,p+16}(G, B, Y) = \sum_{m_\alpha, n_\alpha, Q_I} q^{P_L^2/2}\, \bar{q}^{P_R^2/2}, \tag{D.3}$$

where m_α, n^α take arbitrary integer values, while Q_I take values in the even self-dual lattice Spin(32)/$\mathbb{Z}_2$. To be concrete, the numbers Q_I are either all integer or all half-integer, satisfying in both cases the constraint $\sum_I Q_I =$ even. We will introduce the $(2p + 16) \times (2p + 16)$ symmetric matrix

$$M = \begin{pmatrix} G^{-1} & G^{-1}C & G^{-1}Y^t \\ C^t G^{-1} & G + C^t G^{-1}C + Y^t Y & C^t G^{-1}Y^t + Y^t \\ YG^{-1} & YG^{-1}C + Y & \mathbf{1}_{16} + YG^{-1}Y^t \end{pmatrix}, \tag{D.4}$$

where $\mathbf{1}_{16}$ is the 16-dimensional unit matrix and

$$C_{\alpha\beta} = B_{\alpha\beta} - \frac{1}{2} Y_\alpha^I Y_\beta^I. \tag{D.5}$$

Introduce the $O(p, p + 16)$ invariant metric

$$L = \begin{pmatrix} 0 & \mathbf{1}_p & 0 \\ \mathbf{1}_p & 0 & 0 \\ 0 & 0 & \mathbf{1}_{16} \end{pmatrix}. \tag{D.6}$$

Then the matrix M satisfies

$$M^T L M = MLM = L, \quad M^{-1} = LML. \tag{D.7}$$

Thus, $M \in O(p, p + 16)$. In terms of M the conformal weights are given by

$$\frac{1}{2} P_L^2 = \frac{1}{4}(m^\alpha, n_\alpha, Q_I) \cdot (M - L) \cdot \begin{pmatrix} m^\alpha \\ n_\alpha \\ Q_I \end{pmatrix}, \tag{D.8}$$

$$\frac{1}{2} P_R^2 = \frac{1}{4}(m^\alpha, n_\alpha, Q_I) \cdot (M + L) \cdot \begin{pmatrix} m^\alpha \\ n_\alpha \\ Q_I \end{pmatrix}. \tag{D.9}$$

The spin

$$\frac{1}{2} P_R^2 - \frac{1}{2} P_L^2 = m^\alpha n_\alpha - \frac{1}{2} Q_I Q_I \tag{D.10}$$

is an integer. When $Y = 0$ the lattice sum factorizes:

$$\Gamma_{p,p+16}(G, B, Y = 0) = \Gamma_{p,p}(G, B)\, \bar\Gamma_{\text{Spin}(32)/\mathbb{Z}_2}. \tag{D.11}$$

It can be shown [507] that for some special (nonzero) values $\tilde{Y}_\alpha^I$ the lattice sum factorizes into the (p, p) toroidal sum and the lattice sum of $E_8 \times E_8$:

$$\Gamma_{p,p+16}(G, B, Y = \tilde{Y}) = \Gamma_{p,p}(G', B') \, \bar{\Gamma}_{E_8 \times E_8}. \tag{D.12}$$

Thus, we can continuously interpolate in $\Gamma_{p,p+16}$ between the O(32) and $E_8 \times E_8$ symmetric points.

Finally, the duality group here is $O(p, p + 16, \mathbb{Z})$. An element of $O(p, p + 16, \mathbb{Z})$ is an integer-valued $O(p, p + 16)$ matrix. Consider such a matrix Ω. It satisfies $\Omega^T L \Omega = L$. The lattice sum is invariant under the T-duality transformation

$$\begin{pmatrix} m^\alpha \\ n_\alpha \\ Q_I \end{pmatrix} \rightarrow \Omega \cdot \begin{pmatrix} m^\alpha \\ n_\alpha \\ Q_I \end{pmatrix}, \quad M \rightarrow \Omega M \Omega^T. \tag{D.13}$$

In what follows, we will describe translation orbifold blocks for toroidal CFTs. Start from the $(d, d + 16)$ lattice. We will use the notation $\lambda = (m^\alpha, n_\alpha, Q_I)$ for a lattice vector with its $O(p,p + 16)$ inner product, which gives the invariant square $\lambda^2 = 2m^\alpha n_\alpha - Q_I Q_I \in 2\mathbb{Z}$. Perform a $\mathbb{Z}_N$ translation by $\epsilon/N \notin L$, where ϵ is a lattice vector. The generalization of one-dimensional orbifold blocks (4.21.25) on page 112 is straightforward:

$$Z_{d,d+16}^N(\epsilon)[{}_g^h] = \frac{\Gamma_{p,p+16}(\epsilon)[{}_g^h]}{\eta^p \bar{\eta}^{p+16}} = \frac{\sum_{\lambda \in L + \epsilon \frac{h}{N}} e^{2\pi i g \, \epsilon \cdot \lambda / N} \, q^{p_L^2/2} \, \bar{q}^{p_R^2/2}}{\eta^p \bar{\eta}^{p+16}} \tag{D.14}$$

where $h, g = 0, 1, \ldots, N - 1$. It has the following properties:

$$Z^N(-\epsilon)[{}_g^h] = Z^N(\epsilon)[{}_g^h], \quad Z^N(\epsilon)[{}_{-g}^{-h}] = Z^N(\epsilon)[{}_g^h], \tag{D.15}$$

$$Z^N(\epsilon)[{}_g^{h+1}] = \exp\left[-\frac{i\pi g \epsilon^2}{N}\right] Z^N(\epsilon)[{}_g^h], \quad Z^N(\epsilon)[{}_{g+1}^h] = Z^N(\epsilon)[{}_g^h], \tag{D.16}$$

$$Z^N(\epsilon + N\epsilon')[{}_g^h] = \exp\left[\frac{2\pi i \, gh \, \epsilon \cdot \epsilon'}{N}\right] Z^N(\epsilon)[{}_g^h]. \tag{D.17}$$

Under modular transformations

$$\tau \rightarrow \tau + 1: \quad Z^N(\epsilon)[{}_g^h] \rightarrow \exp\left[\frac{4\pi i}{3} + \frac{i\pi h^2 \epsilon^2}{N^2}\right] Z^N(\epsilon)[{}_{h+g}^h], \tag{D.18}$$

$$\tau \rightarrow -\frac{1}{\tau}: \quad Z^N(\epsilon)[{}_g^h] \rightarrow \exp\left[-\frac{2\pi i \, hg \, \epsilon^2}{N^2}\right] Z^N(\epsilon)[{}_{-h}^g]. \tag{D.19}$$

Under $O(p, p + 16, \mathbb{Z})$ duality transformations it transforms as

$$Z^N(\epsilon, \Omega M \Omega^T)[{}_g^h] = Z^N(\Omega \cdot \epsilon, M)[{}_g^h], \tag{D.20}$$

where $\Omega \in O(p, p + 16, \mathbb{Z})$ and M is the moduli matrix (D.4). The unbroken duality group consists of the subgroup of $O(p, p + 16, \mathbb{Z})$ transformations that preserve ϵ modulo N^2 times a lattice vector.

Appendix E | Toroidal Kaluza-Klein Reduction

In this appendix we will describe the Kaluza-Klein *Ansatz* for toroidal dimensional reduction from 10 to $D < 10$ dimensions. A more detailed discussion can be found in [235,222]. Careted fields will denote the ten-dimensional fields and similarly for the indices. Greek indices from the beginning of the alphabet will denote the $10 - D$ internal (compact) dimensions. Non careted Greek indices from the middle of the alphabet will denote the D noncompact dimensions.

The standard form for the 10-bein is

$$
\hat{e}^{\hat{r}}_{\hat{\mu}} = \begin{pmatrix} e^r_\mu & A^\beta_\mu E^a_\beta \\ 0 & E^a_\alpha \end{pmatrix}, \quad
\hat{e}^{\hat{\mu}}_{\hat{r}} = \begin{pmatrix} e^\mu_r & -e^\nu_r A^\alpha_\nu \\ 0 & E^\alpha_a \end{pmatrix}.
\tag{E.1}
$$

For the metric we have

$$
\hat{G}_{\hat{\mu}\hat{\nu}} = \begin{pmatrix} g_{\mu\nu} + A^\alpha_\mu G_{\alpha\beta} A^\beta_\nu & G_{\alpha\beta} A^\beta_\mu \\ G_{\alpha\beta} A^\beta_\nu & G_{\alpha\beta} \end{pmatrix}, \quad
\hat{G}^{\hat{\mu}\hat{\nu}} = \begin{pmatrix} g^{\mu\nu} & -A^{\mu\alpha} \\ -A^{\nu\alpha} & G^{\alpha\beta} + A^\alpha_\rho A^{\beta,\rho} \end{pmatrix}.
\tag{E.2}
$$

Then the part of the action containing the Einstein-Hilbert term as well as the dilaton becomes

$$
\int d^{10}x \, e^{-2\hat{\Phi}} \sqrt{-\det \hat{G}} \left[\hat{R} + 4\partial_{\hat{\mu}} \hat{\Phi} \partial^{\hat{\mu}} \hat{\Phi} \right]
$$
$$
\to \int d^D x \sqrt{-\det g} \, e^{-2\phi} \left[R + 4\partial_\mu \phi \partial^\mu \phi + \frac{1}{4}\partial_\mu G_{\alpha\beta} \partial^\mu G^{\alpha\beta} - \frac{1}{4} G_{\alpha\beta} F^A{}_{\mu\nu}{}^\alpha F^{\beta,\mu\nu}_A \right]
\tag{E.3}
$$

where

$$
\phi = \hat{\Phi} - \frac{1}{4} \log \left(\det G_{\alpha\beta} \right),
\tag{E.4}
$$

$$
F^A{}_{\mu\nu}{}^\alpha = \partial_\mu A^\alpha_\nu - \partial_\nu A^\alpha_\mu.
\tag{E.5}
$$

The label A in $F^A{}_{\mu\nu}{}^\alpha$ is not an index. It is there to indicate that this is the field strength of the A^α_μ gauge fields originating in the metric.

We now turn to the antisymmetric tensor part of the action:

$$
-\frac{1}{12} \int d^{10}x \sqrt{-\det \hat{G}} e^{-2\hat{\Phi}} \hat{H}^{\hat{\mu}\hat{\nu}\hat{\rho}} \hat{H}_{\hat{\mu}\hat{\nu}\hat{\rho}}
$$
$$
= -\int d^{D}x \sqrt{-\det g}\, e^{-2\phi} \left[\frac{1}{4} H_{\mu\alpha\beta} H^{\mu\alpha\beta} + \frac{1}{4} H_{\mu\nu\alpha} H^{\mu\nu\alpha} + \frac{1}{12} H_{\mu\nu\rho} H^{\mu\nu\rho} \right], \tag{E.6}
$$

where we have used $H_{\alpha\beta\gamma} = 0$ and

$$
H_{\mu\alpha\beta} = e^{r}_{\mu} \hat{e}^{\hat{\mu}}_{\hat{r}} \hat{H}_{\hat{\mu}\alpha\beta} = \hat{H}_{\mu\alpha\beta}, \tag{E.7}
$$
$$
H_{\mu\nu\alpha} = e^{r}_{\mu} e^{s}_{\nu} \hat{e}^{\hat{\mu}}_{\hat{r}} \hat{e}^{\hat{\nu}}_{\hat{s}} \hat{H}_{\hat{\mu}\hat{\nu}\alpha} = \hat{H}_{\mu\nu\alpha} - A^{\beta}_{\mu} \hat{H}_{\nu\alpha\beta} + A^{\beta}_{\nu} \hat{H}_{\mu\alpha\beta}, \tag{E.8}
$$
$$
H_{\mu\nu\rho} = e^{r}_{\mu} e^{s}_{\nu} e^{t}_{\rho} \hat{e}^{\hat{\mu}}_{\hat{r}} \hat{e}^{\hat{\nu}}_{\hat{s}} \hat{e}^{\hat{\rho}}_{\hat{t}} \hat{H}_{\hat{\mu}\hat{\nu}\hat{\rho}} = \hat{H}_{\mu\nu\rho} + \left[-A^{\alpha}_{\mu} \hat{H}_{\alpha\nu\rho} + A^{\alpha}_{\mu} A^{\beta}_{\nu} \hat{H}_{\alpha\beta\rho} + \text{cyclic} \right]. \tag{E.9}
$$

Similarly,

$$
\int d^{10}x \sqrt{-\det \hat{G}}\, e^{-2\hat{\Phi}} \sum_{I=1}^{16} \hat{F}^{I}_{\hat{\mu}\hat{\nu}} \hat{F}^{I,\hat{\mu}\hat{\nu}}
$$
$$
= \int d^{D}x \sqrt{-\det g}\, e^{-2\phi} \sum_{I=1}^{16} \left[\tilde{F}^{I}_{\mu\nu} \tilde{F}^{I,\mu\nu} + 2\tilde{F}^{I}_{\mu\alpha} \tilde{F}^{I,\mu\alpha} \right], \tag{E.10}
$$

with

$$
Y^{I}_{\alpha} = \hat{A}^{I}_{\alpha}, \quad A^{I}_{\mu} = \hat{A}^{I}_{\mu} - Y^{I}_{\alpha} A^{a}_{\mu}, \quad \tilde{F}^{I}_{\mu\nu} = F^{I}_{\mu\nu} + Y^{I}_{\alpha} F^{A,\alpha}_{\mu\nu}, \tag{E.11}
$$
$$
\tilde{F}^{I}_{\mu\alpha} = \partial_{\mu} Y^{I}_{\alpha}, \quad F^{I}_{\mu\nu} = \partial_{\mu} A^{I}_{\nu} - \partial_{\nu} A^{I}_{\mu}. \tag{E.12}
$$

We can now evaluate the D-dimensional antisymmetric tensor pieces using (E.7)–(E.9):

$$
\hat{H}_{\mu\alpha\beta} = \partial_{\mu} \hat{B}_{\alpha\beta} + \frac{1}{2} \sum_{I} \left[Y^{I}_{\alpha} \partial_{\mu} Y^{I}_{\beta} - Y^{I}_{\beta} \partial_{\mu} Y^{I}_{\alpha} \right]. \tag{E.13}
$$

Introducing

$$
C_{\alpha\beta} \equiv \hat{B}_{\alpha\beta} - \frac{1}{2} \sum_{I} Y^{I}_{\alpha} Y^{I}_{\beta}, \tag{E.14}
$$

we obtain from (E.6)

$$
H_{\mu\alpha\beta} = \partial_{\mu} C_{\alpha\beta} + \sum_{I} Y^{I}_{\alpha} \partial_{\mu} Y^{I}_{\beta}. \tag{E.15}
$$

Also

$$
\hat{H}_{\mu\nu\alpha} = \partial_{\mu} \hat{B}_{\nu\alpha} - \partial_{\nu} \hat{B}_{\mu\alpha} + \frac{1}{2} \sum_{I} \left[\hat{A}^{I}_{\nu} \partial_{\mu} Y^{I}_{\alpha} - \hat{A}^{I}_{\mu} \partial_{\nu} Y^{I}_{\alpha} - Y^{I}_{\alpha} \hat{F}^{I}_{\mu\nu} \right]. \tag{E.16}
$$

Define

$$
B_{\mu,\alpha} \equiv \hat{B}_{\mu\alpha} + B_{\alpha\beta} A^{\beta}_{\mu} + \frac{1}{2} \sum_{I} Y^{I}_{\alpha} A^{I}_{\mu}, \tag{E.17}
$$
$$
F^{B}_{\alpha,\mu\nu} = \partial_{\mu} B_{\alpha,\nu} - \partial_{\nu} B_{\alpha,\mu}, \tag{E.18}
$$

we obtain from (E.7)

$$
H_{\mu\nu\alpha} = F^{B}_{\alpha\mu\nu} - C_{\alpha\beta} F^{A,\beta}_{\mu\nu} - \sum_{I} Y^{I}_{\alpha} F^{I}_{\mu\nu}. \tag{E.19}
$$

Finally,

$$B_{\mu\nu} = \hat{B}_{\mu\nu} + \frac{1}{2}\left[A_\mu^\alpha B_{\nu\alpha} + \sum_I A_\mu^I A_\nu^\alpha Y_\alpha^I - (\mu \leftrightarrow \nu)\right] - A_\mu^\alpha A_\nu^\beta B_{\alpha\beta} \tag{E.20}$$

and

$$H_{\mu\nu\rho} = \partial_\mu B_{\nu\rho} - \frac{1}{2}\left[B_{\mu\alpha}F_{\nu\rho}^{A,\alpha} + A_\mu^\alpha F_{a,\nu\rho}^B + \sum_I A_\mu^I F_{\nu\rho}^I\right] + \text{cyclic} \tag{E.21}$$

$$\equiv \partial_\mu B_{\nu\rho} - \frac{1}{2}L_{ij}A_\mu^i F_{\nu\rho}^j + \text{cyclic},$$

where we combined the $36 - 2D$ gauge fields A_μ^α, $B_{\alpha,\mu}$, A_μ^I into the uniform notation A_μ^i, $i = 1, 2, \ldots, 36 - 2D$ and L_{ij} is the $O(10 - D, 26 - D)$-invariant metric (D.6). We can combine the scalars $G_{\alpha\beta}$, $B_{\alpha\beta}$, Y_α^I into the matrix M given in (D.4). In the heterotic case, putting everything together, we obtain the following D-dimensional action:

$$S_D^{\text{het}} = \int d^D x \sqrt{-\det g}\, e^{-2\phi}\left[R + 4\partial^\mu\phi\partial_\mu\phi - \frac{1}{12}H^{\mu\nu\rho}H_{\mu\nu\rho}\right.$$
$$\left. - \frac{1}{4}(M^{-1})_{ij}F_{\mu\nu}^i F^{j\mu\nu} + \frac{1}{8}\text{Tr}(\partial_\mu M\partial^\mu M^{-1})\right]. \tag{E.22}$$

We will also consider here the KK reduction of a three-index antisymmetric tensor $C_{\mu\nu\rho}$. Such a tensor appears in type-IIA string theory and 11-dimensional supergravity. The action for such a tensor is

$$S_C = -\frac{1}{2\cdot 4!}\int d^d x\sqrt{-G}\,\hat{F}^2, \tag{E.23}$$

where

$$\hat{F}_{\mu\nu\rho\sigma} = \partial_\mu\hat{C}_{\nu\rho\sigma} - \partial_\sigma\hat{C}_{\mu\nu\rho} + \partial_\rho\hat{C}_{\sigma\mu\nu} - \partial_\nu\hat{C}_{\rho\sigma\mu}. \tag{E.24}$$

We define the lower-dimensional components as

$$C_{\alpha\beta\gamma} = \hat{C}_{\alpha\beta\gamma}, \quad C_{\mu\alpha\beta} = \hat{C}_{\mu\alpha\beta} - C_{\alpha\beta\gamma}A_\mu^\gamma, \tag{E.25}$$

$$C_{\mu\nu\alpha} = \hat{C}_{\mu\nu\alpha} + \hat{C}_{\mu\alpha\beta}A_\nu^\beta - \hat{C}_{\nu\alpha\beta}A_\mu^\beta + C_{\alpha\beta\gamma}A_\mu^\beta A_\nu^\gamma, \tag{E.26}$$

$$C_{\mu\nu\rho} = \hat{C}_{\mu\nu\rho} + \left(-\hat{C}_{\nu\rho\alpha}A_\mu^\alpha + \hat{C}_{\alpha\beta\rho}A_\mu^\alpha A_\nu^\beta + \text{cyclic}\right) - C_{\alpha\beta\gamma}A_\mu^\alpha A_\nu^\beta A_\rho^\gamma. \tag{E.27}$$

Then,

$$S_C = -\frac{1}{2\cdot 4!}\int d^D x\sqrt{-g}\sqrt{\det G_{\alpha\beta}}\left[F_{\mu\nu\rho\sigma}F^{\mu\nu\rho\sigma} + 4F_{\mu\nu\rho\alpha}F^{\mu\nu\rho\alpha}\right.$$
$$\left. + 6F_{\mu\nu\alpha\beta}F^{\mu\nu\alpha\beta} + 4F_{\mu\alpha\beta\gamma}F^{\mu\alpha\beta\gamma}\right], \tag{E.28}$$

where

$$F_{\mu\alpha\beta\gamma} = \partial_\mu C_{\alpha\beta\gamma}, \quad F_{\mu\nu\alpha\beta} = \partial_\mu C_{\nu\alpha\beta} - \partial_\nu C_{\mu\alpha\beta} + C_{\alpha\beta\gamma}F_{\mu\nu}^\gamma, \tag{E.29}$$

$$F_{\mu\nu\rho\alpha} = \partial_\mu C_{\nu\rho\alpha} + C_{\mu\alpha\beta}F_{\nu\rho}^\beta + \text{cyclic}, \tag{E.30}$$

$$F_{\mu\nu\rho\sigma} = (\partial_\mu C_{\nu\rho\sigma} + 3\text{ perm.}) + (C_{\rho\sigma\alpha}F_{\mu\nu}^\alpha + 5\text{ perm.}). \tag{E.31}$$

Appendix F | The Reissner-Nordström Black Hole

In this appendix we remind the reader of some salient features of static charged black holes in four dimensions.

The relevant fields are the metric and a Maxwell gauge field with action

$$S = \frac{1}{16\pi G} \int d^4x \sqrt{-g}\, R - \frac{1}{16\pi G} \int d^4x \sqrt{-g}\, F_{\mu\nu} F^{\mu\nu}. \tag{F.1}$$

The equations of motion are

$$R_{\mu\nu} - \frac{1}{2} g_{\mu\nu} R = 8\pi G \left[F_{\mu\rho} F_\nu{}^\rho - \frac{1}{4} g_{\mu\nu} F_{\rho\sigma} F^{\rho\sigma} \right], \quad \nabla^\mu F_{\mu\nu} = 0. \tag{F.2}$$

There is a charged and spherically symmetric solution of the equations of motion found by Reissner and Nordström:

$$ds^2 = -f(r)dt^2 + \frac{dr^2}{f(r)} + r^2 d\Omega_2^2, \quad F = \frac{Q}{r^2} dt \wedge dr, \tag{F.3}$$

with

$$f = 1 - \frac{2GM}{r} + \frac{Q^2}{r^2}, \quad d\Omega_2^2 = d\theta^2 + \sin^2\theta\, d\varphi^2. \tag{F.4}$$

It is obvious that the asymptotic Maxwell field is that of an electric charge Q. Setting $Q = 0$ we obtain the Schwarzschild solution that describes a black hole of mass M.

The metric has singularities at the zeros of the function $f(r_\pm) = 0$:

$$r_\pm = GM \pm \sqrt{G^2 M^2 - Q^2}, \tag{F.5}$$

as well as at $r = 0$. It is easy to convince oneself that only $r = 0$ is a singularity of the geometry. Indeed, while the scalar curvature vanishes because the stress tensor source is traceless, a calculation of the Ricci squared invariant gives

$$R_{\mu\nu} R^{\mu\nu} = 4 \frac{Q^4}{r^8}. \tag{F.6}$$

The higher invariants are similar. The other apparent singularities of the metric at $r = r_\pm$, are coordinate singularities and correspond to horizons. We will study them more closely.

Let us start with the known Schwarzschild case $Q = 0$, in which $r_- = 0$. For $r > r_+$, t is the time and the surface $r = $ constant is timelike. The surface $r = r_+$ is a null hypersurface (a horizon) since its normal vector ∂_r is null there. In the interior region, $r < r_+$, it is r that plays the role of time, the surfaces of constant r are spacelike, and thus the (spacelike) singularity at $r = 0$ is unavoidable.

We now go back to the charged case. Let us first consider the case with $Q < GM$. Then $r_+ > r_- > 0$. Since g_{00} vanishes at these two points, then the hypersurfaces $r = r_\pm$ are horizons. The Killing vector associated with time translations, ∂_t, has negative norm for $r > r_+$. This is equivalent to the statement that for $r > r_+$, t is the time, and the surface $r = $ constant is timelike.

In the region $r_- < r < r_+$, r becomes the time and surfaces of constant r are spacelike. Thus once in this region, we will eventually reach $r = r_-$ In the interior region $r < r_-$, however, unlike the Schwarzschild case, things change again. t is now again time, and the surfaces with constant r are timelike. Some geodesics avoid the $r = 0$ singularity. This is a timelike singularity. The null hypersurface $r = r_+$ is known as the *outer horizon* of the Reissner-Nordström black hole while $r = r_-$ is the *inner horizon*.

The case where $Q = GM$ is special. The inner and the outer horizons coincide at $r = r_+ = r_- = GM$. Moreover that hypersurface is null but no longer a horizon. Changing coordinates to

$$\rho = r - Q, \tag{F.7}$$

the metric can be written as

$$ds^2 = -\frac{dt^2}{H^2(\rho)} + H^2(\rho)[d\rho^2 + \rho^2 d\Omega_2^2], \quad H(\rho) = 1 + \frac{Q}{\rho}, \quad A = -H(\rho)dt. \tag{F.8}$$

In these coordinates, the metric is regular at $\rho = 0$. This is the extremal Reissner-Nordström (RN) metric. If we embed the Einstein Maxwell theory in pure $\mathcal{N} = 2_4$ super-gravity (containing apart from these fields two gravitini), then the extremal metric preserved one of the two supersymmetries (see exercise 8.35). This in particular implies the existence of four independent Killing spinors in this background.

At extremality the solution can be easily generalized to a multicenter solution by simply substituting $H(\rho)$ by

$$H_N(\rho) = 1 + \sum_{i=1}^{N} \frac{Q_i}{|\vec{\rho} - \vec{\rho}_i|}. \tag{F.9}$$

You are asked to investigate this solution in exercise 8.36.

The BPS bound for the single-center solution reads

$$Q \leq GM. \tag{F.10}$$

From the point of view of supersymmetric representation theory, as explained in section 11.2 on page 323, violation of the BPS condition implies loss of unitarity. From the geometrical point of view, when $Q > GM$ the solution has no horizon and a naked singularity at $r = 0$. This is an independent view indicating that naked singularities should be avoided.

It is interesting to observe the geometry close to the would-be horizon at $\rho = 0$. The metric simplifies to

$$ds^2 = -\frac{\rho^2}{Q^2}dt^2 + \frac{Q^2}{\rho^2}d\rho^2 + Q^2\,d\Omega_2^2. \tag{F.11}$$

By changing the radial coordinate once again as $\rho \to Q^2/\rho$ we obtain

$$ds^2 = Q^2\,ds_{AdS_2}^2 + Q^2\,d\Omega_2^2, \tag{F.12}$$

where

$$ds_{AdS_2}^2 = \frac{-dt^2 + d\rho^2}{\rho^2} \tag{F.13}$$

is the anti–de Sitter metric in two dimensions, in Poincaré coordinates. Thus, the near-horizon geometry is that of $AdS_2 \times S^2$ with common radius of curvature $|Q|$. This is also known as the Bertotti-Robertson universe.

You are invited in exercise 8.37 on page 217 to show that the near-horizon geometry, viewed as a solution of pure $\mathcal{N} = 2_4$ supergravity, has the maximal possible number of Killing spinors, namely, 8. This is the same number as the trivial solution: flat space and no electric field. In that sense the Reissner-Nordström solution can be thought of as a soliton: it interpolates between two "vacuum" solutions of the theory. At $\rho \to 0$ we obtain the Bertotti-Robertson universe while at $\rho \to \infty$ we obtain flat space.

In general, collapsing matter in four-dimensional gravity forms black holes uniquely characterized by their mass M, various potential $U(1)$ charges Q_i and their angular momentum J. This is known as the "no-hair theorem." There has been considerable debate in the literature over the existence of nonabelian and/or discrete hair (see, for example, [508]). It can be shown that during gravitational collapse, the tails of other global conserved quantities die off exponentially fast in time.

We will also introduce here the higher-dimensional RN solutions. The five-dimensional case is in particularly relevant for the black-hole entropy calculations in string theory, detailed in chapter 12. The metric of of the $(4 + n)$-dimensional RN black hole is

$$ds^2 = -h(r)dt^2 + \frac{dr^2}{h(r)} + r^2 d\Omega_{n+2}^2, \quad h(r) = \left[1 - \left(\frac{r_+}{r}\right)^{n+1}\right]\left[1 - \left(\frac{r_-}{r}\right)^{n+1}\right]. \tag{F.14}$$

The mass and charge are given by

$$M = \frac{(n+2)\Omega_{n+2}}{16\pi\,G_{n+4}}\left(r_+^{n+1} + r_-^{n+1}\right), \quad Q = (r_+r_-)^{\frac{n+1}{2}}\sqrt{\frac{(n+1)(n+2)}{8\pi\,G_{n+4}}}, \tag{F.15}$$

with the volume of the unit sphere $\Omega_{n+2} = 2\pi^{(n+3)/2}/\Gamma\left(\frac{n+3}{2}\right)$.

In this appendix we describe electric-magnetic duality transformations for free gauge fields. We consider a collection of abelian gauge fields in $D = 4$. In the presence of supersymmetry we can write terms quadratic in the gauge fields as

$$\mathcal{L}_{\text{gauge}} = -\frac{1}{8}\text{Im}\int d^4x\sqrt{-\det g}\,\mathcal{F}^i_{\mu\nu}N_{ij}\mathcal{F}^{j,\mu\nu}, \tag{G.1}$$

where

$$\mathcal{F}_{\mu\nu} = F_{\mu\nu} + i^\star F_{\mu\nu}, \quad {}^\star F_{\mu\nu} = \frac{1}{2}\frac{\epsilon_{\mu\nu}{}^{\rho\sigma}}{\sqrt{-g}}F_{\rho\sigma}, \tag{G.2}$$

with the property (in Minkowski space) that ${}^{\star\star}F = -F$ and ${}^\star F_{\mu\nu}{}^{\star}F^{\mu\nu} = -F_{\mu\nu}F^{\mu\nu}$. In components, the Lagrangian (G.1) becomes

$$\mathcal{L}_{\text{gauge}} = -\frac{1}{4}\int d^4x\left[\sqrt{-g}F^i_{\mu\nu}N_2^{ij}F^{j,\mu\nu} + F^i_{\mu\nu}N_1^{ij\,\star}F^{j,\mu\nu}\right] \tag{G.3}$$

with $N = N_1 + iN_2$.

Define now the tensor that gives the equations of motion

$$\mathcal{G}^i_{\mu\nu} = N_{ij}\mathcal{F}^j_{\mu\nu} = N_1 F - N_2{}^\star F + i(N_2 F + N_1{}^\star F), \tag{G.4}$$

The equations of motion can be written in the form

$$\text{Im}\nabla^\mu\mathcal{G}^i_{\mu\nu} = 0, \tag{G.5}$$

while the Bianchi identity is

$$\text{Im}\nabla^\mu\mathcal{F}^i_{\mu\nu} = 0. \tag{G.6}$$

Altogether,

$$\text{Im}\nabla^\mu\begin{pmatrix}\mathcal{G}^i_{\mu\nu}\\\mathcal{F}^i_{\mu\nu}\end{pmatrix} = \begin{pmatrix}0\\0\end{pmatrix}. \tag{G.7}$$

Obviously any $\mathrm{Sp}(2r,\mathbb{R})$ transformation of the form

$$
\begin{pmatrix} \mathcal{G}'_{\mu\nu} \\ \mathcal{F}'_{\mu\nu} \end{pmatrix} = \begin{pmatrix} A & B \\ C & D \end{pmatrix} \begin{pmatrix} \mathcal{G}_{\mu\nu} \\ \mathcal{F}_{\mu\nu} \end{pmatrix},
\tag{G.8}
$$

where A, B, C, D are $r \times r$ matrices $(CA^t - AC^t = 0,\ B^tD - D^tB = 0,\ A^tD - C^tB = 1)$, preserves the collection of equations of motion and Bianchi identities. At the same time

$$
N' = (AN + B)(CN + D)^{-1}.
\tag{G.9}
$$

The duality transformations are

$$
F' = C(N_1 F - N_2{}^\star F) + DF, \quad {}^\star F' = C(N_2 F + N_1{}^\star F) + D^\star F.
\tag{G.10}
$$

In the simple case $A = D = 0,\ -B = C = 1$ they become

$$
F' = N_1 F - N_2{}^\star F, \quad {}^\star F' = N_2 F + N_1{}^\star F, \quad N' = -\frac{1}{N}.
\tag{G.11}
$$

When we perform duality with respect to one of the gauge fields (we will call its component 0) we have

$$
\begin{pmatrix} A & B \\ C & D \end{pmatrix} = \begin{pmatrix} 1 - e & -e \\ e & 1 - e \end{pmatrix}, \quad e = \begin{pmatrix} 1 & 0 & \cdots \\ 0 & 0 & \cdots \\ \cdot & & \cdot \end{pmatrix},
\tag{G.12}
$$

$$
N'_{00} = -\frac{1}{N_{00}}, N'_{0i} = \frac{N_{0i}}{N_{00}}, N'_{i0} = \frac{N_{i0}}{N_{00}}, N'_{ij} = N_{ij} - \frac{N_{i0} N_{0j}}{N_{00}}.
\tag{G.13}
$$

Finally consider the duality generated by

$$
\begin{pmatrix} A & B \\ C & D \end{pmatrix} = \begin{pmatrix} 1 - e_1 & e_2 \\ -e_2 & 1 - e_1 \end{pmatrix},
\tag{G.14}
$$

$$
e_1 = \begin{pmatrix} 1 & 0 & 0 & \cdots \\ 0 & 1 & 0 & \cdots \\ 0 & 0 & 0 & \cdot \\ \cdot & \cdot & \cdot & \end{pmatrix}, \quad e_2 = \begin{pmatrix} 0 & 1 & 0 & \cdots \\ -1 & 0 & 0 & \cdots \\ 0 & 0 & 0 & \cdot \\ \cdot & \cdot & \cdot & \end{pmatrix}.
\tag{G.15}
$$

We will denote the indices in the two-dimensional subsector where the duality acts by $\alpha, \beta, \gamma, \ldots$. Then

$$
N'_{\alpha\beta} = -\frac{N_{\alpha\beta}}{\det N_{\alpha\beta}}, \quad N'_{\alpha i} = -\frac{N_{\alpha\beta}\epsilon^{\beta\gamma} N_{\gamma i}}{\det N_{\alpha\beta}}, \quad N'_{i\alpha} = \frac{N_{i\beta}\epsilon^{\beta\gamma} N_{\alpha\gamma}}{\det N_{\alpha\beta}},
\tag{G.16}
$$

$$
N'_{ij} = N_{ij} + \frac{N_{i\alpha}\epsilon^{\alpha\beta} N_{\beta\gamma}\epsilon^{\gamma\delta} N_{\delta j}}{\det N_{\alpha\beta}}.
\tag{G.17}
$$

Consider now the $\mathcal{N} = 4_4$ heterotic string. The appropriate matrix N is

$$
N = S_1 L + i S_2 M^{-1}, \quad S = S_1 + i S_2.
\tag{G.18}
$$

Performing an overall duality as in (G.11) we obtain

$$N' = -N^{-1} = -\frac{S_1}{|S|^2}L + i\frac{S_2}{|S|^2}M = -\frac{S_1}{|S|^2}L + i\frac{S_2}{|S|^2}LM^{-1}L. \tag{G.19}$$

Thus, we observe that apart from an $S \to -1/S$ transformation on the S field it also affects an $O(6,22,\mathbb{Z})$ transformation by the matrix L, which interchanges windings and momenta of the six-torus.

The duality transformation that acts only on S is given by $A = D = 0, -B = C = L$ under which

$$N' = -LN^{-1}L = -\frac{S_1}{|S|^2}L + i\frac{S_2}{|S|^2}M^{-1}. \tag{G.20}$$

The full $SL(2,\mathbb{Z})$ group acting on S is generated by

$$\begin{pmatrix} A & B \\ C & D \end{pmatrix} = \begin{pmatrix} a\mathbf{1}_{28} & bL \\ cL & d\mathbf{1}_{28} \end{pmatrix}, \quad ad - bc = 1. \tag{G.21}$$

Finally the duality transformation, which acts as an $O(6,22,\mathbb{Z})$ transformation, is given by $A = \Omega, D^{-1} = \Omega^t, B = C = 0$.

Appendix H | Supersymmetric Actions in Ten and Eleven Dimensions

In ten and eleven dimensions, in order to have massless fields with spin not greater than 2 we have to restrict ourselves to $\mathcal{N} \leq 1_{11}$ and $\mathcal{N} \leq 2_{10}$ SUSY. The following are all known space-time supersymmetric string theories. Each of them admits an effective supergravity theory.

- Type-I theory (chiral) with $\mathcal{N} = 1_{10}$ supersymmetry and gauge group O(32).

- Heterotic theories (chiral) with $\mathcal{N} = 1_{10}$ supersymmetry and gauge groups O(32) and $E_8 \times E_8$.

- Type-IIA theory (nonchiral) with $\mathcal{N} = (1, 1)_{10}$ supersymmetry.

- Type-IIB theory (chiral) with $\mathcal{N} = (2, 0)_{10}$ supersymmetry.

The type-IIA supergravity theory is the dimensional reduction of the 11-dimensional $\mathcal{N} = 1_{11}$ Cremmer-Julia-Scherk supergravity.

In the following, we will present their two-derivative effective actions. In particular, we will present the bosonic part of the action as well as the supersymmetry transformations of fermions which are the most useful pieces of data. We also give a guide to references.

We start with some notations and conventions concerning the spin connection and spinor covariant derivatives.

From the vielbein e_μ^a we may construct a one-form as

$$e^a = e_\mu^a dx^\mu. \tag{H.1}$$

The spin connection $\omega^a{}_{b\mu}$ can be constructed out of the metric (or vielbein) in the absence of torsion. The torsionless constraint is

$$\nabla_\mu e_\nu^a = \partial_\mu e_\nu^a + \omega^a{}_{b\mu} e_\nu^b - \Gamma^\rho_{\mu\nu} e_\rho^a = 0, \tag{H.2}$$

from which it follows that

$$\omega^a_{\mu\nu} = \partial_\mu e_\nu^a - \partial_\nu e_\mu^a. \tag{H.3}$$

The spin connection can also be defined as a one-form solving the Maurer-Cartan equation

$$de^a + \omega^a{}_b \wedge e^b = 0, \tag{H.4}$$

where

$$e^a \equiv e^a{}_\mu dx^\mu, \quad \omega^a{}_b \equiv e_b{}^\mu \omega^a{}_{\mu\nu} dx^\nu. \tag{H.5}$$

It is an O(d) connection and the Riemann tensor is its curvature

$$R^{ab}{}_{\mu\nu} = \partial_\mu \omega^{ab}_\nu - \partial_\nu \omega^{ab}_\mu + [\omega_\mu, \omega_\nu]^{ab}. \tag{H.6}$$

Under an infinitesimal Lorentz transformation with parameter $\theta^{ab}(x)$ it transforms as

$$\delta\omega^{ab}_\mu = \partial_\mu \theta^{ab} + [\omega_\mu, \theta]^{ab}. \tag{H.7}$$

It is essential for coupling fermions to gravity via the covariant derivative for spinors

$$\nabla_\mu \psi = \left(\partial_\mu + \frac{1}{4} \omega^{ab}_\mu \Gamma_{ab} \right) \psi. \tag{H.8}$$

H.1 The $\mathcal{N} = 1_{11}$ Supergravity

This supergravity does not appear directly as the low-energy limit of any weakly coupled string theory. It has been conjectured to describe the low-energy limit of strongly coupled IIA theory in ten dimensions as discussed in chapter 11. It is related to type-IIA supergravity by dimensional reduction. It is the supergravity with the maximal possible space-time symmetry.

Here, capital indices refer to 11 dimensions. The relevant fields are the metric g_{MN}, a three-index antisymmetric tensor $\hat{C}_{MNP}$ and a gravitino ψ^M. They form a single representation of the supersymmetry algebra.

The bosonic action is

$$L_{D=11} = \frac{1}{2\kappa^2} \left[R - \frac{1}{2 \cdot 4!} G_4^2 + \frac{1}{(144)^2} \epsilon^{M_1 \cdots M_{11}} G_{M_1 \cdots M_4} G_{M_5 \cdots M_8} \hat{C}_{M_9 M_{10} M_{11}} \right]. \tag{H.9}$$

G_4 is the field strength of $\hat{C}$ defined thus:

$$G_{MNPS} = \partial_M \hat{C}_{NPS} - \partial_N \hat{C}_{PSM} + \partial_P \hat{C}_{SMN} - \partial_S \hat{C}_{MNP}. \tag{H.10}$$

The supersymmetry variation of the gravitino is given by

$$\delta\psi^M = \nabla^M \epsilon + \frac{1}{2 \cdot 3! \cdot 4!} G_{M_1 M_2 M_3 M_4} \left(\Gamma^{MM_1 M_2 M_3 M_4} - 8 G^{MM_1} \Gamma^{M_2 M_3 M_4} \right) \epsilon. \tag{H.11}$$

The full Lagrangian including the fermions as well as the supersymmetry transformation rules can be found in the original paper[1] [509] as well as in [7].

---

[1] There, a slightly different normalization of the three-form is used.

H.2 Type-IIA Supergravity

This is the low-energy limit of the type IIA superstring in ten dimensions. It contains a single supermultiplet of $\mathcal{N} = (1,1)_{10}$ supersymmetry containing the graviton ($g_{\mu\nu}$), an antisymmetric tensor ($B_{\mu\nu}$), a scalar ϕ (dilaton), a vector A_μ and a three-index antisymmetric tensor $C_{\mu\nu\rho}$ as well as a Majorana gravitino (ψ_α^μ) and a Majorana fermion (λ_α). The supergravity action is completely fixed and can be obtained by dimensional reduction of the $\mathcal{N} = 1_{11}$ supergravity.

Upon dimensional reduction, the eleven-dimensional metric gives rise to a ten-dimensional metric, a gauge field and a scalar as follows

$$g_{MN} = \begin{pmatrix} g_{\mu\nu} + e^{2\sigma} A_\mu A_\nu & e^{2\sigma} A_\mu \\ e^{2\sigma} A_\mu & e^{2\sigma} \end{pmatrix}. \tag{H.12}$$

The three-form $\hat{C}$ gives rise to a three-form and a two-form in ten dimensions

$$C_{\mu\nu\rho} = \hat{C}_{\mu\nu\rho} - (\hat{C}_{\nu\rho,11} A_\mu + \text{cyclic}), \quad B_{\mu\nu} = \hat{C}_{\mu\nu,11}. \tag{H.13}$$

The ten-dimensional action can be directly obtained from the eleven-dimensional one using the formulae of Appendix E on page 516. For the bosonic part we obtain,

$$S_{\text{IIA}} = \frac{1}{2\kappa^2} \int d^{10}x \sqrt{g} e^{\sigma} \left[R - \frac{1}{2 \cdot 4!} F_4^2 - \frac{1}{2 \cdot 3!} e^{-2\sigma} H_3^2 - \frac{1}{4} e^{2\sigma} F_2^2 \right]$$
$$+ \frac{1}{4\kappa^2} \int B_2 \wedge dC_3 \wedge dC_3, \tag{H.14}$$

where

$$F_2 = dA, \quad H_3 = dB_2, \quad F_4 = dC_3 - A \wedge H_3. \tag{H.15}$$

This action is not in any standard frame. We can go to the string frame by rescaling the metric as $g_{\mu\nu} \to e^{-\sigma} g_{\mu\nu}$. The ten-dimensional dilaton is $\Phi = 3\sigma/2$. The action becomes

$$S_{\text{IIA},s} = \frac{1}{2\kappa^2} \int d^{10}x \sqrt{g} \left[e^{-2\Phi} \left(R + 4(\nabla\Phi)^2 - \frac{1}{12} H_3^2 \right) - \frac{1}{2 \cdot 4!} F_4^2 - \frac{1}{4} F_2^2 \right]$$
$$+ \frac{1}{4\kappa^2} \int B_2 \wedge dC_3 \wedge dC_3. \tag{H.16}$$

Note that the kinetic terms of the R-R fields A_μ and $C_{\mu\nu\rho}$ do not have dilaton dependence at the tree level, as advocated in section 7.2.1 on page 159.

Alternatively, in the Einstein frame $g_{\mu\nu} \to e^{\Phi/2} g_{\mu\nu}$ the action is

$$S_{\text{IIA,E}} = \frac{1}{2\kappa^2} \int d^{10}x \sqrt{g} \left[\left(R - \frac{1}{2}(\nabla\Phi)^2 - \frac{e^{-\Phi}}{12} H_3^2 \right) - \frac{e^{\Phi/2}}{2 \cdot 4!} F_4^2 - \frac{e^{3\Phi/2}}{4} F_2^2 \right]$$
$$+ \frac{1}{4\kappa^2} \int B_2 \wedge dC_3 \wedge dC_3. \tag{H.17}$$

The supersymmetry variations of the gravitino and the dilatino in the Einstein frame are given by

$$\delta\lambda = \frac{1}{2\sqrt{2}}\nabla_\mu\Phi\Gamma^\mu\Gamma^{11}\epsilon + \frac{3}{16\sqrt{2}}e^{\frac{3\Phi}{4}}F^{(2)}_{\mu\nu}\Gamma^{\mu\nu}\epsilon$$

$$+ \frac{i}{24\sqrt{2}}e^{-\Phi/2}H_{\mu\nu\rho}\Gamma^{\mu\nu\rho}\epsilon - \frac{i}{192\sqrt{2}}e^{\Phi/4}F^{(4)}_{\mu\nu\rho\sigma}\Gamma^{\mu\nu\rho\sigma}\epsilon, \tag{H.18}$$

$$\delta\psi^\mu = \nabla^\mu\epsilon + \frac{1}{64}e^{3\Phi/4}F^{(2)}_{\nu\rho}\left(\Gamma^{\mu\nu\rho} - 14g^{\mu\nu}\Gamma^\rho\right)\Gamma^{11}\epsilon$$

$$+ \frac{1}{96}e^{-\Phi/2}H_{\nu\rho\sigma}\left(\Gamma^{\mu\nu\rho\sigma} - 9g^{\mu\nu}\Gamma^{\rho\sigma}\right)\Gamma^{11}\epsilon$$

$$+ \frac{i}{256}e^{\Phi/4}F^{(4)}_{\nu\rho\sigma\tau}\left(\Gamma^{\mu\nu\rho\sigma\tau} - \frac{20}{3}g^{\mu\nu}\Gamma^{\rho\sigma\tau}\right)\Gamma^{11}\epsilon. \tag{H.19}$$

The rest of the fermionic terms can be obtained as well by the dimensional reduction of the 11-dimensional action in [509]. The derivation of the full action of type-IIA supergravity can be found in [510]. There is a massive version of type-IIA supergravity due to Romans [511]. It can be obtained by a coordinate-dependent compactification of the 11-dimensional theory on S^1. A survey of type-II supergravities can be found in [457].

H.3 Type-IIB Supergravity

Type-IIB supergravity, not surprisingly, is the massless limit of the type-IIB string.

It contains the graviton ($g_{\mu\nu}$), two antisymmetric tensors (B_2, C_2), two scalars Φ, C_0, a self-dual four-index antisymmetric tensor C_4, two Majorana-Weyl gravitini, and two Majorana-Weyl fermions of the same chirality. They form a single representation of the associated supersymmetry algebra. The fields C_0, C_2, and C_4 originate in the R-R sector.

The theory is chiral but anomaly-free, as shown in section 7.9 on page 176. The self-duality condition implies that the field strength F_5 of the four-form is equal to its dual. This equation cannot be obtained from a covariant action. We may however write down an action which gives the type-IIB supergravity equations for all fields except the four-form. The equations have to be supplemented by the self-duality condition.

There is an SL(2,$\mathbb{R}$) global invariance in this theory, which transforms the antisymmetric tensor and scalar doublets (the metric as well as the four-form are invariant in the Einstein frame). We denote by Φ the dilaton that comes from the NS-NS sector and by C_0 the scalar that comes from the R-R sector. Define the complex scalar

$$S = C_0 + ie^{-\Phi}. \tag{H.20}$$

Then, SL(2,$\mathbb{R}$) acts by fractional linear transformations on S and linearly on B_2, C_2

$$S \to \frac{aS+b}{cS+d}, \quad \begin{pmatrix} B_2 \\ C_2 \end{pmatrix} \to \begin{pmatrix} d & -c \\ -b & a \end{pmatrix}\begin{pmatrix} B_2 \\ C_2 \end{pmatrix}, \tag{H.21}$$

where a, b, c, d are real with $ad - bc = 1$. B_2 is the NS-NS antisymmetric tensor while C_2 is the R-R antisymmetric tensor. The equations of motion can be obtained from the following action in the Einstein frame

$$S_{\text{IIB}} = \frac{1}{2\kappa^2} \int d^{10}x \sqrt{-g} \left[R - \frac{1}{2} \frac{\partial S \partial \bar{S}}{S_2^2} - \frac{1}{12} |G_3|^2 - \frac{1}{2 \cdot 5!} F_5^2 \right]$$
$$+ \frac{1}{2i} \int C_4 \wedge G_3 \wedge \bar{G}_3, \tag{H.22}$$

where

$$H_3 = dB_2, \quad F_3 = dC_2, \quad G_3 = i \frac{F_3 + SH_3}{\sqrt{S_2}}, \quad F_5 = dC_4 - C_2 \wedge H_3. \tag{H.23}$$

However, as mentioned earlier, this is not the whole story. The equations of motion derived from (H.22) must be supplemented with the self-duality condition

$$F_5 = {}^\star F_5. \tag{H.24}$$

The action (H.22) and equation (H.24) are SL(2,$\mathbb{R}$) invariant if C_4 is appropriately gauge transformed.

A left and a right Majorana-Weyl spinor ϵ_L and ϵ_R can be put together in a two-component vector of the form $\binom{\epsilon_L}{\epsilon_R}$. We may also use the notation of complex spinors by using $\epsilon_L + i\epsilon_R$. To pass from one notation to the other we use

$$\epsilon^* \leftrightarrow \sigma^3 \epsilon, \quad i\epsilon^* \leftrightarrow \sigma^1 \epsilon, \quad i\epsilon \leftrightarrow -i\sigma^2 \epsilon, \tag{H.25}$$

where σ^i are the Pauli matrices.

In the complex notation, the transformation rules for the fermions are

$$\delta\lambda = i\Gamma^\mu \epsilon^* P_\mu - \frac{i}{24} e^{\Phi/2} \Gamma^{\mu\nu\rho} \epsilon (G_3)_{\mu\nu\rho}, \tag{H.26}$$

$$\delta\psi^\mu = \nabla^\mu \epsilon - \frac{i}{1920} \Gamma^{\mu_1 \cdots \mu_5} \Gamma^\mu \epsilon (F_5)_{\mu_1 \cdots \mu_5} + \frac{1}{96} (\Gamma^{\mu\nu\rho\sigma} - 9g^{\mu\nu}\Gamma^{\rho\sigma}) \epsilon^* (G_3)_{\nu\rho\sigma}, \tag{H.27}$$

where

$$P_\mu = \frac{i}{2} e^\Phi \partial_\mu S. \tag{H.28}$$

The full equations of type-IIB supergravity as well as the supersymmetry transformations can be found in the original papers [512,513], albeit in a somewhat different notation.

H.4 Type-II Supergravities: The Democratic Formulation

We will describe in this section a uniform formulation of the two type-II supergravities in ten dimensions in a way which is symmetric between IIA and IIB and utilises all R-R forms as well as their duals. In particular, it applies also the massive type IIA theory by including the nine-form $C^{(9)}$ and its dual field strength $G^{(0)} = m$ equal to the mass. More details of this formulation can be found in the original reference[2] [514]. We have the extended field content

IIA: $\{g_{\mu\nu}, B_{\mu\nu}, \Phi, C^{(1)}, C^{(3)}, C^{(5)}, C^{(7)}, C^{(9)}, \psi_\mu, \lambda\}$,

IIB: $\{g_{\mu\nu}, B_{\mu\nu}, \Phi, C^{(0)}, C^{(2)}, C^{(4)}, C^{(6)}, C^{(8)}, \psi_\mu, \lambda\}$. (H.29)

In type IIA we have fermions of both chiralities while in IIB $\Gamma^{11}\psi_\mu = \psi_\mu$, $\Gamma^{11}\lambda = -\lambda$.

We may write the following uniform bosonic action:

$$S = \frac{1}{2\kappa^2} \int d^{10}x \sqrt{g} \left[e^{-2\Phi} \left(R - 4(\partial\Phi)^2 + \frac{1}{2} H \cdot H \right) + \frac{1}{4} \sum_{n=0,1/2}^{5,9/2} G^{(2n)} \cdot G^{(2n)} \right], \quad (H.30)$$

where we are using the conventions of appendix B on page 505. n is summed over the integers in IIA and half-integers $\frac{1}{2} \cdots \frac{9}{2}$ in IIB.

We may group all potentials and field strengths together as

$$\mathbf{G} = \sum_{n=0,\frac{1}{2}}^{5,\frac{9}{2}} G^{(2n)}, \quad \mathbf{C} = \sum_{n=0,\frac{1}{2}}^{5,\frac{9}{2}} C^{(2n-1)}. \quad (H.31)$$

The field strengths are given by

$$H = dB, \quad \mathbf{G} = d\mathbf{C} - H \wedge \mathbf{C} + G^{(0)} e^B. \quad (H.32)$$

In the previous equation, one is instructed to extract the part of appropriate form degree. The last term is there only for the massive IIA case. The Bianchi identities are

$$dH = 0, \quad d\mathbf{G} = H \wedge \mathbf{G}. \quad (H.33)$$

We must also impose a set of self-duality conditions that will reduce the number of degrees of freedom to the physical ones.

$$G^{(2n)} = (-1)^{[n]} \star G^{(10-2n)}, \quad (H.34)$$

where $[n]$ stands for the integer part of n. The self-duality conditions involve also the fermion bilinears that we have neglected in (H.34).

The full set of equations of motion involve, the equations of motion stemming from the action (H.30), the Bianchi identities (H.33) and the self-duality conditions (H.34).

For further details the reader is referred to [514].

H.5 $\mathcal{N} = 1_{10}$ Supersymmetry

There are two massless supersymmetry representations (supermultiplets) of $\mathcal{N} = 1_{10}$ supersymmetry. The vector multiplet contains a vector (A_μ) and a Majorana-Weyl fermion (χ_α). The supergravity multiplet contains the graviton ($g_{\mu\nu}$), antisymmetric tensor ($B_{\mu\nu}$) and a scalar ϕ (dilaton) as well as a Majorana-Weyl gravitino (ψ^μ) and a Majorana-Weyl

---

[2] The only difference in conventions is a change of sign for the value of the ϵ-tensor.

fermion (λ). The supergravity multiplet fermions have opposite chirality compared to the gaugini.

The bosonic action is

$$S_{\mathcal{N}=1} = \frac{1}{2\kappa^2} \int d^{10}x \sqrt{-g} \left[R - \frac{1}{2}(\partial\Phi)^2 - \frac{e^{-\Phi}}{12}\hat{H}^2 - \frac{\kappa^2 e^{-\Phi/2}}{2g^2}\text{Tr}[F^2] \right], \tag{H.35}$$

with

$$F = dA - i[A, A] \quad \rightarrow \quad F_{\mu\nu} = \partial_\mu A_\nu - \partial_\mu A_\nu - i[A_\mu, A_\nu], \tag{H.36}$$

and the field strength of the antisymmetric tensor is modified as

$$\hat{H} = dB - \frac{\kappa^2}{g^2}\Omega_3^{YM}, \quad \Omega_3^{YM} = \text{Tr}\left[A \wedge dA - \frac{2i}{3}A \wedge A \wedge A \right], \tag{H.37}$$

and all traces are in the defining (vector) representation.[3] There is also an additional gravitational Chern-Simons contribution to (H.37). Although it is a higher derivative term it is necessary for anomaly cancellation. It is described in section 7.9 on page 176.

The presence of the Chern-Simons contribution makes $B_{\mu\nu}$ transform under gauge transformations so that $\hat{H}$ is gauge invariant,

$$\delta A = d\Lambda - i[A, \Lambda], \quad \delta\Omega_3^{YM} = d\,\text{Tr}[\Lambda\,dA], \quad \delta B = \frac{\kappa^2}{g^2}\text{Tr}[\Lambda\,dA]. \tag{H.38}$$

The supergravity action (H.35) has two *a priori* independent coupling constants κ and g. However they are both dimensionful. There is only one dimensionless coupling constant that can be taken to be g^4/κ^3. Its value can be changed by shifting the dilaton field Φ. In string theory, both κ and g are functions of the string length ℓ_s and the expectation value of the dilaton, namely the string coupling g_s.

The supersymmetric variations of the fermions are also of importance. They are given

$$g\,\delta\chi = -\frac{1}{4}e^{-\frac{\Phi}{4}}\Gamma^{\mu\nu}F_{\mu\nu}\epsilon, \tag{H.39}$$

$$2\sqrt{2}\kappa\,\delta\lambda = -\Gamma^\mu\partial_\mu\Phi\,\epsilon + \frac{e^{-\Phi/2}}{12}(\Gamma^{\mu\nu\rho}\hat{H}_{\mu\nu\rho})\,\epsilon, \tag{H.40}$$

$$\kappa\,\delta\psi_\mu = \nabla_\mu\epsilon + \frac{e^{-\Phi/2}}{96}\left(\Gamma_\mu{}^{\nu\rho\sigma} - 9\delta_\mu{}^\nu\Gamma^{\rho\sigma}\right)\hat{H}_{\nu\rho\sigma}, \tag{H.41}$$

where we have neglected terms bilinear in the fermions in the right-hand sides.

This supergravity theory describes the effective field theory of both the type-I and the heterotic string. To go to the heterotic string frame, we must redefine the metric as $g_{\mu\nu} \rightarrow e^{-\Phi/2}g_{\mu\nu}$ to obtain

$$S_{\text{heterotic}} = \frac{1}{2\kappa^2} \int d^{10}x \sqrt{-g}e^{-2\Phi}\left[R + 4(\partial\Phi)^2 - \frac{1}{12}\hat{H}^2 - \frac{\kappa^2}{2g_h^2}\text{Tr}[F^2] \right]. \tag{H.42}$$

---

[3] The trace can also written in the adjoint representation using $\text{Tr}_{\text{vector}} \rightarrow \frac{1}{30}\text{Tr}_{\text{adjoint}}$ valid for both O(32) and $E_8 \times E_8$.

In all string theories, the gravitational coupling constant in ten dimensions is given by

$$2\kappa^2 = (2\pi)^7 \ell_s^8 g_s^2. \tag{H.43}$$

For the heterotic string the gauge coupling constant is

$$g_h = \frac{2}{\ell_s}\kappa \quad \rightarrow \quad g_h = \sqrt{2}(2\pi)^{7/2}\ell_s^3 g_s. \tag{H.44}$$

To obtain the string-frame effective action for the type-I string we must first take $\Phi \to -\Phi$ in the Einstein action (H.35) and then scale the metric $g_{\mu\nu} \to e^{-\Phi/2}g_{\mu\nu}$ to obtain

$$S_{\text{type I}} = \frac{1}{2\kappa^2}\int d^{10}x\sqrt{-g}\left(e^{-2\Phi}\left[R + 4(\partial\Phi)^2\right] - \frac{1}{12}\hat{H}^2 - \frac{\kappa^2 e^{-\Phi}}{2g_I^2}\text{Tr}[F^2]\right). \tag{H.45}$$

This compatible with the fact that the kinetic terms of the gauge fields come from the disk while the two-form is now coming from the R-R sector. The gauge coupling constant in the type Istring is given by

$$g_I^2 = \sqrt{2}(2\pi)^7\ell_s^6 g_s. \tag{H.46}$$

The action and supersymmetry transformations of $\mathcal{N} = 1_{10}$ supergravity coupled to abelian vector multiplets was determined in [515]. The nonabelian generalization was developed in [516]. A four-fermion term in the action which vanishes in the abelian case, was added in [517]. This is important for supersymmetry breaking via gaugino condensation.

We will review here basic facts about $\mathcal{N}$=1,2 four-dimensional supergravity theories coupled to matter.

I.1 $\mathcal{N}$ = 1$_4$ Supergravity

Apart from the supergravity multiplet, we can have vector multiplets containing the vectors and their Majorana gaugini, and chiral multiplets containing a complex scalar and a Weyl spinor.

There is also the linear multiplet containing an antisymmetric tensor, a scalar and a Weyl fermion. However, this can be dualized into a chiral multiplet, with an accompanying Peccei-Quinn symmetry.[1]

The bosonic Lagrangian can be written as follows:

$$\mathcal{L}_{\mathcal{N}=1} = -\frac{1}{2}R + G_{i\bar{j}}\,D_\mu\phi^i D^\mu\bar{\phi}^{\bar{j}} + V(\phi,\bar{\phi}) + \sum_I \frac{1}{4g_I^2}\mathrm{Tr}[F_{\mu\nu}F^{\mu\nu}]_I + \frac{\theta_I}{4}\mathrm{Tr}[F_{\mu\nu}\tilde{F}^{\mu\nu}]_I. \tag{I.1}$$

ϕ^i are the complex scalars of the chiral multiplets. Supersymmetry requires the manifold of scalars to be Kählerian,

$$G_{i\bar{j}} = \partial_i\partial_{\bar{j}}K(\phi,\bar{\phi}). \tag{I.2}$$

The gauge group $G = \prod_I G_I$ is a product of simple or U(1) factors. The chiral multiplets transform in some representation of the gauge group:

$$\delta\phi^i = Tr[X^i(\phi)\epsilon], \quad D_\mu z^i = \partial_\mu z^i - Tr[A_\mu X^i]. \tag{I.3}$$

X^i are matrix-valued holomorphic Killing vectors. They generate the gauged isometries of the Kähler manifold of chiral multiplets.

---

[1] A Peccei-Quinn symmetry is a translational symmetry of a scalar field, $\phi \to \phi$+constant.

The gauge couplings and θ-angles must depend on the moduli via a holomorphic function f_I,

$$\frac{1}{g_I^2} = \operatorname{Re} f_I(\phi), \quad \theta_I = -\operatorname{Im} f_I(\phi). \tag{I.4}$$

The holomorphic function f_I must be gauge-invariant.[2] The scalar potential V has two contributions $V = V_F + V_D$. V_F comes from the F-terms, where F stands for the auxiliary fields of chiral multiplets. It is determined by a holomorphic function, the superpotential $W(\phi)$:

$$V_F(\phi, \bar{\phi}) = e^K (D_i W G^{i\bar{k}} \bar{D}_{\bar{k}} \bar{W} - 3|W|^2), \tag{I.5}$$

where

$$D_i W = \frac{\partial W}{\partial \phi^i} + \frac{\partial K}{\partial \phi^i} W \equiv \partial_i W + (\partial_i K) W, \tag{I.6}$$

and $G^{i\bar{i}}$ is the inverse of the Kähler metric. The D-term potential originates from the D-terms, where D stands for the auxiliary fields of the vector multiplets. It is given by

$$V_D = \frac{1}{2} \sum_I (\operatorname{Re} f_I)^{-1} \operatorname{Tr}[D_I D_I]. \tag{I.7}$$

The D-terms are the Killing potentials defined by

$$X^i = -iG^{i\bar{k}} \partial_{\bar{k}} D^i. \tag{I.8}$$

The general solution compatible with gauge invariance is

$$D = i(D_i W) X^i. \tag{I.9}$$

For nonabelian factors, K and W are gauge invariant, and (I.9) reduces to

$$D = i\partial_i K X^i. \tag{I.10}$$

However, in the presence of U(1)'s, the superpotential may not be gauge invariant. In this case we have

$$D = i\partial_i K X^i + \xi, \tag{I.11}$$

where ξ is a constant. It is known as the Fayet-Iliopoulos term.

There is an overall redundancy in the data (K, f_I, W). The action is invariant under Kähler transformations

$$K \to K + \Lambda(\phi) + \bar{\Lambda}(\bar{\phi}), \quad W \to W e^{-\Lambda}, \quad f_I \to f_I. \tag{I.12}$$

It seems that this redundancy allows one to get rid of the superpotential. However, this introduces singularities. In particular the metric for the scalars becomes singular.

References for further reading include the standard [519,520] where various $\mathcal{N} = 1$ formulations are presented and [300] where the connection with string effective actions is developed in detail. Another recent and extensive review can be found in [521].

---

[2] There is a possibility where $\operatorname{Im} f_I$ is transforming under gauge transformations. This is the case with anomalous U(1)'s. In this case there are some extra terms in the effective action [518].

I.2 $\mathcal{N} = 2_4$ Supergravity

Apart from the supergravity multiplet we will have a number N_V of abelian vector multiplets and a number N_H of hypermultiplets. There is also an extra gauge boson, the graviphoton residing in the supergravity multiplet.

We pick the gauge group to be abelian. This is without loss of generality since any non-abelian gauge group can be broken to the maximal abelian subgroup by giving expectation values to the scalar partners of the abelian gauge bosons.

We denote the graviphoton by A_μ^0, the rest of the gauge bosons by A_μ^i, $i = 1, 2, \ldots, N_V$, and the scalar partners of A_μ^i as $T^i, \bar{T}^i$. Although the graviphoton does not have a scalar partner, it is convenient to introduce one and later removing by gauging a scaling symmetry.

We introduce the complex coordinates Z^I, $I = 0, 1, 2, \ldots, N_V$, which parameterize the vector moduli space (VMS) $\mathcal{M}_V$. The $4N_H$ scalars of the hypermultiplets parameterize the hypermultiplet moduli space $\mathcal{M}_H$. Supersymmetry requires this to be a quaternionic manifold.[3] The geometry of the full scalar manifold is that of a product $\mathcal{M}_V \times \mathcal{M}_H$.

$\mathcal{N} = 2_4$ supersymmetry implies that the VMS is not just a Kähler manifold. It further satisfies what is known as special geometry. Special geometry eventually leads to the property that the full action of $\mathcal{N} = 2_4$ supergravity (we exclude hypermultiplets for the moment) can be written in terms of one function $\mathcal{F}$, which is holomorphic in the VMS coordinates. $\mathcal{F}(Z^I)$ is called the prepotential. It must be a homogeneous function of the coordinates of degree 2: $Z^I \mathcal{F}_I = 2$, where $\mathcal{F}_I = \frac{\partial \mathcal{F}}{\partial Z^I}$.

The Kähler potential is

$$K = -\log\left[i(\bar{Z}^I \mathcal{F}_I - Z^I \bar{\mathcal{F}}_I)\right], \tag{I.13}$$

which determines the metric $G_{I\bar{j}} = \partial_I \partial_{\bar{j}} K$ of the kinetic terms of the scalars. We can fix the scaling freedom by setting $Z^0 = 1$, and then $Z^I = T^I$ are the physical moduli. The Kähler potential becomes

$$K = -\log\left[2\left(f(T^i) + \bar{f}(\bar{T}^i)\right) - (T^i - \bar{T}^i)(f_i - \bar{f}_i)\right], \tag{I.14}$$

where $f(T^i) = -i\mathcal{F}(Z^0 = 1, Z^i = T^i)$. The Kähler metric $G_{i\bar{j}}$ has the following property:

$$R_{i\bar{j}k\bar{l}} = G_{i\bar{j}}G_{k\bar{l}} + G_{i\bar{l}}G_{k\bar{j}} - e^{-2K} W_{ikm} G^{m\bar{m}} \bar{W}_{\bar{m}\bar{j}\bar{l}}, \tag{I.15}$$

where $W_{ijk} = \partial_i \partial_j \partial_k f$. Since there is no potential, the only part of the bosonic action remaining to be specified are the kinetic terms for the vectors:

$$\mathcal{L}^{\text{vectors}} = -\frac{1}{4} \Xi_{IJ} F_{\mu\nu}^I F^{J,\mu\nu} - \frac{\theta_{IJ}}{4} F_{\mu\nu}^I \tilde{F}^{J,\mu\nu}, \tag{I.16}$$

[3] In the global supersymmetry limit in which gravity decouples, $M_{\text{Planck}} \to \infty$, the geometry of the hypermultiplet space is that of a hyperkähler manifold.

where

$$\Xi_{IJ} = \frac{i}{4}[N_{IJ} - \bar{N}_{IJ}], \quad \theta_{IJ} = \frac{1}{4}[N_{IJ} + \bar{N}_{IJ}], \tag{I.17}$$

$$N_{IJ} = \bar{\mathcal{F}}_{IJ} + 2i\frac{\mathrm{Im}\,\mathcal{F}_{IK}\,\mathrm{Im}\,\mathcal{F}_{JL}Z^K Z^L}{\mathrm{Im}\,\mathcal{F}_{MN}Z^M Z^N}. \tag{I.18}$$

We observe that the gauge couplings, unlike the $\mathcal{N} = 1_4$ case, are not harmonic functions of the moduli.

The self-interactions of massless hypermultiplets are described by a σ-model on a quaternionic manifold (hyperkähler in the global case) A quaternionic manifold must satisfy the following properties:

(i) It must have three complex structures J^i, $i = 1, 2, 3$, satisfying the quaternion algebra

$$J^i J^j = -1\delta^{ij} + \epsilon^{ijk}J^k. \tag{I.19}$$

The metric is Hermitian with respect to each one of them. The dimension of the manifold is $4m$, $m \in \mathbb{Z}$. The three complex structures guarantee the the existence of an $\mathrm{SU}(2)$-valued hyper-Kähler two-form K.

(ii) There exists a principal $\mathrm{SU}(2)$ bundle over the manifold, with connection ω such that the form K is closed with respect to ω

$$\nabla K = dK + [\omega, K] = 0.$$

(iii) The connection ω has a curvature that is proportional to the hyperkähler form

$$d\omega + [\omega, \omega] = \lambda\,K,$$

where λ is a real number. When $\lambda = 0$, the manifold is hyper-Kähler. Thus, the holonomy of a quaternionic manifold is of the form $\mathrm{SU}(2)\otimes H$ while for a hyper-Kähler manifold H, with $H \subset Sp(2m, \mathbb{R})$. The existence of the $\mathrm{SU}(2)$ structure is natural for the hypermultiplets since $\mathrm{SU}(2)$ is the nonabelian part of the $\mathcal{N} = 2_4$ R-symmetry that acts inside the hypermultiplets. The scalars transform as a pair of spinors under $\mathrm{SU}(2)$. When the hypermultiplets transform under the gauge group, then the quaternionic manifold must have appropriate isometries.

$\mathcal{N} = 2_4$ BPS states[4] have masses given by

$$M_{BPS}^2 = \frac{|q_I\,Z^I + p^I\,\mathcal{F}_I|^2}{\mathrm{Im}(Z^I\bar{\mathcal{F}}_I)}, \tag{I.20}$$

where q_I, p^I are the electric and magnetic charges of the state. Further reading can be found in [522–526]. A review on the geometries relevant here can be found in [527]. Quaternionic manifolds and moment maps are described in detail in appendix B of [336].

---

[4] You will find definitions and properties in Appendix J.

Appendix J | BPS Multiplets in Four Dimensions

In this appendix we will restrict our attention to supersymmetry in four dimensions and the associated BPS multiplets. The results however, generalize with very little effort to other dimensions.

BPS states are important probes of non-perturbative physics in theories with extended ($\mathcal{N} \geq 2_4$) supersymmetry. They are special for the following reasons:

- They form multiplets under extended SUSY which are shorter than the generic massive multiplet. Their mass is given in terms of their charges and moduli expectation values.

- At generic points in moduli space they are stable because of energy and charge conservation.

- Their mass formula is exact if one uses renormalized values for the charges and moduli.[1] The argument is that quantum corrections would spoil the relation of mass and charges, and if we assume unbroken SUSY at the quantum level would lead to incompatibilities with the dimension of their representations.

In order to present the concept of BPS states we will briefly review the representation theory of $\mathcal{N}$-extended supersymmetry. A complete treatment can be found in [519]. A general discussion of central charges in various dimensions can be found in [528].

The anticommutation relations are

$$\{Q_\alpha^I, Q_\beta^J\} = \epsilon_{\alpha\beta} Z^{IJ}, \quad \{\bar{Q}_{\dot\alpha}^I, \bar{Q}_{\dot\beta}^J\} = \epsilon_{\dot\alpha\dot\beta} \bar{Z}^{IJ}, \quad \{Q_\alpha^I, \bar{Q}_{\dot\alpha}^J\} = \delta^{IJ} 2\sigma_{\alpha\dot\alpha}^\mu P_\mu, \tag{J.1}$$

where Z^{IJ} is the antisymmetric central charge matrix.

The algebra is invariant under the U($\mathcal{N}$) R-symmetry that rotates $Q, \bar{Q}$. We begin with a description of the representations of the algebra. We first assume that the central charges are zero.

---

[1] In theories with $\mathcal{N} \geq 4_4$ supersymmetry there are no renormalizations of the two-derivative effective action.

Massive representations

We can go to the rest frame $P \sim (M, \vec{0})$. The relations become

$$\{Q_\alpha^I, \bar{Q}_{\dot{\alpha}}^J\} = 2M\delta_{\alpha\dot{\alpha}}\delta^{IJ}, \quad \{Q_\alpha^I, Q_\beta^J\} = \{\bar{Q}_{\dot{\alpha}}^I, \bar{Q}_{\dot{\beta}}^J\} = 0. \tag{J.2}$$

Define the $2\mathcal{N}$ fermionic harmonic creation and annihilation operators

$$A_\alpha^I = \frac{1}{\sqrt{2M}} Q_\alpha^I, \quad A_\alpha^{\dagger I} = \frac{1}{\sqrt{2M}} \bar{Q}_{\dot{\alpha}}^I. \tag{J.3}$$

In order to build the representation, we start with the Clifford vacuum $|\Omega\rangle$, which is annihilated by the A_α^I and we act with the creation operators $A_\alpha^{\dagger I}$. There are $\binom{2\mathcal{N}}{n}$ states at the n-th oscillator level. The total number of states is $2^{2\mathcal{N}} = \sum_{n=0}^{2\mathcal{N}}\binom{2\mathcal{N}}{n}$, half of them being bosonic and half of them fermionic. The spin quantum number follows from symmetrization over the spinorial indices. The maximal spin is the spin of the ground states plus $\mathcal{N}$.

Example. Suppose $\mathcal{N} = 1_4$ and the ground state transforms in the $[j]$ representation of SO(3). Acting with two creation operators, the content of the massive representation is $[j] \otimes ([1/2] + 2[0]) = [j \pm 1/2] + 2[j]$. The two spin-$j$ states correspond to the ground state itself and to the state with two oscillators.

Massless representations

In this case we can go to the frame $P \sim (-E, 0, 0, E)$. The anticommutation relations now become

$$\{Q_\alpha^I, \bar{Q}_{\dot{\alpha}}^J\} = 2 \begin{pmatrix} 2E & 0 \\ 0 & 0 \end{pmatrix} \delta^{IJ}, \tag{J.4}$$

the rest being zero. Since $Q_2^I, \bar{Q}_2^I$ anticommute, they are represented by zero in a unitary theory. We have $\mathcal{N}$ nontrivial creation and annihilation operators $A^I = Q_1^I/2\sqrt{E}, A^{\dagger I} = \bar{Q}_1^I/2\sqrt{E}$, and the representation is $2^\mathcal{N}$-dimensional. It is much shorter than the massive one.

Nonzero central charges

In this case the representations are massive. The central charge matrix can be brought by a U($\mathcal{N}$) transformation to block diagonal form as in (7.9.18), and we will label the real positive skew eigenvalues by Z_m. We assume that $\mathcal{N}$ is even so that $m = 1, 2, \ldots, \mathcal{N}/2$. We will split the index $I \to (a, m)$: $a = 1, 2$ labels the position inside the 2×2 blocks while m labels the blocks. Then

$$\{Q_\alpha^{am}, \bar{Q}_{\dot{\alpha}}^{bn}\} = 2M\delta^{\alpha\dot{\alpha}}\delta^{ab}\delta^{mn}, \quad \{Q_\alpha^{am}, Q_\beta^{bn}\} = Z_n\epsilon^{\alpha\beta}\epsilon^{ab}\delta^{mn}. \tag{J.5}$$

Define the following fermionic oscillators

$$A_\alpha^m = \frac{1}{\sqrt{2}}[Q_\alpha^{1m} + \epsilon_{\alpha\beta}Q_\beta^{2m}], \quad B_\alpha^m = \frac{1}{\sqrt{2}}[Q_\alpha^{1m} - \epsilon_{\alpha\beta}Q_\beta^{2m}], \tag{J.6}$$

and similarly for the conjugate operators. The anticommutators become

$$\{A_\alpha^m, A_\beta^n\} = \{A_\alpha^m, B_\beta^n\} = \{B_\alpha^m, B_\beta^n\} = 0, \tag{J.7}$$

$$\{A_\alpha^m, A_\beta^{\dagger n}\} = \delta_{\alpha\beta}\delta^{mn}(2M + Z_n), \quad \{B_\alpha^m, B_\beta^{\dagger n}\} = \delta_{\alpha\beta}\delta^{mn}(2M - Z_n). \tag{J.8}$$

Unitarity requires that the right-hand sides in (J.8) be non-negative. This in turn implies the Bogomolnyi bound

$$M \geq \max \left[\frac{Z_m}{2} \right].$$

(J.9)

Assume $0 \leq r \leq \mathcal{N}/2$ of the Z_m's to be equal to $2M$. Then $2r$ of the B-oscillators vanish identically and we are left with $2\mathcal{N} - 2r$ creation and annihilation operators. The representation has $2^{2\mathcal{N}-2r}$ states. The maximal case $r = \mathcal{N}/2$ gives rise to the short BPS multiplet whose number of states are the same as in the massless multiplet. The other multiplets with $0 < r < \mathcal{N}/2$ are known as intermediate BPS multiplets.

The BPS properties of representations are reflected in the supertraces of powers of the four-dimensional helicity. Such supertraces appear in loop amplitudes and their supersymmetry-related sum rules are very important for non-renormalization or related properties.

The relation of loop corrections to supertraces was first observed in [529]. General supertraces were computed in [530]. The relationship between B_2 and short multiplets of the $\mathcal{N} = 2_4$ algebra in four dimensions was observed in [303]. It was generalized to different amounts of supersymmetry and used in duality checks in [531].

We will define the helicity supertrace on a supersymmetry representation R as[2]

$$B_{2n}(R) = \mathrm{Tr}_R[(-1)^{2\lambda} \lambda^{2n}].$$

(J.10)

It is useful to introduce the "helicity-generating function" of a given supermultiplet R

$$Z_R(y) = \mathrm{Str}\, y^{2\lambda}.$$

(J.11)

where in the supertrace fermions contribute with an extra minus sign.

For a particle of spin j we have

$$Z_{[j]} = \begin{cases} (-)^{2j} \left(\frac{y^{2j+1} - y^{-2j-1}}{y - y^{-1}} \right) & \text{massive,} \\ (-)^{2j} (y^{2j} + y^{-2j}) & \text{massless.} \end{cases}$$

(J.12)

When tensoring representations, the generating functionals get multiplied,

$$Z_{r \otimes \tilde{r}} = Z_r Z_{\tilde{r}}.$$

(J.13)

The supertrace of the nth power of helicity can be extracted from the generating functional through

$$B_n(R) = \left(y^2 \frac{d}{dy^2} \right)^n Z_R(y)|_{y=1}.$$

(J.14)

For a supersymmetry representation constructed from a spin $[j]$ ground state by acting with $2m$ oscillators we obtain

$$Z_m(y) = Z_{[j]}(y)(1 - y)^m (1 - 1/y)^m.$$

(J.15)

---

[2] In higher dimensions, traces over various Casimirs of the little group have to be considered.

General helicity supertrace formulae for various supersymmetries and representations can be found in [319]. In particular, B_2 distinguishes half-BPS states in theories with $\mathcal{N} = 2_4$ supersymmetry. Concretely, half-BPS multiplets have $B_2 \neq 0$ while (generic) long multiplets have $B_2 = 0$. It turns out that one-loop corrections to the two derivative effective action in $\mathcal{N} = 2_4$ theories are proportional to B_2. Therefore, only half-BPS states contribute.

In this appendix we review the geometry of anti–de Sitter space. This is a widely studied background in string theory and is relevant for the AdS/CFT correspondence. Also relevant is the fact that AdS and flat space have conformal compactifications that are related. In the Euclidean case, the spatial part $\mathbb{R}^n$ can be compactified to S^n by adding a point at infinity. A CFT is naturally defined on S^n. On the other hand Euclidean AdS_{n+1} is conformally equivalent to the $(n+1)$-dimensional disk $\mathcal{D}_{n+1}$. The boundary of compactified Euclidean AdS is therefore compactified Euclidean space. This also holds for Minkowski signature.

K.1 The Minkowski Signature AdS

Anti–de Sitter space AdS_{p+2} of dimension $p+2$ is the hyperboloid defined from flat $(p+3)$-dimensional space with metric

$$ds^2 = -dX_0^2 - dX_{p+2}^2 + \sum_{i=1}^{p+1} dX_i^2 \tag{K.1}$$

via the condition

$$X_0^2 + X_{p+2}^2 - \sum_{i=1}^{p+1} X_i^2 = L^2. \tag{K.2}$$

By construction, the space has isometry $SO(2,p+1)$ and it is homogeneous and isotropic.

We may solve the constraint (K.2) via the parametrization

$$X_0 = L \cosh \rho \cos \tau, \quad X_0 = L \cosh \rho \sin \tau,$$

$$X_i = L \, \sinh \rho \, \Omega_i \quad \left(i = 1, \ldots, p+1, \sum_{i=1}^{p+1} \Omega_i^2 = 1 \right). \tag{K.3}$$

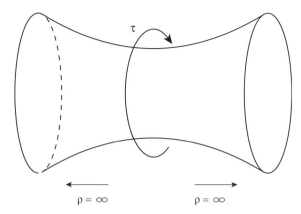

Figure K.1 The space AdS$_{p+2}$ can be constructed as a hyperboloid in $R^{2,p+1}$. It has closed timelike curves in the coordinate τ. To obtain a causal space we must go to the universal cover of the time coordinate.

From their definition, the Ω_i are the standard coordinates of S^p. By substituting this parametrization into (K.1) we obtain the AdS$_{p+2}$ metric as

$$ds^2 = L^2 \left(-\cosh^2 \rho \, d\tau^2 + d\rho^2 + \sinh^2 \rho \, d\Omega_p^2 \right). \tag{K.4}$$

By taking $\rho \in (0, \infty)$ and $\tau \in [0, 2\pi)$ we cover the hyperboloid once. Therefore, ρ, τ, Ω_i are global coordinates of AdS. When we consider the neighborhood of $\rho \simeq 0$ the metric can be approximated by

$$ds^2 \simeq L^2 \left(-d\tau^2 + d\rho^2 + \rho^2 \, d\Omega_p^2 \right) \tag{K.5}$$

and this makes explicit the fact that AdS$_{p+2}$ has topology S$^1 \times \mathbb{R}^{p+1}$. The S^1 is time-like, and indicates that AdS$_{p+2}$ thus defined, has closed timelike curves (see figure K.1). We may obtain a causal space-time by taking the universal cover of the S^1 coordinate. This practically means that we will take $-\infty < \tau < +\infty$. On this cover, there are no closed time-like curves. Here and in most places in the literature AdS$_{p+2}$ stands for this universal cover.

The isometry group SO$(2, p + 1)$ has SO$(2) \times$ SO$(p + 1)$ are the maximal compact subgroup. SO(2) acts by translating the τ coordinate. SO$(p + 1)$ is the transitive symmetry of S^p.

An important property of AdS is its causal structure. To that end, it is convenient to bring the end points of the ρ coordinate to finite values by introducing a new coordinate θ as

$$\tan \theta = \sinh \rho, \quad \theta \in \left[0, \frac{\pi}{2} \right). \tag{K.6}$$

Then, the metric (K.4) takes the form

$$ds^2 = \frac{L^2}{\cos^2 \theta} \left(-d\tau^2 + d\theta^2 + \sin^2 \theta \, d\Omega_p^2 \right). \tag{K.7}$$

Since the causal structure of a space-time does not change by conformal transformations we multiply the AdS metric by $\cos^2 \theta / L^2$ to obtain

$$d\tilde{s}^2 = -d\tau^2 + d\theta^2 + \sin^2 \theta \, d\Omega_p^2. \tag{K.8}$$

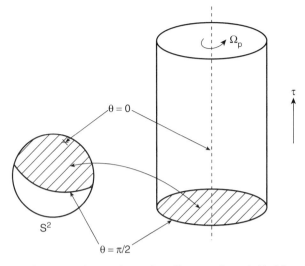

Figure K.2 The space AdS$_3$ can be conformally mapped into half of the Einstein static universe $\mathbb{R} \times S^2$.

This is the metric of the Einstein static universe, with one difference: the θ coordinate takes values in $\left[0, \frac{\pi}{2}\right)$ rather than the full range $[0, \pi)$. The equator $\theta = \pi/2$ is a boundary of the space with the topology of S^p. This is shown in figure K.2 in the case of AdS$_3$. In the case of AdS$_2$ the boundary consists of two points and the coordinate θ takes values in $\left[-\frac{\pi}{2}, \frac{\pi}{2}\right]$.

In general if a space-time is conformal to a space-time that is isomorphic to half of the Einstein static universe then this space-time is called *asymptotically AdS*.

Here the boundary extends in the timelike direction τ and because of this we must specify a boundary condition on $\mathbb{R} \times S^p$ at $\theta = \frac{\pi}{2}$ in order to make the Cauchy problem of AdS well posed.

There is another set of coordinates $(u, t, \vec{x})$ with $ru > 0$, $\vec{x} \in \mathbb{R}^p$, that are useful. They are called Poincaré coordinates. They are defined as

$$X_0 = \frac{u}{2}\left[1 + \frac{1}{u^2}(L^2 + \vec{x}^2 - t^2)\right], \quad X_i = \frac{L\,x^i}{u},$$

$$X_{p+1} = \frac{u}{2}\left[1 - \frac{1}{u^2}(L^2 - \vec{x}^2 + t^2)\right], \quad X_{p+2} = \frac{L\,t}{u}. \tag{K.9}$$

These coordinates cover half of the hyperboloid (K.2). This is shown in figure K.3 for AdS$_2$. The metric in such coordinates is

$$ds^2 = \frac{L^2}{u^2}\left[du^2 - dt^2 + d\vec{x}^2\right]. \tag{K.10}$$

The boundary is at $u = 0$.

In these coordinates, the Poincaré symmetry acting on the $(t, \vec{x})$ coordinates is manifest. Also manifest is the SO(1,1) symmetry that acts as

$$(u, t, \vec{x}) \rightarrow (au, at, a\vec{x}), \quad a > 0. \tag{K.11}$$

This acts as a dilation on the $\mathbb{R}^{1,p}$ coordinates $(t, \vec{x})$.

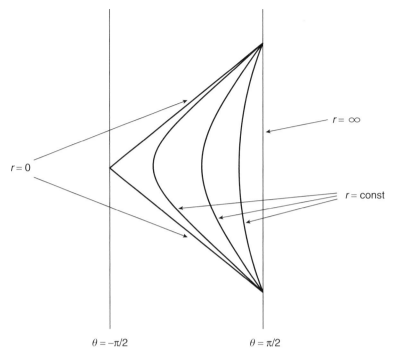

Figure K.3 AdS_2 can be conformally mapped into the strip between $\theta = -\pi/2$ and $\theta = \pi/2$. The modified Poincaré coordinates r, t cover the triangular region.

There is a related metric, which sometimes is also called the Poincaré metric in the literature,

$$ds^2 = L^2 \left[\frac{dr^2}{r^2} + r^2(-dt^2 + d\vec{x}^2) \right]. \tag{K.12}$$

This is related to (K.10) by $r = L^2/u$. In this book we will call the metric in (K.10) the Poincaré metric. Moreover we will call (u, x^μ) the Poincaré coordinates and (r, x^μ) the modified Poincaré coordinates.

A further comment concerns the time-like Killing vectors in the global metric (K.4), ∂_τ and in the Poincaré metric (K.10), ∂_t. ∂_τ has a nowhere vanishing norm proportional to $-\cosh^2 \rho$. On the other hand, the norm of ∂_t, proportional to $-1/r^2$, vanishes at $r = \infty$. This is the (apparent) horizon of the Poincaré coordinates. This is shown in figure K.4 in the AdS_2 case.

K.2 Euclidean AdS

AdS has a global time coordinate τ and therefore the continuation to Euclidean signature is straightforward. This amounts to $\tau \to i\tau_E$ which indicates that Euclidean AdS can be defined as the hyperboloid

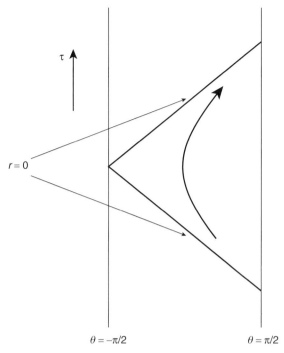

Figure K.4 AdS$_2$ and the timelike Killing vector along t. It becomes null on the horizon $r = 0$.

$$X_0^2 - X_E^2 - \vec{X}^2 = L^2 \tag{K.13}$$

in $\mathbb{R}^{1,p+2}$

$$ds_E^2 = -dX_0^2 + dX_E^2 + d\vec{X}^2. \tag{K.14}$$

We can also obtain the same space by rotating the Poincaré t coordinate as $t \to i\, t$. The metric of Euclidean AdS$_{p+2}$ can be written as

$$\begin{aligned}
ds_E^2 &= L^2 \left(\cos^2 \rho\, d\tau_E^2 + d\rho^2 + \sin^2 \rho\, d\Omega_p^2 \right) \\
&= L^2 \left[\frac{dr^2}{r^2} + r^2 (dt_E^2 + d\vec{x}^2) \right].
\end{aligned} \tag{K.15}$$

We can obtain the Poincaré metric by transforming $r \to 1/u$

$$ds_E^2 = \frac{L^2}{u^2} \left[du^2 + dx_1^2 + \cdots + dx_{p+1}^2 \right]. \tag{K.16}$$

Euclidean AdS$_{p+2}$ is topologically a $(p+2)$-dimensional disk. In the Poincaré coordinates $r = \infty$ is almost all of the boundary. It is topologically an S^{p+1} with a point removed. The full boundary S^{p+1} is recovered by adding the point $r = 0$. This is equivalent to adding the point at infinity, $\vec{x} = \infty$.

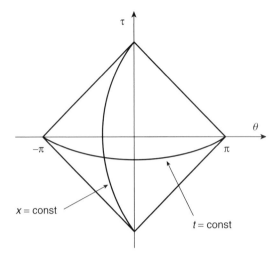

Figure K.5 $\mathbb{R}^{1,1}$ conformally mapped to the interior of the rectangle.

K.3 The Conformal Structure of Flat Space

We will now relate the conformal isometries of AdS space with the standard conformal transformations of flat space. We will start our discussion with the two-dimensional Minkowski space, $\mathbb{R}^{1,1}$, which is simpler to visualize.

The metric is

$$ds^2 = -dt^2 + dx^2, \quad -\infty < t, x < \infty. \tag{K.17}$$

Doing a coordinate transformation

$$\tan u_+ = t + x, \quad \tan u_- = t - x, \quad u_\pm = \frac{\tau \pm \theta}{2}, \tag{K.18}$$

we may rewrite the metric as

$$ds^2 = -\frac{du_+ du_-}{4 \cos^2 u_+ \cos^2 u_-} = \frac{-d\tau^2 + d\theta^2}{4 \cos^2 u_+ \cos^2 u_-}. \tag{K.19}$$

In this fashion, the full Minkowski space has been confined into the compact region $|u_\pm| < \pi/2$ as shown in figure K.5.

The trajectories of photons are insensitive to overall factors of the metric. Therefore the causal structure in the original and the (τ, θ) coordinates is the same. These coordinates are well defined near the conformal boundary. This conformal compactification is useful in order to define asymptotic flatness of a space by comparing the conformal boundary with that of flat space.

The two corners $(\tau, \theta) = (0, \pm\pi)$ correspond to the points at spatial infinity $x = \pm\infty$ in the original coordinates. By identifying them we replace the rectangular image of $\mathbb{R}^{1,1}$ with the space $\mathbb{R} \times S^1$ as shown in figure K.6.

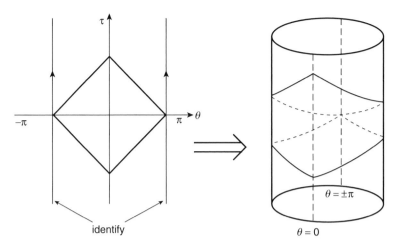

Figure K.6 The identification of $(0, \pi)$ and $(0, -\pi)$ generates a cylinder.

The conformal group of $\mathbb{R}^{1,1}$ is $SO(2,2) \sim SO(2,1) \times SO(2,1)$. Its generators can be written in terms of the $u_\pm$ coordinates as $\partial_\pm$, $u_\pm \partial_\pm$, $u_\pm^2 \partial_\pm$. In particular, the translations along the cylinder can be written as

$$\partial_\tau \pm \partial_\theta = \partial_{\tilde{u}_\pm} = (1 + u_\pm^2)\partial_{u_\pm}. \tag{K.20}$$

We may now discuss higher-dimensional Minkowski space $\mathbb{R}^{1,p}$ with $p \geq 2$. We again start from the flat metric in spherical coordinates

$$ds^2 = -dt^2 + dr^2 + r^2 d\Omega_{p-1}^2 \tag{K.21}$$

and rewrite it in coordinates similar to the above

$$
\begin{aligned}
ds^2 &= -du_+ du_- + \frac{1}{4}(u_+ - u_-)^2 d\Omega_{p-1}^2, \quad u_\pm = t \pm r, \\
&= \frac{1}{\cos^2 \tilde{u}_+ \cos^2 \tilde{u}_-}\left[-d\tilde{u}_+ d\tilde{u}_- + \frac{1}{4}\sin^2(u_+ - u_-)\, d\Omega_{p-1}^2\right], \quad u_\pm = \tan \tilde{u}_\pm, \\
&= \frac{1}{4\cos^2 \tilde{u}_+ \cos^2 \tilde{u}_-}\left[-d\tau^2 + d\theta^2 + \sin^2 \theta\, d\Omega_{p-1}^2\right], \quad \tilde{u}_\pm = \frac{1}{2}(\tau \pm \theta).
\end{aligned} \tag{K.22}
$$

As shown in figure K.7, the full (t,r) plane for a fixed point on the S^{p-1}, is mapped into a triangular region of the (τ, θ) plane. We may now extend the conformally equivalent metric

$$ds^2 = -d\tau^2 + d\theta^2 + \sin^2 \theta\, d\Omega_{p-1}^2 \tag{K.23}$$

from the triangle to the fully extended space

$$0 < \theta < \pi, \quad -\infty < \tau < +\infty. \tag{K.24}$$

The new space has the geometry of $\mathbb{R} \times S^p$, the Einstein static universe as in the AdS case. The points $\theta = 0$ and $\theta = \pi$ correspond to the poles of the S^p. This construction generalizes the conformal embedding of $\mathbb{R}^{1,1}$ to $\mathbb{R} \times S^1$ that we have discussed above.

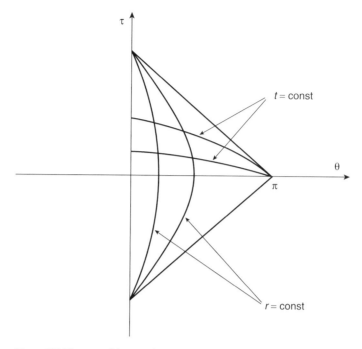

Figure K.7 The map of the (t,r) plane into a triangular region of the (τ, θ) plane.

The subgroup $SO(2) \times SO(p + 1)$ of the conformal group $SO(2,p + 1)$ acts simply on this geometry. The $SO(2)$ generator corresponds to τ translations while $SO(p + 1)$ is the symmetry of S^p. We may therefore start from correlation functions of a conformal field theory in $\mathbb{R}^{1,p}$ and use the $SO(2)$ generator of the conformal group in order to analytically continue them over the whole of the Einstein universe.

K.4 Fields in AdS

We will first consider a massive scalar field in AdS_{p+2} and solve the free equation

$$(\Box - m^2)\phi = 0. \tag{K.25}$$

This can be solved by standard methods. In global coordinates, the solution with well-defined "energy" ω is of the form

$$\phi = e^{i\omega\tau} F(\theta) Y_\ell(\Omega_p), \tag{K.26}$$

with $Y_\ell(\Omega_p)$ the spherical harmonic on S^p, an eigenstate of the Laplacian on S^p with eigenvalue $\ell(\ell + p - 1)$ and

$$F(\theta) = (\sin\theta)^\ell (\cos\theta)^{\Delta_\pm} {}_2F_1(a, b, c; \sin\theta), \tag{K.27}$$

$$a = \frac{1}{2}(\ell + \Delta_\pm - \omega L), \quad b = \frac{1}{2}(\ell + \Delta_\pm + \omega L), \quad c = \ell + \frac{1}{2}(p + 1), \tag{K.28}$$

and

$$\Delta_\pm = \frac{1}{2}(p+1) \pm \frac{1}{2}\sqrt{(p+1)^2 + 4m^2L^2}. \tag{K.29}$$

The energy-momentum tensor is given by

$$T_{\mu\nu} = 2\partial_\mu\phi\partial_\nu\phi - g_{\mu\nu}((\partial\phi)^2 + m^2\phi^2) + \xi(g_{\mu\nu}\Box - \nabla_\mu\nabla_\nu + R_{\mu\nu})\phi^2, \tag{K.30}$$

the last term coming from the coupling of the scalar to the scalar curvature of the background, $\xi \int R \phi^2$. The value of the coupling ξ depends on the particular case. Conformally coupled scalars have $\xi = 1/6$. Note that this coupling in AdS_{p+2}, because the scalar curvature is constant, contributes to the m^2 term in the equation of motion (K.25).

The total energy is

$$E = \int d^{p+1}x \sqrt{-g}\, T_0{}^0, \tag{K.31}$$

and is conserved only if there is no outward flux of energy from the boundary at $\theta = \pi/2$

$$\int_{S^p} d\Omega_p \sqrt{g}\, n_i\, T_0{}^i \,|_{\theta=\pi/2} = 0. \tag{K.32}$$

This translates into the condition

$$(\tan\theta)^p \left[(1 - 2\xi)\partial_\theta + 2\xi \tan\theta\right]\phi^2 = 0 \quad \text{at} \quad \theta = \frac{\pi}{2}. \tag{K.33}$$

This is satisfied by (13.5.2) if a or b in (K.28) is an integer. Requiring the energy ω to be real we obtain

$$|\omega|L = \Delta_\pm + \ell + 2n, \quad n = 0, 1, 2, \ldots. \tag{K.34}$$

This is possible only when $\lambda_\pm$ is real. This gives a lower bound on the mass in AdS_{p+2}

$$m^2L^2 \geq -\frac{1}{4}(p+1)^2, \tag{K.35}$$

known as the Breitenlohner-Freedman (BF) bound. Therefore negative mass-squared scalars are allowed in AdS_{p+2} provided they satisfy this bound.

When $m^2L^2 > -\frac{1}{4}(p-1)(p+3)$, only the Δ_+ solution is normalizable. When $-\frac{1}{4}(p+1)^2 < m^2L^2 < -\frac{1}{4}(p-1)(p+3)$ both solutions are normalizable and there could be different quantizations possible in this case. Which is the correct one depends on other information, for example symmetry considerations.

K.4.1 The wave equation in Poincarè coordinates

In Poincaré coordinates (K.10) the massive wave equation (K.25) reads

$$\frac{u^2}{L^2}\left[\partial_u^2 - \frac{p}{u}\partial_u - \partial_t^2 + \partial \cdot \partial\right]\phi = m^2\phi. \tag{K.36}$$

Fourier transforming in the $\mathbb{R}^{1,p}$ coordinates

$$\phi(u, x) = \int \frac{d^{p+1}q}{(2\pi)^{p+1}}\, \phi(u, q)\, e^{iq\cdot x}, \quad q \cdot x \equiv -q^0 t + \vec{q} \cdot \vec{x}, \tag{K.37}$$

we obtain

$$\left[\partial_u^2 - \frac{p}{u}\,\partial_u - q^2 - \frac{m^2 L^2}{u^2}\right]\phi(u, q) = 0. \tag{K.38}$$

The solution is given in terms of Bessel functions as

$$\phi(u, q) \sim u^{(p+1)/2}\, Z_\nu\!\left(\sqrt{q^2}\,u\right), \quad \nu = \frac{1}{2}\sqrt{(p+1)^2 + 4m^2 L^2}, \tag{K.39}$$

and Z_ν stands for one of the two linearly independent solutions of the Bessel equation, namely, I_ν or K_ν. The two solutions scale as

$$\phi_\pm \sim u^{\Delta_\pm}, \quad \Delta_\pm = \frac{1}{2}(p+1) \pm \frac{1}{2}\sqrt{(p+1)^2 + 4m^2 L^2}. \tag{K.40}$$

In the Euclidean regime, $q^2 \geq 0$ and if we want a function that is regular in the interior of AdS_{p+2}, we must select the K_ν option. It is the one that corresponds to the normalizable mode in Minkowski signature, and also to the λ_- solution in (K.27). Note that, in contract to the global case, the spectrum is now continuous.

K.4.2 The bulk-boundary propagator

In AdS_{p+2}, we are interested in finding solutions of the wave equation (K.25) subject to the boundary condition

$$\phi(u, x)|_{u=0} = u^{4-\Delta}\phi_0(x), \tag{K.41}$$

where $\phi_0(x)$ an arbitrary function on $\mathbb{R}^{1,p}$. A formal way of doing this is via the bulk-boundary propagator K defined as

$$(\Box - m^2)K(u, x; x') = 0, \quad K(u, x; x')|_{u=0} = \delta^{(p+1)}(x - x'). \tag{K.42}$$

Fourier transforming as in (K.37) we translate the boundary condition to

$$K(u, q)|_{u=0} = 1. \tag{K.43}$$

As we argued in section 13.5 on page 413 we must impose the condition very close to the boundary at $u = \epsilon$. We find

$$K_\epsilon(u, x - x') = \left(\frac{u}{\epsilon}\right)^{(p+1)/2} \int \frac{d^{p+1}q}{(2\pi)^{p+1}} \frac{K_\nu(\sqrt{q^2}\,u)}{K_\nu(\sqrt{q^2}\,\epsilon)}\, e^{ip\cdot(x-x')}. \tag{K.44}$$

We may alternatively construct the propagator in configuration space. We will rotate to Euclidean space for convenience. You are asked in exercise 13.18 on page 465 to verify that the function

$$\frac{u^\Delta}{(u^2 + |x - x'|^2)^\Delta}, \quad \Delta(\Delta - p - 1) = m^2 L^2, \tag{K.45}$$

satisfies the massive Laplace equation everywhere. As it will become obvious below we must choose $\Delta = \Delta_+$ the larger of the two roots. Moreover, if $x \neq x'$ it vanishes as $u \to 0$.

This implies that

$$\lim_{u \to 0} \frac{u^{\Delta_+ - \Delta_-}}{(u^2 + |x - x'|^2)^{\Delta_+}} = C_p \delta^{(p+1)}(x - x'), \tag{K.46}$$

$$C_p = \frac{\Gamma\left[\frac{\Delta_+ + \Delta_-}{2}\right] \Gamma\left[\frac{\Delta_+ - \Delta_-}{2}\right] \Omega_p}{2\Gamma[\Delta_+]} = \pi^{\frac{p+1}{2}} \frac{\Gamma\left[\Delta_+ - \frac{p+1}{2}\right]}{\Gamma[\Delta_+]}, \tag{K.47}$$

where are usual Ω_p is the volume of the unit S^p as in (8.8.11) on page 206.

Therefore the normalized propagator is

$$K_\Delta(u, x; x') = \frac{\Gamma[\Delta]}{\pi^{\frac{p+1}{2}} \Gamma\left[\Delta - \frac{p+1}{2}\right]} \frac{u^\Delta}{(u^2 + |x - x'|^2)^\Delta}, \tag{K.48}$$

and the solution of the massive Laplace equation

$$\phi(u, x) = C_p^{-1} \int d^{p+1}x' \frac{u^{\Delta_+}}{(u^2 + |x - x'|^2)^{\Delta_+}} \phi_0(x'), \quad \lim_{u \to 0} \phi(u, x) \simeq u^{\Delta_-} \phi_0(x) + \cdots, \tag{K.49}$$

asymptotes properly at the boundary, proportional to the leading solution. This justifies our choice of branch in the first place.

K.4.3 The bulk-to-bulk propagator

We will concentrate here on AdS$_5$. The propagator is defined as usual as the inverse of the kinetic operator

$$(\Box - m^2) G(u, x; u', x') = u^5 \delta(u - u') \delta^{(4)}(x - x'), \tag{K.50}$$

where we are using Poincaré coordinates. The explicit expression is

$$G_\Delta(u, x; u', x') = -\frac{\Gamma[\Delta]}{2^{\Delta+1} \pi^2 \Gamma[\Delta - 1]} \eta^{-\Delta} {}_2F_1\left(\frac{\Delta}{2}, \frac{\Delta+1}{2}; \Delta - 1; \frac{1}{\eta^2}\right), \tag{K.51}$$

where

$$\eta^2 = \frac{u^2 + u'^2 + (x - x')^2}{2uu'}. \tag{K.52}$$

As one of the bulk points moves to the boundary, the bulk-to-bulk propagator asymptotes to the bulk-to-boundary one

$$\lim_{u \to 0} G_\Delta(u, x; u', x') = -\frac{u^\Delta}{2\Delta - 4} K_\Delta(x; u', x'). \tag{K.53}$$

It is also useful for organizing the AdS perturbation theory to define a new bulk-to-bulk propagator that is zero at the shifted (regularized) boundary at $u = \epsilon$. The form suggested from (K.53) is

$$G_\epsilon(u, x; u', x') = G(u, x; u', x') + \int \frac{d^4 p}{(2\pi)^4} e^{-ip \cdot (x - x')} u^2 u'^2 K_\nu(pu) K_\nu(pu') \frac{I_\nu(p\epsilon)}{K_\nu(p\epsilon)} \tag{K.54}$$

for the AdS$_5$ case. G_ϵ satisfies (K.50) and

$$G_\epsilon(\epsilon, x; u', x') = 0, \quad \partial_u G_\epsilon(\epsilon, x; u', x') = -\epsilon^3 K_\epsilon(u', x'; x). \tag{K.55}$$

Bibliography

[1] S. Weinberg, "Ultraviolet divergences in quantum theories of gravitation," pp. 790–831 in *General Relativity*, ed. S. W. Hawking, and W. Israel, Cambridge University Press, Cambridge, 1979.

[2] E. Alvarez, "Quantum gravity: A pedagogical introduction to some recent results," Rev. Mod. Phys. **61** (1989) 561.

[3] G. Veneziano, "Construction of a crossing—symmetric, Regge behaved amplitude for linearly rising trajectories," Nuovo Cimento A **57** (1968) 190.

[4] L. Alvarez-Gaume and E. Witten, "Gravitational anomalies," Nucl. Phys. B **234** (1984) 269.

[5] M. B. Green and J. H. Schwarz, "Anomaly cancellation in supersymmetric $D = 10$ gauge theory and superstring theory," Phys. Lett. B **149** (1984) 117.

[6] D. J. Gross, J. A. Harvey, E. J. Martinec and R. Rohm, "Heterotic string theory. 1. The free heterotic string," Nucl. Phys. B **256** (1985) 253; "Heterotic string theory. 2. The interacting heterotic string," *ibid.* **267** (1986) 75.

[7] M. Green, J. Schwarz and E. Witten, Superstring Theory, Vols I and II, Cambridge University Press, Cambridge, 1987.

[8] J. Polchinski, *String Theory*, Vols I and II, Cambridge University Press, Cambridge, 1998.

[9] C. V. Johnson, *D-branes*, Cambridge University Press, Cambridge, 2003.

[10] T. Ortin, *Gravity and Strings*, Cambridge University Press, Cambridge, 2004.

[11] R. J. Szabo, *An Introduction to String Theory and D-Brane Dynamics*, Imperial College Press, London, 2004.

[12] B. Zwiebach, *A First Course in String Theory*, Cambridge University Press, Cambridge, 2004.

[13] M. E. Peskin, "Introduction to string and superstring theory: 2," SLAC-PUB-4251; lectures presented at Theoretical Advanced Study Institute in Particle Physics, Santa Cruz, Calif., Jun 23–Jul 19, 1986.

[14] D. Lüst and S. Theisen, "Lectures on string theory," Lect. Notes Phys. **346** (1989) 1.

[15] E. D'Hoker, "TASI lectures on critical string theory," UCLA-92-TEP-30; lectures presented at Theoretical Advanced Study Institute (TASI 92): From Black Holes and Strings to Particles, Boulder, Colo., Jun 3–28 1992.

[16] L. Alvarez-Gaume and M. A. Vazquez-Mozo, "Topics in string theory and quantum gravity," arXiv:hep-th/9212006.

[17] J. Polchinski, "What is string theory?" hep-th/9411028.

[18] H. Ooguri and Z. Yin, "lectures on Perturbative String Theories," arXiv:hep-th/9612254.

[19] E. Kiritsis *Introduction to String Theory,* Leuven University Press, Leuven, 1998, [arXiv:hep-th/9709062].

[20] S. Forste, "Strings, branes and extra dimensions," Fortschr. Phys. **50** (2002) 221 [arXiv:hep-th/0110055].

[21] U. Danielsson, "Introduction to string theory," Rep. Prog. Phys. **64** (2001) 51.

[22] D. Marolf, "Resource letter: The nature and status of string theory," Am. J. Phys. **72** (2004) 730 [arXiv:hep-th/0311044].

[23] J. H. Schwarz and N. Seiberg, "String theory, supersymmetry, unification, and all that," Rev. Mod. Phys. **71** (1999) S112 [arXiv:hep-th/9803179].

[24] L. Brink, P. Di Vecchia, and P. S. Howe, "A locally supersymmetric and reparametrization invariant action for the spinning string," Phys. Lett. B **65** (1976) 471; "A Lagrangian formulation of the classical and quantum dynamics of spinning particles," Nucl. Phys. B **118** (1977) 76.

[25] S. Deser and B. Zumino, "A complete action for the spinning string," Phys. Lett. B **65** (1976) 369.

[26] P. H. Ginsparg, "Applied conformal field theory," arXiv:hep-th/9108028.

[27] R. C. Brower, "Spectrum generating algebra and no ghost theorem for the dual model," Phys. Rev. D **6** (1972) 1655.

[28] P. Goddard and C. B. Thorn, "Compatibility of the dual pomeron with unitarity and the absence of ghosts in the dual resonance model," Phys. Lett. B **40** (1972) 235.

[29] P.A.M. Dirac, "Generalized Hamiltonian dynamics," Cana. J. Math. **2** (1950) 129.

[30] A. J. Niemi, "Pedagogical introduction To BRST," Phys. Rep. **184** (1989) 147.

[31] G. Barnich, F. Brandt, and M. Henneaux, "Local BRST cohomology in gauge theories," Phys. Rep. **338** (2000) 439 [arXiv:hep-th/0002245].

[32] T. Kugo and I. Ojima, "Local covariant operator formalism of nonabelian gauge theories and quark confinement problem," Prog. Theor. Phys. Suppl. **66** (1979) 1.

[33] M. Kato and K. Ogawa, "Covariant quantization of string based on BRS invariance," Nucl. Phys. B **212** (1983) 443.

[34] A. M. Polyakov, "Quantum geometry of bosonic strings," Phys. Lett. B **103** (1981) 207; "Quantum geometry of fermionic strings," Phys. Lett. B **103** (1981) 211.

[35] D. Friedan, "Introduction to Polyakov's string theory," EFI-82-50-Chicago; in Proceedings of Summer School of Theoretical Physics: Recent Advances in Field Theory and Statistical Mechanics, Les Houches, France, Aug 2–Sep 10, 1982.

[36] O. Alvarez, "Theory of strings with boundaries: Fluctuations, topology, and quantum geometry," Nucl. Phys. B **216** (1983) 125.

[37] P. Nelson, "Lectures on strings and moduli space," Phys. Rep. **149** (1987) 337.

[38] N. Marcus and A. Sagnotti, "Tree level constraints on gauge groups for type I superstrings," Phys. Lett. B **119** (1982) 97.

[39] E. Witten, "Interacting field theory of open superstrings," Nucl. Phys. B **276** (1986) 291.

[40] C. B. Thorn, "String field theory," Phys. Rep. **175** (1989) 1.

[41] W. Siegel, "Introduction to string field theory," arXiv:hep-th/0107094.

[42] B. Zwiebach, "Closed string field theory: An introduction," arXiv:hep-th/9305026.

[43] N. E. Mavromatos, "Logarithmic conformal field theories and strings in changing backgrounds," arXiv:hep-th/0407026.

[44] M. Flohr, "Bits and pieces in logarithmic conformal field theory," Int. J. Mod. Phys. A **18** (2003) 4497 [arXiv:hep-th/0111228].

[45] N. Seiberg, "Notes on quantum Liouville theory and quantum gravity," Prog. Theor. Phys. Suppl. **102** (1990) 319.

[46] P. Di Francesco, P. Mathieu, and D. Senechal, *Conformal Field Theory,* Springer, New York, 1997.

[47] S. V. Ketov, *Conformal Field Theory*, World Scientific, Singapore, 1995.

[48] L. Alvarez-Gaume, G. Sierra, and C. Gomez, "Topics in conformal field theory," CERN-TH-5540-89; contribution to Knizhnik Memorial Volume, ed. L. Brink et al., World Scientific, Singapore, 1990.

[49] G. W. Moore and N. Seiberg, "Lectures on Rcft," pp. 1–129 in *Trieste Superstrings 1989*, World Scientific, Singapore, 1990; also pp. 263–362 in *Banff NATO ASI 1989*, Plenum Press, New York, 1990.

[50] M. R. Gaberdiel, "An introduction to conformal field theory," Rep. Prog. Phys. **63** (2000) 607 [arXiv:hep-th/9910156].

[51] J. Fuchs and C. Schweigert, *Symmetries, Lie Algebras and Representations*, Cambridge University Press, Cambridge, 1997.

[52] P. Goddard and D. I. Olive, "Kac-Moody and Virasoro algebras in relation to quantum physics," Int. J. Mod. Phys. A **1** (1986) 303.

[53] M. B. Halpern, E. Kiritsis, N. A. Obers, and K. Clubok, "Irrational conformal field theory," Phys. Rep. **265** (1996) 1 [arXiv:hep-th/9501144].

[54] P. Bouwknegt and K. Schoutens, "W symmetry in conformal field theory," Phys. Rep. **223** (1993) 183 [arXiv:hep-th/9210010].

[55] E. Witten, "Nonabelian bosonization in two dimensions," Commun. Math. Phys. **92** (1984) 455.

[56] A. A. Belavin, A. M. Polyakov and A. B. Zamolodchikov, "Infinite conformal symmetry in two-dimensional quantum field theory," Nucl. Phys. B **241** (1984) 333.

[57] V. S. Dotsenko and V. A. Fateev, "Four point correlation functions and the operator algebra in the two-dimensional conformal invariant theories with the central charge $C < 1$," Nucl. Phys. B **251** (1985) 691.

[58] A. Cappelli, C. Itzykson, and J. B. Zuber, "The ADE classification of minimal and $A_1(1)$ conformal invariant theories," Commun. Math. Phys. **113** (1987) 1.

[59] D. Friedan, E. J. Martinec, and S. H. Shenker, "Conformal invariance, supersymmetry and string theory," Nucl. Phys. B **271** (1986) 93.

[60] N. D. Birrell and P.C.W Davis, *Quantum Fields in Curved Spacetime*, Cambridge Monographs on Mathemetical Physics, Cambridge University Press, Cambridge, 1982.

[61] A. Sevrin, W. Troost, and A. Van Proeyen, "Superconformal algebras in two-dimensions with $N = 4$," Phys. Lett. B **208** (1988) 447.

[62] Z. A. Qiu, "Supersymmetry, two-dimensional critical phenomena and the tricritical Ising model," Nucl. Phys. B **270** (1986) 205.

[63] E. B. Kiritsis, "The structure of $N = 2$ superconformally invariant 'minimal' theories: Operator algebra and correlation functions," Phys. Rev. D **36** (1987) 3048.

[64] W. Lerche, C. Vafa, and N. P. Warner, "Chiral rings in $N = 2$ superconformal theories," Nucl. Phys. B **324** (1989) 427.

[65] N. Warner, "Lectures on $N = 2$ superconformal theories and singularity theory," Lectures presented at Spring School on Superstrings, Trieste, Italy, April 3–14, 1989.

[66] J. H. Schwarz, "Superconformal symmetry and superstring compactification," Int. J. Mod. Phys. A **4** (1989) 2653.

[67] P. S. Aspinwall, "The moduli space of $N = 2$ superconformal field theories," arXiv:hep-th/9412115.

[68] B. R. Greene, "String theory on Calabi-Yau manifolds," arXiv:hep-th/9702155.

[69] T. Eguchi and A. Taormina, "Extended superconformal algebras and string compactifications," Lecture presented at Spring School in Superstrings, Trieste, Italy, Apr 11–22, 1988; "On the unitary representations of $N = 2$ and $N = 4$ superconformal algebras," Phys. Lett. B **210** (1988) 125.

[70] A. Giveon, M. Porrati, and E. Rabinovici, "Target space duality in string theory," Phys. Rep. **244** (1994) 77 [arXiv:hep-th/9401139].

[71] O. Alvarez, T. P. Killingback, M. Mangano, and P. Windey, "String theory and loop space index theorems," Commun. Math. Phys, **111** (1987) 1.

[72] E. Witten, "Elliptic genera and quantum field theory," Commun. Math. Phys. **109** (1987) 525;

[73] W. Lerche, A. N. Schellekens, and N. P. Warner, "Lattices and strings," Phys. Rep. **177** (1989) 1.

[74] V. Schomerus, "Non-compact string backgrounds and non-rational CFT," arXiv:hep-th/0509155.

[75] A. B. Zamolodchikov, " 'Irreversibility' of the flux of the renormalization group in a 2-D field theory," JETP Lett. **43** (1986) 730 [Pis'ma Zh. Eksp. Teor. Fiz. **43** (1986) 565].

[76] A. Cappelli, D. Friedan and J. I. Latorre, "C theorem and spectral representation," Nucl. Phys. B **352** (1991) 616.

[77] A. B. Zamolodchikov, "Conformal symmetry and multicritical points in two-dimensional quantum field theory." Sov. J. Nucl. Phys. **44** (1986) 529 [Yad. Fiz. **44** (1986) 821].

[78] E. Silverstein and E. Witten, "Criteria for conformal invariance of (0,2) models," Nucl. Phys. B **444** (1995) 161 [arXiv:hep-th/9503212].

[79] L. J. Dixon, J. A. Harvey, C. Vafa and E. Witten, "Strings On orbifolds," Nucl. Phys. B **261** (1985) 678; **274** (1986) 285.

[80] L. J. Dixon, D. Friedan, E. J. Martinec, and S. H. Shenker, "The conformal field theory of orbifolds," Nucl. Phys. B **282** (1987) 13.

[81] S. Hamidi and C. Vafa, "Interactions on orbifolds," Nucl. Phys. B **279** (1987) 465.

[82] R. Dijkgraaf, E. P. Verlinde and H. L. Verlinde, "$C = 1$ conformal field theories on Riemann surfaces," Commun. Math. Phys. **115** (1988) 649.

[83] R. Dijkgraaf, C. Vafa, E. P. Verlinde, and H. L. Verlinde, "The operator algebra of orbifold models," Commun. Math. Phys. **123** (1989) 485.

[84] K. S. Narain, M. H. Sarmadi, and C. Vafa, "Asymmetric orbifolds," Nucl. Phys. B **288** (1987) 551; "Asymmetric orbifolds: Path integral and operator formulations," Nucl. Phys. B **356** (1991) 163.

[85] J. L. Cardy, "Conformal invariance and surface critical behavior," Nucl. Phys. B **240** (1984) 514.

[86] J. L. Cardy, "Boundary conditions, fusion rules and the verlinde formula," Nucl. Phys. B **324** (1989) 581.

[87] J. B. Zuber, "CFT, BCFT, ADE and all that," arXiv:hep-th/0006151.

[88] V. B. Petkova and J. B. Zuber, "Conformal boundary conditions and what they teach us," arXiv:hep-th/0103007.

[89] Y. S. Stanev, "Two dimensional conformal field theory on open and unoriented surfaces," arXiv:hep-th/0112222.

[90] R. E. Behrend, P. A. Pearce, V. B. Petkova and J. B. Zuber, "Boundary conditions in rational conformal field theories," [Nucl. Phys. B **579** (2000) 707 [arXiv:hep-th/9908036].

[91] V. Schomerus, "Lectures on branes in curved backgrounds," Class. Quantum Grav. **19** (2002) 5781 [arXiv:hep-th/0209241].

[92] M. R. Gaberdiel and T. Gannon, "Boundary states for WZW models," Nucl. Phys. B **639** (2002) 471 [arXiv:hep-th/0202067].

[93] M. R. Gaberdiel, "Lectures on non-BPS Dirichlet branes," Class. Quantum Grav. **17** (2000) 3483 [arXiv:hep-th/0005029].

[94] A. Lerda and R. Russo, "Stable non-BPS states in string theory: A pedagogical review," Int. J. Mod. Phys. A **15** (2000) 771 [arXiv:hep-th/9905006].

[95] P. Di Vecchia and A. Liccardo, "D branes in string theory. I," NATO Adv. Study Inst. Ser. C. Math. Phys. Sci. **556** (2000) 1 [arXiv:hep-th/9912161]; "D-branes in string theory. II," arXiv:hep-th/9912275.

[96] I. V. Vancea, "Introductory lectures on D-branes," [arXiv:hep-th/0109029].

[97] C. Angelantonj and A. Sagnotti, "Open strings," Phys. Rep. **371** (2002) 1; Erratum, *ibid.* **376** (2003) 339 [arXiv:hep-th/0204089].

[98] P. Goddard, A. Kent, and D. I. Olive, "Virasoro algebras and coset space models," Phys. Lett. B **152** (1985) 88.

[99] E. B. Kiritsis, "Nonstandard bosonization techniques in conformal field theory," Mod. Phys. Lett. A **4** (1989) 437.

[100] S. Mandelstam, "Interacting string picture of dual resonance models," Nucl. Phys. B **64** (1973) 205.

[101] J. H. Schwarz, "Dual resonance theory," Phys. Rep. **8** (1973) 269; "Superstring theory," Phys. Rep. **89** (1982) 223.

[102] S. Mandelstam, "Dual-resonance models," Phys. Rep. **13** (1974) 259.

[103] J. Scherk, "An introduction to the theory of dual models and strings," Rev. Mod. Phys. **47** (1975) 123.

[104] H. Kawai, D. C. Lewellen, and S. H. H. Tye, "A relation between tree amplitudes of closed and open strings," Nucl. Phys. B **269** (1986) 1.

[105] S. R. Coleman and E. Weinberg, "Radiative corrections as the origin of spontaneous symmetry breaking," Phys. Rev. D **7** (1973) 1888.

[106] D. Mumford, "Tata Lectures on Theta," vols. 1–3, Birkhauser, Boston, 1983.

[107] E. D'Hoker and D. H. Phong, "The geometry of string perturbation theory," Rev. Mod. Phys. **60** (1988) 917.

[108] L. Alvarez-Gaume, C. Gomez, G. W. Moore, and C. Vafa, "Strings in the operator formalism," Nucl. Phys. B **303** (1988) 455.

[109] E. J. Martinec, "Conformal field theory on a (super) Riemann surface," Nucl. Phys. B **281** (1987) 157.

[110] H. Sonoda, "Sewing conformal field theories," Nucl. Phys. B **311** (1988) 401; "Sewing conformal field theories. 2," *ibid.* **311** (1988) 417; "Hermiticity and CPT in string theory," *ibid.* **326** (1989) 135.

[111] M. B. Green and J. H. Schwarz, "Infinity cancellations in SO(32) superstring theory," Phys. Lett. B **151** (1985) 21.

[112] C. G. Callan, C. Lovelace, C. R. Nappi, and S. A. Yost, "Adding holes and crosscaps to the superstring," Nucl. Phys. B **293** (1987) 83.

[113] J. Polchinski and Y. Cai, "Consistency of open superstring theories," Nucl. Phys. B **296** (1988) 91.

[114] A. Sagnotti, "Open strings and their symmetry groups," arXiv:hep-th/0208020.

[115] G. Pradisi and A. Sagnotti, "Open string orbifolds," Phys. Lett. B **216** (1989) 59.

[116] P. Hořava, "Background duality of open string models," Nucl. Phys. B **327** (1989) 461; "Strings on worldsheet orbifolds," Phys. Lett. B **231** (1989) 251.

[117] M. Bianchi and A. Sagnotti, "On the systematics of open string theories," Phys. Lett. B **247** (1990) 517; "Twist symmetry and open string Wilson lines," Nucl. Phys. B **361** (1991) 519.

[118] M. Bianchi, G. Pradisi, and A. Sagnotti, "Toroidal compactification and symmetry breaking in open string theories," Nucl. Phys. B **376** (1992) 365.

[119] D. Fioravanti, G. Pradisi, and A. Sagnotti, "Sewing constraints and nonorientable open strings," Phys. Lett. B **321** (1994) 349 [arXiv:hep-th/9311183].

[120] D. C. Lewellen, "Sewing constraints for conformal field theories on surfaces with boundaries," Nucl. Phys. B **372** (1992) 654.

[121] D. Amati, M. Ciafaloni, and G. Veneziano, "Can space-time be probed below the string size?," Phys. Lett. B **216** (1989) 41.

[122] D. J. Gross and P. F. Mende, "The high-energy behavior of string scattering amplitudes," Phys. Lett. B **197** (1987) 129; "String theory beyond the Planck scale," Nucl. Phys. B **303** (1988) 407.

[123] S. H. Shenker, "The strength of nonperturbative effects in string theory," presented at the Cargese Workshop on Random Surfaces, Quantum Gravity and Strings, Cargese, France, May 28–Jun 1, 1990.

[124] D. J. Gross and V. Periwal, "String perturbation theory diverges," Phys. Rev. Lett. **60** (1988) 2105.

[125] P. F. Mende and H. Ooguri, "Borel summation of string theory for Planck scale scattering," Nucl. Phys. B **339** (1990) 641.

[126] S. Weinberg, "The cosmological constant problem," Rev. Mod. Phys. **61** (1989) 1.

[127] E. S. Fradkin and A. A. Tseytlin, "Quantum string theory effective action," Nucl. Phys. B **261** (1985) 1.

[128] L. Alvarez-Gaume, D. Z. Freedman, and S. Mukhi, "The background field method and the ultraviolet structure of the supersymmetric nonlinear sigma model," Ann. Phys. (N.Y.) **134** (1981) 85.

[129] D. H. Friedan, "Nonlinear models in two + epsilon dimensions," Ann. Phys. (N.Y.) **163** (1985) 318.

[130] C. G. Callan, E. J. Martinec, M. J. Perry, and D. Friedan, "Strings in background fields," Nucl. Phys. B **262** (1985) 593.

[131] C. G. Callan and L. Thorlacius, "Sigma models and string theory," in Providence: TASI 88, World Scientific, Singapore, 1989.

[132] A. A. Tseytlin, "Sigma model approach to string theory," Int. J. Mod. Phys. A **4** (1989) 1257.

[133] R. R. Metsaev and A. A. Tseytlin, "Order α' equivalence of the string equations of motion and the σ-model Weyl invariance conditions: Dependence on the dilaton and the antisymmetric tensor," Nucl. Phys. B **293** (1987) 385.

[134] M. T. Grisaru, A. E. M. van de Ven, and D. Zanon, "Two-dimensional supersymmetric sigma models on Ricci flat Kahler manifolds are not finite," Nucl. Phys. B **277** (1986) 388; "Four loop divergences for the N=1 supersymmetric nonlinear sigma model in two-dimensions," *ibid*. **277** (1986) 409.

[135] D. J. Gross and J. H. Sloan, "The quartic effective action for the heterotic string," Nucl. Phys. B **291** (1987) 41.

[136] A. A. Tseytlin, "Exact solutions of closed string theory," Class. Quantum Grav. **12** (1995) 2365 [arXiv:hep-th/9505052].

[137] P. H. Ginsparg and G. W. Moore, "Lectures on 2-D gravity and 2-D string theory," arXiv:hep-th/9304011.

[138] P. Di Francesco, P. H. Ginsparg and J. Zinn-Justin, "2-D gravity and random matrices," Phys. Rep. **254** (1995) 1 [arXiv:hep-th/9306153].

[139] E. Alvarez, L. Alvarez-Gaume, and Y. Lozano, "An introduction to T duality in string theory," Nucl. Phys. Proc. Suppl. **41** (1995) 1 [arXiv:hep-th/9410237].

[140] E. Kiritsis, "Exact duality symmetries in CFT and string theory," Nucl. Phys. B **405** (1993) 109 [arXiv:hep-th/9302033].

[141] E. B. Kiritsis, "Duality in gauged WZW models," Mod. Phys. Lett. A **6** (1991) 2871.

[142] M. Rocek and E. Verlinde, "Duality, quotients, and currents," Nucl. Phys. B **373** (1992) 630 [arXiv:hep-th/9110053].

[143] E. Kiritsis, C. Kounnas, and D. Lust, "A large class of new gravitational and axionic backgrounds for four-dimensional superstrings," Int. J. Mod. Phys. A **9** (1994) 1361 [arXiv:hep-th/9308124].

[144] C. Bachas, "(Half) a lecture on D-branes," arXiv:hep-th/9701019.

[145] A. Van Proeyen, "Tools for supersymmetry," arXiv:hep-th/9910030.

[146] H. Kawai, D. C. Lewellen, and S. H. H. Tye, "Classification of closed fermionic string models," Phys. Rev. D **34** (1986) 3794.

[147] P. H. Ginsparg and C. Vafa, "Toroidal compactification of nonsupersymmetric heterotic strings," Nucl. Phys. B **289** (1987) 414.

[148] J. Polchinski and Y. Cai, "Consistency of open superstring theories," Nucl. Phys. B **296** (1988) 91.

[149] S. Sugimoto, "Anomaly cancellations in type I D9-D9-bar system and the USp(32) string theory," Prog. Theor. Phys. **102** (1999) 685 [arXiv:hep-th/9905159].

[150] N. Berkovits, "Super-Poincaré covariant quantization of the superstring," JHEP **0004** (2000) 018 [arXiv:hep-th/0001035].

[151] N. Berkovits, C. Vafa and E. Witten, "Conformal field theory of AdS background with Ramond-Ramond flux," JHEP **9903** (1999) 018 [arXiv:hep-th/9902098].

[152] N. Berkovits, "Multiloop amplitudes and vanishing theorems using the pure spinor formalism for the superstring," JHEP **0409** (2004) 047 [arXiv:hep-th/0406055].

[153] L. Alvarez-Gaume and P. H. Ginsparg, "The structure of gauge and gravitational anomalies," Ann. Phys. (N.Y.) **161** (1985) 423; Erratum, *ibid.* **171** (1986) 233.

[154] W. Lerche, B. E. W. Nilsson, and A. N. Schellekens, "Heterotic string loop calculation of the anomaly cancelling term," Nucl. Phys. B **289** (1987) 609.

[155] P. Meessen and T. Ortin, "An Sl(2,Z) multiplet of nine-dimensional type II supergravity theories," Nucl. Phys. B **541** (1999) 195 [arXiv:hep-th/9806120].

[156] J. Polchinski, S. Chaudhuri, and C. V. Johnson, "Notes on D-Branes," arXiv:hep-th/9602052.

[157] J. Polchinski, "String duality: A colloquium," Rev. Mod. Phys. **68** (1996) 1245; [arXiv:hep-th/9607050].

[158] J. Polchinski, "TASI Lectures on D-Branes," hep-th/9611050.

[159] C. Bachas, "Lectures on D-branes" hep-th/9806199.

[160] C. V. Johnson, "D-brane primer," arXiv:hep-th/0007170.

[161] M. R. Douglas, "Superstring dualities, Dirichlet branes and the small scale structure of space," arXiv:hep-th/9610041.

[162] W. Taylor, "Lectures on D-branes, gauge theory and M(atrices)," hep-th/9801182.

[163] J. M. Maldacena, "Black holes and D-branes," Nucl. Phys. Proc. Suppl. **61A** (1998) 111 [Nucl. Phys. Proc. Suppl. **62** (1998) 428] [arXiv:hep-th/9705078].

[164] R. I. Nepomechie, "Magnetic monopoles from antisymmetric tensor gauge fields," Phys. Rev. D **31** (1985) 1921.

[165] C. Teitelboim, "Gauge invariance for extended objects," Phys. Lett. B **167** (1986) 63; "Monopoles of higher rank," Phys. Lett. B **167** (1986) 69.

[166] J. Polchinski, "Dirichlet-branes and Ramond-Ramond charges," Phys. Rev. Lett. **75** (1995) 4724 [arXiv:hep-th/9510017].

[167] E. S. Fradkin and A. A. Tseytlin, "Nonlinear electrodynamics from quantized strings," Phys. Lett. B **163** (1985) 123.

[168] A. Abouelsaood, C. G. . Callan, C. R. Nappi, and S. A. Yost, "Open strings in background gauge fields," Nucl. Phys. B **280** (1987) 599.

[169] R. G. Leigh, "Dirac-Born-Infeld action from Dirichlet sigma model," Mod. Phys. Lett. A **4** (1989) 2767.

[170] E. Bergshoeff, M. de Roo, M. B. Green, G. Papadopoulos, and P. K. Townsend, "Duality of type II 7-branes and 8-branes," Nucl. Phys. B **470** (1996) 113 [arXiv:hep-th/9601150].

[171] E. Alvarez, J. L. F. Barbon, and J. Borlaf, "T-duality for open strings," Nucl. Phys. B **479** (1996) 218 [arXiv:hep-th/9603089].

[172] G. W. Gibbons, "Aspects of Born-Infeld theory and string/M-theory," Rev. Mex. Fis. **49S1** (2003) 19 [arXiv:hep-th/0106059].

[173] D. P. Sorokin, "Superbranes and superembeddings," Phys. Rep. **329** (2000) 1 [arXiv:hep-th/9906142].

[174] C. Bachas, "D-brane dynamics," Phys. Lett. B **374** (1996) 37 [arXiv:hep-th/9511043].

[175] G. Lifschytz, "Comparing D-branes to black-branes," Phys. Lett. B **388** (1996) 720 [arXiv:hep-th/9604156].

[176] M. R. Douglas, D. Kabat, P. Pouliot, and S. H. Shenker, "D-branes and short distances in string theory," Nucl. Phys. B **485** (1997) 85 [arXiv:hep-th/9608024].

[177] M. B. Green, J. A. Harvey and G. W. Moore, "I-brane inflow and anomalous couplings on D-branes," Class. Quantum Grav. **14** (1997) 47 [arXiv:hep-th/9605033].

[178] K. Dasgupta, D. P. Jatkar and S. Mukhi, "Gravitational couplings and $\mathbb{Z}_2$ orientifolds," Nucl. Phys. B **523** (1998) 465 [arXiv:hep-th/9707224].

[179] C. P. Bachas, P. Bain and M. B. Green, "Curvature terms in D-brane actions and their M-theory origin," JHEP **9905** (1999) 011 [arXiv:hep-th/9903210].

[180] E. Witten, "Bound states of strings and p-branes," Nucl. Phys. B **460** (1996) 335 [arXiv:hep-th/9510135].

[181] A. A. Tseytlin, "On non-abelian generalisation of the Born-Infeld action in string theory," Nucl. Phys. B **501** (1997) 41 [arXiv:hep-th/9701125]; "Born-Infeld action, supersymmetry and string theory," arXiv:hep-th/9908105.

[182] D. Brecher and M. J. Perry, "Bound states of D-branes and the non-Abelian Born-Infeld action," Nucl. Phys. B **527** (1998) 121 [arXiv:hep-th/9801127].

[183] M. R. Garousi and R. C. Myers, "World-volume interactions on D-branes," Nucl. Phys. B **542** (1999) 73 [arXiv:hep-th/9809100].

[184] A. Hashimoto and W. I. Taylor, "Fluctuation spectra of tilted and intersecting D-branes from the Born-Infeld action," Nucl. Phys. B **503** (1997) 193 [arXiv:hep-th/9703217].

[185] P. Bain, "On the non-Abelian Born-Infeld action," arXiv:hep-th/9909154.

[186] F. Denef, A. Sevrin, and J. Troost, "Non-Abelian Born-Infeld versus string theory," Nucl. Phys. B **581** (2000) 135 [arXiv:hep-th/0002180].

[187] A. Sevrin, J. Troost, and W. Troost, "The non-abelian Born-Infeld action at order F^6," Nucl. Phys. B **603** (2001) 389 [arXiv:hep-th/0101192].

[188] R. C. Myers, "Dielectric-branes," JHEP **9912** (1999) 022 [arXiv:hep-th/9910053].

[189] A. Sen, "A note on marginally stable bound states in type II string theory," Phys. Rev. D **54** (1996) 2964 [arXiv:hep-th/9510229]; "U-duality and intersecting D-branes," *ibid.* **53** (1996) 2874 [arXiv:hep-th/9511026].

[190] C. Vafa, "Gas of D-branes and Hagedorn density of BPS States," Nucl. Phys. B **463** (1996) 415 [arXiv:hep-th/9511088].

[191] P. K. Townsend, "D-branes from M-branes," Phys. Lett. B **373** (1996) 68 [arXiv:hep-th/9512062].

[192] G. Papadopoulos and P. K. Townsend, "Kaluza-Klein on the brane," Phys. Lett. B **393** (1997) 59 [arXiv:hep-th/9609095].

[193] S. Sethi and M. Stern, "D-brane bound states redux," Commun. Math. Phys. **194** (1998) 675 [arXiv:hep-th/9705046].

[194] C. P. Bachas, M. B. Green, and A. Schwimmer, "(8,0) quantum mechanics and symmetry enhancement in type I superstrings," JHEP **9801** (1998) 006 [arXiv:hep-th/9712086].

[195] N. Seiberg and E. Witten, "The D1/D5 system and singular CFT," JHEP **9904** (1999) 017 [arXiv:hep-th/9903224].

[196] A. Giveon and D. Kutasov, "Brane dynamics and gauge theory," Rev. Mod. Phys. **71** (1999) 983 [arXiv:hep-th/9802067].

[197] R. Minasian and G. W. Moore, "K-theory and Ramond-Ramond charge," JHEP **9711** (1997) 002 [arXiv:hep-th/9710230].

[198] E. Witten, "D-branes and K-theory," JHEP **9812** (1998) 019 [arXiv:hep-th/9810188].

[199] P. Hořava, "Type IIA D-branes, K-theory, and matrix theory," Adv. Theor. Math. Phys. **2** (1999) 1373 [arXiv:hep-th/9812135].

[200] D. E. Diaconescu, G. W. Moore, and E. Witten, "E(8) gauge theory, and a derivation of K-theory from M-theory," Adv. Theor. Math. Phys. **6** (2003) 1031 [arXiv:hep-th/0005090].

[201] A. Sen, "Stable non-BPS states in string theory," JHEP **9806** (1998) 007 [arXiv:hep-th/9803194]; "Stable non-BPS bound states of BPS D-branes," *ibid.* **9808** (1998) 010 [arXiv:hep-th/9805019]; "Tachyon condensation on the brane antibrane system," *ibid.* **9808** (1998) 012 [arXiv:hep-th/9805170]; "SO(32) spinors of type I and other solitons on brane-antibrane pair," *ibid.* **9809** (1998) 023 [arXiv:hep-th/9808141].

[202] A. Sen, "Non-BPS states and branes in string theory," arXiv:hep-th/9904207.

[203] A. Lerda and R. Russo, "Stable non-BPS states in string theory: A pedagogical review," Int. J. Mod. Phys. A **15** (2000) 771 [arXiv:hep-th/9905006].

[204] J. H. Schwarz, "TASI lectures on non-BPS D-brane systems," arXiv:hep-th/9908144.

[205] M. J. Duff, R. R. Khuri, and J. X. Lu, "String solitons," Phys. Rep. **259** (1995) 213 [arXiv:hep-th/9412184].

[206] D. Youm, "Black holes and solitons in string theory," Phys. Rep. **316** (1999) 1 [arXiv:hep-th/9710046].

[207] A. W. Peet, "The Bekenstein formula and string theory (N-brane theory)," Class. Quantum Grav. **15** (1998) 3291 [arXiv:hep-th/9712253].

[208] A. W. Peet, "TASI lectures on black holes in string theory," arXiv:hep-th/0008241.

[209] K. S. Stelle, "BPS branes in supergravity," arXiv:hep-th/9803116.

[210] G. T. Horowitz and A. Strominger, "Black strings and P-branes," Nucl. Phys. B **360**, 197 (1991).

[211] A. Strominger, "Heterotic solitons," Nucl. Phys. B **343** (1990) 167; Erratum, *ibid.* **353** (1991) 565.

[212] S. J. Rey, "The confining phase of superstrings and axionic strings," Phys. Rev. D **43** (1991) 526.

[213] C. G. Callan, J. A. Harvey, and A. Strominger, "Worldbrane actions for string solitons," Nucl. Phys. B **367** (1991) 60; "World sheet approach to heterotic instantons and solitons," *ibid.* **359** (1991) 611; "Supersymmetric string solitons," arXiv:hep-th/9112030.

[214] I. Antoniadis, C. Bachas, J. R. Ellis, and D. V. Nanopoulos, "Cosmological string theories and discrete inflation," Phys. Lett. B **211** (1988) 393; "An expanding universe in string theory," Nucl. Phys. B **328** (1989) 117.

[215] C. Kounnas, M. Porrati, and B. Rostand, "On $N = 4$ extended super-Liouville theory," Phys. Lett. B **258** (1991) 61.

[216] A. Sen, "Non-BPS states and branes in string theory," arXiv:hep-th/9904207.

[217] K. Ohmori, "A review on tachyon condensation in open string field theories," arXiv:hep-th/0102085.

[218] W. Taylor and B. Zwiebach, "D-branes, tachyons, and string field theory," arXiv:hep-th/0311017.

[219] A. Sen, "Tachyon dynamics in open string theory," Int. J. Mod. Phys. A **20** (2005) 5513 [arXiv:hep-th/0410103].

[220] L. J. Dixon, V. Kaplunovsky, and C. Vafa, "On four-dimensional gauge theories from type II superstrings," Nucl. Phys. B **294** (1987) 43.

[221] I. Antoniadis, S. Dimopoulos, and A. Giveon, "Little string theory at a TeV," JHEP **0105** (2001) 055 [arXiv:hep-th/0103033].

[222] M. J. Duff, B. E. W. Nilsson, and C. N. Pope, "Kaluza-Klein supergravity," Phys. Rep. **130** (1986) 1.

[223] T. Banks and L. J. Dixon, "Constraints on string vacua with space-time supersymmetry," Nucl. Phys. B **307** (1988) 93.

[224] T. Banks, L. J. Dixon, D. Friedan, and E. J. Martinec, "Phenomenology and conformal field theory or can string theory predict the weak mixing angle?" Nucl. Phys. B **299** (1988) 613.

[225] H. Kawai, D. C. Lewellen, and S. H. H. Tye, "Construction of fermionic string models in four-dimensions," Nucl. Phys. B **288** (1987) 1.

[226] I. Antoniadis, C. P. Bachas, and C. Kounnas, "Four-dimensional superstrings," Nucl. Phys. B **289** (1987) 87.

[227] I. Antoniadis, J. R. Ellis, J. S. Hagelin, and D. V. Nanopoulos, "The flipped $SU(5) \times U(1)$ string model revamped," Phys. Lett. B **231** (1989) 65.

[228] I. Antoniadis, G. K. Leontaris, and J. Rizos, "A three generation $SU(4) \times O(4)$ string model," Phys. Lett. B **245** (1990) 161.

[229] A. Font, L. E. Ibanez, F. Quevedo, and A. Sierra, "The construction of 'realistic' four-dimensional strings through orbifolds," Nucl. Phys. B **331** (1990) 421.

[230] F. Quevedo, "Lectures on superstring phenomenology," arXiv:hep-th/9603074.

[231] D. Bailin and A. Love, "Orbifold compactifications of string theory," Phys. Rep. **315** (1999) 285.

[232] J. T. Giedt, "Heterotic orbifolds," arXiv:hep-ph/0204315.

[233] M. Dine, "Supersymmetry phenomenology (with a broad brush)," arXiv:hep-ph/9612389.

[234] Z. Kakushadze, G. Shiu, S. H. H. Tye and Y. Vtorov-Karevsky, "A review of three-family grand unified string models," Int. J. Mod. Phys. A **13** (1998) 2551 [arXiv:hep-th/9710149].

[235] J. Scherk and J. H. Schwarz, "How to get masses from extra dimensions," Nucl. Phys. B **153** (1979) 61.

[236] R. Rohm, "Spontaneous supersymmetry breaking in supersymmetric string theories," Nucl. Phys. B **237** (1984) 553.

[237] S. Ferrara, C. Kounnas, M. Porrati, and F. Zwirner, "Superstrings with spontaneously broken supersymmetry and their effective theories," Nucl. Phys. B **318** (1989) 75.

[238] E. Kiritsis and C. Kounnas, "Perturbative and non-perturbative partial supersymmetry breaking: $N = 4 \to N = 2 \to N = 1$," Nucl. Phys. B **503** (1997) 117 [arXiv:hep-th/9703059].

[239] I. Antoniadis, E. Dudas and A. Sagnotti, "Brane supersymmetry breaking," Phys. Lett. B **464** (1999) 38 [arXiv:hep-th/9908023].

[240] P. Candelas, G. T. Horowitz, A. Strominger, and E. Witten, "Vacuum configurations for superstrings," Nucl. Phys. B **258** (1985) 46.

[241] P. Candelas, "Lectures on complex manifolds," pp. 1–88 in *Superstrings '87*, ed. L. Alvarez-Gaume *et al.*, World Scientific, Singapore, 1988.

[242] M. Vonk, "A mini-course on topological strings," arXiv:hep-th/0504147.

[243] T. Hubsch, Calabi-Yau Manifolds: A Bestiary for Physicists, World Scientific, Singapore, 1992.

[244] T. Eguchi, P. B. Gilkey, and A. J. Hanson, "Gravitation, gauge theories and differential geometry," Phys. Rep. **66** (1980) 213.

[245] P. Aspinwall, "K3 surfaces and string duality," hep-th/9611137.

[246] A. Strominger, S. T. Yau, and E. Zaslow, "Mirror symmetry is T-duality," Nucl. Phys. B **479** (1996) 243 [arXiv:hep-th/9606040].

[247] K. Hori et al., *Mirror Symmetry*, Clay Mathematics Monographs vol. 1, AMS, Providence, R. I., 2003.

[248] S. Hosono, A. Klemm, and S. Theisen, "Lectures on mirror symmetry," arXiv:hep-th/9403096.

[249] K. Hori and C. Vafa, "Mirror symmetry," arXiv:hep-th/0002222.

[250] M. R. Douglas and G. W. Moore, "D-branes, quivers, and ALE instantons," arXiv:hep-th/9603167.

[251] E. G. Gimon and J. Polchinski, "Consistency conditions for orientifolds and D-manifolds," Phys. Rev. D **54** (1996) 1667 [arXiv:hep-th/9601038].

[252] E. G. Gimon and C. V. Johnson, "K3 orientifolds," Nucl. Phys. B **477** (1996) 715 [arXiv:hep-th/9604129].

[253] G. Aldazabal, A. Font, L. E. Ibanez, and G. Violero, "$D = 4$, $N = 1$, type IIB orientifolds," Nucl. Phys. B **536** (1998) 29 [arXiv:hep-th/9804026].

[254] G. Aldazabal, L. E. Ibãnez, F. Quevedo, and A. M. Uranga, "D-branes at singularities: A bottom-up approach to the string embedding of the standard model," JHEP **0008** (2000) 002 [arXiv:hep-th/0005067].

[255] C. Bachas, "A way to break supersymmetry," arXiv:hep-th/9503030.

[256] R. Blumenhagen, L. Goerlich, B. Kors, and D. Lust, "Noncommutative compactifications of type I strings on tori with magnetic background flux," JHEP **0010** (2000) 006 [arXiv:hep-th/0007024].

[257] C. Angelantonj, I. Antoniadis, E. Dudas and A. Sagnotti, "Type-I strings on magnetised orbifolds and brane transmutation," Phys. Lett. B **489** (2000) 223 [arXiv:hep-th/0007090].

[258] R. Blumenhagen, B. Kors, D. Lust and T. Ott, "The standard model from stable intersecting brane world orbifolds," Nucl. Phys. B **616** (2001) 3 [arXiv:hep-th/0107138].

[259] L. E. Ibanez, F. Marchesano and R. Rabadan, "Getting just the standard model at intersecting branes," JHEP **0111** (2001) 002 [arXiv:hep-th/0105155].

[260] M. Cvetic, G. Shiu and A. M. Uranga, "Chiral four-dimensional $N = 1$ supersymmetric type IIA orientifolds from intersecting D6-branes," Nucl. Phys. B **615** (2001) 3 [arXiv:hep-th/0107166].

[261] E. Kiritsis, "D-branes in standard model building, gravity and cosmology," Phys. Rep. **421** (2005) 105 [arXiv:hep-th/0310001].

[262] R. Blumenhagen, M. Cvetic, P. Langacker, and G. Shiu, "Toward realistic intersecting D-brane models," arXiv:hep-th/0502005.

[263] D. Lust, "Intersecting brane worlds: A path to the standard model?" Class. Quantum Grav. **21** (2004) S1399 [arXiv:hep-th/0401156].

[264] E. Dudas, "Theory and phenomenology of type I strings and M-theory," Class. Quantum Grav. **17** (2000) R41 [arXiv:hep-ph/0006190].

[265] N. Seiberg and E. Witten, "String theory and noncommutative geometry," JHEP **9909** (1999) 032 [arXiv:hep-th/9908142].

[266] M. R. Douglas and N. A. Nekrasov, "Noncommutative field theory," Rev. Mod. Phys. **73** (2001) 977 [arXiv:hep-th/0106048].

[267] C. S. Chu, "Noncommutative open string: Neutral and charged," arXiv:hep-th/0001144.

[268] T. Banks and L. J. Dixon, "Constraints on string vacua with space-time supersymmetry," Nucl. Phys. B **307** (1988) 93.

[269] I. Antoniadis, C. Bachas, D. C. Lewellen, and T. N. Tomaras, "On supersymmetry breaking in superstrings," Phys. Lett. B **207** (1988) 441.

[270] I. Antoniadis, "A possible new dimension at a few TeV," Phys. Lett. B **246** (1990) 377.

[271] E. Kiritsis, C. Kounnas, P. M. Petropoulos, and J. Rizos, "Solving the decompactification problem in string theory," Phys. Lett. B **385** (1996) 87 [arXiv:hep-th/9606087].

[272] E. Witten, "Strong coupling expansion of Calabi-Yau compactification," Nucl. Phys. B **471** (1996) 135 [arXiv:hep-th/9602070].

[273] J. D. Lykken, "Weak scale superstrings," Phys. Rev. D **54** (1996) 3693 [arXiv:hep-th/9603133].

[274] E. Caceres, V. S. Kaplunovsky, and I. M. Mandelberg, "Large-volume string compactifications, revisited," Nucl. Phys. B **493** (1997) 73 [arXiv:hep-th/9606036].

[275] N. Arkani-Hamed, S. Dimopoulos and G. R. Dvali, "The hierarchy problem and new dimensions at a millimeter," Phys. Lett. B **429** (1998) 263 [arXiv:hep-ph/9803315].

[276] I. Antoniadis, N. Arkani-Hamed, S. Dimopoulos and G. R. Dvali, "New dimensions at a millimeter to a Fermi and superstrings at a TeV," Phys. Lett. B **436** (1998) 257 [arXiv:hep-ph/9804398].

[277] I. Antoniadis and C. Bachas, "Branes and the gauge hierarchy," Phys. Lett. B **450** (1999) 83 [arXiv:hep-th/9812093].

[278] C. P. Bachas, "Unification with low string scale," JHEP **9811** (1998) 023; [arXiv:hep-ph/9807415].

[279] L. E. Ibanez, "The second string (phenomenology) revolution," Class. Quantum Grav. **17** (2000) 1117 [arXiv:hep-ph/9911499].

[280] I. Antoniadis and K. Benakli, "Large dimensions and string physics in future colliders," Int. J. Mod. Phys. A **15** (2000) 4237 [arXiv:hep-ph/0007226].

[281] G. G. Ross, "Unified field theories," Rep. Prog. Phys. **44** (1981) 655.

[282] R. Dienes, "String theory and the path to unification: A review of recent developments," Phys. Rep. **287** (1997) 447 [arXiv:hep-th/9602045].

[283] M. Grana, "Flux compactifications in string theory: A comprehensive review," arXiv:hep-th/0509003.

[284] S. B. Giddings, S. Kachru and J. Polchinski, "Hierarchies from fluxes in string compactifications," Phys. Rev. D **66** (2002) 106006 [arXiv:hep-th/0105097].

[285] S. Kachru, R. Kallosh, A. Linde and S. P. Trivedi, "De Sitter vacua in string theory," Phys. Rev. D **68** (2003) 046005 [arXiv:hep-th/0301240].

[286] S. Kachru, R. Kallosh, A. Linde, J. Maldacena, L. McAllister, and S. P. Trivedi, "Towards inflation in string theory," JCAP **0310** (2003) 013 [arXiv:hep-th/0308055].

[287] N. Hitchin, "Generalized Calabi-Yau manifolds," Q. J. Math. Oxford Ser. **54** (2003) 281 [arXiv:math.dg/0209099].

[288] M. Gualtieri, "Generalized complex geometry," arXiv:math.dg/0401221.

[289] J. E. Lidsey, D. Wands and E. J. Copeland, "Superstring cosmology," Phys. Rep. **337** (2000) 343 [arXiv:hep-th/9909061].

[290] M. Gasperini and G. Veneziano, "The pre-big bang scenario in string cosmology," Phys. Rep. **373** (2003) 1 [arXiv:hep-th/0207130].

[291] F. Quevedo, "Lectures on string/brane cosmology," Class. Quantum Grav. **19** (2002) 5721 [arXiv:hep-th/0210292].

[292] E. Kiritsis, C. Kounnas, P. M. Petropoulos, and J. Rizos, "String threshold corrections in models with spontaneously broken supersymmetry," Nucl. Phys. B **540** (1999) 87 [arXiv:hep-th/9807067].

[293] C. Bachas, C. Fabre and T. Yanagida, "Natural gauge-coupling unification at the string scale," Phys. Lett. B **370** (1996) 49 [arXiv:hep-th/9510094].

[294] V. S. Kaplunovsky, "One loop threshold effects in string unification," Nucl. Phys. B **307** (1988) 145; Erratum *ibid*. **382** (1992) 436 [arXiv:hep-th/9205068].

[295] L. J. Dixon, V. Kaplunovsky and J. Louis, "Moduli dependence of string loop corrections to gauge coupling constants," Nucl. Phys. B **355** (1991) 649.

[296] J. P. Derendinger, S. Ferrara, C. Kounnas and F. Zwirner, "On loop corrections to string effective field theories: Field dependent gauge couplings and sigma model anomalies," Nucl. Phys. B **372** (1992) 145.

[297] I. Antoniadis, E. Gava and K. S. Narain, "Moduli corrections to gravitational couplings from string loops," Phys. Lett. B **283** (1992) 209 [arXiv:hep-th/9203071].

[298] I. Antoniadis, E. Gava and K. S. Narain, "Moduli corrections to gauge and gravitational couplings in four-dimensional superstrings," Nucl. Phys. B **383** (1992) 93 [arXiv:hep-th/9204030].

[299] I. Antoniadis, K. S. Narain and T. R. Taylor, "Higher genus string corrections to gauge couplings," Phys. Lett. B **267** (1991) 37.

[300] V. Kaplunovsky and J. Louis, "Field dependent gauge couplings in locally supersymmetric effective quantum field theories," Nucl. Phys. B **422** (1994) 57 [arXiv:hep-th/9402005].

[301] L. E. Ibanez and D. Lüst, "Duality anomaly cancellation, minimal string unification and the effective low-energy Lagrangian of 4-D strings," Nucl. Phys. B **382** (1992) 305 [arXiv:hep-th/9202046].

[302] E. Kiritsis and C. Kounnas, "Infrared regularization of superstring theory and the one loop calculation of coupling constants," Nucl. Phys. B **442** (1995) 472 [arXiv:hep-th/9501020].

[303] J. A. Harvey and G. W. Moore, "Algebras, BPS states, and strings," Nucl. Phys. B **463** (1996) 315 [arXiv:hep-th/9510182].

[304] E. Kiritsis, C. Kounnas, P. M. Petropoulos and J. Rizos, "Universality properties of $N = 2$ and $N = 1$ heterotic threshold corrections," Nucl. Phys. B **483** (1997) 141 [arXiv:hep-th/9608034].

[305] J. J. Atick and A. Sen, "Two loop dilaton tadpole induced by Fayet-Iliopoulos D terms in compactified heterotic string theories," Nucl. Phys. B **296** (1988) 157.

[306] I. Antoniadis, E. Gava, K. S. Narain, and T. R. Taylor, "Topological amplitudes in string theory," Nucl. Phys. B **413** (1994) 162 [arXiv:hep-th/9307158].

[307] M. Bershadsky, S. Cecotti, H. Ooguri and C. Vafa, "Kodaira-Spencer theory of gravity and exact results for quantum string amplitudes," Commun. Math. Phys. **165** (1994) 311 [arXiv:hep-th/9309140].

[308] W. Lerche, B. E. W. Nilsson, A. N. Schellekens and N. P. Warner, "Anomaly cancelling terms from the elliptic genus," Nucl. Phys. B **299** (1988) 91.

[309] M. Abe, H. Kubota and N. Sakai, "Loop corrections to the $E_8 \times E_8$ heterotic string effective lagrangian," Nucl. Phys. B **306** (1988) 405.

[310] E. Kiritsis, "Duality and instantons in string theory," arXiv:hep-th/9906018.

[311] J. J. Atick, G. W. Moore and A. Sen, "Catoptric tadpoles," Nucl. Phys. B **307** (1988) 221.

[312] E. D'Hoker and D. H. Phong, "Two-loop superstrings. VI: Non-renormalization theorems and the 4-point function," Nucl. Phys. B **715** (2005) 3 [arXiv:hep-th/0501197].

[313] I. Antoniadis, C. Bachas, and E. Dudas, "Gauge couplings in four-dimensional type I string orbifolds," Nucl. Phys. B **560** (1999) 93 [arXiv:hep-th/9906039].

[314] E. Poppitz, "On the one loop Fayet-Iliopoulos term in chiral four dimensional type I orbifolds," Nucl. Phys. B **542** (1999) 31 [arXiv:hep-th/9810010].

[315] P. Bain and M. Berg, "Effective action of matter fields in four-dimensional string orientifolds," JHEP **0004** (2000) 013 [arXiv:hep-th/0003185].

[316] I. Antoniadis, E. Kiritsis and J. Rizos, "Anomalous U(1)s in type I superstring vacua," Nucl. Phys. B **637** (2002) 92 [arXiv:hep-th/0204153].

[317] S. Forste and J. Louis, "Duality in string theory," Nucl. Phys. Proc. Suppl. **61A** (1998) 3 [arXiv:hep-th/9612192].

[318] C. Vafa, "Lectures on strings and dualities," hep-th/9702201.

[319] E. Kiritsis, "Introduction to non-perturbative string theory," arXiv:hep-th/9708130.

[320] B. de Wit and J. Louis, "Supersymmetry and dualities in various dimensions," hep-th/9801132.

[321] J. H. Schwarz, "Lectures on superstring and M theory dualities," Nucl. Phys. Proc. Suppl. **55B** (1997) 1 [arXiv:hep-th/9607201].

[322] A. Sen, "An introduction to non-perturbative string theory," hep-th/9802051.

[323] C. M. Hull and P. K. Townsend, "Unity of superstring dualities," Nucl. Phys. B **438** (1995) 109 [arXiv:hep-th/9410167].

[324] E. Witten, "String theory dynamics in various dimensions," Nucl. Phys. B **443** (1995) 85 [arXiv:hep-th/9503124]; "Some comments on string dynamics," arXiv:hep-th/9507121.

[325] P. Horava and E. Witten, "Heterotic and type I string dynamics from eleven dimensions," Nucl. Phys. B **460** (1996) 506 [arXiv:hep-th/9510209].

[326] P. Horava and E. Witten, "Eleven-dimensional supergravity on a manifold with boundary," Nucl. Phys. B **475** (1996) 94 [arXiv:hep-th/9603142].

[327] P. K. Townsend, "Four lectures on M-theory," arXiv:hep-th/9612121.

[328] N. A. Obers and B. Pioline, "U-duality and M-theory," Phys. Rep. **318** (1999) 113 [arXiv:hep-th/9809039].

[329] B. A. Ovrut, "Lectures on heterotic M-theory," arXiv:hep-th/0201032.

[330] A. Sagnotti, "Surprises in open-string perturbation theory," Nucl. Phys. Proc. Suppl. **56B** (1997) 332 [arXiv:hep-th/9702093].

[331] B. S. Acharya and S. Gukov, "M theory and singularities of exceptional holonomy manifolds," Phys. Rep. **392** (2004) 121 [arXiv:hep-th/0409191].

[332] I. Antoniadis and B. Pioline, "Low-scale closed strings and their duals," Nucl. Phys. B **550** (1999) 41 [arXiv:hep-th/9902055].

[333] M. Dine, "TASI lectures on M theory phenomenology," arXiv:hep-th/0003175.

[334] J. Polchinski and E. Witten, "Evidence for heterotic–type I string duality," Nucl. Phys. B **460** (1996) 525 [arXiv:hep-th/9510169].

[335] E. Witten, "Small instantons in string theory," Nucl. Phys. B **460** (1996) 541 [arXiv:hep-th/9511030].

[336] E. Bergshoeff, S. Cucu, T. De Wit, J. Gheerardyn, R. Halbersma, S. Vandoren, and A. Van Proeyen, "Superconformal $N = 2$, $D = 5$ matter with and without actions," JHEP **0210** (2002) 045 [arXiv:hep-th/0205230].

[337] M. J. Duff, R. Minasian and E. Witten, "Evidence for heterotic/heterotic duality," Nucl. Phys. B **465** (1996) 413 [arXiv:hep-th/9601036].

[338] K. Becker, M. Becker and A. Strominger, "Five-branes, membranes and nonperturbative string theory," Nucl. Phys. B **456** (1995) 130 [arXiv:hep-th/9507158].

[339] E. Witten, "Non-perturbative superpotentials in string theory," Nucl. Phys. B **474** (1996) 343 [arXiv:hep-th/9604030].

[340] H. Ooguri and C. Vafa, "Summing up D-instantons," Phys. Rev. Lett. **77** (1996) 3296 [arXiv:hep-th/9608079].

[341] J. A. Harvey and G. W. Moore, "Fivebrane instantons and R**2 couplings in $N = 4$ string theory," Phys. Rev. D **57** (1998) 2323 [arXiv:hep-th/9610237].

[342] M. B. Green and M. Gutperle, "Effects of D-instantons," Nucl. Phys. B **498** (1997) 195 [arXiv:hep-th/9701093].

[343] C. Bachas, "Heterotic versus type I," Nucl. Phys. Proc. Suppl. **68** (1998) 348 [arXiv:hep-th/9710102].

[344] E. Kiritsis, N. A. Obers and B. Pioline, "Heterotic/type II triality and instantons on K3," JHEP **0001** (2000) 029 [arXiv:hep-th/0001083].

[345] W. Lerche, "Introduction to Seiberg-Witten theory and its stringy origin," Nucl. Phys. Proc. Suppl. **55B** (1997) 83 [Fortschr. Phys. **45** (1997) 293] [arXiv:hep-th/9611190].

[346] A. Klemm, "On the geometry behind $N = 2$ supersymmetric effective actions in four dimensions," arXiv:hep-th/9705131.

[347] P. Candelas and X. C. de la Ossa, "Comments on conifolds," Nucl. Phys. B **342** (1990) 246.

[348] A. Strominger, "Massless black holes and conifolds in string theory," Nucl. Phys. B **451** (1995) 96 [arXiv:hep-th/9504090].

[349] B. R. Greene, D. R. Morrison, and A. Strominger, "Black hole condensation and the unification of string vacua," Nucl. Phys. B **451** (1995) 109 [arXiv:hep-th/9504145].

[350] C. Vafa, "Evidence for F-theory," Nucl. Phys. B **469** (1996) 403 [arXiv:hep-th/9602022].

[351] D. R. Morrison and C. Vafa, "Compactifications of F-theory on Calabi–Yau threefolds — II," Nucl. Phys. B **476** (1996) 437 [arXiv:hep-th/9603161].

[352] M. Bershadsky, K. A. Intriligator, S. Kachru, D. R. Morrison, V. Sadov, and C. Vafa, "Geometric singularities and enhanced gauge symmetries," Nucl. Phys. B **481** (1996) 215 [arXiv:hep-th/9605200].

[353] M. Bershadsky, A. Johansen, T. Pantev, V. Sadov, and C. Vafa, "F-theory, geometric engineering and N = 1 dualities," Nucl. Phys. B **505** (1997) 153 [arXiv:hep-th/9612052].

[354] C. Montonen and D. Olive, "Magnetic monopoles as gauge particles?" Phys. Lett. B **72** (1977) 117.

[355] A. Sen, "Strong-weak coupling duality in four-dimensional string theory," Int. J. Mod. Phys. A **9** (1994) 3707 [arXiv:hep-th/9402002].

[356] N. Seiberg, "Exact results on the space of vacua of four-dimensional SUSY gauge theories," Phys. Rev. D **49** (1994) 6857 [arXiv:hep-th/9402044].

[357] N. Seiberg and E. Witten, "Electric-magnetic duality, monopole condensation, and confinement in $N = 2$ supersymmetric Yang-Mills theory," Nucl. Phys. B **426** (1994) 19; Erratum, *ibid*. **430** (1994) 485 [arXiv:hep-th/9407087]; "Monopoles, duality and chiral symmetry breaking in $N = 2$ supersymmetric QCD," *ibid*. **431** (1994) 484 [arXiv:hep-th/9408099].

[358] K. Intriligator and N. Seiberg, "Lectures on supersymmetric gauge theories and electric-magnetic duality," Nucl. Phys. Proc. Suppl. **45BC** (1996) 1 [hep-th/9509066].

[359] L. Alvarez-Gaume and S. F. Hassan, "Introduction to S-duality in N = 2 supersymmetric gauge theories: A pedagogical review of the work of Seiberg and Witten," Fortschr. Phys. **45** (1997) 159 [arXiv:hep-th/9701069].

[360] M. Peskin, "Duality in supersymmetric Yang-Mills theory," arXiv:hep-th/9702094.

[361] M. A. Shifman, "Nonperturbative dynamics in supersymmetric gauge theories," Prog. Part. Nucl. Phys. **39** (1997) 1 [arXiv:hep-th/9704114].

[362] P. Di Vecchia, "Duality in $N = 2, 4$ supersymmetric gauge theories," hep-th/9803026.

[363] A. Bilal, "Duality in $N = 2$ SUSY SU(2) Yang-Mills theory: A pedagogical introduction to the work of Seiberg and Witten," arXiv:hep-th/9601007.

[364] S. Katz, A. Klemm, and C. Vafa, "Geometric engineering of quantum field theories," Nucl. Phys. B **497** (1997) 173 [arXiv:hep-th/9609239].

[365] A. Hanany and E. Witten, "Type IIB superstrings, BPS monopoles, and three-dimensional gauge dynamics," Nucl. Phys. B **492** (1997) 152 [arXiv:hep-th/9611230].

[366] L. Baulieu and I. M. Singer, "Topological Yang-Mills symmetry," Nucl. Phys. Proc. Suppl. **5B** (1988) 12.

[367] D. Birmingham, M. Blau, M. Rakowski, and G. Thompson, "Topological field theory," Phys. Rep. **209** (1991) 129.

[368] M. Marino, "Chern-Simons theory and topological strings," Rev. Mod. Phys. **77** (2005) 675 [arXiv:hep-th/0406005].

[369] A. Sen, "Unification of string dualities," Nucl. Phys. Proc. Suppl. **58** (1997) 5 [arXiv:hep-th/9609176].

[370] C. Vafa and E. Witten, "A one loop test of string duality," Nucl. Phys. B **447** (1995) 261 [arXiv:hep-th/9505053].

[371] B. F. Schutz, A First Course in General Relativity, Cambridge University Press, Cambridge, 1985.

[372] G. 't Hooft, Introduction to General Relativity, Rinton Press, Princeton, N.J.; also http://www.phys.uu.nl/~thooft/lectures/genrel.pdf

[373] S. Carroll, Spacetime and Geometry: An Introduction to General Relativity, Addison-Wesley, Reading, Mass., 2004. An earlier version can be found at arXiv:gr-qc/9712019.

[374] R. M. Wald, General Relativity, Chicago University Press, Chicago, 1984.

[375] P. K. Townsend, "Black holes," arXiv:gr-qc/9707012.

[376] T. Jacobson, "Introductory lectures on black-hole thermodynamics," http://www.glue.umd.edu/~tajac/BHTlectures/lectures.ps

[377] S. W. Hawking, "Particle creation by black holes," Commun. Math. Phys. **43** (1975) 199.

[378] G. W. Gibbons and S. W. Hawking, "Action integrals and partition functions in quantum gravity," Phys. Rev. D **15** (1977) 2752.

[379] L. Susskind, J. Lindesay An Introduction to Black Holes, Information and the String Theory Revolution: The Holographic Universe, World Scientific, Singapore, 2005.

[380] D. N. Page, "Black hole information," arXiv:hep-th/9305040.

[381] S. B. Giddings, "The black hole information paradox," arXiv:hep-th/9508151.

[382] M. Natsuume, "The singularity problem in string theory," arXiv:gr-qc/0108059.

[383] S. R. Das and S. D. Mathur, "The quantum physics of black holes: Results from string theory," Annu. Rev. Nucl. Part. Sci. **50** (2000) 153 [arXiv:gr-qc/0105063].

[384] J. R. David, G. Mandal and S. R. Wadia, "Microscopic formulation of black holes in string theory," Phys. Rep. **369** (2002) 549; [arXiv:hep-th/0203048].

[385] J. M. Maldacena, "Black holes in string theory," arXiv:hep-th/9607235.

[386] K. Sfetsos and K. Skenderis, "Microscopic derivation of the Bekenstein-Hawking entropy formula for non-extremal black holes," Nucl. Phys. B **517** (1998) 179 [arXiv:hep-th/9711138].

[387] J. M. Maldacena and A. Strominger, "AdS(3) black holes and a stringy exclusion principle," JHEP **9812** (1998) 005 [arXiv:hep-th/9804085].

[388] M. Banados, M. Henneaux, C. Teitelboim and J. Zanelli, "Geometry of the (2+1) black hole," Phys. Rev. D **48** (1993) 1506 [arXiv:gr-qc/9302012].

[389] O. Coussaert and M. Henneaux, "Supersymmetry of the (2+1) black holes," Phys. Rev. Lett. **72** (1994) 183 [arXiv:hep-th/9310194].

[390] A. Strominger and C. Vafa, "Microscopic origin of the Bekenstein-Hawking entropy," Phys. Lett. B **379** (1996) 99 [arXiv:hep-th/9601029].

[391] C. G. Callan and J. M. Maldacena, "D-brane approach to black hole quantum mechanics," Nucl. Phys. B **472** (1996) 591 [arXiv:hep-th/9602043].

[392] M. Cvetic and A. A. Tseytlin, "Non-extreme black holes from non-extreme intersecting M-branes," Nucl. Phys. B **478** (1996) 181 [arXiv:hep-th/9606033].

[393] G. T. Horowitz, J. M. Maldacena and A. Strominger, "Nonextremal black hole microstates and U-duality," Phys. Lett. B **383** (1996) 151 [arXiv:hep-th/9603109].

[394] J. M. Maldacena and A. Strominger, "Statistical entropy of four-dimensional extremal black holes," Phys. Rev. Lett. **77** (1996) 428 [arXiv:hep-th/9603060].

[395] C. V. Johnson, R. R. Khuri and R. C. Myers, "Entropy of 4D extremal black holes," Phys. Lett. B **378** (1996) 78 [arXiv:hep-th/9603061].

[396] J. M. Maldacena, A. Strominger and E. Witten, "Black hole entropy in M-theory," JHEP **9712** (1997) 002 [arXiv:hep-th/9711053].

[397] A. Dhar, G. Mandal, and S. R. Wadia, "Absorption vs decay of black holes in string theory and T-symmetry," Phys. Lett. B **388** (1996) 51 [arXiv:hep-th/9605234].

[398] S. R. Das and S. D. Mathur, "Comparing decay rates for black holes and D-branes," Nucl. Phys. B **478** (1996) 561 [arXiv:hep-th/9606185].

[399] J. M. Maldacena and A. Strominger, "Black hole greybody factors and D-brane spectroscopy," Phys. Rev. D **55** (1997) 861 [arXiv:hep-th/9609026].

[400] S. S. Gubser, "Dynamics of D-brane black holes," arXiv:hep-th/9908004.

[401] C. Vafa, "Instantons on D-branes," Nucl. Phys. B **463** (1996) 435 [arXiv:hep-th/9512078].

[402] M. R. Douglas, "Branes within branes," arXiv:hep-th/9512077.

[403] E. Witten, "On the conformal field theory of the Higgs branch," JHEP **9707** (1997) 003 [arXiv:hep-th/9707093].

[404] S. F. Hassan and S. R. Wadia, "Gauge theory description of D-brane black holes: Emergence of the effective SCFT and Hawking radiation," Nucl. Phys. B **526** (1998) 311 [arXiv:hep-th/9712213].

[405] N. Seiberg and E. Witten, "The D1/D5 system and singular CFT," JHEP **9904** (1999) 017 [arXiv:hep-th/9903224].

[406] R. Dijkgraaf, "Instanton strings and hyperKaehler geometry," Nucl. Phys. B **543** (1999) 545 [arXiv:hep-th/9810210].

[407] G. T. Horowitz and J. Polchinski, "A correspondence principle for black holes and strings," Phys. Rev. D **55** (1997) 6189 [arXiv:hep-th/9612146].

[408] A. Dabholkar, F. Denef, G. W. Moore and B. Pioline, "Precision counting of small black holes," arXiv:hep-th/0507014.

[409] A. Sen, "Entropy function for heterotic black holes," arXiv:hep-th/0508042.

[410] S. D. Mathur, "The fuzzball proposal for black holes: An elementary review," Fortschr. Phys. **53** (2005) 793 [arXiv:hep-th/0502050].

[411] H. Lin, O. Lunin, and J. Maldacena, "Bubbling AdS space and 1/2 BPS geometries," JHEP **0410** (2004) 025 [arXiv:hep-th/0409174].

[412] J. L. F. Barbon and E. Rabinovici, "Touring the Hagedorn ridge," arXiv:hep-th/0407236.

[413] C. Vafa and E. Witten, "A strong coupling test of S duality," Nucl. Phys. B **431** (1994) 3 [arXiv:hep-th/9408074].

[414] I. R. Klebanov and E. Witten, "AdS/CFT correspondence and symmetry breaking," Nucl. Phys. B **556** (1999) 89 [arXiv:hep-th/9905104].

[415] N. Drukker, D. J. Gross, and H. Ooguri, "Wilson loops and minimal surfaces," Phys. Rev. D **60** (1999) 125006 [arXiv:hep-th/9904191].

[416] E. Witten, "Anti–de Sitter space and holography," Adv. Theor. Math. Phys. **2** (1998) 253 [arXiv:hep-th/9802150].

[417] E. Witten, "Multi-trace operators, boundary conditions, and AdS/CFT correspondence," arXiv:hep-th/0112258.

[418] O. Aharony, M. Berkooz and E. Silverstein, "Multiple-trace operators and non-local string theories," JHEP **0108** (2001) 006 [arXiv:hep-th/0105309].

[419] O. Aharony, M. Berkooz and B. Katz, "Non-local effects of multi-trace deformations in the AdS/CFT correspondence," arXiv:hep-th/0504177.

[420] I. R. Klebanov and M. J. Strassler, "Supergravity and a confining gauge theory: Duality cascades and χ-SB resolution of naked singularities," JHEP **0008** (2000) 052 [arXiv:hep-th/0007191].

[421] L. Baulieu, "Perturbative gauge theories," Phys. Rep. **129** (1985) 1.

[422] G. 't Hooft, "A planar diagram theory for strong interactions," Nucl. Phys. B **72** (1974) 461; "A two-dimensional model for mesons," *ibid.* **75** (1974) 461.

[423] S. R. Coleman, "1/N," SLAC-PUB-2484, presented at 1979 International School of Subnuclear Physics: Pointlike Structures Inside and Outside Hadrons, Erice, Italy, Jul 31–Aug 10, 1979.

[424] A. A. Migdal, "Multicolor QCD as dual resonance theory," Ann. Phys. (N.Y.) **109** (1977) 365.

[425] S. R. Das, "Some aspects of large-N theories," Rev. Mod. Phys. **59** (1987) 235.

[426] A. V. Manohar, "Large N QCD," arXiv:hep-ph/9802419.

[427] E. Witten, "Baryons in the 1/N expansion," Nucl. Phys. B **160** (1979) 57.

[428] A. M. Polyakov, Gauge Fields and Strings, Harwood Academic Publishers, Chur, Switzerland, 1987.

[429] E. S. Fradkin and M. Y. Palchik, "New developments in D-dimensional conformal quantum field theory," Phys. Rep. **300** (1998) 1.

[430] V. K. Dobrev and V. B. Petkova, "All positive energy unitary irreducible representations of extended conformal supersymmetry," Phys. Lett. B **162** (1985) 127.

[431] S. Ferrara and A. Zaffaroni, "Superconformal field theories, multiplet shortening, and the AdS$_5$/SCFT$_4$ correspondence," arXiv:hep-th/9908163.

[432] G. 't Hooft, "Dimensional reduction in quantum gravity," arXiv:gr-qc/9310026.

[433] L. Susskind, "The world as a hologram," J. Math. Phys. **36** (1995) 6377 [arXiv:hep-th/9409089].

[434] R. Bousso, "The holographic principle," Rev. Mod. Phys. **74** (2002) 825 [arXiv:hep-th/0203101].

[435] J. M. Maldacena, "The large N limit of superconformal field theories and supergravity," Adv. Theor. Math. Phys. **2** (1998) 231 [Int. J. Theor. Phys. 38 (1999) 1113] [arXiv:hep-th/9711200].

[436] S. S. Gubser, I. R. Klebanov, and A. M. Polyakov, "Gauge theory correlators from non-critical string theory," Phys. Lett. B **428** (1998) 105 [arXiv:hep-th/9802109].

[437] E. Witten, "Anti–de Sitter space, thermal phase transition, and confinement in gauge theories," Adv. Theor. Math. Phys. **2** (1998) 505 [arXiv:hep-th/9803131].

[438] O. Aharony, S. S. Gubser, J. M. Maldacena, H. Ooguri, and Y. Oz, "Large N field theories, string theory and gravity," Phys. Rep. **323** (2000) 183 [arXiv:hep-th/9905111].

[439] I. R. Klebanov, "From three-branes to large N gauge theories," arXiv:hep-th/9901018.

[440] M. R. Douglas and S. Randjbar-Daemi, "Two lectures on AdS/CFT correspondence," arXiv:hep-th/9902022.

[441] J. L. Petersen, "Introduction to the Maldacena conjecture on AdS/CFT," Int. J. Mod. Phys. A **14** (1999) 3597 [arXiv:hep-th/9902131].

[442] P. Di Vecchia, "An introduction to AdS/CFT correspondence," Fortschr. Phys. **48** (2000) 87 [arXiv:hep-th/9903007].

[443] I. R. Klebanov, "TASI lectures: Introduction to the AdS/CFT correspondence," arXiv:hep-th/0009139.

[444] J. M. Maldacena, "TASI 2003 lectures on AdS/CFT," arXiv:hep-th/0309246.

[445] E. Alvarez, J. Conde and L. Hernandez, "Rudiments of holography," Int. J. Mod. Phys. D **12** (2003) 543 [arXiv:hep-th/0205075].

[446] A. Salam and E. Sezgin, "Supergravities in Diverse Dimensions, vols. 1 and 2, World Scientific, Singapore, 1989.

[447] J. Sonnenschein, "What does the string / gauge correspondence teach us about Wilson loops?" arXiv:hep-th/0003032.

[448] J. Terning, "Glueballs and AdS/CFT," arXiv:hep-ph/0204012.

[449] R. Apreda, D. E. Crooks, N. J. Evans and M. Petrini, "Confinement, glueballs and strings from deformed AdS," JHEP **0405** (2004) 065 [arXiv:hep-th/0308006].

[450] J. D. Brown and M. Henneaux, "Central charges in the canonical realization of asymptotic symmetries: An example from three-dimensional gravity," Commun. Math. Phys. **104** (1986) 207.

[451] K. Skenderis, "Lecture notes on holographic renormalization," Class. Quantum Grav. **19** (2002) 5849 [arXiv:hep-th/0209067].

[452] S. de Haro, S. N. Solodukhin and K. Skenderis, "Holographic reconstruction of spacetime and renormalization in the AdS/CFT correspondence," Commun. Math. Phys. **217** (2001) 595 [arXiv:hep-th/0002230].

[453] M. Bianchi, D. Z. Freedman and K. Skenderis, How to go with an RG flow," JHEP **0108** (2001) 041 [arXiv:hep-th/0105276]; "Holographic renormalization," Nucl. Phys. B **631** (2002) 159 [arXiv:hep-th/0112119].

[454] I. Papadimitriou and K. Skenderis, "Correlation functions in holographic RG flows," JHEP **0410** (2004) 075 [arXiv:hep-th/0407071].

[455] M. Cvetic and H. H. Soleng, "Supergravity domain walls," Phys. Rep. **282** (1997) 159 [arXiv:hep-th/9604090].

[456] M. Kruczenski, D. Mateos, R. C. Myers and D. J. Winters, "Meson spectroscopy in AdS/CFT with flavour," JHEP **0307** (2003) 049 [arXiv:hep-th/0304032].

[457] A. Paredes, "Supersymmetric solutions of supergravity from wrapped branes," arXiv:hep-th/0407013.

[458] T. Sakai and S. Sugimoto, "Low energy hadron physics in holographic QCD," Prog. Theor. Phys. **113** (2005) 843 [arXiv:hep-th/0412141].

[459] O. Aharony, "The non-AdS/non-CFT correspondence, or three different paths to QCD," arXiv:hep-th/0212193.

[460] F. Bigazzi, A. L. Cotrone, M. Petrini and A. Zaffaroni, "Supergravity duals of supersymmetric four dimensional gauge theories," Riv. Nuovo Cimento **25N12** (2002) 1 [arXiv:hep-th/0303191].

[461] J. M. Maldacena and C. Nunez, "Supergravity description of field theories on curved manifolds and a no go theorem," Int. J. Mod. Phys. A **16** (2001) 822 [arXiv:hep-th/0007018].

[462] J. M. Maldacena and C. Nunez, "Towards the large N limit of pure N = 1 super Yang Mills," Phys. Rev. Lett. **86** (2001) 588 [arXiv:hep-th/0008001].

[463] M. J. Strassler, "The duality cascade," arXiv:hep-th/0505153.

[464] M. J. Strassler, "An unorthodox introduction to supersymmetric gauge theory," arXiv:hep-th/0309149.

[465] S. Kuperstein and J. Sonnenschein, "Non-critical supergravity ($d > 1$) and holography," JHEP **0407** (2004) 049 [arXiv:hep-th/0403254].

[466] O. Aharony, "A brief review of 'little string theories,'" Class. Quantum Grav. **17** (2000) 929 [arXiv:hep-th/9911147].

[467] O. Aharony, M. Berkooz, D. Kutasov and N. Seiberg, "Linear dilatons, NS5-branes and holography," JHEP **9810** (1998) 004 [arXiv:hep-th/9808149].

[468] M. Blau, J. Figueroa-O'Farrill, C. Hull, and G. Papadopoulos, "Penrose limits and maximal supersymmetry," Class. Quantum Grav. **19** (2002) L87 [arXiv:hep-th/0201081].

[469] D. Sadri and M. M. Sheikh-Jabbari, "The plane-wave/super Yang-Mills duality," Rev. Mod. Phys. **76** (2004) 853 [arXiv:hep-th/0310119].

[470] M. Spradlin and A. Volovich, "Light-cone string field theory in a plane wave," arXiv:hep-th/0310033.

[471] A. A. Tseytlin, "Semiclassical strings and AdS/CFT," arXiv:hep-th/0409296.

[472] J. Plefka, "Spinning strings and integrable spin chains in the AdS/CFT correspondence," arXiv:hep-th/0507136.

[473] N. Beisert, "The dilatation operator of N = 4 super Yang-Mills theory and integrability," Phys. Rep. **405** (2005) 1 [arXiv:hep-th/0407277].

[474] J. M. Maldacena, "Eternal black holes in anti–de-Sitter," JHEP **0304** (2003) 021 [arXiv:hep-th/0106112].

[475] J. Maldacena and L. Maoz, "Wormholes in AdS," JHEP **0402** (2004) 053 [arXiv:hep-th/0401024].

[476] J. de Boer, L. Maoz and A. Naqvi, "Some aspects of the AdS/CFT correspondence," arXiv:hep-th/0407212.

[477] L. Randall and R. Sundrum, "A large mass hierarchy from a small extra dimension," Phys. Rev. Lett. **83** (1999) 3370 [arXiv:hep-ph/9905221]; "An alternative to compactification," *ibid*. **83** (1999) 4690 [arXiv:hep-th/9906064].

[478] R. Maartens, "Brane-world gravity," Living Rev. Relativ. **7** (2004) 7 [arXiv:gr-qc/0312059].

[479] V. A. Rubakov, "Large and infinite extra dimensions: An introduction," Phys. Usp. **44** (2001) 871 [Usp. Fiz. Nauk **171** (2001) 913] [arXiv:hep-ph/0104152].

[480] E. Kiritsis, "Holography and brane-bulk energy exchange," arXiv:hep-th/0504219.

[481] M. Claudson and M. B. Halpern, "Supersymmetric ground state wave functions," Nucl. Phys. B **250** (1985) 689.

[482] B. de Wit, J. Hoppe, and H. Nicolai, "On the quantum mechanics of supermembranes," Nucl. Phys. B **305** (1988) 545.

[483] T. Banks, W. Fischler, S. H. Shenker, and L. Susskind, "M theory as a matrix model: A conjecture," Phys. Rev. D **55** (1997) 5112 [arXiv:hep-th/9610043].

[484] L. Susskind, "Another conjecture about M(atrix) theory," arXiv:hep-th/9704080.

[485] K. Becker and M. Becker, "A two-loop test of M(atrix) theory," Nucl. Phys. B **506** (1997) 48 [arXiv:hep-th/9705091].

[486] A. Bilal, "M(atrix) theory: A pedagogical introduction," Fortschr. Phys. **47** (1999) 5 [arXiv:hep-th/9710136].

[487] T. Banks, "Matrix theory," Nucl. Phys. Proc. Suppl. **67** (1998) 180 [arXiv:hep-th/9710231].

[488] D. Bigatti and L. Susskind, "Review of matrix theory," arXiv:hep-th/9712072.

[489] T. Banks, "TASI lectures on matrix theory," arXiv:hep-th/9911068.

[490] W. Taylor, "M(atrix) theory: Matrix quantum mechanics as a fundamental theory," Rev. Mod. Phys. **73** (2001) 419 [arXiv:hep-th/0101126].

[491] N. Ishibashi, H. Kawai, Y. Kitazawa and A. Tsuchiya, "A large-N reduced model as superstring," Nucl. Phys. B **498** (1997) 467 [arXiv:hep-th/9612115].

[492] L. Motl, "Proposals on nonperturbative superstring interactions," arXiv:hep-th/9701025.

[493] R. Dijkgraaf, E. P. Verlinde and H. L. Verlinde, "Matrix string theory," Nucl. Phys. B **500** (1997) 43 [arXiv:hep-th/9703030].

[494] R. Dijkgraaf, E. P. Verlinde and H. L. Verlinde, "Notes on matrix and micro strings," Nucl. Phys. Proc. Suppl. **62** (1998) 348 [arXiv:hep-th/9709107].

[495] C. Crnkovic, P. H. Ginsparg and G. W. Moore, "The Ising model, the Yang-Lee edge singularity, and 2-D quantum gravity," Phys. Lett. B **237** (1990) 196.

[496] V. A. Kazakov, A. A. Migdal and I. K. Kostov, "Critical properties of randomly triangulated planar random surfaces," Phys. Lett. B **157** (1985) 295.

[497] V. A. Kazakov and A. A. Migdal, "Recent progress in the theory of noncritical strings," Nucl. Phys. B **311** (1988) 171.

[498] D. J. Gross and A. A. Migdal, "Nonperturbative two-dimensional quantum gravity," Phys. Rev. Lett. **64** (1990) 127.

[499] E. Brezin and V. A. Kazakov, "Exactly solvable field theories of closed strings," Phys. Lett. B **236** (1990) 144.

[500] M. R. Douglas and S. H. Shenker, "Strings in less than one-dimension," Nucl. Phys. B **335** (1990) 635.

[501] M. R. Douglas, "Strings in less than one-dimension and the generalized K-D-V hierarchies," Phys. Lett. B **238** (1990) 176.

[502] D. J. Gross and N. Miljkovic, "A nonperturbative solution of $D = 1$ string theory," Phys. Lett. B **238** (1990) 217.

[503] I. R. Klebanov, "String theory in two-dimensions," arXiv:hep-th/9108019

[504] J. McGreevy and H. L. Verlinde, "Strings from tachyons: The $c = 1$ matrix reloaded," JHEP **0312** (2003) 054 [arXiv:hep-th/0304224].

[505] T. Takayanagi and N. Toumbas, "A matrix model dual of type 0B string theory in two dimensions," JHEP **0307** (2003) 064 [arXiv:hep-th/0307083].

[506] M. R. Douglas, I. R. Klebanov, D. Kutasov, J. Maldacena, E. Martinec, and N. Seiberg, "A new hat for the $c = 1$ matrix model," arXiv:hep-th/0307195.

[507] P. H. Ginsparg, "Comment on toroidal compactification of heterotic superstrings," Phys. Rev. D **35** (1987) 648.

[508] S. R. Coleman, J. Preskill and F. Wilczek, "Quantum hair on black holes," Nucl. Phys. B **378** (1992) 175 [arXiv:hep-th/9201059].

[509] E. Cremmer, B. Julia and J. Scherk, "Supergravity theory in 11 dimensions," Phys. Lett. B **76** (1978) 409.

[510] I. C. G. Campbell and P. C. West, "$N = 2$ $D = 10$ nonchiral supergravity and its spontaneous compactification," Nucl. Phys. B **243** (1984) 112.

[511] L. J. Romans, "Massive $N = 2a$ supergravity in ten-dimensions," Phys. Lett. B **169** (1986) 374.

[512] J. H. Schwarz, "Covariant field equations of chiral $N = 2$ $D = 10$ supergravity," Nucl. Phys. B **226** (1983) 269.

[513] P. S. Howe and P. C. West, "The complete $N = 2$, $D = 10$ supergravity," Nucl. Phys. B **238** (1984) 181.

[514] E. Bergshoeff, R. Kallosh, T. Ortin, D. Roest and A. Van Proeyen, "New formulations of $D = 10$ supersymmetry and $D8 - -O8$ domain walls," Class. Quantum Grav. **18** (2001) 3359 [arXiv:hep-th/0103233].

[515] E. Bergshoeff, M. de Roo, B. de Wit, and P. van Nieuwenhuizen, "Ten-dimensional Maxwell-Einstein supergravity, its currents, and the issue of its auxiliary fields," Nucl. Phys. B **195** (1982) 97.

[516] G. F. Chapline and N. S. Manton, "Unification of Yang-Mills theory and supergravity in ten dimensions," Phys. Lett. B **120** (1983) 105.

[517] M. Dine, R. Rohm, N. Seiberg, and E. Witten, "Gluino condensation in superstring models," Phys. Lett. B **156** (1985) 55.

[518] L. Andrianopoli, S. Ferrara, and M. A. Lledo, "Axion gauge symmetries and generalized Chern-Simons terms in $N = 1$ supersymmetric theories," JHEP **0404** (2004) 005 [arXiv:hep-th/0402142].

[519] J. Bagger and J. Wess, Supersymmetry and Supergravity, Princeton Series in Physics, 2nd edition, 1992.

[520] T. Kugo and S. Uehara, "Improved superconformal gauge conditions in the $N = 1$ supergravity Yang-Mills Matter System," Nucl. Phys. B **222** (1983) 125.

[521] P. Binetruy, G. Girardi, and R. Grimm, "Supergravity couplings: A geometric formulation," Phys. Rep. **343** (2001) 255 [arXiv:hep-th/0005225].

[522] E. Cremmer, C. Kounnas, A. Van Proeyen, J. P. Derendinger, S. Ferrara, B. de Wit, and L. Girardello, "Vector multiplets coupled To $N = 2$ supergravity: Superhiggs effect, flat potentials and geometric structure," Nucl. Phys. B **250** (1985) 385.

[523] B. de Wit, P. G. Lauwers, and A. Van Proeyen, "Lagrangians of $N = 2$ supergravity-matter systems," Nucl. Phys. B **255** (1985) 569.

[524] S. Cecotti, S. Ferrara and L. Girardello, "Geometry of type II superstrings and the moduli of superconformal field theories," Int. J. Mod. Phys. A **4** (1989) 2475.

[525] L. Andrianopoli, M. Bertolini, A. Ceresole, R. D'Auria, S. Ferrara, P. Fre, and T. Magri, "$N = 2$ supergravity and $N = 2$ super Yang-Mills theory on general scalar manifolds: Symplectic covariance, gaugings and the momentum map," J. Geom. Phys. **23** (1997) 111 [arXiv:hep-th/9605032].

[526] J. Louis and K. Foerger, "Holomorphic couplings in string theory," Nucl. Phys. Proc. Suppl. **55B** (1997) 33 [arXiv:hep-th/9611184].

[527] B. de Wit and A. Van Proeyen, "Isometries of special manifolds," arXiv:hep-th/9505097.

[528] L. Andrianopoli, R. D'Auria and S. Ferrara, "U-duality and central charges in various dimensions revisited," Int. J. Mod. Phys. A **13** (1998) 431 [arXiv:hep-th/9612105].

[529] T. L. Curtright, "Charge renormalization and high spin fields," Phys. Lett. B **102** (1981) 17.

[530] S. Ferrara, C. A. Savoy, and L. Girardello, "Spin sum rules in extended supersymmetry," Phys. Lett. B **105** (1981) 363.

[531] C. Bachas and E. Kiritsis, "F^4 terms in $N = 4$ string vacua," Nucl. Phys. Proc. Suppl. **55B** (1997) 194 [arXiv:hep-th/9611205].

Index